2004 NSTI Nanotechnology Conference and Trade Show
NSTI Nanotech 2004
Volume 3

An Interdisciplinary Integrative Forum on
Nanotechnology, Biotechnology and Microtechnology

March 7-11, 2004
Boston Sheraton Hotel
Boston, Massachusetts, USA
www.nsti.org

Nano Science and Technology Institute
Boston • Geneva • San Francisco

Artwork provided by Prof. Eric J. Heller, http://www-heller.harvard.edu/

TransportXVII (cover):

Two-dimensional electron flow in a semiconductor heterostructure. Electrons were launched from the upper center into a weakly random potential, the randomness caused by positively charged donor atoms in the "delta layer" above the 2 dimensional electron gas. Trajectories were launched evenly over 180 degrees, using strict overwrite of successive trajectories. This gives a three dimensional hidden surface effect to the caustics (cusps) which has much the same topology as an erosion landscape. Color is assigned according to the direction of the trajectory.

2004 NSTI Nanotechnology Conference and Trade Show

NSTI Nanotech 2004

Volume 3

An Interdisciplinary Integrative Forum on
Nanotechnology, Biotechnology and Microtechnology

March 7-11, 2004
Boston, Massachusetts, USA
http://www.nsti.org

NSTI Nanotech 2004 Joint Meeting

The 2004 NSTI Nanotechnology Conference and Trade Show includes:
7th International Conference on Modeling and Simulation of Microsystems, MSM 2004
4th International Conference on Computational Nanoscience and Technology, ICCN 2004
2004 Workshop on Compact Modeling, WCM 2004

NSTI Nanotech 2004 Proceedings Editors:

Matthew Laudon
mlaudon@nsti.org

Bart Romanowicz
bfr@nsti.org

Nano Science and Technology Institute
Boston • Geneva • San Francisco

Nano Science and Technology Institute
One Kendall Square, PMB 308
Cambridge, MA 02139
U.S.A.

Nano Science and Technology Institute Order Number PCP04020393
ISBN 0-9728422-9-2
ISBN 0-9728422-5-X (www)
ISBN 0-9728422-6-8 (CD-ROM, Vol. 1-3)

Additional copies may be ordered from:

Nano Science and Technology Institute
Publishing Office
One Kendall Sq., PMB 308
Cambridge, MA 02139, USA
http://www.nsti.org

Printed in the United States of America

Nano Science and Technology Institute
Boston • Geneva • San Francisco

Sponsors

Nano Science and Technology Institute

Intel Corporation

Motorola

Texas Instruments

Ciphergen Biosystems, Inc.

iMediasoft Group

Frontier Carbon Corporation

Veeco Instruments

Hitachi High Technologies America, Inc.

Racepoint Group, Inc.

International SEMATECH

FEI Company

Keithley

Saint-Gobain High Performance Materials

Zyvex Corporation

Accelrys, Inc.

ANSYS, Inc.

Atomistix

PolyInsight

Georgia Department of Industry & Trade

Swiss Business Hub USA

Australian Government - Invest Australia

State of Bavaria, Germany - The United States Office For Economic Development

m+w zander

Engis Corporation

COMSOL

Engelhard Corporation

Nanonex, Inc.

nanoTITAN

Tegal Corporation

Umech Technologies

MEMSCAP

Swiss House for Advanced Research and Education SHARE

Basel Area Business Development

Greater Zürich Area

Development Economic Western Switzerland

Location:Switzerland

Nanoworld AG

Nanosensors

Nanofair

Burns, Doane, Swecker & Mathis, L.L.P.

Mintz Levin Cohn Ferris Glovsky and Popeo, PC

Jackson Walker L.L.P.

Nano Science and Technology Institute
Boston • Geneva • San Francisco

Supporting Organizations

American Physical Society

IEEE Electron Devices Society

IEEE Boston Section

The American Ceramic Society

ASME Nanotechnology Institute

NANOPOLIS

The Global Emerging Technology Institute, Ltd.

Nanotechnology Researchers Network Center of Japan

TIMA-CMP Laboratory, France

Squire, Sanders & Dempsey

Boston University

Massachusetts Technology Collaborative

CIMIT (Center for Integration of Medicine & Innovative Technology)

Draper Laboratory

International Scientific Communications, Inc

DARPA (Defense Advanced Research Projects Agency)

Swiss Federal Institute of Technology

Ibero-American Science and Technology Education Consortium

Applied Computational Research Society

Media Sponsors

Technology Review

iMediasoft Group

Taylor & Francis

NanoApex

Institute of Physics

nanotechweb.org

Photonics Spectra

R&D Magazine and Micro/Nano

Earth and Sky Radio Series

Nano Nordic Network

Table of Contents

Nano Photonics, Optoelectronics and Imaging

Nanoscale Electronics and Quantum Devices

Atomic and Mesoscale Modelling of Nanoscale Phenomena

Nano Devices and Systems

Carbon Nano Structures and Devices

Surfaces and Films

Nano Particles and Molecules

Commercial Tools, Processes and Materials

NSTI Nanotech 2004 Program Committee

TECHNICAL PROGRAM CO-CHAIRS

Matthew Laudon	*Nano Science and Technology Institute, USA*
Bart Romanowicz	*Nano Science and Technology Institute, USA*

TOPICAL AND REGIONAL SCIENTIFIC ADVISORS AND CHAIRS

Nanotechnology

Matthew Laudon	*Nano Science and Technology Institute, USA*
Philippe Renaud	*Swiss Federal Institute of Technology, Switzerland*
Mihail Roco	*National Science Foundation, USA*
Wolfgang Windl	*Ohio State University, USA*

Biotechnology

Robert S. Eisenberg	*Rush Medical Center, Chicago, USA*
Srinivas Iyer	*Los Alamos National Laboratory, USA*

Pharmaceutical

Kurt Krause	*University of Houston, USA*

Microtechnology

Narayan R. Aluru	*University of Illinois Urbana-Champaign, USA*
Bernard Courtois	*TIMA-CMP, France*
Anantha Krishnan	*Defense Advanced Research Projects Agency, USA*
Bart Romanowicz	*Nano Science and Technology Institute, USA*

Semiconductors

David K. Ferry	*Arizona State University, USA*
Andreas Wild	*Motorola, Germany*

NANOTECHNOLOGY CONFERENCE COMMITTEE

M.P. Anantram	*NASA Ames Research Center, USA*
Phaedon Avouris	*IBM, USA*
Xavier J.R. Avula	*Washington University, USA*
Wolfgang S. Bacsa	*Université Paul Sabatier, France*
Roberto Car	*Princeton University, USA*
Franco Cerrina	*University of Wisconsin - Madison, USA*
Murray S. Daw	*Clemson University, USA*
Alex Demkov	*Motorola, USA*
Toshio Fukuda	*Nagoya University, Japan*
David K. Ferry	*Arizona State University, USA*
Sharon Glotzer	*University of Michigan, USA*
William Goddard	*CalTech, USA*
Gerhard Goldbeck-Wood	*Accelrys, Inc., UK*
Niels Gronbech-Jensen	*UC Davis and Berkeley Laboratory, USA*
Karl Hess	*University of Illinois at Urbana-Champaign, USA*
Charles H. Hsu	*Maximem Limited, Taiwan*
Hannes Jonsson	*University of Washington, USA*
Anantha Krishnan	*Defense Advanced Research Projects Agency, USA*
Alex Liddle	*Lawrence Berkeley National Laboratory, USA*
Chris Menzel	*Nano Science and Technology Institute, USA*
Stephen Paddison	*Los Alamos National Laboratory, USA*
Sokrates Pantelides	*Vanderbilt University, USA*
Philip Pincus	*University of California at Santa Barbara, USA*
Joachim Piprek	*University of California at Santa Barbara, USA*
Serge Prudhomme	*University of Texas at Austin, USA*
Nick Quirke	*Imperial College, London, UK*
PVM Rao	*Indian Institute of Technology, Delhi, India*
Philippe Renaud	*Swiss Federal Institute of Technology of Lausanne, Switzerland*
Robert Rudd	*Lawrence Livermore National Laboratory, USA*
Douglas Smith	*University of San Diego, USA*
Clayton Teague	*National Nanotechnology Coordination Office, USA*
Dragica Vasilesca	*Arizona State University, USA*
Arthur Voter	*Los Alamos National Laboratory, USA*
Phillip R. Westmoreland	*University of Massachusetts Amherst, USA*
Wolfgang Windl	*Ohio State University, USA*

| Gloria Yueh | *Midwestern University, USA* |
| Xiaoguang Zhang | *Oakridge National Laboratory, USA* |

BIOTECHNOLOGY CONFERENCE COMMITTEE

Dirk Bussiere	*Chiron Corporation, USA*
Amos Bairoch	*Swiss Institute of Bioinformatics, Switzerland*
Stephen H. Bryant	*National Institute of Health, USA*
Fred Cohen	*University of California, San Francisco, USA*
Daniel Davison	*Bristol Myers Squibb, USA*
Andreas Hieke	*Ciphergen Biosystems, Inc., USA*
Leroy Hood	*Institute for Systems Biology, USA*
Sorin Istrail	*Celera Genomics, USA*
Srinivas Iyer	*Los Alamos National Laboratory, USA*
Brian Korgel	*University of Texas-Austin, USA*
Kurt Krause	*University of Houston, USA*
Daniel Lacks	*Tulane University, USA*
Mike Masquelier	*Motorola, Los Alamos NM, USA*
Atul Parikh	*University of California, Davis, USA*
Andrzej Przekwas	*CFD Research Corporation, USA*
George Robillard	*BioMade Corporation, Netherlands*
Jonathan Rosen	*Center for Integration of Medicine and Innovative Technology, USA*
Tom Terwilliger	*Los Alamos National Laboratory, USA*
Michael S. Waterman	*University of Southern California, USA*

MICROSYSTEMS CONFERENCE COMMITTEE

Narayan R. Aluru	*University of Illinois Urbana-Champaign, USA*
Xavier J. R. Avula	*University of Missouri-Rolla, USA*
Stephen F. Bart	*Nano Science and Technology Institute, USA*
Bum-Kyoo Choi	*Sogang University, Korea*
Bernard Courtois	*TIMA-CMP, France*
Robert W. Dutton	*Stanford University, USA*
Gary K. Fedder	*Carnegie Mellon University, USA*
Elena Gaura	*Coventry University, USA*
Steffen Hardt	*Institute of Microtechnology Mainz, Germany*
Lee W. Ho	*Corning Intellisense, USA*
Eberhard P. Hofer	*University of Ulm, Germany*
Michael Judy	*Analog Devices, USA*
Yozo Kanda	*Toyo University, Japan*
Jan G. Korvink	*University of Freiburg, Germany*
Anantha Krishnan	*Defense Advanced Research Projects Agency, USA*
Mark E. Law	*University of Florida, USA*
Mary-Ann Maher	*MemsCap, France*
Kazunori Matsuda	*Naruto University of Education, Japan*
Tamal Mukherjee	*Carnegie Mellon University, USA*
Andrzej Napieralski	*Technical University of Lodz, Poland*
Ruth Pachter	*Air Force Research Laboratory, USA*
Michael G. Pecht	*University of Maryland, USA*
Marcel D. Profirescu	*Technical University of Bucharest, Romania*
Marta Rencz	*Technical University of Budapest, Hungary*
Siegried Selberherr	*Technical University of Vienna, Austria*
Sudhama Shastri	*ON Semiconductor, USA*
Armin Sulzmann	*Daimler-Chrysler, Germany*
Mathew Varghese	*The Charles Stark Draper Laboratory, Inc., USA*
Dragica Vasilesca	*Arizona State University, USA*
Gerhard Wachutka	*Technical University of München, Germany*
Jacob White	*Massachusetts Institute of Technology, USA*
Thomas Wiegele	*BF Goodrich Aerospace, USA*
Wenjing Ye	*Georgia Institute of Technology, USA*
Sung-Kie Youn	*Korea Advanced Institute of Science and Technology, Korea*
Xing Zhou	*Nanyang Technological University, Singapore*

CONFERENCE OPERATIONS MANAGER

| Sarah Wenning | *Nano Science and Technology Institute, USA* |

NSTI Nanotech 2004 Proceedings Topics

2004 NSTI Nanotechnology Conference and Trade Show

NSTI Nanotech 2004, Vol. 1, ISBN 0-9728422-7-6

- Bio Nano Systems and Chemistry
- Bio Nano Analysis and Characterization
- Bio Molecular Motors
- Ion Channels
- Bio Nano Computational Methods and Applications
- Bio Micro Sensors and Systems
- Micro Fluidics and Nanoscale Transport
- MEMS Design and Application
- Smart MEMS and Sensor Systems
- Micro and Nano Structuring and Assembly
- Wafer and MEMS Processing

NSTI Nanotech 2004, Vol. 2, ISBN 0-9728422-8-4

- Advanced Semiconductors
- Nano Scale Device Modeling
- Compact Modeling
- Circuits
- System Level Modeling
- MEMS Modeling
- Modeling Fundamental Phenomena in MEMS
- Computational Methods and Numerics

NSTI Nanotech 2004, Vol. 3, ISBN 0-9728422-9-2

- Nano Photonics, Optoelectronics and Imaging
- Nanoscale Electronics and Quantum Devices
- Atomic and Mesoscale Modelling of Nanoscale Phenomena
- Nano Devices and Systems
- Carbon Nano Structures and Devices
- Nano Composites
- Surfaces and Films
- Nano Particles and Molecules
- Characterization and Parameter Extraction
- Micro and Nano Structuring and Assembly
- Commercial Tools, Processes and Materials

Message from the Program Committee

The Nano Science and Technology Institute is proud to present the 2004 Nanotechnology Conference and Trade Show (Nanotech 2004). The charter of the Nanotech conference, and its numerous sub-conferences, remains the same since its original conception in 1997. The Nanotech provides for a single interdisciplinary integrative community, allowing for core scientific advancements to disseminate into a multitude of industrial sectors and across the breadth of traditional science and technology domains converging under Nanotechnology, Biotechnology and Microtechnology.

The Nanotech Program Committee makes every effort to provide a scientifically outstanding environment, through its review and ranking process. The Nanotech received a total of 656 abstracts for the 2004 event. All abstracts submitted into the Nanotech are reviewed and scored by a minimum of three (3) expert reviewers. Of the 656 submitted abstracts, 25% were accepted as oral presentations, 40% were accepted as poster presentations resulting in a 35% rejection rate. In addition to the regular program there are a number of invited sessions, panels, overview presentations and tutorials provided for completeness. The Nanotech program committee thanks the hard work of all of this year's reviewers (116 active reviewers in total). This grassroots self-review process by the nano, bio and micro technology communities is a source of pride for the Nanotech conference. This process is in place to provide for a yearly evolution and technical validation of the Nanotech conference content. We thank the authors for submitting their latest work, making a meeting of this caliber possible.

We hope the reader will find the papers assembled in these proceedings rewarding to read, and that the conference continues to foster further advances in this fascinating and multi-disciplinary field. Although the Nanotech conference makes every effort to be as comprehensive as possible, due to the rapid advancements in science and industry, there will inevitably be under-represented sections of this event. We look to you, as the true science and business leaders of small-technologies, to assist us in identifying the needs of our participants so that we can continue to grow the content of this event to best serve this community.

We would like to take this opportunity to thank the many individuals who have worked so hard to make this meeting happen, and to welcome the new members of the program committee. Conferences of this scope are possible only because of the continuing interest and support of the community, expressed both by their submission of papers of high quality and by their attendance. The Nanotech 2004 program committee is grateful to all keynote speakers, authors and session chairs for contributing to the success of the event. We are also indebted to the foundations, agencies and companies whose financial contributions made this meeting possible. We encourage your feedback and participation. Additionally, if you have an interest in conference or session organizational assistance, please contact the conference manager and we will attempt to accommodate. Information concerning next year's conference is posted online at URL: http://www.nsti.org. We look forward to seeing you again next year, and thank you for your continuing support and participation.

Technical Program Co-chairs, Nanotech 2004

Matthew Laudon, Nano Science and Technology Institute, USA

Bart Romanowicz, Nano Science and Technology Institute, USA

NSTI
Nano Science and Technology Institute

Ultimate Limits to Optical Displacement Detection in Nanoelectromechanical Systems

T. Kouh[*], D. Karabacak[*], D. H. Kim[**] and K. L. Ekinci[*]

[*]Aerospace and Mechanical Engineering, Boston University, Boston, MA, USA, tkouh@bu.edu
[**]On leave from Seoul National University of Technology, Seoul 139-743, Korea, dhkim63@bu.edu

ABSTRACT

We describe an optical technique for the detection of displacement in Nanoelectromechanical Systems (NEMS). The technique is based upon path stabilized optical interferometry. We evaluate the effectiveness of this technique in NEMS by detecting displacements from a series of doubly clamped beam resonators with decreasing dimensions. Our measurements and analyses indicate that the technique tends to become less effective beyond the optical diffraction limit.

Keywords: nanoelectromechanical systems, optical interferometry, displacement detection

1 INTRODUCTION

Nanoelectromechanical Systems (NEMS) are among the most promising manifestations of the emerging field of nanotechnology [1]. These are electromechanical systems with dimensions in the deep submicron — mostly operated in their resonant modes. In this size regime, NEMS come with extremely high resonance frequencies, diminished active masses, tolerable force constants and high quality (Q) factors of resonance. These attributes collectively make NEMS suitable for a multitude of technological applications such as ultra-fast actuators, sensors, and high frequency signal processing components.

There exist fundamental and technological challenges to NEMS optimization. One of the remaining challenges to developing technologies based upon NEMS is a robust, sensitive and broadband displacement detection method for sub-nanometer displacements. Most mainstay displacement sensing techniques used in the domain of Microelectromechanical Systems (MEMS) are not scaleable into the domain of NEMS — necessitating the development of new techniques to realize the full potential of NEMS.

Optical detection, including path stabilized interferometry and Fabry-Perot interferometry, has been used extensively in the domain of MEMS. Using path stabilized interferometry, for instance, shot noise limited displacement sensitivities, $\sqrt{S_x} \sim 10^{-6}$ nm/$\sqrt{\text{Hz}}$ are routinely attainable on objects — such as AFM cantilevers — with cross sections much larger than the diffraction limited optical spot [2]. Unfortunately, this conventional approach fails for objects with smaller cross sections [3]. The signal reflected from a NEMS device, for instance, will

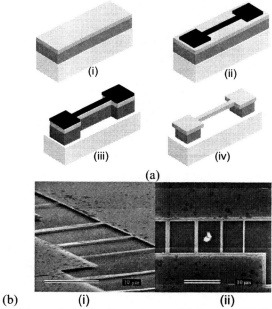

Figure 1: (a) Surface nanomachining of NEMS. A semiconductor heterostructure with structural (top) and sacrificial (middle) layers on top of a substrate (bottom) is patterned using electron beam lithography and selective etch processes. (b) Scanning electron micrographs of silicon doubly-clamped beams.

have a complicated diffraction form — most likely reducing the above-mentioned sensitivity.

In this work, our main objective is to explore the applicability of optical interferometry to NEMS displacement detection — first, by experimentally investigating optical displacement detection in silicon NEMS and then by developing a theoretical picture.

2 EXPERIMENT

Now, we turn to a detailed description of the experimental procedure. We first illustrate the method for fabrication of suspended single crystal silicon structures, and then describe the optical measurements.

The fabrication process is illustrated in Figure 1 (a). The starting material for device fabrication is a silicon-on-insulator (SOI) wafer with a 219 nm-thick silicon device layer on top of 396-nm-thick oxide layer. Fabrication begins by defining large area contact pads by optical lithography. A 40-nm-thick layer of Cr is then evaporated and, subsequently, standard liftoff is carried out with acetone. Samples are then coated with a bi-layer PMMA (polymethyl methacrylate) resist prior to patterning by electron beam lithography. After resist exposure and development, 40 nm of Cr is evaporated on the samples, followed by liftoff in acetone. The pattern in the Cr metal mask is then transferred to the oxide layer beneath it by anisotropic reactive ion etching (RIE). We use a plasma of CF_4 and O_2 at a pressure of 300 mTorr with respective flow rates of 50 sccm and 5 sccm, and a microwave power of 300 W. The etch rate under these conditions is ~6 nm/s. The vertically-etched structures are then released by a controlled selective isotropic etching of the underlying oxide layer using hydrofluoric acid (HF). Figure 1 (b) shows scanning electron micrographs of completed NEMS devices.

The measurements on fabricated devices are carried out inside an ultrahigh vacuum (UHV) chamber using a path stabilized optical interferometer. The block diagram of this set up is displayed in Figure 2 (a). After fabrication, the devices are introduced into the main chamber through a load lock and transferred onto a single-axis linear translator. The motion of the translator brings the devices towards an optical view-port as shown in the Figure 2 (b) — allowing probing of the devices by a laser beam.

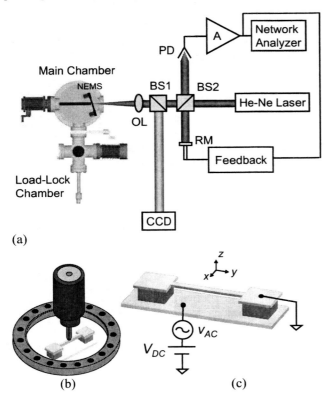

(a)

(b)   (c)

Figure 2: (a) Schematic diagram of the optical measurement setup and the UHV chamber. (b) Close up of the NEMS device in the vicinity of the quartz optical view-port. (c) Schematic diagram of the electrostatic actuation.

The whole optical interferometer is mounted on a XYZ translation stage. The interferometer comprises various beam-splitters (BS), a reference mirror (RM) and a photo-detector (PD) as shown in Fig. 2 (a). Coherent light from a He-Ne laser with wavelength, $\lambda \approx 635$ nm is split into two beams — the first beam traveling along a reference path and a second beam used as a probe for NEMS displacement detection. A feedback circuit and a mirror mounted on a piezoelectric actuator control the length of the reference path. The probe beam used for NEMS displacement detection is focused on the device by a 50X objective lens (OL) with numerical aperture, $NA=0.5$. The light reflecting from the NEMS is collected by the same lens.

The probe beam reflected from the NEMS device and the beam in the reference path form an interference pattern on the photodetector. The optical path length that the probe beam travels and consequently, the intensity of the interference pattern on the photodetector change as the NEMS device displaces out of plane (in the z-direction). The intensity variations on the photodetector are measured by a network analyzer.

The NEMS devices are actuated electrostatically [4] and their displacements are detected using a network analyzer as shown in Figure 2 (c). The electrostatic force, F_e, exerted on the doubly clamped beam as a function of the applied voltage, V, can be written as $F_e = -\frac{1}{2}\frac{\varepsilon_0 wL}{(d-z)^2}V^2$ where ε_0 is the permittivity of free space, d is the initial gap between the beam and the substrate, z is the displacement of the beam, w and L are the width and length of the beam, respectively. In general, the excitation voltage, V includes both V_{DC} and v_{AC} components.

3 RESULTS

We now turn to a discussion of our experimental results. Figure 3 shows the out-of-plane fundamental resonance frequencies of a family of doubly-clamped silicon beams, with rectangular cross sections and different aspect ratios (length/width). The beams are 10 µm long and 200 nm thick with widths of 1000 nm, 750 nm, 500 nm and 250 nm. There is a metallization layer of 1 nm Cr and 1.5 nm Au atop the beams. This particular family of devices yields out-of-plane resonant frequencies between 16.5 MHz and 17.1 MHz. Quality (Q) factors measured were in the range $300<Q<3000$ at room temperature.

The fundamental resonance frequency, $\omega_0/2\pi$, of a doubly-clamped beam of length, L, and thickness, t, varies linearly with the geometric factor t/L^2 according to the simple relation $\omega_0/2\pi = 1.03(\sqrt{E/\rho})t/L^2$ where E is the Young's modulus and ρ is the mass density. The measured frequencies are within 4% of the theoretical predictions.

The inset in Figure 3 shows the rms displacement of the center of the 1000 nm-wide beam as a function of V_{DC}

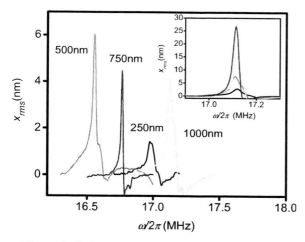

Figure 3: Frequency spectrum of four suspended silicon beams of 10 μm-long (L) and 200 nm-thick (t) with different widths (w) of 1000 nm, 750 nm, 500 nm and 250 nm. The inset shows the frequency spectrum of the silicon beam of 10 μm × 200 nm × 1000 nm as a function of DC driving amplitude, V_{DC}. The lowest curve corresponds to 1 V and V_{DC} is increased by 1V at fixed AC voltage of 800mV. All the measurements are done at $2×10^{-9}$ Torr.

while v_{AC} is fixed at 800 mV. With increasing V_{DC}, the beam is driven into larger rms amplitude as described above.

4 ANALYSIS

The purpose of the numerical analysis is to gain an understanding of the ultimate limits to optical interferometric displacement detection in NEMS. The measured signal is the photodetector current that flows as a result of the intensity modulations — arising from the NEMS motion — in the interference pattern formed upon the photo-detector.

Obviously, the dimensions of the NEMS beams are on the order of the wavelength of the light used for detection. This necessitates a detailed modeling of the behavior of the electromagnetic (EM) field in the near field region of the NEMS — as many of the analytical solutions based on far-field approximations fail in this size regime.

Our numerical analysis is for a 2-dimensional case. The computation is based on solving Maxwell's Equations for a source free medium with monochromatic light, in the region of focus around the resonator using a finite element method. The laser beam is modeled as an in-plane Gaussian beam of transverse electric (TE) polarization as shown in Figure 4 (a) [5]. The finite element EM field solutions are obtained for the domain, shown in Fig. 4 (a) along the path of the Gaussian beam.

The resonator is modeled to be within the focused region and the surfaces are assumed perfectly reflective.

The center location of the beam is then displaced at various amplitudes to simulate the resonator motion. The displacement of the resonator affects the electric field pattern reflected back to the photodetector. A representative intensity modulation on the photo-detector is shown in Figure 5 (a). Here, the peak intensity is observed to drop significantly as the beam moves between two displacement maxima, x_{max} and $-x_{max}$. The intensity variations thus determined are integrated across the first light fringe for x_{max} and $-x_{max}$ for different rms amplitudes of vibration. Figure 5 (b) shows the weakly parabolic relationship — obtained from multiple simulations — between rms vibration amplitude and the power oscillation observed on the detector.

At temperature, T, the spectral density of the thermomechanical displacement fluctuations of the resonator is given by $S_x = 4k_B TQ / m\omega_o^3$ where k_B is the Boltzmann constant, Q is the quality factor, m is the effective mass of the resonator, and ω_o is the resonant frequency [6]. For our silicon beams of dimensions 10 μm × 200 nm × 1000 nm with $Q\sim1000$ at room temperature,

(a)

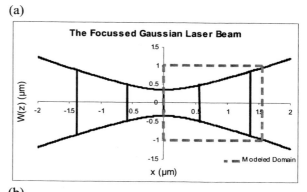

(b)

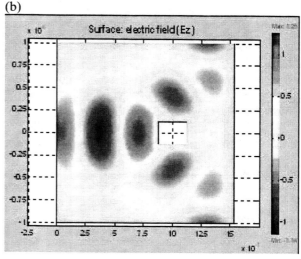

Figure 4: (a) The numerical computation domain modeled in the focus region of the Gaussian laser beam. (b) Electric field in the near-field region of the beam cross section, at stationary position (dimensions in μm).

(a)

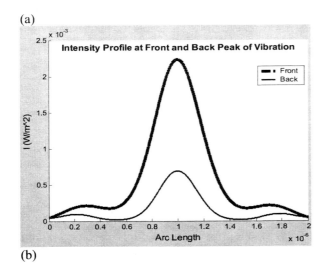

(b)

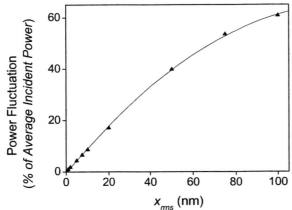

Figure 5: (a) Intensity profile change of interference of the reflected and reference beams, for a vibration range of ± 100 nm. (b) Change in power incident on photodiode based on amplitude of vibration of resonator, for a beam of 250 nm thickness and width.

$\sqrt{S_x} \approx 0.1\text{pm}/\sqrt{\text{Hz}}$. We were not able to observe these fluctuations in our optical measurements.

Our displacement sensitivity in this work is determined by the various noise sources in the detection circuit. Figure 6 shows the circuit diagram of the photodetector. The noise sources are shown in lighter shades. The shot noise current spectral density, S_i^{shot} is given by $S_i^{shot} = \eta e^2 P_o / \hbar\omega$ where η is the quantum efficiency of the photodetector, e is the electronic charge, $\hbar$ is the Planck's constant, ω is the frequency of the photon, and P_o is the average power incident on the photodetector [7]. For the typical values of $\eta \sim 10\%$, $\hbar\omega \sim 3\times10^{-19}$J, and $P_o \sim 10\mu$W, $S_i^{shot} \approx 0.1\text{pA}^2/\text{Hz}$. The dark current noise current spectral density, S_i^D is $2eI_D$ where I_D is the dark current. For our photodetector, $I_D \sim 2.5$ nA and the corresponding $S_i^D \approx 10^{-3}\,\text{pA}^2/\text{Hz}$. The amplifier noise in this work is $S_i^a \approx 44\text{pA}^2/\text{Hz}$ [8]. These noise current spectral

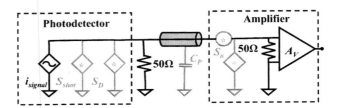

Figure 6: Circuit representation of the detection circuit. The gray elements indicate the noise sources in the circuit such as shot noise, dark current noise, and amplifier noise.

densities determine our sensitivity limits as $S_x = S_i / |\partial i_{signal}/\partial x|^2$. In our optical path stabilized interferometer, $\partial i_{signal} / \partial x \sim 1\text{nA/nm}$. This gives the total displacement noise spectral density from the photodetector and the amplifier as $S_x^{total} = S_x^{shot} + S_x^D + S_x^a$. This corresponds to the displacement sensitivity of $\sqrt{S_x^{total}} \approx 7\text{pm}/\sqrt{\text{Hz}}$.

5 CONCLUSIONS

We have presented experimental and theoretical studies on optical displacement detection in silicon NEMS using a path stabilized interferometer. We measured out of plane fundamental resonance of silicon NEMS with dimensions as small as 10 μm × 200 nm × 1000 nm in UHV at room temperature.

The circuit analysis of the optical displacement detection scheme shows that the sensitivity of the path stabilized optical interferometer is limited by the amplifier noise.

We gratefully acknowledge support from the NSF under grants 210752, 216274 and 324416.

REFERENCES

[1] A. N. Cleland, M. L. Roukes, Applied Physics Letters 69, 2653, 1996.

[2] T.R. Albrecht, P. Grütter, D. Rugar, D.P.E. Smith, Ultramicroscopy 42, 1638, 1992.

[3] D. W. Carr and H. G. Craighead , J. Vac. Sci. Technol. B 15, 2760, 1997.

[4] B. Choi and E. G. Lovell, Journal of Micromechanics and Microengineering, 7, 24, 1997.

[5] FEMLAB® Computation Package, Comsol AB.

[6] A.N. Cleland, "Foundations of Nanomechanics", Springer, 2003.

[7] J. W. Wagner, Physical Acoustics 19, 201, 1990.

[8] FEMTO Messtechnik GmbH, Paul-Lincke-Ufer 34, D-10999 Berlin, Germany.

Local optical field in the neighborhood of structured surfaces:

phase singularities and Talbot effect

B. Levine[*], M. Caumont[*], C. Amien[**], B. Chaudret[**], B. Dwir[***] and W.S. Bacsa[*]

[*] Laboratoire de Physique des Solides, Université Paul Sabatier, UMR-CNRS
118 route de Narbonne, Toulouse 31062, France, bacsa@lpst.ups-tlse.fr
[**] Laboratoire de Chemie de Coordination, UP-CNRS, 205, route de Narbonne, 31077 Toulouse, amiens@lcc-toulouse.fr
[***] Laboratory of Nanostructures, Swiss Federal Institute of Technology Lausanne, Lausanne 1015, Switzerland, benjamin.dwir@epfl.ch

ABSTRACT

We have probed the local optical field in the vicinity of structured surfaces to observe optical coherence effects. We have used a sub-wavelength sized optical fiber probe to map the local field of micrograjings and nanometer sized gold particles. While probing the middle field region, the ratio of the aperture size to the object-probe distance is smaller (a/h =1) than in the nearfield region (a/h=10). In the middle field one probes mostly the transverse field component and probe induced effects are reduced. We have observed self-imaging (Talbot effect) of the micrograting structure in the direction of the incident beam and apparent phase singularities from the deposited gold particles in the direction of the reflected beam.

Keywords: local optical field, structured surfaces, middle field, Talbot effect, phase singularities

1 INTRODUCTION

Structured surfaces with dimensions comparable or smaller than the wavelength of light are increasingly used as parts of functional units in integrated devices. Lens based optical microscopes rely on ray optics and ignore the wave aspect and the lateral resolution is limited by diffraction. Scanning optical probe techniques with apertures smaller than the wavelength circumvent the diffraction limit of lens based systems. Nearfield optics (1) aims to use the enhanced longitudinal local field using an optical probe in the proximity of surfaces. The transmission of optical waves through the aperture of the probe is in general strongly reduced as the aperture size drops below the size of the wavelength. Apertures in the size range of /4 are typically used. To reach the nearfield the distance to the surface is often ten times smaller than the aperture size although the lateral resolution is given by the aperture size. The small probe-substrate distance makes it difficult to image surfaces with height variations in the size range of the probe aperture.

We note that the illuminating beam which strikes the surface overlaps with that on the surface scattered field within its vicinity. The overlap of propagating beams gives rise to lateral and vertical standing waves with wave fronts perpendicular to the direction of counter-propagating waves (2-4). The optical probe is used to detect the local optical field of the standing waves or interferograms. The standing waves are clearly related to the surface structure and can give information about the surface at variable distances from the surface. To obtain high lateral resolution a smaller distance is still needed but the distance can be larger than in nearfield optics by one order of magnitude. We denote the distances ranging from 0.5 m to 100 m as the middle field. We report here local optical field measurements in the proximity of agglomerated nano-sized gold particles on a polished silicon wafer and micrometer sized optical gratings etched into GaAs. We also discuss the different types of observed standing waves displaced in the incident or reflected beam direction.

Figure 1 shows standing wave patterns recorded at two different distances (100 m, 10 m) from the surface. The optical probe and surface have been illuminated by a laser beam (wavelength: 669nm, 10mW, s-polarized) and the local field is detected through a metal coated optical fiber probe (Nanonics Imaging Ltd., aperture size *100nm*). The gold particles 2-3nm in size have been deposited on a polished Si wafer from a suspension and the solvent was evaporated off. Concentric diffraction fringes around the agglomerated gold particles are observed (Fig.1) as well as diagonal fringes. The incident and reflected beam of a flat surface form standing waves parallel to the surface. They become visible as parallel fringes when the image plane is tilted with respect to the surface. The diagonal fringes are due to the tilt of the image plane with respect to the surface and can be corrected by tilting the image plane with the piezo-electric scanner (CP-R Digital Instruments). The tilt angle is given by the fringe spacing (5). The fact that the diagonal fringes are not exactly straight lines indicates that the image plane is not only tilted but slightly deformed due to the movement of the piezo-electric tube scanner. The substrate is not displaced in a perfectly parallel manner. The imaging of optical standing waves on flat surfaces in turn can be used to verify the symmetric scanner movement. We find that the concentric fringes depend on the size of the island and distance from the surface. They are also influenced by the angle of incidence and the polarization of the incident beam. The bottom image (fig.1) shows the same region at considerably smaller distances to the surface. The first concentric fringe is smaller, displaced, and the same diagonal fringes seen in the top image have much smaller amplitude. The maximum amplitude of the standing wave oriented parallel to the surface is constant. The lower contrast of the parallel fringes in the lower image hence shows that concentric fringes around the island are much larger and have a higher amplitude. We conclude that with smaller distance to the surface the concentric fringes get considerably more intense and show more details of the island.

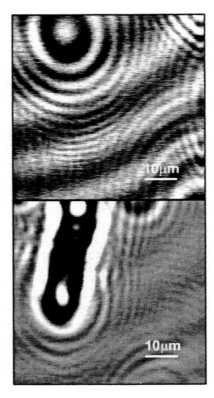

Figure 1

Standing wave patterns of agglomerated gold islands at two different distances (top: 100 m, bottom: 10 m) from the surface.

It can be shown that the displacement of the image with changing object-image distance is related to the angle of incidence and the image recorded in the middle field can be represented as a linear superposition of the direct and diffracted image (5). The relative contribution of the two parts depends on the image height.

When illuminating the islands of gold particles not with s polarized light but with light polarized at 45 deg in the direction of the incident beam direction we observe spiral shaped standing waves irrespective of the shape of the island at sufficiently large distances from the surface (Fig 2). Phase singularities as observed in the spiral shaped standing waves are observed in a number of wave phenomena (6,7). It has been suggested that phase singularities could be used to trap particles (8). Phase singularities have also been observed in the near-field on top of optical wave guide structures (9). Phase singularities in their most simple form can be created through the superposition of three plane waves (10). The 45deg polarization of the incident beam gives an additional field component in the direction parallel to plane of incidence and can in principle explain the observed spiral formation. A detailed explanation of the observed contrast is in progress and will be given elsewhere. So far we have assumed that the optical probe is not sensitive to polarization. But the transmission of light through the aperture of the optical fiber probe is expected to be sensitive to the polarization of the local optical field. The induced dipole at the probe edge, oriented perpendicular to the plane of incident makes the largest contribution. The induced dipole is oriented parallel to the probe edge for s-polarized light and has a larger emission amplitude in the plane of incident in which also falls the axis of the optical fiber probe.

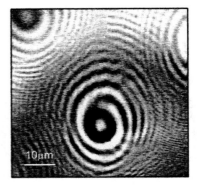

Figure 2

Spiral shaped standing waves of island of gold particles (distance 100 m).

This is consistent with what we observe experimentally for the optical fiber used in the experiment. The signal for p-polarized light is by a factor two smaller than for s-polarized light. The recorded signal is hence dominated by contributions of the local field with s-polarization. Figure 1 and 2 show also fringes from the neighboring island. While the spiral formation in the standing wave has been observed for the metal islands we will now show a particularly interesting type of standing wave formation on micro-gratings.

The micro-gratings have been fabricated by electron beam lithography on a GaAs substrate. The shown micro grating has an elliptical shape (size: 80x50 m) where the edge (10 m) has the grating structure etched into the substrate. Figure 3 shows a 60 m scan at large distance (*100 μm*) from the surface in the middle field region. Concentric diffraction fringes around the grating pattern as well as displaced fringes of the shape of the micro-grating are observed in the direction of the incident beam (angle of incidence 45deg) where the periodic structure from the grating is reproduced. This self-imaging of the grating at macroscopic distances of optical gratings has been first reported by Talbot (11) in the 19th century. In general, a self-focused image can be observed at periodic distances from the grating. The Talbot effect has been later explained theoretically by Lord Rayleigh (12). The fact that the scattered waves from the grating are periodic and they have at the same time a fixed phase relationship has the consequence that the image is reproduced at characteristic distances from the surface. This characteristic distance (d_{Talbot}) depends on the periodicity of the object (a) and the wavelength (of the incident beam:

$$d_{Talbot} \quad \frac{\lambda}{1 \quad \sqrt{1 \quad \frac{\lambda^2}{a^2}}} \qquad (1)$$

We observe here the Talbot image of an optical grating in the vicinity of the surface and in the middle field region. Its appearance indicates that the surface is at a distance which is a multiple of the Talbot distance d_{Talbot}.

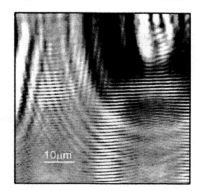

Figure 3

Talbot effect of a micro-grating recorded in the middle field region.

In case the distance is known the appearance of the Talbot image gives information about the wavelength. This can be exploited in spectrometer designs. While the diffracted wave around the object is displaced in the direction of the reflected wave for a non-normal incident beam, the Talbot image is displaced in the direction of the incident beam. Talbot self-imaging has been exploited at macroscopic distances in wavefront sensing and in transform spectrometer designs (13-14).

CONCLUSION

We have observed the local optical field in the middle field region and the overlap zone of a non-normal incident and the reflected beam from aggregated gold particles and microgratings. We find that the recorded standing waves or interferogram of the gold particles contains apart of diffraction fringes, phase singularities when illuminated with the incident beam polarized at 45deg. Images recorded from GaAs microgratings show displaced self-focused images in the direction of the incident beam attributed to the Talbot effect in the vicinity of structured surfaces.

REFERENCES

[1] D. Courjon, Near-Field microscopy and Near-Field Optics, Imperial College Press (2003)
[2] N. Umeda, Y. Hagashi, K. Takayanagi, Appl. Opt. 31 (1992) 4515
[3] A. Kramer, T. Hartmann, S.M. Stadler, R. Guckenberger, Ultramicroscopy 61 (1995) 191
[4] W.S. Bacsa, A. Kulik, Appl. Phys. Lett 70 (1997) 3509
[5] B. Levine, A. Kulik, W.S. Bacsa, Phys. Rev. B 66, 233404, 2002.
[6] J.F. Nye, M.V. Berry, Proc. Roy. Soc. London A 336, 165, 1974
[7] I. Freund, N. Shvartsman, V. Freilikher, Opt. Comm. 101 (193) 247
[8] K.T. Gahagan, G.A. Swartzlander, Opt. Lett. 21 (1996) 827
[9] M.L. Balistreri, J.P. Korterik, L. Kuipers, N. F. van Hulst, Phys. Rev. Lett. 85 (2000) 294
[10] W. Braunbek, Zeitschrift für Naturforschung 6a (1951) 1
[11] W. H.F. Talbot, Philos. Mag. **9**, No. IV, 401, 407 1836.
[12] Lord Rayleigh, Phil. Mag. 11, 196, 1881
[13] L. Kung, A. Bhatnagar, D.A.B. Miller, IEEE Laser and Electro-Optical Society 2000 Annual Meeting, Rio Grande, Puerto Rico (November 13-16, 2000), paper TuU2
[14] A.W. Lohmann, in Proc. of Conference on Optical Instruments and Techniques, London (1961) 58

High Resolution Backside Imaging and Thermography using a Numerical Aperture Increasing Lens

M. S. Ünlü[*], S. B. Ippolito[*], M. G. Eraslan[*], S. A. Thorne[*], A. Vamivakas[*], B. B. Goldberg[*], and Y. Leblebici,

[*]Boston University, Boston, MA, selim@bu.edu
[**]Swiss Federal Institute of Technology, Lausanne, Switzerland

ABSTRACT

Nanoscale imaging of defects in ICs is a great current technological challenge as IC feature sizes continue to shrink. We have developed novel techniques based on a Numerical Aperture Increasing Lens (NAIL) to study semiconductors at very high spatial resolution. The NAIL is placed on the surface of a sample and its convex surface effectively transforms the NAIL and the planar sample into an integrated solid immersion lens. Addition of the NAIL to a standard microscope increases the NA by a factor of square of the index n, to a maximum of NA = n. In silicon, the NA is increased by a factor of 13, to NA = 3.6. The spatial resolution improvement laterally is about a factor of 4 while longitudinally it is a factor of 12.5 corresponding to an overall reduction of the volume of interrogation by a factor of 50.

Subsurface solid immersion microscopy can be applied to thermal imaging of blackbody radiation at IR wavelengths. We have designed, built, and demonstrated the use of a subsurface solid immersion microscope with capability for confocal imaging in 3-5μm wavelength range and demonstrated a resolution of 1.4μm, representing the highest resolution subsurface thermography to date.

Keywords: thermal imaging, high-resolution, failure analysis, solid immersion lens.

1 NUMERICAL APERTURE INCREASING LENS

The semiconductor industry has continued to advance at the rapid pace of Moore's Law through continued shrinking of the physical dimensions of the semiconductor devices, which are already at the nanoscale and will soon approach the atomic scale. New materials and devices at nanoscale herald a revolutionary age for science and technology, provided we can observe the detailed operation and discover and utilize the underlying principles. An important topic in nano-optics is optical microscopy and spectroscopy. Optical spectroscopy provides a wealth of information on structural and dynamical properties of materials, especially when combined with high-resolution microscopy because the spectral features can be spatially resolved. However, there are fundamental limitations of conventional microscopy. In case of imaging objects with optical fields propagating to the far-field, the basic constraint is the diffraction of light, which limits standard optical microscopy to a spatial resolution comparable to the wavelength of light. For imaging objects through a substrate, which is opaque for short wavelengths, this limitation becomes more stringent. Reducing the wavelength or increasing the collected solid angle can improve the spatial resolution of surface microscopy. We have recently developed novel techniques based on a Numerical Aperture Increasing Lens (NAIL) to study semiconductors at very high spatial resolution. [1,2] The NAIL is placed on the surface of a sample and its convex surface effectively transforms the NAIL and the planar sample into an integrated solid immersion lens. Addition of the NAIL to a standard microscope increases the NA by a factor of square of the index n, to a maximum of NA = n. In silicon, the NA is increased by a factor of 13, to NA = 3.6. Figure 1 shows inspection images of Si circuits fabricated by 180nm and 130nm technologies, displaying the striking improvement provided by the NAIL technique. Using an optimized confocal system we demonstrated lateral spatial resolution of approximately 200 nm. [1] The spatial resolution improvement laterally is about a factor of 4. One of the important features of NAIL microscopy is improved

Figure 1. Qualitative comparison of the images displays the NAIL technique's striking improvement over state-of-the-art resolution.

light collection efficiency (scales with the square of NA), particularly important in the study of quantum dots as well as a variety of semiconductor failure analysis modalities including thermal imaging.

2 THERMAL EMISSION MICROSCOPY

Thermal emission microscopy is a non-contact optical microscopy technique that collects mid-infrared photons emitted to image the spatial distribution of temperature in a sample. The spatial distribution of temperature within a sample can be calculated, because the optical power emitted by the sample is a function of its local temperature. The optical power per unit area emitted by an object is proportional to its absolute temperature to the fourth power (Stephan Boltzmann's Law). Thermal emission microscopes are important tools in failure analysis of Si integrated circuits (ICs). Current Si IC technology has many opaque metal layers and structures fabricated above semiconductor devices, thereby hindering topside microscopy of the buried devices in their final state. Therefore, microscopy through the backside or substrate of a Si IC is often preferred. We demonstrate the improvement the NAIL yields in thermal emission microscopy of Si ICs. The theoretical lateral spatial resolution limit is 2.5 µm (~5µm for best commercial systems) for conventional thermal emission microscopes operating at wavelengths up to 5 µm. Current Si IC technology has reached submicron process size scales, well beyond the spatial resolution capability of conventional thermal emission microscopy.

The confocal scanning thermal emission microscope we built for this measurement consists of the elements shown in Fig. 2. The thermal test sample has an Al line and pads fabricated on a Si substrate. Joule heating the Al line generates a spatial distribution of temperature in one lateral direction and the longitudinal direction narrow enough to demonstrate a significant improvement in spatial resolution. However, without an accurate thermal model of the exact temperature distribution, the resulting best spatial resolution is unknown. The sample is flip chip bonded to a printed circuit board for connection and mounting on the xyz scanning stage. The computer controls the stage that scans the sample and NAIL, while acquiring the signal voltage. The sample is driven by a 900Hz sine wave from the signal genera-tor, while the lockin measures the amplitude of the second harmonic. The NAIL has a radius of curvature of 1.61 mm and a center thickness of 1.07 mm, optimized for the sample substrate thickness of 1 mm. The mid-infrared achromatic objective lens has NA = 0.25, resulting in the NAIL microscope having NA = 3.06. A cold mirror reflects the near-infrared wavelengths to an InGaAs camera for visual inspection and transmits the mid-infrared wavelengths to a cooled 50 µm diameter InSb detector, for thermal emission microscopy. [3,4]

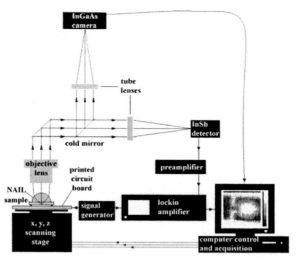

Figure 2. NAIL Confocal scanning thermal emission microscope configuration.

Figure 3 shows the resolution improvement with NAIL technique. The lower panel compares the inspection images taken by the InGaAs camera of the Al line, patterned to be 1µm wide. The thermal emission image taken by the InSb detector of the Joule heating is shown in the upper row. These images are taken at best focus. The optical power emitted due to the Joule heating follows Stephan Boltzmann's Law. Figure 4 shows a linecut in the lateral direction of the thermal emission image in Fig. 3(upper right). The full-width-at-half-maximum (FWHM) of the signal is 1.6 µm. The signal represents a convolution of the NAIL microscope line spread function and the finite spatial distribution of thermal emission in the sample. Deconvolution of the actual linewidth results in a spatial resolution of 1.4 µm representing a significant improvement over conventional thermal emission microscopy.

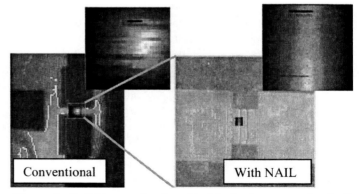

Figure 3. Solid immersion lens blackbody thermography. Left is a standard backside visual image (bottom), with blackbody image (top) showing >5µm resolution. On right, visual NAIL image clearly showing 200nm wires separated by 4µm. Red image in NAIL blackbody, showing 1.3µm resolution. The greater heating toward the center is due to electromigration and reduced thermal conductance away from the contacts

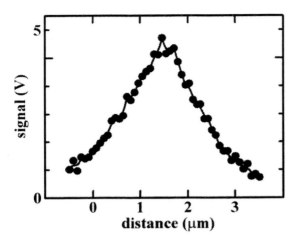

Figure 4. Lateral linecut of image in Fig. 3. The FWHM of the signal is 1.6 μm.

To evaluate the longitudinal spatial resolution we take successive images at different defocus distances in the longitudinal direction (z) and obtained a longitudinal spatial resolution of ~7μm. In comparison, the ultimate limit of longitudinal spatial resolution in conventional thermal emission microscopy is 18μm.

3 CONCLUSIONS AND FUTURE WORK

In summary, we have demonstrated drastic resolution improvements utilizing Numerical Aperture Increasing Lens technique. The application of the NAIL technique to subsurface thermal emission microscopy of Si integrated circuits demonstrates improvements in the amount of light collected and in both the lateral and longitudinal spatial resolutions, well beyond the limits of conventional thermal emission microscopy. The spatial resolution improvement laterally is about a factor of 4. Theoretically, overall reduction of the volume of interrogation is a factor of 50. We have designed, built, and demonstrated the use of a subsurface solid immersion microscope with capability for confocal imaging in 3-5μm wavelength range and demonstrated a lateral resolution of 1.4μm along with a focal depth of ~7μm. To the best of our knowledge, both of these values correspond to the best resolution in subsurface thermography to date.

Currently, we are building upon these experimental results and exploring the ultimate limitations of NAIL microscopy in thermal imaging. The diffraction of light theoretically limits the lateral spatial resolution to 0.73 μm and the longitudinal spatial resolution to 1.5 μm, for thermal emission microscopes with a NAIL, operating at free space wavelengths up to 5 μm. To eliminate the experimental limitations due to large thermal sample, we are building a system where a UV laser is focused on the front surface of the substrate (double-side polished Si wafer) to form a thermal image smaller than 500 nm in diameter.

REFERENCES

[1] S. B. Ippolito, B. B. Goldberg, and M. S. Ünlü, "High-resolution Subsurface Microscopy Technique," *Appl. Phys. Lett.*, Vol. 78(26), pp. 4071-4073 (2001).

[2] B. B. Goldberg, S. B. Ippolito, L. Novotny, Z. Liu, and M. S. Ünlü, "Immersion Lens Microscopy of Photonic Nanostructures and Quantum Dots," *IEEE J. Selected Topics in Quantum Electron.*, Vol. 8(5), pp. 1051-1058, (2002).

[3] S. B. Ippolito, S. A. Thorne, M. G. Eraslan, B. B. Goldberg, Y. Leblebici, and M. S. Ünlü, "High spatial resolution subsurface thermal emission microscopy," Proc. IEEE/LEOS Annual Meeting, Tucson, AZ, October 2003.

[4] S. Thorne, S. B. Ippolito, M. Eraslan, B. Goldberg, M.S. Ünlü, and Y. Leblebici, "High Resolution Backside Thermography using a Numerical Aperture Increasing Lens," Proceedings of 29th International Symposium for Testing and Failure Analysis, 2-6 November 2003, Santa Clara, CA

An ultra-fast scheme for sample-detection in dynamic-mode Atomic Force Microscopy

Deepak R. Sahoo, Abu Sebastian and Murti V. Salapaka

Department of Electrical and Computer Engineering, Iowa State University, Ames, IA, 50011.
{deepak, abuseb, murti}@iastate.edu.

ABSTRACT

In typical dynamic mode operation of atomic force microscopes steady state signals like amplitude and phase are used for detection and imaging of material. In these methods, high quality factor of the cantilever results in high resolution, but low bandwidth and vice versa. In this paper we present a methodology that exploits the deflection signal during the transients of the cantilever motion. The principle overcomes the limitations on the trade off between resolution and bandwidth present in existing methods and makes it independent of the quality factor. Experimental results provided corroborate the theoretical development.

Keywords: atomic force microscopy, state-space model, state observer, hypothesis testing

1 INTRODUCTION

Atomic force microscopes [1] (AFMs) utilize a cantilever to image and manipulate sample properties at the nano-scale. Cantilevers have been utilized in biological sciences to perform remarkable feats such as cutting DNA strands [2] and monitoring RNA activity [3]. On a similar note there are impressive proposals on using cantilever based nanoprobes to interrogate cell dynamics that will have significant impact on human health. At present extensive research is being carried out to develop tip-based data storage devices [4] using cantilevers as sensors and actuators for high density data storage in the range of 10^{12} bits/in^2 with data read-write rates of the order of 100 kHz. These events often have time-scales in the micro-second or nano-second regimes. Current technology does not meet the aforementioned high precision and bandwidth requirements.

The cantilever is often operated in the dynamic-mode due to its gentle nature on the sample [5]. In this mode of operation, a cantilever with high quality factor is employed for high resolution. However, due to large settling time, the steady state signals (like the demodulated amplitude or phase) of the cantilever response are slow and therefore the corresponding methods have a smaller bandwidth. Using active Q control the bandwidth can be increased [6], [7], [8]; however the trade off between bandwidth and resolution remains inherent [8]. The existing methods do not utilize the cantilever model and

do not exploit the deflection signal during the transient state of the cantilever.

In this paper we present a new principle that harnesses the transient part of the cantilever dynamics. As in steady state methods, high quality factors result in high resolution; however in the method presented, the bandwidth is largely independent of the quality factor and is determined by the resonant frequency of the cantilever. As is seen later it also provides advantages with respect to resolution; particularly of events that have very small time scales.

2 SAMPLE-DETECTION

The transient signal based detection method relies on identifying the first mode model of the cantilever-dynamics and consequently construction of an observer that provides an estimate of the cantilever-dynamics. The resulting architecture facilitates the detection of tip-sample interaction force during the transient state of the cantilever as described below.

2.1 Model of the cantilever-dynamics

When the cantilever is forced sinusoidally near its first resonance frequency, its dynamic response is well described by the first mode model given by:

$$\dot{x} = \overbrace{\begin{bmatrix} 0 & 1 \\ -\omega_0{}^2 & -\frac{\omega_0}{Q} \end{bmatrix}}^{A} x + \overbrace{\begin{bmatrix} 0 \\ 1 \end{bmatrix}}^{B} (\eta + w),$$

$$y = \overbrace{\begin{bmatrix} 1 & 0 \end{bmatrix}}^{C} x + v,$$
(1)

where the cantilever states $x = [p\ v]^T$, Q, ω_0, η, w, y and v denote the cantilever-tip position (p) and velocity ($v = \dot{p}$), the quality factor, the first resonant frequency, the thermal-noise, external forces acting on the cantilever, the deflection signal and the photo-diode noise respectively. The cantilever model described above can be identified precisely using its thermal-noise response [9]. The cantilever can be imagined to be a system that takes in the thermal-noise η, the dither signal g and the tip-sample interaction force $\phi(x)$ as inputs (in which case $w = \phi + g$) and produces the photodiode signal y as the output.

2.2 Observer based state-estimation

The construction of the observer (see Figure 1) is based on the cantilever-model and power spectral densities of the noise. The observer dynamics is given by:

$$\dot{\hat{x}} = A\hat{x} + Bg + L(y - \hat{y}); \quad \hat{x}(0) = \hat{x}_0, \qquad (2)$$
$$\hat{y} = C\hat{x},$$

and associated state estimation error ($\tilde{x} := x - \hat{x}$) dynamics is given by:

$$\dot{\tilde{x}} = (A - LC)\tilde{x} + B\eta - Lv; \quad \tilde{x}(0) = x(0) - \hat{x}(0). \qquad (3)$$

The observer mimics the dynamics of the cantilever (given in Equation (1)) by utilizing the correcting term $L(y - \hat{y})$ where L is the gain of the observer and $y - \hat{y} := e$ is the error in estimating the deflection signal. The error $\tilde{x}$ between the estimated state $\hat{x}$ and the actual state x of the cantilever, when no noise terms are present ($\eta = v = 0$) is only due to the mismatch in the initial state of the observer and the cantilever (see Equation (3)). The error $\tilde{x}$ goes to zero when the real part of all the eigenvalues of the matrix $(A - LC)$ are negative. Since the pair (A, C) is *observable* for the cantilever model (i.e. $rank([A \ CA]^T) = 2$ when a second order model is assumed) the eigenvalues of the matrix $A - LC$ can be placed anywhere by appropriately choosing L [10]. Thus the error signal e due to initial condition mismatch can be reduced to zero and in principle arbitrarily fast by suitably choosing L. When there is a change in the tip-sample interaction the cantilever dynamics is effected. This introduces an error in tracking which evolves according to the cantilever-observer dynamics as given by Equation (3). Also, when the change in the tip-sample potential persists, the observer by utilizing its input y may track the altered cantilever state. It can also be shown that in the presence of noise sources η and v the error signal e is a zero-mean stationary process. Thus the error signal e shows the signature of the change in the tip-sample behavior (buried in noise) immediately after the change is introduced. The error e may recover its zero mean nature even when the interaction change persists. This is in contrast to the steady state methods where the information is available not in the initial part but after the cantilever has come to a steady state. The Kalman observer [11] can be employed for optimal tracking in which case the error process (also known as the *innovation*) has zero mean and is white during perfect tracking.

2.3 Estimation-error characterization

The error profile due to a tip-sample interaction change can be better characterized if a model of the effect of the tip-sample interaction change on the cantilever-motion is available. We assume that the sample's influence on the cantilever tip is approximated by an impact condition where the tip-position and velocity instantaneously

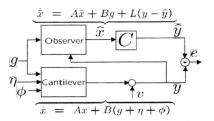

$$\dot{x} = Ax + B(g + \eta + \phi)$$

Figure 1: The observer estimates the state to be $\hat{x}$ in presence of thermal noise η and photo-diode noise v. The actual state is x. By a choice of the observer gain L the error e in the state estimate goes to zero when the cantilever is freely oscillating. When the cantilever is subjected to the sample force ϕ, its dynamics is altered whereas the observer dynamics remains the same. This is registered as a nonzero value in the error e.

assume a new value (equivalent to resetting to a different initial condition). This is satisfied in most typical operations because in the dynamic mode, the time spent by the tip under the sample's influence is negligible compared to the time it spends outside the sample's influence [12]. The assumption is also corroborated by experimental results provided later.

The error dynamics (characterized in the Laplace domain from Equation(3)) is given by,

$$e(s) = \frac{\eta(s) + (s^2 + \frac{\omega_0}{Q}s + \omega_0^2)v(s) + (s + \frac{\omega_0}{Q})\nu_1 + \nu_2}{s^2 + (\frac{\omega_0}{Q} + l_1)s + (\omega_0^2 + l_2 + \frac{\omega_0}{Q}l_1)}, \qquad (4)$$

where $[\nu_1, \nu_2]^T$ is the initial condition reset due to change in tip-sample interaction and $L = [l_1 \ l_2]^T$ is the gain of the observer that must satisfy the stability criterion: $(\frac{\omega_0}{Q} + l_1) > 0$ and $(\omega_0^2 + \frac{\omega_0}{Q}l_1 + l_2) > 0$.

From Equation (4) it can be seen that the tracking bandwidth is characterized by,

$$B \propto \frac{\omega_0}{Q} + l_1. \qquad (5)$$

Since the choice of the gain term l_1 is independent of the quality factor Q, the tracking bandwidth of the observer is effectively decoupled from Q.

From Equation(3), it can be shown that the signal to noise ratio in the error signal e due to thermal noise,

$$SNR_{th} \propto (\omega_0^2 + \frac{\omega_0 l_1}{Q} + l_2)\nu_1^2 + (\frac{\omega_0}{Q}\nu_1 + \nu_2)^2. \quad (6)$$

Thus by increasing values of l_1 and l_2, SNR_{th} and bandwidth B increase. The signal to noise ratio due to photo-diode noise is given by,

$$SNR_v = \frac{\int_0^B \frac{\omega^2\nu_1^2 + (\frac{\omega}{Q}\nu_1 + \nu_2)^2}{(\omega^2 - \omega_0^2 - l_2 - \frac{\omega_0 l_1}{Q})^2 + \omega^2(\frac{\omega_0}{Q} + l_1)^2}d\omega}{\int_0^B \frac{[(\omega^2 - \omega_0^2)^2 + (\frac{\omega\omega_0}{Q})^2]R}{(\omega^2 - \omega_0^2 - l_2 - \frac{\omega_0 l_1}{Q})^2 + \omega^2(\frac{\omega_0}{Q} + l_1)^2}d\omega}, \qquad (7)$$

where the photo-diode noise v is assumed to be white with noise power equal to R. It can be seen that SNR_v

decreases with increasing values of l_1 and l_2. Therefore the bandwidth constraint in the detection scheme is mainly imposed by the photo-diode noise. It is evident that a desired tradeoff between signal to noise ratio and bandwidth can be obtained by an appropriate choice of l_1 and l_2 that is independent of Q. This provides considerable flexibility when compared to the existing steady state methods where Q determines the bandwidth. For typical cantilever parameters and ambient conditions, the Kalman design yields a bandwidth $B \gg \omega_0/Q$ and the innovation process carrying the signature of tip-sample interaction has a zero mean and white component. Note that the observer gain l_1 can be chosen large enough so that the cantilever state is tracked within a couple of cycles of the dither forcing. This shows that the optimal bandwidth is primarily dictated by the resonant frequency ω_0 of the cantilever.

2.4 Hypothesis-testing based detection

The sample detection problem is formulated by considering a discretized model of the cantilever (given in Equation(1)) and the impact model for the tip-sample interaction, as described by,

$$
\begin{aligned}
x(i+1) &= Fx(i) + Gg(i) + G_1\eta(i) + \delta_{\theta,i+1}\nu, \\
y(i) &= Hx(i) + v(i); \ i \geq 0,
\end{aligned} \tag{8}
$$

where $\delta_{i,j}$ denotes the dirac delta function, θ denotes the time instant when the tip-sample impact occurs and ν signifies the magnitude of the impact. It is assumed that the thermal noise and the photodiode noise are white and uncorrelated. As indicated before, given this statistics, the optimal observer is a Kalman observer [11]. With an observer having gain K (the discrete-time equivalent of L), the innovation sequence $e(i)$ is given by [13],

$$
e(i) = \Upsilon(i;\theta)\,\nu + e^w(i), \tag{9}
$$

where $\Upsilon(i;\theta) = [H; \ H(F-KH); \ \cdots \ H(F-KH)^{i-\theta}]$ and $e^w(i)$ is the innovation sequence when $\nu = 0$. $\Upsilon(i;\theta)$ is a dynamic profile with unknown arrival time θ. When there is no change in tip-sample interaction (i.e. $\nu = 0$) the innovation sequence has zero mean and is white [13]. When there is a change in tip-sample interaction the innovation sequence becomes nonwhite and is sum of a zero-mean and white sequence $e^w(i)$ and $\Upsilon(i;\theta)\nu$ with θ and ν unknown.

Thus the objective of detecting a change in tip-sample interaction is translated to the task of detecting the dynamic profile $\Upsilon(i;\theta)\nu$ in a zero mean white sequence. This problem can be cast in hypothesis testing framework as,

$$
\begin{aligned}
H_0 &: \ Y_i = e^w(i), \ i = 1, 2, \ldots, n; \\
H_1 &: \ Y_i = \Upsilon(i;\theta)\,\nu + e^w(i), \ i = 1, 2, \ldots, n;
\end{aligned}
$$

where the observed data $Y_i = e(i)$ is the innovation sequence. The dynamic profile is detected by using a likelihood ratio test [13], [14] and a decision signal is obtained.

3 EXPERIMENTAL RESULTS

The advantages of the new methodology are well demonstrated in the following experiment performed using a Digital Instruments multi-mode AFM. A cantilever with first resonance frequency $f_0 = 70.1 \ kHz$ and quality factor $Q = 180$ was forced at f_0 to an amplitude of $80 \ nm$. A $0.5 \ V$ pulse train having $1 \ ms$ time period and duty cycle of 50% was applied to the piezo-scanner holding an HOPG (Highly Oriented Pyrolytic Graphite) sample. Each pulse applied to the piezo generated a sample profile (see Figure 2) having four peaks separated by approximately $100 \ \mu s$. The sample was brought close to the cantilever so that the tip would interact with the four peaks in the sample profile. Since the settling time of the cantilever is in the order of $Q/f_0 \approx 2.57 \ ms$, the cantilever was interacting with the peaks in the sample profile during its transient state and it never recovered the steady state during the experiment. From the amplitude profile of the deflection signal (see Figure 2), it is not possible to detect the four peaks in the sample profile. Since the steady state data based signals are slowly varying, it can be argued that corresponding methods fail to detect the small time-scale (high bandwidth content) profiles in the sample that may arise during scanning.

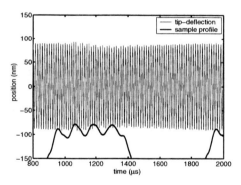

Figure 2: Cantilever-tip deflection data with respect to approximate sample position is shown. Note that from the amplitude profile of the deflection signal the four peaks in the sample profile are not discernible.

Observe that the peaks are easily discernible in the innovation sequence (see Figure 3(b)). When the cantilever is not interacting with the sample (until $\approx 950 \ \mu s$) the innovation sequence has zero mean and is white. As soon as it encounters the first peak in the sample profile ($\approx 950 \ \mu s$) the innovation sequence becomes nonwhite and dynamic profile is detected. Between the first and the second peaks in the sample profile, the innovation sequence recovers the zero mean and white nature until the second peak appears ($\approx 1050 \ \mu s$). Overlapping dynamic profiles may appear in the innovation sequence ($\approx 1050 \ \mu s$) due to multiple hits with the sample in consecutive cycles. The likelihood ratio (see Figure 3(c)) increases significantly when the dynamic profile is present in the innovation sequence. The peaks are detected within 2 cycles. The overlapping dynamic profiles are detected as a single event (second peak) as shown by the detection signal (see Figure 3(e)). Note that the cantilever has not reached steady state and is in transient during the entire experiment.

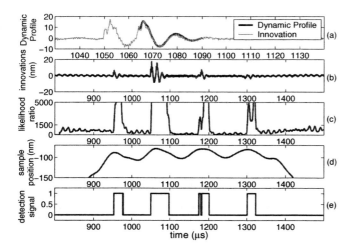

Figure 3: (a) The dynamic profile buried in innovation sequence, (b) the innovation sequence, (c) likelihood ratio, (d) sample profile and (e) the detection signal are shown when the cantilever is interacting with the sample during its transient state. The four peaks are detected by the appearance of dynamic profile in the innovation sequence and it being captured by likelihood ratio as shown by the detection signal.

The dynamic profile (see Figure 3(a)) persists for approximately $25\mu s$ ($\approx 2/f_0$ seconds) which is captured within a data window of size M=128 with a 5 MHz sampling. The dynamic profile is detected in $23.94\mu s$ ($\approx 2/f_0$ seconds) of its inception (with threshold ϵ=1681.3 corresponding to a false alarm rate of $P_F = 0.1\%$ and detection probability $P_D = 90\%$ for a minimum step size to detect ν=0.25 nm). To ensure at least one hit with cantilever the sample has to be present for more than 1 cycle ($1/f_0$ seconds) of the cantilever oscillation. A good estimate of the bandwidth is $f_0/4$ Hz=17.5 kHz. The experiment demonstrates a detection-bandwidth $\approx$ 10 kHz. This is considerably large as compared to the cantilever's natural bandwidth as determined by $f_0/Q \approx$ 390 Hz. Note that high quality factor of the cantilever does not limit the bandwidth in the proposed scheme. It is evident from the innovation sequence and the likelihood ratio that the cantilever interactions with the peaks in the sample profile are not uniform. However by feeding back the demodulated amplitude signal to the sample positioner and the cantilever this issue can be effectively addressed.

This research is supported by NSF Grant NSF ECS-0330224 to Prof. Murti V. Salapaka.

REFERENCES

[1] G. Binnig, C.F. Quate, and C. Gerber. Atomic force microscope. *Physical Review Letters*, 56(9):930–933, 1986.

[2] J. H. Hoh, R. Lal, S. A. John, J. P. Revel, et al. Atomic force microscopy and dissection of gap junctions. *Science*, 253:1405–1408, 1991.

[3] S. Kasas, N. H. Thomson, B. L. Smith, H. G. Hansma, and others. Escherichia coli rna polymerase activity observed using atomic force microscopy. *Biochemistry*, 36(3):461–468, 1997.

[4] H. J. Mamin, R. P. Ried, B. D. Terris, and D. Rugar. High-density data storage based on the atomic force microscope. *Proceedings of the IEEE*, 87(6):1014–1027, June 1999.

[5] Wisendanger. *Scanning Probe Microscopy and Spectroscopy*. Cambridge University Press, 1994.

[6] T. Sulchek, R. Hseih, J. D. Adams, G. G. Yaralioglu, S. C. Minne, C. F. Quate, J. P. Cleveland, and D. M. Adderton. High-speed tapping mode imaging with active q control for atomic force microscopy. *Applied Physics Letters*, 76:1473–1475, 2000.

[7] Tomás Rodríguez and Ricardo García. Detailed analysis of forces influencing lateral resolution for q-control and tapping mode. *Applied Physics Letters*, 79:135–137, 2001.

[8] Rainer D. Jäggi, Alfredo Franco-Obregòn, Paul Studerus, and Klaus Ensslin. Detailed analysis of forces influencing lateral resolution for q-control and tapping mode. *Applied Physics Letters*, 79:135–137, 2001.

[9] M. V. Salapaka, H. S. Bergh, J. Lai, A. Majumdar, and E. McFarland. Multimode noise analysis of cantilevers for scanning probe microscopy. *Journal of Applied Physics*, 81(6):2480–2487, 1997.

[10] Chi-Tsong Chen. *Linear System Theory and Design*. Oxford University Press, 1999.

[11] Ali H. Sayed Thomas Kailath and Babak Hassibi. *Linear Estimation*. Prentice Hall, NJ, 2000.

[12] M. V. Salapaka, D. Chen, and J. P. Cleveland. Linearity of amplitude and phase in tapping-mode atomic force microscopy. *Physical Review B.*, 61, no. 2:pp. 1106–1115, Jan 2000.

[13] Alan S. Willsky and Harold L. Jones. A Generalized Likelihood Ratio Approach to the Detection and Estimation of Jumps in Linear Systems. *IEEE Transactions on Automatic Control, pp. 108-112*, Feb. 1976.

[14] Steven M. Kay. *Fundamentals of Statistical Signal Processing, Detection Theory, vol.II*. Prentice Hall, PTR, 1993.

Nano-Scale Effects in GaN-based Light-Emitting Diodes

Joachim Piprek and Shuji Nakamura

College of Engineering, University of California, Santa Barbara, CA 93106-9560, USA, piprek@ieee.org

ABSTRACT

We here investigate the effects of built-in polarization on the properties of non-symmetric InGaN quantum wells. The impact of these nano-scale effects on the performance of blue light emitting diodes is analyzed utilizing advanced numerical simulation.

Keywords: wurtzite semiconductors, GaN, light-emitting diodes, built-in polarization, InGaN quantum well, nano-photonics

1 INTRODUCTION

Breakthrough developments in nitride semiconductor technology have triggered a worldwide surge in research on GaN-based light emitting devices. Blue and green nitride light-emitting diodes (LEDs), for instance, enable more efficient and less costly full-color outdoor displays. Compact ultraviolet AlGaN/GaN light sources are currently under development for applications in solid-state lighting, short-range communication, and bio-chemical detection. GaN-based laser diodes have great potential in a number of applications such as optical data storage and printing.

Light is generated in these devices by spontaneous or stimulated recombination of quantum well electron-hole pairs. However, the nano-scale physics of recombination mechanisms in GaN-based quantum wells is still not fully understood. Polarization, quantum well non-uniformities, and exciton effects may have a major influence [1].

2 POLARIZATION

In c-face nitride materials, both spontaneous and piezoelectric polarization were found to be much stronger than in other III-V compounds [2]. Different alloy compositions show different polarization and net charges remain at the alloy interfaces. For ternary nitride alloys grown on GaN, Fig. 1 gives the net interface charge density as well as the resulting electrostatic field within the alloy. The built-in polarization field significantly affects the properties of quantum wells. The wider the quantum well, the more separated the electrons and holes, and the smaller the optical gain and spontaneous emission. The transition energy is reduced by the built-in field, leading to a red-shift of the emission wavelength. However, with increasing carrier injection into the quantum well, charge screening is expected to reduce polarization field effects.

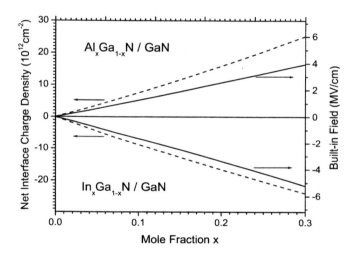

Fig. 1: Density of interface polarization charges (dashed) and resulting electrostatic field (solid) for ternary alloys grown on GaN [2].

3 DEVICE

As a practical example, we here investigate single-quantum-well blue light-emitting diodes as described in [3]. These devices exhibit an output power of 4.8 mW at 20 mA injection current and 3.1 V forward voltage. The low voltage results in a relatively high power efficiency of 7.7% (ratio of light power to electrical power). The external quantum efficiency is as high as 8.7% despite a large dislocation density. Fluctuations of the Indium composition within the InGaN quantum well are assumed to localize carriers in radiative centers and to prevent them from recombining nonradiatively at dislocations [1]. The device was grown by metal organic chemical vapor deposition (MOCVD) on c-face sapphire. Layer sequence and band diagram are shown in

Fig. 2. The 2-nm-thick strained $In_{0.2}Ga_{0.8}N$ quantum well is sandwiched between n-InGaN and p-AlGaN barriers. The p-barrier acts as blocking layer in order to reduce electron leakage from the quantum well into the p-doped regions.

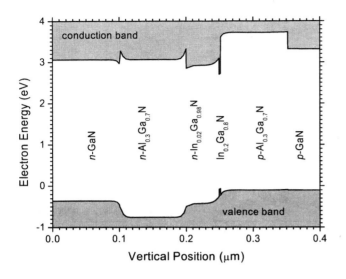

Fig. 2: Energy band diagram of our LED.

4 MODEL

In order to simulate nano-scale effects in the InGaN quantum well and their impact on the device performance, we here employ the simulation software APSYS which self-consistently includes drift and diffusion of electrons and holes, built-in polarization and thermionic emission at hetero-interfaces, as well as spontaneous and defect related recombination of carriers within the quantum well. The polarization charges are considered partially screened by charged interface defects so that only half of them contribute to the built-in polarization field [4]. Schrodinger and Poisson equations are solved iteratively in order to account for the quantum well deformation with changing device bias. Spontaneous emission of photons by electron-hole recombination within the quantum well is calculated using a free-carrier model and assuming non-parabolic valence bands. In the quasi three-dimensional simulation of the LED, current crowding and self-heating effects are taken into account. Further details of the model as well as a discussion of material parameters can be found elsewhere [5].

5 QUANTUM WELL

The effect of the net polarization charges on the quantum well is shown in Fig. 3 including screening by free carriers.

Without polarization, the internal electrical field of the pn-junction is positive and the holes accumulate at the p-side of the quantum well. With polarization, the electrical field is strongly negative within the quantum well and the holes accumulate at the n-side. In both cases, the peak of the electron distribution is near the center of the quantum well. Due to the more non-symmetric quantum barriers, there are less confined electrons without polarization. Built-in polarization allows for more quantum levels.

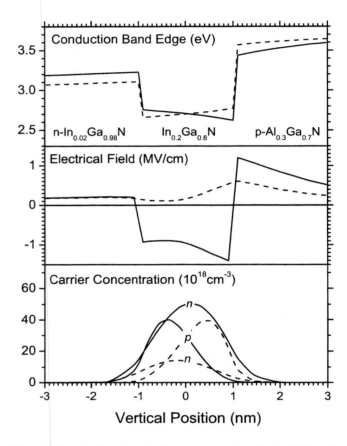

Fig. 3: Conduction band edge, internal electrical field, and carrier profiles with (solid) and without (dashed) built-in polarization for our non-symmetric LED quantum well.

6 EMISSION SPECTRUM

Calculated LED spectra exhibit a significant reduction of the peak emission intensity due to the polarization field (Fig. 4). These spectra represent the radiation from the entire device including current crowding and self-heating. The maximum temperature increase of about 80 K is observed at the edge of the device where the current crowding is strongest. Lorentzian broadening with an intraband scattering time of 0.1 ps gives a full-width at half-maximum (FWHM) of 17 nm for the spectrum without polarization. Due to the additional confined quantum states, built-in polarization results in a broader line shape with FWHM = 26 nm. This result is close to the measured line broadening of 25 nm [3]. Uniform current injection at room temperature gives smaller FWHM values in both cases [5]. Built-in polarization red-shifts the emission peak by only 3 nm due to the opposite tilt of the quantum well in both cases (cf. Fig. 3).

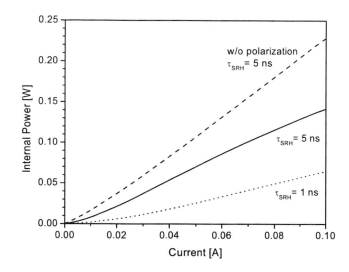

Fig. 5: Calculated internal LED light power vs. injection current for a 300μm x 300μm device (with polarization: solid and dotted lines, without polarization: dashed line, τ_{SRH} gives the non-radiative carrier lifetime within the quantum well).

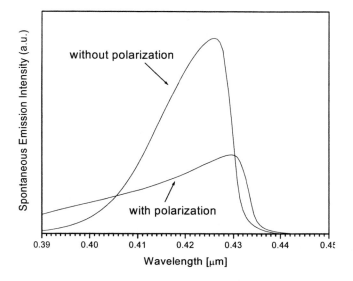

Fig. 4: Emission spectrum calculated with and without polarization charges.

8 CONCLUSION

Utilizing self-consistent numerical simulation, we were able to demonstrate the strong effect of built-in quantum well polarization on LED characteristics.

7 LIGHT POWER

The light power generated inside the LED is plotted in Fig. 5 as function of the injection current. It is strongly affected by the non-radiative carrier lifetime τ_{SRH} inside the quantum well. This number is unknown for our device but it is typically considered to be on the order of a few nanoseconds. We therefore show the results for two different lifetimes (solid and dotted curves), illustrating the strong impact of this parameter. With τ_{SRH} = 5ns, the maximum internal quantum efficiency is η_{int} = 48% while τ_{SRH} = 1ns results in η_{int} = 22%. Without built-in polarization, the LED would emit a significantly higher light power with η_{int} = 78% (dashed line in Fig. 5) due to the better overlap of electron and hole wave function within the quantum well (Fig. 3) and the correspondingly higher spontaneous recombination rate (Fig. 4).

The measured light emission is much lower due to inefficient light extraction from the LED. Most of the photons emitted from the quantum well are unable to leave the device due to total internal reflection. For τ_{SRH} = 5ns, the measured external efficiency of 8.7% gives a light extraction efficiency of η_{opt} = 18% for this example device.

REFERENCES

[1] S. Nakamura and S. F. Chichibu, *Introduction to Nitride Semiconductor Blue Lasers and Light-Emitting Diodes,* Taylor & Francis, London, 2000.

[2] F. Bernardini, V. Fiorentini, and D. Vanderbilt, Phys. Rev. B 56, R10024 (1997); Phys. Rev. B 63, 193201 (2001).

[3] S. Nakamura, M. Senoh, N. Iwasa, and S. Nagahama, Appl. Phys. Lett., vol. 67, pp. 1868–1870 (1995).

[4] C. A. Flory and G. Hasnain, IEEE J. Quantum Electronics 37, 244 (2001).

[5] J. Piprek, *Semiconductor Optoelectronic Devices – Iintroduction to Physics and Simulation,* Academic Press, San Diego, 2003.

A Minnow, An E. Coli and Ubiquinone:
The story of Rectified Brownian Motion

Ronald F. Fox

School of Physics, Georgia Institute of Technology
Atlanta, Georgia, 30332-0430
ron.fox@physics.gatech.edu

ABSTRACT

Rectified Brownian motion provides a mechanistic alternative to the *power stroke* model for the function of motor proteins and other enzyme complexes. ATP is the source of metabolic free energy for motor proteins such as kinesin and myosin V. In the power stroke models, the energy of ATP hydrolysis causes chemo-mechanical energy conversion although the precise manner by which this occurs has yet to be identified. In the rectified Brownian motion model, the energy released by ATP hydrolysis causes an irreversible conformational switch in the ATP binding protein that results in release of the motor protein head from the track along which it moves, and Brownian motion provides the power for the head to move to a new binding site. Asymmetric boundary conditions for this diffusive motion result in a directed motion on the average.

Keywords: rectified Brownian motion, power stroke, ubiquinone, kinesin, Langevin equation

1. THE CENTRAL PROBLEM

Motors proteins move along protein tracks and carry cargo from one cellular region to another. Kinesin moves along microtubule tracks and myosin V moves along actin tracks. How these proteins function is the central question. The mechanism may be much more general and also apply to rotary enzyme complexes, ATPases, polymerases, membrane transporters and other molecular systems. Two different mechanisms vie for recognition, the power stroke model and the rectified Brownian motion model. Andrew Huxley [1] introduced the rectified Brownian motion model many years ago but at a time when the detailed macromolecular structure of muscle fibers was not well determined. In the absence of an appreciation of the robustness of thermal energy at the macromolecular scale, the power stroke model is usually put forth and tends to dominate text book presentations [2]. The present author has addressed this dichotomy in the recent past and clearly favors the rectified Brownian motion mechanism [3,4]. The purpose of this paper is to make the case for rectified Brownian motion more widely known and to argue why it is a natural mechanism in the context of robust thermal energy. Moreover, this mechanism suggests how thermal energy can be constructively used in the design of nanotechnological devices.

2. THERMAL ENERGY

In order to appreciate the importance of thermal energy at the sub-cellular level, it helps to look at three examples that represent very different scales. These examples are a minnow, an E. Coli and ubiquinone.

The minnow is 16 cm long and has a cross-sectional diameter of 4 cm. It has a mass of 134 gm. Using its fins it can swim up to 100 cm/s. Throughout this paper we will assume that the ambient temperature is 25 °C. The viscosity of its environment is one centi-poise. The Reynolds number for the swimming motion is 80,000 which is very large. Normally this set of values would complete our description of the minnow in motion. However, for comparative purposes let us consider its thermal motion as well. The thermal velocity of the minnow's center of mass is 1.75×10^{-8} cm/s. This is ten orders of magnitude smaller than its swimming speed. It is a random motion that causes the center of mass to diffuse in addition to its secular, swimming motion. In one second the root-mean-square displacement caused by this diffusion is 3.7 nm. This is to be compared to the 100 cm it swims in one second. Thus, the thermal motion is entirely negligible for the minnow. The Reynolds number for this thermal motion is 1.4×10^{-6}, a very small value that will be characteristic of all examples to follow. The secular power expended by swimming is 5.96×10^{-4} W. This shows how efficiently the little fish can swim. Its thermal power is 5.7×10^{-23} W, 19 orders of magnitude smaller than the swimming power and showing in another way how insignificant the thermal energy considerations for the minnow really are.

An E. Coli is a bacterium that is two microns long with a cross-sectional diameter of one micron. It has a mass of 2×10^{-12} gm. Using its flagella, it too can swim. It is capable of "runs" of about one second duration, interspersed with "tumbles" of about 0.1 second duration. The runs have top speeds of 2×10^{-3} cm/s. The tumbles reorient the direction of motion in an effectively random way. The viscosity of its environment is 2.7 centi-poise. Because of the very small mass, the thermal velocity for the E. Coli is 0.14 cm/s, much larger than its secular, swimming speed. Its secular Reynolds number is 1.5×10^{-6} and its thermal Reynolds number is 10^{-3}, each of which is very small. What this means physically is that for the E. Coli inertia is of no importance and its motion is dominated by viscosity. This is graphically exhibited by the fact that if it is moving at top speed and the flagella are suddenly turned off then the E. Coli will come to a complete stop after moving a distance of only 1.3×10^{-10} cm. This remarkable result was noted by Howard Berg [5]. Clearly, the E. Coli world is dominated by thermal energy and viscosity. While the thermal velocity is nearly 100 times greater than the swimming velocity, it does not persist in a single direction for any appreciable length of time like in the case of a secular run. Instead it causes a diffusion of the center of mass of the E. Coli that amounts to a root-mean-square displacement of half a micron in one second. In that same one second the secular motion moves the E. Coli 20 microns. This example lends some credence to the idea of a power stroke in that it shows how the E. Coli is able to overcome a very robust thermal environment and achieve a secular motion in spite of all the thermal agitation. Moreover it can do this at the relatively low power of 1.23×10^{-17} W compared to the thermal power of 1.92×10^{-13} W.

Ubiquinone is a ubiquitous molecular species that is found in aerobic bacterial membranes and in the organelles, mitochondria and chloroplasts. It plays a central role as an intermediate in electron transport chains. These systems of membrane embedded protein complexes are made up of a predominately iron-sulfur protein complex and a predominately cytochrome complex that are coupled together by diffusive shuttling of ubiquinone between the two complexes. In bacteria, iron-sulfur proteins reduce oxidized ubiquinone near the surface of the membrane that is adjacent to the interior of the cell. Cytochromes oxidize the reduced ubiquinone near the surface of the membrane that is adjacent to the external environment. Since ubiquinone oxidation and reduction involves a pair of electrons and a pair of protons, the location of the reduction near the inside and the location of the oxidation near the outside results in protons being translocated from inside the cell to outside the cell while electrons are passed along the electron transport chain. This process is a paradigm for rectified Brownian motion. Ubiquinone has a molecular weight of 862 when oxidized (864 when reduced). This makes its mass 1.44×10^{-21} gm. This is nine orders of magnitude smaller than the mass of E. Coli and 23 orders of magnitude smaller than the mass of the minnow. It moves in the lipid interior of the membrane where the viscosity is 25 centi-poise. It has a spherical conformation with a radius of 0.75 nm. Its motion inside the membrane lipid interior can be described by the Langevin equation, a stochastic differential equation [6]. It has no fins or flagella and therefore is unable to "swim" through the membrane. Instead the only source of motion available to ubiquinone is thermal motion. The relaxation time for the Langevin equation for ubiquinone is 4×10^{-15} s. This is such a short time that the Langevin description may be replaced by an equivalent diffusion process for all times long compared with this short relaxation time. The ubiquinone diffusion constant is 1.2×10^{-7} cm^2/s. For a membrane with a thickness of 8 nm, the expected time for ubiquinone to cross the membrane thickness is 2.8×10^{-6} s. This is nine orders of magnitudes longer than the relaxation time. Thus, the ubiquinone motion in the membrane is in the extreme limit of diffusion for the Langevin equation. Therefore, in this case, rectified Brownian motion is described by diffusion with asymmetric boundary conditions.

The asymmetry of the boundary conditions for diffusion is the key to rectified Brownian motion. This situation should not be confused with "Brownian ratchets" in which an asymmetric saw-tooth potential that is usually oscillated plays a central role [7]. For ubiquinone the boundary conditions are produced by the non-equilibrium concentrations of electron donors on the one side and of electron acceptors on the other. Specifically, the reduced form of the electron donor is kept in excess over the oxidized form by metabolism, as is the oxidized form of the electron acceptor compared to its reduced form. As long as metabolism is operating, these disequilibria are maintained and asymmetric boundary conditions for ubiquinone diffusion function. The result is that on the average there is a non-zero flux of reduced ubiquinone (UQH$_2$) from the inside surface of the membrane to the outside surface, and a non-zero flux of oxidized ubiquinone (UQ) from the outside surface back to the inside surface. These two fluxes create a ubiquinone cycle. The physical motion of the ubiquinone molecule is provided by the thermal agitation. The thermal power associated with this motion is 3 micro-Watts. While this may at first seem small, compare it with the secular power of the minnow.

This description can be made quantitative by explicitly using the diffusion equation. Let $f(x,t)$ denote the probability density at time t for reduced UQH$_2$ and let $g(x,t)$ denote the probability density at time t for oxidized UQ. The inside surface of the membrane is located at $x = 0$ and the outside surface is located at $x = d$. In steady state, it is expected that the probability density for UQH$_2$ at the inside surface, denoted by Q_{in}^r, and the probability density for UQH$_2$ at the outside surface, denoted by Q_{out}^r, satisfy $Q_{in}^r > Q_{out}^r$, because UQH$_2$ is produced at the inside surface and is converted at the outside surface. Similarly, in steady state it is expected that the probability density for UQ at the inside surface, denoted by Q_{in}^o, and the probability density for UQ at the outside surface, denoted by Q_{out}^o, satisfy $Q_{out}^o > Q_{in}^o$, because UQ is produced at the outside surface and is converted at the inside surface. The reduced species satisfies the diffusion equation

$$\frac{\partial}{\partial t} f(x,t) = D \frac{\partial^2}{\partial x^2} f(x,t)$$

with the boundary conditions at steady state given by

$$f_{SS}(0) = Q_{in}^r \quad and \quad f_{SS}(d) = Q_{out}^r$$

where the subscript SS denotes the steady state values. Similarly, the oxidized species satisfies the diffusion equation

$$\frac{\partial}{\partial t} g(x,t) = D \frac{\partial^2}{\partial x^2} g(x,t)$$

with the boundary conditions at steady state given by

$$g_{SS}(0) = Q_{in}^o \quad and \quad g_{SS}(d) = Q_{out}^o$$

These equations are easily solved and have the steady state solutions

$$f_{SS}(x) = f_{SS}(0) - \frac{x}{d}\left(f_{SS}(0) - f_{SS}(d)\right)$$

$$g_{SS}(x) = g_{SS}(0) - \frac{x}{d}\left(g_{SS}(0) - g_{SS}(d)\right)$$

The probability currents, or fluxes, are defined by

$$-D\frac{\partial}{\partial x} f_{SS}(x) = \frac{D}{d}\left(f_{SS}(0) - f_{SS}(d)\right) > 0$$

$$-D\frac{\partial}{\partial x} g_{SS}(x) = \frac{D}{d}\left(g_{SS}(0) - g_{SS}(d)\right) < 0$$

wherein the left-hand sides define the fluxes in the manner that is standard for diffusion, and the right-hand sides are the results for the particular steady state solutions given above. The inequalities result from the boundary conditions. The meaning of these fluxes is simple, the reduced species goes from 0 to d and the oxidized species goes from d to 0, thereby creating the ubiquinone cycle.

3. MOTOR PROTEINS

The preceding considerations set the stage for a discussion of the function of motor proteins. In this presentation, the focus will be on kinesin, a motor protein that moves along microtubule tracks. Kinesin is a rather large molecule with a molecular weight around 500-600 kD [8]. It is made up of two heavy chains and two light chains. The heavy chain contains the head, i.e. the motor unit, that is comprised of about 340-350 amino acid residues. This region has a rough size of 7.5 nm x 4.5 nm x 4.5 nm. The head makes direct contact with the microtubule. The bulk of the heavy chain, called the neck, is a dimerized alpha-helical coiled coil. The total length of the complete kinesin is around 100 nm. The light chains are associated with the end of the molecule that binds the load. All of the catalytic activity is in the heavy chains. The ATP binding and hydrolysis takes place on the catalytic core of the heads. The neck is attached to the catalytic cores by segments of 15 amino acids called neck linkers.

The motion of kinesin is processive, i.e. many sequential steps occur before kinesin is completely released from the microtubule track. In one cycle ATP is bound, hydrolyzed and released while the trailing head detaches from the microtubule, moves forward to the next binding site and reattaches. The issue here is whether ATP powers a *power stroke* or whether ATP facilitates *rectified Brownian motion*. In the former view, the energy released by hydrolysis of the γ-phosphate of ATP causes a conformation change in the kinesin that results in the movement of one head by a distance of 16 nm in order for this head to move from one binding site to the next. In the latter view, ATP hydrolysis causes an irreversible switch of the bound head to the unbound state, and the movement of the head from one binding site to the next is caused by Brownian motion. No mechanism for the power stroke has been supported by experiment to date.

From the rectified Brownian motion point of view, it is important to note that the motion of the detached head is in the diffusion regime for reasons that parallel the situation for ubiquinone. Kinesin heads are larger than ubiquinone but only by about 40-fold, and they are much smaller than an E. Coli. The diffusion time for a kinesin head to move 16 nm, in the absence of a load, is 1.7 x 10^{-6} s. This is much faster than the chemical reaction steps for ATP hydrolysis and release, that are longer than milliseconds in duration. Thus, the head diffusion is far and away the fast step in the process [4]. In fact, phosphate release is the slowest step according to biochemical assays [9].

The motion of a kinesin head can be modeled as a mean first passage time process with a reflecting boundary condition at one end and a absorbing boundary condition at the other end. These boundary conditions reflect detachment and rebinding respectively. The Langevin equation for this process is reduced to a Langevin equation for the equivalent diffusion process and the backward Fokker-Planck equation for this process is used to solve for the first passage time distribution [4]. The mean first passage time as a function of load can be determined analytically but the first passage time distribution itself must be obtained numerically [4]. For diffusion constant D, a distance d and a load c, the solution for the mean first passage time, T(d), is

$$T(d) = \frac{1}{c}\left\{\frac{D}{c}\left(\exp\left[\frac{c}{D}d\right] - 1\right) - d\right\}$$

This formula has an exponential character that implies that while the value for no load is 1.7 x 10^{-6} s, the value for a load of 5 pN is 4.3 s, more than a million times longer. Indeed, 5-6 pN is the size of the stall load according to measurements made using laser tweezers [10]. In calculating detailed load-velocity profiles the head diffusion step requires use of the entire first passage time distribution rather than a mere insertion of the mean first passage time into the rate formula. This is a consequence of the broad, exponential tail in the mean first passage time distribution function [4]. The results obtained agree quantitatively with experimental results and the use of the entire distribution moves the stall force from 5 pN to 6 pN for ATP concentrations of 2.0 mM. Changes in the ATP concentration change the load-velocity profile by altering the maximum velocity and the stall force, so long as these changes are below the saturation concentration of roughly 2.0 mM. For example, an ATP concentration of 8 μM gives a maximum velocity of only 80 nm/s and a stall force of 5 pN whereas for a 2.0 mM concetration the maximum velocity is 700 nm/s and the stall force is 6 pN. That these results are quantitatively consistent with the measured values is strong support for the model.

Evolutionary evidence also exists for the rectified Brownian motion model. In the rectified Brownian motion model ATP binding and hydrolysis cause the kinesin head to switch from microtubule binding to release. In the power stroke model this conformation change also results in the complete translocation of the head from the old binding site to the new one. In the rectified Brownian motion model diffusion of the head creates the translocation. Is there a precedent for nucleotide stimulated switching activity? The answer is yes. The G-proteins, that use GTP in place of ATP, are such switches. These proteins are central to the second messenger mechanism of hormone action [2]. Study of the amino acid sequences of the nucleotide binding sites for G-proteins, kinesin heads and myosins has revealed strong similarities. It has been proposed [8] that there was an ancestor protein that gave rise to G-protein switches on the one hand and to kinesin heads and myosin heads on the other. If so, this strongly suggests that kinesin and myosin heads function as switches too. The importance of nucleotide hydrolysis in these systems is that it ensures irreversibility of the switch rather than being a source of energy for a power stroke. The energy require to move a kinesin head is supplied by Brownian motion, and the directed, processive motion of kinesin results from asymmetric boundary conditions for the head diffusion.

REFERENCES

[1] A. F. Huxley, Prog. Biophys. Chem. **7**, 255, (1957).
[2] D. Voet and J. G. Voet, *Biochemistry* (Wiley, New York, 1995).
[3] R. F. Fox, Phys. Rev. E **57**, 2177 (1998).
[4] R. F. Fox and M. H. Choi, Phys. Rev. E 63, 051901 (2001)
[5] H. C. Berg, *Random Walks in Biology*, (Princeton University Press, Princeton, N. J., 1993).
[6] C. W. Gardiner, *Handbook of Stochastic Methods* (Springer-Verlag, Berlin, 1983).,
[7] R. D. Astumian, Science **276**, 917 (1997).
[8] R. D. Vale and R. A. Milligan, Science **288**, 88 (2000).
[9] S. P. Gilbert, M. L. Moyer and K. A. Johnson, Biochemistry **37**, 792 (1998); M. L. Moyer, S. P. Gilbert and K. A. Johnson, *ibid.* **37**, 800 (1998).
[10] K. Svoboda and S. Block, Cell **77**, 773 (1994).

Multiscale (Nano-to-Micro) Design of Integrated Nanobio Systems

J. W. Jenkins[1], M. Hagan[2], A. Chakraborty[2], S. Sundaram[1]

[1]CFD Research Corporation, 215 Wynn Drive, Huntsville, AL 35805
[2]Department of Chemical Engineering, 209 Gilman Hall, UC Berkeley, CA 94720

ABSTRACT

The design of devices that incorporate both nanoscale and continuum scale physics is challenging due to the nature of simulation techniques required for each scale. This paper presents a hierarchical approach where micro/mesoscopic models of molecular phenomena are coupled to continuum device descriptions. This method is developed and demonstrated with reference to DNA hybridization on a microcantilever platform. A macroscopic (Langmuir) model of DNA hybridization is coupled to a mesoscopic description that captures the effect of critical parameters such as salt concentration and graft density. Bridging of the two models was achieved by way of a stochastic Master Equation which was transformed into a Fokker-Planck equivalent that can then be solved in a PDE framework.

Keywords: Microfluidics, nanoscale modeling, DNA hybridization.

1 INTRODUCTION

Biological systems can often be thought of as nanoscale systems that leverage molecular interactions to perform specific tasks. Integrated nano-bio systems have emerged as strong candidates for single molecule level detection, genomic sequencing, and the harnessing of naturally occurring biomotors [1]. Research efforts directed at the problem of integrating nanotechnology and biology to form integrated nano-bio systems is rapidly becoming a priority.

In this context, there is a growing need for novel design tools that can incorporate sufficiently detailed models of the nanoscale molecular phenomena with overarching transport and related effects in a microfluidic devices, in a computationally feasible framework. Currently no models are available that can perform this task. This void is sought to be filled by the proposed research, whose innovative aspects include

- Development of a novel multi-scale simulation tool for integrated nanosystem design, analysis and optimization
- Multi-tiered modeling approach consisting of (a) microscopic/mesoscopic models with stochastic nature and (b) device scale, deterministic continuum models

Our overall objective is to develop a generalized, multi-scale, multiphysics CFD (continuum)-based design software where nanoscale effects of biosystems are accurately, efficiently and seamlessly integrated with transport-based models carrying device-level information

1.1 DNA Hybridization on Microcantilevers

The general techniques applicable to multi-scale modeling of nano-bio systems is developed and demonstrated in the context of a sample problem: the design and characterization of a DNA sensing microcantilever system. Recent experiments show that the adsorption of biomolecules on one surface of a microcantilever generates surface stresses that cause the cantilever to deflect [2]. These experiments are performed by immobilizing a group of probe molecules on one side of the microcantilever beam. A solution of target molecules that can bind to the immobilized probes is then introduced. The experiments reveal that when the target binds to the probe a surface stress is generated and the tip of the microcantilever is deflected. The deflection can be monitored using an optical technique. One of the most interesting result from these experiments is that the identity and concentration of the target molecule can be related to the deflection of the microcantilever.

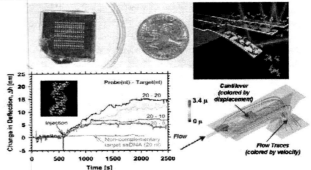

Figure 1: Model of a chemomechanical cantilever-based sensor. The figure shows several aspects of the project; the experimental cantilever array, some experimental data from DNA hybridization, and simulation results to date [2].

As shown in Figure 1, the microcantilevers can be created in an array with a different probe molecule immobilized on the target molecules. A fluid sample can then be screened for various target molecules based on the deflection of each cantilever. The ultimate goal is to create devices based on this technology that can be used in medical diagnostics and screening. The design of medical devices based on microcantilevers poses a challenge because of the scales (molecular to continuum) involved in the problem. Design of devices containing integrated nano-bio components can benefit from the help of accurate simulation tools in much the same way as the design of microfluidic devices have benefited. Currently a large stumbling block in the development of simulation tools for integrated nano-bio devices is the lack of adequate simulation methods capable of handling nanoscale physics, device level physics, and the coupling of the two scales.

2 MACROSCALE DNA BINDING MODEL

In many cases the hybridization of DNA from solution onto a surface with a complimentary strand is modeled using Langmuir kinetics. Using this model the kinetics are modeled as a second order binding rate with a first order debinding rate.

$$\frac{dC_{PT}}{dt} = k_a C_P^{Bulk}\left(\sigma - C_{PT}\right) - k_d C_{PT} \qquad (1)$$

where C_{PT} is the surface concentration of hybridized probe/target pairs, C_P^{Bulk} is the concentration of probe DNA in solution, σ is the graft density of probes on the surface, and the rate constants are given by k_a and k_d. This type of a rate equation can be easily incorporated into a CFD modeling framework, making it an attractive choice for use in hybridization modeling. Using this model, quantitative estimates of DNA hybridization in the microcantilever system have been obtained and reported elsewhere [2]. Figure 2 illustrates an example simulation of DNA hybridization onto a paddle cantilever.

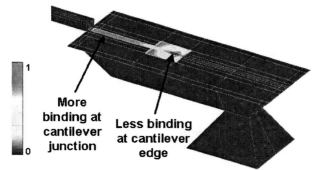

More binding at cantilever junction

Less binding at cantilever edge

Simulated DNA Binding On Rigid μCantilever under Flow Conditions (3D)

Figure 2: 3D model of the binding of DNA to a paddle type cantilever. The legend is a normalized amount of binding.

However this macroscopic model has serious shortcomings such as the fact that it does not allow for the variation in binding rate with salt concentration, graft density, and DNA chain length. All of these parameters are known to have a strong impact on the hybridization kinetics. A microscopic models that more accurately describes the functional dependence of the hybridization rate on these fundamental parameters is needed.

3 MICROSCALE DNA BINDING MODEL

We consider the system of single stranded DNA hybridizing with a complimentary DNA strand immobilized

on the microcantilever surface (termed the "brush" layer). This system can be modeled by appealing to methods used for characterizing surface immobilized polyelectrolyte polymer systems. A brief overview of the theory given below (more details may be found in [3]). The main goal of the polyelectrolyte theory is to predict the free volume distribution of the monomers in the direction normal to the immobilized surface. The free volume distribution arises due to the conformations taken on by the surface immobilized polymers. This theory is valid under the conditions of sufficiently high immobilized probe density (>0.01 polymers/nm^2).

A description of the effective binding rate begins with the calculation of electrostatic potential within the DNA brush layer. In order to obtain the electrostatic potential we will need to solve the Poisson equation for the system. The solution to the Poisson equation can be broken into two contributions, one for inside of the brush layer and one for outside of the brush layer in the bulk solution. These equations are given below in equations 2 and 3. In the equations below, y is the direction normal to the surface.

$$\frac{d^2\Psi^i}{dy^2} = -\frac{4\pi}{\varepsilon}\left[-2C_{salt}Ze\sinh\left(\frac{Ze}{kT}\Psi^i\right) + \frac{fZ_s e}{v_s}\Phi(y)\right] (2)$$

$$\frac{d^2\Psi^o}{dy^2} = -\frac{4\pi}{\varepsilon}\left[-2C_{salt}Ze\sinh\left(\frac{Ze}{kT}\Psi^o\right)\right] \qquad (3)$$

Where the subscripts "i" and "o" denote the electrostatic potential inside and outside of the brush layer. Also, C_{salt} is the electrolyte concentration, v_s is the volume of a polymer segment. The potentials are subject to the following continuity boundary conditions at the edge of the brush layer (denoted by y*):

$$\Psi^i\left(y^*\right) = \Psi^o\left(y^*\right)$$

$$\frac{d\Psi^i\left(y^*\right)}{dy} = \frac{d\Psi^o\left(y^*\right)}{dy} \qquad (4)$$

$$\Psi^o\left(\infty\right) \to 0$$

In order to obtain an equation for the free volume distribution, $\Phi(y)$, we first need to define an expression for the total potential energy for the system of immobilized polyelectrolyte polymer strands. The reader is referred to reference [3] for details. From the potential energy expression, a differential equation for the free volume distribution can be derived and is given below.

$$\frac{d\Phi}{dy} = \frac{r\left(\frac{d\Psi^i}{dx}\right) - 2\beta y}{\left(1-\Phi\right)^{-1} - 2\chi} \qquad (5)$$

Where r is a dimensionless number for the average segment charge, and β is a potential energy coupling parameter. The other boundary conditions for equations 2-3 and 5 are given below:

$$\Phi\left(y^*\right) = 0$$

$$\frac{d\Psi^i}{dx}(0) = -\frac{\kappa q_{surf}}{2ZeC_{salt}} \qquad (6)$$

$$\frac{d\Psi^i}{dx}\left(y^*\right) = -2\sinh\left(\frac{Ze}{2kT}\Psi^i\left(y^*\right)\right)$$

Where y^* is the ultimate brush height, κ is the Debye length, q_{surf} is the surface charge distribution. The first condition is to ensure that the volume fraction distribution goes to zero at y^*. The second condition is to account for the affect of a charged surface on the electrostatic potential. The final condition is to ensure that the value and derivative of the electrostatic potential match at y^*.

The main problem with solving equations this system of equations subject to 4 and 6 is that y^* is not known ahead of time, and must be determined by a constraint equation. The constraint equation is a balance equation on the number of segments contained in the brush volume given below in equation 7.

$$\sigma N v_s = \left(\frac{1}{\kappa}\right)\int_0^{y^*} \Phi(t)dt \qquad (7)$$

Where N is the number of segments in the polymer chain, σ is the density of chains (chains/nm^2), and the rest of the symbols have already been defined.

The solution to the above equations gives the free volume distribution, which can be used to compute the effective molecular binding rate in the brush layer, as has been shown in [4],

$$k_{eff}(t) = \frac{k_a^{Bulk}\sqrt{L}}{\sigma C_{Bulk}}\int_0^{y^*} C_{Target}(y,t)\Phi(y,t)f_{max}dy \qquad (8)$$

where L is the total chain length, k_a^{Bulk} is the bulk hybridization rate, C_{Target} is the target molecule concentration, and f_{max} is the maximum binding fraction defined by physical constraints of the polymer molecules.

The above expression provides a microscopic model that captures the impact of parameters such as salt concentration, brush density etc. This detail can be directly communicated up to the continuum level by means of the Langmuir model described earlier in equation 1. However, this communication is better handled by a "bridging equation" based on a master equation description of the hybridization process. More details on the exact nature of this multi-scale information cascade can be found in [5].

4 MESOSCALE BINDING MODEL

In general the binding rate will be a function of time, since during the binding process DNA chains are being integrated into he brush layer changing the free volume distribution. The hybridization process inside of the brush layer can be described using a Master Equation such as.

$$\frac{dP(n,t)}{dt} = k_{eff}(n-1,t)P(n-1,t) + k_{rev}P(n+1,t)$$
$$- \left[k_{eff}(n,t) + k_{rev}\right]P(n,t) \qquad (9)$$

Here P denotes the probability of finding "n" DNA chains hybridized in the brush layer at time "t". In a well characterized limit [6], the Master Equation can be cast in a Fokker-Planck form to give the same probability function as output. The FPE is a partial differential equation with a continuous variable, unlike the Master Equation which describes probability in discrete states. The Fokker-Planck Equation offers a unique advantage of being based on partial differential equation, which facilitates integration with the continuum approach. At CFDRC we have successfully coupled a FPE-based description of biomolecular events with a continuum-based convective-diffusive-reactive treatment of biosystems (CFD-ACE+) [5]. Our main goal is to be able to simulate nano-resolved, biomolecular reactions in complex, spatially inhomogeneous systems with convective-diffusive transport.

The master equation (equation 9) can be converted into a Fokker-Planck Equation by taking the immobilized area, A, as the largeness parameter. The FPE is used to obtain solutions for the probability distribution function P(n,t) in a continuous sense by describing the rate of change of P(n,t) in terms of a gradient of probability flux and source terms S. The flux term consists of a drift D_1 and diffusion coefficient D_2. In general the drift and diffusion coefficients are functions of the local conditions in the simulation and must be solved simultaneously with the momentum and mass transport equations. The implementation in the CFD-ACE+ software package breaks the flux of probability into a real space term and an auxiliary variable, making the probability space 4 dimensional as given in equation 10.

$$\frac{\partial f}{\partial t} + \nabla_x \cdot \left[uf - \nabla_x(Df)\right] +$$
$$\nabla_n \cdot \left[D_1 f - \nabla_n(D_2 f)\right] = S \qquad (10)$$

The drift and diffusion coefficients for the particular problem are relatively simple expressions given below.

$$D_s\left(\frac{n}{A},t\right) = k_{eff}\left(\frac{n}{A},t\right) + (-1)^n k_{rev} \qquad (11)$$

Equations 10 and 11 assumes that the effective binding rate is a function of the amount of DNA that has already hybridized. This will change the free volume distribution and ultimately change the binding rate. This equation also assumes that the debinding rate k_{rev} is constant, and does not vary with time.

The solution of the Fokker-Planck equation using the drift and diffusion coefficients given in Equation 14 can be solved using CFD-ACE+ FPE solver. The solutions of the FPE above give the distribution of bound DNA molecules as a function of time. The first moment of the distribution is the number of hybridized DNA molecules. The temporal change in the first moment is the local binding rate. Solutions to this equation will give binding rates that use the molecular binding rates taken from polyelectrolyte brush theory, and can be incorporated into the surface chemistry solver in CFD-ACE+.

5 EXAMPLE RESULTS

In the example calculations presented here, the ratio of the effective binding rate given in equation 8 to the bulk solution binding rate are given as a function of probe density. The rate ratio is given for two DNA chain lengths of 20 and 40 base pairs from a chain density of 0.01 to 0.15 chains/nm^2. The scaling of the rate ratio is seen to approach $\sigma^{-1.8}$ as the grafted probe density increases. This is in agreement with [4]. For this calculation, the background electrolyte concentration was held constant at 0.50 mol/L. This result is shown in Figure 3.

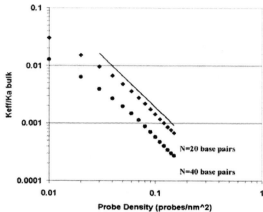

Figure 3: Functional dependence of the effective hybridization rate relative to the free solution binding rate with probe density. The calculations were performed for chains of length 20 and 40 base pairs. The line in the figure is to illustrate the $\sigma^{-1.8}$ scaling with probe density.

Figure 4 shows the equilibrium coverage of immobilized probe DNA molecules. The equilibrium coverage is the fraction of immobilized DNA molecules that have hybridized with a DNA chain out of solution. Figure 4 shows that the equilibrium coverage decreases dramatically with an increase in probe density, and that the equilibrium coverage is dependent on the chain length.

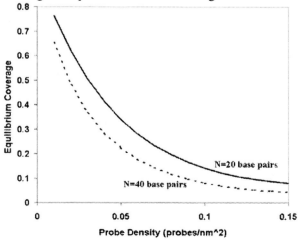

Figure 4: Functional dependence of the equilibrium coverage with probe density. The calculations were performed for chains of length 20 and 40 base pairs.

6 SUMMARY

Progress in our effort to incorporate nanoscale models into continuum level design simulations of microfluidics devices has been summarized. A macroscopic (Langmuir) model of DNA hybridization on a microcantilever was described and demonstrated. Following the approach of [4], a mesoscopic model that captures the effect of critical parameters (such as salt concentration and probe density) was adapted. The model showed consistent behavior in the variation of the effective binding rate with the probe density and in the calculated coverage. Information from the mesoscopic model was communicated to deterministic, continuum models by way of a stochastic Master Equation. The Master Equation was transformed into a Fokker-Planck equivalent that can then be solved in four-dimensional space (along with transport of DNA in the physical domain). Fully coupled, device level calculations with the multi-scale approach are underway and will be reported in a future communication.

ACKNOWLEDGEMENT

The authors gratefully acknowledge support for this work under DARPA SIMBIOSYS program (F30602-01-02-0540, Program Monitor: Dr. Anantha Krishnan).

REFERENCES

[1] Oster, Wang. Nature **396** (1998)
[2] M. Yue, H. Lin, D. Dedrick, S. Satyanarayana, A. Majumdar, A. Bedekar, J. Jenkins. JMEMS Accepted for publication [1] G. Wu, et. al., PNAS 98, 1560-64. 2001
[3] S. Misra, S. Varanasi, P. Varanasi. Macromolecules **22** (1989) 4173-4179
[4] M. Hagan, A. Chakraborty. J. Chem. Phys. (submitted 2003).
[5] J.W. Jenkins and S. Sundaram, ICCN 2003.
[6] van Kampen. Can. J. of Phys. **39** (1961) 551.

Stochastic Analysis of Particle -Pair Transport in Brownian Ratchet Device

Renata Retkute, James P Gleeson

Department of Applied Mathematics, University College Cork
Cork, Ireland, r.retkute@ucc.ie

ABSTRACT

The motion of particle pairs in a Brownian ratchet device is studied using Langevin simulations. A Lennard-Jones interaction between the particles is added to a standard spring-bead model. The effects of such interaction upon a recent model for Brownian motors [3] is investigated, with emphasis on the steady-state current.

Keywords: electrophoresis, Brownian ratchet, Langevin equation, DNA

1 INTRODUCTION

Eletrophoresis is a technique used to separate polyelectrolyte strands with different lengths [1]. One of its fields of application in nanotechnology is use in separation of DNA [2]. Clusters of particles can undergo net transport on a potential energy that is externally driven to fluctuate between several states in Brownian ratchet device.

In [3], a model for coupled Brownian motors, inspired by the motion of individual two-headed molecular motors on cytoskeletal filaments was proposed. The motors were modelled as two elastically coupled Brownian particles, each moving in a flashing ratchet potential. With a view to modelling ratchet-separation devices for DNA, we examine in this paper the effect of a Lennard-Jones interaction [4] between the particle pair, in addition to the existing spring force.

Using Langevin simulations we obtain currents as a function of noise strength, the equilibrium separation of the particles and the rate of switching between potentials. The outline is as follows. In Sec. 2 we introduce the Brownian ratchet mechanism. The model together with type of potential is discussed in Sec. 3. The numerical results are presented in Sec. 4.

2 BROWNIAN RATCHET

Noise induced transport has been recently become a widely studied area [5]–[7]. Much effort has been made to understand the dynamics of Brownian ratchets in the presence of stochastic forcing. The fluctuations and a broken symmetry are sufficient prerequisites for molecular transport to occur.

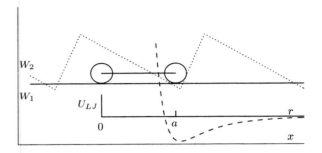

Figure 1: A simple Brownian ratchet device. Particles, with center at coordinate x, are subject to one of the two-state flashing ratchet potential, constant potential, W_1 and an asymmetric saw-tooth potential, W_2, which switches periodically with period τ. The interaction between particles, that are at the distance r apart, is modelled by the Lennard-Jones potential U_{LJ}, which has a minimum at equilibrium distance a.

Hammond et al [8] described the development and use of an integrated electrode array (IDEA) device for transportation of DNA based on a Brownian ratchet mechanism. The ratchet like potential is generated by applying a voltage difference to a series of pattern electrodes. The traps vanish and the particles undergo Brownian motion after the electrodes are discharged. When applying an ac electric field, because of the difference in the electrophoretic mobilities it is possible to observe a directional motion with shorter clusters moving faster then longer ones. This allows a separation of polymers with different lengths.

Material models of DNA frequently use elastic coupling between neighboring particles [9]. In order to investigate the effects of more realistic interaction between particles, we begin by adding a Lennard-Jones interaction to the ratchet model for Brownian motors described in [3].

3 MODEL

We consider two overdamped Brownian particles coupled through a spring of spring constant k and equilibrium length a. Graphical representation of the model is given in Figure 1. The excluded volume interaction between two particles, that are at the distance r apart,

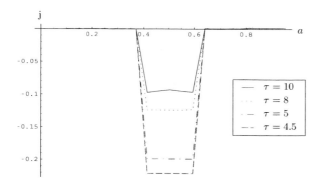

Figure 2: Deterministic case: the current j vs an equilibrium distance a for $\epsilon = 0.1$ and $k = 0.8$.

is modelled by the Lennard-Jones potential:

$$U_{LJ}(r) = 4\epsilon\left(\left(\frac{\sigma}{r}\right)^{12} - \left(\frac{\sigma}{r}\right)^{6}\right), \qquad (1)$$

The potential of Eq. (1) is stiff for small distance r, and for strength of interaction $\epsilon = 0$ model would correspond to the standard spring-bead model [3]. We choose the value of parameter σ such that the minimum of Lennard-Jones potential is at the equilibrium distance, e.g. $\sigma = a \cdot 2^{-1/6}$.

Following [3], particles are subject to one of the two-state flashing ratchet potential which switches periodically with period τ. $W_j(x)(j = 1, 2)$ defines the potential in state j at point x. W_1 is a flat potential and we choose the following asymmetric potential W_2:

$$W_2 = U\left(\frac{1}{2}\sin\left(\frac{2\pi x}{L}\right) + \frac{1}{8}\sin\left(\frac{4\pi x}{L}\right)\right), \qquad (2)$$

where U and L represent depth and period of potential, respectively. We set $L = 1$ and $U = 1$. The span of this potential is about 1.1 and its ratio of downhill region to the uphill region is around $1/4$.

The equations of motion of the particles are

$$\gamma\dot{x}_1 = -z_1(t)\frac{\partial W_2(x_1)}{\partial x_1} + k((x_2 - x_1) - a)$$
$$+ \frac{\partial U_{LJ}(x_2 - x_1)}{\partial x_1} + \sqrt{D}\xi_1, \qquad (3)$$
$$\gamma\dot{x}_2 = -z_2(t)\frac{\partial W_2(x_2)}{\partial x_2} - k((x_2 - x_1) - a)$$
$$- \frac{\partial U_{LJ}(x_2 - x_1)}{\partial x_2} + \sqrt{D}\xi_2, \qquad (4)$$

where x_i denote position of particle i^{th}. $\xi_i(t)$ denotes white noise with zero mean and correlation given by $\langle\xi_i(t)\xi_j(s)\rangle = \delta(t - s)\delta_{ij}$. D represents the intensity of fluctuations. We set the physical scales of the problem by putting the friction constant γ to unity.

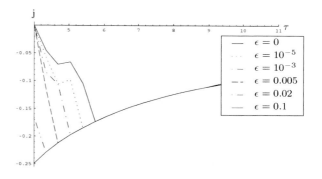

Figure 3: Deterministic case: the current j vs the potential switching period τ for $a = 0.5$ and $k = 0.8$.

The z_i are periodic functions with period τ, given by $z_1(t) = 1$, $z_2(t) = 0$ for $0 \le \tau < \tau/2$ and $z_1(t) = 0$, $z_2(t) = 1$ for $\tau/2 \le t < \tau$.

The quantity of our interest is the current, which we define by:

$$j = \frac{\langle x(T) - x(t_0)\rangle}{T - t_0}, \qquad (5)$$

where $\langle\rangle$ denotes the ensemble average.

We obtain currents as a function of noise strength D, the equilibrium separation of the particles a, strength of Lennard-Jones potential ϵ and the period of switching between potentials τ.

4 RESULTS AND DISCUSSION

4.1 Deterministic case

Particles in an asymmetric potential can drift on average in one direction even when operated at zero noise level, e.g. when $D = 0$. The phase space of the system can be either periodic or diffusive, depending on the value of the control parameter.

For $\epsilon > 0$, direct current is possible if the equilibrium separation of the particles is larger that the smaller arm of the potential $W_2(x)$, L_{min}, and smaller then the longer arm of the potential $W_2(x)$, L_{max}. Lengths L_{min} and L_{max} given by:

$$L_{max} = \frac{2L}{\pi}\arcsin\left[\frac{1}{2}\sqrt{1 + \sqrt{3}}\right], \qquad (6)$$
$$L_{min} = L - L_{max}. \qquad (7)$$

For period of potential equal to one particles move away from the initial coordinates if the equilibrium separation satisfies: $0.38 < a < 0.62$. This is investigated in Figure 2, which shows the dependance of current on the equilibrium distance and period switching. Zero value of current corresponds to closed orbits in the phase space of the system.

In Figure 3 we plot current j as function of τ for $a = 0.5$

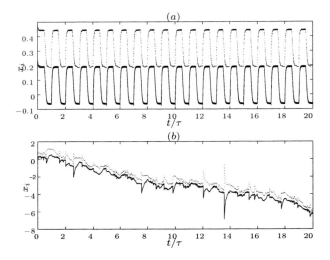

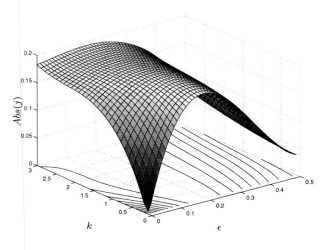

Figure 4: Position of particles for (a) deterministic case $D = 0$ and (b) stochastic case $D = 0.01$, for $a = 0.25$, $\epsilon = 0.1$, $\tau = 5$, $k = 0.8$. The current arises as a result of the presence of noise in the system.

Figure 5: Stochastic case: the absolute value of current j vs spring constant k and intensity of interaction ϵ, for $D = 0.01$, $\tau = 5$ and $a = 0.5$.

and $k = 0.8$, for different values of ϵ. As the switching period increases, current for different values of ϵ does not differ significantly. For the case where $\epsilon = 0$, the values of the equilibrium distance for which a direct current occurs depends on switching period τ [10]. The length of the window for nonzero current is smaller for smaller values of τ, although the dependance is not monotonic, see, for example the $\epsilon = 10^{-5}$ curve in Figure 3.

4.2 Stochastic case

We have already discussed in Section 4.1 that for zero-noise case, the transport of particles occurs only when the equilibrium distance a satisfies the condition $L_{min} < a < L_{max}$. The addition of thermal fluctuations to the system permits net motion (non-zero current) even for values of a outside the deterministic limits. Figure 4 shows the position of particles for (a) deterministic case $D = 0$ and (b) stochastic case $D = 0.01$, for $a = 0.25$, $\epsilon = 0.1$, $\tau = 5$, $k = 0.8$. The phase space of the particle pair is a closed orbit for the deterministic case, whilst for the stochastic case the particles are moving in the negative direction.

Since the model under consideration has seven parameters, we have analyzed the dependance of current only on two parameters. In our calculations we fixed $\tau = 5$ and $a = 0.5$. For convenience, we have plotted absolute values of current.

Figure 5 shows the absolute value of current j as a function of spring constant k and intensity of interaction ϵ, for $D = 0.01$. Note $\epsilon = 0$ corresponds to the case studied in [3]. For any fixed value of ϵ, introducing stronger coupling between the particles causes current to increase

initially, and for $k \simeq 2.5$ the currents saturates. On other hand, for any fixed value of k, the current has a maximum in the range of ϵ from 0.1 to 0.3. The location of the peak in $k - \epsilon$ space is governed by the strength of the Lennard-Jones potential.

Figure 6 shows the absolute value of current j vs spring constant k and intensity of noise D, for $\epsilon = 0.1$. The maximum of the current is at $D = 0$. On increasing the noise strength and keeping spring constant k small, random hopping in both directions dominates the ratchet mechanism and the absolute value of the current tends to zero. However, for any fixed value of noise intensity D, the current increases monotonically with increasing k.

Our preliminary results from exploring a subset of parameter space indicate that the addition of Lennard-Jones interaction to a standard spring-bead model can have important effect upon current. A fuller understanding of particle-particle interaction as well as particle-ratchet interaction is required for modelling of DNA transport and separation devices.

5 ACKNOWLEDGMENT

This work was supported by the Science Foundation Ireland through Investigator Award 02/IN.1/IM062.

REFERENCES

[1] G.W. Slater, S. Guillouzic, M.G. Gauthier, "Theory of DNA electrophoresis (1999-2002 1/2)," Electrophoresis, 23, 3791-3816, 2002.

[2] J. S. Bader, R. W. Hammond, S. A. Henck, "DNA Transport by a Micromachined Brownian Ratchet Device," PNAS, 96, 13165-13169, 1999.

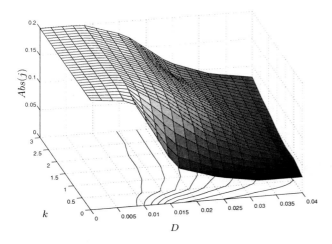

Figure 6: Stochastic case: the absolute value of current j vs spring constant k and intensity of noise D, for $\epsilon = 0.1$, $\tau = 5$ and $a = 0.5$.

[3] D. Dan, A.M. Jayannavar and G.I. Menon, "A biologically inspired ratchet model of two coupled Brownian motors," Physica A, 318, 40-47, 2003.

[4] D. Ceperlay and M.H. Kalos, "Computer simulations of dynamics of a single polymer chain," Phys Rev Lett, 45, 313-316,1978.

[5] B. Yan, R.M. Miura, Y. Chen, "Direction Reversal of Fluctuation-induced Biased Brownian Moion on Distorted Ratchets," J. theor. Biol., 210, 141-150, 2001.

[6] H. Linke, T.E. Humphrey, P.E. Lindelof, "Quantum Ratchets and Quantum Heat Pumps," Appl.Phys. A, 75, 237-246, 2002.

[7] P. Reimann, P. Hänggi, "Introduction to the Physics of Brownian Motors," Appl. Phys. A, 1-10, 2002.

[8] R.W. Hammond, J.S. Bader, S.A. Henck, M.W. Deem, G.A. McDermott, J.M. Bustillo, J. M. Rothberg, "Differential transport of DNA by a rectified Brownian motion device," Electrophoresis, 21, 74-80, 2000.

[9] Z. Csahok, F. Family, T. Viscek, "Transport of Elastically Coupled Particles in an Asynnetric Period Potential," Physical Review E, 55, 5179-5183, 1997.

[10] A. Igarashi, S Tsukumoto, H. Goko, "Transport properties of elastically coupled Brownian motors," Physical Review E, 64, 0519081-0519085, 2001.

Superposition behavior in Babinet s effect on the Au nano particles patterns

S.-C. Wu, C.-F. Chen, W.-C. Chao, and H.-L Chen

National Nano Device Laboratories, 1001-1 Ta-Hsueh Road, Hsinchu,
Taiwan, R. O. C., scwu@ndl.gov.tw

ABSTRACT

The individual Babinet's effect of two types of complementary patterns can be added together on a composed pattern mixed with both patterns under the FTIR (Fourier Transform Infer-Red) spectroscopy measuring. Herein the added spectrum can show the superposition behavior in Babinet's effect. These two types of complementary patterns are Split Ring Resonator (SRR) and wire array (CLS Capacitively Loaded Strip) respectively. The mixed pattern is combined with SRR and wire array, which is so called the LHM (Left Handed Meta-material) pattern. Those samples are fabricated on Si wafer using standard integrated circuit (IC) processes. The metal conductors on the pattern surface are made of gold nano particles. Positive sample is deposited the Au nano particles on the periodic conducting parts of the pattern. Negative one is a reverse way that empties the conducting part and filled the surrounding space with Au particles.

Keywords Babinet's effect, left handed meta-material, nano Au particles, transmission enhancement

1 INTRODUCTION

Babinet's effect [1] is that the diffraction pattern of an aperture is the same as the pattern of an opaque object of the same shape illuminated in the same manner. The original principal stated by Babinet, which is in scalar form, and applied at large distances from the scattering screen where the energy densities of the electric and magnetic fields of the wave at any point are equal. In addition, the two spectra of complementary patterns are out of phase at each point on a detector. The electromagnetic Babinet principle for diffraction fields behind complementary perfectly conducting objects [2] relies on the duality of Maxwell's equations as well. Accordingly, the electromagnetic field at one position can be obtained by adding up those fields coming from different directions. Thus, the quantity of Babinet's effect on one mixed pattern can be added by each amount on those composed patterns. It is the goal of this study to demonstrate this superposition behavior by analyzing the FTIR spectra at LHM (Left Handed Meta-material) combined pattern. Especially, the superposition behavior of Babinet's effect is seldom reported since Babinet presented his observation. Interestingly, except Babinet effect, the optical characteristics of sub-wavelength transmission [3], doubly negative permeability and permittivity effects [4], and Wood's anomaly [5] were also observed in those spectra. The LHM is a specific pattern, which indicates the unique characteristics of a negative refraction index [4]. The pattern is a combination of the Split Ring Resonator (SRR) at $\mu < 0$ and the wire grating (CLS Capacitively Loaded Strip) pattern with $\varepsilon < 0$ [6].

Currently, novel properties of metal nano particles attract attention of researchers all over the world. For instance, Au nano particles are of specific behaviors and unique properties [7], and very useful in bios, optics and nano-electronics applications. Accordingly, we try to include Au nano particles into this LHM fabrication study and wish to investigate some new effects herein. Fortunately, the transmittance-enhanced behavior is found out obviously from the spectrum. The surface plasmon resonated effect [8] of Au nano particle on top pattern surface is supposed one of the attribution factors. Simultaneously, the combining effects at both the response of metamaterial pattern and the collective surface plasmon effect of the nano particles will be evaluated.

Some results of SRR and wire array investigation have been presented in other journals [9-11]. This work addresses only the superposition behavior in Babinet effect of LHM array obtained using FTIR. The description in this phenomenon will be presented following up the experiment. Section 2 states the experiment and characterization details. In Section 3, we report the main results and discuss the experimental data. Section 4 draws the conclusions.

2. Experiment

The patterned samples were fabricated using standard IC (integrated-circuit) processes. The substrate was a 6" double polished n-type silicon wafer. The wafer was cleaned by RCA cleaning. Reference [10] describes negative and positive patterning of samples. For positive one, the conducting parts in the pattern are filled with Au metal. Therefore, in the negative samples, the space along the conducting parts is filled with the gold particles. The conducting parts are thus empty and not covered by Au film; only SiO_2 remains on the Si substrate. Photographs were taken using SEM (Scanning Electronic Microscopy) instruments. Figure 1 shows the schematic graph of three types of patterns herein. The thickness of Au was ~20nm while the thickness of the Si substrate was ~0.7mm. The horizontal distance between the two LHM units was 3.2 μm and the vertical distance was 6.8 μm. Typically, the area of an entire LHM array sample is about 4 mm X 4 mm. An entire sample has $1\ 0\ X200 = 30,000$ LHM units. The FTIR equipment used herein was QS 300 produced by Bio-Rad Laboratories, Inc. The range of wavenumbers of the light source was 4000~400 cm^{-1} (The wavelengths ranged from 2.5

to 25 μm). In this experiment the incident light was perpendicular to the pattern surface. Each of spectrums was obtained using 16scans. The time resolution was 2 sec and the scanning length was 30 sec.

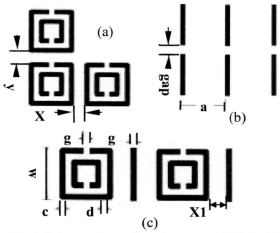

Fig.1 Schematic graphs of pattern (a) SRR (b) wire array. (c) LHM

c	d	g	w	X	Y	gap	X1	a
1	1.2	1.84	10.48	3.2	6.8	4.4	3.2	13.68

Table 1 Dimensions of all patterns. (unit : μm)

3. Results and Discussion

Figure 2 shows the SEM photograph of the positive and negative LHM arrays with 1000x and 35,000x magnifications, respectively.

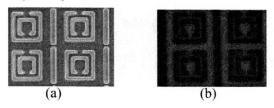

Fig. 2 SEM photographs of the complementary LHM patterns, (a) positive, (b) negative.

Figure 3 (a)-(c) display the FTIR transmittance spectra at 400~4000 cm^{-1}, corresponding to three types of patterns. From

Fig. 3 (a), there are over 15 positions showing the Babinet's effect. Notably, whereon 5 points in this spectrum are below wave number 750 cm^{-1}. But no Babinet's effect can be observed at same region of wave number in spectrum of SRR pattern, shown as Fig. 3 (b). Additionally, only 4 points on the SRR spectrum exhibit Babinet effect, which all above wave number 750 cm^{-1}.

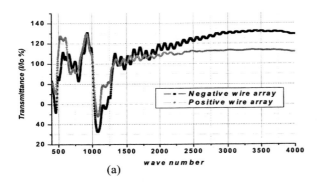

(a)

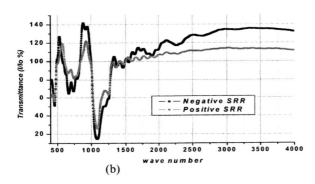

(b)

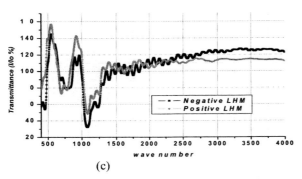

(c)

Fig. 3 The FTIR spectra of the Babinet's effect on the complementary SRR (a), wire array (b), and (c) LHM patterns.

However, there are more than 10 positions after wave number 750 cm^{-1} can show Babinet's behaviors in spectrum of wire array. Figure 3 (c) shows the spectrum of LHM, in which Babinet's behaviors are obvious in many regions. The range of wave numbers is from 696~710 cm^{-1} (14.1~14.4 µm), to 2926 cm^{-1} (3.4 µm). Babinet's behavior is clearly exhibited in over 25 positions, shown as Fig. 3 (c). Interestingly, only one position remained in this spectrum under wave number 750 cm^{-1} demonstrates the Babinet phenomenon. It illuminates the added result from superposing both spectra of two patterns during longer wavelength regime. Fig. 4 zooms into the SRR spectrum at 1200~2800 cm^{-1} to show clearer the relationship between the Babinet's behavior of complementary patterns. In this region, there are over 20 positions showing Babinet effect and some of them even occurring at same positions. For instance, at around 700, 1220, 1393, 1444, 1530, and 1800 cm^{-1}, the phase difference between waveforms is almost equal to π. Accordingly, it seems that the waveform of wire array spectrum dominates in shorter wavelength region and that of SRR does in longer range.

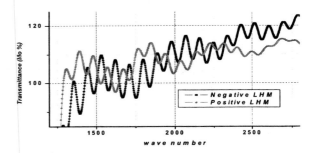

Fig. 4. Magnified spectrum of FTIR data that clearly shows the Basinet's effect in the 1200~2800 cm^{-1} regime.

Meanwhile, the transmittance (T) values can take to be $T_p (\lambda) = T_0 (\lambda) \pm \Delta T (\lambda)$ and $T_n (\lambda) = T_0 (\lambda) \pm \Delta T (\lambda)$ for that of positive and negative LHM patterns, respectively. $T_p (\lambda)$, and $T_n (\lambda)$ are the peak or valley transmittance values of the positive and negative patterns in the regions of Babinet's effects, and T_0 is the approximate transmission of the undiffracted light, which is considered to be the average value of $T_p (\lambda)$+ $T_n (\lambda)$. Some T_0 values exceed 100%, probably because of the effect of transmission enhanced on both types of patterns. Where $\Delta T (\lambda)$, i.e., the difference between T_n and T_0, or T_p and T_0, is derived from the difference in diffraction from opaque obstacles in both LHM patterns. Notably, the complementary behaviors of the spectra are clearer than those of SRR and wire array pattern. It is seldom reported that Babinet's behavior is still exhibited under a sophisticated and periodic patterned array and even the superposition characteristics. Especially the incident wavelength is also compatible with the size of array. The annihilation effect of precise overlapping of two reciprocal patterns is thus observed. This finding is being investigated and will be discussed in a later study.

SEM observes that the dispersion of Au nano particles on the conducting surface in array pattern [10] is like the cluster in a semicontinuous metal. Accordingly, the Au nano particles could play a role of transmission enhancing, which is same as the percolation-enhanced nonlinear scattering (PENS) effect [12]. It is believed that would be a marked effect when polarized coherent light source is used.

4. Conclusion

Complementary LHM patterns were fabricated to include Au nano particles by standard IC processes. More than 25 Basinet's effects were observed over a wide range of wavelengths. Notably, those effects are coincident with the superposition of both spectra that correspond to SRR and wire pattern. Interestingly, some complementary waveforms arise even at the same positions. This result has seldom been reported and should be studied further. Whether the complementary spectra could be added together to yield a flat response was also investigated. The enhanced factor of the negative array exceeds that of the positive one, although the

surface of the former is covered with more gold particles. Therefore the collective localized electric field, enhanced by Au nano particles, can combine with the surface plasmon to increase transmittance. Therefore further investigation is needed, particularly in the field distributed over the conductor in both types of LHM array.

Acknowledgement: The author would like to thank the National Science Council of the Republic of China, Taiwan for financially supporting this research under Contract No. NSC92-2215-E-492-009.

Reference

[1] M. Babinet, Memories d'optique meteorologique, C. R. Acad. Sci. 4, 638, 1837.

[2] E. T. Copson, An integral equation method for solving phase diffraction problems, Proc. R. Sdoc. London Sect. A 186, 100-110, 1948

[3] T. W. Ebbesen, H. J. Lezec, H. F. Ghaemi, T. Thio, and P. A. Wolff, *Nature*, 391, 667-669, 1998.

[4] R. A. Shelby, D. R. Smith, and S. Schultz, Science, 292, 77-79, 2001.

[5] R. W. Woods, Anomalous Diffraction Gratings, Phys, Rev. 48, 928-936, 1935

[6] J. B. Pendry, A. J. Holden, W. J. Stewart, and I. Youngs, Phys. Rev. Lett. 76, 4773-4776, 1996

[7] X. M. Lin, R. Parthasarathy, and H. M. Jaeger, Appl. Phys. Lett. 78, 1915-1917, 2001

[8] M. B. Sobnack, W. C. Tan, N. P. Wanstall, T. W. Preist, and J. R. Sambles, Phys. Rev. Lett. 80, 5667-5670, 1998

[9] A-Chuan Hsu, et. al, PIERS, 586, 2003.

[10] S.-C. Wu, et al, First international Meeting on Applied Physics, 906-907, 2003.

[11]A-Chuan Hsu, et. al., accepted by Jap. J. Appl. Phys.

[12] A.K. Sarychev, V. A. Shubin, and V. M. Shalaev, Phys. Rev. E, 59 (6) (1999) 7239-7242.

Quantum Dots/Conductive Polymer Nanocomposite

Hsueh-Shih Chen[*], Ho-Chang Huang[**] and Chien-Ming Chen[***]

[*]Union Chemical Laboratories, Industrial Technology Research Institute
321, Kuang-Fu Rd. Section 2, Hsinchu, Taiwan, sean@itri.org.tw
[**] Union Chemical Laboratories, Industrial Technology Research Institute
321, Kuang-Fu Rd. Section 2, Hsinchu, Taiwan
[***] Union Chemical Laboratories, Industrial Technology Research Institute
321, Kuang-Fu Rd. Section 2, Hsinchu, Taiwan, jeremychen@itri.org.tw

ABSTRACT

The nanocomposite composed of CdSe quantum dots and conductive polymer (OC1C10-PPV) was investigated. The spherical 3.8-nm functionalized CdSe quantum dots were synthesized by a colloidal chemical method. The conductive polymer OC1C10-PPV was synthesized from 1-methoxyl-4-(3,7dimethyl-octanoxyl) benzene and 4-methoxylphenol. The quantum dots were mixed with OC1C10-PPV before and after polymerization. It was found quantum dots will affect the polymerization of OC1C10-PPV. The photoluminescence (PL) peaks from CdSe and OC1C10-PPV was 590 nm (FWHM~26 nm) and 555 nm (FWHM~50 nm), respectively. Photoluminescence spectra of nanocomposite displayed mixed characters of quantum dots and OC1C10-PPV.

Keywords: conductive polymer, nanoparticle, quantum dots

1 INTRODUCTION

There has been a great effort to develop luminescent materials such as quantum dots (QDs) in the past decade. The quantum confinement effect (QCE) has been demonstrated to exist on many types of materials while their dimension reduces to nanometer scale. For semiconductive QDs, the wavelength of the electron is close to the diameter of the crystallite so the matter wave of the carrier will be confined to form standing wave like a particle in a box. The confinement effect for the electron leads to a split of energy band edge, and therefore electronic, optical or magnetic properties of the QDs are different to those of bulk material [1]. For most of the semiconductors, the discrete band edge causes higher band gap energy (E_g). The E_g of the semiconductive QDs blueshifts to higher energy when the dimension of crystallite reduced [2-3]. Consequently, scientists can modify the band structure of materials by the change of particle size. Moreover, the optical properties of semiconductors can be altered by particle size [4-5].

The colloidal semiconductive QDs used as luminescent materials have been devoted recently. Alivisatos et al. used semiconductive colloidal QDs as biological tagging for diagnostics and visualization. It has significant advantages compared to existing organic dyes such as a high luminescent efficiency, reduced photobleaching, multi-color labeling, parallel screening, infrared labels, blood diagnostics, and bio-compatible [6].

The luminescent conjugate polymer or conductive polymer also attracts much attention recently [7]. The advantage of the polymer-based product is the simple process. However the reliability of polymer is a challenge for the commercial applications. Some inorganic filler have been introduced into polymer to improve the thermal or optical properties [8]. On the other hand, the composite composed of luminescence QDs and luminescent conductive polymer is a new material for optical applications. The mechanisms of the material synthesis and luminescence are not clearly. In this study, the QDs/conductive polymer which QDs mixed with conductive polymer before/after polymer polymerization was investigated.

2 EXPERIMENTAL

Cadmium oxide, selenium, trioctylphosphine (TOP), trioctylphosphine oxide (TOPO) and hexadcylamine (HDA) are supplied from Aldrich. 4-methoxyphenol was purchased from TCI. Acetic acid, tetrahydrofuran, n-hexane was obtained from Tedia. Paraformaldehyde and sodium hydroxide was purchased from Showa. All solvents were purified before it was used.

2.1 Preparation of CdSe QDs

The CdSe QDs were synthesized from organometallic precursors via thermal decomposition. The selenium powder was dissolved in trioctylphosphine under nitrogen to form TOPSe stock solution. The cadmium oxide was mixed with the fatty acid and heated at ~100℃ in a flask free of water and oxygen. TOPO and HDA were then added into the flask at 320 ℃. After the clear liquid appeared, the TOPSe stock solution was injected into the flask. The ratio of cadmium to selenium is 1.2. The reaction time was 10 minutes. The slurry was washed by methanol and toluene for several times to separate TOPO-capped CdSe QDs.

2.2 Synthesis of OC1C10-PPV

1-methoxyl-4-(3,7dimethyl-octanoxyl) benzene was synthesized from 3,7-dimethyl-1-bromooctane and 4-methoxylphenol in alcohol at 75°C.

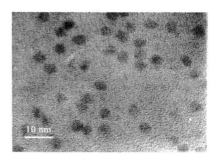

1-methoxyl-4-(3,7dimethyl-octanoxyl) benzene was transferred to 4-methoxyl-2,5-chloromethyl-(3,7dimethyl-octanoxyl) benzene 75 °C in mixture of acetic acid, hydrochloric acid and paraformaldehyde. Then 4-methoxyl-2,5-chloromethyl-(3,7dimethyl-octanoxyl) benzene polymerized to OC1C10-PPV in THF at room temperature. OC1C10-PPV was orange with naked eye.

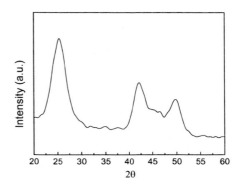

2.3 Preparation of QDs/OC1C10-PPV

For Sample A, the CdSe QDs were dispersed to OC1C10-PPV in toluene (QDs = 5 wt %). An orange solution was obtained after it stirred in dark at room temperature for 3 hours.

For Sample B, QDs added into the precursor (4-methoxyl-2,5-chloromethyl-(3,7dimethyl-octanoxyl) benzene) before polymerization and then polymerized to OC1C10-PPV (QDs = 5 wt %). An orange solution was obtained after the reaction.

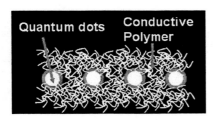

Figure 1 QDs/conductive polymer nanocomposite

2.4 Characterization

The size of QDs was analyzed by transmission electron microscope (JEOL JEM-2010). Photoluminescence spectra were measured by Hitachi F-4500 fluorescence spectrophotometer. The molecular weight distribution was determined by gel permeation chromatography (Waters, GPC).

3 RESULTS AND DISCUSSION

Monodispersed CdSe QDs were obtained without any size selection. The size of CdSe QDs is about 3.8 nm, as shown in Fig. 2. The XRD patterns for the CdSe QDs are shown in Fig. 3. The peaks are broad since the crystallites are in nanoscale.

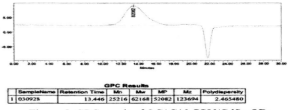

Figure 2 TEM image of the CdSe QDs. The darker spots were CdSe QDs.

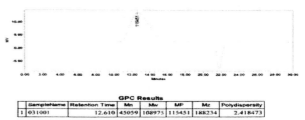

Figure 3 XRD of CdSe QDs.

GPC Results

SampleName	Retention Time	Mn	Mw	MP	Mz	Polydispersity
1 031001	12.610	45059	108975	115451	188234	2.418473

Figure 4 GPC result of OC1C10-PPV

GPC Results

SampleName	Retention Time	Mn	Mw	MP	Mz	Polydispersity
1 030928	13.446	25216	62168	52082	123694	2.465480

Figure 5 GPC result of OC1C10-PPV/CdSe QDs

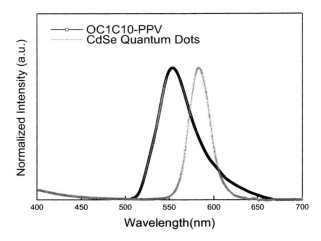

Figure 6 PL from OC1C10-PPV and CdSe quantum Dots, respectively.

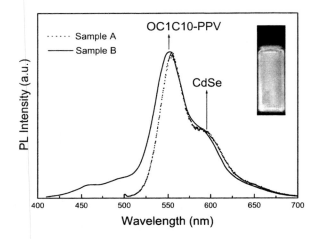

Figure 7 PL from QDs/OC1C10-PPV. The content of QDs is 5 wt%. The picture is the specimen under UV excitation.

From GPC measurement, the molecular weight and polydispersity of OC1C10-PPV were confirmed to 115,451 and ~2.42, respectively, as shown in Fig. 4. The molecular weight and polydispersity of QDs/OC1C10-PPV (Sample A) were similar to those of OC1C10-PPV. It displayed that the addition of QDs with OC1C10-PPV does not affect the molecular weight or distribution of OC1C10-PPV. On the other hand, if the QDs were added into the precursor before OC1C10-PPV polymerization, the molecular weight of QDs/OC1C10-PPV decreased and has a wider distribution than that of OC1C10-PPV, as shown in Fig. 5. The phenomenon implies the CdSe QDs maybe disturb the polymerization of OC1C10-PPV.

PL spectra of QDs and OC1C10-PPV were shown in Fig. 6. The peak positions of OC1C10-PPV and CdSe QDs were 555 nm (FWHM~50 nm) and ~590 nm (FWHM~26 nm), respectively. The PL spectra of QDs which mix with OC1C10-PPV before and after polymerization were shown in Fig. 7. The PL spectra displayed mixed characters of QDs and OC1C10-PPV. The PL from Sample B has a wider band than that of Sample A. The phenomenon indicates the molecular weight distribution of sample B is larger than that of Sample A, and implies the QDs influence the OC1C10-PPV polymerization and broaden the molecular weight distribution. The result is consistent with the previous data measured from GPC.

4 CONCLUSION

The CdSe QDs/OC1C10-PPV nanocomposite was fabricated. The QDs were mixed with OC1C10-PPV before and after polymerization. It was found that the molecular weight of QDs/OC1C10-PPV was similar to that

of OC1C10-PPV while QDs directly mixed with OC1C10-PPV. However, QDs will decrease the molecular weight and broaden the molecular weight distribution of QDs/OC1C10-PPV while QDs were added into the precursor before polymerization of OC1C10-PPV. The PL peaks from CdSe and OC1C10-PPV was 590 nm (FWHM~26 nm) and 555 nm (FWHM~50 nm), respectively. PL spectra of nanocomposite displayed mixed characters of QDs and OC1C10-PPV.

ACKNOWLEDGEMENTS

The authors would like to thank the *Industrial Technology Research Institute* of the R.O.C., Taiwan for financially supporting this research.

REFERENCES

[1] A. P. Alivisatos, Science 271, 933, 1996.
[2] T. Takagahara, and K. Takeda, Phys. Rev. B, 46 15578, 1992.
[3] D. J. Norris, A. Sacra, C. B. Murray, and M. G. Bawendi, Phys. Rev. Lett., 72, 2612, 1994.
[4] M. B. Jr., M. Moronne, P. Gin, S. Weiss, A. P. Alivisatos, Science, 281, 2013, 1998.
[5] C. Tsai, K. H. Li, D. S. Kinosky, R. Z. Qian, T. C. Hsu, J. T. Irby, S. K. Banerjee, A. F. Tasch, and Joe C. Campbell, Appl. Phys. Lett., 60 , 1700, 1992.
[6] M. B. Jr., M. Moronne, P. Gin, S. Weiss, A. P. Alivisatos, Science, 281, 2013, 1998.
[7] Yong Cao, Ian D. Parker, Gang Yu, Chi Zhang and Alan J. Heeger, Nature, 397, 414, 1999.
[8] Y. K. Kim, K. Y. Lee, O. K. Kwon, D. M. Shin, B. C. Sohn, J. H. Choi, Synthetic Metals, 111, 207, 2000.

Modeling of Nanoelectronic and Quantum Devices[*]

D. K. Ferry, R. Akis, M. J. Gilbert, and G. Speyer

Department of Electrical Engineering and Center for Solid State Electronics Research
Arizona State University, Tempe, AZ 85287-5706

ABSTRACT

The semiconductor industry is constantly pushing towards ever smaller devices and it is expected that we will see commercial devices with gate lengths less than 10 nm within the next decade. Such small devices have active regions that are smaller than relevant coherence lengths, so that full quantum modeling will be required. In addition, novel new structures, such as molecules, may represent the active regions in such small devices. Here we outline a fully quantum mechanical approach to the modeling of coherent transport in ballistic structures. Examples of an SOI MOSFET and a molecule are presented.

Keywords: quantum transport, semiconductors, molecules, electron states, MOSFETs

1 INTRODUCTION

Almost 25 years ago, the prospects of making very small transistors was discussed, and a suggested technique for a 25 nm gate length, Schottky source-drain device, was proposed [1]. At that time, it was suggested that the central feature of transport in such small devices would be that the micro-dynamics could not be treated in isolation from the overall device environment (of a great many similar devices). Rather, it was thought that the transport would by necessity be described by quantum transport and that the array of such small devices on the chip would lead to considerable coherent many-device interactions. While this early suggestion does not seem to have been fulfilled, as witnessed by the quite normal behavior of today's research devices [2,3], there have been many subsequent suggestions for treatment via quantum transport [4-8]. There is ample suggestion that the transport will not be normal, but will have significant coherent transport effects and this, in turn, will lead to quantum behavior.

In this paper, we will review a full quantum formulation of the three dimensional transport, which is coupled to a three dimensional Poisson solver, in order to treat the coherent transport in small devices. In the next section, we will outline the basic approach, and then treat two example cases—a small silicon-on-insulator MOSFET, and a molecule attached to two conductors. Only the basic introduction to this approach is given, as more details are contained in the references and in a recent review article that is scheduled for publication [9].

2 A QUANTUM TRANSPORT FORMULATION

There have been many suggestions for different quantum methods to model ultra-small semiconductor devices [10-12]. However, these approaches are uaually quasi-two-dimensional, as the length and the depth are modeled rigorously, while the third dimension (width) is usually included through the assumption that there is no interesting physics in this dimension. Other simulation proposals have simply assumed that only one sub-band in the orthogonal direction is occupied, therefore making higher-dimensional transport considerations unnecessary. These may not be valid assumptions, especially as we approach devices whose width is comparable to the channel length, both of which may be less than 10 nm.

It is important to consider all the modes that may be excited in the source (or drain) region, as this may be responsible for some of the interesting physics that we wish to capture. In the source, the modes are three dimensional (3D) in nature, even in a thin SOI device. These modes are propagated from the source to the channel, and the coupling among the various modes will be dependent upon the details of the total confining potential at each point along the channel. Moreover, as the doping and the Fermi level in short-channel MOSFETs increases, we can no longer assume that there is only one occupied sub-band. Hence, we use a full 3D quantum simulation, based on the use of recursive scattering matrices [13-16].

Consider the Schr dinger equation in three dimensions:

$$-\frac{\hbar^2}{2}\left[\frac{1}{m_x}\frac{d^2}{dx^2}+\frac{1}{m_y}\frac{d^2}{dy^2}+\frac{1}{m_z}\frac{d^2}{dz^2}\right] + V(x,y,z) = E \quad (1)$$

Here, it is assumed that the mass is constant, in order to simplify the equations (for nonparabolic bands, the reciprocal mass enters between the partial derivatives). We have labeled the mass corresponding to the principle coordinate axes. In silicon, these take on the values of m_L and m_T as appropriate. We then choose to implement this on a finite difference grid with uniform spacing a. Therefore, we replace the derivatives appearing in the discrete Schr dinger equation with finite difference representations of the derivatives. The Schr dinger equation then reads

[*] Work supported in part by the Office of Naval Research.

$$t_x \left[\Psi_{i+1,\cdot,k} + \Psi_{i-1,\cdot,k} \right]$$
$$+ t_y \left[\Psi_{i,\cdot+1,k} + \Psi_{i,\cdot-1,k} \right]$$
$$+ t_z \left[\Psi_{i,\cdot,k+1} + \Psi_{i,\cdot,k-1} \right] \qquad (2)$$
$$+ \left[V_{i,\cdot,k} - 2t_x - 2t_y - 2t_z \right] \Psi_{i,\cdot,k}$$
$$= E \Psi_{i,\cdot,k},$$

where t_x, t_y and t_z are the hopping energies

$$t_x = \frac{\hbar^2}{2m_x a^2},$$

$$t_y = \frac{\hbar^2}{2m_y a^2}, \qquad (3)$$

$$t_z = \frac{\hbar^2}{2m_z a^2}.$$

Each hopping energy corresponds with a specific direction in the silicon crystal. The fact that we are now dealing with three sets of hopping energies is quite important.

There are other important points that relate to the hopping energy. The discretization of the Schrödinger equation introduces an artificial band structure, due to the periodicity that this discretization introduces. As a result, the band structure in any one direction has a cosinusoidal variation with momentum eigenvalue (or mode index), and the total width of this band is $4t$. Hence, if we are to properly simulate the real band behavior, which is quadratic in momentum, we need to keep the energies of interest below a value where the cosinusoidal variation deviates significantly from the parabolic behavior desired. For practical purposes, this means that $E_{max} \ll t$. The smallest value of t corresponds to the longitudinal mass, and if we desire energies of the order of the source-drain bias ~ 1 V, then we must have $a \le 0.2$ nm. That is, we must take the grid size to be comparable to the Si lattice spacing

With the discrete form of the Schrödinger equation defined, we now seek to obtain the transfer matrices relating adjacent slices in our solution space. For this, we will develop the method in terms of slices [9]. This is modified here by the two dimensions in the transverse plane. We begin first by noting that the transverse plane has $N_y \times N_z$ grid points. Normally, this would produce a second-rank tensor (matrix) for the wave function, and it would propagate via a fourth-rank tensor. However, we can re-order the coefficients into a $N_y N_z \times 1$ vector, so that the propagation is handled by a simpler matrix multiplication. Since the smaller dimension is the z direction, we use N_z for the expansion. Now, equation (2) can be rewritten as a matrix equation as, with s an index of the distance along the x direction,

$$H(s) + T_x(s+1) + T_x(s-1) - EI = (s). \qquad (4)$$

Here, I is the unit matrix, E is the energy to be found from the eigenvalue equation, and

$$H = \begin{pmatrix} H_0(\mathbf{r}) & \tilde{t}_z & \cdots & 0 \\ \tilde{t}_z & H_0(\mathbf{r}) & \cdots & \cdots \\ \cdots & \cdots & \cdots & \tilde{t}_z \\ 0 & \cdots & \tilde{t}_z & H_0(\mathbf{r}) \end{pmatrix}, \qquad (5)$$

$$T_x = \begin{pmatrix} \tilde{t}_x & 0 & \cdots & 0 \\ 0 & \tilde{t}_x & \cdots & 0 \\ \cdots & \cdots & \cdots & \cdots \\ 0 & 0 & \cdots & \tilde{t}_x \end{pmatrix}. \qquad (6)$$

The dimension of these two super-matrices is $N_z \times N_z$, while the basic Hamiltonian terms of (5) have dimension of $N_y \times N_y$, so that the total dimension of the above two matrices is $N_y N_z \times N_y N_z$. In general, if we take k and s as indices along y, and η and v as indices along z, then

$$\tilde{t}_{z,\eta v} = t_z \delta_{\eta v}, \quad \tilde{t}_{y,k} = t_y \delta_k, \quad \tilde{t}_{x,ss} = t_z \delta_{ss}, \qquad (7)$$

and

$$H_0(\mathbf{r}) = \begin{pmatrix} V(s,1,\eta) & W & t_y & \cdots & 0 \\ t_y & V(s,2,\eta) & W & \cdots & 0 \\ \cdots & & \cdots & \cdots & t_y \\ 0 & & 0 & t_y & V(s,N_y,\eta) & W \end{pmatrix} \qquad (8)$$

The quantity W is $2(t_x + t_y + t_z)$.

With this set of matrices, the general procedure follows that laid out in the previous work. One first solves the eigenvalue problem on slice 0 at the end of the source (away from the channel), which determines the propagating and evanescent modes for a given Fermi energy in this region. The wave function is thus written in a mode basis, but this is immediately transformed to the site basis, and one propagates from the drain end, using the scattering matrice iteration

$$\begin{pmatrix} C_1(s+1) & C_2(s+1) \\ 0 & 1 \end{pmatrix} = \begin{pmatrix} 0 & 1 \\ 1 & (T_x)^{-1}[EI - H] \end{pmatrix}$$
$$\times \begin{pmatrix} C_1(s) & C_2(s) \\ 0 & 1 \end{pmatrix} \begin{pmatrix} 1 & 0 \\ P_1(s) & P_2(s) \end{pmatrix} \qquad (9)$$

The dimension of these matrices is $2N_y N_z \times 2N_y N_z$, but the effective propagation is handled by submatrix compuations, through the fact that the second row of this equation sets the iteration conditions

$$C_2(s+1) + P_2(s) = C_2(s)(T_x)^{-1}[EI - H], \qquad (10)$$
$$C_1(s+1) + P_1(s) = P_2(s)C_1(s).$$

At the source end, $C_1(0) = 1$, and $C_2(0) = 0$ are used as the initial conditions. These are now propagated to the N_x slice, which is the end of the active region, and then onto the $N_z + 1$ slice. At this point, the inverse of the mode-to-site transformation matrix is applied to bring the solution back to the mode representation, so that the transmission coefficients of each mode can be computed. These are then summed to give the total transmission and this is used in a version of the Landauer formula to compute the current through the device (there is no integration over the transverse modes, only over the longitudinal density of states and energy).

3   AN SOI MOSFET

In the SOI MOSFET under consideration, we have an oversized source and drain region which are doped to 3 10^{19} cm^{-3} n-type. The dimensions of the source and the drain are 18 nm wide, 10 nm long and 6 nm high, corresponding to the thickness of the silicon (SOI) layer. The source and drain of the device have been given an exaggerated size to exacerbate the interaction of the modes excited in the source with the constriction present at the source-channel interface. The channel of our device is a p-type region. The channel is 10 nm in length, 6 nm in height and 8 nm in width. In the z direction of our device, the gate oxide has a thickness of 2 nm. Further, we have assumed that the oxide in this device is perfect in so much as the oxide does not have any spatial variation in thickness or charges present. Below the silicon layer, lies the buried oxide layer. This is 10 nm in thickness. This is large enough to insure that there is no leakage from the silicon layer, but also small enough to not waste computational resources simulating a region where no interesting physics is present. As discussed above, the x direction is along the channel, and the z direction is normal to the top gate. The simulation is carried out at low temperature. For this structure, we find that the threshold voltage is about 0.45 V, which does not vary much with doping in the channel, as there are only 2-3 dopants in the channel region. In Fig. 1, we illustrate the role of the impurities by plotting the local potential in the x-y plane, at the center of the channel. Here, both donors and acceptors are treated as discrete entities, and the potential clearly illustrates the local potential variations. The position of the dopant atoms has a significant effect on the resultant device characteristics. This is a result of the interference that the potential spikes produce. Dopants that are positioned closer to the source of the device have a greater effect on the threshold voltage than do dopants positioned further down the channel due to increased interaction with the waves incident at the source-channel interface, causing additional reflections.

In Fig. 2, we plot the density in a vertical cut through the x-z plane at the center of the channel, for a gate voltage of 1 V. Here, the source-drain bias is only 10 mV, but the conductance through the channel consists of one full mode propagating from the source to the drain [17]. However, looking at the figure, one does not draw this conclusion. Rather, the self-consistent potential has created quantum dots within the channel region, presumably due to quantum reflections at the source- and drain-interfaces with the

channel, and it appears that the conductance is supported by resonant tunneling into and through this quantum dot. The actual position of the dot is bias dependent, and also depends upon the details of the location of the impurities in the channel.

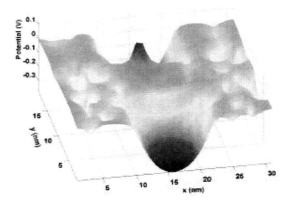

Fig. 1 The local potential in an x-y plane through the center of the channel. The gate voltage is 1 V, and the source-drain potential is 10 mV.

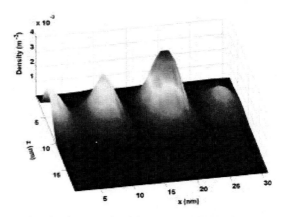

Fig. 2 density in a vertical (x-z) plane through the center of the channel. The gate is at the top, and the bottom region of no density is the SOI layer. Here, the gate bias is 1 V, and the source-drain bias is 10 mV.

The fact that the transport properties can be dramatically affected not only by the position in the xy plane, but by their position in the z direction as well, adds additional importance to such simulations. The positions of the dopants in the source and the drain can cause pools of electron density to form. This leads to noticeable variation in the density distribution in the source and, particularly, in the drain. This illustrates the growing importance of the mechanisms of coherence in ultra-short devices.

4   TRANSPORT IN A MOLECULE

As can be seen in eqn. (5), we need to know the Hamiltonian (energy spectrum) of the molecule in its configuration in order to use the recursive scattering matrix approach. The first-principles program, FIREBALL 2000, a local atomic orbital density functional theory (DFT) based method in the local-density approximation (LDA), was used to calculate the Hamiltonian employed by the transfer matrix code [18]. We have first applied our approach to a xylyldithiol molecule connected to Au leads. Stretching of the molecule, corresponding to pulling the leads apart, has been simulated to compare with recent experiments [19]. In addition, FIREBALL 2000 was also used to calculate the Hellman-Feynman forces used to drift the xylyldithiol atoms upon stretching. In order to preserve the periodicity of the unit cell, the gold atoms were left fixed in these simulations, although the dynamics of the gold atoms are believed to be important in the stretching. Molecules were initially attached in the hollow-site configuration.

Our calculations agree within an order of magnitude with experimental calculations of the molecular conductance and indicate an interesting trend in the conductance upon stretching, with an apparent resonance for ~0.2 nm. Orbital plots [20] help explain this phenomenon. As the molecule is stretched, orbitals near the Fermi level change in degree of localization. At the resonance, there is a conductance enhancement due to the planarization of the molecule leading to enhanced coupling between gold states and molecular states. This is evident in the LUMO-like orbital, as shown in Fig. 3. We also note the effect of charge transfer at the metal-molecule interface with applied bias, agreeing well with other theoretical observations [21].

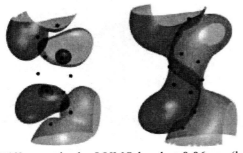

Fig. 3 Difference in the LUMO level at 0.06 nm (left) and 0.2 nm (right) stretch of the xylyldithiol molecule. The dots are atomic positions.

5 CONCLUSIONS

Coherent transport is becoming much more important in real semiconductor devices as the gate length is reduced into the nanometer regime. Hence, the role of coherent effects in device operation become important, and the control of decoherence within the source and drain assumes more importance [22]. We have developed a series of fully three dimensional quantum transport models, coupled with three dimensional Poisson solvers, to investigate such coherent transport in small systems. This has been illustrated here with an SOI MOSFET, and a molecule attached to two gold leads. These approaches are quite general and provide an alternative approach to other methods which have recently appeared in the literature [10-12, 23].

REFERENCES

[1] J. R. Barker and D. K. Ferry, Sol.-State Electron 23, 531, 1980.

[2] R. Chau, 2001 Silicon Nanoelectronics Workshop, Kyoto, Japan, June 10-11, 2001.

[3] B. Doris, M. Ieong, T. Kanarsky, Y. Zhang, R. A. Roy, O. Dokumaci, Z. Ren, F.-F. Jamin, L. Shi, W. Natzle, H.-J. Huang, J. Mezzapelle, A. Mocuta, S. Womack, M. Gribelyuk, E. C. Jones, R. J. Miller, H.-S. P. Wong, and W. Haensch, 2002 International Electron Device Meeting Technical Digest. New York: IEEE, 2002:267-270.

[4] M. Fischetti, J Appl Phys. 83, 270, 1998.

[5] K. Likharev, In: Morko H, ed. Advanced Semiconductor and Organic Nano-Technique. New York: Academic, 2002.

[6] R. Venugopal, Z. Ren, S. Datta, M. S. Lundstrom, and D. Jovanovic, J. Appl. Phys. 92, 3730, 2002.

[7] K. Natori, J. Appl. Phys. 76, 4879, 1994.

[8] M. J. Gilbert, R. Akis, and D. K. Ferry, J. Comp. Electron. *In press.*

[9] D. K. Ferry, R. Akis, M. J. Gilbert, and S. M. Ramey, to be published in *Silicon Nanoelectroncs*, S. Oda and D. K. Ferry, Eds., *in preparation.*

[10] F. G. Pikus, and K. K. Likharev, Appl. Phys. Lett.; 71, 3661, 1997.

[11] S. Datta, Superlatt Microstuct. 28, 253, 2000.

[12] J. Knoch, B. Lengeler, and J. Appenzeller, IEEE Trans. Elec. Dev. 49, 1212, 2002.

[13] T. Usuki, M. Takatsu, R. A. Kiehl, and N. Yokoyama, Phys. Rev. B, 50, 7615, 1994.

[14] T. Usuki, M. Saito, M. Takatsu, R. A. Kiehl, and N. Yokoyama, Phys. Rev. B, 52, 8244, 1995.

[15] R. Akis, D.K. Ferry, and J. P. Bird, Phys. Rev. B, 54, 17705, 1996.

[16] M. J. Gilbert, S. N. Milicic, R. Akis, and D. K. Ferry, IEEE Trans. Nanotechnolog., *In press.*

[17] M. J. Gilbert and D. K. Ferry, Superlatt. Microstruc., *in press.*

[18] O. F. Sankey and D. J. Niklewski, Phys.Rev. B 40, 3979, 1989.

[19] B. Xu and N. J. Tao, Science 301, 1221, 2003.

[20] G. Speyer, R. Akis, and D. K. Ferry, Superlatt. Microstruc., *in press.*

[21] Y. Xue and M. A. Ratner, Phys Rev. B 68, 115406, 2003.

[22] D. K. Ferry and J. R. Barker, Appl. Phys. Lett. 74, 582, 1999.

[23] M. Brandbyge, J.-L. Mozos, P. Ordejon, J. Taylor, and K. Stokbro, Phys. Rev. B 65, 165401, 2002.

A sub-40nm nanostructured La$_{0.7}$Sr$_{0.3}$MnO$_3$ planar magnetic memory

T. Arnal[*], M. Bibes[*], Ph. Lecoeur[*], B. Mercey[**], W. Prellier[**] and A.M. Haghiri-Gosnet[*]

[*] Institut d'Electronique Fondamentale, IEF/ UMR 8622, Université Paris Sud,
Bâtiment 220, 91405 Orsay Cedex, France, anne-marie.haghiri@ief.u-psud.fr
[**] CRISMAT – Laboratoire de Cristallographie et Sciences des Matériaux,
6 Boulevard du maréchal Juin, 14050 Caen Cedex, France, bernard.mercey@ismra.fr

ABSTRACT

A single-step nanolithography planar process, which allows generating the core-element of a spin-polarized magnetic memory in the fully spin-polarized La$_{0.7}$Sr$_{0.3}$MnO$_3$ (LSMO) manganite, is reported. Taking benefit of the proximity effects due to backscattered electrons, a conventional electron-beam patterning process at 30 KeV has been optimized to generate sub-50 nm-wide nanokinks in the magnetic microbridge. The best layout for the nanokinks, the electron beam patterning parameters and the results of the ion beam etching (IBE) for transferring these nanopatterns in the magnetic oxide are reported.

Keywords: spintronics, manganite, nanolithography, MRAM.

1 INTRODUCTION

Due to their large spin polarisation [1], ferromagnetic oxides, such as La$_{0.7}$Sr$_{0.3}$MnO$_3$ (LSMO), Fe$_3$O$_4$ and CrO$_2$, are promising materials for spin-electronics and can be used as injectors of spin-polarised electrons in magnetic memories. Also, their peculiar transport properties across interfaces, like tunnel junctions [2-3] and domain walls [4] allow an enhancement of the magneto-resistive response at low fields in memories applications.

The introduction of a nanometric constriction inside a thin ferromagnetic film has several effects, such as the pinning at this constriction of a domain wall (DW) [5]. Upon crossing a thin DW, the spin of the electrons cannot maintain a parallel alignment with the local magnetization, which can induce a large increase of the resistance [6]. If a magnetic field is applied or a pulsed current is injected, the DW can move and thus disappear. A large MR effect can be observed and gigantic values should be recorded in a half-metal. The challenge is thus to pattern and to etch these highly polarized half-metallic oxides in a sub-50 nm range.

We report on the optimization of the single-step nanolithography planar process that allows generating the core-element of a spin-polarized magnetic memory in La$_{0.7}$Sr$_{0.3}$MnO$_3$ (LSMO). Comparatively to conventional vertical tunnel junctions, this approach is easier for growth, since it is based on a single layer of LSMO. The best geometry for the nanokinks will be first discussed based on dynamic micromagnetic simulations of the magnetization reversal under the application of magnetic field. This paper will then focus on the planar process, that allows to create nanokinks in the oxide film using a conventional single-step 30 KeV electron beam lithography (EBL) (nanokinks and bridge are patterned at the same process level). It will be demonstrated that sub-40 nm wide nanokinks can be patterned in routine in La$_{0.7}$Sr$_{0.3}$MnO$_3$ microbridge by taking advantages of the EBL proximity effects.

2 PRINCIPLE OF THE DEVICE

The concept of the planar memory is schematically described in the Figure 1-a). At the center of the bridge, two couples of straight nanokinks isolate a single magnetic monodomain that exhibits a high shape anisotropy along the direction of the field. Magnetization will thus rotate at higher field values in this central part generating a bistable MR response for intermediate fields.

For proper operation of the device, the whole track should be free of undesired pinning centers like grain-boundaries. Therefore, this study requires first epitaxial LSMO thin films of high cristallinity. The epitaxial 40nm-thick LSMO films have been grown on (100)-oriented

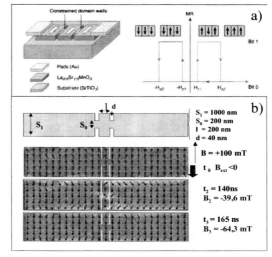

Figure 1 – a) principle of the planar memory with the MR response during reversal of the magnetization, b) micromagnetic simulations showing the pinning of two domain walls at the nanokinks for intermediate fields.

SrTiO₃ (STO) substrates using an original low-pressure Pulsed Laser Deposition (PLD) system (λ=248nm) at a reduced substrate temperature of 630 °C and under a low oxygen/ozone pressure of 5×10^{-3} mbar. The ozone concentration in the O_2/O_3 mixture ensures a full oxidation of the structure during growth. Details on the films deposition will be published elsewhere [7]. From our microstructural study, such epitaxial films appear free of any defect with a perfect "cube-on-cube" epitaxy, with an average grain-size larger than 1 μm. The surface roughness, deduced from AFM measurements, is about 0.2nm, i.e. half one unit cell. This exceptional flatness confirms the 2D step edge growth mode observed *in-situ* on the streaky RHEED patterns.

Dynamical micromagnetic simulations have been performed in order to determine the best geometry for both the kinks and the central domain. The Landau-Lifschitz equation has been resolved using the software *Oommf*. The thin film parameters were previously deduced from both SQUID and magneto-optical Kerr effects (MOKE) measurements. Note that the easy axes for magnetization are <110> and <1‾10> in the plane [8] and that the bridge is aligned along one of this axis. At 4K, the film parameters are: the saturation moment M_S = 5800 A/m (3.67 μB/f.u), the stiffness constant A = 1.84 10^{-12} J/m , the anisotropy coefficient K_1 = -3 10^5 erg/cm³ (K_2 = 0 with the <001> direction as a pure hard axis) and the damping coefficient α = 0.5. The pinning of DW occurs only if the lateral thickness **d** of the kink is lower than 40 nm, as well as if the central domain is smaller than 200 nm (see Fig. 1 – b)).

3   EXPERIMENTAL

A first level for large Cr/Au contact pads and alignment cross-marks is elaborated using UV lithography. Pads are aligned along <110>.

During the second step, the bridge and the nanokinks are patterned simultaneously. This high resolution nanopatterning is performed using a 30KeV SEMFEG microscope equipped with a beam-scanning system and coupled with a computer-assisted design software ELPHY (from Raith company). Exposures on the single 300nm-thick layer of 950 k molecular weight polymethyl - polymethacrylate (PMMA) organic resist were carried out with a beam diameter of approximately 1 nm and with a minimal beam current of 13 pA. Note that the adjustment of focus and the correction of astigmatism were performed by observing at high magnification a small contamination dot of several nanometers in diameter. The current was measured using a small Faraday cell placed near the sample. The exposure time per pixel was fixed at 15 μs and the pixel-to-pixel distance was of 13.7 nm, corresponding to a writing speed lower than 1 nm/s. All the electronic doses have been normalized to the value of D = 1 = 140 μC/cm². A conventional development of 2.5 minutes was performed in a mixture 1:3 of methyl-isobutyl-ketone (MIBK) / isopropylic alcohol (IPA). No ultra-sonic agitation [9] was used for this conventional process and the samples are gently rinsed in pure IPA during 2 min.

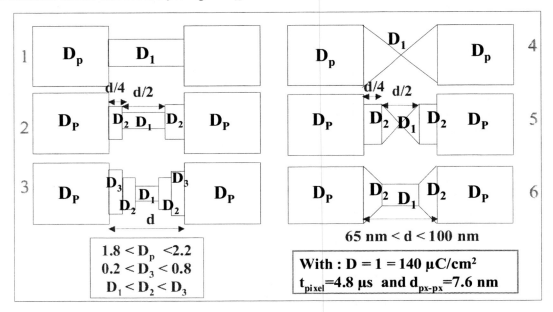

Figure 2 – Design of the six different geometries for the elaboration of the nanokinks (d is the width of the kink pattern). D_P and D_l are respectively the electronic doses of the bridge and of the kink. Note that for a better understanding, the dimensions of the patterns are not been reproduced at the exact scale: the width d of the nanokinks has been widely broadened.

The Al metallic mask for etching is evaporated and the nanopatterns are transferred to the LSMO film by an argon ion beam etching (IBE). During IBE, the sample holder was rotating and tilted at 45° to produce vertical profiles. The voltage and the current density of ions were fixed respectively at 500 V and 0.5 mA/cm^2. Finally, small Pt contacts lines are patterned on top of the nanostructured bridge in a 4-points configuration. The devices are inspected using a SEM imaging mode at a high magnification (> x50000) to precisely determine the final size of the nanokinks.

4   RESULTS AND DISCUSSION

Six different geometries have been tested for generating the nanokinks. These different designs are presented in the Figure 2. For these resolution tests, the total length of the bridge and its width are respectively 5 μm and 1 μm. The lateral thickness **d** of the kink is varied in the range 65 nm - 100 nm. In the geometries n°1, 2 and 3 (see figure 2), the kink has been cut out with lateral stripes. To take benefit of the proximity effects observed between the two large adjacent rectangles of the bridge, electron doses were decreased in these lateral stripes ($D_1 < D_2 < D_3$). For the geometries n°4, 5 and 6, the nanokinks has been divided using both rectangular and triangular-shape patterns.

Let us first consider the simplest geometry n°1 that is the conventional one. Because of the proximity effects of the 1μmx2.5μm rectangles, the width of the kink after development is around 40nm for a $d_{pattern}$ = 65 nm. To get a kink with the exact designed length (0.35 μm), doses for both the rectangles and the kink should be fixed at the optimal values D_p = 2 (280 μC/cm^2) and $D_l \sim$ 0.4 (56 μC/cm^2). Under these conditions, the kink is 40 nm – wide. This corresponds to a 38% lateral reduction, which traduces the strong importance of the proximity effects, due to backscattered electrons from LSMO surface under PMMA resist.

The kink has the appearance of a small resist bridge in suspension over the LSMO surface (Fig.3a-geometry n°1). The fragile nanobridge will act as a mask during the Al evaporation allowing the transfer towards a kink during the Al lift-off process. The shape of the fragile nanobridge has been then optimized using the six different geometries. Figure 3-a) gives tilted SEM images of the kink which allows a direct observation of the resist nanobridge profiles. It is shown that the geometry n°5 produces the strongest bridge: the nanobridge exhibits long feet for a better strength and preserves the desired 0.35 μm length. The strength of such "geometry n°5" nanobridges has been confirmed during the Al lift-off process, since the lateral width of the transferred in Al kink approaches the measured value after development.

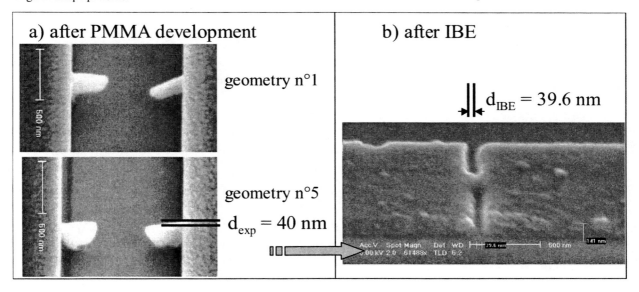

Figure 3 – Optimization of the shape of the nanobridge: a) tilted SEM images of the PMMA nanobridge after conventional development of the resist on top of the LSMO. The geometry n°5 appears to be the best way to get a strong bridge with large feet – b) SEM image of the geometry n° 5 nanokink that has been transferred in the LSMO thin film using IBE at 500 Volts.

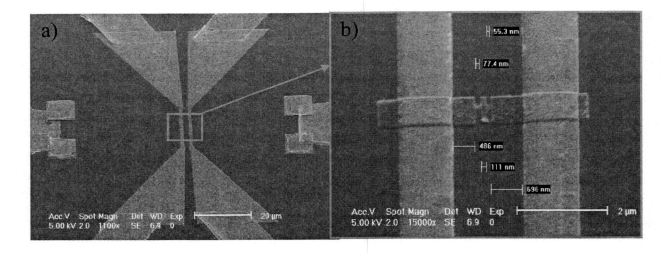

Figure 4 – a) a general top SEM view of the device after the last level of contact pads and b) a zoom of the central part of the nanostructured LSMO bridge. Note the two couples of kinks with lateral d values of d= 55 nm to 75 nm.

The majority of manganite nanodevices are ordinarily etched using conventional Ar ion beam etching (IBE) after an Al lift-off process. Al is preferred due to its high reactivity with oxygen. The bilayer Al / Al₂O₃ offers the best selectivity of about 2 for IBE of manganites. To avoid removal of the Al mask in solutions, which always attack the manganite film, the etching time is larger than the required time for a total Al etching. The kink transferred in LSMO by IBE is presented in the Figure 3-b) for the optimal geometry (n°5). No lateral reduction is observed between the PMMA development and the IBE process: the kinks exhibit the same width value of ~40 nm than the one of the nanobridge after development. This important result confirms first that both our lift-off and IBE processes are fully optimized for preserving lateral dimensions. Figure 4 gives both an overview of one device with its pads after the third contact level, as well as a zoom of the central domain with ~50nm-wide nanokinks.

5 CONCLUSION

Due to an optimized electron-beam pattern design, 40 nm-wide kinks have been successfully transferred in LSMO thin film, from a standard EBL process at 30 KeV. Proximity effects produce an interesting lateral reduction of the nanokink of around 40%.

The use of negative resist with a sufficiently high resolution will be tested in the future, in order to avoid the small over-etching for Al. The suppression of Al in the process should prevent the top surface of the manganite of a small reduction of oxygen. Moreover, the negative resist

mask can leave on the device after the complete process for further etching to narrow the constrictions.

REFERENCES

[1] J.H.Park, E.Vescovo, H.J.Kim, C.Kwon, R.Ramesh, T.Venkatesan, Nature 392, 794, 1998.

[2] M. Bowen, M. Bibes, A. Barthélémy, J.-P. Contour, A. Anane, Y. Lemaitre, A. Fert, App. Phys. Lett 82, 233, 2003.

[3] Y Lu, X W Li, G Q Gong, G Xiao, A Gupta, Ph Lecoeur, J-Z Sun, Y Y Wang, V P Dravid, Phys. Rev. B 54, R8357, 1996.

[4] J Wolfman, A-M Haghiri-Gosnet, B Raveau, C Vieu, E Cambril, A Cornette, H Launois, J. Appl. Phys. 89, 6955, 2001.

[5] P. Bruno, Phys. Rev. Lett. 83, 2425, 1999.

[6] M. Viret, Y. Samson, P. Warin, A. Marty, F. Ott, E. Sondergard, O. Klein, C. Fermon, Phys. Rev. Lett. 85, 3962, 2000.

[7] Ph. Lecoeur and B. Mercey, to be published.

[8] A.M. Haghiri-Gosnet, J. Wolfman, B. Mercey, Ch. Simon, Ph. Lecoeur, M. Korzenski, M. Hervieu, R. Desfeux, G. Baldinozzi, J. Appl. Phys. 88, 4257, 2000.

[9] S. Yasin, D.G. Hasko, H. Ahmed, App. Phys. Lett 78, 2760, 2001.

Discontinuous Gold Films for Nanocell Memories

David P. Nackashi[*], Neil H. Di Spigna[*], David A. Winick[*], Christian J. Amsinck[*]
Long Cheng[**], James M. Tour[**], Paul D. Franzon[*]

[*]Department of Electrical and Computer Engineering
North Carolina State University, Raleigh NC, 27695, paul_franzon@ncsu.edu
[**]Departments of Chemistry and Computer Science
Center for Nanoscale Science and Technology
Rice University, Houston TX, 77005, tour@rice.edu

ABSTRACT

An important component to the nanocell, among other self-assembled networks, is the fabrication of a framework by which molecular elements can be interconnected. This framework must be nanometric in scale, created in a material suitable for attachment chemistries and remain electrically discontinuous until molecular attachment. Utilizing the Volmer-Weber mechanism by which gold grows on silicon dioxide surfaces, nanometric islands of gold are fabricated to provide this framework. Using standard photolithography techniques, the regions where these islands are located are well defined. A two-layer photoresist stack is developed that prevents edge shorting around the boundaries of each region. The discontinuous gold films fabricated in this study are repeatable, offer a fill factor of 63%, and are easily patterned down to the one-micron scale.

Keywords: molecular electronics, Volmer-Weber, self-assembled, integration

1 INTRODUCTION

Since a moleculer mechanism was first proposed as a basis for electronic circuits [1], self-assembly of these circuits has remained a primary goal for most molecular electronic technologies [2]. By utilizing specific chemical reactions, molecules can be attached to specific regions on a semiconductor or metal surface. Although photolithographic technologies limit lateral resolutions, molecular self-assembly, when combined with photolithography, offers the potential to scale circuit architectures below these limits. The nanocell [3] is a device that proposes to use chemical self-assembly followed by post fabrication training to realize electronic circuits. The post fabrication training step reduces the demand for nanometer-scale layout and integration.

An important component to the nanocell is the fabrication of a framework by which molecular elements can be interconnected. This network must be: 1) fabricated using gold, or some other material appropriate for sulfur attachment chemistry, 2) electrically discontinuous, 3) patternable to allow molecular self-assembly only in well defined areas, and 4) be repeatable and robust to the environment. Previous research [4] has shown that gold, when evaporated in thin layers, grows discontinuously according to the Volmer-Weber mechanism. Using this technique, a process was developed to repeatably grow and pattern discontinuous gold films as a backbone for molecular interconnection.

2 FABRICATION

As previously shown [4], a thin layer of gold can be sputtered onto a thermally grown silicon dioxide film, which results in the formation of discontinuous islands.

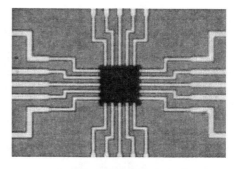
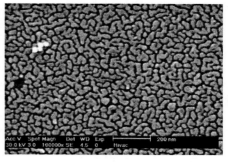

Figure 1: The nanocell device (top) and the center region of the discontinuous gold film (bottom).

Rather than growing as a continuous film, gold tends to exhibit three dimensional island formation during the initial stages of growth, known as the Volmer-Weber mechanism. This growth mechanism results from strong gold-gold interactions, coupled with relatively weak gold-silicon dioxide interactions [5].

2.1 Device Fabrication

The nanocell device (Figure 1) consists of a series of metallic leads that come into close proximity. Variations in numbers of leads and gap distances have been fabricated, including gaps under 1 micron. The substrate is a 1000 angstrom thick silicon dioxide film, thermally grown on a (100) p-type, 1-10 ohm-cm silicon wafer. The leads are formed by depositing 100 angstroms of titanium for adhesion, followed by 1500 angstroms of gold, patterned using a liftoff process.

2.2 Discontinuous Film Growth

The silicon dioxide surfaces, patterned with the Ti/Au leads, are first solvent cleaned with acetone followed by a methanol bath. Afterwards, the samples undergo a dehydration bake of 150°C for 10 minutes prior to the photolithography step used to pattern the films. After the photoresist stack is patterned with the nanocell regions defined, the samples are quickly placed under vacuum in the E-beam system.

The discontinuous gold films are deposited using E-beam evaporation, and patterned using a liftoff photolithography process. The evaporation system uses a Thermionics, Inc. 5-position electron gun for evaporation, and a Sycon, Inc. STM-100 quartz crystal microbalance for measuring film thicknesses. The base pressures for all evaporations are $5*10^{-6}$ Torr, and rates are held constant at 1.0 angstrom/second during the evaporation for controllability. Evaporations were timed and the average thickness was monitored using the Sycon thickness monitor. Comparisons between samples evaporated with different measured thicknesses were made. Endpoints were determined by monitoring the average thickness on the Sycon monitor, which takes into account any small perturbations in the evaporation rate. Since the thickness reported by the monitor represents an average, rather than an absolute, these values are used only as a reference for experimentation.

2.3 Edge Defects

One observation made during early experiments is the tendency for the gold islands to agglomerate against the edges of the photoresist used to pattern the regions of discontinuous film within the nanocell. Easily seen using SEM microscopy (Figure 2), this agglomeration against the step edges results in the formation of electrical shorting paths between any two patterned leads. These shorting paths render a device useless, since they cannot be corrected and the low-resistive path prevents the measurement of any current passing through the highly resistive paths within an assembled nanocell. The frequency of devices that exhibit edge shorts is related to the perimeter of the nanocell region in which the discontinuous film is patterned, and the orientation of the sample in the evaporation system. This makes reasonable sense, because the tendency of the gold to agglomerate into the resist edges should be greater in the direction of the evaporation flux.

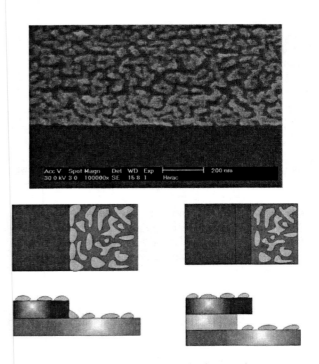

Figure 2: Edge shorts (top) and photoresist process to minimize them (bottom).

3 CHARACTERIZATION

3.1 Morphology

It is understood that the evaporation method used to deposit the gold films significantly contributes to the morphology of the gold islands themselves. Previous studies using sputter deposition [4, 5] result in small, more colloidal gold island formation. Using E-beam evaporation, the average area and perimeter of each gold island tends to be greater, an effect that might be explained by the lower mobility of evaporated gold versus sputtered gold. A study is planned that will experimentally compare sputtered verses evaporated gold targets, and derive a quantitative analysis of island area, perimeter and fill factor.

To statistically study the discontinuous gold films, a custom image analysis software routine was developed using LabVIEW (National Instruments, Inc.) Shown below (Figure 3) is a portion of an SEM image taken of a Sycon measured 40 angstrom thick film, evaporated in approximately 40 seconds. The scale bar was separated from the image to remove any inconsistencies during analysis. By correlating the scale bar to the number of pixels, the resulting image area was determined to be 1200nm x 707nm, or 848,588nm^2. A summary of the data collected is shown below (Table 1). Calculations to determine the average gap distances are underway, but the average is estimated to be on the order of 10nm.

Island Analysis	
Image Area	848,588 nm^2
Number of Particles	196
Average Size	2713 nm^2
Fill Factor	0.63

Table 1: Island analysis for 40 angstrom measured deposition condition, image from Figure 3

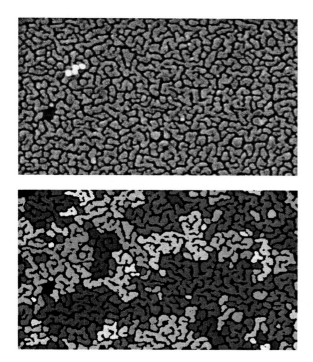

Figure 3: A section (504x297 pixels, or 1200x707nm) of the SEM image (above), and after analyzed using the particle counter routine (below)

Although the average particle size was determined to be 2713 nm^2, gold islands ranged in size from 50nm^2 to just over 10,000nm^2. The majority of the particles (72%) however, were determined to be in the 225nm^2 to 1700nm^2 range. The presence of only a few large area particles contributes to the resulting average area calculation.

The contribution of these relatively few, large islands is also important to the performance of the nanocell network. The gap distances between gold islands formed under the above conditions are too large for direct molecular attachment. When used as a seed layer, however, these islands can be bridged using molecularly passivated gold colloids (Figure 4) or nanorods [6]. The larger islands can allow for low attachment densities, requiring only a few connections be made to complete a circuit between two leads [6]. Although the nature of these connections is not conclusively determined, nanometallic filament formation is suspected between the gold nanorods and the gold islands. More experimentation is under way using islands fabricated from refractory metals, such as palladium, to minimize electromigration and isolate the current path within the nanocell.

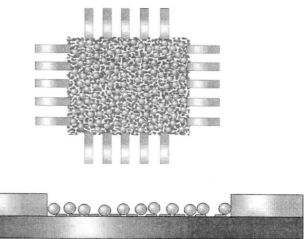

Figure 4: Representation of an assembled nanocell using metallic colloids attached to a discontinuous gold film.

3.2 Electrical Characterization

Three separate experiments were used to determine the electrical discontinuity of the gold films. First, electrical measurements were taken using a Magne-tron Instruments M-700 4-point probe system. For the sample analyzed above, the sheet resistance was out of range, indicating a >10 G-ohm sheet resistance. This is the expected result for a discontinuous film. Second, a series of 1500 angstrom thick gold pads were shadow masked on an unpatterned discontinuous film sample, formed using the above conditions. Distances between the pads ranged from 100 microns to 1 millimeter. Currents measured were in the pA range at 20V, also indicating open circuits. Finally, the discontinuous gold films were patterned on a substrate containing patterned gold probes, with gap distances from 10 microns down to under 1 micron in length. Currents measured from these devices indicated either open circuits (pA range at 20V) or direct shorts (approximately 100 ohm resistances.) The shorts were much more prevalent in the gap distances less than 5 microns. After SEM analysis, it was determined that the cause of these shorting paths was from the edge defects previously explained (Figure 2). Using the two-layer photoresist process, the number of shorts was greatly reduced, and confined to the gaps less than 3 microns.

4 CONCLUSION

Discontinuous gold films have been fabricated and patterned, resulting in a large fill factor and average gap distances of approximately 10nm. The perimeter and area of these islands make this technique desirable as a seed layer for molecular attachment, using gold colloids or nanorods passivated with thiol-terminated molecules. This process offers an alternative to previously researched seed layers fabricated from the self-assembly of gold colloids, which often suffer from lack of density, patternability and reproducibility. The discontinuous gold films fabricated in this study are repeatable, offer a fill factor of 63%, and are easily patterned down to the one-micron scale.

Acknowledgement. Financial support came from the Defense Advanced Research Projects Agency.

REFERENCES

[1] A. Aviram, M. A. Ratner, "Molecular Rectifiers," Chem. Phys. Lett., **29**, pp277-283, 1974.

[2] R. StanleyWilliams, P. J. Kuekes, "Molecular Nanoelectronics," IEEE Int. Sym.On Circuits and Systems, pp. 5-7, 2000.

[3] J. M. Tour, W. L. V. Zandt, C. P. Husband, L. S. Wilson, P. D. Franzon, and D. P. Nackashi, "Nanocell Logic Gates for Molecular Computing," IEEE Transactions on Nanotechnology, **1**, pp100-109, 2002.

[4] F. Parmigiani, G. Samoggia, G.P.Ferraris, "Optical Properties of sputtered gold Clusters," J. Appl. Phys., **57**, pp. 2524-2528, 1985.

[5] Paulson, C.G., Friedberg, A.L., "Coalescence and Agglomeration of Gold Films," Thin Solid Films, **5**, p47, 1970.

[6] James M. Tour, Long Cheng, David P. Nackashi, Yuxing Yao, Austen K. Flatt, Sarah K. St. Angelo, Thomas E. Mallouk and Paul D. Franzon, "Nanocell Electronic Memories," **125**, pp.13279-13283, 2003.

Valence-band Energies of GaAs/AlGaAs and InGaAs/InP V-groove [1$\bar{1}$0] Quantum Wires

B. Lassen*, L. C. Lew Yan Voon*,**, R. Melnik* and M. Willatzen*

*Mads Clausen Institute, University of Southern Denmark,
Grundtvigs Allé 150, DK-6400 Sønderborg, Denmark, benny@mci.sdu.dk
**Department of Physics, Worcester Polytechnic Institute,
100 Institute Road, Worcester, MA 01609, USA, llew@wpi.edu

ABSTRACT

Comparison between the Burt-Foreman and Luttinger-Kohn valence-band Hamiltonians have been performed for realistic V-groove GaAs/AlGaAs and InGaAs/InP quantum wires. Significant differences in band structure is only found for InGaAs/InP quantum wires.

Keywords: quantum wires, V-groove, $k \cdot p$ theory, valence band, band structure

1 INTRODUCTION

High quality V-groove GaAs/AlGaAs quantum wires (QWR's) of nanometer cross-section have been grown by low pressure organometallic chemical vapor deposition techniques [1]. The structural parameters and quality were initially ascertained using conventional and high resolution transmission electron microscopy, and by low-temperature cathodoluminescence [1]. Subsequently, low temperature polarized photoluminescence and polarized photoluminescence excitation (PLE) experiments were carried out and, together with a $k \cdot p$ calculation, it was deduced that the optical transitions involved one-dimensional excitons and that a strong suppression of the band-edge absorption occured [2], [3].

The calculation was crucial in labeling the nature of the structures in the optical spectra. In the original papers [2], [3], a one-band parabolic effective-mass equation was used for the conduction states and a four-band Luttinger-Kohn Hamiltonian was used for the valence states. Excitonic corrections were *ad hoc* accounted for by rigidly redshifting the theoretical single-particle transition energies by 10-20 meV for each structure investigated. Excellent agreement was thus obtained between theory and low-incident-angle linearly polarized PLE spectra (on planarized structures in order to minimize surface grating effects) for the 2.5 nm structure (the dimension being the thickness of the crescent region of the QWR), but not so for the other two structures (5 nm and 1.5 nm QWR's). We note that Stier and Bimberg had carried out an eight-band calculation [4] but concentrated on piezoelectric effects.

The goal of this work is to investigate whether the discrepancy between the theory and experiment in [3] is band-structure related. Specifically, we address the issue of whether the Luttinger-Kohn (LK) Hamiltonian is appropriate for this QWR structure. The theoretically correct Hamiltonian has been shown to be the so-called Burt-Foreman (BF) one [5], [6]. Foreman had shown that the two can give very different heavy hole (hh) effective masses for InGaAs/GaAs (001) quantum wells (QW's). Meney *et al.* [7] had also shown that the four-band BF Hamiltonian gives different band structures compared to the LK one for 10 nm GaAs/AlAs (001) QW's, especially for intermediate wave vectors, with energy differences of up to 10 meV. Similar conclusions were recently obtained for the intersubband optical matrix elements [8]. We recently implemented the BF Hamiltonian for a triangular GaAs/AlAs (001) QWR with wire axis in the [001] direction and showed that there are differences of up to 10 meV for a baseline of 10 nm [9]. The V-groove QWR's grown have a wire axis along the [1$\bar{1}$0] direction. It remains to prove that there can be a significant difference between the BF and LK calculations for such a structure and geometry. We, therefore, derived the BF Hamiltonian for a QWR along the [1$\bar{1}$0] direction and carried out calculations for both GaAs/AlGaAs and InGaAs/InP QWR's. V-groove QWR's of InGaAs with {111}A InP facets and on (001) oriented InP:S substrates have also been grown by low-pressure MOCVD [10].

2 GEOMETRY OF QWR's

The actual geometry of the V-groove QWR's is fairly complicated and, thus, complete information is not available from the published literature. The most detailed description can be found in [1]. The overall structure is that of a crescent-shaped QWR with QW sidearms and accompanied by two vertical QW's that have an inverted Y structure (see Fig. 1 of [1]). Here we report comparisons for the 2.5 nm QWR for which the experimental shape is available [2] (Fig. 1).

3 THEORY

We have derived the BF Hamiltonian for a QWR with [1$\bar{1}$0] wire axis. We will take as starting matrix the D matrix of Stravinou and van Dalen [11]. One can break down the process of obtaining the Hamiltonian in the new rotated coordinate system into three steps.

The first step involves rotating the basis functions of the three-band Hamiltonian. The second involves replacing the unprimed (unrotated) wave vector by the primed (rotated) one. The third involves re-expressing the matrix in terms of the coupled angular momentum JM states. One can give a succinct presentation of the procedure. Given

$$x'_j = U_{ji}x_i, \ k'_j = U_{ji}k_i, \tag{1}$$

and the original unrotated three-band Hamiltonian matrix

$$D_{\alpha\beta}(\mathbf{k}) = \sum_{ij} D_{\alpha\beta}^{ij}k_ik_j,$$

then, the first two steps lead to

$$D'_{\alpha'\beta'}(\mathbf{k}') = \sum_{\alpha\beta}\sum_{ij}\sum_{i'j'} U_{\alpha'\alpha}U_{\beta'\beta}U_{i'i}U_{j'j}D_{\alpha\beta}^{ij}k_{i'}k_{j'}. \tag{2}$$

This results in the matrix given in Table 1. A, B, C_1 and C_2 are as defined in [11]:

$$
\begin{aligned}
A &= 1 - 6\sigma - 12\delta = -(\gamma_1 + 4\gamma_2), \\
B &= 1 - 6\pi = -(\gamma_1 - 2\gamma_2), \\
C_1 &= 6\delta - 6\sigma = 1 + \gamma_1 - 2\gamma_2 - 6\gamma_3, \\
C_2 &= 6\pi = 1 + \gamma_1 - 2\gamma_2,
\end{aligned}
\tag{3}
$$

where σ, π, δ are the Foreman parameters [6] reflecting the interaction of the valence states with atomic-like s, p, d states, and the γ_i's are the usual Luttinger parameters. The final step of converting to JM states will not be written out explicitly. One significant difference between the BF and LK Hamiltonians is the presence of an additional interaction between light-hole states of opposite spin and denoted by the matrix element C; a similar difference is present in the quantum-well Hamiltonian [6], [11].

4 ANALYSIS

Analysis of the resulting Hamiltonian reveals a number of interesting results.

4.1 Zone-center Energies

First, there can be a difference between LK and BF energies even at the Brillouin zone center. This is a new result not present in previous studies on quantum wells. However, at the zone center of the first Brillouin zone, all the off-diagonal matrix elements of the BF and LK Hamiltonians are the same. Furthermore, the Luttinger parameters between GaAs and $Al_{0.3}Ga_{0.7}As$ are

Table 1: Stravinou and van Dalen Hamiltonian in $[1\bar{1}0]$ rotated LS basis.

$$
D'(\mathbf{k}') = \frac{\hbar^2}{2m_0}
\begin{pmatrix}
|X'\rangle & |Y'\rangle & |Z'\rangle \\
k'_x A k'_x + k'_y B k'_y + k'_z B k'_z & k'_x C_1 k'_y - k'_y C_2 k'_x & k'_x C_1 k'_z - k'_z C_2 k'_x \\
k'_y C_1 k'_x - k'_x C_2 k'_y & \frac{1}{2}[k'_y(A+B+C_1-C_2)k'_y + k'_z(A+B-C_1+C_2)k'_z + k'_x B k'_x] & \frac{1}{2}[k'_y(A-B+C_1-C_2)k'_z + k'_z(A-B-C_1+C_2)k'_y] \\
k'_z C_1 k'_x - k'_x C_2 k'_z & \frac{1}{2}[k'_z(A-B+C_1-C_2)k'_y + k'_y(A-B-C_1+C_2)k'_z] & \frac{1}{2}[k'_y(A+B-C_1-C_2)k'_y + k'_z(A+B+C_1-C_2)k'_z + k'_x B k'_x]
\end{pmatrix}
$$

not very different, but there are substantial differences between $In_{0.53}Ga_{0.47}As$ and InP (Table 2). Together, these suggest that there might only be a small difference between the BF and LK energies for GaAs/AlGaAs but larger for InGaAs/InP QWR's.

Table 2: Luttinger parameters.

	γ_1	γ_2	γ_3
GaAs [12]	6.85	2.10	2.90
AlAs [12]	3.45	0.68	1.29
InAs [13]	20.4	8.3	9.1
InP [13]	4.95	1.65	2.35
$In_{0.53}Ga_{0.47}As$	14.03	5.39	6.19

4.2 Boundary Conditions

The standard technique is to integrate the differential equations across an interface. In the process, contributions arise from, for example, $\widehat{k}'_i \mathcal{A}(\mathbf{r}')f(\mathbf{r}')$ terms, where $\mathcal{A}(\mathbf{r}')$ are position-dependent material parameters (within LK theory, they are related to the Luttinger parameters) and $f(\mathbf{r}')$ is the envelope function. For quantum wells, this corresponds to the term $\widehat{k}'_z \mathcal{A}(z')f(z')$; when applied to the interaction between heavy-hole and light-hole states of the same spin, it only leads to π and δ contributions using the BF Hamiltonian [11], but it also includes the σ contribution if the LK Hamiltonian is used. In the QWR case, the same matrix element contains the term $\widehat{k}'_x C_1 f(z')$ which includes the σ contribution; this is the s-like coupling term absent from previous implementations of the BF theory, but which is present in the LK theory [6]. One can similarly analyze the other matrix elements. Overall, one expects that, at finite k_z the difference between BF and LK might be reduced.

5 NUMERICAL RESULTS

The above analysis was tested with a numerical implementation of the problem using a finite-element algorithm for solving the coupled partial differential equations.

For GaAs/AlGaAs, we obtained energy differences of $< 1\,\mathrm{meV}$ for the small wave vector values relevant to the experimental data (Fig. 2). We further verified that this is due to the wire orientation by finding a small difference of $< 1\,\mathrm{meV}$ for the same triangular wire we used in [9] (for which there were differences of order $10\,\mathrm{meV}$ between BF and LK energies when oriented along [001]) but now oriented along [1$\bar{1}$0]. This confirms that the effective difference between the BF and LK theories is reduced for the [1$\bar{1}$0]-oriented wire compared to [001].

The sum squared of the envelope-function components for the lowest electron and hole states are given in Fig. 1, showing the confinement of these states inside the QWR. The electron state was obtained using a one-band model.

We have also redone the above calculations for a lattice-matched $In_{0.53}Ga_{0.47}As$/InP QWR (Fig. 2). This system is chosen because V-groove QWR's based upon these materials have also been made [10], though the experimental structures are probably all strained. It is also interesting because there is a larger difference in the Luttinger parameters. Indeed, this translates into a correspondingly larger difference in LK and BF results (Fig. 2).

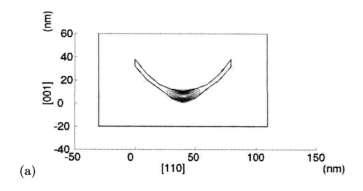

(a)

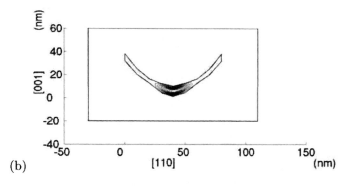

(b)

Figure 1: Geometry of embedded wire and first (a) electron, (b) hole probability density for the GaAs/AlGaAs QWR.

6 CONCLUSIONS

It is, therefore, concluded that the discrepancy between theory and experiment for GaAs/AlGaAs V-groove QWR's is not due to the use of the Luttinger-Kohn Hamiltonian. Possible reasons for the differences are excitonic binding energies, differences (both in geometry and composition) between the real and modeled structures, and the Luttinger parameters are not exactly known. On the other hand, InGaAs/InP QWR's have a better potential of revealing differences between the Luttinger-Kohn and the Burt-Foreman Hamiltonians.

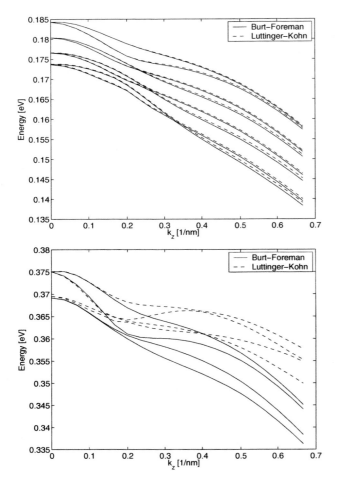

Figure 2: Valence band structure for 2.5 nm wire using BF (lines) and LK (crosses) Hamiltonians. Top (bottom) for GaAs/AlGaAs (InGaAs/InP).

ACKNOWLEDGMENTS

This work was supported by an NSF CAREER award (NSF Grant No. 9984059) and a Balslev award (Denmark).

REFERENCES

[1] A. Gustafsson, F. Reinhardt, G. Biasiol, and E. Kapon, "Low-pressure organometallic vapor deposition of quantum wires on V-grooved substrates," Appl. Phys. Lett. **67**, 3673 (1995).

[2] F. Vouilloz, D. Y. Oberli, M.-A. Dupertius, A. Gustafsson, F. Reinhardt, and E. Kapon, "Polarization anisotropy and valence band mixing in semiconductor quantum wires," Phys. Rev. Lett. **78**, 1580 (1997).

[3] F. Vouilloz, D. Y. Oberli, M.-A. Dupertius, A. Gustafsson, F. Reinhardt, and E. Kapon, "Effect of lateral confinement on valence-band mixing and polarization anisotropy in quantum wires," Phys. Rev. B **57**, 12378 (1998).

[4] O. Stier and D. Bimberg, "Modeling of strained quantum wires using eight-band $k \cdot p$ theory," Phys. Rev. B **55**, 7726 (1997).

[5] M. G. Burt, "The justification for applying the effective-mass approximation to microstructures," J. Phys. Cond. Matter **4**, 6651 (1992).

[6] B. A. Foreman, "Effective-mass Hamiltonian and boundary conditions for the valence bands of semiconductor microstructures," Phys. Rev. B **48**, 4964 (1993).

[7] A. T. Meney, B. Gonul, and E. P. O'Reilly, "Evaluation of various approximations used in the envelope-function method," Phys. Rev. B **50**, 10893 (1994).

[8] C. Galeriu, L. C. Lew Yan Voon, and M. Willatzen, "Modelling intersubband transitions in the valence bands of quantum-well structures," In H. Sigg, editor, *Proc. Workshop on Intersubband Transitions in Quantum Wells* (2003).

[9] B. Lassen, R. Melnik, M. Willatzen, and L. C. Lew Yan Voon, "Differences between Luttinger-Kohn and exact envelope function approaches for quantum-wire electronic bandstructures," In H. Sigg, editor, *Proc. Workshop on Intersubband Transitions in Quantum Wells* (2003).

[10] M. Kappelt, M. Grundmann, A. Krost, V. Trck, and D. Bimberg, "InGaAs quantum wires grown by low pressure metalorganic chemical vapor deposition on InP V-grooves," Appl. Phys. Lett. **68**, 3596 (1996).

[11] P. N. Stravinou and R. van Dalen, "Operator ordering and boundary conditions for valence-band modeling: Application to [110] heterostructures," Phys. Rev. B **55**, 15456 (1997).

[12] *Landolt-Börnstein, Numerical Data and Functional Relationships in Science and Technology*, volume 22a of *New Series, Group III* (Springer-Verlag, Berlin, 1982).

[13] S. A. Stoklitsky, Q. X. Zhao, P. O. Holtz, B. Monemar, and T. Lundström, "Optical intervalence-subband transitions in strained p-type $In_{1-x}Ga_xAs$/InP quantum wells," J. Appl. Phys. **77**, 5256 (1995).

Spin-orbit Interaction and Energy States in Nanoscale Semiconductor Quantum Rings

Yiming Li

Department of Nano Device Technology, National Nano Device Laboratories
Microelectronics and Information Systems Research Center, National Chiao Tung University
P.O. Box 25-178, Hsinchu 300, TAIWAN, ymli@mail.nctu.edu.tw

ABSTRACT

We study the effect of spin-orbit iteration on the electron energy spectra in tree-dimensional (3D) nanoscale semiconductor quantum rings. Ultrasmall InAs quantum ring embedded in GaAs matrix is numerically solved with the effective one electronic band Hamiltonian, the energy- and position-dependent electron effective mass approximation, and the spin-dependent Ben Daniel-Duke boundary conditions. The multishift QR algorithm is implemented in the nonlinear iterative method for solving the corresponding nonlinear eigenvalue problem. It is found that the spin-dependent boundary conditions lead to a spin-splitting of the electron energy states with non-zero angular momentum. The splitting is strongly dependent on the ring dimension. Meanwhile, it is larger than that of quantum dot and demonstrates an experimentally measurable quantity for relatively small quantum rings.

Keywords : nanoscale semiconductor quantum rings, spin-orbit interaction, mathematical modeling, computer simulation, multishift QR algorithm.

1 INTRODUCTION

Semiconductor quantum nanostructures, such as quantum dots, quantum rings , and quantum molecules in recent years have been of a great interest from experimental and theoretical points of view [1]–[20], [22], [23]. Semiconductor quantum rings have been fabricated and studied recently [13]–[17]. They possess very interesting physical properties including, such as far-infrared spectrum and magnetic effects. Unusual optical and magnetic properties can be controlled by morphological changes during the fabrication of nanostructure and by the number of electrons which are bounded in a quantum ring. Therefore, they are very attractive for potential applications in nanoelectronics and optics. It is known that the electron spin plays an important role in the manipulation of energy states and modify the intrinsic property of structures. Spintronics is currently a fascinating branch in electronics. Study of the spin-dependent energy spectra is an essential element for the development of semiconductor spintronics. In semiconductor spintronic device, the carrier's generation-recombination and transport can be controlled by electron spin polarization and the electron charge. It becomes necessary to study the spin-dependent electron confinement for quantum nanostructures in the development of semiconductor spintronics. It has been known that the spin-orbit interaction impacts the energy and electronic properties for III-V semiconductors [18]–[20]. However, no clear description of the spin-orbit interaction on ultra-small nanoscale quantum rings can be drawn from the literature.

We in this paper investigate the effect of spin-orbit interaction [1]–[5] on the electron energy states in nanoscale semiconductor quantum rings. The effective one-band Hamiltonian approximation with the spin-dependent Ben Daniel-Duke boundary conditions is formulated and solved numerically. Most of calculations of the electron spectra in semiconductor quantum nanostructures were done within different 1D approximations. The confinement potential in the radius direction often was approximated by a parabolic potential and in the height direction was taken to be the infinite outside the quantum ring. We for the first time adopt a realistic hard-wall (of finite height) 3D confinement potential that is induced by real discontinuity of the conduction band at the edge of the quantum ring. To solve the corresponding 3D effective one band Schrödinger equation, the multishift QR algorithm is implemented in the nonlinear iterative method for solving the corresponding nonlinear eigenvalue problem. The QR algorithm for solving the nonsymmetric eigenvalue problem is one of the jewels in the crown of matrix computations. With the multishift QR algorithm [21], it is possible to reduces the cost of simulation time up to $1 - 2$ orders of magnitude. The nonlinear iterative method was successfully developed by us for the simulation of semiconductor quantum nanostructures [5], [12], [17]. The spin-dependent boundary conditions mainly come from a difference between the spin-orbit interaction parameters in the quantum ring and the semiconductor environment matrix. Due to significant spin-orbit interaction in the nonsimply connected torus topology, experimentally measurable spin splitting can be observed in InAs/GaAs quantum ring. The spin splitting depends on the variations of geometric (dot- and ring-liked) structures. They are dominated by the inner radius, base radius, and height of the quantum ring. Under zero magnetic fields, it is found that quantum ring can produce about 2 meV spin splitting of excited electronic states which is substantially larger than that of quantum dot (~1 meV).

This article is organized as follows. Section 2 introduces the mathematical model and the simulation technique. Section 3 describes the results illustrating the effect of the spin-orbit interaction on the electron energy spectra for ultra-small InAs/GaAs quantum rings. Section 4 draws conclusions.

2 MATHEMATICAL MODEL AND SIMULATION METHOD

Consider the electrons are confined in three-dimensional quantum dot structures and apply an effective one electronic band Hamiltonian, we have [5], [12], [17], [18]

$$\hat{H} = \hat{H}_0 + \hat{V}_{so}(\mathbf{r}), \qquad (1)$$

where $\hat{H}_0$ is the system Hamiltonian without spin-orbit interaction and $V_{so}(\mathbf{r})$ indicates the spin-orbit interaction for the conduction band electrons. The expression for $\hat{H}_0$ is as follows:

$$\hat{H}_0 = -\frac{\hbar^2}{2}\nabla_{\mathbf{r}}\left(\frac{1}{m(E,\mathbf{r})}\right)\nabla_{\mathbf{r}} + V(\mathbf{r}), \qquad (2)$$

where $\nabla_{\mathbf{r}}$ stands for the spatial gradient, $m(E,\mathbf{r})$ is the energy and position dependent electron effective mass

$$\frac{1}{m(E,\mathbf{r})} = \frac{P^2}{\hbar^2}[\frac{2}{E + E_g(\mathbf{r}) - V(\mathbf{r})}$$
$$+\frac{1}{E + E_g(\mathbf{r}) + \Delta(\mathbf{r}) - V(\mathbf{r})}]. \qquad (3)$$

In (3), $V(\mathbf{r})$ is the confinement potential, $E_g(\mathbf{r})$ and $\Delta(\mathbf{r})$ stand for the position dependent band gap and the spin-orbit splitting in the valence band, respectively. P in (3) is the momentum matrix element. The spin-orbit interaction for the conduction band electrons $V_{so}(\mathbf{r})$ is given by [2], [5], [12], [17]

$$\hat{V}_{so}(\mathbf{r}) = i\nabla\beta(E,\mathbf{r}) \cdot [\hat{\boldsymbol{\sigma}} \times \nabla], \qquad (4)$$

where $\beta(E,\mathbf{r})$ is the spin-orbit coupling parameter and $\hat{\boldsymbol{\sigma}} = \{\boldsymbol{\sigma_x}, \boldsymbol{\sigma_y}, \boldsymbol{\sigma_z}\}$ is the vector of the Pauli matrices. The energy and position dependent $\beta(E,\mathbf{r})$ has the form

$$\beta(E,\mathbf{r}) = \frac{P^2}{2}[\frac{1}{E + E_g(\mathbf{r}) - V(\mathbf{r})}$$
$$-\frac{1}{E + E_g(\mathbf{r}) + \Delta(\mathbf{r}) - V(\mathbf{r})}]. \qquad (5)$$

For those quantum ring systems that have sharp discontinuity on the conduction band interfaces between

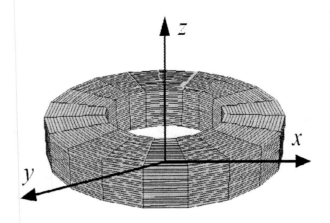

Figure 1: A 3D plot of the nanoscale semiconductor quantum ring.

the quantum ring (material 1) and semiconductor matrix (material 2), the hard-wall confinement potential is

$$V(\mathbf{r}) = \begin{cases} 0, & \mathbf{r} \in \text{material 1} \\ V_0, & \mathbf{r} \in \text{materia l } 2, \end{cases} \qquad (6)$$

where V_0 is the structure band offset. Combining the Hamiltonian in equations (1), (2), (4), and taking an integration of this Schrödinger equation with respect to the direction perpendicular to the system interface, the spin dependent Ben Daniel-Duke boundary conditions for the electron wave function $\Psi(\mathbf{r})$ is written as follows:

$$\Psi_{\text{material }1}(\mathbf{r}_s) = \Psi_{\text{material }2}(\mathbf{r}_s)$$
$$\left\{\frac{\hbar^2}{2m(E,\mathbf{r}_s)}\nabla - i\beta(E,\mathbf{r}_s)[\hat{\boldsymbol{\sigma}} \times \nabla]\right\}_n \Psi(\mathbf{r}_s) = C_0, \qquad (7)$$

where C_0 is some constant, $\mathbf{r}_s$ denotes the position of the system interface. We note that the expressions of electron effective mass in (3), spin-orbit coupling parameter in (5), and the equations of Ben Daniel-Duke boundary condition in (7) are all energy and position dependent relationships in this study.

We now consider the quantum ring as shown in Fig. 1, with the inner radius R_{in}, radius R_0 and the thickness z_0 in the cylindrical coordinate (R, ϕ, z). The origin of the system is at the center of the structure and the z axis is chosen along the rotation axis. Since the system is cylindrically symmetric, the wave function can be represented as

$$\Psi(\mathbf{r}) = \Phi(R, z)\exp(il\phi), \qquad (8)$$

where $l = 0, \pm1, \pm2, ...$ is the electron orbital quantum number and the original model remains a two-dimensional problem in (R, z) coordinate. From Eqs. (1)-(6) and (8), we obtain equations

$$-\frac{\hbar^2}{2m_1(E)}\left(\frac{\partial^2}{\partial R^2} + \frac{\partial}{R\partial R} + \frac{\partial^2}{\partial z^2} - \frac{l^2}{R^2}\right)\Phi_1(R,z) =$$

$$E\Phi_1(R,z), \ \forall(R,z) \in \text{material 1} \qquad (9)$$

$$-\frac{\hbar^2}{2m_2(E)}\left(\frac{\partial^2}{\partial R^2} + \frac{\partial}{R\partial R} + \frac{\partial^2}{\partial z^2} - \frac{l^2}{R^2}\right)\Phi_2(R,z)$$

$$+V_0\Phi_2(R,z) =$$

$$E\Phi_2(R,z), \ \forall(R,z) \in \text{material 2}. \qquad (10)$$

For the same reasons that the problem is symmetry along the z axis, the spin-dependent boundary conditions in (7) are given by

$$\Phi_1(R,z) = \Phi_2(R,z), z = f(R) \text{ and} \qquad (11)$$

$$\frac{1}{m_1(E)}\left\{\frac{\partial\Phi_1(R,z)}{\partial R} + \frac{df(R)}{dR}\frac{\partial\Phi_1(R,z)}{\partial R}\right\}_{z=f(R)}$$

$$-\frac{1}{m_2(E)}\left\{\frac{\partial\Phi_2(R,z)}{\partial R} + \frac{df(R)}{dR}\frac{\partial\Phi_2(R,z)}{\partial R}\right\}_{z=f(R)}$$

$$+\frac{2\boldsymbol{\sigma}(\beta_1 - \beta_2)}{\hbar^2}\frac{l}{R_0}\Phi_1(R_0,z) = 0, \qquad (12)$$

where $z = f(R)$ is the generating contour of the quantum ring on $\{R,z\}$ plane, σ refers to the electron spin polarization along the z direction. The electron energy state and wave function in this model have a tightly coupled and complicated relationships with the parameters of quantum ring and the electron angular momentum. The solution of the equations $(9) - (10)$ together with the boundary conditions $(11) - (12)$ is solved with the proposed numerical method.

The energy dependence of the electron effective mass and spin-orbit coupling parameter complicates the analytical solution [5], [12], [17], [22], [23]. The finite volume discretized Schrödinger equation leads to a nonlinear algebraic eigenvalue problem and is solved with the nonlinear iterative method to calculate all bounded energy levels. To solve the corresponding matrix eigenvalue problem efficiently, we use the multishift QR method. Computationally, this method converges monotonically and is highly cost effective in the computer simulation of 3D quantum rings. The nonlinear iterative method is outlined as:

Step 1. Set initial energy E_0;

Step 2. Compute electron effective mass m;

Step 3. Compute spin-orbit coupling parameter β;

Step 4. Solve the Schrödinger equation; and

Step 5. Update the newer computed energy and back to Step 2.

The iteration is terminated when the computed energy is convergent to a specified tolerance error. To obtain the complete numerical solution of the Schrödinger equation in Step 4, the Schrödinger equation is discretized with the finite volume method. The discretized Schrödinger equation with its boundary conditions leads to a generalized algebraic eigenvalue problem. The eigenvalues of the problem are computed with the multishift QR method. The key idea of the multishift QR method is to introduce carefully chosen perturbations to reveal deflations that are not yet evident on the subdiagonal. In our experience, the proposed nonlinear iterative method converges monotonically. The cost of simulation time can be reduced about 1.5 orders of magnitude.

The energy spectrum of the quantum ring is a set of discrete energy states that is formed and numerated by a set of numbers (n, l, σ), where n is the nth solution of the problem with a fixed l and σ. For the same value of n, the parallel (antiparallel) orbital momentum, and spin, the energy states still have two-fold degenerate (well-known Kramers degeneracy). But nth states with antiparallel orbital momentum and spin are separated from the nth state with parallel orbital momentum and spin. For cylindrical quantum rings, a conventional notation nL_σ for the electron energy states is adopted, where $L = S, P, D, ...$ denotes the absolute value of l, and $\sigma = \pm 1$ refers to the electron spin directions corresponding to the electron angular momentum direction. For all calculations we choose the lowest energy states $(n = 1)$.

3 RESULTS AND DISCUSSION

In the calculation of the electron energy spectra for InAs/GaAs quantum ring we choose the semiconductor band structure parameters for InAs as follows. The energy gap is $E_{1g} = 0.42$ eV and the spin-orbit splitting $\Delta_1 = 0.48$ eV. The value of the nonparabolicity parameter $E_{1p} = 3m_0P_1^2/\hbar^2 = 22.2$ eV, where m_0 is the free electron effective mass. For GaAs, $E_{2g} = 1.52$ eV, $\Delta_2 = 0.34$ eV, and $E_{2p} = 24.2$ eV. The band offset is taken as $V_0 = 0.77$ eV. The spin splitting effect is obviously zero for the lowest energy state $1S_{\pm 1}$. The dependence of the $1P$ energy level splitting

$$\Delta E_{1P} = E_{1P_{+1}} - E_{1P_{-1}}$$

on the ring size is shown in Fig. 2. Our calculation demonstrates significant spin splitting for ultra-small semiconductor quantum rings. The splitting is strongly dependent on the ring radius and decreases

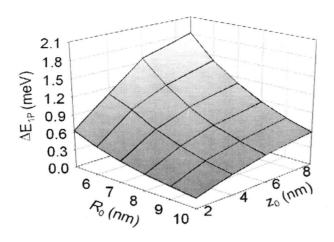

Figure 2: Spin splitting of $|l| = 1$ states for InAs/GaAs quantum rings with different base radii and heights, where the inner radius is chosen as 2 nm.

when the radius increases. At the same time for quantum rings with relatively small thickness the spin splitting is small. This is a direct result of electron wave function tunneling into the barrier along z-direction and energy dependence of the electron effective mass and spin-orbit coupling parameters. To clarify the result we have compared the "weight" of electron wave function inside and outside the quantum ring [5], [12], [17]. For the quantum ring with small thickness the electron "spreads" out of the quantum ring, the energy level properties are controlled by band parameters of GaAs matrix. Under this situation an effective difference of spin-orbit coupling parameters is smaller then $\beta_1(E = 0) - \beta_2(E = 0)$. When z_0 increases the difference also increases and then becomes z-independent. It makes the spliting effect lager for lager z_0.

Table 1: The $1P$ energy level splitting vs. R_{in}

R_{in} (nm)	2	4	6	8	10
ΔE_{1P}	0.81	0.72	0.66	0.58	0.47

The energy splitting for the state $1P$ depending on the ring size is defined as ΔE_{1P} and is shown in Fig. 2, where the ring's inner radius is 10 nm [13]–[17]. Our approach demonstrates a significant spin splitting (~2 meV) for ultrasmall quantum ring. It is larger than that of quantum dot (~1 meV) which was reported in our works [5], [12], [17]. For the small InAs/GaAs quantum ring ($z_0 = 2$ nm and $R = 6$ nm), spin splitting of $|l| = 1$ states with different inner radii R_{in} is summarized in Tab. 1. It reports the variation of ΔE_{1P}, it increases when the inner radius decreases.

4 CONCLUSIONS

The spin-orbit interaction play an important role in the formation of electron energy states in nanoscale quantum nanostructures and lead to a significant modification of the electron energy spectrum under zero magnetic field. We have studied the effect of the spin-orbit interaction on the electron energy states for ultrasmall semiconductor quantum rings. We found the spin-orbit interaction can significantly modify the electron energy spectrum of InAs/GaAs semiconductor quantum rings. Under zero magnetic fields, ultrasmall InAs/GaAs quantum ring produces 2 meV spin splitting of excited electronic states which is substantially larger than that of quantum dot (~1 meV).

5 ACKNOWLEDGMENTS

This work is supported in part by the TAIWAN NSC grants: NSC - 92 - 2112 - M - 429 - 001 and NSC - 92 - 815 - C - 492 - 001 - E. It is also supported in part the grant of the Ministry of Economic Affairs, Taiwan under contract No. 91 - EC - 17 - A - 07 - S1 - 0011.

REFERENCES

[1] 1. A. Emperador et al., Phys. Rev. B 68, 115312, 2003.

[2] S. Tarucha et al., Physica E 3, 112, 1998.

[3] G. Dresselhaus, Phys. Rev. 100, 580, 1955.

[4] Y.A. Bychkov et al., J. Phys. C 17, 6039, 1984.

[5] Y. Li et al., Eur. Phys. J. B 28, 475, 2002.

[6] D. Bimberg, Semicond. 33, 951, 1999.

[7] J. Akinaga et al., IEEE Trans. Nanotech. 1, 19, 2002.

[8] C.M. Hu et al., Phys. Rev. B 60, 7736, 1999.

[9] D. Richards et al., Phys. Rev. B 59, R2506, 1999.

[10] A.V. Moroz et al., Phys. Rev. B 60, 14272, 2000.

[11] E. Silveira et al., Physica E 2, 929, 1998.

[12] Y. Li et al., Proc. MSM (2001) p. 562.

[13] U. F. Keyser et al., Phys. Rev. Lett. 90, 196601, 2003.

[14] A. Fuhrer et al., Nature 413, 822, 2001.

[15] A. Lorke et al., Phys. Rev. Lett. 84, 2223, 2000.

[16] J.M. Garcia et al., App. Phys. Lett. 71, 2014, 1997.

[17] Y. Li Int. J. Mod. Phys. C 14, 501, 2003.

[18] E. A. de Andrada e Silva et al., Phys. Rev. B 55, 16 293, 1997.

[19] A. G. Mal'shukov and K. A. Chao, Phys. Rev. B 61, R2413, 2000.

[20] P. N. Racec et al., Phys. Rev. B 56, 3595, 1997.

[21] K. Braman et al., SIAM J. Matrix Anal. Appl. 23, 948, 2002.

[22] Y. Li et al., J. Comput. Elec. 1, 227, 2002.

[23] Y. Li et al., Comput. Phys. Commun. 141, 66, 2001.

Fault Detection and Diagnosis Techniques for Molecular Computing

Mehdi B. Tahoori[1] and Subhasish Mitra[2]

[1]Northeastern University, Boston, MA USA, mtahoori@ece.neu.edu
[2]Intel Corp, Sacramento, CA USA, subhasish.mitra@intel.com

ABSTRACT

Fault detection and diagnosis techniques that are essential for defect tolerance, fault tolerance and self-repair of molecular computing systems are discussed. These techniques enable robust molecular computing system design protected from manufacturing defects and run-time errors in the underlying hardware.

Keywords: molecular electronics, test, diagnosis, BIST, defect-tolerance, fault-tolerance, self-repair.

1 INTRODUCTION

Nano-scale devices and molecular electronics promise to overcome the fundamental physical limitation of lithography-based silicon VLSI technology. It is projected that molecular electronics can achieve density of 10^{12} devices per cm^2 and operate at THz frequencies. Researchers have demonstrated several successful nano-scale electronic devices including carbon nano-tubes [Iijima 91] [Bachtold 01][Fuhrer 00][Rueckes 00], silicon nano-wires [Kamins 00] [Huang 01], single electron devices, and quantum dot cells [Tougaw 94]. Regular programmable architectures for nano-scale devices that are conceptually similar to *field programmable gate arrays* (FPGAs) are currently being investigated by several researchers [Dehon 03a][Dehon 03b][Beckett 03][Butts 02][Goldstein 01] [Rueckes 00][Ziegler 02]. Unlike FPGAs where logic functions are realized by *look-up tables* (LUTs), logic functions in these architectures are based on *programmable logic arrays* (PLAs).

Several earlier publications stressed the need for defect and fault tolerance for circuits realized by nano-scale devices [Butts 02] [Collier 99] [Dehon 03a][Goldstein 02]. *Defect-tolerance* techniques are generally used to tolerate manufacturing defects and are mainly useful for yield enhancement purposes [Siewiorek 92]. In the context of traditional systems, repair of memory defects using spare rows and columns is an example of defect tolerance. Some publications such as [Butts 02] observed that nano-scale devices are more vulnerable to temporary errors and permanent faults during system operation compared to conventional CMOS devices. Fault-tolerance techniques are used to tolerate errors occurring during normal operation. Fault-tolerance techniques include masking (e.g., Triple Module Redundancy), concurrent error detection, recovery (e.g., rollback and roll-forward recovery) and self-repair [Siewiorek 92, Pradhan 96].

Thorough testing and precise high-resolution location of the failing resource in a defective part are keys to successful implementations of defect and fault tolerance. Thorough manufacturing testing is required to identify a defective manufactured part. During system operation, periodic testing is required to identify a defective system component with a permanent fault. High-resolution diagnosis is essential to precisely locate a defective resource so that efficient repair can be performed. Since a molecular computing system is expected to have close similarities with reconfigurable systems (such as those implemented using FPGAs), testing and diagnosis techniques for such systems can be tailored to be efficient in the context molecular systems. In this paper, we discuss the applicability of test and diagnosis techniques originally developed for FPGAs to molecular computing systems.

The rest of this paper is organized as follows. Defect and fault tolerance techniques for reconfigurable systems are discussed in Sec. 2. Application-independent test and diagnosis techniques are presented in Sec. 3. These techniques are mainly useful for thorough manufacturing testing and defect-tolerance purposes. Application-dependent test and diagnosis techniques are discussed in Sec. 4. These techniques are used for fault-tolerance and self-repair, and to some extent for defect tolerance. Section 5 concludes this paper.

2 DEFECT AND FAULT TOLERANCE IN RECONFIGURABLE SYSTEMS

The Termac project at HP-labs is an example of the use of defect tolerance in a reconfigurable system [Culbertson 97]. In this project, defect tolerance is achieved by applying thorough testing and diagnosis to identify defective (unusable) resources from defect-free (usable) resources, and an efficient algorithm to map an entire design to the usable resources.

As a practical industrial effort for defect tolerance, Xilinx has announced Easy-Path solution [Xilinx Easy Path]. Due to manufacturing defects, some FPGA parts cannot be used for all possible applications. For example, if a programmable switch is permanently off, any design which uses that particular switch will become faulty. In the regular manufacturing flow, these parts result in yield loss. However, these chips may still be usable for some designs that do not use the defective resources. By testing the resources of an FPGA with respect to a specific design to be implemented on it, these previously marked "faulty" chips can be used and sold to customers. These FPGAs, which are good only for a few particular design and do not have general programmability of typical FPGAs are called *application-specific FPGAs* (ASFPGAs). ASFPGAs are profitable for relatively large volume designs which have been completely finalized, i.e. the final placed and routed version is fixed. The main challenge in this flow from defect tolerance perspective is to thoroughly test the FPGA chip with respect to a particular application and to identify the defective resources.

The ROAR (Reliability Obtained by Adaptive Reconfiguration) project at the Stanford Center for Reliable Computing is an example of the design of a self-repairing reconfigurable architecture based on dual FPGAs with embedded "soft" micro-controllers [Mitra 04]. The difference between the ROAR project and the Teramac project is in the "on-line" aspect of error detection, recovery and self-repair addressed by the ROAR project.

The ROAR project achieves fault-tolerance in reconfigurable systems using on-line error detection during system operation also called *concurrent error detection* (CED), very fast fault location and quick recovery from temporary failures and fast repair of the system from permanent faults. Unlike conventional fault-tolerant systems where the *Field Replaceable Unit* (FRU) is a chip or a board, the FRU for an *adaptive computing system* (ACS) is a logic or a routing resource; thus, there the system can be repaired taking advantage of the fine granularity of the FRU without any standby spare hardware.

Concurrent error detection (CED) is a major component of any dependable system. CED techniques detect errors during system operation. The major objective of any CED technique is to guarantee system data integrity. By data integrity we mean that the system either produces correct outputs or produces an error signal when incorrect outputs are produced. Various CED techniques using in the ROAR project are discussed in [Saxena 03]. For the system to recover from errors after error detection, any system with CED must be equipped with sufficient capabilities to locate the faulty part and perform recovery.

There are two types of failures that can affect system operation, temporary and permanent failures. When an error is detected, the system first assumes that the error is temporary and performs temporary error recovery. This is because, the general belief is that temporary failures are more common compared to permanent faults. However, if the error still persists after a pre-determined number of recoveries, the system assumes that the error is due to a permanent failure. After a permanent fault is detected and the faulty part is located, permanent fault recovery schemes that reconfigure the system to replace the faulty part with originally unused resources are used to repair the system. The reader is referred to [Mitra 04] for a comprehensive description of temporary error recovery and permanent fault repair techniques used in the ROAR project.

Abramovici et al. also presented a roving STAR approach for fault detection and location during system operation [Abramovici 99]. One concern with the roving STAR approach is the performance and availability degradation of the system because of the moving of the STAR, even when failures are not present.

The purpose of Fig. 1 is to show where the various test and diagnosis techniques discussed in this paper fit in. Application-independent test and diagnosis are used after manufacturing mainly for identifying defective parts and also for defect tolerance techniques such as those used in the Teramac project. Application-dependent test and diagnosis is used for fault tolerance during system operation and also for defect-tolerance purposes such as the Xilinx EasyPath flow.

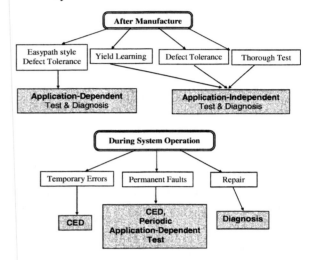

Figure 1 Test and diagnosis for defect and fault tolerance

3 APPLICATION-INDEPENDENT TEST & DIAGNOSIS

New manufacturing process techniques and steps for nano-scale devices that are conceptually different from conventional lithography-based process techniques may result in new failure mechanisms not completely understood today. Various fault models such as stuck-at, bridging, stuck-open, transition, delay and bridging faults are used for conventional CMOS logic testing [Abramovici 90]. Among them, single stuck-at fault model seems to be the most effective, in terms of detection of defective parts although it is not very accurate [McCluskey 00]. The industry uses transition fault model for detecting delay defects. It's not known whether more sophisticated fault models will be required for molecular technologies. Therefore, test techniques for such systems shouldn't be restricted to a particular fault model, but very comprehensive to cover all logical fault models. Some examples of such tests are exhaustive or pseudo-exhaustive testing in which detection of all faults in a given class (for example, all combinational faults inclusive of all stuck-at faults) is guaranteed [McCluskey 84]. The major drawback of pseudo-exhaustive testing is the test length and test time.

Due to reprogrammability, a reconfigurable chip can be configured in an incredibly large number of ways by the user. From manufacturing testing point of view, it must be ensured that the part is functional under all these configurations, as the actual user configuration is not known during production test. Application-independent testing not only consists of developing and applying test vectors, as in conventional application-specific integrated circuits (ASICs), but also of generating and configuring the chip with a set of test configurations, i.e. similar to FPGA testing. The chip must be configured into a test configuration prior to test vector application, in which some resources are programmed (activated) as the resource under test (RUT) and possibly appropriate test circuitry is implemented on a different portion of the chip. The main challenge in application-independent testing is to achieve a high fault coverage with respect to an appropriate fault list, while the number of test configurations is minimized. Note that the total test time is dominated by loading test configurations rather than application of test vectors.

In order to manage the complexity of testing millions of reconfigurable devices on chip, different type of resources (i.e. programmable logic, programmable memory, and programmable interconnect) are tested in different set of test configurations.

Most of the area of a programmable chip is dedicated to interconnects. Interconnect faults consist of permanent on and off faults for programmable switches, and open and bridging faults for wires. Testing interconnect resources is performed by configuring the device into a number of *wires under test* (WUTs). The logic resources are configured as transparent logic, identity function, followed by flip-flops. This improves controllability and observability, as well as the diagnostic resolution. Delay testing for interconnects ensures the performance of the mapped designs. Automatic interconnect test configuration generation techniques are presented in [Tahoori 03a].

Since the logic resources in reconfigurable molecular computing is based on PLA (regular AND and OR planes), easily testable PLA structures and corresponding test sets can be exploited to fully test the logic blocks. Special fault models for PLAs, PLA test and design for testability techniques are discussed in [Khakbaz 82, McCluskey 86].

High resolution diagnosis is required to identify defective resources such as programmable switch, wire, or logic cell. In the Teramac project, this was done by considering only those resources that are used in the passing test configurations to be fault-free. No conclusion is made for resources in the failing test configurations [Culbertson 97]. Such a simple approach can be effective for manufacturing processes with low defect densities. For high defect densities, more sophisticated diagnosis techniques such as those discussed in [Abramovici 00, Tahoori 02] may be required.

4 APPLICATION-DEPENDENT TEST & DIAGNOSIS

In application-dependent testing, the resources are tested only with respect to a particular mapped design. In other words, unused resources are not tested. This test typically consists of fewer test configurations and hence shorter test time compared to application-independent test.

Similar to an application-independent test approach, logic and interconnect resources can be tested separately. In this specific strategy, the interconnect is first tested by modifying the logic block configuration and preserving the interconnect configuration. Then, logic blocks are tested by modifying the interconnect configuration and preserving the logic block configuration. . In the first phase, only the configuration of logic blocks are changed, and the interconnect configuration is *not* modified. Hence, no extra placement and routing is required for test configuration generation. In the second phase, the configuration of original used logic blocks is preserved and configuration of interconnects and unused logic blocks are changed to exhaustively test all used logic blocks. A *Built-in Self Test* (BIST) architecture can be embedded in this test configuration in order to generate the test vectors and observe the outputs in order to generate Go/NoGo signal [Tahoori 03b][Tahoori 04a].

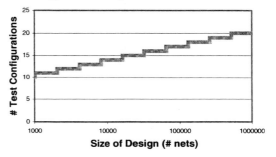

Figure 2 Test configurations for 100% coverage

By implementing special logic functions in the used logic resources, all possible interconnect faults can be detected in very few test configurations. For example, if all used logic resources are configured as AND function, by applying all-1 pattern as the test vector, any stuck-at-0 fault in interconnects will be detected. As presented in [Tahoori 03b], the required number of test configurations is logarithmic to the number of nets in the design. This is shown in Figure 2. For example, for 10^{12} programmable resources, at most 40 test configurations are required to detect all possible stuck-at, open, and bridging faults. This approach can also be extended to generate very few test configurations for thorough testing of defects affecting the interconnects. The technique in [Tahoori 04b] uses only two to four configurations to verify the timing of all paths in the design in order to guarantee the required system performance.

An extended version of the test configurations used for application-dependent test can be applied for diagnosis. The idea is to activate different sets of faults in each test configuration so that the failing pattern for the total set of test configurations uniquely identifies faulty resources. The number of required test configurations is still logarithmic to the size of the design. For example, for 10^{12} devices, at most 120 configurations are required to uniquely identify any stuck-at, open, or bridging fault.

5 CONCLUSION

Thorough testing and precise diagnosis techniques discussed in this paper can enable design of molecular systems with defect and fault tolerance. The discussed techniques are effective for reconfigurable hardware such as FPGAs. Future work includes a thorough (probably simulation-based) study of the implementation and improvement of these techniques in a molecular computing system.

REFERENCES

[Abramovici 00] M. Abramovici, C. Stroud, "BIST-Based Detection and Diagnosis of Multiple Faults in FPGAs," Proc. of Int'l Test Conf., 2000.

[Abramovici 99] M. Abramovici, C. Stroud, C. Hamilton, C. Wijesuriya and V. Verma, "Using Roving Stars for On-line Testing and Diagnosis of FPGAs," Proc. Intl. Test Conf., pp. 973-982, 1999.

[Abramovici 90] M. Abramovici, M. Breuer and A. Friedman, "Digital Systems Testing and Testable Design," IEEE Press, 1990.

[Bachtold 01] A.Bachtold, P.Harley, T.Nakanishi, C.Dekker, "Logic Circuits with Carbon Nanotube Transistors", Science vol 294, pp. 1317-1320, 2001.

[Beckett 03] P. Beckett, "Exploiting multiple functionality for nano-scale reconfigurable systems", Proc. ACM Great Lakes Symposium on VLSI, pp. 50-55, 2003.

[Butts 02] M.Butts, A. DeHon, S.C.Goldstein, "Molecular Electronics: Devices, Systems and Tools for Gigagate, Gigabit Chips", Proc. Int'l Conf. on Computer-Aided Design, pp. 443--440,2002.

[Collier 99] C.P.Collier, E.W.Wong, M.Belohradsky, F.M.Raymo, J.F.Stoddart, P.J.Kuekes, R.S.Williams, J.R.Heath, "Electronically Configurable Molecular-Based Logic Gates", Science, vol 285, pp. 391-394, 1999.

[Culbertson 97] Culbertson, W., R. Amerson, R. Carter, P. Keukes and G. Snider, "Defect-Tolerance in the Teramac Custom Computer," Proc. Intl. Symp. Field Programmable Custom Computing Machines (FCCM), pp. 116-124, 1997.

[Dehon 03a] A. DeHon, "Array-Based Architecture for FET-Based, Nanoscale Electronics", IEEE Trans. on Nanotechnology, Volume 2, Number 1, Pages 23--32, Mar 2003.

[Dehon 03b] A. DeHon, "Stochastic Assembly of Sublithographic Nanoscale Interfaces", IEEE Trans. on Nanotechnology, Volume 2, Number 3, Pages 165--174, September 2003.

[Fuhrer 00] M.S.Fuhrer, J.Nygard, L.Shih, M.Forero, Y.Yoon, M.S.C.Mazzoni, H.J.Choi, J.Ihm, S.G.Louie, A.Zettl, P.L.McEuen, "Crossed Nanotube Junctions", Science vol 288, pp. 494-497, 2000.

[Goldstein 02] S.C. Goldstein, D.Rosewater, "Digital Logic Using Molecular Electronics", Proc. IEEE Int'l Solid-State Circuits Conf., 2002.

[Goldstein 01] S.C.Goldstein, M.Budiu, "NanoFabrics: Spatial Computing using Molecular Electronics" Proc. Int'l Symp. on Computer Architecture, 2001.

[Iijima 91] S. Iijima, "Helical Microtubules of Graphitic Carbon," Nature, vol. 354, pp. 56, 1991.

[Kamins 00] T.I. Kamins, R.S.Williams, Y. Chen, Y.-L. Chang, and Y.A. Chang, "Chemical vapor deposition of Si nanowires nucleated by TiSi2 islands on Si," Applied Physics Letters, vol. 76, no. 562, 2000.

[Khakbaz 82] J. Khakbaz, and E.J. McCluskey, "Concurrent Error Detection and Testing for Large PLA's," Joint Special Issue on VLSI, IEEE Trans. on Electron Devices, pp. 756-764 and IEEE J. of Solid-State Circuits, pp. 386-394, Apr. 1982.

[McCluskey 00] E. J. McCluskey, and C. W. Tseng, "Stuck-Fault Tests vs. Actual Defects," Proc. Intl. Test Conf., 2000, pp. 336-343.

[McCluskey 86] E. J. McCluskey, Logic Design Principles, Prentice-Hall Inc., Englewood Cliffs, N.J., 1986.

[McCluskey 84] E. J. McCluskey, "Verification Testing - A Pseudoexhaustive Test Technique," IEEE Trans Comp., pp.541-546, 1984.

[Mitra 04] Mitra, S., W.J. Huang, N. Saxena, S.Y. Yu and E.J. McCluskey, "Dependable Reconfigurable Computing: Reliability Obtained by Adaptive Reconfiguration," ACM Trans. Embedded Computing Systems, To appear.

[Rueckes 00] T.Rueckes, K.Kim, E.Joselevich, G.Y.Tseng, C.Cheung, C.M.Lieber, "Carbon Nanotube-Based Nonvolatile Random Access Memory for Molecular Computing", Science, vol 289, pp. 94-97, 2000.

[Huang 01] Y.Huang, X.Duan, Y.Cui, L.J.Lauhon, K.Kim, C.M.Lieber, "Logic Gates and Computation from Assembled Nanowire Building Blocks", Science vol 294, pp. 1313-1317,2001.

[Pradhan 96] D. K. Pradhan, "Fault-Tolerant Computer System Design", Prentice Hall, 1996.

[Saxena 03] N. Saxena, S. Mitra, C. Zeng and E. J. McCluskey, "Concurrent Error Detection and Design Diversity In Reconfigurable Computing - New Opportunities," IEEE Design and Test of Computers, 2003.

[Siewiorek 92] Siewiorek, D. P., and R. S. Swarz, "Reliable Computer Systems: Design and Evaluation," 2nd Edition, Digital Press, 1992.

[Tahoori 02] M.B.Tahoori, "Diagnosis of Open Defects in FPGA Interconnects," Proc. IEEE Int'l Conf. on Field-Programmable Technology, pp. 328-331, 2002.

[Tahoori 03a] M. B. Tahoori, S. Mitra, "Automatic Test Configuration Generation for FPGA Interconnect Testing," Proc. IEEE VLSI Test Symp., 2003.

[Tahoori 03b] M. B. Tahoori, "Application-Dependent Interconnect Testing of FPGAs", Proc. IEEE Symp. On Defect and Fault Tolerance in VLSI, pp. 409-416, 2003.

[Tahoori 04a] M. B. Tahoori, E. J. McCluskey, M. Renovell, P. Faure, "A Multi-Configuration Strategy for an Application Dependent Testing of FPGAs", to appear in Proc. VLSI Test Symp., 2004.

[Tahoori 04b] M. B. Tahoori, S. Mitra, "Thorough Delay Testing of Designs on Programmable Logic Devices," in preparation.

[Tougaw 94] P.D. Tougaw and C.S. Lent, "Logical Devices Implemented Using Quantum Cellular Automata," Applied Physics, Vol 75(3), pp. 1818-1825, 1994.

[Ziegler 02] M.M.Ziegler, M.R.Stan, "Design and analysis of crossbar circuits for molecular nanoelectronics", Proc. IEEE Int'l Conf. on Nanotechnology, 2002.

[Xilinx EasyPath] Xilinx Easy Path Solution, http://www.xilinx.com, 2002.

Translating the Integration Challenges to Molecular Device Requirements - Analysis of Scaling Constraints in Molecular Random Access Memories

C. Amsinck[*], N. Di Spigna[*], D. Nackashi[*] and P. Franzon[*]

[*]Department of Electrical and Computer Engineering
North Carolina State University, Raleigh, NC, 27695, paul_franzon@ncsu.edu

ABSTRACT

Integrating molecular memory devices into large scale arrays is a key requirement for translating the miniature size of molecular devices into ultradense memory systems. This in turn imposes constraints on the individual molecular memory devices. A circuit theory approach is used to derive a general parameterized memory circuit model, from which quantitative relationships between the device on:off ratio, noise margin and memory size are studied. Assuming a small interconnect impedance and a reasonable noise margin, a 7:1 on:off ratio would be sufficient for a 4kbit memory, while a 16kbit memory would require a 13:1 ratio. Parasitic impedances become significant in architectures employing molecular interconnect, and full-scale memory circuit simulations are presented as a case study. This way, trends for the impact of all system parameters on system scalability are examined.

Keywords: molecular electronics, random access memories, scalability, device integration, molecular memory

1 MOLECULAR RAM SYSTEM

Translating the size advantage of molecular memory devices into ultradense memory systems requires large-scale integration. In this paper, we examine how the molecular device characteristics affect the scalability of these devices into large-scale random access memory arrays. A general resistive m by n crossbar is shown in Figure 1.

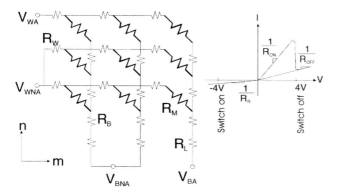

Figure 1: Molecular Random Access Memory and molecular current-voltage characteristic

The horizontal (word) lines have resistance R_W, the vertical (bit) lines have resistance R_B. These represent the interconnect resistance. A load resistor R_L is connected to the bitlines, and the resistors at the crosspoints, R_M, represent the molecular memory device. The value of R_M is either R_{ON} or R_{OFF}, depending on the state of the device. The inset current-voltage (I-V) characteristic, based on [1-3] shows the molecular memory behavior, exhibiting voltage-controlled switching between two conductivity states, with resistance R_{ON} or R_{OFF}.

The accessed memory cell is at the crosspoint of the accessed w̲ordline, with voltage V_{WA} applied, and the accessed b̲itline, which is biased to V_{BA}. V_{WNA} and V_{BNA} are applied at the remaining n̲on-a̲ccessed w̲ordlines and b̲itlines, respectively.

Ignoring interconnect resistance, the voltage across the load resistor is always determined by the voltage divider formed between the load resistance and the device resistance. Thus, the difference between the voltages across the load resistor in the two states, which is the memory noise margin, is given as:

$$\Delta V = (V_{WA} - V_{BA})(\frac{R_L}{R_L + R_{ON}} - \frac{R_L}{R_L + R_{OFF}}) \qquad (1)$$

The optimal load resistance value can be obtained by setting the derivative of equation (1) equal to zero, which yields that R_L should be the geometric mean of R_{ON} and R_{OFF}.

2 DEVICE ON:OFF SCALABILITY

If negligible interconnect resistance (R_W and R_B) can be assumed, circuit transformations based on the superposition principle for linear circuits can be applied on an arbitrarily large memory array of the type shown in Figure 1. The m−1 bitlines terminating in V_{BNA} can be merged into a single bitline with all resistors divided by m−1, since they are all in parallel. The n−1 wordlines terminating in V_{WNA} can be expressed as a single wordline with all resistors divided by n−1, since they are also all in parallel. This results in the circuit template shown in Figure 2, where an arbitrarily large array can be analyzed with a circuit consisting of 6 resistors, whose values change with array size. From this template, the impact of individual parameters on scalability can be predicted qualitatively.

The assumption of negligible interconnect resistance is likely to be accurate in the case where the architecture relies on lithographically defined metal wires, such as in [4, 5]. Even nanoscale metallic wires have impedances that should be negligible compared to the device resistance. Most metals have bulk resistivity on the order of at least $10^{-6}\Omega m$, so lithographically defined nanowires (pitch=133nm, width=40nm, height=8nm [5]) should have resistance on the order of 100-1000Ω. This is still several orders of magnitude lower than the resistance of molecular devices [1,2,6,7]. However, one cannot assume negligible resistances in architectures employing molecules as part of the interconnect structure, which will be analyzed in the latter part of this work. In either case, capacitive and inductive parasitic effects are neglected, since operating speed performance is not examined in this study.

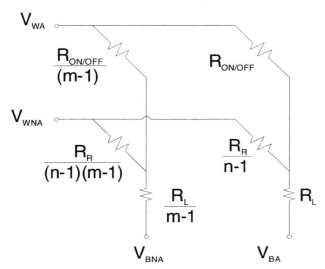

Figure 2: Circuit Template parameterized by RAM size

In order to quantitatively explore the relationship between the memory size, the noise margin and the memory device's on:off ratio, k, the memory operations were defined as given in table 1.

Parameter	Reset	Write	Read
V_{WA} (V)	6	-3	3
V_{WNA} (V)	-1	-1	-1
V_{BA} (V)	0	2	0
V_{BNA} (V)	0	0	0

Table 1: Applied voltages for memory operations

Furthermore, it was assumed that the device's reverse-bias resistance is equal to the off-state resistance, i.e. $R_R=R_{OFF}$, and that $R_L = \sqrt{R_{ON}R_{OFF}}$. Then, the noise margin can be derived by using superposition analysis on the circuit template in Figure 2. Summing up the contributions of all the voltage sources, an expression for the output voltage, V_{OUT}, across R_L can be derived.

$$V_{OUT} = \frac{R_L R_R V_{WA}}{(n-1)(R_L + \frac{R_R}{n-1})(\frac{R_L R_R}{(n-1)(R_L + \frac{R_R}{n-1})} + R_M)}$$
$$- \frac{R_L V_{BA}}{R_L + \frac{R_M R_R}{(n-1)(R_M + \frac{R_R}{n-1})}}$$
$$+ \frac{R_L R_M V_{WNA}}{(R_L + R_M)(\frac{R_L R_M}{R_L + R_M} + \frac{R_R}{n-1})} \quad (2)$$

By taking the difference between V_{OUT} for $R_M=R_{ON}$ and $R_M=R_{OFF}=k*R_{ON}$, the noise margin can be derived:

$$NM = \frac{(k-1)(n-1)(V_{WLA} - V_{WLNA}) + (k^{\frac{3}{2}} - \sqrt{k})(V_{WLA} - V_{BLA})}{(k - 1 + \sqrt{k} + n)(\sqrt{k} + n)} \quad (3)$$

Figure 3 graphically explores the relationships between the memory size (given as the number of memory words), the noise margin (as a fraction of the applied bias between the word- and bitline during a read operation, as defined in table 1), and the ratio between the resistances of the two molecular conductivity states (on:off ratio).

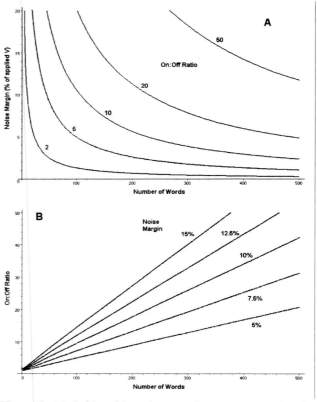

Figure 3: (a) Achievable noise margin vs. memory size for various on:off ratios (b) Required on:off ratio vs. memory size for various noise margins

For a 64x64 memory, a on:off ratio of 7:1 is sufficient for a noise margin of 10% of the applied voltage. A 13:1 ratio would allow for a 128x128 (16kbit) RAM circuit with a similar noise margin. Based on these results in conjunction with on:off ratios cited in existing literature [1-3,5,6], it is thus estimated that achieving sufficient on:off ratio of molecular memory devices will not be the most critical challenge towards building molecular memories, as other issues such as device reliability, reproducibility and architectural fault tolerance will likely impose more strict constraints on the RAM scalability.

3   INTERCONNECT SCALABILITY

In order for molecular memory to take full advantage of molecular dimensions and achieve the greatest possible integration density, a physical architecture must be developed which employs molecules not only for the memory devices, but for interconnect as well. In this scenario, molecular "wires" will likely exhibit resistances that are no longer negligible compared to the molecular memory devices [7]. The fraction of voltage dropped across the load resistor cannot exceed that of a voltage divider formed with the molecular device as well as $m*R_W$ and $n*R_B$. Thus, if R_W and/or R_B becomes significant, the noise margin diminishes quickly with increasing memory size. Full-scale simulations were performed using commercial circuit simulation software [8], since lumped models were found to have insufficient accuracy. This also prohibits using a modified version of the circuit template, since accurate results would only be obtained for a limited range of parameters. Thus, a case study was performed on a sample molecular RAM system. The system parameters, summarized in table 2, represent an attempt to start from a realistic estimate for what a fully molecular RAM architecture might look like, based on current experimental results [1,2,3,4,5].

Parameter	Value
R_{ON} (Ω)	10e6
R_{OFF} (Ω)	100e6
R_R (Ω)	1e9
R_W (Ω)	100e3
R_B (Ω)	100e3

Table 2: Default parameters chosen for case-study of fully molecular RAM

For a 16x16 RAM, the noise margin of a read operation, which is the most critical, is about 32% of the applied wordline-bitline bias of 3V. As shown in Figure 4, the noise margin decreases rather rapidly as the memory size is increased.

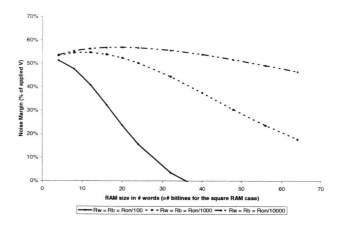

Figure 4: Scalability of fully molecular RAM for different system parameters

The solid line representing the default parameters shows that increasing the array size beyond a 32x32 causes the noise margin to become zero. The impact of the parasitic wire impedance becomes so severe that reading an "off" molecular device close to the edge of the array results in a higher voltage at the output than reading an "on" device far from the edge. Sizing the load resistor becomes nontrivial when significant parasitic degradation from the interconnect resistance is present – the geometric mean of the two molecular conductivity states in no longer the optimal value. Heuristically, an analysis similar to the case without wire resistance can be performed if only the accessed wordline and bitline is considered, which leads to:

$$R_L = \sqrt{(mR_W + nR_B + R_{ON})(mR_W + nR_B + R_{OFF})} \quad (4)$$

For a 16x16 array with the default parameters, the load resistance is 36.9MΩ. In order to avoid obtaining misleading results, R_L was fixed at this, albeit nonideal, value, so that the impact of memory scaling was measured directly rather than the secondary impact of changing the load resistance being incorporated into the results as well. As a result, the dashed lines in Figure 4, representing more ideal interconnect, show a better noise margin for a 16x16 memory than for a 2x2.

Figure 4 also shows the correlation between achievable memory size and the necessary interconnect conductivity. Reasonable noise margins are only achievable for large arrays if the interconnect is several orders of magnitude more conductive than the devices. It is important to note that for the assumed molecular characteristic, the wordline impedance plays a much more significant role than the bitline impedance. The current at each crosspoint along the wordline is divided between the input resistance looking into the memory device, and the input resistance looking into the remaining wordline. On the bitline, however, this current division occurs according to the ratio between the input resistance looking back at the wordline through the reverse-biased device, and the input resistance of the remaining bitline. Since the reverse-biased device has a much higher resistance (R_R, which is assumed to be at least as large as R_{OFF}) than the forward biased device (which can be as low as R_{ON}), the input resistance of the wordline, and thus R_W, has to be much lower than the input resistance of the bitline, and thus R_B. This means that for scaling this type of molecular memory, with this particular I-V characteristic, to large memory arrays, it is preferable to build very asymmetrical arrays, where n>>m, and/or focus on making R_W<<R_B in the physical architecture.

For the case of square arrays with the fixed load resistance, if the number of memory elements in the array exceeds the interconnect-to-device conductivity ratio R_W/R_{ON} by more than a factor of 4, unacceptably low noise margins result.

4 CONCLUSION

The scalability of molecular electronic random access memories based on molecular memory elements relying on binary information storage as two conductivity states was analyzed. For the case of small interconnect resistance, such as would be the case in a lithographically defined crossbar, a circuit template was presented that allows accurate prediction of the scalability based on the molecular on:off ratio. It was found that molecular memory devices with a reliable and repeatable 7:1 on:off ratio can scale up to a 64x64 array in a low-parasitic physical architecture. For the case of significant interconnect resistance, such as would be present in a fully molecular memory system employing molecular "wires", the additional interconnect resistance parameters make analysis significantly more complex. For this scenario, a case study was performed for a sample molecular memory system with parameters chosen based on estimates of what might be realizable based on existing literature. It was found that interconnect impedance has a rather significant impact on system scalability. For the molecular memory device model we employed, the wordline resistance presents a much more restrictive limitation than the bitline resistance, suggesting that constructing square memory arrays is disadvantageous, and memory arrays where the total wordline resistance is much smaller than the total bitline resistance results in improved scalability.

Acknowledgement: Financial support came from the Defense Advanced Research Projects Agency.

REFERENCES

[1] M. A. Reed, J. Chen, A. M. Rawlett, D. W. Price and J. M. Tour, "Molecular random access memory cell," *App. Phys. Lett.*, vol. 78, pp. 3735-3737, Jun. 4 2001

[2] C. P. Collier, E. W. Wong, M. Belohradsky, F. M. Raymo, J. F. Stoddart, P. J. Kuekes, R. S. Williams, J. R. Heath: "Electronically Configurable Molecular-based Logic Gates" *Science*, vol. 285, pp. 391-394, Jul 16, 1999

[3] A. Bandyopadhyay, A. J. Pal: "Large conductance switching and memory effects in organic molecules for data-storage applications" *Appl. Phys. Lett.*, vol. 82, pp. 1215-1217, Feb 24, 2003

[4] Y. Chen, D. A. A. Ohlberg, X. Li, D. R. Stewart, R. S. Williams, J. O. Jeppesen, K. A. Nielsen, J. F. Stoddart, D. L. Olynick, E. Anderson: "Nanoscale molecular-switch devices fabricated by imprint lithography" *Appl. Phys. Lett.*, vol. 82, pp. 1610-1612, March 10, 2003

[5] Yon Chen, Hun-Young Jung, Douglas A. A. Ohlberg, Xuema Li, Duncan R. Stewart, Jan O. Jeppesen, Kent A. Nielsen, J. Fraser Stoddart, R. Stanley Williams: "Nanoscale molecular-switch crossbar circuits", *Nanotechnology*, vol. 14, no. 4, pp. 462-468, Apr. 2003

[6] J. Chen, M. A. Reed, A.M. Rawlett and J.M. Tour, "Large On-Off Ratios and Negative Differential Resistance in a Molecular Electronic Device," *Science*, vol. 286, pp. 1550-1552, Nov. 19, 1999

[7] A. S. Blum, J. C. Yang, R. Shashidhar, B. Ratna: "Comparing the conductivity of molecular wires with the scanning tunneling microscope" *Appl. Phys. Lett.*, vol. 82, pp. 3322-3324, May 12, 2003

[8] Avant! Corporation: Star-HSPICE 2001.2 and AvanWaves 2001.2

A Schrödinger-Poisson Solver for Modeling Carbon Nanotube FETs

D.L. John[†], L.C. Castro[†], P.J.S. Pereira, and D.L. Pulfrey

Department of Electrical and Computer Engineering
The University of British Columbia
Vancouver, BC V6T 1Z4, Canada

ABSTRACT

We present details of a coupled Schrödinger-Poisson solver for modeling quantum transport effects in carbon nanotube field-effect transistors. The Poisson solution is effected using a two-dimensional finite difference algorithm in a coaxial structure with azimuthal symmetry. The Schrödinger solution is implemented by the scattering matrix method, and the resultant, spatially unbounded wavefunctions, defined on the nanotube surface, are normalized to the flux computed by the Landauer formula. The solver illustrates the need for detailed modeling of the nanotube due to the impact of interference effects and evanescent modes on the carrier profiles. Non-equilibrium carrier distributions are presented for particular cases.

Keywords: carbon nanotubes, modeling, nanoelectronics, transistors, quantum transport

1 INTRODUCTION

Carbon nanotubes [1] are attracting great interest for their use in nanoscale electronic devices. Recent modeling efforts of carbon nanotube field-effect transistors (CNFETs) have been successful in examining the sub-threshold behaviour of these devices through a simple solution to Laplace's equation [2], [3], while the above-threshold behaviour has been modeled using bulk device concepts [4], [5]. Accurate CNFET modeling requires a self-consistent solution of the charge and local electrostatic potential. In order to properly treat such quantum phenomena as tunneling and resonance, the charge is computed via Schrödinger's equation. Owing to the presence of metal-semiconductor interfaces, we also account for the penetration of evanescent wavefunctions from the metal into the energy gap of the nanotube.

We deal specifically with the coaxial geometry of the CNFET shown in Fig. 1. The device consists of a semiconducting carbon nanotube surrounded by insulating material (relative permittivity ϵ_{ins}) and a cylindrical, wrap-around gate contact. The source and drain contacts terminate the ends of the device. The device dimensions of note are the gate radius, R_g, the nanotube

---

[†]These authors contributed equally to this work.

radius, R_t, the insulator thickness $t_{\text{ins}} = R_g - R_t$, and the device length, L_t.

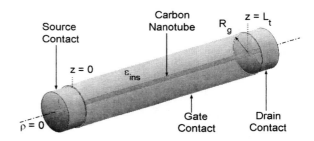

Figure 1: Coaxial CNFET model geometry.

In this closed, metallic cylinder system, Poisson's equation, restricted to just two dimensions by azimuthal symmetry, is

$$\frac{\partial^2 V}{\partial \rho^2} + \frac{1}{\rho}\frac{\partial V}{\partial \rho} + \frac{\partial^2 V}{\partial z^2} = -\frac{Q}{\epsilon}, \tag{1}$$

where $V(\rho, z)$ is the potential within the outer cylinder, and Q is the charge density. It must be noted that, although the solution of Eq. (1) encompasses the entire volume of the device, we are primarily concerned with the longitudinal potential profile along the surface of the tube, hereafter labeled $V_{CS}(z) \equiv V(R_t, z)$, since knowledge of this potential is required for carrier transport calculations.

We treat the nanotube as a quasi-one-dimensional conductor, and the linear carrier density is then computed via the time-independent Schrödinger equation given by

$$\frac{\partial^2 \Psi}{\partial z^2} = -\frac{2m}{\hbar^2}(E - U)\Psi, \tag{2}$$

where $\Psi(z, E)$ is the wavefunction of a carrier with total energy E and effective mass m, traveling in a region with local effective potential $U(z)$. While Q may include sources such as trapped charge within the dielectric, we neglect any charge other than that of electrons and holes on the nanotube.

2 SOLUTION METHOD

We require a solution to Eq. (1) with $Q = Q(V)$. Convergence for this non-linear system is achieved with the Picard iterative scheme, whereby iteration $k + 1$ is given by

$$V_{k+1} = V_k - \alpha \mathcal{L}^{-1} r_k \,,$$

$$r_k = \mathcal{L} V_k + Q(V_k) \,,$$

where r_k is the residual of the k-th iteration, $0 < \alpha \leq 1$ is a damping parameter, and $\mathcal{L}$ represents the linear, differential operator allowing Eq. (1) to be written as $\mathcal{L} V = -Q$.

2.1 Potential

The boundary conditions for V are given by

$$V(R_g, z) = V_{GS} - \phi_G/q \,, \tag{3}$$

$$V(\rho, 0) = -\phi_S/q \,, \tag{4}$$

$$V(\rho, L_t) = V_{DS} - \phi_D/q \,, \tag{5}$$

$$\frac{\partial V}{\partial \rho}(0, z) = 0 \,, \tag{6}$$

where $\phi_{G,S,D}$ represent the work functions of the gate-, source-, and drain-metallizations, respectively, and V_{GS} and V_{DS} are the gate- and drain-source voltages. Due to the discontinuity in ϵ across the nanotube surface, we must also apply the usual matching condition

$$\epsilon_{\text{ins}} \frac{\partial V}{\partial \rho}\bigg|_{R_t^+} - \epsilon_t \frac{\partial V}{\partial \rho}\bigg|_{R_t^-} = -\frac{q(p-n)}{2\pi R_t} \,,$$

where p and n are the one-dimensional hole and electron carrier densities.

The solution to Eq. (1) was obtained via the finite difference technique, implemented by discretizing the spatial domain and using central differencing to generate a linear system of equations, for some known Q, and subject to the boundary conditions specified by Eqs. (3)–(6). Finite differencing was chosen over an FFT-Green's function approach due to its flexibility in modeling more complex structures. The singularity at $\rho = 0$ was addressed by applying l'Hôpital's Rule to the offending term, yielding

$$\frac{1}{\rho} \frac{\partial V}{\partial \rho} \simeq \frac{\partial^2 V}{\partial \rho^2} \,.$$

The amount of energy band bending in the vacuum level, along the length of the nanotube, is given by $E_{vac}(z) = -qV_{CS}(z)$, since we assume that the local electrostatic potential rigidly shifts the nanotube band structure. The potential energies seen by electrons and holes in the nanotube are

$$U_e(z) = E_{vac}(z) - \chi_{CN} \,, \tag{7}$$

$$U_h(z) = -U_e(z) + E_g \,, \tag{8}$$

where E_g and χ_{CN} are, respectively, the nanotube band-gap and electron affinity.

2.2 Charge

Having established a solution for the potential and its relation to the energy band structure, we now determine the carrier concentration. In our system, the charge density is given by

$$Q = \frac{q(p-n)}{2\pi R_t} \frac{\delta(\rho - R_t)}{\rho} \,,$$

where $\delta(\cdot)/\rho$ is the Dirac delta function in cylindrical coordinates, and $n(z)$ and $p(z)$ are computed via Eq. (2), where the nanotube effective mass is obtained from the tight-binding approximation of the band structure, and is the same for both electrons and holes due to symmetry [1]. Only the first, doubly-degenerate band is included in the calculations presented herein. The potential energy, U, for each carrier type is specified by Eqs. (7)–(8), given a potential profile $V_{CS}(z)$.

We solve Eq. (2) using the scattering-matrix method in which a numerical solution is propagated by cascading 2×2 matrices [6]. We find that the use of piecewise constant potentials (plane-wave solutions) are preferable to piecewise linear potentials (Airy function solutions) due to the considerable reduction of simulation time without an appreciable increase in the error. Matching of the wavefunction and its derivative on the boundary between intervals n and $n+1$, assuming a constant effective mass, is performed via the usual relations

$$\Psi_n = \Psi_{n+1} \,,$$

$$\frac{\partial \Psi_n}{\partial z} = \frac{\partial \Psi_{n+1}}{\partial z} \,.$$

In order to completely specify the wavefunction, we require two boundary conditions. In the contacts, the wavefunction at a given energy is of the form

$$\Psi = \begin{cases} A e^{ik_S z} + B e^{-ik_S z} & , \quad z < 0 \,, \\ C e^{ik_D z} + D e^{-ik_D z} & , \quad z > L_t \,, \end{cases}$$

where k_S and k_D are the wavevectors in the source and drain contacts, respectively, and A, B, C, and D are constants. As an example, noting that an analogous calculation may be performed for the drain by exchange of variables, we now illustrate source injection. For this case, $D = 0$ for all energies. In addition, we expect that the Landauer equation [7] will hold for the flux, and must be equal to the probability current. For the transmitted wave, this yields

$$\frac{2q}{\pi\hbar} f_S T = \frac{q\hbar}{m} k_D |C|^2 \,, \tag{9}$$

where the pre-factor of 2 accounts for the aforementioned band degeneracy, f_S is the Fermi-Dirac carrier distribution in the source, and T is the transmission probability specified by

$$T = \frac{k_D |C|^2}{k_S |A|^2} \,.$$

Simple manipulation yields the normalization condition

$$|A|^2 = \frac{2m}{\pi\hbar^2}\frac{f_S}{k_S}. \qquad (10)$$

At any given energy, multiplication of the unnormalized wavefunction by a constant satisfies Eq. (10).

Including source and drain injection components, the normalized wavefunctions yield the total carrier densities in the system,

$$n(z) = \int_{\mathcal{E}_e}^{\infty} \left(|\Psi_{e,S}|^2 + |\Psi_{e,D}|^2 \right) \, \mathrm{d}E,$$

$$p(z) = \int_{\mathcal{E}_h}^{\infty} \left(|\Psi_{h,S}|^2 + |\Psi_{h,D}|^2 \right) \, \mathrm{d}E,$$

where $\mathcal{E}_{e,h}$ is taken to be the bottom of the band, for either electrons or holes, in the appropriate metallic contact, and corresponds to the bottom of the band in the metal. In practice, the integrals are performed using adaptive Romberg integration, where repeated Richardson extrapolations are performed until a predefined tolerance is reached [8]. We find that an adaptive integration method is a necessity for convergence, in order to properly capture Ψ, which is typically highly-peaked in energy for propagating modes. Alternatively, one could employ a very fine discretization in energy, however the Romberg method allows for the mesh size to change based on the requirements of the integrand, and results in a much improved simulation time.

3 RESULTS

We now present results for a CNFET with a (16, 0) nanotube ($R_t \approx 0.63$ nm; $E_g \approx 0.62$ eV), $L_t = 20$ nm, $t_{\mathrm{ins}} = 2.5$ nm, and $\epsilon_{\mathrm{ins}} = 25$. All work functions are taken to be 4.5 eV unless otherwise noted, and $\chi_{CN} = 4.2$ eV. The nanotube is presumed to have a free-space relative permittivity $\epsilon_t = 1$ [9], and $\mathcal{E}$ was taken to be 5.5 eV below the metal Fermi level, as a rough estimate [10].

In equilibrium, i.e., for $V_{DS} = 0$, we obtain reasonable agreement for the carrier concentrations away from the contacts with that computed using equilibrium statistics [2]. Out of equilibrium, however, interference effects influence the carrier distributions throughout the device. Fig. 2 shows the carrier distributions for $V_{GS} = 0.5$ V as a function of position and V_{DS}, and Fig. 3 shows the corresponding conduction band edges for $V_{DS} = 0$ and 0.4 V.

Under a positive gate bias, the band bending results in an increase in the electron concentration throughout the device as more propagating modes are allowed in the channel. As V_{DS} is increased, this concentration is considerably reduced in the mid-length region. Evanescent modes dominate the carrier concentrations near the end contacts, thus impacting on the local potential. Due to

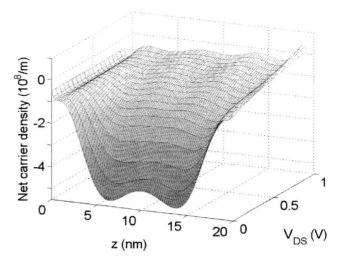

Figure 2: Net carrier density, $p(z) - n(z)$, for the model device as a function of position and V_{DS}.

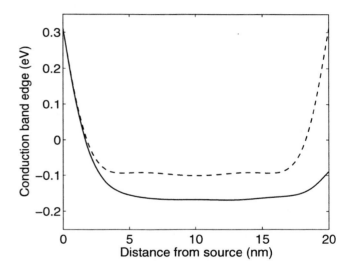

Figure 3: Conduction band edges for the model device with $V_{GS} = 0.5$ V, and $V_{DS} = 0$ (dashed) and 0.4 V (solid). Energies are with respect to the source Fermi level.

the exponential dependence of the transmission probability on the barrier shapes, the flux is significantly modified if these modes are neglected.

We note, also, that it is important to allow for the full inclusion of quantum mechanical reflection for the thermionic component of the flux. Often, carriers above the barrier are assumed to have a transmission probability near unity. However, this approximation does not hold in general, as Fig. 4 shows, wherein the significant reflection is due to $\mathcal{E}$ being much lower than the conduction band edge. The effect is most important for devices where the metal-nanotube work function difference yields a negative barrier, shown in Fig. 4(a). Here, a classical treatment would considerably overestimate the Landauer flux, a function of T, for energies in the vicinity of the Fermi level.

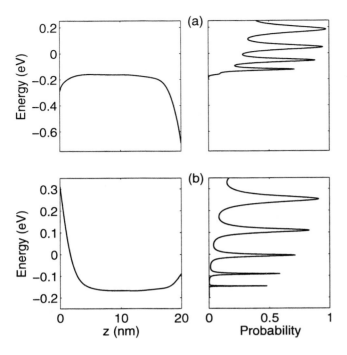

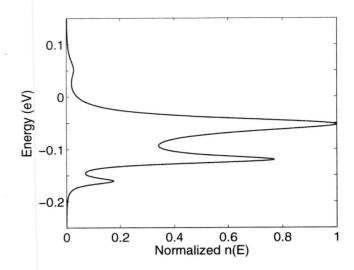

Figure 5: Source-originated electron concentration at $L_t/2$, normalized to its maximum value. $V_{GS} = 0.5\,\text{V}$, $V_{DS} = 0.4\,\text{V}$, and $\phi_S = \phi_D = 3.9\,\text{eV}$. Energies are with respect to the source Fermi level.

3. for devices dominated by thermionic emission, a full solution of Schrödinger's equation is still required in order to account for significant reflection above the barriers.

Figure 4: Conduction band edges and transmission probabilities for electrons at $V_{GS} = 0.5\,\text{V}$ and $V_{DS} = 0.4\,\text{V}$: (a) $\phi_S = \phi_D = 3.9\,\text{eV}$ and (b) $\phi_S = \phi_D = 4.5\,\text{eV}$. Energies are with respect to the source Fermi level.

Finally, the present Schrödinger-Poisson method allows for explicit calculation of the carrier distribution functions, as shown in Fig. 5. The result is in marked contrast to a previous self-consistent model [5] that utilized quasi-equilibrium distribution functions to calculate the non-equilibrium carrier concentrations. Moreover, while the model provided in Ref. [4] yields more appropriate non-equilibrium carrier distributions, it is not equipped to account for the resonant peaks illustrated here.

4 CONCLUSIONS

From this work on the modeling of CNFETs with a coupled Schrödinger-Poisson solver, we conclude that:

1. equilibrium statistics are not adequate in describing the carrier distributions in energy;

2. consideration of the evanescent modes is crucial for the accurate simulation of devices where transport is dominated by tunneling through the interfacial barriers;

REFERENCES

[1] R. Saito, T. Takeya, T. Kimura, G. Dresselhaus, and M.S. Dresselhaus, Phys. Rev. B, 57, 4145, 1998.

[2] D.L. John, L.C. Castro, J.P. Clifford, and D.L. Pulfrey, IEEE Trans. Nanotechnol., 2, 175, 2003.

[3] S. Heinze, M. Radosavljević, J. Tersoff, and Ph. Avouris, Phys. Rev. B, 68, 235418, 2003.

[4] L.C. Castro, D.L. John, and D.L. Pulfrey, Proc. IEEE COMMAD, 303–306, 2002.

[5] J.P. Clifford, D.L. John, and D.L. Pulfrey, IEEE Trans. Nanotechnol., 2, 181, 2003.

[6] D.Y.K. Ko and J.C. Inkson, Phys. Rev. B, 38, 9945, 1988.

[7] D.K. Ferry and S.M. Goodnick, Transport in Nanostructures, Cambridge University Press, New York, 1997.

[8] L.W. Johnson and R.D. Riess, Numerical Analysis, Addision-Wesley, Don Mills, Ontario, 1977.

[9] F. Léonard and J. Tersoff, Appl. Phys. Lett., 81, 4835, 2002.

[10] N.W. Ashcroft and N.D. Mermin, Solid State Physics, Harcourt College Publishers, New York, 1st edn., 1976.

Wetting and Hydrophobicity of Nanoscale Systems with Impurities

P. Gonnet, U. Zimmerli, J.H. Walther, T. Werder and P. Koumoutsakos

Institute of Computational Science, ETH Zurich, Switzerland
[gonnetp|zimmerli|walther|werder|petros]@inf.ethz.ch

ABSTRACT

Molecular dynamics simulations are performed to study the influence of surface and fluid impurities on water-carbon interfaces. In order to quantify these interactions we consider both the canonical problem of wetting of a doped flat graphitic surface by a water droplet with impurities and the influence of static dipole moments perpendicular to the wall of a carbon nanotube on its wetting behavior. As model fluid impurities we consider aqueous solutions of KCl with molar concentrations up to 1.8 M. The contact angle is found to decrease weakly with increasing ionic concentration, from 90° at 0 M to 81° at 1.8 M concentration, and with increasing dipole moments across the nanotube wall from 109° to 93°. The influence of solid impurities, modeled by H and OH groups, is found to be more significant, yet it is dependent on the partial charge distribution on the carbons near the doping site.

1 INTRODUCTION

Carbon nanotubes are envisioned as key components for the development of nanoscale devices, such as sensors and actuators. The method of choice to investigate their behavior and the interaction with their environment are molecular dynamics (MD) simulations. Since many of the applications of carbon nanotubes are targeting biological systems, aqueous environments are of highest interest.

MD simulations depend crucially on the underlying interaction potentials and so the calibration of these potentials is essential for obtaining quantitative results in agreement with experiments. In [1], Gogotsi *et al.* proposed, based on electron microscopy studies, that the interior of a multi-walled carbon nanotube (MWCNT) was hydrophilic. Our own studies, however, based on fitting carbon-water potentials to experimental observations of nano-droplets on a graphite surface [2] and *ab initio* and DFT calculations [3], showed that graphite surfaces — and therefore also the interior of a MWCNT — were, in fact, hydrophobic. A later study by Gogotsi *et al.* [4] attributed the observed hydrophilic behavior to impurities on the MWCNT interior.

Given the importance of impurities in real-world carbon graphite interactions, we wish to establish a model for which these can be effectively simulated using molecular dynamics. This model comprises three types of improvements:

In recent density functional theory (DFT) studies it was found that carbon nanotubes exhibit significant dipole moments across their surface due to their curvature [5]. We validate these results and investigate their influence on the water carbon nanotube interaction by performing MD simulations of water droplets in carbon nanotubes. Improvements to current molecular dynamics potentials for the water-carbon nanotube interaction are discussed.

Next, we consider the effect of surface impurities — H and OH groups — on the wetting behavior of water on a graphite surface. We show that the effect of adding partial charges to a graphite surface greatly depends on how the partial charges are distributed on the carbon atoms.

Finally, we consider the influence of fluid impurities on the contact angle. As model fluid impurities we consider aqueous solutions of KCl with molar concentrations up to 1.8 M. Quantum chemistry calculation are performed to derive pair potentials for the ion-graphite interactions.

2 METHODS

All simulations were performed using the parallel molecular dynamics code FASTTUBE [2], [6], [7]. All simulations are carried out with an integration time step of 2 fs and a cutoff distance of 1.0 nm for the Lennard Jones potentials. The water model used in this study is the rigid extended Simple Point Charge potential SPC/E [8].

The graphite surface was modeled as two staggered, rigid hexagonal carbon sheets with an inter-layer spacing of 3.4 Å. The carbon nanotube in the dipole simulations was modeled either as described in Walther *et al.* [6], using a Morse potential for the bonds, a harmonic angle potential and a twofold torsion potential or it was modeled rigidly.

Given a particular water model, the wetting properties of the fluid-solid interface are determined solely by the water-carbon interaction potentials as demonstrated in [7]. The present study employs the potential model of [7], which involves a 12-6 Lennard-Jones potential

$$U_{12-6}(r_{ij}) = 4\varepsilon_{CO} \left[\left(\frac{\sigma_{CO}}{r_{ij}} \right)^{12} - \left(\frac{\sigma_{CO}}{r_{ij}} \right)^{6} \right],$$

acting between the carbon and oxygen atoms of the water. The parameters of the potential are $\sigma_{CO} = 3.19\,\text{Å}$ and $\varepsilon_{CO} = 0.392\,\text{kJ}\,\text{mol}^{-1}$.

The potassium-water interactions used in the fluid impurity simulations are described by a 12-6 Lennard-Jones interaction between the potassium and the oxygen atom of the water and by a Coulomb potential between all charges. The ionic charges for potassium and chloride are $q_K = +1e$ and $q_{Cl} = -1e$, respectively. The Lennard-Jones parameters are $\sigma_{KO} = 3.26026\,\text{Å}$ and $\varepsilon_{KO} = 0.482331\,\text{kJ}\,\text{mol}^{-1}$, cf. Borodin $et\ al.$ [9]. The potentials governing the chloride-water interaction are taken from Smith $et\ al.$ [10] with a 12-6 Lennard-Jones potential and a Coulomb interaction. The parameters for these potentials are $\sigma_{ClO} = 3.550\,\text{Å}$, $\varepsilon_{ClO} = 0.406415\,\text{kJ}\,\text{mol}^{-1}$.

The potassium-potassium interaction is described by a 12-6 Lennard-Jones and a Coulomb interaction. The parameters for the Lennard-Jones potential are obtained by inverting the Lorentz-Berthelot mixing rules with the values for σ_{KO} and ε_{KO} from Borodin $et\ al.$ [9], and with the SPC/E oxygen-oxygen values. This results in $\sigma_{KK} = 3.3545\,\text{Å}$ and $\varepsilon_{KK} = 0.35782\,\text{kJ}\,\text{mol}^{-1}$. The chloride-chloride interactions are obtained from Ref. [10] and consist of a 12-6 Lennard Jones interaction with $\sigma_{ClCl} = 0.49623\,\text{Å}$ and $\varepsilon_{ClCl} = 0.19384\,\text{kJ}\,\text{mol}^{-1}$, a Coulomb interaction, and a polarization term to model the dipole interaction with $D_{ClCl} = -0.61036\,\text{kJ}\,\text{mol}^{-1}\text{nm}^{4}$. Finally, the KCl interactions were estimated by Smith $et\ al.$ [10], and consist of a 12-6 Lennard-Jones potential with $\varepsilon_{KCl} = 0.26336\,\text{kJ}\,\text{mol}^{-1}$ and $\sigma_{KCl} = 0.43663\,\text{Å}$ [10], a polarization term with $D_{KCl} = -0.3404\,\text{kJ}\,\text{mol}^{-1}\text{nm}^{4}$, and Coulomb interactions. In the fluid impurities simulations we included a quadrupole term between the ions and the carbon atoms, as described in [11].

The partial charges of the chemisorbed hydrogen and hydroxyl groups are estimated from $ab\ initio$ MP2 calculations of iso-butane $((CH_3)_3CH)$ and tert-butanol using the Gaussian98 package with the 6-311G(d,p) basis set. The partial (Mulliken) charge for the chemisorbed hydrogen atom is $0.108\,e$ and that of the oxygen and hydrogen of the OH groups are $-0.3836\,e$ and $0.2240\,e$ respectively. Besides this Coulombic interaction, the chemisorbed hydrogens interact through a 12-6 Lennard-Jones potential with the oxygen atoms of the water molecules. The corresponding parameters of $\varepsilon_{OH} = 0.051098\,\text{kJ}\,\text{mol}^{-1}$ and $\sigma_{OH} = 2.633\,\text{Å}$ are taken from the AMBER96 [12] force field.

From the MD simulation trajectories, the location of the equimolar dividing water surface is determined within every single horizontal layer of the binned drop.

μ_C [Debye]	tube model	ϵ_{CO} kJ mol^{-1}	contact angle
0.0	flex.	0.3135	111.0°
0.0	rigid	0.3135	109.0°
0.0	rigid	0.3920	98.4°
0.03273	rigid	0.3920	95.4°
0.06546	rigid	0.3920	92.9°
0.32732	rigid	0.3920	62.7°

Table 1: Influence of the dipole moment per carbon atom μ_C across the nanotube shell on the contact angle of a water droplet inside a carbon nano tube.

Second, a circular best fit through these points is extrapolated to the graphite surface where the contact angle θ is measured.

3 CARBON NANOTUBE DIPOLE

The simulation setup chosen to investigate the dipole effect on the water carbon nanotube interaction is similar to the one presented by Werder $et\ al.$ [2]. A (64,0) single wall carbon nanotube with a diameter of 5 nm, is used and filled with a droplet of 2112 water molecules. The system is equilibrated and the contact angle of the water droplet within the carbon nanotube is assessed in order to investigate the water carbon nanotube interaction. The initial setup was an initially flat 2.37 nm thick drop, which corresponds to 12 layers of randomly orientated water molecules. The system is equilibrated for 200 ps with a Berendsen thermostat at a temperature of 300 K. Statistics are then collected during 200 ps with samples taken every 0.04 ps.

For the carbon water interaction, two different Lennard-Jones type potentials between the carbon and the oxygen atoms are used. To ensure consistency with the simulations presented in [2], validation runs are performed with the interaction potential presented by Bojan and Steele [13], with $\epsilon_{CO} = 0.3135$ and $\sigma_{CO} = 0.319$. Production runs are performed with the interaction potential suggested by Werder $et\ al.$ [7].

The dipole arising due to the curvature of the nanotube is modeled using two charges of $\pm 0.5\,e$ each. The charges are placed in radial direction inside and outside the carbon nanotube around each carbon atom in order to reproduce the desired dipole moment, as predicted by Dumitrica $et\ al.$. In order to estimate the maximal effect of the dipole in multi-walled carbon nanotubes twofold and tenfold dipole moments of the predicted ones are considered. In these cases, the charges are taken to be 1 and 5 e respectively.

From these simulations (Table 1) we can conclude that although the dipole effect is small, it still changes the carbon nanotube interaction with its environment, causing a change in contact angle of at least 3°.

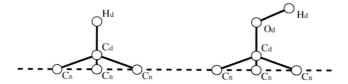

Figure 1: H and OH doping of carbon atoms. C_d is the doped carbon atom and the C_n are the three neighboring carbons.

	Coverage	Distributed	Simple
H	5%	97.09°	88.27°
	10%	98.84°	82.95°
	15%	100.96°	77.91°
	18%	100.97°	75.15°
OH	5%	107.20°	–
	10%	110.56°	–
	15%	99.34°	–
	18%	90.84°	–
none	0%	95.14°	

Table 2: Contact angles for the 16 different surface impurities simulations and the control simulation. The missing contact angle for the simple charges and OH-doping mean that the graphite surface was wet completely (no drop discernible).

4 SURFACE IMPURITIES

Based on our work in [7], we study the effect of surface impurities on the contact angle of a water nano-droplet. The impurities studied were H and OH groups [14] bound to single carbon atoms on the graphite surface (Figure 1).

The simulations consist of 2000 SPC/E water molecules placed in a regular cubic lattice ($20 \times 20 \times 5$ molecules) 3 Å above a graphite bilayer with a separation of 3.4 Å. The doping sites are chosen randomly such that no two are separated by less than three bonds in the graphite surface. In the first half of the 200 ps equilibration time, the water is coupled to a Berendsen thermostat at a temperature of 300 K, whereas in the second half of the equilibration and for the sampling a constant energy simulation is performed.

Two sets of simulations were run with eight cases of H and OH doping each with 5%, 10%, 15% and 18% of the surface carbons doped with two different charge models each, resulting in 16 simulations. A control case without doping was also run to evaluate the contact angle on a pristine graphite surface.

The two charged models differ in the way charges are distributed in the graphite plane. In the first, distributed charge model, the doped carbon atom (C_d) has a charge of -0.2270 e and the neighboring carbons (C_n) of 0.0395 e when doped with hydrogen and of -0.0345 e and 0.0647 e respectively when doped with OH. In the second, simple charge model, the C_n are charge neutral whereas the doped carbon has a charge of -0.1084 e when doped with hydrogen and -0.1596 e when doped with OH.

The contact angles resulting from the different simulations (Table 2) demonstrate that the wetting behavior of a graphite surface depends principally on the charge distribution between the carbon atoms, where the simple charge model causes significantly smaller contact angles.

5 FLUID IMPURITIES

The water molecules are initially placed on a regular lattice. For the KCl simulations two different initial configurations for the water are considered: a $4.2 \times 4.2 \times 3.0$ nm ("cubic") layout and a $5.2 \times 5.2 \times 2.0$ nm ("flat")

configuration. The ions are placed within the water lattice, replacing the water molecule at that location. For each case, the number of molecules in the water lattice is adjusted to 2000 water molecules. The ionic concentration ranges from 0.0 (reference) to 1.8 M, the latter corresponding to 64 ions pairs, with intermediate concentrations represented by 4, 8, 16, 32, and 48 pairs, respectively. One simulation was also performed with a droplet in vacuum to study the effect of the graphite surface (case 16).

Two validation studies were performed on the pristine system (zero ionic concentration) to assess the influence on the contact angle of the initial conditions (cases 1 and 2), and on the numerical treatment of the Coulomb interaction (cases 1 and 3). As demonstrated in Table 3, starting from a cubic or flat configuration of the water has little effect on the equilibrated contact angle, with a deviation less than 0.3%. Clearly, for pristine systems with an atomistically smooth surface the advancing and receding contact angle are equal. The effect of using the Smooth Particle Mesh Ewald (SPME) method versus a smooth truncation (STC) is also small with deviations less than 1.4%, cf. Table 3, and confirms the result obtained in [7], that the contact angles from simulations using a truncation of the Coulomb interaction at 1.0 nm are independent of the cutoff. The total potential energy of the three system is -44.61, -44.66, and -43.09 kJ mol^{-1} (cases 1–3) reflecting the convergence of the simulations. The 3% lower cohesive energy obtained from the simulation using truncation is consistent with the observed (1.4%) lower contact angle as compared with the SPME simulations.

Snapshots from these simulations (cases 4 and 15), cf. Figure 2, confirm that the potassium ions have moved to the graphite surface, leaving the chloride solvated in the vicinity of the liquid-vapor interface.

The complete precipitation of potassium ions onto the graphite surface (Figure 2b) persists at higher concentrations until the surface saturates. For the particu-

case	N_C	N_{KCl}	θ(deg)	IC	method
1	15048	0	90.99	cubic	SPME
2	15048	0	91.20	flat	SPME
3	15048	0	89.73	cubic	STC
4	15048	4	91.83	cubic	SPME
5	15048	4	88.99	flat	SPME
6	15048	4	88.99	flat	STC
7	15048	8	90.77	cubic	SPME
8	15048	8	89.23	flat	SPME
9	15048	16	90.12	cubic	SPME
10	15048	16	92.92	flat	SPME
11	15048	32	88.99	cubic	SPME
12	15048	32	87.46	flat	SPME
13	15048	48	85.22	cubic	SPME
14	15048	48	85.42	flat	SPME
15	15048	64	81.41	cubic	SPME
16	0	4	-	cubic	SPME

Table 3: Overview of the MD simulations of droplets of aqueous solutions of KCl on graphite. N_C is the number of carbon atoms, N_{KCl} the number of ion pairs, and θ the contact angle. The method refers to the numerical treatment of the Coulomb interaction: Smooth Particle Mesh Ewald (SPME) or Smooth Truncation (STC).

lar droplet size considered this appears to be at concentrations exceeding approximately 1 M. At higher concentrations, K-Cl clusters form at the liquid-vapor interface as shown in Figure 2d. The precipitated potassium attracts the chloride towards the graphite surface, and the simulations indicate that K-Cl surface crystals form at concentrations exceeding ≈ 0.5 M. As the chloride binds to the potassium at the surface, the electrostatic attraction is reduced, leaving the remaining chloride at the liquid-vapor interface to form nano-crystals at higher concentrations.

The change in contact angle with the increasing ionic concentration is small. Moreover, at this droplet size, the accuracy of the simulations does not allow us to measure the expected initial increase in surface tension which would lead to an increase in the contact angle. However, as the ions precipitate onto the surface, the contact angle is reduced from an initial 90° to 82° at a concentration of 1.8 M.

6 CONCLUSIONS

The impurities and system properties studied in this work all have non-negligable qualitative effects on the wetting behaviour of carbon surfaces and should therefore be taken into account when realistic models are to be extablished. The quantitative effects, however, are strongly dependent on details of the model and its parametrisation. These details are the subject of ongoing research.

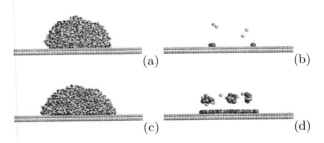

Figure 2: The equilibrated structure of the dilute (0.11 M) and concentrated (1.8 M) droplets of KCl on a graphite surface. The dilute droplet (a) reveals a contact angle of $\approx 90°$, and a precipitation of the potassium ions onto the surface (b, same as a but without water molecules). In the concentrated drop, the ions precipitate onto the graphite surface and some forms crystals at the liquid-vapor interface (d).

REFERENCES

[1] Y. Gogotsi, J. A. Libera, A. Güvenç-Yazicioglu, and C. M. Megaridis. *Appl. Phys. Lett.*, 79(7):1021–1023, 2001.

[2] T. Werder, J. H. Walther, R. Jaffe, T. Halicioglu, F. Noca, and P. Koumoutsakos. *Nano Letters*, 1(12):697–702, 2001.

[3] U. Zimmerli, M. Parrinello, and P. Koumoutsakos. *in preparation*, 2003.

[4] Y. Gogotsi, N. Naguib, and J. A. Libera. *Chem. Phys. Lett.*, 365:354–360, 2002.

[5] T. Dumitrică, C. M. Landis, and B. I. Yakobson. *Chem. Phys. Lett.*, 360(1-2):182–188, 2002.

[6] J. H. Walther, R. Jaffe, T. Halicioglu, and P. Koumoutsakos. *J. Phys. Chem. B*, 105:9980–9987, 2001.

[7] T. Werder, J. H. Walther, R. L. Jaffe, T. Halicioglu, and P. Koumoutsakos. *J. Phys. Chem. B*, 107:1345–1352, 2003.

[8] H. J. C. Berendsen, J. R. Grigera, and T. P. Straatsma. *J. Phys. Chem.*, 91:6269–6271, 1987.

[9] O. Borodin, R. L. Bell, Y. Li, and G. D. Bedrov, D. Smith. *Chem. Phys. Lett.*, 336:292–302, 2001.

[10] G. D. Smith, R. L. Jaffe, and H. Partridge. *J. Phys. Chem. A*, 101(9):1705–1715, 1997.

[11] F. Y. Hansen and L. W. Bruch. *Phys. Rev. B*, 51(4):2515–2536, 1995.

[12] W. D. Cornell, P. Cieplak, C. I. Bayly, I. R. Gould, K. M. Merz, Jr., D. M. Ferguson, D. C. Spellmeyer, T. Fox, J. W. Caldwell, and P. A. Kollman. *J. Am. Chem. Soc.*, 117:5179–5197, 1995.

[13] M. J. Bojan, A. V. Vernov, and W. A. Steele. *Langmuir*, 8:901–908, 1992.

[14] H. P. Boehm. In Pierre Delhaés, editor, *Graphite and Precursors*, chapter 7, pages 141–178. Gordon and Breach Science Publishers, 2001.

First-principles molecular-dynamics simulations of a hydrous silica melt: Hydrogen diffusion mechanisms and electronic properties

Markus Pöhlmann*/**, Helmut Schober*, Magali Benoit***, and Walter Kob***
* Institute Laue-Langevin, 38042 Grenoble Cedex 9, France, poehlman@ill.fr, schober@ill.fr
** Physik-Department E13, Technische Universität München, 85747 Garching, Germany,
mpoehlma@ph.tum.de
*** Laboratoire des Verres, Universite Montpellier II, 34095 Montpellier, France
magali@ldv.univ-montp2.fr, kob@ldv.univ-montp2.fr

ABSTRACT

We study a sample of liquid silica containing 3.84 wt.% H_2O with *ab initio* molecular dynamics simulation in its liquid state at temperatures of 3000 K and 3500 K. At these temperatures the liquid can be brought into equilibrium after several picoseconds. Hence we are able to investigate possible diffusion mechanisms for hydrogen atoms. It turns out that intermediate states in the liquid play a decisive role for the diffusion of hydrogen. Quenches of selected configurations to ambient temperature allow us to study the electronic structure of the material. In particular we find electronic states in the band gap of amorphous silica. The correlation of these electronic states with the structural intermediate states that are present in our system and which we make responsible for the hydrogen diffusion has already been discussed for a long time.

Keywords: hydrous silica, *ab initio* molecular dynamics, hydrogen diffusion, electronic structure

1 Introduction

The design of appropriate dielectrics for semiconductor devices has been one of the challenging tasks in micro electronics over the past decades [1], [2]. This research was driven by the requirement of high storage capacity at low dielectric constant. With progressing miniaturization in forthcoming microchip generations the dielectric constant of the oldfashoined dielectric silicon dioxide (k=3.9 to 4.2) has to be decreased. The development of dielectrics with ultra low dielectric constants is inevitable. Silica (SiO_2) replacements are still under (partly discouraging) investigation [1]. In addition, since device manufacturing is a highly cost-driven business, replacement materials have to be easily and cheaply synthesizable. Appropriate candidates are not yet found and therefore silica is still the material of choice for oxide gates.

One of the reasons for the degradation of dielectric properties of silica are impurities that produce charges in the bulk material and/or at the silica silicon interface. Hydrogen and/or water once introduced as passivant for interface traps or at other stages of the fabrication process was made responsible for various instabilities of the dielectric SiO_2. Such instabilities or defects like Si-O dangling bonds are naturally electrically active and hence degrade the insulating properties. Generally, water is dissolved partly to SiOH groups when added to silica at elevated temperatures, a balance that follows the equation

$$-Si - O - Si - \ + \ H_2O \ \longleftrightarrow \ 2(SiOH)$$

Many other reactions that set off hydrogen atoms, hydrogen molecules or oxonium ions have been proposed [3]. However, the formation of the electrical active centers in combination with the hydrogen and water diffusion has, according to ref. [3], never been understood.

At this point, the analytic power of molecular dynamics computer simulations comes into play. Unfortunately, they so far struggled with the difficulty of finding appropriate potentials describing all possible water silica adducts (water molecules, SiO-H groups, Si-H groups, oxonium ions,..). This problem can be overcome with the use of recently developed *ab initio* techniques. Especially *ab initio* Car-Parrinello molecular dynamics techniques [4] have already been successfully applied to a considerable number of silica systems [5].

2 The Method

Here we present an *ab initio* molecular dynamics approach in order to understand the dissolution and diffusion of water in bulk amorphous silica. In a first stage we studied a melt of SiO_2 with 3.84 wt.% H_2O (30 SiO_2 units and 4 H_2O units). The calculations were stably driven over a time of 25 ps with the Car-Parrinello code CPMD [6] in the liquid state at temperatures of 3000 K and 3500 K. The electronic system was treated with a density functional approach in a general gradient approximation using the PBE functional [7], [8] and for the core electrons we used Troullier-Martins pseudopotentials [9]. The technical details of the simulation are given elsewhere [10]. At these high temperatures the liquid attains equilibrium after the mentioned time. The diffusion process of hydrogen is, for the first time, directly observable in the time window present days *ab initio* molecular dynamics simulations can offer. The high water concentration chosen in this model allows us to obtain a statistically significant picture of the transition states that are involved in the hydrogen diffusion. The question whether an extrapolation of the diffusion processes to lower temperatures (especially below the glass transition temperature) is possible remains to be addressed. However, quenching the equilibrated liquid to ambient temperature allows us comparisons with experimental data on the structure. Due to the *ab initio* character of the simulation, the electronic structure of the material is accessible in particular.

3 Structure

The structure of the liquid has been analyzed regarding commonly discussed quantities like the radial pair distributions, angle distributions, coordination numbers and Q^n species [10]. These quantities indicate that the water is dissolved and forms SiOH groups with the SiO_2 matrix. Hence the silica tetrahedral network is partially broken. Free water molecules, free hydrogen atoms or molecules and SiH units are not present. The nature of the OH bonds is highly covalent. The Si-OH groups constitute "dead-end-pieces" in the structure having a high radial mobility. This radial mobility facilitates the approach of an SiOH oxygen to a second silicon atom and therefore the formation of oxygen tri-clusters of the form $\frac{Si}{Si} > OH$ (bridging hydroxyl group). Since the threefold coordinated oxygen atoms violate the stoichiometry, the system aims for a re-compensation of this stoichiometry violation. Such a charge re-compensation can only be performed by one fold coordinated oxygen atoms. In fact the detailed analysis of the structure revealed the presence of SiO dangling bonds. Water like structures are also found in the liquid but occur much less than the dangling bonds and the $\frac{Si}{Si} > OH$ bridging hydroxyl groups. Those water molecules are not free but mostly attached to a silicon atom in the form $\frac{H}{H} > OSi$ which constitutes another oxygen tri-cluster. A snapshot of the structure at 3500 K including oxygen tri-clusters and SiO dangling bonds is displayed in Fig. 1.

4 Hydrogen Diffusion

In this section we analyze the mechanisms of hydrogen transport in the liquid. In particular we investigate the contribution of the intermediate states displayed in Fig. 1. Since we know from Section 3 that the hydrogen atoms are accommodated in the form of OH groups, the question of hydrogen transport reduces to that of the formation, the rupture and the motion of OH groups. Since we find two oxygen coordinated hydrogen atoms and almost no zero oxygen coordinated hydrogen atoms, it becomes also evident that hydrogen atoms do not drift

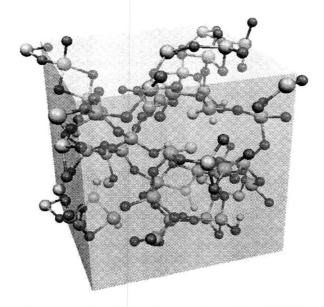

Figure 1: Snapshot of the structure at 3500 K (silicon green, oxygen red, hydrogen white). The structure contains intermediate states like SiO dangling bonds (yellow) and oxygen tri-clusters (orange).

freely through the silica network if an OH bond is broken. Furthermore, in almost all cases, a second OH bond (of the same hydrogen atom) is formed before the rupture of the initial OH bond. Analyzing the nearest neighbor environments of the oxygen atom of the initial OH bond and of the oxygen atom of the formed OH bond, we find the following balances:

$$SiOH + O < \frac{Si}{Si} \longleftrightarrow SiO + HO < \frac{Si}{Si} \quad (1)$$

$$\frac{Si}{Si} > OH + O < \frac{Si}{Si} \longleftrightarrow \frac{Si}{Si} > O + HO < \frac{Si}{Si} \quad (2)$$

$$SiOH + HOSi \longleftrightarrow SiO + \frac{H}{H} > OSi \quad (3)$$

$$SiO < \frac{H}{H} + O < \frac{Si}{Si} \longleftrightarrow SiOH + HO < \frac{Si}{Si} \quad (4)$$

Obviously, as already anticipated above, the SiO dangling bonds and the two oxygen tri-clusters $\frac{Si}{Si} > OH$ (the bridging hydroxyl group) and $\frac{H}{H} > OSi$ (the water containing tri-cluster) constitute the acceptor and donator states for hydrogen transfer reactions. Hydrogen atoms can be transfered to the under saturated SiO dangling oxygen (reaction (1) and (3)) as well as saturated oxygens as the bridging oxygen $\frac{Si}{Si} > O$ as indicated in reactions (1), (2) and (4) and an hydroxyl group oxygen (reaction (3) and (4)). Note that, by a hydrogen transfer to the latter one, the formation of a water-like unit of the form $SiO <^H_H$ occurs.

In Fig. 2 we present the probabilities of the decay products for the OH ruptures in reactions (1) to (4). Note that equations (1)-(4) are balances and hence forward and back reactions are counted. According to Fig. 2, the bridging hydrogen group $^{Si}_{Si}> OH$ is the donator with the highest probability, followed by the SiOH unit and the water containing tri-cluster $SiO <^H_H$.

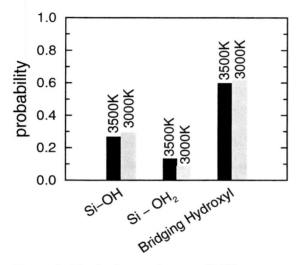

Figure 2: The hydrogen donators SiOH groups, water tri-cluster $SiO<^H_H$ and bridging hydroxyl groups $^{Si}_{Si}>OH$ and their relative contribution to hydrogen release.

Apart from the hydrogen diffusion reactions associated to OH ruptures, we have also evidence for the transport of OH units. These hydroxyl units can be transfered from one silicon atom to another involving again a bridging hydroxyl group as intermediate state. The corresponding equation reads as follows:

$$SiOH + Si \longleftrightarrow {}^{Si}_{Si}> OH \longleftrightarrow Si + HOSi \qquad (5)$$

5 Electronic Structure

As already discussed above, the major concern associated to hydrogen in silicon technology is the creation of electrical active deficiencies like SiO dangling bonds (according to reactions (1) and (3)). The insulating properties of bulk silica are due to its band gap of ≈ 9 eV . It is well known that hydrogen (atomic or molecular), molecular water, and also SiOH groups do not possess electronic states in the SiO_2 band gap [11]. If electronic properties are degraded, the degradation has to be due to the intermediate states like the dangling bonds. Hence it is mandatory to investigate the influence of these states on the electronic structure of the

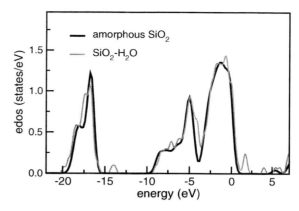

Figure 3: Electronic density of states (edos) of a quenched hydrous configuration comprising two SiO dangling bonds compared to the edos of amorphous silica.

material with particular attention to the band gap. It has already been proposed that the SiO dangling bond created by reactions (1) and (3) produces a gap state at around 2 eV [11].

Performing ultra rapid quenches from the liquid structure to 300 K, we can freeze configurations with SiO dangling bonds and other intermediate states and study their electronic properties. We note that a significant underestimation of the band gap is usual in DFT [12]. Nevertheless it is recognized in literature that the electronic states are well represented in pure silica, even if the DFT band gap is not greater than 5 eV [13]. In Fig. 3 we present the electronic density of states (edos) of a typical frozen configuration (including SiO dangling bonds that are stabilized by weak hydrogen bonds) compared to the one of bulk amorphous silica. We note that the water addition does not alter the electronic properties of silica dramatically. Nevertheless band gap states emerge. The determination of the maximally-localized Wannier functions (Wannier centers) [14] allows a description of the charge distribution in terms of well-defined and localized functions. For instance by a projection of the Wannier functions onto the Kohn-Sham density of states, we could indeed assign the peak at 2 eV in Fig. 3 to the electronic states associated to the SiO dangling bonds.

The most important question for the dielectric properties of the material is of course whether the transition states are electrically charged and if this charge is mobile. A well known tool to investigate this in quantum chemistry are the Hirshfeld charge density deformations [15]. These charges represent, in principle, the difference of the charge density of the bonded atom with respect to the one of the free atom. Integration of the Hirshfeld charge density over the (atomic) volume gives the total electronic charge of an atom Q_i. The addition of the nuclear charge Z_i yields the net atomic charge q_i.

These net atomic charges are given in Table 1. Table

species	q_i(SiO$_2$-H$_2$O)	q_i(SiO$_2$)
Si	$+0.220 \pm 0.031$	$+0.218 \pm 0.010$
bridging oxygen	-0.101 ± 0.026	-0.109 ± 0.007
SiOH oxygen	-0.116 ± 0.021	-
SiO dangling O	-0.277 ± 0.046	-
bridging hydroxyl O	$+0.023 \pm 0.000$	-
H	$+0.038 \pm 0.030$	-

Table 1: Average Hirshfeld net atomic charges (average $\pm$ sigma) for the hydrous silica system and pure silica [13] . For hydrous silica the average was taken over the atoms of the same kind of one configuarion quenched to 300 K.

1 shows that the net atomic charges for the silicon and bridging oxygen atoms in the hydrous sample are very similar to those of pure silica from Ref. [13]. On the other hand, the charges of the different oxygen species in the hydrous sample differ considerably. In particular the SiO dangling bonds exhibit a net charge that is more than twice as high as the bridging oxygens' one. It becomes evident that these bonds constitute indeed electrical active centers in the network. In contrast, the tri-cluster is undersaturated with electrons and exhibits a positive charge.

6   Summary and Perspectives

In this work, we have studied the principle hydrogen diffusion mechanisms in hydrous silica and the important intermediate states for this diffusion were revealed. Especially the existence of the long-time predicted SiO dangling bond that is generated due to a hydrogen release could be confirmed. Quenches of configurations from the liquid to ambient temperature allowed conclusions on the electronic structure. It becomes evident that water addition and the resulting defects in the network are indeed responsible for states in the band gap and hence degrade the insulating properties of the material.

Several dopants to silica that should suppress hydrogen diffusion have been proposed [2]. The most prominent among them are fluorine and nitrogen. We point out that the presented method of investigation will be applicable also to these systems and that the suppression of hydrogen diffusion under parallel surveillance of the electronic properties could be studied.

REFERENCES

[1] R. D. Miller. Science 286, 421-423, 1999.

[2] P. Balk. J. Non-Cryst. Solids 187, 1-9, 1995.

[3] E. H. Poindexter. J. Non-Cryst. Solids 187, 257-263, 1995.

[4] R. Car, M. Parrinello. Phys. Rev. Lett. 55, 2471, 1985

[5] A. Pasquarello. Curr. Opinion in Sol. State Mat. Sci. 5, 503-508, 2001.

[6] CPMD Version 3.3, J. Hutter, A. Alavi, T. Deutsch, M. Bernasconi, S. Goedecker, D. Marx, M. Tuckerman, and M. Parrinello. MPI für Festkörperforschung and IBM Zürich Research Laboratory 1995-99.

[7] W. Kohn and L. Sham. Phys. Rev A 140, 1133, 1965

[8] J. P. Perdew, K. Burke, and M. Ernzerhof. Phys. Rev. Lett. 77, 865, 1996

[9] N. Troullier and J. L. Martins. Phys. Rev. B 43, 1993, 1991

[10] M. Pöhlmann, M. Benoit, W. Kob. Submitted to Phys. Rev. B

[11] J. Robertson. The physics and technology of amorphous SiO$_2$. R. Devine (ed.) pp 91-101. Plenum Press New York, 1988.

[12] R. W. Godby, M. Schlüter, L. J. Sham. Phys. Rev. B 37, 10 159 (1988)

[13] M. Benoit, S. Ispas, P. Jund, and R. Jullien. Eur. Phys. J. B 13 631, 2000; J. Sarnthein, A. Pasquarello, and R. Car. Phys. Rev. Lett. 74, 4682-4685 (1995)

[14] P. L. Silvestrelli, N. Marzari, D. Vanderbilt, M. Parrinello. Solid State Comm. 107, 7, 1998

[15] F. L. Hirshfeld. Theor. Chim. Acta 44, 129, 1977

Dynamic, Transport, and Mechanical Properties of Polymer Nanocomposites and Nanocomposite Solid Polymer Electrolytes

O. Borodin,* G. D. Smith,* D. Bedrov,* S. G. Bardenhagen** and J. Nairn*

* Department of Materials Science and Engineering, 122 S. Central Campus Drive, Rm. 304, University of Utah, Salt Lake City, UT 84112
** Los Alamos National Laboratory, Los Alamos, NM 87545

ABSTRACT

Molecular dynamics simulations have been performed on PEO-based polymer nanocomposites and nanocomposite polymer electrolytes. MD simulations revealed that the PEO density is significantly perturbed by TiO_2 surfaces, PEO and ion dynamics in the vicinity of nanoparticles was found to be intimately related to the structure of the solid nanoparticle surfaces, polymer-surface interactions and the nanoparticle surface area in contact with the polymer.

Molecular dynamics simulations were coupled with the material point method calculations in order to allow prediction of mechanical properties of nanocomposites at length and time scales orders of magnitude larger than those accessible by molecular dynamics simulations.

Keywords: nanocomposites, molecular dynamics simulations, material point method

1 INTRODUCTION

Viscoelastic properties of polymer nanocomposites cannot be adequately described by conventional composite theories that do not take into account the size of particles and particle polymer interactions [1-2], indicating a need for fundamental understanding of the effect of nanoparticles on the structure and dynamics of the interfacial polymer. Molecular dynamics (MD) simulations are well suited for the investigation of structural and dynamic properties on the nanometer length scale. Coarse-grained MD simulations have provided valuable insight into factors responsible for alteration of polymer matrix properties with addition of interfaces; in particular, they found that the strength of nanoparticle – polymer interaction and nanoparticle specific surface area were the most important factors controlling properties of the interfacial polymer[1]. Addition of nanoparticles with attractive interactions led to the decreased polymer dynamics and increased viscosity, whereas addition of nanoparticles with the repulsive, or excluded volume, interactions led to increased polymer dynamics and decreased viscosity. These effects were found to scale linearly with the specific surface area of nanoparticles and not with the volume fraction was found for conventional composites. Another important factor controlling interfacial polymer dynamics is the structure of

the surface, where interfacial polymer motion is strongly correlated to the surface structure [3].

Application of MD simulations to nanocomposites is limited to systems with a few nanoparticles because of high computational cost. To extend the length scale from the nanoscale to mesoscale and thus enable prediction of dynamic-mechanical properties of polymer nanocomposites with tens and hundreds of nanoparticles, we have coupled atomistic MD simulations with the material point method (MPM). This method is an extension of the particle-in-cell method that uses two representations of the continuum, one based on material points and the other based on a computational grid.

In this contribution we discuss influence of TiO_2 nanoparticles on poly(ethylene oxide) (PEO) and PEO/LiBF$_4$ dynamics and structure and show the path for coupling MD simulations with MPM calculations in order to obtain homogenized mechanical properties of polymer nanocomposites and nanocomposite polymer electrolytes.

We focus on the PEO/TiO$_2$ system because PEO is a major component of the PEO-based solid polymer electrolytes, such as PEO/LiBF$_4$/TiO$_2$, which are potential candidates for use in the solvent-free secondary lithium batteries. Addition of ceramic nanoparticles to polymer electrolytes, initially done to enhance mechanical properties, has been also found to improve anodic stability, cyclability, sometimes conductivity and cation transfer number of polymer electrolytes without causing any observed degradations.

2 PEO-T$_i$O$_2$ INTERACTIONS

MD simulation studies of the interface between poly(ethylene oxide) (PEO) and TiO$_2$ have been performed at 423 K using a quantum chemistry based force field. MD simulations revealed that the PEO density is significantly perturbed by TiO$_2$ surfaces, forming layers of highly dense polymer (compared to the bulk melt) that persisted up to 15 Å from the surface. Conformational and structural relaxations of the interfacial PEO were found to be dramatically slower than those of bulk PEO as shown in Figure 1. These effects are attributed to intrinsic slowing down of PEO dynamics and increased dynamic heterogeneity of the interfacial polymer. The surface structure and electrostatic interactions between PEO and TiO$_2$, rather than the increased polymer density at the TiO$_2$

surface, determine the nature of PEO relaxation at the TiO_2 interface.

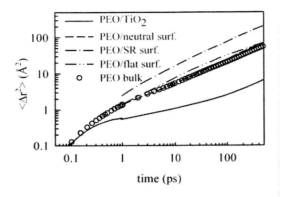

Figure 1. Mean-square displacements for the interfacial PEO backbone atoms next to TiO_2, flat, neutral, soft-repulsive surfaces and bulk PEO.

MD simulations of PEO between flat (structureless) surfaces with interactions modified to match PEO density profiles to those of the PEO between the atomistic TiO_2 surfaces revealed no significant differences between the structural and conformation relaxation of the interfacial PEO and the PEO bulk (see Figure 1), indicating that the PEO density near surfaces has only a minor influence on interfacial PEO dynamics and that perturbation of the interfacial PEO dynamics near TiO_2 surfaces is largely due to atomistic structure of the TiO_2 surfaces and not densification of interfacial PEO.

Elimination of the Coulomb interactions between PEO and TiO_2 significantly increased the structural and conformational relaxation of the interfacial PEO, demonstrating that these interactions are responsible for the slow down of PEO dynamics. Moreover, relaxation of the interfacial PEO in the PEO/neutral surface system is similar to that in bulk PEO suggesting that the surface with the TiO_2 structure and no Coulomb interaction with PEO do not significantly perturb PEO dynamics. Reduction of the repulsive interaction between the neutral surface and PEO further facilitated relaxation of the interfacial PEO in agreement with the results polymer films simulations in vacuum.

We conclude that in addition to such intrinsic parameters as torsional potentials, the polymer-surface interactions and surface structure (at least for attractive polymer surface interactions) are two fundamental factors that determine interfacial polymer relaxations in PEO/TiO_2 nanocomposites.

3 PEO/LiBF$_4$/TiO$_2$ NANOCOMPOSITES

Addition of a TiO_2 nanoparticle to PEO/$LiBF_4$ solid polymer electrolyte resulted in formation of the highly structured layer with the thickness of 5-6 Å as seen from Figure 2 that has more than an order of magnitude slower mobility than that of bulk PEO/$LiBF_4$. The PEO and ions in the layers extending from 6Å to 15 Å from the TiO_2 nanoparticle also revealed some structuring and reduced dynamics, whereas the PEO/$LiBF_4$ located further than 15 Å was basically unaffected the presence of the TiO_2 nanoparticle. Both cation and anion tended to form a region with the increased concentration in the interfacial layers extending from 5 Å to 15 Å. No ions were dissolved by the first interfacial layer of PEO. Addition of the nanoparticle with the soft-repulsion interactions with PEO resulted in formation of the PEO interfacial layer with reduced PEO density but increased ion concentration. The PEO and ion mobility in the interfacial layer next to the soft-repulsive nanoparticle were higher than those of bulk PEO/$LiBF_4$ by 20-50%, whereas conductivity of nanocomposite electrolyte with the soft-repulsive particle increased only by 10%.

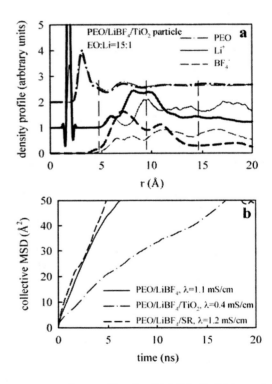

Figure 2. Density profiles for PEO/$LiBF_4$/TiO_2 and collective mean-square displacements for PEO/$LiBF_4$ with TiO_2 and soft-repulsive particles embedded in polymer electrolytes

4 COUPLING OF MD SIMULATIONS AND MPM CALCULATIONS

In the previous sections, we have shown that the nature of the polymer-nanoparticle interactions and the structure of the nanoparticle surface significantly perturb structural, transport and dynamic properties of polymers and polymer electrolytes. In this section we will demonstrate that mechanical properties extracted from MD simulations can be used in MPM calculation in order to obtain homogenized properties of nanocomposites at much larger length-scales.

MD simulations have been performed on three systems: (a) pure polymer and (b) polymer nanocomposite consisting of an elastic cylinder embedded in a polymer matrix with the polymer having the same interactions with the cylinder as with itself (neutral system), (c) polymer nanocomposite consisting of an elastic cylinder embedded in a polymer matrix with the polymer interacting with the cylinder twice as strong as with itself (attractive system). The system is shown in Figure 3.

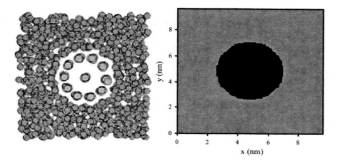

Figure 3. Snapshot of the MD simulation system consisting from a cylinder in a polymer Rcyl=2.1 nm, box size=9.615 nm (left) and setup of the MPM run (right).

Shear modulus G(t) has been extracted from MD simulations and approximated with a sum of 3 exponents. Bulk modulus (K) of the polymer was assumed to be time independent. Elastic properties of the cylinder were also measured in MD simulations. The properties ($K_{polymer}$, $G(t)polymer$ and K_{cyl}, G_{cyl}) extracted from MD simulations were used in MPM simulations of the composite.

A number of stress relaxation, creep, and dynamic-mechanical experiments have been performed on the pure materials and composites. Stress relaxation experiments are the most efficient and allow property extraction and are accurate to within 1 % for pure polymer MPM calculations. Stress on the material points have been found to exhibit large oscillations during MPM calculations. However, averaging over material points reduced oscillations for the composite and virtually eliminated them for the pure viscoelastic material. The velocity damping function had to be added to the particle equation of motions to significantly reduce the oscillations, while leaving the average (smoothed) G(t) behavior of the composite unchanged. A series of computational experiments indicated that the shear modulus for the composite not only increases in magnitude and shows significantly longer relaxation times, but the shape of the relaxation becomes more stretched.

MPM calculation of the neutral system using properties of the pure polymer and elastic material from MD simulations predicted shear modulus of the composite in good agreement with the one obtained from MD simulations without any adjustment to the polymer or nanoparticle properties. In order to reproduce in MPM calculations the shear modulus G(t) from MD simulations of the nanocomposite with the attractive polymer-nanoparticle interactions, we introduced a layer of polymer around a nanoparticle that has relaxation two orders of magnitude slower than the bulk polymer and the width of 1.7 nm corresponding to approximately 1.7 layers of polymer. This model was able to reproduce MD simulations results reasonably well as shown in Figure 4. We are currently working on extending this model to systems with more than one nanoparticle and investigating the ability of this model to capture the dependence of shear modulus for attractive nanocomposites on the size of nanoparticles.

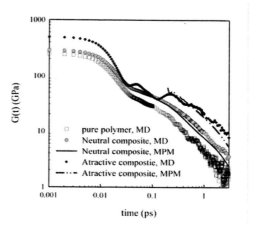

Figure 4. Shear modulus of bulk polymer, neutral and attractive nanocomposites from MD simulations and MPM calculations.

ACKNOWLEDGMENT

The authors are indebted to NASA and DOE for financial support through grants NAG3 2624 and 34000 A respectively.

REFERENCES

[1] G. D. Smith, D. Bedrov, L. Li, O. Byutner J. Chem. Phys. 117, 9478, 2002.
[2] S. C. Glotzer, Nature Materials, 2, 713, 2003.
[3] G. D. Smith, D. Bedrov, O. Borodin Phys. Rev. Lett. 90, 26103, 2003.

Hybrid atomistic-continuum fluid mechanics

T. Werder, J. H. Walther, E. Kotsalis, P. Gonnet and P. Koumoutsakos

Institute of Computational Science, ETH Zürich
Hirschengraben 84, 8092 Zürich, Switzerland
{werder,walther,kotsalis,gonnetp,koumoutsakos}@inf.ethz.ch

ABSTRACT

We outline a multiscale algorithm for the simulation of dense fluids that couples molecular dynamics and continuum fluid dynamics. The coupling between the two models is realized by a common overlap domain and by the alternating Schwarz method. We prefer this method to those based on direct flux exchange, since the accurate estimation of the fluxes requires much more statistics. The boundary conditions on the atomistic region are imposed using stochastic walls to thermalize and confine the molecular dynamics system, and the USHER algorithm to insert new atoms. The forcing of the continuum is achieved using appropriate source terms in the momentum equations. To test the algorithm, we study equilibrium liquid argon systems and the flow of liquid argon past a carbon nanotube.

Keywords: multiscale algorithm, molecular dynamics, fluid dynamics

1 Introduction

The study of fluid mechanics at the nanoscale level [1] has received wide attention due to its importance for the understanding and development of biosensors. From the computational side, classical molecular dynamics (MD) have so far been the most useful tool to characterize phenomena such as wetting, hydrophobicity and boundary conditions. One system of particular interest with regard to biosensors is carbon nanotubes in water, which has recently been studied in some detail [2], [3]. Despite the success of molecular dynamics, its limitations in terms of accessible length and time scales are currently of the order of $10\,\text{nm}^3$ and $10\,\text{ns}$. In order to allow the computational analysis of nanoscale systems integrated in microfluidic environments, a multiscale approach is indispensable for the following two reasons. First, even with special purpose parallel hardware and state of the art algorithms, it remains prohibitively expensive to reach the microscale using classical MD simulations. Second, the amount of data that such a computation would generate is intractable.

For both dilute [4], and dense fluids [5]–[7] different algorithms exist to couple atomistic and continuum simulations. Here, we present an algorithm to impose arbitrary, non-periodic boundary conditions on MD simulations of dense fluids, cf. Fig. 1. We show how this algorithm couples standard molecular dynamics codes with commercial computational fluid dynamics packages. As a first application we consider a flow of liquid argon around a carbon nanotube (CNT), where only the CNT and its interaction with the nearby argon are modeled using MD, cf. Fig. 4, and where the continuum part is described by the isothermal, incompressible, Navier-Stokes (NS) equations.

2 The computational method

The central part of any hybrid scheme is the coupling of the atomistic and continuum regions. One can distinguish between two fundamental types of coupling, the former aim at matching the fluxes of mass, momentum, and energy at the interface [6], [7] and, the others are based on overlapping cells where the mass density ρ, the mean velocity $\boldsymbol{u}$, and the temperature T are imposed. The latter schemes have two main advantages over flux based schemes, namely that they decouple time scales [5] and that densities are substantially easier to sample than fluxes. This can be quantified by the number of atomistic samples M_q required to achieve a statistical error E_q in a quantity q measured in a domain of volume V [8], where E_q is defined as $E_q = \sigma(\bar{q})/\bar{q} = \sigma(q)/\sqrt{M}\bar{q}$, and where $\bar{q}$ and $\sigma(q)$ denote the average and standard deviation of q. Recently, estimates were given for the number of samples needed for the mean velocity, the density, and the pressure [8]

$$M_u = \frac{k_B T_0}{\bar{u}_x^2} \frac{1}{\rho V E_u^2}, \qquad (1)$$

$$M_\rho = \frac{\kappa_T k_B T_0}{V E_\rho^2}, \qquad (2)$$

$$M_P = \frac{\gamma k_B T_0}{P_0^2 \kappa_T} \frac{1}{V E_P^2}. \qquad (3)$$

To illustrate Eqns. (1)-(3), consider the following numerical example: Assume we measure $\boldsymbol{u}$, ρ, and the pressure P of bulk water in a cell of volume $1\,\text{nm}^3$ at atmospheric conditions ($P = 1\text{bar}$, $T = 293\,\text{K}$) and with a ratio of specific heats $\gamma = 1$, an isothermal compressibility of $\kappa_T = 48.95 \cdot 10^{-6}\,\text{bar}^{-1}$, and a mean velocity

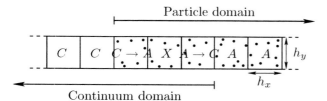

Figure 1: One dimensional sketch of a computational domain. There are pure continuum (C) and particle cells (A), as well as overlap cells $C \rightarrow A$, X, and $A \rightarrow C$, where both descriptions are valid. The Schwarz iteration by turns imposes the values measured in the $C \rightarrow A$ cells on the atomistic region and the values measured in the $A \rightarrow C$ cells on the continuum region until the solution converges in X.

$\bar{u}_x = 100\,\mathrm{ms}^{-1}$. Then, to obtain a fractional error of 5% the number of required (*statistically independent*) samples amounts to $M_u \approx 10^2$, $M_\rho \approx 1$, and $M_P \approx 10^8$. We note that the pressure is particularly expensive to sample and that we therefore favour density based schemes, where the pressure does not need to be computed or imposed directly.

2.1 The Schwarz algorithm

The algorithm is based on the Schwarz iteration scheme [5], i.e., the computational domain consists of continuum cells C, particle cells A, and overlap cells X, where both descriptions are valid, cf. Figs. 1 and 2. The following iteration is run until the solution is converged in the overlap region.

Step 1) Run an MD simulation until a steady state is reached, with $(\rho, \boldsymbol{u}, T)$ imposed in the $C \rightarrow A$ cells.

Step 2) Solve the continuum equations with the steady state solution of the MD in the $A \rightarrow C$ cells as boundary conditions.

2.2 The atomistic part

The main difficulties in imposing the continuum field values on the atomistic system in the $C \rightarrow A$ domain are the introduction and removal of particles but also the altered dynamics of the molecules in the $C \rightarrow A$ region due to the "missing particles" exterior to the atomistic domain. For a monatomic Lennard-Jones fluid, the first problem is solved using the USHER algorithm [9]. For the second problem, we use the notion of stochastic boundary condition introduced to simulate the contact of an MD system with a thermal reservoir [10]. A generic molecular dynamics time step, where $(\rho, \boldsymbol{u}, T)$ are given in the $C \rightarrow A$ cells, reads:

1. *Force computation.* Compute forces according to the pair potential U and the wall force $F_\mathrm{w}(x)$.

2. *Velocity update.* Update the particle velocities using the leap-frog scheme.

3. *Andersen thermostat.* Choose particles randomly in the $C \rightarrow A$ cells with probability ν. Reset their instantaneous velocities $\widetilde{u}$ according to a biased Maxwell-Boltzmann distribution with a mean velocity $\boldsymbol{u}$ and temperature T.

$$f(\widetilde{u}_k) = \sqrt{\frac{m}{2\pi k_b T}} \exp\left(-\frac{(\widetilde{u}_k - u)^2}{2 k_b T}\right). \quad (4)$$

4. *Position update.* Update the particle positions using the leap-frog scheme.

5. *Stochastic wall.* The boundary between $C \rightarrow A$ and C cells is formed by a stochastic wall W [10]. The normal velocity of particles k that hit W is sampled from the probability density [12]

$$f(\widetilde{u}_k) = \frac{m}{k_b T}|u_k| \exp\left(-\frac{(\widetilde{u}_k - u)^2}{2 k_b T}\right), \quad (5)$$

and the tangential components are sampled from Eq. (4).

6. *Mass flux.* We consider an incompressible system and ensure mass conservation by collecting and redistributing the particles that leave the atomistic domain. Note that the particles are not periodically mapped, but they are re-introduced in the cell with the lowest instantaneous density using the USHER algorithm [9]

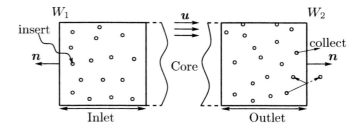

Figure 2: Schematic representation of an atomistic simulation domain with non-periodic boundary conditions. In the core region an NVE molecular dynamics simulation is performed, while the inlet and outlet are thermostatted at a temperature T and with mean velocity $\boldsymbol{u}$. W_1 and W_2 denote the positions of the stochastic walls. The outward normal $\boldsymbol{n}$ points toward the continuum region.

2.3 The continuum part

The continuum system is described by a finite volume discretization of the isothermal, incompressible, Navier-Stokes (NS) equations in the entire computational domain including the atomistic region. By the complete overlay, we avoid explicit internal interfaces in the continuum system and instead utilize body forces to impose the average molecular velocity field. We note that the external boundary conditions of the continuum system are not limited to cyclic conditions, but can take any for the flow solver admissible form.

The x-component of the discrete momentum equation reads

$$A_P^x u_P^x + \sum_i^N A_i^x u_i^x + S_1^x = 0, \qquad (6)$$

where the sum includes the contributions from the N neighboring finite volumes. The coefficients (A_i) in Eq. (6) contain convective and diffusive fluxes across the cell faces which determine the total change of the cell velocity field. The central coefficient is given by $A_P = \sum_i A_i + S_2$, where S_1 and S_2 denote the contribution from body forces per unit volume within the cell, e.g., $B^x = S_1^x + S_2^x u_P^x$. For incompressible flow, an equation for the pressure is derived using the SIMPLE method [13]. The velocity field is imposed by choosing $S_1 = 10^{30}$ and $S_2 = -10^{30}$.

Here, with the assumptions of an isothermal, incompressible fluid, we solely need to extract the mean velocity from the atomistic domain. Since in general, the macroscopic time scale τ_C on which $\boldsymbol{u}$ changes is much larger than the microscopic time scale τ_P, $\boldsymbol{u}$ is simply taken to be the time averaged mean velocity. If $\tau_P/\tau_C \approx 1$, then a statistically relevant ensemble average can be reached through multiple realizations of a short simulation with perturbed initial conditions.

3   Results

To test the proposed algorithm, we have interfaced a Navier-Stokes solver (StarCD [14]), with an in-house molecular dynamics code (FASTTUBE [15]). We consider liquid argon systems in equilibrium and flowing past a carbon nanotube. The argon interacts through a Lennard-Jones pair potential

$$V(r) = 4\epsilon_{\alpha\beta}\left[\left(\frac{\sigma_{\alpha\beta}}{r}\right)^{12} - \left(\frac{\sigma_{\alpha\beta}}{r}\right)^6\right], \qquad (7)$$

where $\epsilon_{\alpha\beta}$ and $\sigma_{\alpha\beta}$ denote the Lennard-Jones parameters for the species α and β. For the argon-argon interaction, we use coefficients of $\epsilon_{\mathrm{ArAr}} = 1.0\,\mathrm{kJ\,mol}^{-1}$ and $\sigma_{\mathrm{ArAr}} = 0.34\,\mathrm{nm}$. The carbon nanotube is modeled as a rigid structure which interacts with the argon through a Lennard-Jones potential with coefficients $\epsilon_{\mathrm{CAr}} = 0.57\,\mathrm{kJ\,mol}^{-1}$ and $\sigma_{\mathrm{CAr}} = 0.34\,\mathrm{nm}$.

3.1   Equilibrium systems

To validate the proposed algorithm, we consider an equilibrium system of 1589 argon atoms at 300 K in a

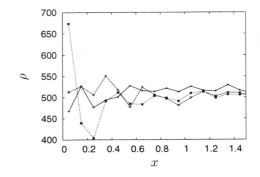

Figure 3: Density profiles extracted from equilibrium atomistic simulations with different boundary conditions (BC) at $x \leq 0$: Periodic BC ($\diamond$), periodic BC with a stochastic wall ($+$), and non-periodic with a stochastic wall ($\square$).

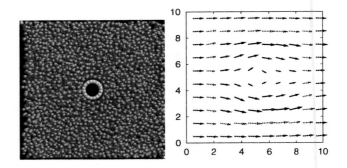

Figure 4: Flow of argon around a carbon nanotube. Left: A snapshot of the atomistic system. Right: The converged velocity field in the atomistic domain. The continuum domain is not shown.

computational box of dimensions $5 \times 5 \times 5\,\mathrm{nm}$. The y and z-directions are treated periodically, while for the x-direction, the following variants were considered: (i) periodic boundary conditions (PBC) as a reference solution, (ii) PBC but with a stochastic wall (SW), and (iii) a non-periodic system with SW. Figure 3 displays for these different cases the mass density along the x-axis (only leftmost part shown) measured in bins of thickness 0.05 nm and averaged over 500 snapshots (taken every 0.2 ps). We note that the stochastic wall alone does not disturb the averaged local density field. However, for the non-periodic system (iii), we observe a layering of the liquid near the wall. Toward the bulk system, the layering decays quickly and is not expected to affect the bulk properties.

3.2   Flow of liquid argon around a carbon nanotube

For the study of a flow around a carbon nanotube, we choose the size of the finite-volume cells to match the size and location of the bins used in the molecular

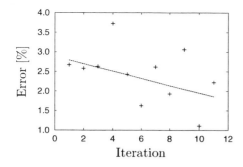

Figure 5: Convergence of the velocity field imposed on the edge of the atomistic domain. The error measure is defined in Eqn. (8). The solid line is a linear fit to the data.

system for the averaging and for the imposition of the boundary conditions. In the present study we use cells of size $1.0 \times 1.0 \times 4.26$ nm to secure a sufficient averaging and forcing of the molecular system. The continuum mesh consists of $30 \times 30 \times 1$ mesh points. The boundary conditions of the continuum system are chosen to be periodic in y and z. A flow is imposed along the x-direction with an inlet velocity of $u_x = 100$ ms^{-1} at the left boundary and an outlet boundary condition at the right boundary. The atomistic domain extends over $10 \times 10 \times 4.26$ nm (including the $C \to A$ cells) and contains 5600 argon atoms. Finally, the carbon nanotube is of chirality (16,0) and oriented along the z-axis. A snapshot of the atomistic domain is given in Fig. 4.

In every iteration of the atomistic domain, we let the system adapt to the new boundary conditions for 20 ps, and we subsequently sample the system during 80 ps.

We quantify the convergence through the measure

$$e^k = \frac{1}{N} \sum_{i=1}^{N} \frac{|\boldsymbol{u}_i^k - \boldsymbol{u}_i^{k-1}|}{|\boldsymbol{u}_i^{k-1}|} \qquad (8)$$

where $\boldsymbol{u}_i^k$ is the mean velocity imposed on the i'th of N $C \to A$ cells in the k'th iteration. In Fig. 5, we show this measure as a function of the iteration number. The fractional error in sampling the velocity in the atomistic region is estimated to be $\approx 2.5\%$. Therefore we conclude that the solution has nearly converged after the first iteration, cf. Fig. 5.

4 Conclusion and Outlook

We have presented a hybrid multiscale algorithm for the coupling of atomistic and continuum systems. It is based on the Schwarz domain decomposition technique, the Usher algorithm for particle insertions, and stochastic walls. Future work includes the extension of the algorithm to treat water, to find optimal parameters for the overlap size and finally to include adaptivity of the scheme.

REFERENCES

[1] P. Koumoutsakos, U. Zimmerli, T. Werder, and J. H. Walther, ASME/SPIE handbook of Nanotechnology in press (2003).

[2] T. Werder *et al.*, J. Phys. Chem. B **107**, 1345 (2003).

[3] J. H. Walther, T. Werder, R. L. Jaffe, and P. Koumoutsakos, Phys. Rev. E accepted (2003).

[4] A. L. Garcia, J. B. Bell, W. Y. Crutchfield, and B. J. Alder, J. Comput. Phys. **154**, 134 (1999).

[5] N. G. Hadjiconstantinou, J. Comput. Phys. **154**, 245 (1999).

[6] E. G. Flekkøy, G. Wagner, and J. Feder, Europhys. Lett. **52**, 271 (2000).

[7] R. Delgado-Buscalioni and P. V. Coveney, Phys. Rev. E **67**, 046704 (2003).

[8] N. G. Hadjiconstantinou, A. L. Garcia, M. Z. Bazant, and G. He, J. Comput. Phys. **187**, 274 (2003).

[9] R. Delgado-Buscalioni and P. V. Coveney, J. Chem. Phys. **119**, 978 (2003).

[10] G. Ciccotti and A. Tenenbaum, J. Stat. Phys. **23**, 767 (1980).

[11] E. Matteoli and G. Ali Mansoori, J. Chem. Phys. **103**, 4672 (1995).

[12] R. Tehver, F. Toigo, J. Koplik, and J. R. Banavar, Phys. Rev. E **57**, R17 (1998).

[13] S. V. Patankar and D. B. Spalding, Int. J. Heat Mass Transfer **15**, 1787 (1972).

[14] STAR-CD Version 3.1A. Manual, 1999.

[15] J. H. Walther, R. Jaffe, T. Halicioglu, and P. Koumoutsakos, J. Phys. Chem. B **105**, 9980 (2001).

Impact Induced Desorption of Large Molecular Structures from Graphitic Substrates

R.P.Webb and K.J.Kirkby

Advanced Technology Institute, University of Surrey, Guildford, GU2 7XH, UK,
R.Webb@eim.surrey.ac.uk

ABSTRACT

We use molecular dynamics computer simulation to show how the impact of a C60 molecule on a graphite surface can cause an acoustic wave across the surface of the substrate that is strong enough to desorbed a neighboring C_{60} molecule that has already been adsorbed. This could have potential implications for experimentalists attempting to grow fullerene based structures on graphite like materials using an energetic deposition technique. It suggests that unlike normal growth conditions were sticking probabilities are the principle concern an extra problem of deposition induced desorption may further slow a growth process.

Keywords: Computer Simulation, Fullerene Deposition, Graphite

1 INTRODUCTION

The phenomenon of sputtering has been well known and understood for over a century now and has been put to great uses in areas such as Secondary Ion Mass Spectrometry (SIMS) and Focused Ion Beam (FIB) lithography techniques, and been the cause of some minor technological problems employing ion beams by causing premature saturation effects in implantation and undesired surface morphology changes in many other applications. In the majority of cases these problems have been overcome relatively simply.

Currently many new applications are being considered which involve the positioning of large fullerene based objects on surfaces to create both active and passive devices. The investigation presented here uses Molecular Dynamics computer simulation [1] to highlight a potential pit-fall that might cause some initial problems when attempting to build these systems using an energetic deposition mechanism particularly if the substrate material is a layered material such as graphite.

We have employed full many body potentials with an additional long range component to model the Van de Waals forces [2] to demonstrate the behaviour of the surrounding surface after a relatively gentle impact of fullerenes (single cage fullerenes – C_{60} – have been used as well as multiple cage - C_{300} and C_{840}) onto a graphite (hopg) surface. It is seen that the impact, whilst leaving the molecule intact can cause an acoustic wave on the surface of the graphite. Previous work [3] has shown that this acoustic wave can cause substantial ejection of benzene molecules adsorbed onto the surface of the substrate. Here we use the same technique and show that the same acoustic wave can cause desorption of a large molecules as well.

2 SIMULATION METHOD

The precise details of the computer simulation technique used in this investigation have been given elsewhere [1,4] hence only a brief summary outlining the important components will be given here. The classical molecular dynamics simulation technique is employed solving Newton's Laws of Motion for many thousands of atoms obeying a realistic many body interaction potentials. The interaction potential used is the Brenner hydrocarbon potential [5,6] with an additional long range term as described by Beardmore and Smith [7]. The elastic properties of this potential have been investigated more fully by Smith and Christopher [8] in a simulation of nano-indentation for diamond, graphite and fullerite films. In their simulations they find good agreement with experiment. In the results presented here the elastic behavior of the graphite surface is quite key in determining the exact results and so it is important that this behavior is modeled well. However it can be expected that the phenomenon discussed here is likely to occur in any layered material, like graphite, to a lesser or greater extent.

The long-range term provides the interlayer bonding term for graphite and the Van der Waals attraction of molecules to the graphite surface. The binding energy of clusters to the graphite surface is calculated by moving the particle close to the surface and away to beyond the cut-off of the long range potential. The potential energy of a fullerene C_{60} molecule is plotted as a function of distance from the surface and shown in figure 1. The binding energy is found to be 0.47eV to the graphite surface at an equilibrium distance of 6.3 Å above the surface, the distance being measure from the center of the cage structure. Previous simulations [9] of binding and equilibrium distances of benzene molecules in this way have found good agreement with experimental values. Similar results are found for C_{300} structures with a binding energy of 0.94eV at 10 Å above the surface. Cell sizes contain about 100,000 atoms depending upon the over-layer and projectile simulated. The lateral edges of the simulation cell are treated as periodic but the surface and bottom are treated as free boundaries. The simulations are run for typically 4ps, if they are run for much longer then reflections from the periodic boundaries can cause noticeable effects to the results.

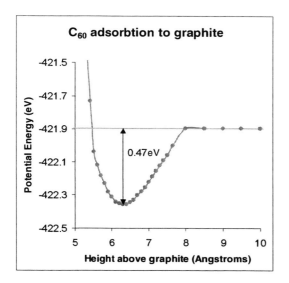

Figure 1: Potential Energy of C60 as a function of height above graphite surface.

3 RESULTS

The simulations have been run to observe the effects of the impact from a fullerene molecule on a graphite surface on which a fullerene molecule has been adsorbed. As described above the impact at energies of 200eV (that is an equivalent energy of 3.33eV per atom) can cause the propagation of acoustic waves across the surface which have sufficient energy to cause the ejection of small benzene molecules that have previously been adsorbed to the surface. Can the same mechanism cause the ejection of larger adsorbed molecules?

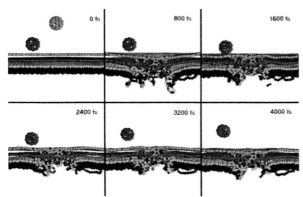

Figure2: The energetic impact of a 4keV C_{60} on a graphite surface – see text for discussion.

The figure above shows that an energetic impact of a C_{60} molecule can cause the desorption of a neighboring C_{60} molecule. The figure shows the ejection of a nearby adsorbed C_{60} molecule (drawn in red) caused by the impact of a 4keV C_{60}^{+} molecule (drawn in green). In the initial frame the "red" C_{60} is bound to the surface and the impacting "green" C_{60} is about to make impact with the surface. In this figure the atoms between the "viewing point" and the impact point have been removed from the drawing, but are present in the simulation, to allow the observer a clearer view as to what happens at the impact site and close to the desorbed molecule. In each frame the time after the start of the simulation in fs is shown. It can be seen that the energetic C_{60} impacts the surface and breaks through the surface of the graphite and destroys itself and causes an acoustic wave to propagate across the surface from the impact site. This wave then interacts with the adsorbed C_{60} giving it enough energy (at about 2400fs) to overcome the binding between it and the surface.

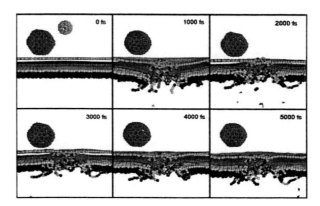

Figure 3: The energetic impact of a 4keV C_{60} near a C_{300} adsorbed molecule

In figure 3 we show a similar set of results for a similar energy impact, but this time near a C_{300} adsorbed cluster. The effect is very similar despite the increased mass of the cluster and the higher binding. The impacting C_{60} creates an acoustic wave in the surface of the graphite with enough strength to cause the C_{300} cluster to be desorbed from the surface. The mechanism is very similar, the initial impact causes the surface to move downwards – away from the adsorbed molecule – pulling it with it. The surface then recovers and springs back, hitting the adsorbed molecule and giving it enough energy to escape the surface binding potential.

These examples have been modeled at a higher energy (4keV for the C_{60} or 66.67eV per atom) than would be used in a deposition system it does demonstrate a potential problem of such a technique. Also, whilst the overall energy is relatively high the energy per incoming atom is lower than is often used in a deposition system It should also be noted that fullerene impacts of 250eV which are enough to "soft-land" a fullerene on the surface also create acoustic waves [10]. We have investigated a number of impacts at 250eV at a range of distances away from the adsorbed fullerene molecule and the general results can be compared in figure 4 below.

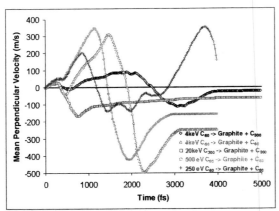

Figure 4: Perpendicular Velocities of "adsorbed" molecule for various fullerene impacts

Figure 4 shows the mean perpendicular velocity of all the atoms in the adsorbed molecule (either a C_{300} cluster of a C_{60} molecule) as a function of time during the simulation. The same effect can be seen here as were described above from the atom positions. Looking at the red curve on the figure (the 4keV C_{60} impact near an adsorbed C_{60}). The perpendicular velocity of the adsorbed molecule show that the molecule is first pulled towards the surface (+ve direction) and then pushed away from the surface (-ve direction), this is repeated with more energy and at about 2400fs the molecule is give enough energy to to escape the surface, which it eventually does at about 3ps. A very similar effect is seen (orange curve) when the energy of the impacting molecule is dropped to 500eV (8.3eV per atom). When the adsorbed molecule is changed for a C_{300} cluster (blue curve) then the story is very similar but the velocities are lower. If we change the impacting particle to a C_{300} cluster as well then the adsorbed cluster is desorbed from the surface with the first vibration of the surface. However in the case of the impact by a 250eV C_{60} (brown curve), an energy at which soft-landing occurs, it can be seen that the adsorbed molecule undergoes the same behavior except that when the molecule is pushed away from the surface it does not acquire enough velocity to escape the surface and remains bound. However, the velocity on the second bounce is substantially higher as it comes closer to the surface and at 4ps it does appear to be about to start moving away from the surface and the curve gives a hint that in the following time-steps the molecule might well shake itself free of the surface. Unfortunately the simulation time beyond 4ps is untrustworthy and it is possible that reflections from the boundaries could cause ejection of the molecule at this point. Simulations on larger systems are now underway to investigate if impacts at these low energies do indeed cause ejection as well or if this is just a side effect of the boundary conditions of the simulation.

4 CONCLUSIONS

Although at the present time we have not been able to observe coupling of the wave with sufficient energy to a neighboring fullerene to cause desorption for low energy impacts it has been demonstrated that it does occur for larger energies and that there is some indication from the simulations that there could well be a problem for low energies at which soft-landing occurs. It is felt that this could, therefore potentially, be a real problem and will almost certainly contribute to a reduction on the experimentally observed sticking probability in this kind of system. This investigation is very much preliminary at this stage.

REFERENCES

[1] DE Harrison Jr., Crit. Rev. Solid State Mater. Sci. 14, S1, 1989
[2] R Smith, K Beardmore, Thin Solid Films, 272, p255, 1996
[3] M Kerford, RP Webb, Nucl. Instrum. Meth. B, 180, p44, 2001
[4] R Smith, DE Harrison Jr., BJ Garrison, Phys. Rev. B40, p93, 1989
[5] DW Brenner, Phys. Rev. B46, p1948, 1992
[6] DW Brenner, Phys. Rev. B42, p9458, 1990
[7] R Smith, K.Beardmore, Thin Solid Films 272, p255, 1996
[8] D Christopher, R Smith, Nucl. Instrum. Meths. 180, p117, 2001
[9] RP Webb, M Kerford, E Ali, M Dunn, L Knowles, K Lee, J Mistry, F Whitefoot, Surfacd & Interface Analysis 31, p297, 2001
[10] R Smith, RP Webb, Proc. R. Soc. Lond. A, 441, p495, 1993

Multiscale modeling in Nanotechnology

A. Maiti[*]

[*]Accelrys Inc., 9685 Scranton Road, San Diego, CA 92121

ABSTRACT

Technologically important nanomaterials come in all shapes and sizes. They can range from small molecules to complex composites and mixtures. Depending upon the spatial dimensions of the system and properties under investigation, computer modeling of such materials can range from first-principles Quantum Mechanics, to Forcefield-based Molecular Mechanics, to mesoscale simulation methods. In this contribution we illustrate the use of all three modeling techniques through a number of recent applications: (1) carbon nanotubes (CNTs) as nano electromechanical sensor (NEMS) devices; (2) mesoscale modeling of polymer-nanotube composites; and (3) mesoscale diffusion of drug molecules across cell membranes.

Keywords: nanotubes, nanocomposites, mesoscale modeling, membranes, diffusion.

1 INTRODUCTION

As with any new technology, Nanotechnology has many challenges to overcome, typically associated with control and precision at the nanoscale. Some of the challenges include: device integration (interconnect failure, addressability issues), growth and synthesis (difficulty in size-dispersion control, requirement of novel assembly techniques, presence of structural defects), contact resistance (necessity of atomic-level structural precision at junctions), functionalization (challenges with chemical inertness) and doping (non-uniformity of dopant levels). Computer modeling is a great approach to surmounting some of the above obstacles because it often provides deeper insight into the system properties as a function of size/shape, structural defects, added functional groups, and system surroundings. However, the simulation method and the resolution details depend upon the system size and the properties to be investigated. Thus, to study electronic structure one has to take recourse to a Quantum mechanical method, either First-Principles or semi-empirical. To study electronic transport, one has to use a code based on the non-equilibrium Green's function (NEGF) formalism. To investigate macro-molecular structures like DNA, proteins or large nanotubes, one would employ classical molecular mechanics based on force fields, while investigations of phenomena at larger length and time-scales would require the use of mesoscale and possibly finite-elements methods.

This work illustrates the use of several of the above simulation techniques in three different application areas: (1) CNTs as NEMS devices; (2) polymer-nanotube composites; and (3) drug diffusion across cell membranes.

2 CNT-BASED ELECTROMECHANICAL SENSORS

Interest in the application of carbon nanotubes as electromechanical sensors got a significant boost from the pioneering experiment of Tombler et al. [1], in which the middle part of the segment of a metallic CNT suspended over a trench was pushed with an Atomic Force Microscope (AFM) tip. Beyond a deformation angle of $\sim 13°$ the electrical conductance of the tube dropped by more than two orders of magnitude. The effect was found to be completely reversible, i.e., through repeated cycles of AFM-deformation and tip removal, the electrical conductance displayed a cyclical variation with constant amplitude. An interesting explanation was put forward by O(N) tight-binding calculations [2], which show that beyond a critical deformation several C-atoms close to the AFM tip become sp^3-coordinated. This leads to the tying up of π-electrons into localized σ-states, which would explain the large drop in electrical conductance.

Considering the significance of the above result, it was important to carry out an independent investigation using first-principles QM. Unfortunately, the smallest models required to simulate the AFM-deformation of a CNT typically involve a few thousand atoms, which makes first-principles QM simulations unfeasible. This necessitated a combination of first-principles DFT and classical molecular mechanics. Bond reconstruction, if any, is likely to occur only in the highly deformed, non-straight part of the tube close to the AFM-tip. For such atoms (~ 100-150 atoms including AFM-tip atoms) a DFT-based QM description was used, while the long and essentially straight part away from the AFM tip was described accurately using the Universal Forcefield (UFF) [3], which had previously been used in CNT simulations [4].

Because of known differences in the electronic response of zigzag and armchair tubes to mechanical deformation, the simulations were performed on a (12, 0) zigzag and a (6, 6) armchair tube, each consisting of 2400 atoms. The AFM tip was modeled by a 6-layer deep 15-atom Li-needle normal to the (100) direction, terminating in an atomically sharp tip. To simulate AFM-tip-deformation, the Li-needle was initially aimed at the center of a hexagon on the bottom-side of the middle part of the tube. The Li-needle tip was then displaced by an amount δ toward the tube along the needle-axis, resulting in a deformation angle $\theta = tan^{-1}(2\delta/l)$, l being the unstretched length of the tube. At each end of the tube, a contact region defined by a unit cell plus one atomic ring (a total of 36 and 60 atoms for the armchair and the zigzag tube respectively) was then fixed and the whole tube relaxed with the UFF, while constraining the needle atoms as well. The contact region atoms were fixed in order to simulate an ideal undeformed semi-infinite carbon nanotube lead, and to ensure that all possible contact modes are coupled to the deformed part of the tube. Following the UFF relaxation, a cluster of 132 C-atoms for the (6, 6) tube, and a cluster of 144 C-atoms for the (12, 0) tube were cut out from the middle of the tubes. These clusters, referred to below as the *QM clusters* (plus the 15 Li-tip atoms) were further relaxed with Accelrys' DFT-code DMol^3 [5], with the end atoms of the cluster plus the Li-tip atoms fixed at their respective classical positions.

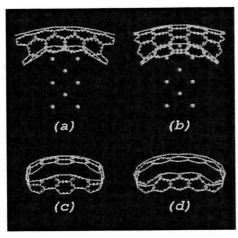

Figure 1. DMol^3-relaxed Li-tip-deformed QM clusters for: (a) the (6, 6) armchair (132 C-atoms); and (b) the (12, 0) zigzag (144 C-atoms), in side views. The deformation angle is 25° for both tubes. Figs. (c) and (d) are respective views along the tube length, with the Li-tip hidden for clarity.

Fig. 1 displays the tip-deformed QM-cluster for (6, 6) and (12, 0) tubes at the highest deformation angle of 25° considered in these simulations. Even under such large deformations, there is no indication of sp^3 bonding, and the structure was very similar to what was observed for a (5, 5) tube in a previous work. The absence of sp^3 coordination is inferred based on an analysis of nearest-neighbor distances of the atoms with the highest displacements, *i.e.*, the ones closest to the Li-tip. *Although for each of these atoms the three nearest neighbor C-C bonds are stretched to between 1.45-1.75 Å, the distance of the fourth neighbor, required to induce sp^3 coordination is greater than 2.2 Å for all tubes in our simulations.*

Following the structural relaxation of the CNTs, the transmission and conductance were computed using the recursive Green's function formalism [6]. A nearest-neighbor sp^3 tight-binding Hamiltonian in a non-orthogonal basis was chosen, and ideal semi-infinite contacts assumed

at both ends. These simulations indicate that the conductance remains essentially constant for the (6, 6) armchair tube up to deformation as large as 25°. However, for the (12, 0) zigzag tube the conductance drops significantly, by two orders of magnitude at 20°, and 4 orders of magnitude at θ=25° [7]. Since sp^3 coordination could be ruled out, what could be the cause for such a large conductance drop in the experiment of ref. [1]? Also, why did the armchair tube display no significant drop in conductance even up to large angles of deformation?

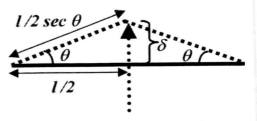

Figure 2. Schematic diagram representing deformation with an AFM tip. The tip-deformed tube undergoes a tensile strain. The deformation angle θ is related to the tip displacement (δ) and non-deformed tube-length (l) by $\theta = tan^{-1}(2\delta/l)$.

A simple explanation emerges if one zooms out from the middle of the tube and looks at the profile of the whole tube under AFM-deformation. One immediately discovers an overall stretching of the tube under AFM-deformation, as indicated schematically in Fig. 2. The effect of tensile stretch on CNT conductance has been theoretically studied in great detail [8]. An important result from such studies is that the *rate of change* of bandgap as a function of strain depends on the CNT chiral angle θ, more precisely as proportional to $cos(3\theta)$. Thus, stretched armchair tubes ($\theta = 30°$) do not open any bandgap, and always remain metallic. On the other hand, a metallic (3n, 0) zigzag tube ($\theta = 0$) can open a bandgap of ~100 meV when stretched by only 1%. This bandgap increases linearly with strain, thus transforming the CNT into a semiconductor at a strain of only a few percent. In general, all metallic tubes with $n_1 - n_2 = 3n$ will undergo the above metal-to-semiconductor transition, the effect being the most pronounced in metallic zigzag tubes. An experiment as in ref. [1] is thus expected to show a decrease in conductance upon AFM-deformation for all metallic CNTs except the armchair tubes. Researchers are also beginning to explore the electromechanical response of a squashed CNT, where sp^3 coordination is a possibility.

In addition to the above results for metallic CNTs, theory also predicts that for semiconducting tubes ($n_1 - n_2 \neq 3n$), the bandgap can either increase (for $n_1 - n_2 = 3n -2$) or decrease (for $n_1 - n_2 = 3n -1$) with strain. These results have recently prompted more detailed experiments on a set of metallic and semiconducting CNTs deformed with an AFM-tip [9], as well as on CNTs under experimental tensile stretch [10]. Commercial applications from such work could lead to novel pressure sensors, transducers, amplifiers, and logic devices [11].

3 CNT-POLYMER NANOCOMPOSITES

CNTs are found to possess remarkable mechanical and electrical properties. As discussed in the previous section, the CNTs can be metallic or semiconducting depending upon their chirality. The metallic CNTs are much more conducting than copper, and can carry very high density of electrical currents. Mechanically, the CNTs are stronger than steel yet being much lighter. One would like to utilize such unique properties of CNTs by enhancing mechanical and electrical properties of polymers through dispersing CNTs into a polymeric matrix. Such composites have potential applications not only as structural materials [12] but also for functional applications that make use of their conductivity, electromagnetic interference shielding [13] and optoelectronic properties [14].

The overall properties of CNT-polymer composite material depend on how uniformly the CNTs are dispersed within the polymeric matrix, how effective the interfacial bonding between the two systems is, and the way in which the above properties depend on the nanotube diameter, chirality, and functionalization of the CNT surface with organic functional groups.

There have been a few atomistic studies of the interaction of specific polymers with small segments of isolated CNTs [15]. However, extension of such studies to composites with several CNTs, each a few nanometers long, is extremely difficult because of: (1) a huge number of atoms involved; (2) long time-scales associated with typical dispersion processes. To surmount these challenges, we have adopted a mesoscale approach, in which groups of atoms are lumped into single "beads".

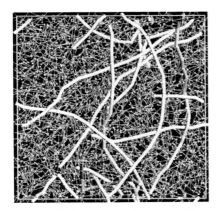

Fig. 3. Snapshot from a DPD simulation of CNTs well dispersed in an amorphous polymeric matrix.

The mesoscale simulations were performed using the code DPD (Dissipative Particle Dynamics) [16]. DPD simulates soft spherical beads interacting through a simple pair-wise potential, and thermally equilibrated through a Langevin thermostat. The total force on each DPD particle (bead) is expressed as a sum of three pair-wise additive terms [17]: (1) a conservative force, which is taken to be a soft repulsion; (2) a dissipative force, proportional to the relative velocity of the beads; and (3) a random force, necessary to maintain the system temperature. For polymeric systems, one has to include an additional interaction due to "springs" connecting the monomeric "beads". While the dissipative and random forces act in unison as a thermostat for the simulation, it is the conservative soft repulsive force that embodies the essential chemistry of the system. Ideally, one would like to derive the conservative force from detailed atomistic interactions. In 1997, Groot and Warren [17] made an important contribution on this front by establishing a relation between a simple functional form of the conservative repulsion in DPD and the Flory-Huggins χ-parameter theory.

The scaling of the repulsive interaction as a function of the bead-size, and its relation to surface tension in strongly segregated systems were recently analyzed by Maiti and McGrother [18]. This work was used to derive the interaction parameters for a CNT-polymer nanocomposite system. In addition, extra "spring" and "angle"-dependent terms were introduced in order to reproduce the Young's modulus and bending rigidity of the CNTs, which were computed from atomistic simulations. Details will be described elsewhere. Fig. 3 is a snapshot from a DPD simulation on a nanocomposite containing a volume fraction of 5% CNT and 95% polymer. The solubility parameters of the CNT and the polymer were close enough in this simulation, which resulted in a uniform dispersion of the CNT within the polymer. Such mixing is necessary for good mechanical properties of the composite. On the other hand, if the solubility parameter of the CNT differs from that of the polymer by a critical amount, it results in bundles of CNTs segregating out of the polymer, as illustrated in Fig. 4. Ongoing simulations are exploring the possibility of aligning the CNTs within the polymer matrix [[19], [20]] through the application of an external shear. Investigations of the effect of attaching organic functional groups to the CNTs are also underway.

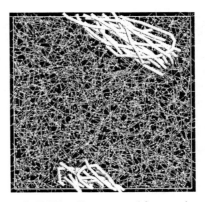

Fig. 4. Mesoscale CNT bundles segregated from a polymeric matrix.

Finally, we are exploring possibilities of importing such mesoscale data into a Finite-elements code [21] that would yield important mechanical and electrical properties of the system as a function of the mesoscale morphology.

4. DRUG DIFFUSION THROUGH CELL-MEMBRANE

DPD and other mesoscale modeling codes have so far been primarily used in the study of simple organic liquids, oil/water/surfactant systems, polymer blends, polymeric micelles, and so on. However, since most biological systems are organic in nature, such codes can be extended to the study of biological processes as well. Here we illustrate a novel application of DPD by simulating the diffusion of an Aspirin molecule across a cell membrane.

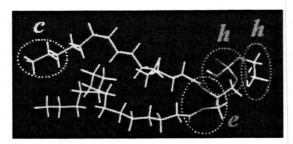

Fig. 5. DPD Bead representation of phospholipids PE.

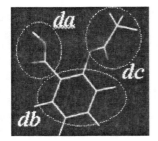

Fig. 6. DPD Bead representation of Aspirin.

A cell membrane typically consists of a lipid bilayer (phospholipids, glycolipids, cholesterol), transmembrane proteins (structural, transport, and channel proteins), and receptors (sugar, protein, etc.). To start with, we chose just a bilayer of the phospholipid phosphatidylethanolamine (PE) [22], whose atomic structure and corresponding mesoscopic beads are shown in Fig. 5. The atomic structure and meso-representation of the drug Aspirin is shown in Fig. 6. The hydrocarbon c-beads are hydrophobic, while the "head group" h-beads are hydrophilic. The connectivity of the lipid is described as $c_5e_1[h_2]c_5$ (where [] denotes a branching out), while that of the drug is simply $da_1db_1dc_1$. We have attempted a full atomic-scale simulation of the diffusion of Aspirin through the PE-bilayer. However, nothing significant happens in terms of diffusion within a few nanoseconds of simulation times that we are limited to. On the other hand, in a mesoscale simulation (see Fig. 7 for a snapshot), the soft repulsion allows a basic time-step of ~ 100 ps. This fact, coupled with an order of magnitude reduction in the number of beads (as compared to the number of atoms), and much simpler functional form of the DPD interaction (compared to the complex analytical form of typical force fields), effectively leads to *milliseconds* of simulation time within days on a single processor of a typical workstation. Preliminary results indicate trans-membrane diffusion of Aspirin within a few microseconds.

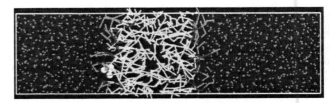

Fig. 7. Mesoscale representation of a three-bead Aspirin diffusing through a PE bilayer. The drug molecule is shown in ball and stick (yellow) representation for clarity. Red beads represent water.

To our knowledge, this is the first mesoscale simulation of drug diffusion across membranes. Current investigations include: (1) studying relative diffusion rates of different drugs; (2) effect of cholesterol content in the membrane; (3) diffusion of hydrophobic drugs encased in a micelle, where processes like endocytosis is a possibility.

ACKNOWLEDGEMENTS: The author would like to acknowledge collaborations with M. P. Anantram and Alexei Svizhenko (NASA), and Simon McGrother, James Wescott, and Paul Kung (Accelrys).

REFERENCES

[1] T. W. Tombler *et al.*, Nature 405, 769, 2000.
[2] L. Liu *et al.*, Phys. Rev. Lett. 84, 4950, 2000.
[3] A. K. Rappe *et al.*, J. Am. Chem. Soc. 114, 10024, 1992.
[4] N. Yao & V. Lordi, J. Appl. Phys. 84, 1939, 1998.
[5] http://www.accelrys.com/mstudio/dmol3.html.
[6] A. Svizhenko *et al.*, J. Appl. Phys. 91, 2343, 2002.
[7] A. Maiti, A. Svizhenko, and M. P. Anantram, Phys. Rev. Lett. 88, 126805, 2002.
[8] L. Yang and J. Han, Phys. Rev. Lett. 85, 154, 2000.
[9] E. D. Minot *et al.*, Phys. Rev. Lett. 90, 156401, 2003.
[10] J. Cao *et al.*, Phys. Rev. Lett. 90, 157601, 2003.
[11] A. Maiti, Nature Materials (London) 2, 440, 2003.
[12] D. Qian *et al.*, Appl. Phys. Lett. 76, 2868, 2000.
[13] J. Sandler *et al.*, Polymer 40, 5967, 1999.
[14] H. Ago *et al.*, Adv. Mater. 11, 1281, 1999.
[15] S. B. Sinnott et al., Carbon 36, 1, 1998; Y. Hu *et al.*, Composites Science & Technology 63, 1663, 2003; M. in het Panhuis *et al.*, J. Phys. Chem. B 109, 478, 2003.
[16] http://www.accelrys.com/mstudio/ms_modeling/dpd.html
[17] R. D. Groot and P. B. Warren, J. Chem. Phys. 107, 4423, 1997.
[18] A. Maiti and S. McGrother, J. Chem. Phys. (in press).
[19] R. Haggenmueller *et al.*, Chem. Phys. Lett. 330, 219, 2000.
[20] C. Bower *et al.*, Appl. Phys. Lett. 74, 3317, 1999.
[21] A.A. Gusev Macromolecules 34, 3081, 2001.
[22] R. D. Groot and K. L. Rabone, Biophysical Journal 81, 725, 2001.

Kinetics of microdomain structures in multi-phase polymer-liquid crystalline materials

S. K. Das[*], A. D. Rey[**]

Department of Chemical Engineering, McGill University
3610 University Street, Montreal, Quebec H3A 2B2, Canada.
[*]susanta.das@staff.mcgill.ca
[**]alejandro.rey@mcgill.ca

ABSTRACT

A mesoscopic kinetic model for phase separation in the presence of liquid crystalline order has been formulated, and solved using high performance numerical methods. The thermodynamic phase diagram on temperature-polymer concentration plane indicates the presence of coexistence regions between isotropic and liquid crystalline phases. These regions are itself partition by the phase separation spinodal and the phase ordering spinodal. We characterize the morphologies following temperature quenches in the phase diagram. The scenario is completely different from isotropic mixing since the continuous phase exhibits liquid crystalline ordering. Microdomains of the dispersed phase induce long- and short-range forces affecting the kinetics of the separation and the emerging structures. Presence of topological defects and elastic distortions around the microdomains formed during the phase separation dominate the morphology. The free energy of the system establishes dynamics and correlations of the morphological structures.

Keywords: multiphase, phase separation, microdomain structures, topological defects, phase diagram

1 INTRODUCTION

Multiphase polymer-liquid crystal blends are new multifunctional materials with unique electro-optical properties. The formation process is driven by thermodynamic instabilities, and the emerging microstructures reflect the curvature elasticity of the liquid crystalline phase. Such multiphase polymer dispersed liquid crystal (PDLC) makes a new composite material with unique physical properties that originate from the orientational ordering of the liquid crystal. Mechanical and electro-optical properties of this system are primarily determined by the collective behaviors of these binary mixtures. Because when flexible polymers are introduced into LC's the electro-optical properties of the system are considerably affected due to the deformation of the nematic director field, which can result in nontrivial collective behaviors, leading to the formation of spatially modulated structures. Depending on the time scale that controls these processes, a rich variety of morphologies have been observed [1-2]. Phase separation of such systems can be induced either through a thermal quench [3-4] or through polymerization [5]. Because of the number of nonequilibrium processes involved, however, there is a little theoretical understanding of the factors that control the domain morphology. A Cahn-Hilliard framework that allows composition and orientational density to evolve in a coupled fashion as functions of position and time following a temperature quench was performed [3]. Their framework includes the orientational density's second-order tensorial nature [6] where free energy of the system contains orientational density's three term gradient expansions. But details morphological structures, especially the free energy profiles of the system, characterization of morphological structures, phase separation and phase transition mechanism, topological defect structures, etc., remain unclear yet.

In this paper we present a nonlocal dynamical model focusing on the interplay between phase-separation and phase ordering kinetics in mixtures of short, liquid crystals (rigid rods) and long, flexible polymers, as a first step towards the rational design and control of the microdomain morphology. Here we consider fully nonlocal model without resorting to the three term gradient expansions of Landau-type [6] while derived free energy of the system. Computationally, this is challenging because it would require evaluating multiple convolutions at each moment in time. The advantage of our system is that we can calculate two order parameters (conserved and non-conserved) solving two coupled time-dependent equations together from a microscopic model of polymers and liquid crystals without loosing any information of order parameters.

2 MODEL FORMULATION

2.1 Thermodynamic phase diagram

In this section we represent the free energy to construct the static phase diagrams. According to [7], the free energy density of polymer-liquid crystal mixtures can be written as;

$$f(\varphi, S) = \beta \Delta \, F_{mix}/N_T = f^{(i)} + f^{(n)}, \tag{1}$$

$$f^{(i)} = \frac{\varphi_I}{n_I}\ln\varphi_I + \frac{\varphi_A}{n_A}\ln\varphi_A + \chi\varphi_I\varphi_A \, , \qquad (2)$$

$$f^{(n)} = \frac{1}{2}\left(\frac{\Gamma_0}{n_A}\right)\varphi_A^{\,2}\,S^2 - \frac{\varphi_A}{n_A}\ln\left(\frac{I_0\left(\Gamma_0\varphi_A S\right)}{2}\right)\bigg] \, , \qquad (3)$$

$$\varphi_I + \varphi_A = 1, \ \Gamma_0 = \left(\chi_a + 5/4\right)n_A, \ \beta = 1/k_b T \, , \qquad (4)$$

$$I_0\left(\Gamma_0\varphi_A S\right) = \int_{-1}^{1}\exp\left(\frac{3}{2}\Gamma_0\varphi_A S\left(x^2 - \frac{1}{3}\right)\right)dx \, , \qquad (5)$$

where T is the absolute temperature, k_b is the Boltzmann constant, S is the 'scalar' orientational order parameter of the liquid crystals, χ is the Flory-Huggin's interaction parameter, the terms χ_a and $\left(5/4\right)n_A$ in Γ_0 indicates the orientation-dependent attractive interactions between the mesogens and excluded volume interactions between mesogenic molecules respectively, n_I and n_A are the number of segments on the isotropic (monomer or polymer) component and the number of segments (axial ratios) on the mesogen respectively, and φ_I and φ_A are the corresponding volume fractions, respectively. The first two terms in the right hand side of equation (2) represent the entropy of mixing. For thermodynamical reasons, the entropy of mixing must be dominant at high temperatures and so we can introduce the temperature parameter, τ, defined by $1/\tau = \chi = U_0/k_b T$, where U_0 controls the miscibility of the two species in the isotropic phase. The two terms on the right side of equation (3) represent the free energy change due to the alignment of the liquid crystals.

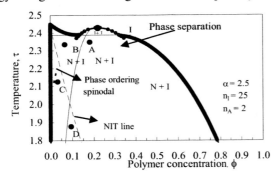

Fig. 1 Phase diagram of the binary mixtures of polymer and liquid crystals on the temperature–concentration plane for $n_I = 25$, $n_A = 2$, $\alpha = 2.5$.

As shown by De Gennes and Prost [6] equations (1-5) predicts the emergence of a stable nematic phase when $\Gamma_0\varphi_A = 4.55$. Using the following definitions:

$$\Gamma_0 = \left(\chi_a + 5/4\right)n_A, \ \alpha = \chi_a/\chi, \ \chi = 1/\tau, \qquad (6)$$

where $\alpha = \chi_a/\chi$ represents the relative strength of interactions, and τ is the reduced temperature, it is found

that the threshold $\Gamma_0\varphi_A = 4.55$ gives the following concentration dependence of the reduced nematic-isotropic transition (NIT) temperature:

$$\tau_{NI}\left(\varphi_A\right) = \frac{\alpha\,n_A\,\varphi_A}{4.55 - 1.25\,n_A\,\varphi_A} \, . \qquad (7)$$

For simplicity, we denote the isotropic component composition, φ_I, by ϕ $\left(\phi \equiv \varphi_I\right)$ in the phase diagram (Fig 1).

2.2 Kinetic equations

The dimensionless total free energy of the system consists of the bulk free energy and a nonlocal free energy that controls the cost of gradients in composition and orientational density, in the absence of surface terms and external fields, can be expressed as [8]

$$\tilde{F} = \tilde{T} \times \oint_{\tilde{v}}\left(\tilde{f}^h + \frac{1}{\tilde{T}}\tilde{f}^g\right)d\tilde{v} \, , \qquad (8)$$

$$\tilde{f}^h = \left[\frac{\tilde{\varphi}}{n_I}\ln\tilde{\varphi} + \frac{\left(1-\tilde{\varphi}\right)}{n_A}\ln\left(1-\tilde{\varphi}\right) + \chi\tilde{\varphi}\left(1-\tilde{\varphi}\right)\right.$$
$$\left. + \frac{3}{4}\left(\frac{\tilde{\Gamma}_0}{n_A}\right)\left(1-\tilde{\varphi}\right)^2\tilde{Q}:\tilde{Q} - \frac{\left(1-\tilde{\varphi}\right)}{n_A}\ln\left(\frac{\tilde{I}_0}{2}\right)\right] , \qquad (9)$$

$$\tilde{f}^g = \left[\frac{1}{2\tilde{D}}\times\left(\tilde{\nabla}\tilde{\varphi}\right)^2 + \frac{\tilde{R}}{\tilde{D}}\times\left(\tilde{\partial}_i\tilde{\varphi}\right)\left(\tilde{\partial}_j\tilde{Q}_{ij}\right)\right.$$
$$\left. + \frac{\tilde{G}}{2\tilde{D}}\times\left(\tilde{\partial}_k\tilde{Q}_{ij}\right)^2 + \frac{\tilde{P}}{2\tilde{D}}\times\left(\tilde{\partial}_i\tilde{Q}_{ik}\right)\left(\tilde{\partial}_j\tilde{Q}_{jk}\right)\right] , \qquad (10)$$

where

$$\tilde{I}_0 = \int_0^{2\pi}\int_0^\pi exp\left[\frac{3}{2}\tilde{\Gamma}_0\left(1-\tilde{\varphi}\right)\tilde{Q}:\left(\sigma\sigma - \frac{\delta}{3}\right)\right.$$
$$\left. sin^2\theta d\theta d\omega\right] . \qquad (11)$$

The dimensionless governing equations of the system becomes [8]

$$\frac{\partial\tilde{\varphi}}{\partial\tilde{t}} = \left[\tilde{D}\times\tilde{T}\times\tilde{\nabla}^2\left(\frac{\partial\tilde{f}^h}{\partial\tilde{\varphi}}\right) - \tilde{\nabla}^4\tilde{\varphi} - \tilde{R}\times\tilde{\nabla}^2\left(\tilde{\nabla}\cdot\tilde{\nabla}\tilde{Q}\right)\right] , \qquad (12)$$

$$\left[\frac{\partial\tilde{Q}}{\partial\tilde{t}}\right]^{[s]} = \left[-\tilde{T}\times\tilde{D}\times\tilde{E}\times\frac{\partial\tilde{f}^h}{\partial\tilde{Q}} + \tilde{E}\times\tilde{R}\times\tilde{\nabla}\cdot\tilde{\nabla}\tilde{\varphi}\right.$$
$$\left. + \tilde{E}\times\tilde{G}\times\tilde{\nabla}^2\tilde{Q} + \tilde{E}\times\tilde{P}\times\tilde{\nabla}\left(\tilde{\nabla}.\tilde{Q}\right)\right]^{[s]} \qquad (13)$$

where $\tilde{t}$ is the dimensionless time, $\tilde{D}$ is the dimensionless diffusion parameter, $\tilde{E}$ is the phenomenological constant, $\tilde{R}$ is the coupling parameter, $\tilde{G}$, and $\tilde{P}$ represents dimensionless Frank elastic parameters respectively and

$\tilde{Q}$ is a second rank symmetric and traceless tensor [9]. Dimensionless Eqs. (8)-(13) are solved by a high performance numerical scheme with periodic boundary condition [8].

3 RESULTS AND DISCUSSIONS

A typical phase diagram of the system on the temperature-

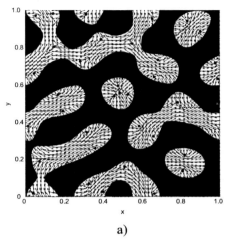

a)

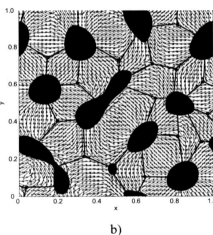

b)

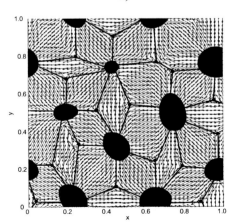

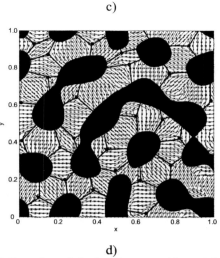

c)

d)

Fig. 2 Snapshot of the local composition of the system at a late time step following a quench to; a) point A, b) point B, c) point C and d) point D for $n_I = 25$, $n_A = 2$, $\alpha = 2.5$, $\tilde{D} = 1000$, $\tilde{E} = 1.0$, $\tilde{R} = 0.2$, $\tilde{G} = 0.1$, and $\tilde{P} = 0.1$. Black corresponds to isotropic polymer and white corresponds to pure liquid crystals (LCs). The arrows represent the local nematic director, and defects are marked with small solid circles.

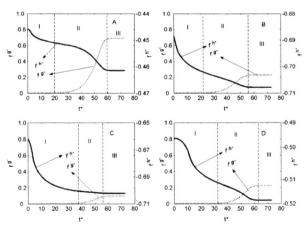

Fig. 3 Free energy profiles of the system following a quench to; a) point A, b) point B, c) point C and d) point D for $n_I = 25$, $n_A = 2$, $\alpha = 2.5$, $\tilde{D} = 1000$, $\tilde{E} = 1.0$, $\tilde{R} = 0.2$, $\tilde{G} = 0.1$, and $\tilde{P} = 0.1$. Dimensionless homogeneous energy, f^{h^*}, Dimensionless gradient energy, $f^{g^*} = 325 \times \tilde{f}^g$, and Dimensionless time, $t^* = 10^5 \times \tilde{t}$.

concentration plane is shown in Fig. 1 which is computed with $n_I = 25$, $n_A = 2$, and $\alpha = 2.5$. In the figure, ϕ, denotes the isotropic component composition (polymer concentration). The coexistence (binodal) curve of the phase equilibrium is derived by a double tangent method.

Details procedure for computing binodal and spinodal curves for such a system is documented in the work [8]. We study the morphology following four quenches from the isotropic, homogeneous phase into the isotropic-nematic (IN) coexistence region below the triple point line. Four regions are indicated by filled circles and denoted by points A, B, C and D respectively in the phase diagram (Fig. 1).

Fig. 2 represents the snapshot of compositional order of the system at points A, B, C and D respectively. In Fig. 2(a), mass matrix phase is isotropic and droplet phase is nematic. In Fig. 2(a) we can see that a pair of topological defects forms inside each microdomain due to the presence of repulsive Peach-Koehler forces. In our system the repulsive force naturally arises from interaction via the elastic deformation of liquid crystal [1]. We can see from the Fig. 2(a) that orientation inside the droplet is perpendicular implying strong normal anchoring of liquid crystal molecules at the droplet boundary. Nematic droplets must develop defects because the LCs wants to be parallel to each other and parallel to the droplet interface too. This result agrees quite well with the results reported by Lapena et.al [3] (see Fig. 3, [3]). In Fig. 2(b)-2(d), mass matrix phase is nematic and microdomain phase is isotropic. In Fig. 2(b)-2(d), isotropic microdomains suspended into the nematic matrix are surrounded by the topological defects. One interesting feature of the defect lattice is its topology. Solid lines represent the interconnection between defect cores and isotropic microdomains. Defect structures form cellular polygonal networks that are mostly four-sided and the side of each polygon ends either at the droplet or at another defect. Most of the defects are +1/2 disclinations. Some of them are +1 disclinations which eventually split into two +1/2 disclinations as can be seen from Fig. 2(b) and 2(c). In the case of point C (see Fig. 2(c)), microdomians are almost positionally ordered whilst they form fabrillar networks for the case of point D (see Fig. 2(d)).

To get better understanding of underlying physics in phase separation processes, we calculated free energy profiles at each of the quenching positions of the system. Fig. 3 represents the dimensionless homogeneous and gradient energy as a function of dimensionless time following quenches to the point A, B, C and D. In Fig. 2, we can clearly see three distinct regimes, namely initial time lag regime (I), growth/relaxation regime (II) and the plateau regime (II). In the II regime, the free energy shows growth in the gradient energy and decrease in homogeneous energy indicating that phase separation and phase ordering spinodal decomposition (SD) drives the system to be unstable, leading to the breakdown of the interconnected domains and formation of isotropic microdomains or fabrillar networks. In the crossover regime III, a plateau regime corresponding to the onset of the breakdown of the interconnected structure (see Fig. 2) appears. The plateau is quite pronounced in both of the energy profiles, which indicates that phase separation get saturated and signals a transition from early stage to intermediate stage of phase separation.

4   CONCLUSION

A nonlocal mesoscopic dynamic model for multiple phase separation, based on a tensor theory, in the presence of liquid crystalline order has been formulated, and solved using high performance numerical methods. We characterized the emerging morphologies following four temperature quenches into the physically meaningful regions of phase diagram. Phase separations from temperature quenches of isotropic binary mixtures start with the formation of small domains that grow and coarsen as time elapses which leads to polydisperse dispersions of growing microdomains that eventually phase separate macroscopically. It has been found that ordering dramatically affects morphology. Topological defects arise due to the elastic distortions around the microdomains formed during the phase separation. Defect structures form cellular polygonal networks that are mostly four-sided and the side of each polygon ends either at the droplet or at another defect. The free energy of the system establishes the dynamics and correlation of the morphological structures. Formation of interconnected (bicontinuous) networks or microdomins depends on whether ordering or phase separation is the initially dominant process. Compared to the experimental and numerical results available in the literature, our simulation results may able to provide new insights into the understanding of new emerging microdomain tropological defect morphology in liquid crystalline materials.

REFERENCES

[1] J.W. Doane, Liquid Crystals: Applications and Uses, ed. B. Bahadur, World Scientific, NJ, 1990.

[2] P.S. Drzaic, Liquid crystal dispersions, World Scientific, Singapore, 1995.

[3] A.M. Lapena, S. C. Glotzer, S. A. Langer, and A.J. Liu, Phys. Rev. E 60(1), R29, 1999.

[4] M. Graca, S. A. Wieczorek, and R. Holyst, Macromolecules, 36, 6903-691, 2003.

[5] G.P. Crawford and S. Zumer, Liquid Crystals in Complex Geometries Formed by Polymer and Porous Networks, Taylor & Francis, London, 1996.

[6] P. G. de Gennes and J. Prost, The Physics of Liquid Crystals, Oxford University Press, NY, 1993.

[7] S. K. Das and A. D. Rey, Computational Modelling of Multi-Phase Equilibria of Mesogenic Mixtures. J. Comp. Mat. Sci. 2003, (In Press).

[8] S. K. Das and A. D. Rey, A nonlocal dynamic model of multi-phase inhomogeneous mixtures of flexible polymer and liquid crystal materials, J. Chem. Phys., 2003, (to be appear).

[9] A.D. Rey and T. Tsuji, Macromol. Theory Simul. 7, 623-639, 1998.

A Carbon Nanotube-Based Sensor for Measuring Forces at the Cellular Scale

C. Roman, F. Ciontu, B. Courtois

TIMA Laboratory, 46. Av. Félix Viallet, 38031 Grenoble, France,
{cosmin.roman, florin.ciontu, bernard.courtois}@imag.fr

ABSTRACT

In this paper we perform a theoretical study of a potential design for a carbon-nanotube device able to transduce forces developed at the scale of basic cellular processes into current variations. The proposed device transduces the deflection of a carbon nanotube upon the application of an external perturbation into a variation of the currents at the ends of a second nanotube, perpendicular to the former. The first stage of this study consists of an assessment of the sensitivity of the devices with forces in the tens of pNs, developed typically at the cellular scale. In the second stage, we focus on the transduction of the deflection of the cantilever into an electrical signal.

Keywords: carbon nanotubes, nano-mechanical cantilever, biosensors, molecular modeling,

1 INTRODUCTION

The last few years have brought nanotech applications closer to the realm of reality mainly due to considerable progress in fabricating nanostructures with controlled properties. Carbon nanotubes followed this path with huge advances in synthesis techniques as well as functionalization [1], solubility and selection [2]. Given this trend it is reasonable to extrapolate that in several years there is a possibility of having nanotubes with well controlled properties at significantly lower costs. Aside from this, an essential aspect that allows envisioning the design of carbon-nanotube based structures is the very good correlation between theoretical models and experimental data.

Bio-sensing is one application domain offering some clear opportunities to transpose scientific advances related to nanotubes into applications.

In this paper we theoretically investigate a potential design for a carbon-nanotube device able to transduce forces developed at the scale of basic cellular processes into current variations.

Measuring cell forces is one possible application for this device. Also, compared to microscopic cantilevers carbon nanotube based devices would have the advantage of a far better scale compatibility with the elementary biological processes. The scale compatibility is the first of the four requirements for the future generation of bio-sensors, followed by the need for label-free detection, scalability to allow massive parallelization and sensitivity of detection range.

The remainder of the paper is structured as follows. In Section 2 we explain the functioning principle of the sensor. Section 3 contains the details on the calculations performed to characterize the mechanics of the sensor. The transduction of mechanical movement into a current variation is presented in Section 4. Finally, we present our conclusions in Section 5.

2 FORCE SENSOR OPERATION PRINCIPLE

The proposed device transduces the deflection of a carbon nanotube upon the application of an external perturbation into a variation of the currents at the ends of a second nanotube, perpendicular to the former.

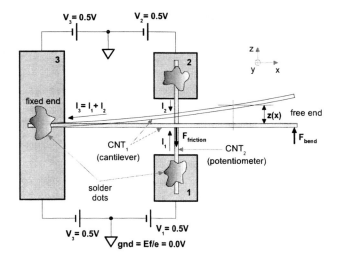

Figure 1: Schematic representation of the principle of operation of a carbon nanotube-based force sensor, including electronic biasing.

The schematic of the device is depicted in Figure 1. The transversal tube is locked down at both its ends while the longitudinal tube is fixed at one end and free standing at the other. A deformation force will elastically bend the cantilever, shifting the position of the cross-junction along the former tube. If the tubes are electrically biased with the aid of some metallic leads, then currents start to flow through each of the three branches. Nevertheless the conductance ratio of the two transversal branches is greatly affected by the position of the cross-junction. As can be observed from Figure 1, one of the two branches gets shorter as the other gets longer. If the transversal tube is not ballistic then the currents will change their ratio accordingly, that is higher currents will flow though the shorter branches. Current imbalance can be measured with differential front-end electronics, yielding a measure of the cantilever's deflection, which is further multiplied with the spring constant of the system to obtain the applied force.

The junction's integrity is maintained mainly by van der Waals forces and additionally by hydrophobic-hydrophilic effects if the system is immersed. An off-plane force, i.e. a force along the y axis, could compromise the coupling of the tubes or even break them apart.

Basically the sensor is a molecular potentiometer who's actuation could be performed for instance by cell motility or by any other molecular phenomena. A sensor of this kind has the same functionality with a lateral force microscope. However it has the advantage of allowing massive parallelization.

3 MECHANICS OF NANOTUBES AND CELL MEMBRANES

Modeling the mechanics of the device was performed by using molecular dynamics and required the *ab initio* parameterization of the force-field in order to cope with the heterogeneity of a system including both carbon nanotubes and molecules. The cantilever's oscillation modes have been studied under perturbations induced by forces of tens of pN, a range compared for instance to those developed by cells during motility related processes.

We opted for the latter one as implemented in the freely-available program NAMD [3]. This force-field is faster then ab initio and semi-empirical methods allowing simulations with more than 10^5 atoms for time intervals of about 1ns. Compared with Brenner potentials that describe typically carbon-carbon interactions, the CHARMM force field was parameterized for a large spectrum of molecules, notably for amino-acids and phospholipids. This advantage becomes obvious when simulating the sensor, or just part of it, in contact with a cellular membrane.

3.1 Ab initio force-field parameterization

Empirical force fields in CHARMM's class were previously used to model carbon-nanotubes [4] but their main task was to account for hydrophobic-hydrophilic effects somehow neglecting the mechanics of the tubes under important deformations. Accurate experimental information about Young's modulus and Poisson's ratio of carbon nanotubes is still missing from literature, besetting the parameterization procedure to rely on ab initio calculations. All quantum mechanical computations in this section were performed with Siesta [5], within the density functional theory (DFT).

The first set of simulations were conducted in order to obtain statistics on bond lengths, angles, Urey-Bradley (UB) and improper dihedrals values as required by CHARMM force field. This was achieved for three different nanotubes, one armchair (5,5), one zig-zag (8,0) and one chiral (6,3) having approximately the same characteristic lengths. A variable-cell relaxation of the tubes, considered infinitely long, was performed subject to a convergence criteria of residual forces less than 0.01eV/A. We used a LDA Hamiltonian, an integration grid cutoff of 60Ry, a double zeta (DZ) atomic orbital basis set with an energy shift of 160meV or equivalently a confinement/cutoff radius of 2.85Å. The final values for r0, θ0, r0UB and RvdW are presented in Table 1.

k	189,580693	r_0	1,43257520454
k_θ	115,723856	θ_0	118,891824301
k_{UB}	22,699013	r_{0UB}	2,46731009022
intra-tube			
ε_{vdW}	-0,105262	R_{vdW}	4,000000
inter-tube			
ε_{vdW}	-0,070000	$R_{vdW}/2$	1,992400

Table 1: CHARMM force-field parameters

The second phase of the parameterization procedure consisted in fitting the spring constants k, k_θ, k_{UB} and the Lennard-Jones well-depth ε_{vdW} against energy versus strain curves as obtained with Siesta. Since our calculations are similar with those performed in Reference 6 and rely on the same code, we took into account their calculated Poisson's ratio of 0.14 when preparing strained tubes for relaxation. As opposed to the same reference we extended the study to strains in the range [-10,10]% with a step of 1%, in order to obtain well-behaved parameters even at large deformations. The system under study was a (5, 5), 5 cells long carbon nanotube. For each strain we took the already relaxed structure of this tube, modified its length to $l_0(1+\Delta l/l_0)$ and radius to $r_0(1-v\Delta l/l_0)$, to accelerate the forthcoming relaxation. After constraining the boundary atoms in planes perpendicular to the tube's axis we re-minimize the energy.

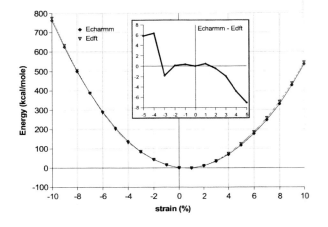

Figure 2: Energy vs. strain curves as obtained with Siesta and NAMD. The inset shows the error around the origin

To obtain positive spring constants and a negative well-depth we used Lagrange multipliers within the goal function, defined as the mean square of energy differences as calculated with Siesta and NAMD at different strains (see Figure 2), but not before shifting their energy minima to zero.

3.2 Force sensor simulation

In the first MD simulation, the cantilever measuring 36nm was pushed upon with a constant force of 10pN equally distributed between its ten terminal atoms while keeping the other end of the tube fixed.

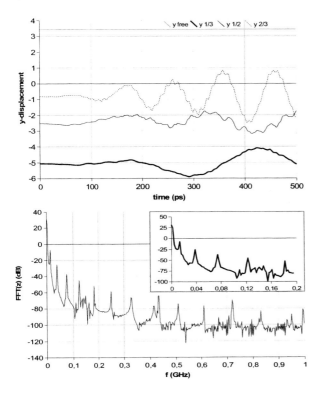

Figure 3: Displacement along y axis. (Bottom) Spectrum of the z deflection for one atom belonging to the sensor's tip

Turning on the force at the initial simulation time was similar to applying a step function stimulus, simultaneously exciting all the frequency modes of the system.

Three different positions of the (5,5) transverse tube, measuring 20 nm, were chosen to study the influence of the friction; at one third, at half and at two thirds from either edge of the cantilever. The simulation step was of 1fs, the total simulation time was of 0.5ns and proved to be sufficient for capturing at least one period of the cantilever's fundamental mode. A supplementary relaxation was performed before running the system in order to minimize the van der Waals interaction energy.

There should be in principle no important deflection along y since the applied force is constrained in the x-z plane. However Figure 3(top) revealed a different situation.

Even if initially constant, the cantilever's deflection starts to oscillate with increasing amplitude. A closer examination confirmed that the motion of the cantilever is stick-and-slip like due to rapid fluctuations in the van der Waals potential of the underlying tube. As we will see in the next section this spurious movement will greatly influence the charge transport through the junction as the latter is extremely sensitive to the inter-tube distance.

Heating is more pronounced with the transverse tube approaching the sensor's tip and it explains the smearing of the cantilever's spectrum observed in Figure 3(bottom). We plan further studies of this system in the presence of a Nose-Hover thermostat that could minimize the artifacts induced by thermal effects.

4 SENSOR ELECTRONIC TRANSPORT

The transduction of the cantilever's deflection into an electrical signal has been investigated in the framework provided by Landauer-Buttiker theory with a tight-binding description of the system.

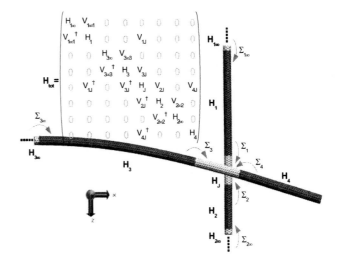

Figure 4: System's real-space partitioning and Hamiltonian.

Modeling the charge transport through the carbon-nanotube cross-junctions is a delicate task. Since the junction is maintained, by non-covalent bonding, it is relatively free to move in the x-z plane. Thus, thermal fluctuations of the junction's position will hugely influence the behavior of the device this effect to obtain the non-zero temperature behavior of this device and need to be addressed specifically.

4.1 Model description

The computation scheme is very similar with the one described in Reference 7. We used a Tight-Binding Hamiltonian including only π orbitals but as opposed to typical calculations we included the cosine factor like in the Slater-Koster scheme to account for the anisotropic inter-tube coupling. An exponential decay was considered as well in order to limit the interaction range between non-covalently bonded atoms of the two distinct tubes.

Infinite pristine (5,5) CNTs were placed at the end of each the first three regions (denoted with $H_{1...3}$) of the sensor's tubes to simulate the effect of electron reservoirs.

The transverse nanotube was "doped" by modifying the on-site energy ε_0 from zero to a random value equally distributed within [-1eV,1eV]. Leads and the tubes they contact were set to an electrochemical potential of ±0.5eV.

In order to accelerate the computation of the Green's functions we partitioned the system like in Figure 4. After completing the Hamiltonian's matrix elements a fast, self-energy based elimination method was used to invert an otherwise large system matrix. Self-energies were propagated backwards from leads to junction as illustrated in the same figure. Practically one single large inversion was performed to invert for the junction Green's functions, corresponding to a Hamiltonian that is not too sparse as opposed to the other regions. We used the Fisher-Lee relation and the Landauer-Büttiker formula, as described in Reference 8 to obtain the interest, thermally-smeared conductances, i.e. G_{12}, G_{13} and G_{23}.

4.2 Simulation results

Simulations at zero temperature with a longitudinal tube deflected in different positions showed that the device behaves non-monotonically. Universal conductance fluctuations play an important role, especially in the case of the sharp dopant distribution. Even more important are the atomic details of the cross-junction. Displacements as low as half an Angstrom give rise to important fluctuations in G13 and G23 as can be observed in Figure 5.

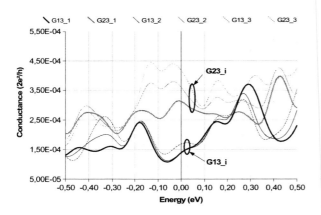

Figure 5: Smeared conductance between leads 3, 2 and 1 varies strongly at small perturbations of the cantilever.

Because relaxing the sensor in molecular dynamics is too slow to obtain enough intermediary positions of the cantilever, we took samples from the dynamical trajectory of the system as obtained in Section 3. Around a given junction position we took other twenty five closely located

sites. The length of the distribution interval was of ~2Å, consistent with thermal displacement fluctuations as know from the classical cantilever theory.

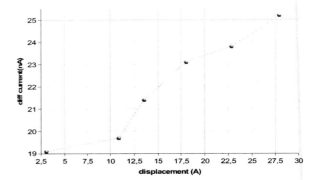

Figure 6: Deflection current dependence.

After computing the conductances and currents we convoluted the obtained values with a thermal-smearing function yielding the curve in Figure 6. This second smearing is considered here for to randomize the phase of electrons that tunnel through a fluctuating junction, and should not be confounded with the smearing applied to obtain the smooth conductance curves of Figure 5, accounting for the broadening of the Fermi-Dirac distribution with temperature.

5 CONCLUSIONS

After investigating the carbon nanotube-based device proposed in Section 2 by modeling phenomena characteristic to different scales we evaluate positively its suitability for measuring forces specific to the cellular scale. We note however several challenges posed by the coupling of phenomena belonging to different scales like the influence of the inter-tubes distance, typically modeled though molecular dynamics, and the variation of the currents computed though quantum transport calculations. Future work will have to take into account more thoroughly and systematically the influence of thermal noise on the functioning of the device.

REFERENCES

[1] G. della Torre et al., Nanotechnology 14, 765-771, 2003
[2] M. Zheng et al., Science 302, 1545-1548, 2003
[3] L. Kale et al., J. Comp. Phys. 151, 283-312, 1999
[4] W.H. Noon et al., J.P. Ma. Chemical Physics Letters, 355, 445-448, 2002
[5] J.M. Soler et al., Int. J. Quantum Chem. 65, 453, 1997
[6] D.S. Portal et al., Phys. Rev. B 59, 12678, 1999
[7] M.P. Anantram and T.R. Govindan, Phys. Rev. B 58, 4882, 1998
[8] S. Datta, Cambridge University Press, 1995

Influence of the Number of Washing on the Characteristic of Nanocrystalline Copper Oxide Powders

I. Ramli*, E. N. Muhamad*, A.H. Abdullah*, Y.H. Taufiq-Yap*, S.B. Abd. Hamid**

*Combinatorial Technology and Catalysis Centre, Department of Chemistry, Universiti Putra Malaysia, 43400 UPM Serdang, Selangor, Malaysia.
**Combinatorial Technology and Catalysis Centre, University of Malaya, 50603 Lembah Pantai, Kuala Lumpur, Malaysia.

ABSTRACT

The influence of the number of washing on the copper oxide physico-chemical properties was investigated by carrying out precipitation of copper (II) nitrate trihydrate ($Cu(NO_3)_2 \cdot 3H_2O$) with an aqueous solution of sodium hydroxide. The structural evolution of the desired CuO phase, via an intermediate $Cu(OH)_2$ phase, was investigated using XRD, FTIR, total surface area (BET method) and SEM. An investigation on the mobility of oxygen species was carried out using transient technique i.e. temperature programmed reduction (TPR). Results show that the BET surface area increased gradually with the number of washing from 8.5 to 15.9 $m^2 g^{-1}$. While, the total amount of oxygen atom being reduced is roughly the same. This indicates the ease of reducibility of the oxide when the precursor is washed several times. XRD pattern of these oxides gave well crystalline CuO with main peaks appeared at $2\theta = 35.5, 38.7$ and $48.7°$. The powders obtained are of uniform size distribution, finely grained with an average size of 20 nm.

Keywords: copper oxide, nanocrystalline, precipitation, washing effect, temperature programmed reduction

1 INTRODUCTION

The catalytic properties of heterogeneous catalysts are strongly affected by every step of the preparation together with the quality of raw materials. The choice of a laboratory method for preparing a given catalyst depends on the physical and chemical characteristics desired in the final composition. Bulk catalysts are normally comprised of active substance is generally achieved by precipitation [1] with main purpose of formation of very small crystallite. Washing of the solids is then required to remove the mother liquor (usually water) completely, to eliminate impurities and to exchange certain undesirable or useless ions for others that are easily decomposable by calcinations.

This study was carried out with purpose to prepare catalysts for oxidation reaction especially propylene oxidation. As been reported that the simple oxide Cu_2O [2] and supported mixed copper oxides [3] exhibit significant activity and selectivity in the partial oxidation of propylene to acrolein and has been chosen as a model catalysts for the determination of the involved reaction mechanisms. Thus, research focuses on the influence of number of washing on the copper oxide physico-chemical properties was investigated.

2 EXPERIMENTAL

Copper oxides were prepared using precipitation method under ambient temperature. An aqueous solution of NaOH (Analytical UNIVAR Reagent 97 %) was added continuously into aqueous solution of $Cu(NO_3)_2 \cdot 3H_2O$ (Hamburg Chemical GmbH 99 %) with constant stirring until a pH is reached at ~ 13.5. The precipitate is allowed to age for 120 minutes. After ageing the precipitate was recovered by filtration and washed with 30 ml of cold distilled water. The washing process is repeated for 60, 90, 120, 150 and 180 ml of distilled water. All the samples were dried in oven at 373 K for 12 h. Six series of copper oxide catalysts were obtained and calcined in air at 723 K for 4 h and designated as CuO1, CuO2, CuO3, CuO4, CuO5 and CuO6, where number 1 to 6 indicating number of washing. The catalysts were then characterised by using XRD, FTIR, BET surface area measurement, SEM and TPR in H_2.

3 RESULTS AND DISCUSSION

X-ray diffraction (XRD) patterns of the precursors prepared through this method showed that the precipitation occur mainly via $Cu(OH)_2$ phase (Figure 1). Calcination of the synthesised samples at 723 K display a high and intense XRD patterns indicating that all the samples are highly crystalline (Figure 2). As the washing volume increases, the CuO phase peaks become apparent which consist of three main peaks appeared at $2\theta = 35.3, 38.7$ and $48.7°$, correspond to ($^-111$), (111) and ($^-202$) planes, respectively.

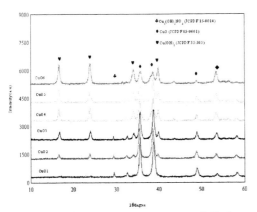

Figure 1: XRD patterns of CuO precursors

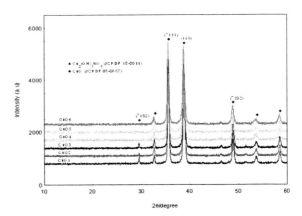

Figure 2: XRD patterns of CuO catalysts

Catalysts	FWHM $_{-111}$	FWHM $_{111}$	$t_{-111}{}^a$ / (Å)	$t_{111}{}^a$ / (Å)
CuO1	0.3003	0.3618	275.32	230.70
CuO2	0.3191	0.3768	259.08	221.51
CuO3	03121	0.3805	264.91	219.37
CuO4	0.3309	0.3890	249.83	214.55
CuO5	0.3605	0.4260	229.35	195.94
CuO6	0.3696	0.4478	223.66	186.37

a crystallite size by means of *Scherrer's* formula: $t\ (Å) = (0.89\ x\ \lambda)\ /\ \beta_{hlk}\ cos\ \theta_{hlk}$

Table 1: XRD data of CuO Catalysts

The Fourier transform infrared spectroscopy (FTIR) spectra of CuO catalysts were also recorded (Figure 3). Main feature recorded in samples CuO1, CuO2 and CuO3 show sharp band at 1384 cm^{-1} which is referred to as copper hydroxynitrate band [4]. This is probably due to insufficient decomposition of nitrate during calcinations. Few times of washing is also thought to contribute to the above observations. This occurrence may be avoided when using a large amount of water to wash the precipitates, thereby the suspension may thoroughly mix to remove the impurities [5]. The band at region 1636 cm^{-1} is associated with the bending mode of OH groups of adsorbed water. The CuO vibration bends which are found to appear at 575, 500 and 460 cm^{-1} cannot be observed clearly because it is shifted due to the existence of OH group.

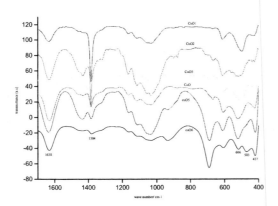

Figure 3: FTIR Spectra of CuO Catalysts

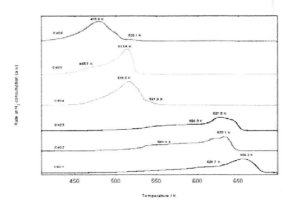

Figure 4: TPR Profile of CuO Catalysts

NSTI-Nanotech 2004, www.nsti.org, ISBN 0-9728422-9-2   Vol. 3, 2004

Peaks[a]	T_{ONSET} / K	T_{MAX} / K	Reduction activation energy, E_r/kJ mol^{-1}	Oxygen atom removed from catalysts (mol g^{-1})	Oxygen atom removed from catalysts (atom g^{-1})	Coverage (atom cm^{-2})
CuO1						
1	445	626.8	104.81	4.81×10^{-3}	2.90×10^{21}	3.41×10^{16}
2		656.2	109.72	6.18×10^{-3}	3.72×10^{21}	4.38×10^{16}
Total oxygen atom removed				**1.04×10^{-2}**	**6.62×10^{21}**	**7.79×10^{16}**
CuO2						
1	425	561.5	93.89	2.73×10^{-3}	1.64×10^{21}	1.39×10^{16}
2		633.1	105.87	8.71×10^{-3}	1.24×10^{21}	4.44×10^{16}
Total oxygen atom removed				**1.14×10^{-2}**	**6.88×10^{21}**	**5.83×10^{16}**
CuO3						
1	444	605.3	101.22	5.07×10^{-3}	3.05×10^{21}	2.70×10^{16}
2		627.8	104.98	5.53×10^{-3}	3.33×10^{21}	2.95×10^{16}
Total oxygen atom removed				**1.06×10^{-2}**	**6.38×10^{21}**	**5.65×10^{16}**
CuO4						
1	424	515.6	86.21	9.67×10^{-3}	5.83×10^{21}	4.79×10^{16}
2		541.9	90.61	6.93×10^{-4}	4.17×10^{20}	3.42×10^{15}
Total oxygen atom removed				**1.04×10^{-2}**	**6.25×10^{21}**	**5.13×10^{16}**
CuO5						
1	415	468.7	80.50	3.13×10^{-3}	1.88×10^{21}	1.28×10^{16}
2		513.5	85.86	6.15×10^{-3}	3.70×10^{21}	2.52×10^{16}
Total oxygen atom removed				**9.28×10^{-3}**	**5.58×10^{21}**	**3.80×10^{16}**
CuO6						
1	370	478.9	80.08	8.78×10^{-3}	5.29×10^{21}	3.34×10^{16}
2		520.1	86.97	3.13×10^{-4}	1.89×10^{20}	1.19×10^{15}
Total oxygen atom removed				**9.09×10^{-3}**	**5.48×10^{21}**	**3.46×10^{16}**

Table 2: Temperature Programmed Reduction (TPR)
Total number of O_2 removed from the copper oxide catalysts by reduction in H_2/Ar stream (5% H_2)

The oxides have also been characterised by TPR with hydrogen as the reducing gas (5 % H_2-Ar). See Figure 4. The figure show that the first peak maximum of the reduction profile is significantly reduced from ~ 650 to 470 K as the number of washing is increased. The total amount of oxygen atom being reduced was calculated and listed in Table 2. As the number of washing increases, the total amount of oxygen atom being removed is also increase.

This result shows the increment of ease of reducibility of the oxides as the number of washing is increases. Meanwhile, the total surface area of the CuO catalysts is found climbing with the increasing number of washing. This result is suggested due to the smaller crystallite size synthesised. The crystallite size is calculated from the Scherrer's formula (Table 1).

(a)

(b)

Figure 5: SEM Images for (a) CuO1 and (b) CuO6

Figure 5 display scanning electron micrograph (SEM) of two of the calcined samples, i.e CuO1 and CuO6. Both images show needle-like structure. The type of morphologies for the crystallites could be detected from the XRD is triclinic crystallite (tenorite).

4 CONCLUSION

The CuO oxides were demonstrated to have enhanced in surface area, thus smaller crystallite size when it were washed several times. From the XRD and FTIR spectrums, there is still small amount of nitrate remain in the samples probably due to insufficient amount of water used to wash the precipitate. On the other hand, TPR profiles showed that the peak of the reduction temperature shifted to lower temperature indicating the ease of reducibility of the oxide as the number of washing increases.

5 ACKNOWLEDGEMENTS

This project was financially supported by COMBICAT Research Group, MALAYSIA.

REFERENCES

[1] M.Campanati, G. Fornasari and A. Vaccari. Catal. Today, 77, 301, 2003.
[2] M. M. Bettahar, G. Costentin, L. Savary and J. C. Lavalley, Appl. Catal. A, 145, 12, 1996.
[3] C. Cheng Hien, W. P. Chuang and T. J. Huang, Appl. Catal. A, 131, 73, 1995.
[4] J. L. Li and T. Inui, Appl. Catal. A, 137, 110 1996.
[5] C. Perego and P. Villa, Catal. Today, 34, 285 1997.

Solid Modeling of Nanoscale Artifacts

P.V.M. Rao, M. Sharma and R. Venugopal

Department of Mechanical Engineering, Indian Institute of Technology Delhi,
New Delhi - 110016, INDIA, pvmrao@mech.iitd.ernet.in

ABSTRACT

Solid modeling involves creation and manipulation of complete and unambiguous mathematical representations of 3-D objects. The purpose of such a representation is to provide tools for visualization, calculation of geometric properties and realization by design and manufacturing processes. Existing methods of solid modeling namely constructive solid geometry (CSG), boundary representation (B-rep) and decomposition methods are found to be inadequate for representing nanoscale artifacts as they do not account for arrangement of atoms and molecules in the form of lattice structures, lattice planes and lattice constants. The proposed work is a step in the direction of bridging this gap. The paper presents a new solid modeling scheme for representing nanoscale artifacts. The applicability and effectiveness of the new representation is demonstrated by taking many examples including devices such as nanotubes.

Keywords: solid modeling, crystalline structure, constructive solid geometry, boundary representation, spatial enumeration

1 INTRODUCTION

Solid modeling involves the creation and manipulation of complete, unambiguous mathematical representations of 3-D objects. The purpose of such a representation is to provide tools for visualization, graphical user interaction and calculation of geometric properties needed at various stages of product life cycle. Geometric model of an object also forms a basis for integrating various life cycle activities, particularly product design and manufacturing. Existing methods of solid modeling can be classified into one of three major categories: constructive solid geometry (CSG), boundary representation (B-rep) and decomposition methods [1].

Constructive solid geometry (CSG) is a solid modeling method that combines simple solid primitives to build more complex models using Boolean operators: *union, difference* and *intersection* [2]. The resulting model is a procedural model stored in the mathematical form of a binary tree where leaf nodes are solid primitives, correctly sized and positioned, and each branch node is a Boolean operator. In a CSG representation, geometry and topology of an object is stored implicitly. This representation is not convenient to use for nanoscale artifacts, as it would be impractical to build and model an artifact as a Boolean combination of individual atoms / molecules.

In a B-rep scheme, [3] the solid objects are represented as unions of their boundaries or enclosing surfaces. The enclosing surfaces can include planar polygons, quadrics and free-form surface patches. In this scheme topological and geometric information are explicitly defined. Topological information provides the relationships among vertices, edges/curves and faces/patches. In addition to connectivity, topological information also includes orientation of edges and faces etc. Geometric information usually consists of equations of the edges/curves and faces/patches. B-Rep too has limitations in representation of nanoscale artifacts as it is difficult to identify a group of atoms / molecules that can define unique boundary surface of an object. Moreover, any such representation will have a few hundreds and thousands of boundary atoms / molecules to represent an object.

In decomposition methods, a solid is decomposed into a collection of adjoining, non-intersecting solid primitives. Depending on the shape and parameterization of primitives, decomposition schemes of solid modeling primarily have three general forms: spatial enumeration, cell decomposition and Octree encoding [4]. Spatial enumeration is a direct approach to solid modeling representation in which a solid is modeled as a collection of identical volume cells. Correct object representations are easy to maintain but difficult to create due to the simplistic structure. Cell decomposition is a generalization of spatial enumeration in which objects are represented as a collection of simple primitives, which are not necessarily required to have the same size or shape. A special case of cell decomposition/spatial enumeration modeling is *voxel based modeling* in which object is modeled as a collection of cubical cells of same size which are located in a fixed grid in a 3D discrete space.

The octree encoding [5, 6] is similar to both spatial enumeration and cell decomposition in that, objects are represented as a collection of fundamental primitive solids or cells. However, the representation is made more efficient by organizing the cells in an eight-array tree. Cells that are partially full are further subdivided into cells. However the storage required increases exponentially as the tree depths increase.

The above discussed decomposition methods of solid representation are used to model solid objects whose dimensions can go as small as a few micrometers [7]. The same representations have limitations in modeling nanoscale artifacts as the cell and grid sizes reach dimensions comparable to inter-atomic distances. However, among the three major solid modeling schemes discussed above, the decomposition methods are closest in terms of representing arrangement of individual atoms and molecules with which any 3D object is made of. It was felt decomposition methods with some modifications could be used as potential tools for modeling nanoscale artifacts.

2. GEOMETRY OF CRYSTALLINE SOLIDS

Solids can be broadly classified as crystalline and amorphous solids. In a crystalline solid, the arrangement of atoms is in periodically repeating fashion, whereas no such regularity of arrangement is found in amorphous solids. Such an arrangement of atoms in a crystalline solid is referred as space lattice. The space lattice can be defined by referring to a unit cell. The unit cell is defined as the smallest unit which, when repeated in space indefinitely will generate a space lattice. Thus the geometry of crystalline solids can be modeled as an array of unit cells in three dimensions in which every point has surroundings identical to that of every other point in the array. There exist 14 different types of space lattices, known as Bravais Lattices, which belong to seven different crystal systems. For example one of the seven crystal systems is cubic crystal system is defined by three mutually perpendicular translation vectors, which are equal in magnitude. The magnitude of this vector is fixed for a given material and can be defined in terms of lattice constant For example the lattice constant for Aluminum is 0.321 nm (nano meters), and that of Iron is 0.287 nm. The three space lattices in the cubic crystal system namely Simple Cubic (SC), Body Centered Cubic (BCC) and Face Centered Cubic (FCC) are shown in the Table 1.

The crystalline solid looks different when cut at different orientations. System of Miller indices is the universally accepted system of indices that has been developed to describe the orientation of crystallographic planes and crystal faces relative to crystallographic axes. Miller Indices are a symbolic vector representation for the orientation of an atomic plane in a crystal lattice and are defined as the reciprocals of the fractional intercepts, which the plane makes with the crystallographic axes (x, y and z). Miller indices are represented by a set of 3 integer numbers <hkl>. For example if x, y and z intercepts of a plan are 2, 1 and 3 respectively, the Miller indices are <362> obtained by multiplying the reciprocal of x, y and z intercept values (½, 1 and ↓) to convert them into integers. Table 2 shows typical planes and the corresponding Miller indices for cubic crystal system.

Lattice Structure	Effective number of atoms per unit cell	Figure
Simple Cubic (SC) Atoms at Eight corners of unit cell	1	
Body Centered Cubic (BCC) Atoms at Eight corners and at the center of unit cell	2	
Face Centered Cubic (FCC) Atoms at Eight corners and at center of six faces of unit cell	4	

Table 1: Cubic Crystal System

Miller Indices	<100>	<110>	<111>
Illustration			

Table 2. Miller Indices for Cubic Crystal System

3. MODELING OF NANOSCALE OBJECTS

The new solid modeling representation proposed for nanoscale artifacts is discussed in this section. The proposed representation has some similarities with decomposition models discussed earlier. In the present scheme a solid object is represented as a number of spherical primitives. These spheres are located in a grid like structure defined by one of the 14 Bravais lattice structures. Center of sphere corresponds with grid point and the radius of the sphere depends on the lattice constant. This type of model can be considered as a special case of decomposition scheme, with primitive as an atom with voids between it and surrounding atoms.

Every lattice structure has a template with fixed number of cells and their relative arrangement which when extended in lattice directions as an array, gives bulk of solid which we call as an extended structure. For example a 3D object in a cubic crystal structure can be represented as a three-dimensional array, the three directions of the array correspond to three crystallographic directions. Table 3 shows an extended structure for three different types of lattice structures of cubic system.

Number of cells in template	Lattice Type	Illustration
1	Simple Cubic (SC)	
2	Body Center -ed Cubic (BCC)	
4	Face Center -ed Cubic (FCC)	

Table 3. Extended structures of Cubic Crystal System

In order to construct solid model an object in the present scheme, first a boundary representation of the same is obtained. Next the size of array of an extended structure is chosen such that the object to be represented can be made to completely lie inside the structure. All the grid positions, which are contained in the object being represented, are binary coded 1 or else 0.

The generic representation of a solid in the present scheme has following format:

< t, a, l >
< Δ l_i * c binary data>

Where t corresponds to type of lattice structure, a corresponds to lattice constant, l is vector whose elements represent size of array along different crystallographic directions and c correspond to number of unit cells in template corresponding to lattice structure t. Figure 3 shows views of a 3D object in B-Rep and proposed scheme.

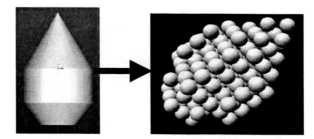

Figure 3. B-Rep and proposed model for a 3D object

Orientation of the object with respect to crystallographic direction is important parameter in the present system of modeling. In other words different orientations of the object with respect to given extended structure will yield different representations for the same object. This is true with other decomposition models such as octree encoding or voxel based modeling. Among the infinite orientations possible for the object with respect to crystallographic directions, some of the orientations have special significance from manufacturing point of view. Here the orientation of the object or object plane is be represented using a vector as an input. Such a representation has direct relation with Miller indices for crystal planes discussed earlier.

In the present work only a few of the fourteen Bravais lattice structures are discussed here. However extension of concept to other lattice structures is non-trivial.

4. RESULTS & CONCLUSIONS

The proposed scheme of solid modeling is used to construct and model many nanoscale artifacts. Tables 4 and 5 show results of modeling and visualization for two typical nanoscale artifacts, a stepped hollow shaft and a bracket. In the two cases the results of visualization are shown for same size of artifact but for different lattice structures or for different orientations. Solid modeling scheme discussed in the present work can be used for visualization of nanoscale artifacts and manipulation of the same leading to designs with preferred atomic terraces and orientations. Secondly it can be used for manufacturing of nanoscale artifacts, which require preferred atomic planes for atomic level manipulations as it is done using STM probes [8].

Lattice & Orientation	Illustration
Simple Cubic <100>	
Face centered Cubic <100>	

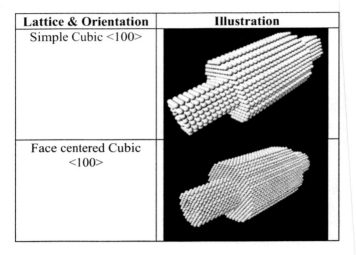

Table 4. Solid model of a Stepped Hollow Shaft

Lattice & Orientation	Illustration
Body Centered Cubic <100>	
Body Centered Cubic <110>	
Body Centered Cubic <111>	

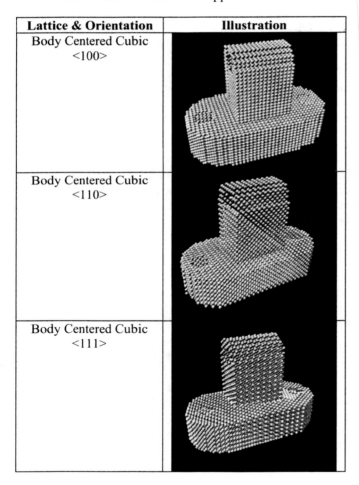

Table 5. Solid model of a Bracket

In our present work modeling of only crystalline solids with lattice structures belonging to cubic system are dealt in detail. The model can be generalized for other lattice structures. Fig. 4 show a nanotube modeled using present scheme, which has hexagonal structure.

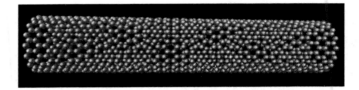

Figure 4. B-Rep and proposed model for a nanotube

A different approach is needed when modeling those objects, which consist of molecules and bonds in stead of objects made of single type of atoms. The real life artifacts are more complex in terms of geometry having crystal defects, dislocations and grain boundaries. We are presently extending our work along these lines and results if which will be communicated through another paper.

REFERENCES

[1] A.A.G. Requicha and J.R. Rossignac, "Solid Modeling and Beyond", IEEE Computer Graphics and Applications, 12, 31, 1992.
[2] A.A.G. Requicha and H.B. Voelcker, "Historical Summary and Contemporary Assessment", IEEE Computer Graphics and Applications, 2, 9, 1982.
[3] M. Mantyla, "An Introduction to Solid Modelling", Computer Science Press, Maryland, USA, 1988.
[4] D.W. Storti, M.A. Ganter and C. Nevrinceanu, "Tutorial on Implicit Solid Modeling", Mathematica Journal, 2, 70, 1992.
[5] D. Meagher, "Geometric Modeling Using Octree Encoding", Computer Graphics and Image Processing, 19, 129, 1982.
[6] D. Ayala, P. Brunet, R. Juan, and I. Navazo, "Object Representation by Means of Non-minimal Division Quadtrees and Octrees, ACM Transactions on Graphics, 4, 41, 1985.
[7] A. Perrin, V. Ananthakrishnan, F. Gao, R. Sarma, and G.K. Ananthasuresh, "Voxel-Based Heterogeneous Geometric Modeling for Surface Micromachined MEMS", Proceedings of the 2001 International Conference on Modeling and Simulation of Microsystems, Hilton Head Island, South Carolina, U.S.A., 136, 2001.
[8] P.V.M. Rao, C.P. Jensen and R.M. Silver, "A Generic Model for STM Tip Geometry Measured with FIM, Paper Accepted for Publication in Journal of Vacuum Science & Technology - B.

Modeling and Simulations of Tubular Nanowires

A. Buldum

Department of Physics, The University of Akron
Akron, OH, 44325, buldum@physics.uakron.edu

ABSTRACT

Here, we present theoretical investigations on tubular nanowires. In order to have a better understanding of the physical properties of the nanowires in different sizes and to have a better comparison with the other nanowire forms a systematic study is performed. Atomic models of tubular aluminum nanowires in different radii and thickness are created. To obtain typical atomic structures of the tubes, molecular dynamics (MD) simulation method is used and annealing - quenching simulations are performed. For tubes in large diameters, periodic boundary conditions are modified and angular boundary conditions are used. An important question related with the tubular nanowires is their structural stability. Energetics and stability of these nanowires and their dependence on the radius and thickness are studied.

Keywords: Nanowires, Coating, Molecular Dynamics

1 INTRODUCTION

Metallic nanowires have attracted great interest due to their interesting low-dimensional physics and due to possible future technological applications. Nanowires and contacts having radius in the range of the Fermi wavelength λ_F have been fabricated and have shown novel electronic and mechanical properties[1-3]. Ultrathin nanowires suspended between two metal electrodes have been produced [4-9]. Previous theoretical investigations include studies on quantum point contacts [10-13] as well as the structural, vibrational and electronic properties of metallic nanowires [14-17]. Most of these investigations were related with atomic scale nanowires. Recently, metallic tubular nanowires (nanotubes or tubes) in different radii and lengths were created using physical vapor deposition techniques. Polymeric nanofibers were coated with aluminum and then the polymer cores were removed that leaved tubular nanowires of coating material [18]. The smallest inner radii were around 10 nm and the approximate thickness of the tubes was controlled by the sputtering process. The tubes did not collapsed after the removal of the polymer core. Beside these tubular nanowires, epitaxial core-shell nanowire heterostructures [19] and single-walled nanotube (SWNT) coatings with tin oxide [20] or different metallic nanoparticles such as Au, Ti, Ni Pd, Al were reported [21-23].

Here, we present theoretical investigations on the structural properties of the tubular aluminum nanowires. In order to have a better understanding of the properties of the nanowires in different sizes and to have a better comparison with the other nanowire forms a systematic study is performed. Atomic models of tubular aluminum nanowires in different diameters and thickness are created. To obtain typical structures, molecular dynamics (MD) simulation method is used and annealing - quenching simulations are performed. For tubular nanowires in large diameters, periodic boundary conditions are modified. An important question related with the tubular nanowires is their structural stability. Energetics and stability of these nanowires and their dependence on the diameter and thickness are studied. In section two, the details of the models are given. Section three is for the presentation of the results on the structure and stability of the tubes and it is followed by our conclusions in section four.

2 MODEL

Atomistic simulations, which provide the methodologies for detailed microscopic modeling, are powerful and widely used tools in physics, chemistry and materials science. Directly from the atomic nature of the system, they offer prediction of mechanical and electronic properties of materials. In order to study the structural properties of tubular aluminum nanowires, an atomic model of a slice or angular region of the tube is created. A well-tested state-of-the art embedded-atom-type interaction potential [24] is used for the Al-Al atom interactions. As the radii of the tubes can be quite large, angular portions of tubes in larger diameters are simulated using modified periodic boundary conditions. For atomic displacements perpendicular to the axis of the tubes, angular boundary conditions are used. On the other side, regular periodic boundary conditions are employed for the atomic displacements along the axis of the tubes. Thus, the tubes were modeled as if they are perfectly cylindrical and infinitely long. Then, molecular dynamics simulations were performed and structures are relaxed at different constant temperature ranging from 100 K to 300 K using velocity scaling algorithm. The time steps are taken as 2.0 fs. In addition, to obtain typical atomic structures, annealing and quenching simulations are performed.

3 RESULTS AND DISCUSSION

In order to simulate tubular aluminum nanowires that have similar geometrical features with the experiments [18], tubes that have 10-25 nm radii and 1-5nm thicknesses are selected. In figure 1, atomic structure of a tube in 17 nm radius and in 5 nm thickness is shown. The simulation cell includes to a 30° portion of the tube that contains 6112 atoms. The initial structure is created from an fcc Al crystal structure with (111) surface facing radially outwards. The positions of the atoms that are in a certain radial distance interval to the tube's axis (in 0.5 nm shell) are fixed. As it can be seen in the figure, the tube's structure can be considered as a poly-crystalline nanostructure. After relaxing the tube's atomic structure at constant T=100K and 300K temperatures, we found that the tube preserved its shape and only the atoms in or close to the boundary regions had significant displacements. Further increasing temperature of the structure to 500K did not alter this result.

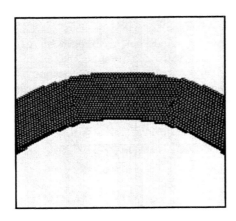

Figure 1: Top view of a region of an aluminum tube with radius 17 nm and thickness 5 nm after equilibration at T=100K.

In order to investigate the structural stability of these tubular nanowires a systematic simulation study is performed. Atomic structures of the tubes in different radii and thickness are equilibrated at T=100K. When the inner shells of the tubes are kept rigid, the structures of the tubes preserved their selves and only relatively small atomic displacements were observed. This simulation case corresponds to the existence of support for the thinner tubes such as nanofibers at the core region or it corresponds to the relaxation of only outer surfaces or shells of thicker tubular structures. On the other side, when all the atoms of the tubes were relaxed, significant structural changes occured especially in the case of thinner tubes with thickness less then 2.5 nm (for 17 nm radius).

Calculation and analysis of potential energy of nanostructures provide important information related with the nanostructures' stability. In figure 2, variation of potential energy of the tubes as functions of tube thickness and radius are shown. In these calculations, the atoms in the inner 0.5 nm shell of the tubes were fixed and the rest of the atoms were allowed to move. After equilibration of the atomic structures, further MD simulation runs were performed and average atomic potential energy values were obtained. As it can be seen in figure 2(a), for a fixed tube radius (17 nm), average potential energy of the atoms decreases with the increase of the tube thickness. This is directly related with the surface to volume ratio of these tubular nanostructures. Relaxing both inner and outer surfaces of the tubes resulted with relatively lower potential energy values due to significant deformations in the case of thinner tubes. In figure 2(b), average potential energy values of the atoms as a function of tube's radius is shown for constant tube thickness of 3.0 nm. As the tube's radius increases, the average potential per atom is found to be decreasing, however, to have a better understanding of the overall variation of potential energy of the tubes with radius, a more systematic study is required and the two limiting cases (solid cylinder and completely flat slab) must be taken into account.

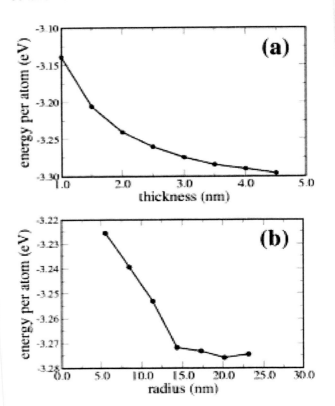

Figure 2: (a) Average potential energy per atom as a function of tube's thickness for a fixed radius of 17 nm. (b) Average potential energy per atom vs. tube's radius for a fixed tube thickness of 3.0 nm.

Transmission electron microscope (TEM) images of tubular nanowires observed in the experiments showed that the surfaces of the tubes may not be smooth and may contain islands or asperities depending on growth conditions [18]. To study the energetics of the tubes with asperities, two atomic models of tubes with the same inner radii (17nm) and the number of atoms (7364 atoms in 30° region) are created. One of the models contains a pyramidal asperity that has 364 asperity atoms and 2.6 nm height. In the other model, the asperity atoms are distributed on the

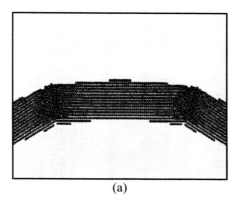

(a)

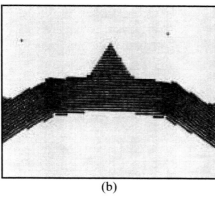

(b)

Figure 3: Top views of tubular nanowires (a) without asperities and (b) with asperities. Both structures contain same number of atoms and relaxed at constant T=300 L temperature.

outer surface of the tube. Both structures are equilibrated at constant T=300 K temperature. In figure 3, the top views of the tubes' atomic structures are shown. It is found that the potential energy difference between two structures is 59 eV and the energy of the structure with no asperity is lower. Thus, smooth tube surfaces are energetically more favorable. On the other side, asperities or islands that are grown may exist depending on their sizes, as their atoms may not diffuse on the tubes' surfaces at relatively low temperatures.

4 CONCLUSIONS

Theoretical investigations on tubular nanowires are presented. In order to have a better understanding of the physical properties of the nanowires in different sizes and to have a better comparison with the other nanowire forms a systematic study is performed. To obtain typical structures, molecular dynamics (MD) simulation method is used and annealing - quenching simulations are performed. For tubular nanowires in large diameters, periodic boundary conditions are modified. Energetics and stability of these tubes and their dependence on the radius and thickness are studied. It is found that, as increase in tubes' radius or thickness results with decrease in the potential energy. Although they were observed in TEM images, islands or asperities on the surface of the tubes are not energetically favorable.

REFERENCES

[1] J. K. Gimzewski and M. E. Welland (ed) "Ultimate Limits of Fabrication and Measurements", NATO ASI Series, Vol. 292 Kluwer, Dordrecht, 1995.

[2] P. A. Serena and N. Garcia (ed) "Nanowires", NATO ASI Series, Kluwer, Dordrecht, 1997.

[3] I. P. Batra, A. Buldum and S. Ciraci, J. Phys:Condens Matter, 13, R537, 2001 (and the references there in).

[4] J. Gimzewski and R. M ller, Phys. Rev. B, 41, 2763, 1987.

[5] N. Agra t, J. G. Rodrigo, and S. Vieira, Phys. Rev. B, 47, 12 345, 1993.

[6] L. Olesen et al, Phys. Rev. Lett. 72, 2251, 1994.

[7] J. M. Krans et al, Nature, 375, 767, 1995.

[8] H. Ohnishi, Y. Kondo, and K. Takayanagi, Nature, 395, 780, 1998.

[9] Y. Kondo and K. Takayanagi, Science, 289, 606, 2000.

[10] S. Ciraci and E. Tekman, Phys. Rev. B, 40, R11969, 1989.

[11] T. N. Todorov and A. P. Sutton, Phys. Rev. Lett., 70, 2138, 1993.

[12] M. Brandbyge, K. W. Jacobsen and J. K. N rskov, Phys. Rev. B, 52, 8499; Phys. Rev. B, 56, 14956, 1997.

[13] H. Mehrez and S. Ciraci, Phys. Rev. B, 56, 12632, 1997.

[14] O. Gulseren, F. Ercolessi and E. Tosatti, Phys. Rev. Lett., 80, 3775, 1998.

[15] E. Tosatti et al, Science, 291, 288, 2001.

[16] B. Wang, S. Yin, G. Wang, A. Buldum and J. Zhao, Phys. Rev. Lett., 86, 2046, 2001.

[17] P. Sen, S. Ciraci, A. Buldum and I. P. Batra, Phys. Rev. B, 64, 195420, 2001.

[18] W. Liu, M. Graham, E. A. Evans and D. H. Reneker, J. Mater. Res., 17, 3206, 2002.

[19] L. J. Lauhon, M. S. Gudiksen, D. Wang and C. M. Lieber, Nature, 420, 57, 2002.

[20] W. Q. Han and A. Zettl, Nano Lett. 3, 681, 2003.

[21] Y. Zhang, N. Franklin, R. Chen, and H. Dai, Chem. Phys. Lett., 331, 35, 2000.

[22] J. Kong, M. Chapline and H. Dai, Adv. Mater. 13, 1384, (2001).

[23] H. Choi, M. Shim, S. Bangsaruntip and H. Dai, J. Am. Chem. Soc. 124, 9058, 2002.

[24] F. Ercolessi and J. B. Adams, Europhysics Letters, 26, 583 (1994).

Orientation Effects of Elastic-Plastic Deformation at Surfaces: Nanoindentation of Nickel Single Crystals

Oyeon Kum

Department of Chemistry, Clemson University, Clemson, SC 29634

ABSTRACT

Orientation effects in nanomechanical properties at the surfaces with molecular dyanmics were observed as a function of indenter size and indenter speed in three crystal orientations: $\langle 100 \rangle$, $\langle 110 \rangle$, and $\langle 111 \rangle$. The force vs displacement curves for indentation follows the Hertzian solution for elastic deformation of $F = kd^{1.5}$. However, the force fitted constant k showed a dependency on indenter size, velocity, and crystal orientations. The results of dislocation nucleations in different orientations also showed anisotropy: Stacking faults in the $\langle 100 \rangle$; deep partial dislocations in the $\langle 110 \rangle$; shallow partial dislocations followed by the stacking faults in the $\langle 111 \rangle$.

Keywords: nanoindentation, anisotropy, indentation curve, dislocation nucleation

1 INTRODUCTION

Large-scale simulations of single crystals, nickel, using Morse-type pair potentials, the analytical form of embedded atom method (EAM) potential, and the tabular form of the EAM potential were recently performed in bulk under shock compressed conditions [1], [2] to investigate anisotropy of fcc single crystal deformation. However, knowledge of the mechanical behavior associated with the contact of small volumes under the surface is also important from both scientific and technological viewpoints. The development of nanoindentation techniques such as interfacial force microscopy or atomic-force microscopy provide easy experimental tools to investigate the detailed mechanism of deformation during indentation at a very small scale. The results of such research contribute to the understanding of microscopic fracture mechanics to develop device miniaturization and computer disk drives.

Modern nanoindentation techniques have allowed for the measurement of load versus indentation depth curves of small volumes where the contact radius is less than 100 nm [3]. These small volumes are available sizes for large-scale molecular dynamics simulations to make direct comparisons between experiments and simulations. Such comparison may solve the discrepancies among different models and experiments. For example, a typical nanoindentation curve exhibits abrupt bursts in displacement at constant load separated by regions of positive slope, whereas a classical curve shows a relatively smooth positive slope. Such discrepancies are postulated to be driven from the discrete nucleation of dislocation loops below the indenter, but the exact mechanisms are difficult to investigate experimentally and still under discussion [4].

In this paper, I investigate the orientation effects of nickel single crystals in (100), (110), and (111) surfaces using molecular dynamics. The indenter velocity, crystal orientations, indenter sizes were important parameters for anisotropy shown in indentation curves. This study showed the atomistic insight into anomalous dislocation nucleation of single crystals at surfaces in three crystal orientations.

2 MODEL

Because a typical dislocation separation is the order of μm for well-annealed metals, the area under the nanoindenter should behave close to that of a perfect single crystal–dislocation free. In this paper, I used Voter's tabular form of embedded atom model (EAM) potential [5] to describe nickel single crystals. Previous atomistic calculations have studied indentation and retraction using EAM potentials [6], or other semiempirical potential models [7], as well as first-principles methods [8]. These calculations showed strong bonding between the indenter tip and the surface due to the large energy of adhesion between two clean surfaces. This leads to a jump to contact upon approach and necking between the tip and surface during retraction. In experiments, the tip-surface adhesion interaction will be dramatically reduced because the tip and surface are not atomically clean or surfaces are passivated by the addition of an alkanethiol layer which prevents bonding between the tip and the surface [9].

The model discussed in this work is tailored to address the passivated surfaces by using a strongly repulsive potential to describe the interactions between the indenter and the metal surfaces. Each atom in the indented material interacts with the idealized spherical indenter via the repulsive potential,

$$V(r) = \epsilon(\sigma/r)^{\alpha}, \qquad (1)$$

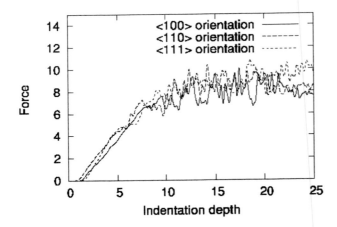

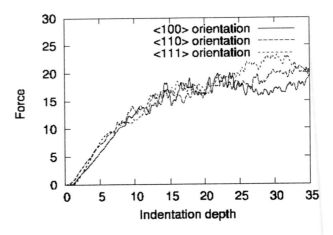

Figure 1: Indentation force vs displacement curves in the three orientations. The indenter velocity is about 670 m/sec and its diameter is 4.0 nm. Force unit is 43.59042 nN and length unit is $\mathring{A}$.

Figure 2: Indentation force vs displacement curves in the three orientation directions. The indenter velocity is about 670 m/sec and its diameter is 6.0 nm. See Fig. 1 caption for units.

where ϵ is energy, σ is the indenter radius, and r is the distance between atom and the center of the indenter. α is the power of the potential. α was chosen to be 150 in the simulations to guarantee the strong repulsive force between indenter and surface. As is customary, I used reduced (or dimensionless) quantities to specify various physical parameters. Energy unit is eV, unit of length $\mathring{A}$, and unit of mass atomic unit of nickel mass. Thus, the time unit $t_0 = (m\sigma^2/\epsilon)^{1/2}$ and force unit is 43.59042 (nN).

3   RESULTS AND DISCUSSION

Simulations were performed for three indenter diameters (4.0, 5.0, and 6.0 nm) and two indenter speeds (about 670 and 67 m/sec). The indenter approaches to the cold (zero temperature) fcc nickel single crystals oriented in the $\langle 100 \rangle$, $\langle 110 \rangle$, and $\langle 111 \rangle$ directions. The total number of atoms involved are 400 000, 397 600, and 397 440 for the $\langle 100 \rangle$, $\langle 110 \rangle$, and $\langle 111 \rangle$ orientations, respectively.

Figure 1 shows indentation force vs displacement depth curves for three crystal orientations with indenter diameter of 4.0 nm and at the indenter velocity of about 670 m/sec. The typical features of curves for three orientations are the elastic deformations followed by the plastic deformations and all curves have displacement excursions which correspond to the initiation of plastic deformation (yield points). Overall, the indentation curves follow the Hertzian solution for elastic deformation which is a simple power law of $F = k\ d^{1.5}$. The values of k, corresponding to indentation modulus or micro-hardness, are 0.32, 0.40, and 0.37 in the $\langle 100 \rangle$, $\langle 110 \rangle$, and $\langle 111 \rangle$ orientations, respectively.

The force vs displacement curves for three crystal

orientations with indenter diameter of 6.0 nm at the indenter velocity of about 670 m/sec are shown in Fig. 2. Larger diameter of indenter produces higher micro-hardness. The values of k are 0.45, 0.61, and 0.63 in the $\langle 100 \rangle$, $\langle 110 \rangle$, and $\langle 111 \rangle$ orientations, respectively. The anisotropy is also shown clearly in the different orientation directions. Compared with Fig. 1, indenter size can be considered as an important parameter of anisotropy.

Figure 3 shows indentation force vs displacement curves for three orientations at the indenter velocity of about 67 m/sec with the indentor diameter of 5.0 nm. Compared with the Fig. 1 and Fig. 2, slow indenter velocity resulted in different microscopic mode of indentation curves in elastic deformation. Each orientation showed different excursion point: $\langle 110 \rangle$ orientation showed the weakest yield strength and in the $\langle 100 \rangle$ direction, the yield strength was the strongest. In the $\langle 111 \rangle$ orientation, it was about average of the two. The results suggest the different mode of dislocation nucleation, requiring high energy for stacking faults and relatively low energy for partial dislocations. The fitted k were 0.24, 0.30, and 0.27 in the $\langle 100 \rangle$, $\langle 110 \rangle$, and $\langle 111 \rangle$ orientations, respectively. These were different from those of the same diameter of 5 nm with velocity of about 670 m/sec which were not shown here. This result also suggests the anisotropy due to velocity differences.

To study the dislocation nucleation, the centrosymmetry parameter is used which is defined as follows [10]:

$$P = \sum_{i=1,6} |\mathbf{R}_i + \mathbf{R}_{i+6}|^2, \qquad (2)$$

where $\mathbf{R}_i$ and $\mathbf{R}_{i+6}$ are the vectors or bonds corresponding to the six pairs of opposite nearest neighbors in the fcc lattice. The 12 nearest-neighbor vectors for each

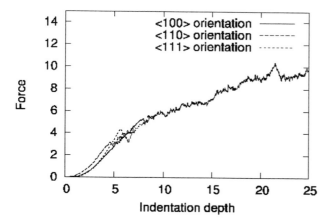

Figure 3: Indentation force vs displacement curves in the three orientation directions. The indenter velocity is about 67 m/sec and its diameter is 5.0 nm. See Fig. 1 caption for units.

atom are first determined in an undistorted bulk fcc lattice with the orientation of the slab. The analogous set of 12 vectors for each atom in the distorted lattice $\mathbf{R}_i$ is then generated by finding those neighbors in the distorted lattice with vectors closest in distance to the undistorted nearest-neighbor vectors. It is possible that this set will contain duplicates or non-nearest neighbors if a given atom has fewer than 12 nearest neighbors or a very distorted local environment. Each "equal and opposite" pair of vectors is added together, then the sum of the squares of the six resulting vectors is calculated. This final number is a measure of the departure from centrosymmetry in the immediate vicinity of any given atom and is used to determine if the atom is near a defect [10].

The centrosymmetry parameter, P is useful to distinguish partial dislocations and stacking faults [10]. The range of values from 0.5 to 4.0 represents partial dislocations and the range between 4.0 and the value of the surface atoms is defined as stacking faults [10]. It is zero for atoms in a perfect nickel crystal lattice, 24.78 $Å^2$ for surface atoms in the $\langle 100 \rangle$ orientation, 30.88 $Å^2$ and 6.11 $Å^2$ for surface atoms in the $\langle 110 \rangle$ orientation, 18.59 $Å^2$ for surface atoms in the $\langle 111 \rangle$ orientation. Note that the $\langle 110 \rangle$ orientation has two different surface atom values because two layers of surface atoms have different numbers of neighbors. These values assume that the nickel nearest neighbor distance does not change in the vicinity of the defects.

Figure 4 shows early dislocation nucleations in three crystal orientations. For the $\langle 100 \rangle$ orientation, dislocation nucleation starts with stacking faults in the (111) plane. This slip plane agreed with that of the shocked fcc single crystal deformation [1]. Table 1 shows the

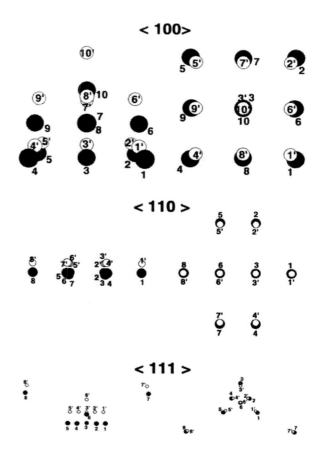

Figure 4: Dislocation nucleations at the time of 28.5, 20.0, 46.0, and at the force of 2.604, 2.363, and 6.734 in Fig. 2 for three crystal orientations, $\langle 100 \rangle$, $\langle 110 \rangle$, and $\langle 111 \rangle$ from top to bottom, respectively. White circles are undistorted atoms and black circles are corresponding distorted atoms. The indenter is moving with constant velocity from top to bottom in the vertical direction (z-axis). Left-handed column shows the active view similar to the plane normal to x-axis and right-handed column is the top view. 10, 8, and 8 atoms are involved in the deformation for $\langle 100 \rangle$, $\langle 110 \rangle$, and $\langle 111 \rangle$ orientations, respectively. Atoms with P>0.5 are selected. The atomic number and its P value are in the Table 1. Prime number is the corresponding undistorted atom number. The atom size in the picture is proportional to the deformed area in the crystal. The centrosymmetry parameter is useful to discriminate different deformation modes, but it fails to provide information on the Burgers vectors of dislocations. Burgers vectors are not shown clearly in the picture.

Table 1: The centrosymmetry, P values at the dislocation nucleation in three crystal orientations. The atom number corresponds to that in Fig. 4.

Atom #	$\langle 100 \rangle$	$\langle 110 \rangle$	$\langle 111 \rangle$
1	4.81	0.82	0.57
2	4.81	0.63	0.55
3	5.32	0.67	0.95
4	4.81	0.63	0.55
5	4.81	0.63	0.57
6	4.75	0.67	0.72
7	4.75	0.63	0.51
8	4.75	0.82	0.51
9	5.75		
10	14.00		

centrosymmetry values for all atoms in Fig. 4. These results suggest that the dislocations propagated non-symmetrically, and they evolved into the mixed modes with stacking faults and partial dislocations. For $\langle 110 \rangle$ direction, dislocation nucleation started early at the smallest displacement and propagated through in the mode of only partial dislocations. This slip direction agreed with that of the shocked fcc single crystal deformation [1]. For $\langle 111 \rangle$ orientation, the initiation of the dislocation nucleation took the longest time among the three orientations and propagated through with mixed modes of partial dislocations and stacking faults. The centrosymmetry parameter is useful to discriminate the modes of deformation but does not provide helpful information on the Burgers vectors of dislocations. The slip vector defined in the other paper [11] is known to be useful for calculating Burgers vectors of dislocations. It is now under investigation.

4 CONCLUSIONS

I have studied molecular dynamics nanoindentation simulations at surfaces for nickel single crystals with a system size of about 400 000 atoms and with Voter's tabular form of EAM potential. To simulate the elastic-plastic deformation during indentation on a *passivated* surface, a hard-sphere like indenter described by a strong repulsive potential interacted with the fcc metal surface. Anisotropy parameters at the surfaces were investigated as a function of indenter size and indenter velocity for three crystal orientations: $\langle 100 \rangle$, $\langle 110 \rangle$, and $\langle 111 \rangle$. In general, the load-displacement curves followed the Hertzian solution for elastic deformation which is a simple power law of $F = k\, d^{1.5}$ for all orientations, sizes, and velocities of the indenter. However, the indentation curves showed somewhat complicated dependence on the indenter velocity as well as indenter size in the three crystal orientations. These dependencies were quantified by the micro-modulus, k.

In summary, discrete yield phenomena were different in different orientations. Thus, the results showed the dependency of anisotropic elastic-plastic deformation on indenter size, speed and crystal orientations. The centrosymmetry parameter discriminated the mode of deformation. For $\langle 100 \rangle$ orientation, dislocation nucleation starts at the mode of stacking faults in the two (111) surfaces. However, the dislocation propagation occurred non-symmetrically and soon evolved into the mixed modes of stacking faults and partial dislocations. For $\langle 110 \rangle$ direction, dislocation nucleation occurred in the mode of partial dislocation. For $\langle 111 \rangle$ orientation, the dislocation nucleation took the longest time among the three orientations and started at the mode of partial dislocation. However, the geometry was different from that of $\langle 110 \rangle$ direction. In this orientation, two dislocation modes, partial dislocations and stacking faults, occurred immediately and almost simultaneously. The results for slip plane and its direction agreed with those observed in the shocked nickel single crystals simulations.

ACKNOWLEDGMENTS

Arthur F. Voter is thanked for the kindness of giving his package of EAM potential models for fcc metals and their alloys. Ethan Ballard is thanked for his proof reading.

REFERENCES

[1] Oyeon Kum, J. Appl. Phys., 93, 3239, 2003.

[2] Oyeon Kum, Nanotech2003, 2, 538, 2003.

[3] S.G. Corcoran, R.J. Colton, E.T. Lilleodden, and W.W. Gerberich, Phys. Rev. B, 55, R16057, 1997.

[4] A. Gouldstone, H.J. Koh, K.Y. Zeng, A.E. Giannakopoulous, and S. Suresh, Acta. mater., 42, 2277, 2000.

[5] A.F. Voter, Los Alamos Unclassified Technical Report #LA-UR93-3901, 1993.

[6] J. Belak, D.B. Boercker, and I.F. Stowers, MRS Bull., 18, 55, 1993.

[7] D.W. Brenner, S.B. Sinnott, J.A. Harrison, and O.A. Shenderova, Nanotechnology, 7, 161, 1996.

[8] R. Pérez, M.C. Payne, and A.D. Simpson, Phys. Rev. Lett., 75, 4748, 1995.

[9] P. Tangyunyong, R.C. Thomas, J.E. Houston, T.A. Michalske, R.M. Crooks, and A.J. Howard, Phys. Rev. Lett., 71, 3319, 1993.

[10] C.L. Kelchner, S.J. Plimpton, and J.C. Hamilton, Phys. Rev. B, 58, 11085, 1998.

[11] J.A. Zimmerman, C.L. Kelchner, P.A. Klein, J.C. Hamilton, and S.M. Foiles, Phys. Rev. Lett., 87, 165507-1, 2001.

Modeling of PDMS - Silica Nanocomposites

J. S. Smith*, G. D. Smith** and O. Borodin***

Department of Materials Science and Engineering, University of Utah,
122 S. Central Campus Dr., Rm. 304,
Salt Lake City, UT USA, * jsmith@cluster2.mse.utah.edu,
** gsmith@gibbon.mse.utah.edu, *** borodin@eng.utah.edu

ABSTRACT

A hydrogen bonding pathway between polydimethyl-siloxane (PDMS) and hydroxyl groups on a silica surface was studied using quantum chemistry calculations of disiloxane and hexamethyldisiloxane molecules with small silica clusters. A newly developed classical force field for PDMS was developed for atomistic molecular dynamics simulation studies of PDMS – silica nanocomposites to determine the effect of these interactions on the dynamics and structure of PDMS. A three nanometer silica particle (β-crystobalite) with (111) surface hydroxyl group density of 4.8 OH groups/nm^2 was simulated in a PDMS melt in the temperature range of 300 to 500K. The density and structure of PDMS chains near the silica surface were strongly influenced by the hydrogen bonding interaction which is not properly represented in other current force fields. Residence time correlation analysis confirmed that PDMS oxygen – silica surface hydrogen atom dynamics were consistent with polymer hydrogen bonding.

Keywords: nanocomposites, quantum chemistry, molecular dynamics, polymers, modeling

1 INTRODUCTION

There is an extensive and rapidly growing literature describing the properties of filled and nanocomposite polymer systems. In addition, there has been much effort to model the constitutive behavior of such systems, and PDMS – silicone systems have been perhaps the second most studied system next to rubber – carbon black. Recent simulations have investigated the effects of particle shape, interaction, and surface structure on various systems [1,2,3]. In order to correctly model these systems the underlying mechanisms of polymer – particle interaction must be investigated.

In the PDMS – silica system experiments have determined that the PDMS chains near the surface are constrained and their motions are much slower compared to bulk PDMS. Experiments suggest that this behavior is due to hydrogen bonding between the silica surface hydroxyl atoms and the PDMS chains, or covalent bonding of certain chains to the surface [4]. Recent quantum chemistry and simulations investigating PDMS – silica interfaces concluded that hydrogen bonding was not possible, but failed to suggest an alternate explanation for the polymer dynamics [5]. We have undertaken an ab initio quantum chemistry investigation of likely hydrogen bonding geometries between PDMS oligomers and hydroxyl bearing silica clusters in order to develop a classical force field to correctly describe PDMS – silica interactions. Molecular dynamics simulations have been initiated to investigate the particle polymer interface, understand underlying mechanisms such as hydrogen bonding, and generate useful mechanical properties for use in constitutive models.

2 PDMS – SILICA INTERACTION

2.1 Quantum Chemistry

In order to investigate the intermolecular bonding between PDMS and silica surfaces with hydroxyl groups, the binding energies of representative molecules were calculated using the Gaussian 98 package [6]. The binding energy is defined as the difference in the energy of the complex and the sum of the energies of the isolated molecules. The smallest molecules were disiloxane (DS, SiH_3OSiH_3) and silanol (H_3SiOH) shown in their minimum energy geometry in Figure 1. While frozen in their minimum energy geometries, the molecules were moved closer together and farther apart along the bonding path and Hartree-Fock (HF) and Møller-Plesset second order perturbation theory (MP2) energies were calculated with 6-311G(2df), aug-cc-pvDz, and aug-cc-pvTz basis sets.

Additional calculations were done on a larger trisiloxysilanol cluster (TSS, $(H_3SiO)_3SiOH$, see Figure 2) at the aug-cc-pvDz level where the MP2 binding energy was found to be -5.18 kcal/mol, with a DS oxygen to silanol oxygen separation, r_{OO} = 2.943 angstroms (O-H distance ~2.0 angstroms). This binding energy is 3.2 to 4.2 kcals stronger than that reported in the literature [5]. The hexamethyldisiloxane molecule (HMDS, $(CH_3)_3SiO-Si(CH_3)_3$, see Figure 3) has a hydrogen bonding approach path with TSS which had a binding energy of -7.08 kcal/mol at the 6-31G(2d) level of theory.

This demonstrates the feasibility of hydrogen bonding even in the presence of additional methyl groups and suggests that the level of theory (basis set) is very important in determining the binding energy, although it limits the size of molecules that can be examined. The additional binding energy of ~1.8 kcal/mol is similar to the binding energy of methyl groups to a silica surface found in the literature on much larger molecules [5]. The variations of binding energy with separation are shown in Figure 4.

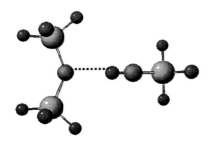

Figure 1: Hydrogen bonding paths between disiloxane and silanol.

Figure 2: Hydrogen bonding path between disiloxane and the TSS cluster.

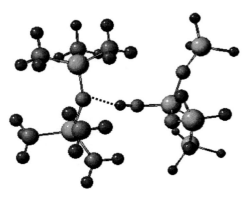

Figure 3: Hydrogen bonding path between hexamethyldisiloxane and a TSS cluster.

2.2 Force Field Methodology

The interaction between the particle surface and PDMS chains in this treatment is of a purely nonbonded nature. Parameters for the PDMS – PDMS and silica – silica atomic intra- and intermolecular interactions are published elsewhere [7,8]. The total binding energy, $U^{BIND}(r)$ of an ensemble of PDMS molecules and particles represented by coordinate vector **r**, includes contributions from electrostatic, van der Waals (VDW) and hydrogen bonding, U^{HB}, interactions.

$$U^{BIND}\left(r_{ij}\right) = \sum_{i<j} \frac{q_\alpha q_\beta}{4\pi\varepsilon_0 r_{ij}} + A_{\alpha\beta}\exp\left(-B_{\alpha\beta}r_{ij}\right) - \frac{C_{\alpha\beta}}{r_{ij}^6} + U^{HB}\left(r_{ij}\right) \quad (1)$$

The electrostatic and VDW energies are represented by columbic and Buckingham equations where r_{ij} is the distance between the ith particle atom and j^{th} polymer atom of type α and β respectively. The equation for hydrogen bonding is presented in section 2.4. In order to correctly represent the PDMS – hydroxylated silica interaction the following methodology was carried out. First the partial charges, q, for the atom of each representative molecule were calculated. Second, it was determined that the Buckingham potential parameters, A, B and C from the newly parameterized PDMS force field would be used to represent the PDMS polymer – silica particle interactions [7]. Finally an appropriate hydrogen bonding function is parameterized which represents the both the relatively strong and short ranged nature of this interaction [9].

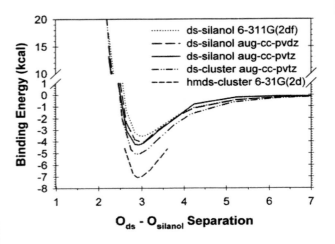

Figure 4: The variation of binding energy with molecular separation showing the effect of level of theory and molecular structure.

2.3 Charge Parameterization

Electrostatic charges were calculated from minimization of the $(\varphi_i^{QC} - \varphi_i^{FF})^2$ objective function for the DMTS molecule, where φ_i^{QC} and φ_i^{FF} are the electrostatic potentials on a grid point i from quantum chemistry calculations and from the force field respectively [10]. The grid points within one van der Waals (VDW) radius of each atom were excluded from the calculation because such separations are highly unlikely in MD simulations and points any farther than 4.0 Å from any atom were also excluded. The VDW radii were 2.0, 2.0, 2.5 and 1.8 angstroms, for Si, O, C, and H, respectively.

2.4 Hydrogen Bond

The excess binding energy between the disiloxane – silanol molecule was fit at the 6-311G(2df) level with a special attractive hydrogen bonding function [9]:

$$U^{HB}(r_{ij}) = \sum_{i<j} - A_{OH}^{HB} \exp(-B_{OH}^{HB} r_{ij}) \qquad (2)$$

where r_{ij} is the atomic separation of the silica surface hydroxyl's hydrogen atom and PDMS' oxygen atom with hydrogen bonding parameters A_{OH}^{HB} and B_{OH}^{HB}. Equation 2 was fit using A and B parameters of 5942 and 3.649 respectively. These same parameters fit the higher level aug-cc-pvDz results for the DS – TSS energies as shown in Figure 5. Also shown in Figure 5 is the binding energy calculated using the consistent force field (CFF) parameters of Sun and Rigby [11] which does not capture the hydrogen bonding energy as reported recently in the literature [5].

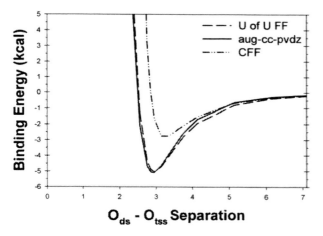

Figure 5: Fit of the binding energy between disiloxane and a TSS cluster using the current U of U force field and the CFF force field.

3 SIMULATIONS AND PROPERTIES

3.1 Simulation Methodology

MD Simulations of an ensemble of 40 PDMS molecules, $MD_{19}M$ (20 repeat units each and 1571 g/mol), and a single silica nanoparticle were performed at 500 K. All simulations were carried out with the *Lucretius* [12] MD simulation package using a Nose-Hoover thermostat [13] and barostat [14] to control the temperature and pressure. The particle-mesh Ewald (PME) technique [15] was used to treat all electrostatic interactions. A multiple time step reversible reference system propagator algorithm [14] was employed with a time step of 0.5 fs for bond, bend and torsional motions, a 1.0 fs time step for all VDW and real electrostatic interactions within a sphere of radius 6.0

Å, and a 2.0 fs time step for nonbonded interactions in the shell between radii of 6.0 and 10.0 Å and for the reciprocal space PME calculations.

For the initial investigations, a force field potential was used which combined the newly derived nonbonded parameters and partial charges to describe PDMS and PDMS – silica interactions and the CFF internal parameters [11] to describe the bond, bend and dihedral energies of PDMS. This parameter combination will certainly affect the thermodynamic and dynamic behavior of the PDMS chains but should give a qualitative indication of the effect of the PDMS polymer interaction with a hydroxylated silica surface and a reasonable comparison with previous works using the unaltered CFF force field.

A freely rotating ß-cristobalite crystal (approximately, 27.8, 24.7, and 29.6 angstroms thick) surrounded by PDMS melt was simulated at 500 K. This crystal was fully hydroxylated on the 111 surface to a density of 4.8 OH groups/nm². The intra- and inter-atomic interactions of only silica atoms were represented by a CFF type force field from the literature [8]. After box size equilibration in an NPT ensemble, NVT runs were then carried out for an additional 2.4 ns. The crystal was then fixed at the center of the box with faces parallel to the periodic boundaries so that the distance between crystal faces was approximately 20 to 25 angstroms and NVT production runs were carried out for an additional 1.9 ns.

3.2 PDMS – Silica Properties

Although qualitative, initial density profiles of PDMS layers near the hydroxylated (111) silica surfaces show increased structure closer to the hydroxyl atoms while the bare surfaces show simply a broad first layer peak with fewer undulations. The PDMS constituent atom arrangement next to the surface also shows that the surface hydroxyl groups induce strong ordering in the atom peaks and that the PDMS oxygen atom peak moves closer to the hydroxylated surface as it participates in hydrogen bonding whereas the bare surface is predominantly bordered by the PDMS methyl groups.

The first coordination shell of PDMS' oxygen atoms around the silica surface hydroxyl's hydrogen atoms is 2 angstroms according to the radial distribution function. Using this distance to define O-H hydrogen-bonding pairs we can calculate mean hydrogen bonding residence time correlation functions:

$$R_1(t) = \langle H(0) \bullet H(t) \rangle \qquad (3)$$

$R_2(t)$ = <average time until Oxygen atom exits the hydroxyl's H atom coordination shell> (4)

where $H(t) = 1$ if the oxygen atom is in the first coordination shell of the hydroxyl group's hydrogen atom and $H(t) = 0$ otherwise. The difference between the residence time correlation functions $R_1(t)$ and $R_2(t)$ is that the $R_2(t)$ function probes the time that each oxygen atom spends in the hydroxyl group's hydrogen atoms coordination shell until it first exits regardless of any subsequent reentries, whereas the $R_1(t)$ function probes the total time the oxygen atom spends in a particular coordination shell no matter how many times it leaves and reenters. The difference between the residence time correlation functions suggest that hydrogen bonds form, break, and reform many times before the PDMS oxygen atom finally leaves the surface hydroxyl's hydrogen atom.

The mean squared displacement (MSD) of the PDMS backbone silicon and oxygen atoms were examined to see the effect of the additional hydrogen bonding near the hydroxyl covered silica surface. The MSD of the PDMS surface layer was an order of magnitude less after 800 ps when hydroxyl groups are present than for PDMS near bare silica surfaces. These results support the assertion that silica surfaces in the PDMS – silica nanocomposites slow down the PDMS chains by means of hydrogen bonding at surface hydroxyls. This mechanism certainly factors in to the mechanical behavior of these composites.

4 CONCLUSIONS

An ab initio quantum chemistry investigation of disiloxane or hexamethyldisiloxane with silanol or a larger silica cluster revealed hydrogen bonding configurations with a minimum binding energy of -5.2 kcal/mol at the MP2/aug-cc-pvDz level of theory. From this data a quantum chemistry based PDMS - silica force field has been developed that includes a hydrogen bonding function which shows transferability between levels of theory and to other polymer systems.

Molecular dynamics simulations of PDMS - silica systems at 500 K qualitatively showed increased density and structure near the hydroxyl covered (111) SiO_2 surface. PDMS oxygen atoms were drawn near to the surface to participate in hydrogen bonds contrary to previous simulation studies in the literature [5]. Residence time correlation analysis showed that hydrogen bonding atom pairs remained coordinated for some time forming, breaking and reforming hydrogen bonds. This hydrogen bonding model is also supported by an order of magnitude decrease in the mean squared displacement of PDMS backbone atoms next to a hydroxylated surface as compared with the same bare surface. Further studies are underway with the recently completed PDMS force field [7] and the results will be presented at the conference.

ACKNOWLEDGEMENT

This work was supported by Los Alamos National Laboratory under contract 79277001032F.

REFERENCES

[1] G. Smith, D. Bedrov and O. Borodin, PRL, 90, 226103, 2003.
[2] O. Borodin, G. Smith, R. Bandyopadhyaya and O. Byutner, Macromolecules, 36, 7873, 2003.
[3] D. Bedrov, G. Smith and J. Smith, J. Chem. Phys., 119, 10438, 2003.
[4] V. Litvinov, H. Barthel and J. Weis, Macromolecules, 35, 4356, 2002.
[5] M. Tsige, T. Soddemann, S. Rempe, G. Grest, J. Kress, M. Robbins, S. Sides, M. Stevens and E. Webb III, J. Chem. Phys., 118, 5132, 2003.
[6] M. Frisch et. al., GAUSSIAN 98, revision A.7, Gaussian, Inc., Pittsburg, PA, 1998.
[7] J. Smith, G. Smith and O. Borodin, manuscript in preparation, 2003.
[8] J. Hill and J. Sauer, J. Phys. Chem., 98, 1238, 1994.
[9] G. Smith, O. Borodin and D. Bedrov, J. Comp. Chem., 23, 1480, 2002.
[10] C. Breneman and K. Wiberg, J. Comput. Chem., 11, 361, 1990.
[11] H. Sun and D. Rigby, Spectrochemica Acta Part A, 53, 1301, 1997.
[12] Lucretius, lucretius.mse.utah.edu
[13] S. Nose, in: M. Meyer and V. Pontikis, (Eds.), Computer Simulations in Materials Science, Kluwer academic Pub., Netherlands, 21,1991.
[14] G. Martyna, M. Tuckerman, D. Tobias and M. Klein, Mol. Phys., 87, 1117, 1996.
[15] M. Deserno and C. Holm, J. Chem. Phys., 109, 7678, 1998.

Optical Absorption Modeling of Arbitrary Shaped Nanoparticles

P. Mullin[*], G. Kozlowski[*], P.G. Kozlowski[*], J.D. Busbee[**] and J.G. Jones[**]

[*]248 Fawcett Hall, Wright State University
3640 Colonel Glenn Hwy, Dayton, OH 45435, USA
Pmullin@aosepc.com
Gregory.Kozlowski@wright.edu

[**]AFRL/MLMR Bldg 653, 2977 P Street, Suite 13
WPAFB, OH 45433
John.Busbee@wpafb.af.mil
John.Jones@wpafb.af.mil

ABSTRACT

Limitations of a method for modeling the optical absorption spectra of nanosize particles is explored. Computer software has been developed to predict absorption peaks based on the particle's geometry. The software breaks the particle's surface into small sites and builds a matrix describing site to site surface polarization interactions. From this matrix the particle's normal mode resonances are determined.

With the method it is possible to simulate nearly any geometrical shape. Cubes with cut corners or edges, such as chemically fabricated silver particles are possible. Long, thin bars are also candidates for simulation. Results dependence on surface resolution is explained.

Keywords: absorption, spectra, simulation, nanoparticle

1 INTRODUCTION

Mie theory and other models have shown good accuracy in predicting the optical response in fields of spherical nanoparticles. These methods are however severely limited in their application to some laboratory experiments. Real particles, of silver or MgO for example, are far from spherical. They are typically cubes or many times cubes with the corners severed. In the case of our current experiments the particle fields may even include rods many times longer than the average cube edge. What is desired is a method of predicting response to particles of any arbitrary shape.

An analytic computer model to predict the absorption portion of optical response in fields of nanoparticles is investigated. In the results presented here particles are assumed to be nanosized cubes. However the method is applicable to nearly any continuous shape without holes. Particles are also assumed to be significantly smaller than the wavelength of incident light. The model considers a single particle and its interaction with incident light,

particle to particle interaction is not modeled. Therefore simulation of experiments with fields of particles is limited to low density, widely spaced particles. The method focuses on surface polarization resonant modes set up with the incident light. These resonant modes are a function of the incident light wavelength, particle geometry and material. The model treats the cube surface as an array of small discrete surface patches with properties similar to the bulk material.

A key assumption is that only surface polarization resonant modes are required to account for complete optical absorption spectra. R. Fuchs [1] has shown that this assumption is valid by correlating calculated results with experiment. Fuchs used the method to predict absorption in fields of cubic MgO and NaCl particles.

2 PATCH MODEL

Only surface mode polarization is considered in the model. Transverse and longitudinal modes are assumed to be insignificant. Conceptually an electric field, $\vec{E}_0$ is applied to the particle. With $\vec{r}$ just inside the surface polarization is related to susceptibility of the base material by:

$$P(\vec{r}) = \chi(\omega)\vec{E}(\vec{r}) \tag{1}$$

And, where $\hat{n}$ is the normal at the surface it follows that polarization at one point is effected by the entire surface and the incident $\vec{E}_0$ field. Polarization inside the particle is assumed to be zero.

$$\frac{p(\vec{r})}{\chi(\omega)} = -2\pi p(\vec{r}) + \int_S \frac{\hat{n}(\vec{r}) \bullet (\vec{r} - \vec{r}')}{|\vec{r} - \vec{r}'|^3} p(\vec{r}')dS' + \vec{E}^0 \bullet \hat{n}(\vec{r}) \tag{2}$$

The first term on the right side of Eq. (2) is a solid angle approximation to account for the polarization at the site of $\vec{r}$ on the surface.

And in discrete form with the surface divided into i equal area patches:

$$[\chi^{-1}(\omega) + 2\pi]p_i = \sum R_{ij}p_j + \vec{E}^0 \cdot \hat{n}_i \qquad (3)$$

The polarization at each patch is p_i. R_{ij} is representative of each patch's interaction with each other patch individually. It is a two dimensional matrix where both indices ranges are the total number of patches on the particle. Therefore R_{ij} contains the particle's geometry dependant information. R_{ij} is used in the model to determine resonant peak locations and magnitudes.

$$R_{ij} = \frac{\hat{n}_i \bullet (\vec{r}_i - \vec{r}_j)}{|\vec{r}_i - \vec{r}_j|^3} \Delta S_j \qquad (4)$$

Obviously because of the denominator the diagonal elements of Rij must be set equal to zero. The local effect of a patch is represented by the 2π term in Eq. (3).

The polarization Eq. (3) includes the particle geometry in R_{ij}, the material properties in the frequency dependent susceptibility and the local patch approximation. Given a susceptibility value for a specific material Eq. (3) generates a system of equations with one equation and one variable for each patch. Solving the system of equations yields polarization for the entire surface.

Determining if the particular polarization pattern is a significant resonant mode requires calculating the expectation value for the complete particle.

$$\langle \chi \rangle = \frac{|\vec{M}|}{v|\vec{E}_0|} \qquad (5)$$

With a uni-directional $\vec{E}_0$ field the total resultant induced dipole moment, $\vec{M}$, is related to the average susceptibility by Eq. (5). In Eq. (5) v is simply the volume of the particle.

The total dipole moment may be determined by summing over the polarization at each surface patch.

$$M_1 = \sum_i p_i x_{1i} s_i \qquad (6)$$

Where M_1 is the induced dipole moment along one axis direction and x_{1i} is the component of $\vec{r}$ parallel to that axis.

Significant absorption peaks will be characterized by spikes in the imaginary part of the total particle susceptibility expectation value. To find resonant modes the real part of $\chi(\omega)$ is varied from 0 to –1 in small steps. At each step Eq. (3) is solved for the complete surface polarization. The dipole moment and total expectation value of susceptibility are calculated. Peaks indicate at what values of $\chi(\omega)$ resonant modes occur. In general the wider the peak the more significant the resonant mode.

An example of the susceptibility spectrum is shown below in Fig. 1.

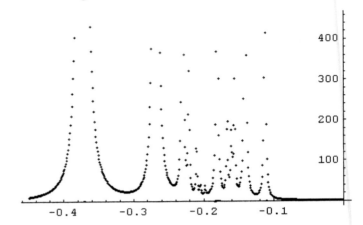

Figure 1: Susceptibility Peaks of a Cube

3 RESOLUTION DEPENDENCE

Of particular interest is the models dependence on the size of each patch (or overall resolution of the surface). As the resolution is increased computer processing time increases exponentially so there are pratical limits to the total number of patches.

Figs. 2-5 show normalized resonant peak distributions for a simple cube. Various patch resolutions are shown to illustrate changes in the resultant peaks. Fig. 2, with only 9 patches per cube side was generated in just a few minutes of computer time. Note the low frequency peak that does not show up in higher resolution calulations. This low frequency peak is diminished in Fig. 3 with 16 patches per face. The peak is probably a resonant mode reacting with the patch structure as opposed to the geometry of the cube shape.

Peaks show closer grouping in Fig. 4 with 25 patches per side. Fig. 4 took about one hour to fully process. Finally Fig. 5 ran almost four hours without yielding results greatly different from Fig. 4. For a simple cube 36 patches per side appears to be adequate. However, as more complex geometries are simulated higher resolutions may be required. For example, a cube with cut corners will need to be simulated with patch size small enough to accurately represent the cut face.

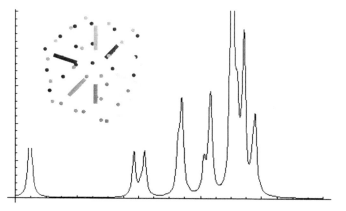

Figure 2: 9 Patches per side

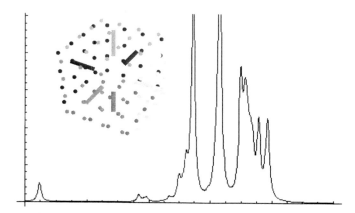

Figure 3: 16 Patches per side

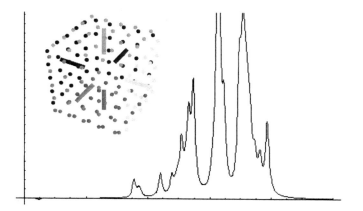

Figure 4: 25 Patches per side

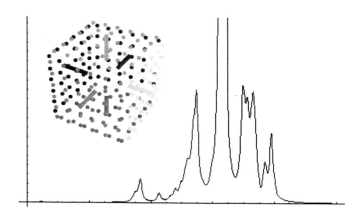

Figure 5: 36 Patches per side

(Processing times are approximate running Mathematica on a Pentium 4 desktop computer)

4 POLARIZATION MODES

As explained in section 2 each absorption peak is associated with a particular resonant mode surface polarization. Once the peaks are found graphics representing the polarization can be generated.

Fig. 5 shows one cube face for the eight most prominent resonant modes. The widest peak results from the mode labeled A to the smaller H. The spectrum for a cube shows about 10 peaks total, two are small enough to be ignored. The first two peaks represent the majority of the response. The spike associated with A is twice the size of B and B in turn is twice the size of C.

5 SOFTWARE

The software model has been developed in Mathematica version 4.2. The first section of the worksheet is dedicated to definition of the particle. The shape is defined by first entering an X, Y, and Z length and a patch resolution. A cube (that can be rectangular) is generated by the program. More complex shapes are achieved by defining planes to cut the cube. An unlimited number of planes may be defined by inputting a normal vector and a point on the plane. The cube is cut with the part in the positive vector direction being discarded. In this fashion cubes with corners or edges cut off can easily be input. Diamond, parallelogram and other shapes are also possible. The program outputs a graphic of the shape, as seen in Figs. 2-5 for verification.

Once the shape is input the program handles all calculations. The large matrix R_{ij} is generated first. The X, Y, and Z coordinates of each patch are recorded in another matrix for reference later.

A range and step size can then be defined for generating the graph of total particle susceptibility. The program loops through solving Eq. (3) for each step in material susceptibility.

Peaks are then picked out by hand and polarization plot like those in Fig. 6 may be generated.

The plots in Figs. 2-5 are generated by applying a realistic material susceptibility. The susceptibility data can be calculated from the standard single oscillator expression or taken from published experimental data such as [3].

REFERENCES

[1] R. Fuchs , Phys. Rev. B, 11, 1732, 1975.
[2] P. Mulvaney, MRS Bulletin, 26, 1009, 2001.
[3] P.B. Johnson and R.W. Christy, Phys. Rev. B, 6, 4370, 1972.

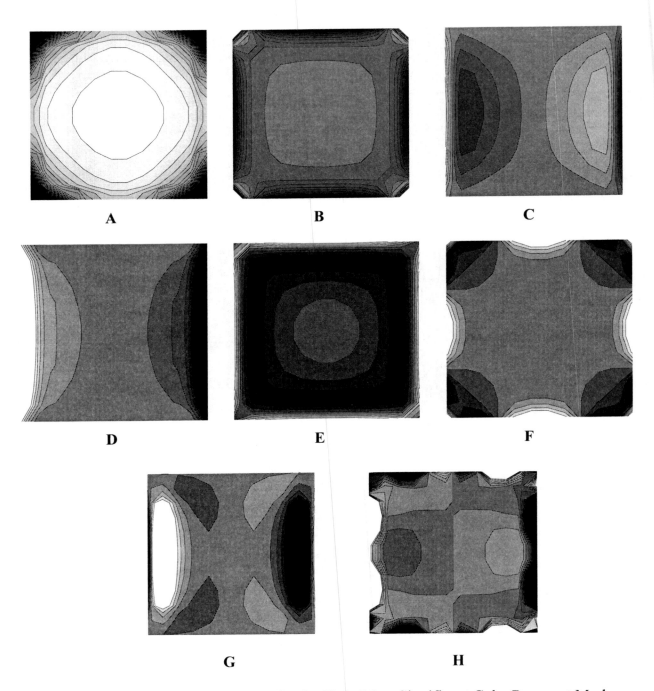

Figure 6: Surface Polarization Patterns for the Eight Most Significant Cube Resonant Modes

Thin Film Instability and Nanostructure Formation: a Molecular Dynamics Study

M. Han[*], J.S. Lee[**], S.H. Park[***] and Y.K. Choi[****]

[*]Micro Thermal System Research Center, Seoul National University, Korea, bard2@snu.ac.kr
[**]School of Mechanical and Aerospace Engineering, Seoul National University, Korea
[***]Department of Mechanical and System Design Engineering, Hongik University, Korea
[****]School of Mechanical Engineering, Chung-Ang University, Korea

ABSTRACT

One of the recent studies showed that the structures can be self-constructed by controlling the instability of thin fluid film of a nano-scale [S.Y. Chou and L. Zhuang, , *J. Vac. Sci. and Technol. B*, **17**, 3197-3202 (1999)]. The Molecular dynamics simulation is used to probe the phenomena. And two sources of instability are investigated: the normal temperature gradient and the long-range attractive potential by the wall. In the first case, the wall temperature is controlled so that the film maintains its temperature gradient. It is observed that the growth rate decreases in a monotonic fashion. The effect of the surface tension gradient, rather than evaporation, may drive the growth. Another source of instability is considered that is the interaction between the upper wall and the fluid film through a potential originating from the interaction between ion and non-polar molecule. The solid potential induces the formation of one or more vertical structures. This may result from the distortion of the pressure distribution.

Keywords: molecular-dynamics simulation, thin film, nanostructure

1 INTRODUCTION

Recently it is the subject of intense researches to build small structures of a scale less than 100 nm. Manufacturing the nanostructures with efficiency and low cost has great significance in the application of the emerging small-scale technologies; the good examples are the electronic circuits integrated in an increasingly smaller dimension or the nano-electro-mechanical systems. One of the recent studies showed that the structures can be self-constructed by controlling the instability of thin fluid film of a nano-scale[1,2]. The processes involve several physical elements, which may determine the characteristics like the structure arrangements and formation time. Previous studies on the subject rely on the instability theory of continuous media based on the long-wave approximation [3]. In this study, the Molecular dynamics simulation is used to probe the thin fluid film on a substrate and its dynamical behavior during destabilization and structure formation. Two sources of instability are investigated: one is the temperature gradient

applied in the normal direction and the other is an attractive potential by the wall in the distance.

2 RESULTS

2.1 Simulation Methods

Non-equilibrium molecular dynamics simulation is conducted [4]. The simulation system consists of the argon atoms confined by two insulated platinum walls. The argon atoms are equilibrated to form a thin fluid film on a platinum wall. The size of the computational domain is W*2.4*29.4 nm^3, where that of width, W, is varied from 2.4 to about 136 nm. The depth is set to be a smallest size possible, 7.06 nm, to minimize the state variation in the direction as in the Hele-Shaw Cell. The argon atom interacts with other argon and platinum atoms through 12-6 Lennerds-Jones potential.

$$u(r) = 4\varepsilon \left[\left(\frac{\sigma}{r} \right)^{12} - \left(\frac{\sigma}{r} \right)^{6} \right] \tag{1}$$

u is the interaction energy between two molecules that are separated by the distance of r when they have the energy parameter ε and intermolecular diameter σ. And the interaction is cut off in the distance of 1.02 nm. The walls are of a f.c.c. lattice in which the platinum atom interacts with others with a harmonic potential. The temperature of the system is controlled by the particles embedded in the farthest rows in the walls from the inside. They observe the Langevin Equation whose parameters are given according to a known phonon characteristics [5].

2.2 Instability by temperature gradient

The temperature of the wall the thin film is on is controlled to induce the instability. And the film maintains its temperature gradient of about 150.4 degrees per nanometer in a relatively constant manner except for an initial time (~0.5 ps), when it increases up to about 225.2 degrees per nanometer. These are about the maximum possible amount in the given system. A more amount of gradient induces the evaporation at the solid-liquid interface. The growth rate of the fluctuations in the early stages (0.3<

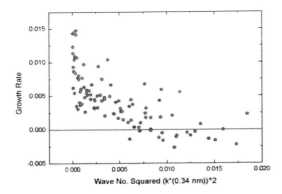

Figure 1. Growth rate of fluctuations vs. wave no. squared

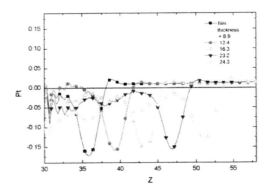

Figure 3. Tangential pressure distribution according to the interfacial postion (ε_{lr} =1; The data are smoothed with 10 points by Savitsky and Golay's method. All scales are reduced in Argon's units)

t <0.75 ps) is observed. According to the continuum theory based on the long-wave approximation, the growth rate has a parabolic behavior with a peak with respect to the wave number squared. In the measured results in Fig. 1, it decreases in a monotonic fashion. Two thing may contribute to this discrepancy: one is from the fact that the major group of waves in the figure are not appropriate for the long-wave approximation. The other is that the amplitude of capillary waves in the nano-scale may increase during the initial stage. And the growth of the lower wave numbers continue in the later while that of the higher quickly decays. This instability may mainly come from the surface tension gradient, due to the temperature, with respect to the interfacial position. The vapor-coil effect from the evaporation may not contribute any noticeable effect. Finally, even with the large temperature gradient applied, any fluctuation does not grow up to an enough height to build a structure with the wall in the distance.

2.3 Instability by solid potential

Another source of instability is considered in the context of the structure formation. The interaction of a longer range than that of Van der Waals kind between the upper wall and the fluid film may have the significant effect on the film evolution. This contribution is implemented as a solid potential acting on the fluid atoms. The potential relevant for the interaction between ion and non-polar molecule is integrated in the solid domain to give the following expression.

$$potential(z_o) = \varepsilon_{lr} \frac{\pi}{z_o} \rho \tag{2}$$

ε_{lr} is the proportionality that represents the strength of the interaction. z_o is the distance between the argon atom and the solid surface. ρ is the mean number density of the solid. The coulomb interactions among fluid atoms are assumed to of a negligible influence compared to the Van der Waals. This case simulate the system where the upper wall is composed of ionic atoms and induces the dipoles on the non-polar fluid atoms. In the given values of ε_{lr} (=0.5~2), it is observed that the solid potential induces the formation of one or more vertical structures. (Fig. 2) This may be the direct result of the distortion of the pressure distribution. (Fig. 3) As the position of the liquid-vapor interface gets close to the upper wall, the tensional stress (the negative pressure) in the direction tangential to the interface become stronger. This phenomenon is usually

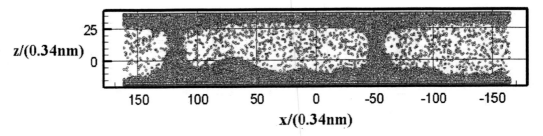

Figure 2. Vertical structure formation by solid wall potential ($\varepsilon_{l\_}$=1, t=0.4 ns)

treated as the concept of "disjoining pressure" in the continuum model.

The time that it takes for the fluctuation to reach the upper wall for the first time tends to decrease as the strength of the potential increases. (Fig. 4) The linearly decreasing dependence of the slope in the figure corresponds to the prediction of the continuum model that is based on the long-wave approximation. On the other hand, it can be observed that the spacing of neighboring structures is inversely proportional to the strength. However, more quantitative result for the trend is not pursued because of the limitation in the computational size of the problem.

3 CONCLUSION

In this work, the instability of thin fluid film and the nanostructure formation is studied by the Molecular dynamics simulation. Liquid argon film is located on the platinum substrate and two physical conditions are applied to induce the vertical structures that are observed in the literatures. The primary findings are summarized as follows

1) When a large thermal gradient is applied to the fluid film of a nano size, the instability that is observed in the larger scale is still realizable. And the surface tension gradient due to the temperature, that is, the thermocapillary effect, was the major source in the simulation.

2) It is shown that the interaction energy of the solid whose range is longer that the Van der Waals is very effective in inducing the instability of the thin film and vertical structures formation between the solid walls. The formation time is inversely proportional to the strength of the solid potential.

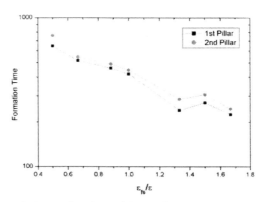

Figure 4. Formation time of the vertical structure vs. the relative strength of the solid potential

4 ACKNOWLEGEMENT

The authors gratefully acknowledge that this work is supported by Micro Thermal Research Center.

REFERENCES
[1] S.Y. Chou and L. Zhuang, J. Vac. Sci. Technol. 17, 3197, 1999.
[2] E. Schaffer, T. Thurn-Albrecht, T.P. Russell, and U. Steiner, Nature 403, 874, 2000.
[3] L. Wu, and S.Y. Chou, Appl. Phys. Lett. 82, 3200, 2003.
[4] D.C. Rapaport, "The Art of Molecular Dynamics Simulation," Cambridge, 1995.
[5] M.P. Allen and D.J. Tildesley, "Computer Simulation of Liquids," Oxford, 1987.

Design and Development of New Nano-Reinforced Bonds and Interfaces

Y. Sun[*], S.A. Meguid[*], K.M. Liew[**] and L.S. Ong[**]

[*]Department of Mechanical and Industrial Engineering
University of Toronto, 5 King's College Road
Toronto, Ontario M5S 3 G8, CANADA
meguid@mie.utronto.ca

[**]School of Mechanical and Production Engineering
Nanyang Technological University
Nanyang Avenue, Singapore 639798

ABSTRACT

This investigation focuses on the design and development of nano-reinforced adhesively-bonded high performance joints. The influence of homogeneous dispersion of carbon nanotubes and alumina nanoparticles on the cohesive strength, interfacial properties and mechanical behaviour of epoxy adhesives is examined. Mechanical tests and optical measurements are conducted, using carbon fibre/epoxy laminate and aluminium alloy adherends with high performance epoxy adhesives, to evaluate the optimum weight percentage of the nanofillers to be added to induce maximum enhancement in strength and strain energy release rate. The results reveal that the introduction of nanofillers into the epoxy adhesive leads to a dramatic improvement in the debonding and shear strength of the interface without sacrificing its toughness. The presence of nanoparticles plays a major role in determining the fracture toughness of the interface and is typically associated with a significant improvement in the delamination resistance of the joint.

Keywords: nano-reinforced interfaces; carbon nanotubes; alumina nanopowder.

1. INTRODUCTION

Nanomaterials possess high strength and hardness, and excellent ductility and toughness. Undoubtedly, more attention is being paid to the applications of nanomaterials nowadays. Recent works ([1]-[4]) show that the bulk properties of materials can be improved by adding nanoparticles. Gadakaree [5] improved the mechanical properties of fibre-reinforced composites by improving the toughness of the glass matrix with micro-scale particle fillers. Liu and Xu [6] et al. studied the influence of the addition of nano TiN on TiC grains and showed that a wide distribution of nano TiN at the interface of TiC/TiC grains leads to improvement of mechanical properties when the nano additions are below 6 wt.%. Ramanan Krishnamoorti [7] studied the shear properties of polymer-layered silicate nano composites. Sklep [8] successfully prepared a series of polymer/clay nano composites with the layered silicate clays as the inorganic dispersed phase via intercalation polymerisation. Young and Wang ([9]-[10]) also studied the effect of filler particles in rigid reinforced resin composite systems and observed the toughness improvement. The influence of nanofillers on interfacial properties and bonding region in two-phase materials is not well understood [11]. The authors are unaware of any work that deals with the change in fracture toughness and energy release rate of composite interfaces as a result of dispersing nanoparticles into epoxy adhesives.

The work described herein is devoted to the characterization of the tensile and shear properties, fracture toughness and energy release rate of adhesively bonded interfaces reinforced by nanoparticles. SEM and TEM were also used to elucidate some of the findings.

2. EXPERIMENTAL DESCRIPTION

2.1 Materials and Sample Preparation

Unidirectional carbon fabric (G1157/EFP 100) provided by Hexcel Fabrics was formed into laminate for one of the substrates by autoclave. The carbon fabric has a nominal weight per unit area of 290 g/m^2, and its laminate has a fibre volume fraction of 55 %. The respective ultimate tensile, compressive and shear strength of the laminate are 1200 MPa, 155 MPa, and 57 MPa. The tensile modulii E_{11}= 180 GPa and E_{22}= 12 GPa, shear modulii $G_{12} = G_{13} = 28$ GPa, and Poisson's ratio = 0.263. Aluminium 6061-T6 is used as another substrate, and is chosen for its excellent joining characteristics, good acceptance of coatings, and a good combination of relatively high strength, good workability and high corrosion resistance. It has tensile yield strength of 275 MPa, and an elastic modulus of 69 GPa.

A two-component paste epoxy adhesive (Dexter Hysol EA 9330) is used in this study. This adhesive system possesses high peel strength and excellent environmental resistance, and bonds to a variety of substrates.

Nanofillers

Single-walled carbon nanotubes (SWNTs) were used as fillers in the epoxy adhesive, and they are synthesized by the electric arc technique. Fig. 1 shows the SEM image of a single wall carbon nanotubes. It can be seen that the nanotubes have diameters between 30-60 nm and are a few hundred nanometers long.

Another filler is alumina nanofibre powder, which is approximately 2-4 nm in diameter and a few hundred nanometers in length. The surface area of the alumina nanopowder ranges from 300-700 square meters per gram. Fig. 2 shows TEM images of the fibre powder. As can be seen, they are agglomerated in the longitudinal direction. It can also be seen that the alumina nanofibres have aspect ratios greater than 20 and many of the fibres appear to be hundreds of nanometers long.

Preparation of adherends
The two dissimilar aluminium and carbon fibre adherends were machined from their respective sheets. The carbon fibre reinforced laminates were fabricated by autoclave at specified pressure and temperature. The surface was sand-blasted, cleaned by chemical solvent and dried before bonding. Carbon nanotubes and alumina nanopowders were respectively dispersed into the above-mentioned epoxy adhesive. The dispersion was carried out by stirring the mixture for 30 minutes at 50°C, and then adding and blending the remaining components of the adhesive. This ensured the homogeneous dispersion of the nanofillers into the adhesive. Fig. 3 shows typical micrographs of the epoxy resin with uniformly dispersed carbon nanotubes; the weight concentration in these two cases was 2.5%. Additional nano mixtures with nanofillers at nominal weight percentages of 1.5%, 5%, 7.5%, 10%, 12.5% and 15% were also prepared. The substrates were then bonded together and carefully cured at a controlled room temperature for 7 days.

2.2 Details of Testing

ASTM standard 4541-95 was adhered to during mechanical testing. The tests were carried out using an electro-hydraulic servo-controlled machine at a crosshead speed of 1.0 mm/min. It is worth noting that the load co-axiality was ensured using special crosshead adapters and strain gauged samples. Single-lap tests were carried out as per the ASTM D3162-92 standard.

Double Cantilever Beam (DCB) specimens were used to characterise Mode I interfacial strain energy release rate G and fracture toughness G_{Ic}. The specimen contains a non-adhesive polytetrafluoroethylene insert on the interfacial plane, which serves as a delamination initiator. Opening forces are applied to the DCB by means of upper and lower spherical joints bonded to one end of the specimen. The tests were conducted under displacement control on an Instron universal testing. Cross head motion was set at 0.50 mm/min. The specimens are 135 mm long and 25 mm wide, and the test procedure followed ASTM D5528-94a. During the test, a curve of the applied load versus opening displacement is recorded and instantaneous delamination front locations are marked on the chart at different intervals

of delamination growth. Modified Beam Theory (MBT) was used to calculate the energy release rate, such that G_I:

$$G_I = 3P\delta / [2b (a + \Delta)] \tag{1}$$

Where: P is load, δ is load point displacement, b is specimen width, a is delamination length and Δ is a correction factor for rotation at the delamination front, determined experimentally by generating a least squares plot of the cube root of compliance.

3. RESULTS AND DISCUSSIONS

Tensile and shear specimen
Testing results are recorded and compared for three cases: EANP (Epoxy Adhesive with alumina nanopowder), EANT (Epoxy Adhesive with carbon nanotube) and EA (Epoxy Adhesive). Figs. 4(a) and 4(b) show the improvement in Young's modulus and ultimate tensile strength for the cases involving different weight fractions of homogeneously dispersed nanofillers. The increase continues with the increasing weight percentage of the nanofillers, up to a limit of 10%, above which the properties degrade to below those of the EA. These results indicate the sensitivity of the epoxy to the concentration of the dispersed nanofillers.

Fig. 4(c) shows the resulting shear strength for the cases involving different concentrations of dispersed alumina nanopowders and carbon nanotubes. Analogous to the tensile tests, the results also reveal the sensitivity of the shear properties to the concentration of the nanofillers. An increase in the weight fraction of the nanofillers beyond 7-8% results in a reduction in the shear properties of the adhesive.

Energy release rate
Fig. 5(a) shows the energy release rate for the three different cases examined: EA, EANT (2.5% of nanotubes) and EANP (2.5% nanopowder). Compared with bonds containing nanofillers (EANT and EANP), the energy release rate of a joint composed solely of epoxy adhesive (EA) has lower values for G_I for the same delamination length. For nano-reinforced interfaces, the energy release rate G_I is higher, and the improvement is quite significant after the delamination length reaches 75mm. The energy release rate for the carbon nanotube case is on average 16% higher than that of corresponding to pure epoxy adhesive, at a delamination length of 95mm. The interfaces reinforced by alumina nanopowder show even better results at this delamination length, with a 34% improvement in G_I over the unreinforced case.

Effect of weight percentage of nanofillers on G_{IC}
Fig. 5(b) shows the normalised energy release rate G_{IC} for different weight fractions of homogeneously dispersed nanofillers for DCB specimens bonded with prepreg glass reinforced laminate and aluminium sheet. The energy release rates show significant improvement with increasing weight percentage of nanofillers and achieve the best improvement when the weight percentage of fillers is

between 2-5%. However, for percentages above 10%, the properties degrade to below the EA, indicative of the sensitivity of the epoxy adhesive to the concentration of the dispersed nanofillers. The use of alumina nanopowder appears generally more beneficial than the use of nanotubes in the reinforcement of interfaces.

It is worth noting that at a given weight percentage, the presence of nanoparticles plays a major role in determining the strength and toughness of the interface. The phenomenon can be explained as follows: The nanofillers are characterized by large surface areas per unit gram. As the number of adhesively joined sites increase, the cohesive strength of the epoxy increases leading to a higher mechanical strength of the interface. The experimental results show that there is a limit to the number of dispersed nanofillers beyond which a drop in the properties is observed. Once the nanoparticles fully fill the gaps and porosities and all contact points are established, the additional particles cannot interact effectively within the epoxy adhesive and consequently poor matrix infiltration occurs. It is also believed that agglomeration of the nanoparticles may act as failure initiation sites, which results in lowering the strength and toughness of the adhesive.

4. CONCLUSIONS

The influence of the nanofillers in a special epoxy adhesive for the purpose of increasing bond strength and toughness is studied. Two different types of nanofillers were used: carbon nanotubes and alumina nanopowder. The work concentrated on the experimental determination of the tensile, shear and delamination properties of the nano-reinforced interface. The results reveal that at a given weight percent, the presence of nanoparticles plays a major role in determining the strength of the interface. This is attributed to the large surface area and the consequent interlocking of the nanofillers with the epoxy adhesive. Experimental results also show that there is a limit to the number of dispersed nanofillers beyond which a drop in the properties is observed.

REFERENCES:
[1] Niihara K., and Nkahira A., Advanced Structural Inorganic Composites, Elsevier Scientific Publishing Co. Trieste, Italy, 1990.
[2] Zhang Z., and Cui Z., Nanotechnology and Nanomaterials, Defence Technology Press. Beijing, 2001.
[3] Kim J.K., and Mai Y.W., Engineered Interfaces in Fibre Reinforced Composites, Chapter 7. New York, Elsevier, 1998.
[4] Hodzic A., Kalyanasundaram S., Lowe A.E., Stachurski Z.H., Composite Interfaces, 4, 375,1991.
[5] Gadakaree K.P., J. Mater. Sci. 27, 3827,1992.
[6] Liu N., Xu Y.D., Li H., Li G.h. and Zhang L.D., Journal of the European Ceramic Society, 22, 2409-2414, 2002.

[7] Krishnamoorti R., Vaia R.A., and Giannelis E.P., Chem. Mater. 8, 1728-1734,1996.
[8] Sklep, Advanced polymer/clay nanocomposites, report of state key laboratory of engineering plastics, Chinese Academy of Sciences, 2000.
[9] Spanoudakis J, and Young RJ, Journal of Materials Science, 19, 473-486, 1984.
[10] Wang S.Z, Sabit A. and Bor Z. J, Composites Part B: Engineering, 28, 215-231, 1997.
[11] Hodzic A., Stachurski Z.H., Kim J.K., Polymer, 41, 6895-6905, 2000.

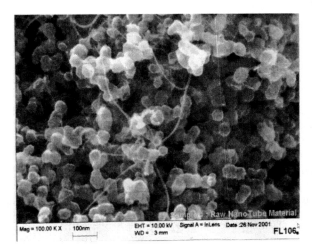

Fig. 1 SEM images of single wall carbon nanotubes

Fig. 2 TEM micrographs of alumina nanofibre powder

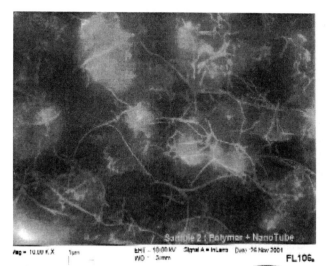

Fig. 3 SEM image of carbon nanotubes dispersed into epoxy adhesive

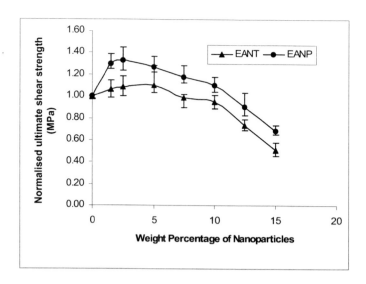

Fig. 4(c) Normalized ultimate shear strength for different weight fractions of nanofillers.

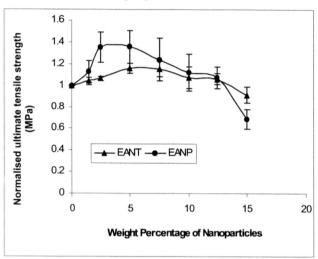

Fig. 4(a) Normalized ultimate tensile strength for different weight fractions of nanofillers.

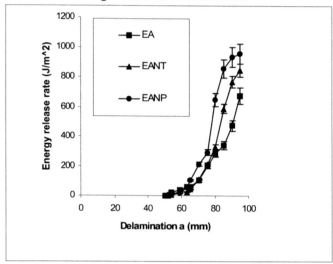

Fig. 5(a) Energy release rate vs. delamiantion

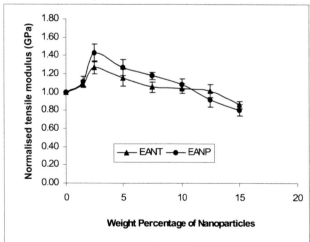

Fig. 4(b) Normalized tensile modulus for different weight fractions of nanofillers.

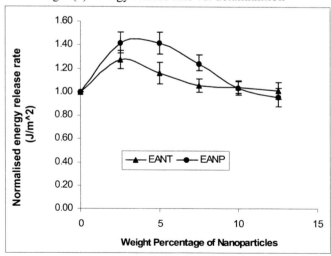

Fig. 5(b) Normalized energy release rate for different weight fractions of nanofillers.

Numerical Simulation of electronic properties in quantum dot heterostructures

B. Vlahovic, I. Filikhin, V. M. Suslov, and K. Wang
Department of Physics, North Carolina Central University
Durham, NC 27707

ABSTRACT

A confined structure in all three dimensions leads to carrier's discrete energy level spectrum in quantum dots. This property has profound impact on many applications, such as single electron transistors, quantum dot laser, high efficiency photovoltaic cells, information storage etc. A finite element method is utilized to model the residual stress distribution. The effect of residual stress on the electronic and optical properties is studied. This is accomplished by incorporating both the valence subbands and the strain-induced potential field into Schr dinger equation. A finite-difference method was applied to solve the equation system. The density of states is obtained from the spectrum of the eigenstates. The discrete eigenstate distributions for both with and without residual stress are compared. The effect of the quantum dot size and geometry to the energy state distribution is discussed.

Keywords: Numerical Modeling, Quantum Dots, Residual Stress, Electronic State

INTRODUCTION AND METHODS

Quantum dots have been the research topics of many publications because of their unique electronic properties. Many research papers involving both experiments and theoretical modeling have been published since the early 1990s. Publications on quantum dots prior to 1998 have been reviewed by Bimberg et al. [1].

Quantum dots are usually fabricated by growing nanometer sized materials on various substrates. The differences in material's properties between the quantum dots and the substrate material, such as lattice parameters and thermal expansion coefficients, will generate residual stress in the quantum dots. The residual stress affects the electrical, optical properties, and device lifetime. The theoretical simulations in quantum dots can help to understand various effects that will have impacts on the electronic properties. Many approaches have been applied in the quantum dot simulations. Besides the dependence on the material properties, the electronic properties of quantum dots also strongly depend on the quantum dot size and geometry. The complex geometry of quantum dots and the anisotropy of materials often cause simulation challenges. Numerical methods, such as finite element and finite difference method, are avenues to model the distribution of the lattice mismatch and thermal residual stress, their effects on the electronic, transport, and photovoltaic properties in quantum dots. Johnson et al. [2,3,4] have applied finite element modeling to simulate the residual stress distribution and its effects on electrical and optical properties of quantum dots.

Due to the strong dependence of energy levels on the geometry and sizes, the quantum dots are perfect candidates for high efficiency photovoltaic cells. Solar cells with 63% efficiency have been predicted by theoretical work [5,6]. Recently, we have developed numerical computer programs for simulation of residual stress distributions and its effects on the electrical, transport, and optical properties in the quantum dots [7,8]. A finite element method has been utilized to model the residual stress distribution. The effects of lattice mismatch, thermal expansion, and geometry of the sample have been systematically studied.

A finite difference method was applied to model the electronic properties. The effects of residual stresses on the electronic and optical properties are also studied. This was accomplished by incorporating both the valence subbands and the strain-induced potential field into the time-independent Schr dinger equation

$$\sum_{\beta=1}^{4} H_{k.p}^{\alpha\beta} \psi^{\beta} + V^{\alpha\beta} \psi^{\beta} = E \psi^{\alpha} \tag{1}$$

where ψ^{α} is the wave function in subband a, E is the energy, $H_{k.p}^{\alpha,\beta}$ is the **k.p** Hamiltonian operator, and $V^{\alpha\beta}$ is a potential.

$$V^{\alpha\beta} = V_C^{\alpha\beta} + V_\varepsilon^{\alpha\beta} \tag{2}$$

where V_C is due to the valence-band alignment of material at a given position in device and V is strain-induced potential. Finite-difference method was then applied to solve the equation system[8]. Description of the algorithm can be found in our earlier paper[8]. The density of states are obtained from the spectrum of the eigenstates in the numerical solutions.

RESULTS

Calculations have been performed on a series of quantum dots with different size and geometry. Figure 1 provides an example of the strain distribution in a stand alone Ge quantum dot grown on a Si substrate. Growth temperature is assumed to be 1000K. The diameter of the quantum dot is 20 nm. For the system with axial symmetry, only three components of the strain tensor are needed. As it is shown, the residual strain is mainly concentrated at the interface and the edge of the quantum dot.

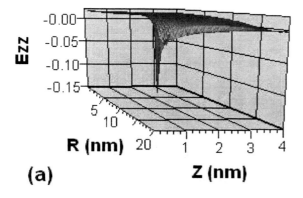

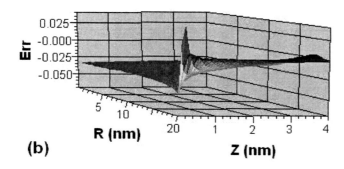

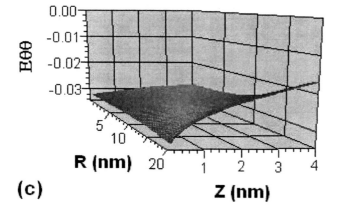

Figure 1. Distribution of residual strain in Ge Quantum Dot

Figure 2 provides the distribution of energy levels for cone shape quantum dots with different diameters. The discrete distributions of the energy levels and the effects of quantum dot size can be clearly observed

We have also studied the effects of different geometries. The energy level distributions for two quantum dots with different shape are provided. Figure 3(a) is for a conical quantum dot with a diameter of 40 nm and a height of 4

nm. Figure 3(b) is for a pyramidal shape studied by Johnson et al. [3]. Their results are redrawn here for comparison. The impact of residual stress can be clearly observed. The geometry of the quantum dot has shifted the energy distribution significantly.

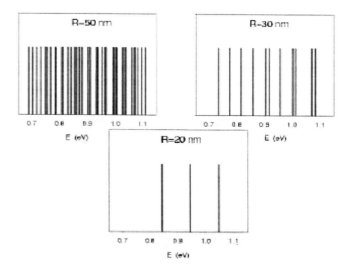

Figure 2. Distribution of energy levels for cone shape quantum dots with diameters of 40nm, 60nm and 100nm

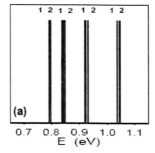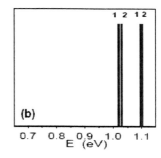

Figure 3. Effects of strain on the confined energies for cylindrical and pyramidal quantum dots. (a) is for cylindrical geometry and (b) is for pyramidal geometry. 1s indicate the energy level without the effects of strain and 2s when the strain potential is taken into account.

Figure 4 illustrates the effects of residual strain on the wave function for energy level E_{13} of a cylindrical Ge quantum dot with a radius of 20 nm. Where the first index "1" in E_{13} would indicate the first energy level or ground state in z direction and the second index "3" would indicate the third energy level or the second excited state in r direction, if z and r variables would be possible to be separated. In our case they are not separable because of the residual strain potential. However, the strain potential is a perturbation of our problem defined by equation (1) - (2) and in this approach the nomenclature is appropriate. The wave function displayed in Figure 4(c) is for the quantum dot with strain field included in the calculation. Figure 4(a) and 4(b) are the wave function distribution viewed from the angle along the radius direction inside the quantum dots. Figure 4(a) is the wave function without the effect of the strain and figure 4(b) is the one with strain effects. The differences can be clearly observed in figure 4.

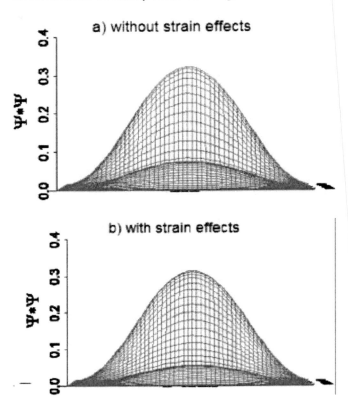

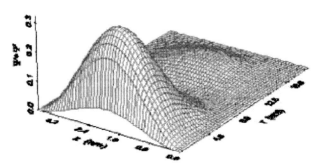

Figure 4. Square wave functions for a cylindrical quantum dot.

CONCLUSIONS

The time independent 3D Schrödinger equation is solved using finite element method. The effects of the shape and quantum dot size on the electron wave function and electron energy level distribution inside quantum dots are demonstrated for conical, pyramidal and cylindrical quantum dots structures. The comparison of wave functions and eigenstates for calculations with and without the strain potential demonstrate the significance of the residual stress impact. The effects of size, confinement, and thermal stress which is related to the material composition can be seen directly from the energy level distribution. Since the density of the states can be obtained from the spectrum of the eigenstates, the effect on the optical properties and charge transport can be estimated. The impact of the considered parameters is significant, consequently the accurate calculations should be made when specific device is to be considered for potential application.

REFERENCES

[1] D. Bimberg, M. Grundmann, and N. N. Ledentsov, Quantum Dot Heterostructures, John Wiley & Sons, New York, 1999.
[2] Johnson, H. T., L. B. Freund, C. D. Akyüz, and A. Zaslavsky, Journal of Applied Physics, 84, 3714-3725(1998).
[3] Johnson, H. T. and L. B. Freund. International Journal of Solids and Structures, 38, 1045-1062(2001).
[4] Freund, L. B. and H. T. Johnson, Journal of the Mechanics and Physics of Solids, 49, 1925-1935(2001).
[5] A. Luque and A. Marti, Phys. Rev. Lett. 78, 5014-5017(1997).
[6] M. Yamaguchi and A. Luque, IEEE Trans. Electron Devices 46(10), 2139-2144(1999
[7] B. Vlahovic, K. Wang, V.M. Suslov, S.L. Yakovlev, C. Otieno, A. Soldi, and J. Dutta, Residual Stress Modeling In Quantum Dots, Proceedings of the 2003 Nanotechnology Conference, Nanotech 2003.
[8] B. Vlahovic, V. Suslov, and K. Wang, Numerical Simulations of Residual Stresses and Their effects on Electronic Properties in Quantum Dots, Technical Proceedings of the 2003 Nanotechnology Conference, Nanotech 2003.

Molecular Dynamics (MD) Calculation on Ion Implantation Process with Dynamic Annealing for Ultra-shallow Junction Formation

Ohseob Kwon, Kidong Kim, Jihyun Seo, and Taeyoung Won

Department of Electrical Engineering, School of Engineering, Inha University
253 Yonghyun-dong, Nam-gu, Incheon, Korea 402-751, E-mail: kos@hsel.inha.ac.kr

ABSTRACT

In this paper, we report a molecular dynamics (MD) simulation of the ion implantation for nano-scale devices with ultra-shallow junctions. In order to model the profile of ion distribution in nanometer scale, the molecular dynamics with a damage model has been employed while the kinetic Monte Carlo (KMC) diffusion model was used for the dynamic annealing between cascades. The distribution of dopants during the ion implantation was calculated from the MD approach. The calculation has been performed for B with energies down to 100eV and dose 1×10^{14} ions/cm^2. The B, As, and Ge implant has been simulated with the energies of 0.5, 1, 2, 4, 8, and 16 keV and with dose 1×10^{14} ions/cm^2 into Si <100>, respectively.

Keywords: molecular dynamics, kinetic monte carlo, ion implantation, diffusion, dynamic annealing

1 INDRODUCTION

Ion Implantation induced dopant and damage distribution is of major process among the silicon technology. A need for predictive range profile in the area of ion implantation modeling puts much emphasis on accurate physical modeling of the implant process for nano-scale devices. Monte Carlo methods based on BCA (Binary Collision Approximation) have been used in the past with describing many physical mechanisms. However, although the multiple interactions must be taken into account for ultra-low energy ion implantation, BCA has shortcomings because the basic assumption is the binary collision in which many factors for ion implantation is difficult to be illustrated realistically. Despite molecular dynamics (MD) approach requires too excessive computation time, an accurate distribution of dopants can be expected quite during the ion implantation can be calculated by using appropriate functions for inter-atomic potentials [1,2]. In addition, it is not clear whether ion-beam induced annealing plays a role under ion implantation. In this paper, we report our study on the ion implantation process for ultra-low energy by using MD method for describing dynamic annealing for dopant redistribution and defect recombination within a recoil cascade.

2 SIMULATION MODEL

In order to model the ultra shallow junction, a recoil interaction approximation was applied for ion implantation (RIA). The Ziegler-Biersack-Littmark (ZBL) potential model, Eq. (1), has been used for the interaction among atoms. In order to model the electronic stopping power, the density functional theory by Echenique *et al* [3] was implemented in this work. Furthermore, the Firsov model was employed in order to model the energy loss during the inelastic collisions [4].

$$E_{ZBL} = V(r_{ij}) = \frac{Z_1 Z_2 e^2}{R_{ij}} \sum_{k=1}^{4} c_k \exp(-d_k \frac{r_{ij}}{a}), \qquad (1)$$

The distribution of the concentration of dopants was calculated using the environment-dependent inter-atomic potential (EDIP). The critical step is performed such as following steps for the simulation of ion implantation process using MD; a) the initial configuration including position and velocity of recoil atom is defined. b) the potential is calculated between recoil atom and lattice atoms. Finally, c) the next configuration is calculated.

The lattice atoms may be annealed by ion bombardment during ion implantation. The ion-beam induced annealing affects dopant redistribution and defect recombination within a recoil cascade. The dynamic annealing during ion implantation is effective as much as annealing after implantation. The dynamic annealing may be explained the defect recombination within a recoil cascade, the stabilization of defects when overlapped by subsequent collision cascades, and the results in a lower overall number of defects. For the consideration of dynamic annealing during ion implantation and an accurate model for diffusion of intrinsic point defects (I, V) and impurities (B) in ion-implanted silicon, our approach is based upon the coupled calculation of MD and KMC. From the atomic distribution during the ion implantation from MD simulation, the dynamic development (hopping) of impurities and defects are calculated using KMC. The KMC simulation is interactively performed with the results of the MD simulation.

In this type of KMC model, point defects and dopants are treated at an atomic scale while they are considered to diffuse in accordance with the reaction rates, which are given as input parameters [5]. The input parameters can be obtained from either first-principles calculations or classical MD simulation, or experimental data. Especially, the formation of clusters and extended defects, which usually control the annealing kinetics after ion implantation, should be minimized in the range of low dose in an effort to create dilute concentrations of I and V. Therefore, a simple kick-out mechanism has been tested and a good agreement with the experimental data [6] was verified in this condition. However, a more recent model, interstitialcy mechanism, is preferred when compared to the traditional kick-out mechanism by *ab initio* molecular dynamics [7].

An atomistic diffusion mechanism involving fast-migrating intermediate species of the form is proposed. The reactions $X_s + I \Leftrightarrow X_m$,and $I + V \Leftrightarrow 0$ are essentially diffusion limited, with capture radius of second neighbor atomic length (3.84Å) and direction of particle migration is limited to six neighbor sites. Here, X_s is the immobile substitution impurity, which through reaction with a self-interstitial (I) forms a fast-migrating species X_m which diffuse at a rate D_m. The diffusion rate is form in Arrhenius type ($D_0 exp(-E_{act}/KT)$). KMC simulations are performed by using the damage profiles and defects distribution from the MD simulations. The KMC is an event-driven technique, i.e., simulate events at random with probabilities according to the corresponding event rates. In this way, it self-adjusts the reasonable time step as the simulation proceeds.

3   RESULTS AND DISCUSSION

Fig. 1 shows the calculation results for B implant with the energies of 1, 3, and 5keV with the dose of $1 \ 10^{14} ions/cm^2$ into Si<100>. The tile and rotation angle is 0'. In case of 1keV, the peak is shown near the surface to the energy of 5keV, but on the other hand the end of range is deeper than the energy of 1keV. The mean range of ion implantation with the energy of 1keV is 27Å because the most part of implantation energy is lost due to the bombardment with surface atoms. The max range of ion implantation with the energy of 5keV is 870Å, which is about 4.3 times deeper than the ion implantation with the energy of 5keV. The difference may be explained that the surface is more amorphized, followed by the more energy loss than the ion implantation with the energy of 1keV. In case of Fig. 2, B implant with the energy of 5keV with the dose of $1 \ 10^{14} ions/cm^2$ and $1 \ 10^{15} ions/cm^2$ into Si<100>, small dose dependence is only shown. However, if there is more dose difference, the results similar with Fig.1 may be shown.

Fig. 3 shows the simulation results for B implantation with for energies down to 100eV below 1keV, and Fig. 4 shows the mean range and the sputter energy per ion. In Fig.3, the most atoms are within the range of 100Å. The differences of peak values are shown as the implantation energy increases. In Fig. 4, the linear difference of mean range below the energy of 1keV is shown as the implantation energy increases. The effect of surface amorphization is little more than above the range of implantation energy of 1keV.

Fig.5, 6, and 7 were performed on a Si target of <110> crystal direction at a temperature of 300K. Dopants and damage profiles were simulated for B, As, and Ge ions with the dose of $1 \ 10^{14} ions/cm^2$. Fig.5 shows the simulation results with the energy of 0.5, 1, 2, 4, 8, 16keV B implant. In the case of Fig. 6 and Fig. 7, the implant ions are As and Ge, respectively. As boron ion dose increases, local damage accumulation affects the dopant distribution more appreciably in the case of ultra-low energy ion implantation. In other words, the channeling tail drops very steeply with the increase of the amount of dose. In the case of As, the channeling tail drops very steeply with depth. This phenomenon is due to the large atomic mass of As.

Fig. 8 shows the simulation results with the energy of 8keV B implant into Si after 500 and 4,000 B ions. The KMC simulation is performed by using 3D dopant distribution and defects. Fig. 9 shows the result of B implantation with/without the dynamic annealing. In Fig. 6, the mean range with/without the dynamic annealing is 719.4 ±13.3Å and 241.8±4.4Å, respectively. The peak value is decreased and the concentration at the tail increases due to the effect of dynamic annealing. Table 1 illustrates the parameters of EDIP for Si.

4   CONCLUSIONS

The simulation of low-energy ion implantation for B, As, and Ge has been performed to energy range as low as 100eV by using the MD method. In case of B implantation, the characteristics of the mean range, max range, and sputtered atoms is investgated, followed by showing the difference of simulation results as the range of implantation energy. In the below range of implantation energy of 1keV with dynamic annealing, the more accurate potential for describing the mechanism of ultra-low energy ion implantation may be needed.

ACKNOLEDGEMENT

This work was supported partly by the Ministry of Information & Communication (MIC) of Korea through Support Project of University Information Technology Research Center (ITRC) Program supervised by IITA (Institute of Information Technology Assessment).

REFERENCES

[1] J. Nord, *et al,* Phys. Rev. B (2002).

[2] J. Peltola *et al.*, Nucl. Instr. Meth. Phys. Res. B, (2002).

[3] P. M. Echenique *et al.*, Appl. Phys. A 71, 503 (2000).

[4] P. Keblinski *et al,* Phys. Rev. B 66, 4104 (2002).

[5] Martin Jaraiz, Atomistic simulations in Materials Processing.

[6] N.E.B. Cowern *et al*, Phys Rev Lett Vol 69 (1992).

[7] Paola Alippi *et al*, Phys Rev B, Vol 64 (2001).

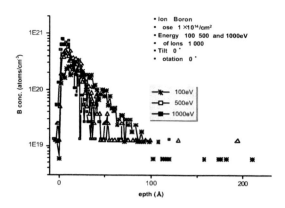

Fig. 3 A plot showing the simulation results for B implant with the energies of 100, 500, and 1,000eV and the dose of 1 10^{14}ions/cm^2 into Si<100>.

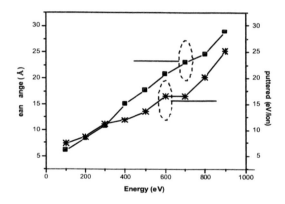

Fig. 4 A plot showing the mean range and the sputtered atoms for B implant with the energies of 100, 500, and 1,000eV and the dose of 1 10^{14}ions/cm^2 into Si<100>.

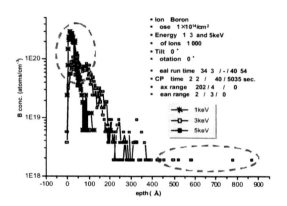

Fig. 1 A plot showing the simulation results for B implant with the energies of 1, 3, and 5keV and the dose of 1 10^{14}ions/cm^2 into Si<100>.

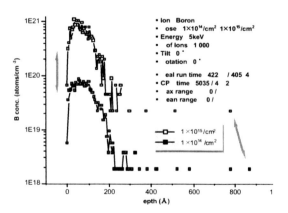

Fig. 2 A plot showing the simulation results for B implant with the dose of 1 10^{14}ions/cm^2 and 1 10^{15}ions/cm^2 and the energy of 5keV into Si<100>.

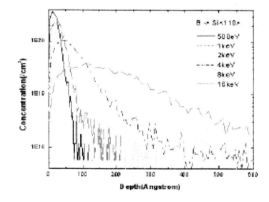

Fig. 5 A plot showing the simulation results for B implant with the energies of 0.5, 1, 2, 4, 8, and 16keV and the dose of 1 10^{14}ions/cm^2 into Si<110>.

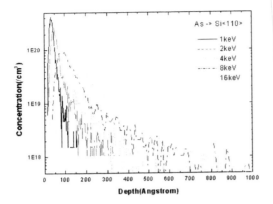

Fig. 6 A plot showing the simulation results As implant into Si with energies of 1, 2, 4, 8, and 16keV and the dose of 1×10^{14} ions/cm^2.

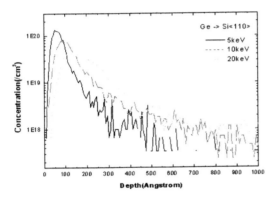

Fig. 7 A plot showing the simulation results for Ge implant into Si with the energies of 5, 10, and 20keV and the dose of 1×10^{14} ions/cm^2.

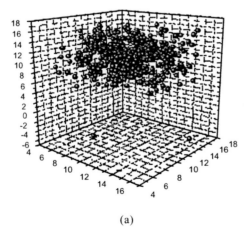

(a)

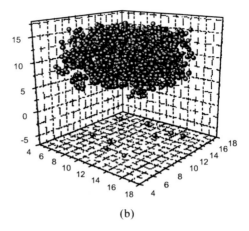

(b)

Fig. 8 Plots showing the simulation results with the energy of 5keV B implant into Si; (a) after 500 B ions and (b) after 4,000 B ions.

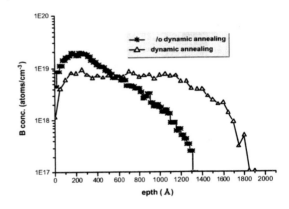

Fig. 9 A plot showing the simulation result with the energy of 5keV and the dose of 1×10^{14} ions/cm^2 B implant into Si with/without the dynamic annealing.

Table 1. Parameters of EDIP for Si.

Parameter	Value	Parameter	Value
A	7.9821730 eV	?	1.1247945 Å
a	3.1213820 Å	μ	0.6966326
	1.4533108 eV	?	1.2085196
Q_0	312.1341346	s	0.5774108 Å
	3.1083847	?	0.2523244
B	1.5075463 Å	ß	0.0070975
c	2.5609104 Å		

2D Quantum Mechanical Device Modeling and Simulation: Single and Multi-fin FinFET

Kidong Kim, Ohseob Kwon, Jihyun Seo and Taeyoung Won

Department of Electrical Engineering, School of Engineering, Inha University
253 Yonghyun-dong, Nam-gu, Incheon, Korea 402-751
Phone: +82-32-860-8686 Fax:+82-32-875-7436 E-mail: kkd@hsel.inha.ac.kr

ABSTRACT

A two-dimensional quantum mechanical modeling has been performed to simulate a nano-scale FinFET by obtaining the self-consistent solution of coupled Poisson and Schrödinger equations. Calculated current-voltage (I-V) curves are carefully compared with experimental data to verify the validity of our theoretical work. The transconductance (G_{mmax}=380) is optimized through varying the Si-fin thickness (T_{fin}) from 10nm to 75nm. In order to ascertain the current drivability of FinFET, we investigated the dependence on the number of fins. The electron distributions for single and multiple fins FinFETs are reported with several gate voltage V_g=1.5V, -0.3V, 1.5V. In addition, calculated I_d-V_g curve of single fin FinFET is also compared with three and five fins FinFET. From these simulation results, the mechanism of the formation of channel and high current drivability of multiple fins FinFET can be understood.

Keywords: FinFET, double gate MOSFET, quantum mechanical modeling and simulation, coupled Poisson and Schrödinger equations, high current drivability

1 INTRODUCTION

Recently, a double-gate (DG) structure has attracted a great deal of attention for the application of sub-40nm MOSFET. Among the proposed variations of DG MOSFETs, a self-aligned double-gate MOSFET structure including FinFET is one of the most attractive devices to implement a nano-scale planar MOSFET [1-4]. In order to optimize the structure of FinFET, it is necessary to undertake a two-dimensional (2-D) quantum mechanical (QM) simulation due to the inherent quantum effects on the electronic properties of nano-scale semiconductor devices. To fulfill the numerical simulation of nano-scale structures such as FinFET, we need to get a self-consistent solution of the coupled Poisson and Schrödinger equations.

In this paper, two-dimensional quantum-mechanical (QM) simulation of FinFET in a self-consistent manner is reported. We compare the current-voltage (I-V) characteristics with the experimental data. The simulation of multi-fin FinFETs has been performed to analyze the high current drivability of multi-fin FinFETs [1]. The electron densities of single fin and three-fin FinFETs are also demonstrated.

2 NUMERICAL MODEL FOR COMPUTER SIMULATION

For nano-scale device simulation, the nonlinear Poisson and Schrödinger equation should be solved in a simultaneous manner.

$$\nabla \varepsilon(r) \nabla \Phi(r) = -\rho(r) \tag{1}$$

$$\frac{\hbar^2}{2} \nabla \frac{1}{m^*(r)} \nabla \psi(r) + V(r)\psi(r) = E\psi(r) \tag{2}$$

where, ε is the dielectric constant, Φ the electrostatic potential, ρ the total charge density, ψ the wave function, $\hbar$ the Planck's constant divided by 2π, m^* the effective mass, V the potential energy, and E the energy. Here, one of the most important parameter in these equations is the quantum electron density as follows:

$$n(r) = \frac{1}{\pi}\left(\frac{2m^*(r)k_B T}{\hbar^2}\right)^{\frac{1}{2}} \sum_n \psi_n^2(r) F_{-1/2}\left(\frac{E_F - E_n(r)}{k_B T}\right) \tag{3}$$

Here, k_B is the Boltzmann's constant, T the temperature, E_F the Fermi level, and the Fermi-Dirac integrals of order k. These integrals are defined as follows:

$$F_k(\eta) = \frac{1}{\Gamma(k+1)} \int_0^\infty \frac{\varepsilon^k d\varepsilon}{e^{\varepsilon - \eta} + 1}, \quad k \geq -1 \tag{4}$$

and also have the following property

$$\frac{d}{dx} F_k(\eta) = F_{k-1}(\eta), \qquad k \leq -1 \tag{5}$$

The continuity equation for current density is given by

$$divj(r) = 0 \tag{6}$$

To obtain quantum mechanical solutions, we have to employ an iterative procedure. The first step we have to undertake is to calculate an electric potential in Equation (1). With the potential value initiated, the program calculates a built-in potential from Equation (1) with the Newton's method. Thereafter, the program determines self-consistent solutions of Equations (1), (2) and (6). The Newton's method has been employed with a constraint that should satisfy a certain error criteria [3].

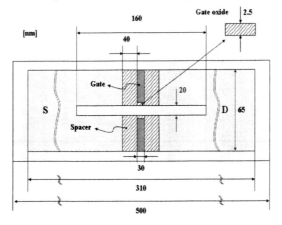

Figure 1: A schematic diagram illustrating the top view of the cross section of the single fin FinFET under this study.

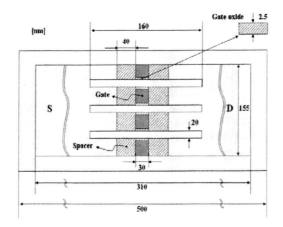

Figure 2: A schematic diagram illustrating the top view of the cross section of the three-fin FinFET under this study.

Figs. 1 and 2 show a top view of the cross section of the single and multiple fins FinFET used in this work. We employed a finite difference method (FDM) as a method of numerical analysis. As a test vehicle for verifying the validity of our numerical simulator, we chose an n-channel FinFET because the n-FinFET shows relatively good short-channel performance down to a gate-length of 17nm [4]. We investigated the short channel effect with a simple drift-diffusion model, which seems to be all right with the level of the driving current.

3 SIMULATION RESULTS

To verify the validity of our simulation, we compared our calculations with the experimental results of Digh Hisamoto et al [1] and Jakub Kedziersk et al [5]. Figs. 3 and 4 demonstrate the comparison of the typical current-voltage (I-V) characteristics for FinFET with L_{eff}=30nm and T_{si}=10nm, 20nm. The I_d-V_g curves of N-channel FinFETs are shown in Fig. 3 in a condition of V_d=0.1V and 1.5V, respectively. The calculated value of the subthreshold swing (S) is 74.58mV/dec at V_d=1.5V.

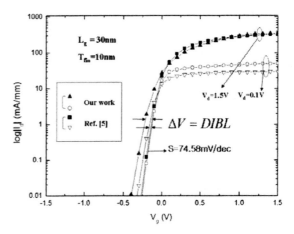

Figure 3: A plot showing the I_d-V_g curves for n-FinFET with L_g=30nm, T_{fin}=10nm which are compared with experimental data.

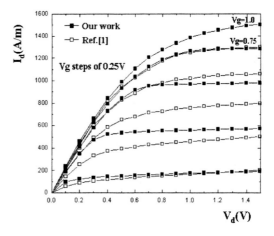

Figure 4: A plot showing the I_d-V_d curves for n-FinFET with L_g=30nm, T_{fin}=20nm which are compared with experimental data.

The I_d-V_d curves are also shown in Fig. 4. In spite of the low channel doping concentration, the subthreshold leakage current is well suppressed. Furthermore, there seems to be no kink effect, which comes from effect of floating body.

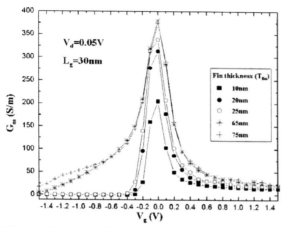

Figure 5: A plot showing the dependence of transconductance (G_m) on the Si-fin Thickness (T_{fin}). G_{mmax}=380 at T_{fin}=65nm, G_{mmax}=375 at T_{fin}=75nm.

Fig. 5 shows a functional dependence of transconductance (G_m) at V_d=0.05V on the thickness of Si-fin. The simulation reveals that G_m increases as the width of Si-fin increases. However, the value of G_m is found to be maximum at 65nm of Si-fin width. This is because as long as the Si-fin width increases the parasitic resistance, it also increases the carrier mobility. However, the charge centroid is reduced [1]. As a result of this reason, we can obtain the optimal Si-fin width.

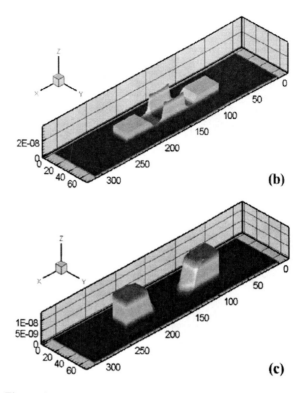

(b)

(c)

Figure 6: A plot showing the electron densities of single fin FinFET with L_g=30nm, T_{si}=20nm, V_d=0.1V. (a) V_g=1.5V, (b) V_g=0.3V, (c) V_g=-1.5V.

Figs. 6 and 7 demonstrate the distribution of electrons at several gate voltage (a) V_g=1.5V, (b) V_g=0.3V, (c) V_g=-1.5V. These figures exhibit how the channels of FinFET are formed, high current drivability of multi-fin FinFET is good because of the formation of multiple channels, and the electron density is high at S/D regions as the gate voltage decrease.

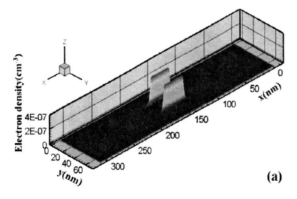

(a)

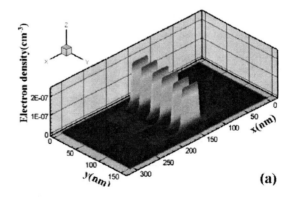

(a)

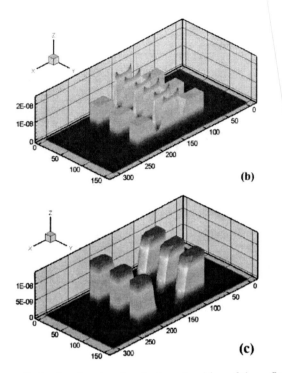

(b)

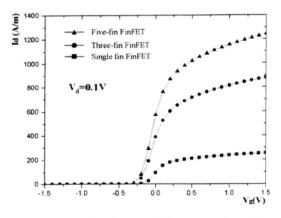

(c)

Figure 7: A plot showing the electron densities of three-fin FinFET with L_g=30nm, T_{si}=20nm, V_d=0.1V. (a) V_g=1.5V, (b) V_g=0.3V, (c) V_g=-1.5V.

Figure 8: A plot showing the high current drivability of multi-fin FinFET which is proportional to the number of fins.

Finally, high current drivability of multi-fin FinFETs is shown in Fig. 8. The current of three-fin and five-fin FinFETs is about three and five times larger than single fin FinFET. This is because the multi-fin FinFET is designed to obtain larger channel width than single fin FinFET. These multi-fin devices are good structure for applying in self-aligned and quasi-planner structures like FinFET.

4 CONCLUSION

In this paper, 2D (two dimension) numerical modeling and simulations for N-channel FinFET were reported. The optimization of Si-fin width was also demonstrated through extracting the transconductance of different structures. We also demonstrated the high current drivability of multi-fin FinFETs with optimization. We note that the current drivability of multi-fin FinFET is proportional to the number of fins and multi-fin structure is suitable for self-aligned and quasi-planner devices.

5 ACKNOWLEDGEMENT

This work was supported partly by the Ministry of Information & Communication (MIC) of Korea through Support Project of University Information Technology Research Center (ITRC) Program supervised by IITA, and partly by Ministry of Science and Technology (MOST) through Tera-Nano (TND) Program supervised by KISTEP. The author (K. Kim) would like to express special thanks to Prof. P. Vogl and Stefan of WSI (Walter Schottky Institute) for fruitful discussion.

REFERENCES

[1] D. Hisamoto, Wen-chin Lee, Jakub Kedzierski, Hideki Takeuchi, Kazuya Asano, Charles Kuo, Erik Anderson, Tsu-Jae King, Jeffrey Bokor and Chenming Hu, "FinFET-A self-aligned double-gate MOSFET scalable to 20nm," IEEE Trans. Electron devices, vol.47, p.2320 (2000).

[2] A. Svizhenko, M. P. Anantram, T. R. Govindan and B.Biegel, "Two-dimensional quantum mechanical modeling of nanotransistors," J. Appl, Phys, vol.91, no.4, p.2343 (2002).

[3] S. E. Laux, A. Kumar and M. V. Fischetti, "QDAME simulation of 7.5nm double-gate Si nFETs with Differing Acess geometries," IEDM Tech. Dig., p.715 (2002).

[4] X. Huang, Wen-Chin Lee, Charles Kuo, Digh Hisamoto, Leland Chang, Jakub Kedzierski, Erik Anderson, Hideki Takeuchi, Yang-Kyu Choi, Kazuya Asano, Vivek Subramanian, Tsu-Jae King, Jeffrey Bokor and Chenming Hu, "Sub-50nm FinFET: PMOS," IEDM Tech. Dig., p.67 (1999).

[5] J. Kedzierski, David M. Fried, Edward J. Nowak, Thomas Kanarsky, Jed H. Rankin, Hussein Hanafi, W. Natzle, Diane Boyd, Ying Zhang, Ronnen A. Roy, J. Newbury, Chienfan Yu, Qingyun Yang, P. Saunders, Christa P. Willets, A. Johnson, S. P. Cole, H. E. Young, N. Carpenter, A. Rakowski, Beth Ann Rainey, Peter E. Cottrell, Meikei Ieong and H. –S. Philip Wong, "High-performance symmetric-gate and CMOS-compatible V_t asymmetric-gate FinFET devices", IEDM Tech. Dig., p.437 (2001).

Kinetic Monte Carlo Modeling of Boron Diffusion

in Si Crystalline Materials

Jihyun Seo, Chi-Ok Hwang, Ohseob Kwon, Kidong Kim, and Taeyoung Won

Department of Electrical Engineering, School of Engineering, Inha University
253 Yonghyun-dong, Nam-gu, Incheon, Korea, E-mail : sjh@hsel.inha.ac.kr

ABSTRACT

In this paper, we report a lattice-free kinetic Monte Carlo (KMC) result of boron diffusion at low temperatures 450℃ and 550℃ with vacancy + interstitialcy mechanism or vacancy + kick-out one with dilute self-interstitials (I) and vacancies (V) created in a B-doped marker layer (4×10^{11}B/cm^2 per marker) by Si implantation (50keV, 10^{11}/cm^2). In this type of KMC model, point defects and dopants are treated at an atomic scale while they are considered to diffuse in accordance with their event rates, which are given as input parameters from *ab initio* calculations or experimental data. Especially, the formation of clusters and extended defects, which usually control the annealing kinetics after ion implantation, is to be minimized in the range of low doses in an effort to create dilute concentrations of I and V. Therefore, simple vacancy and kick-out or interstitialcy mechanisms without interstitial clusters or extended defects are tested in these conditions and both are in a good agreement with the SIMS data. However, in these dilute concentrations vacancy mechanism plays a dominant role in B diffusion in place of the usual kick-out or interstitialcy mechanism in boron enhanced diffusion.

Keywords: Atomistic diffusion modeling and simulation, kinetic Monte Carlo, Boron diffusion, Interstitialcy mechanism, Kick-out mechanism.

1 INTRODUCTION

In nano-scale semiconductor process simulation, an atomic scale model is essential because of the destined failure of continuum approach in small nano scales. Molecular dynamics (MD), the most in-detailed description at an atomic level is not adequate in thermal annealing simulations after ion implantation due to that MD can not span the long time of thermal annealing. Therefore, kinetic Monte Carlo (KMC) method for simulating the thermal annealing has been preferred.

In this paper, we report a model for diffusion of Boron at low temperatures with vacancy and interstitialcy or kick-out mechanism [4,6,7] with Si self-interstitials (I) and vacancies (V) created in a B-doped marker layer by Si implantation [3]. For an accurate atomic-scale model for diffusion of intrinsic point defects (I, V) and impurities (B)

in ion-implanted silicon, we use a lattice-free KMC method. In this type of KMC model, point defects and dopants are treated at an atomic scale while they are considered to diffuse in accordance with their event rates, which are given as input parameters from *ab initio* calculations or experimental data. Especially, the formation of clusters and extended defects, which usually control the annealing kinetics after ion implantation, is minimized in the range of low doses to create dilute concentrations of I and V. In these conditions, the formation of clusters and extended defects are ignored to study only boron diffusion mechanisms.

2 KINETIC MONTE CARLO IMPLEMENTATION

In KMC, a physical system, which consists of many possible events, evolves as a series of independent event occurrings. All events have their own event rates. We consider thermally activated events in thermal annealing simulation after ion implantation. If the probability of next event to occur is independent of the previous history, and the same at all times. The transition probability is a constant. Then the process is a so-called Poisson process. To derive the time dependence, consider a single event with a uniform transition probability r. Let f the transition probability density, which gives the probability rate at which the transition occurs at time t. The change of $f(t)$ over some short time interval dt is proportional to r, dt and f because f gives the probability density that the physical system still remains at time t.

$$df(t) = -rf(t)dt \qquad (1)$$

And the solution is given by with boundary conditions.

$$f(t) = re^{-rt}, f(0) = r \qquad (2)$$

Therefore, we update the simulation time with ($t = t + \Delta t$) according to event rates as follows because an ensemble of independent Poisson processes will behave as one large Poisson process:

$$\Delta t = -\frac{\ln u}{R} \qquad (3)$$

Here, u is the random number and R the total sum of all possible event rates. We select an event according to the event rates and KMC is proper to simulate non-uniform time evolution processes.

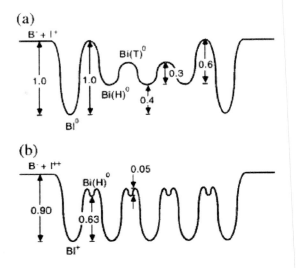

Figure 1: Boron diffusion paths and energy barriers taken from [6], (a) kick-out mechanism and (b) interstitialcy mechanism.

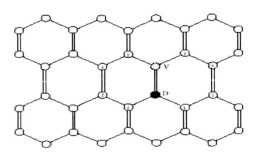

Figure 2: Boron vacancy diffusion mechanism taken from [8] : The vacancy can go around the six member ring of atoms and approach the dopant from a different direction. After the dopant hops into the now empty site the dopant vacancy pair completes one diffusion step.

We implement an atomistic diffusion mechanism involving fast-migrating intermediate species of the form. The reactions $X_s + V \Leftrightarrow XV_m$, $X_s + I \Leftrightarrow X_m$,and $I + V \Leftrightarrow 0$ are essentially diffusion limited, with capture radius of second neighbor atomic length (3.84Å) and direction of particle migration is limited to six neighboring sites. Here X_s is the immobile substitution impurity, which through reaction with a vacancy or a Si self-interstitial (I) forms a fast-migrating species XV_m and X_m, which diffuse at a rate D_m. Ordinary, the diffusion rate is in Arrhenius form ($D_0 \exp(-E_b/kT)$). A simulation box of $1000 \times 1000 \times 2500$ Å^3 (about 120 (B), 750 (I, V) particles) is used. The width of the simulation box is much smaller than that of the implantation window. Thus, the periodic boundary condition is applied to the box in the lateral direction. Figure 1 shows the Boron diffusion paths and energy barriers of kick-out and interstitialcy mechanisms. Figure 2 shows Boron vacancy diffusion mechanism (ring mechanism) [8]. In addition, all the necessary reactions and parameters are given in Table 1 and Table 2.

Table 1 : Reaction equations

Common diffusion mechanism	
I + V $\Leftrightarrow$ 0 Bs + V $\Leftrightarrow$ BV	
Interstitialcy mechanism	Kick-out mechanism
Bs + I $\Leftrightarrow$ BI	Bs + I $\Leftrightarrow$ BI BI $\Leftrightarrow$ Bi

Table 2 : Parameters for simulation [9], [10].

Events	V_0 (10^{15}/s)	E_b (eV)
I migration	0.1	0.9
V migration	0.0025	0.43
BI migration	0.1	0.68
BI $\to$ B + I	0.01	0.9
BV migration	0.01	1.0
BV $\to$ B + V	0.01	0.5

3 SIMULATION RESULTS

Figure 3 (a) and (b) show boron dopant profiles in MBE-grown marker layer with of about 7 nm (4×10^{11}B/cm^2 per marker) modified by 50 keV Si ion implantation with the dose of 1×10^{11}/cm^2 using MARLOWE code. The white circles represent the initial distribution of boron and the black circles the distribution of boron after the thermal annealing for 15 min at 450℃, 550℃, respectively. Figure 5 (a) and (b) shows the boron distribution using vacancy and kick-out mechanism. The two cases of boron diffusion profiles are quite close to SIMS data (asterisk symbols). A significant fraction of B atoms in the shallow marker are displaced and migrate a considerable distance ($\sim 10^2$ nm) from their initial positions.

In the interstitialcy mechanism, boron with one interstitial (BI) is only mobile species. It can diffuse into silicon lattice via switching BI (tetrahedral sites, T) to Bi (hexagonal sites, H) and Bi to BI. The BI can also break up into substitution B and one mobile interstitial. But, the BI is not mobile particle in kick-out mechanism. In this case, BI can only break up (BI$\Leftrightarrow$B+I) or kick B out of the silicon lattice (BI$\Leftrightarrow$Bi, kick-out mechanism). Thus, boron-interstitial (Bi) can diffuse into interstitial sites (H, T).

Figures 4 and 6 represent the time evolution of the number of mobile particles and immobile ones during the thermal annealing. In early stage of diffusion, vacancy mechanism plays a dominant role because of the frequent V migration. The considerable differences are not detected between interstitialcy and kick-out mechanism results in our simulations. These simulation results are caused by the similarity of the total energy barriers between kick-out and interstitialcy mechanisms. In addition, in these dilute concentrations of I and V, the vacancy mechanism plays a dominant role in B enhanced diffusion due to the higher migration frequency of vacancies.

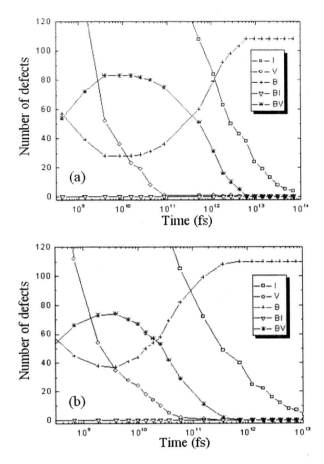

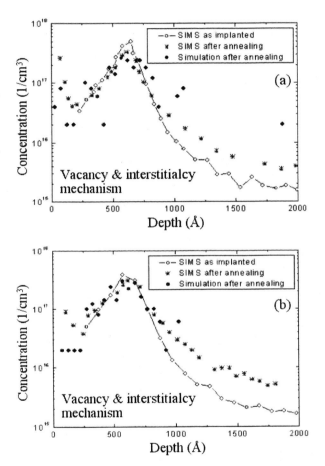

Figure 3: Dopant profiles of MBE-grown boron marker modified by 50 keV Si ion implantation to dose of 1×10^{11} /cm^2 [3] using vacancy and interstitialcy mechanism. (a) 450℃, (b) 550℃ after 15min annealing

Figure 4: Time evolution of the number of particles in boron MBE-grown marker layer modified by 50 keV Si ion implantation to dose of 1×10^{11} /cm^2 using vacancy and interstitialcy mechanism. (a) 450℃, (b) 550℃ during the annealing

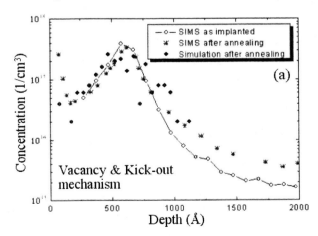

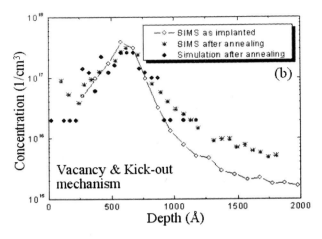

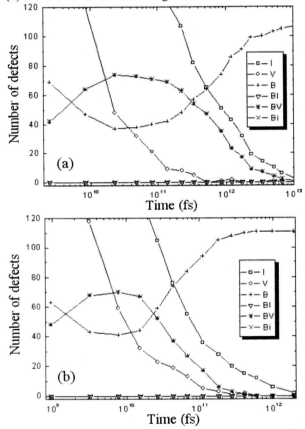

Figure 5: Dopant profiles of boron MBE-grown marker modified by 50 keV Si ion implantation to dose of 1×10^{11} /cm² [3] using vacancy and kick-out mechanism. (a) 450℃, (b) 550℃ after 15 min annealing.

Figure 6: Time evolution of the number of particles in boron MBE-grown marker layer modified by 50 keV Si ion implantation to dose of 1×10^{11} /cm² using vacancy and kick-out mechanism. (a) 450℃, (b) 550℃ during the annealing

4 CONCLUSION

In this paper, we implement a KMC simulation for B migration via vacancy and interstitialcy mechanism or kick-out mechanism at low temperatures, 450 and 550℃. Also, to ignore the formation of clusters and extended defects and to study only Boron diffusion mechanisms, low dose experimental data are used.

In our simulations, we find that there is no essential B diffusion difference between vacancy + kick-out and vacancy + interstitialcy mechanisms. This result can be inferred from the fact that the total energy barriers of kick-out and interstitialcy mechanisms are similar. In addition, in these dilute concentrations of I and V, in place of the usual kick-out or interstitialcy mechanism vacancy mechanism plays a dominant role in B diffusion due to the frequent migration of vacancies.

ACKNOLEDGEMENT

This work is supported by the Ministry of Information & Communication (MIC) of Korea through ITRC program (Support Project of University Information Technology Research Center Program supervised by Institute of Information Technology Assessment in Korea) and partly by Ministry of Science and Technology (MOST) through Nano-device Modeling program supervised by Korea Institute of Science and Technology Evaluation and Planning (KISTEP).

REFERENCES

[1] N.E.B. Cowern, G.F.A. van de Walle, D.J. Gravesteijn, and C. J. Vriezema, Phys. Rev. Lett 67, 212 (1991) .

[2] A.E. Michel, Nucl. Instrum. Methods, Phys. Res, Sect. B 37/38, 379 (1989).

[3] N.E.B. Cowern, G.F.A. van de Walle, P. C. Zalm, and D. J. Oostra, Phys Rev Lett Vol 69 (1992).

[4] Paola Alippi, L. Colombo, P. Ruggerone, A. Sieck, G. Seifert and Th. Frauenheim, Phys Rev B, Vol 64 (2001).

[5] M.T. Robinson and J.D. Plummer, Appl. Phys. Lett. 56, 1787 (1990).

[6] B. Sadigh, T.J. Lenosky, S.K. Theiss, M.-J. Caturla, T.D. de la Rubia and M.A. Foad, Phys. Rev. Lett. 83, 4341-4344 (1999).

[7] W. Windl, M.M. Bunea, R. Stumpf, S.T. Dunham and M.P. Masquelier, Phys. Rev. Lett., 83, 4345-4348 (1999)

[8] S. Chakravarthi, Ph.D. thesis, Boston University (1995).

[9] C.S. Nichols, C.G. van de Walle and S. T. pentelides, Phys. Rev. B Vol. 40(8), 5484 (1989).

[10] H. Oka, K. Suzuki, Z. Jinyu, M. Yu, R. Huang, FUJITSU. Tech. J., 39,1 p.128-137 (June 2003)

A Study of the Threshold Voltage Variations for Ultrathin Body Double Gate SOI MOSFETs

Chien-Shao Tang[1], Shih-Ching Lo[2], Jam-Wem Lee[3], Jyun-Hwei Tsai[2] and Yiming Li[3,4,*]

[1]Departmenet of Electrophysics, National Chiao Tung University, Hsinchu 300, TAIWAN
[2]National Center for High-Performance Computing, Hsinchu 300, TAIWAN
[3]Department of Nano Device Technology, National Nano Device Laboratories, Hsinchu 300, TAIWAN
[4]Mircoelectronics and Information Systems Research Center, National Chiao Tung University, Hsinchu 300, TAIWAN
[*]P.O. Box 25-178, Hsinchu 300, TAIWAN, ymli@faculty.nctu.edu.tw

ABSTRACT

Silicon on insulator (SOI) devices have been of great interest in these years. In this paper, simulation with density-gradient transport model is performed to examine the variation of threshold voltage (V_{TH}) for double gate SOI MOSFETs. Different thickness of silicon (Si) film, oxide thickness, channel length and doping concentration are considered in this work. According to the numerical study, both drift-diffusion (DD) and density gradient (DG) models demonstrate that the thickness of Si film greatly affects the threshold voltage (5 ~ 15 % variation). It is found that the thickness of Si film decreases, V_{TH} variation increases; and the dependence relation is nonlinear. Therefore, this effect must be taken into account for the realization of double gate SOI ULSI circuit.

Keywords: double-gate devices, ultrathin body, quantum mechanical effects, threshold voltage, modeling and simulation.

1 INTRODUCTION

Double gate SOI devices are more and more attractive for sub-50 nm ultra-large scaled integrated (ULSI) circuits manufacturing because of their inherent suppression of short-channel effects (SCEs), high transconductance and ideal subthreshold swing (S-swing) [1-9]. The double gate SOI devices, owing to a difficulty of manufacturing uniformity, suffer fluctuations of silicon and gate oxide thin films. This is important for design and fabrication of the double gate SOI devices, in particular for the applications of ULSI circuits. From the fabrication point of view, the double gate SOI manufacturing may have a 10 % thin film thickness fluctuation. Therefore, threshold voltage variation resulting from this fluctuation becomes an important problem and should be subject to further investigation when exploring important information for circuit designer and device engineer. Besides the thickness discrepancy in Si thin film, the correlations between the gate oxide thickness variations to threshold voltage changes should also be examined. Considering these two non-uniformity properties for the double gate SOI device, we probably can ensure the ability for double gate SOI ULSI circuit design and manufacturing. Various works have recently been proposed for these devices [1-16].

In this work, we computational examine the fluctuation effects of Si film thickness on the threshold voltage variation. In the numerical simulation of the threshold voltage variation, the classical and quantum correction transport models, DD and DG models are considered and solved numerically [10-14]. Addition to the thickness variation effects, the doping concentration, oxide thickness, and channel length are simulated systemically to discuss the threshold voltage variation. Subjecting to the comprehensive numerical studies, the complete conclusion and suggestion are drawn.

2 COMPUTATIONAL MODELS

Classical and quantum models are employed to compare the quantum effects on the threshold voltage variation. DD and DG models are considered in this work. Firstly, the three governing equations of DD model are listed in the following. Equation (1) shows the Poisson equation:

$$\nabla \varepsilon \cdot \nabla \psi = -q(p - n + N_D - N_A), \tag{1}$$

where ε is the electrical permittivity, q is the elementary electronic charge, n and p are the electron and hole densities, and N_D and N_A are the number of ionized donors and acceptors, respectively. The other two are continuity equations, they are presented in the equations (2) and (3), respectively:

$$q\frac{\partial n}{\partial t} - \nabla \cdot \mathbf{J_n} = -qR, \tag{2}$$

$$q\frac{\partial p}{\partial t} + \nabla \cdot \mathbf{J_p} = -qR, \tag{3}$$

$\mathbf{J}_n$ and $\mathbf{J}_p$ are the electron and hole current densities while μ_n and μ_p are the electron and hole mobility. Moreover, it should be noticed that ϕ_n and ϕ_p are the electron and hole quasi-Fermi potentials.

It is known that in comparing with the classical simulation, the quantum mechanical model explores different deviation quantitatively. In principle, the Schrödinger equation coupled with classical model is the most accurate way to solve the carrier concentration. However, it is not only computationally expensive but also difficult in simulation of multi-dimensional cases. Consequently, researchers devote their efforts on developing of quantum correction models. Those models are claimed to have a similar result as quantum mechanically calculated one but requires the same computation cost as that of the classical calculation. To include quantization effects in a classical device simulation, a simple approach is to introduce an additional potential, such as quantity Λ, in the classical density formula, which reads:

$$n = N_C \exp\left(\frac{E_F - E_C - \Lambda}{k_B T}\right), \quad (4)$$

where N_C is the conduction band density of states, E_C is the conduction band energy, and E_F is the electron Fermi energy. It is difficult to describe all quantum mechanical effects in terms of a variable Λ. A carefully quantum correction should be taken into considerations for avoiding any under- or over-estimations. Among the quantum correction models, DG model is a well-known approximation [15]. We chose it in this work and Λ is:

$$\Lambda = -\frac{\gamma \hbar^2}{12m} \frac{\nabla^2 \sqrt{n}}{\sqrt{n}}, \quad (5)$$

where $\hbar$ is the reduced Planck constant, m is the density of states mass, and γ is a fitting factor.

The computing procedure is briefly described as follows. Firstly, the stop criteria, mesh, output variables and simulation models are chosen. If DD model is chosen, Poisson equation is solved iteratively until the result converges. If DG model is chosen, the modified potential should be added in Poisson equation and solved. Then, continuity equations are solved. After all equations converge, we check the convergence of whole system. If the whole system converges, the simulation is terminated. Otherwise, we perform the outer loop iteration until it converges. This scheme produces a self-consistent result.

3 RESULTS AND DISCUSSION

In our numerical studies, the influence of oxide thickness (T_{OX} = 1.5, 2, 3 and 4 nm), channel length (L_G = 35, 65, 90 and 130 nm), Si film thickness (T_{Si} = 20, 40 and 65 nm) and doping concentration (N_A = 1×10^{17}, 5×10^{17}, 1×10^{18}, and 5×10^{18} cm^{-3}) on the threshold voltage for double gate SOI n-typed MOSFETs is simulated using ISE-DESSIS ver. 8.0.3 [17]. Each factor is examined sequentially. The definition of threshold voltage employed in this study is the Gm maximum method. The method firstly find out the gate voltage at the maximum of Gm, then make a tangent line of the drain current – gate voltage (I_{DS}-V_{GS}) curve at the gate voltage. Finally, extrapolated intercept of the tangent line to the V_{GS}-axis; the extrapolation is defined as V_{TH}.

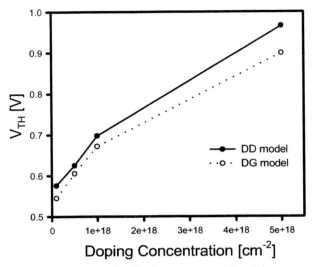

Figure 1: V_{TH} vs. the doping concentration. They are computed with DD (solid line) and DG (dot line) models.

The influence of doping concentration on V_{TH} is discussed firstly. Different doping concentrations of a 65 nm double gate SOI NMOSFET with T_{OX} = 1.5 nm and T_{si} = 40 nm are simulated with DD and DG models, shown in Fig. 1. It is found that V_{TH} increases as doping concentration increases. Therefore, a heavy doped substrate reduces V_{TH} roll-off caused by SCE. The variation is almost linear. An increase of 10^{18} cm^{-3} in doping concentration induces a 0.13 V shift on V_{TH}. From Fig. 1, the V_{TH} variations of the heavy doping devices are smaller than that of light doping ones. In comparing the extreme cases of the scenario, we observe a 30 % variation of V_{TH}.

Next, the threshold voltage affected by Si film thickness is discussed. A 65 nm double gate SOI NMOSFET with T_{OX} = 1.5 nm and N_A = 5×10^{17} cm^{-3} is explored and presented in Fig. 2. Both DD and DG models demonstrate that the thicknesses of the Si film will greatly affect the threshold voltage of the double gate devices. We note that the decrease of T_{Si} results in a significant increase on V_{TH}. For example, a 10-nm-thin T_{Si} shifts V_{TH} up to 0.02 V (~ 4% variation). This phenomenon is mainly caused from the shield-effect that retards the SCE. Thus, employing a thin Si film of double gate SOI NMOSFET will enhance the scaling down ability of semiconductor devices. From the fabrication point of view, the manufacturing may have a 10 % fluctuation in thin film thickness; accordingly, a 0.01 V variation of V_{TH} should be carefully estimated for manufactured devices.

Oxide thickness scaling effects is shown in the following paragraph. Figure 3 exhibits the simulated results of a 65 nm double gate SOI NMOSFET. T_{Si} and N_A are 40 nm and 5×10^{17} cm^{-3}, respectively. Since the gate oxide thickness scaling results in an oxide capacitance increases. Additionally, V_{TH} decreases when the oxide capacitance increases. This implies that the change of V_{TH} is linearly dependent on T_{OX}. From our simulation results, a nanometer decrement of T_{OX} will cause a 0.014 V (~ 2.2 %) threshold voltage decrement. The extreme cases of the simulated scenario are about 7 % variation of V_{TH}.

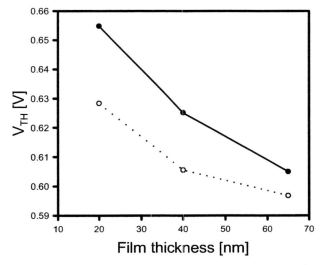

Figure 2: V_{TH} vs. T_{Si}. The notations are the same with Fig. 1.

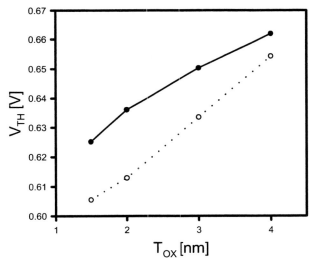

Figure 3: V_{TH} vs. T_{OX}. The notations are the same with Fig. 1.

Figure 4 plots the double gate SOI NMOSFET simulated results of V_{TH} vs. the channel length. $T_{Si} = 40$ nm, $T_{OX} = 1.5$ nm, and $N_A = 5 \times 10^{17}$ cm^{-3} are chosen in this

examination. We observe that the shield-effect strongly reduces drain-induced barrier lowering (DIBL) that roll-off of V_{TH} does not so sensitive to channel length scaling. Take a realistic number for example; a 10 nm reduction on the channel length induces 0.04 V (~7 %) decreasing of V_{TH}. Furthermore, Fig. 4 suggests that the threshold voltage lowering effects would become serious when the channel length scaled down.

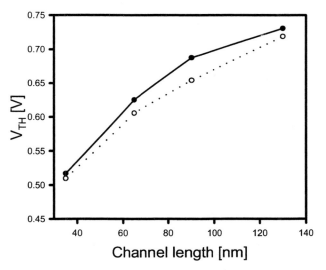

Figure 4: Simulated curves of V_{TH} vs. the channel length for DD and DG models.

In summary, from Figs. 1 to 4, the quantum effect on the V_{TH} variation is explored. According to these figures, DD and DG models predict similar V_{TH} variation under different device characteristics. However, DD model overestimates V_{TH} for about 4 %, which is about 0.03 ~ 0.05 V. Those facts are strongly related to electron distributions difference between classical and quantum theories. Therefore the quantum effect should be included in nanoscale semiconductor devices simulations.

To clarify the fluctuation of Si film thickness related V_{TH} variation, the DG simulation is performed for different film thicknesses and doping concentrations. Those results are shown in Fig. 5 that higher V_{TH} variation will be concluded for a thinner film thickness (~ 5 % for $T_{si} = 20$ nm); moreover the variation is nonlinear dependency. Therefore, this effect must be taken into account for the thinner silicon film double gate SOI ULSI circuit.

From the standpoint of devices physics, problems should be overcome are SCE and nonscalability of the oxide thickness scaled down to about 1.5 nm. The practical limit does not allow us to scale down devices feature and keep a constant ratio between channel lengths to oxide thickness. The shield-effect of double gate SOI NMOSFET minimizes the short channel effects and makes drain current improvement with device scalability enhancement. In addition, a manufacturing error exists during the devices fabrication. According to the results, V_{TH} depends on T_{Si},

T_{OX}, N_A and L_G. The relationship among them can be expressed as nonlinear function (6)

$$V_{TH} = f(N_A, -T_{Si}, T_{OX}, L_G),\qquad(6)$$

where f should be subject to further formulation. As mentioned above, V_{TH} fluctuates severely in the extreme small devices. Therefore, all effect should be considered in the fabrication of nanoscale double gate SOI NMOSFETs.

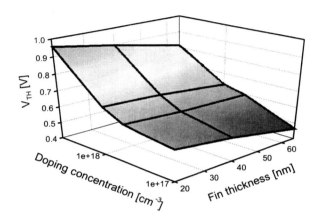

Figure 5: A three-dimensional plot of V_{TH} with respect to the doping levels and film thickness.

4 CONCLUSIONS

Double gate SOI devices with ultra-thin silicon and gate oxide films have been investigated on the process related V_{TH} variation. We would like to conclude that double gate SOI MOS structures are promise in preventing SCE. However, the film thickness, channel length, oxide thickness, and doping profile must be carefully optimized to reduce the influence of V_{TH} variation, especially for nanoscale devices. From the fabrication point of view, we have to pay attention to the implementation of film thickness, doping profile, geometry, and its corresponding circuit design for the challenge of threshold voltage variation.

5 ACKNOWLEDGEMENTS

This work is supported in part by the TAIWAN NSC grants: NSC − 92 − 2215 - E − 429 − 010, NSC − 92 − 2112 − M − 429 - 001, and NSC − 92 - 815 − C − 492 − 001 - E. It is also supported in part by the grant of the Ministry of Economic Affairs, Taiwan under contract No. 91 − EC − 17 − A − 07 - S1 - 0011.

REFERENCES

[1] N. Sano et al., "Device modeling and simulations toward sub-10 nm semiconductor devices," IEEE T. Nanotech. 1, 63, 2002.

[2] M. Masahara et al., "15-nm-thick Si Channel Wall Vertical Double-gate MOSFET," IEDM Tech. Dig. 949, 2002.

[3] S. E. Laux et al., "QDAME Simulation of 7.5 nm Double-gate Si nFETs with Differing Access Geometries," IEDM. Tech. Dig. 715, 2002.

[4] G. Pei et al., "FinFET design considerations based on 3-D simulation and analytical modeling," IEEE Trans. Elec. Dev. 49, 1411, 2002.

[5] T. Schulz et al., "Planar and vertical double gate concepts," Solid-State Elec. 46, 985, 2002.

[6] S.H. Tang et al., "Comparison of short-channel effect and offstate leakage in symmetric vs. asymmetric double gate MOSFETs," Proc. IEEE Int. SOI Conf. 120, 2000.

[7] M. Mouis and F. N. Genin, "A. Poncet Gate Control in Ultra-short Channel Double-gate MOSFETS Accounting for 2D and Quantum Confinement Effects," Proc. Device Res. Conf., Indiana, USA, 195, 2001.

[8] J. Walczak and B. Majkusiak, "The remote roughness mobility resulting from the ultrathin SiO2 thickness nonuniformity in the DG SOI and bulk MOS transistors," Microelec. Eng. 59, 417, 2001.

[9] G. Baccarani and S. Reggiani, "A Compact Double-Gate MOSFET Model Comprising Quantum-Mechanical and Nonstatic Effects," IEEE Trans. Elec. Dev. 46, 1656, 1999.

[10] M. G. Ancona et al., "Density-Gradient Analysis of MOS Tunneling," IEEE Trans. Elec. Dev. 47, 2310, 2000.

[11] A. Asenov et al., "The Use of Quantum Potentials for Confinement and Tunnelling in Semiconductor Devices," J. Comput. Elec. 1, 503, 2002.

[12] Y.–K. Choi et al., "Investigation of Gate-Induced Drain Leakage (GIDL) Current in Thin Body Devices: Single-Gate Ultra-Thin Body, Symmetrical Double-Gate, and Asymmetrical Double-Gate MOSFETs," Jpn. J. Appl. Phys. 42, 2073, 2003.

[13] Y. Li et al., "A Novel Parallel Approach for Quantum Effect Simulation in Semiconductor Devices," Int. J. Modelling & Simulation 23, 94, 2003.

[14] Y. Li et al., "A Practical Implementation of Parallel Dynamic Load Balanceing for Adaptive Computing in VLSI Device Simulation," Engineering with Computers 18, 124, 2002.

[15] Y. Li, "A Computational Efficient Approach to the Numerical Solution of the Density-Gradient Equations for Ultra-Thin Oxide MOS devices," WSEAS Trans. Circuits 1, 1, 2002.

[16] Y. Li et al., "Modeling of Quantum Effects for Ultrathin Oxide MOS Structures with an Effective Potential," IEEE Trans. Nanotech. 1, 238, 2002.

[17] DESIS-ISE TCAD Release 8.0.3, ISE integrated Systems Engineering AG, Switzerland, 2003.

Single Wall Carbon Nanotube-Based Structural Health Sensing Materials

A. Neal Watkins[*], JoAnne L. Ingram[**], Jeffrey D. Jordan[***], Russell A. Wincheski[***], Jan M. Smits[****], and Phillip A. Williams[*****]

[*]NASA Langley Research Center, MS 493, Hampton, VA 23681, a.n.watkins@larc.nasa.gov
[**]Swales Aerospace, Inc.
[***]NASA Langley Research Center
[****]Lockheed-Martin
[*****]National Research Council Postdoctoral Associate

ABSTRACT

Single wall carbon nanotube (SWCNT)-based materials represent the future aerospace vehicle construction material of choice based primarily on predicted strength-to-weight advantages and inherent multifunctionality. The multifunctionality of SWCNTs arises from the ability of the nanotubes to be either metallic or semi-conducting based on their chirality. Furthermore, simply changing the environment around a SWCNT can change its conducting behavior. This phenomenon is being exploited to create sensors capable of measuring several parameters related to vehicle structural health (i.e. strain, pressure, temperature, etc.) The structural health monitor is constructed using conventional electron-beam lithographic and photolithographic techniques to place specific electrode patterns on a surface. SWCNTs are then deposited between the electrodes using a dielectrophoretic alignment technique. Prototypes have been constructed on both silicon and polyimide substrates, demonstrating that surface-mountable and multifunctional devices based on SWCNTs can be realized.

1   INTRODUCTION

Since their discovery in 1991 by Iijima [1], carbon nanotubes have attracted a growing amount of attention due to their remarkable strength and conductive properties [2]. Single wall carbon nanotubes (SWCNTs) are one of the strongest known materials with a Young's modulus of approximately 1 TPa (about 10 times that of steel) making it a material of choice for the next generation of structural materials. SWCNTs can exhibit either semiconducting or metallic behavior based on the chirality of the tube, and ballistically transport electrons while maintaining their spin state down the length of the tube [2]. Due to these properties, SWCNTs have been used to fabricate several nanoscale devices, such as field-effect transistors [3,4] and molecular logic devices. Sensors with SWCNTs as the active element have also been constructed to detect a variety of gaseous analytes [5,6].

Because of these unique mechanical and electrical properties, SWCNTs seem to an ideal choice for future aerospace vehicle construction material. Not only does using advanced materials incorporating SWCNTs lead to more efficient, lighter aerospace designs, but the inherent multifunctionality could lead to designs which are "self-sensing"; sensor skins capable of probing the environment around the vehicle could be designed as part of the vehicle itself. Several advantages to this approach are readily apparent. First, the added cost of adding external sensors and devices is alleviated, as is the intrusiveness of the added sensors to the performance of the vehicle (i.e. additional weight, change in the flow characteristics, etc.) Second, since SWCNT-based devices can be made extremely small, a variety of sensors can be placed in a very small are and conceivably in places on a vehicle not currently amenable to adding larger, external devices.

This work describes the construction and testing of prototype sensors based on SWCNTs for measuring parameters related to vehicle structural health, specifically strain. The sensors were constructed using conventional lithographic techniques to pattern electrodes on both silicon and flexible plastic substrates. SWCNTs were then deposited and aligned using a dielectrophoretic technique [7-9], and the resulting SWCNTs created electrical bridges between the electrodes. It has been demonstrated using an atomic force microscope (AFM) tip that the conductance through a SWCNT can change with an applied strain [10, 11], and this is the mechanism for the proposed sensor.

2   EXPERIMENTAL

2.1   Materials

Purified SWCNTs produced using the HiPCO process [12] were purchased from Carbon Nanotechnologies, Inc., and used as received. Toluene (ACS reagent grade, 99.5%), hydrofluoric acid (HF, 48% in water), sulfuric acid (H_2SO_4, ACS reagent grade), and hydrogen peroxide (H_2O_2, 30%) were purchased from Aldrich and used as received. Rapid-curing polyimide precursors solution (PI2525) was purchased from HD Microsystems and used as received. Oxidized silicon wafers (500 nm SiO_2 thickness) were purchased from TTI Silicon.

2.2 Sensor Fabrication

Before use, the oxidized silicon wafers were cleaned by immersion in a solution of 70:30 H_2SO_4:H_2O_2 for 30 minutes at $100°C$ ("piranha" solution, *Caution: strong oxidizer, may explode*) to remove any organic contaminants present. After cleaning, the wafers were rinsed with de-ionized water four times, ethanol four times, and finally dried under a stream of nitrogen.

Sensor construction on silicon wafers was accomplished with conventional photo- and electron-beam lithography to deposit circuit elements followed by SWCNT deposition. First, a layer of photoresist was deposited on the surface by spin coating at 3000 rpm for 30 sec. Large circuit elements (e.g. bond pads, wires, etc.) were defined via illumination with UV radiation through a custom designed photomask. Following development of the photoresist, an adhesion layer of chromium or titanium (10 nm) and a thin layer of gold (20 nm) were evaporated onto the surface. Two sets of electrodes were then defined using electron-beam lithography. The first set of electrodes is used to deposit and align the SWCNTs (alignment electrodes). For the alignment electrodes, a thin layer of poly(methyl methacrylate) (PMMA) was deposited onto the pattern by spin coating at 3500 rpm for 30 seconds. The electrodes were then defined using an electron beam. The typical width of each electrode was approximately 1 m, and the separation between the electrodes was 3 m. Deposition and alignment of the SWCNTs was accomplished using a dielectrophoretic technique. First, a suspension of nanotubes in toluene (approximately 1 mg/mL) was prepared by ultrasonication for 24 hours. Because the concentration of nanotubes in the suspension was small, only a slight discoloration occurs. Several drops of the suspension were then deposited on the sample to cover the surface. Two spring-loaded pins were then used to make contact with the pads and an AC voltage was applied to the pads to generate an electric field. A brief period of time ranging from tens of seconds to several minutes was used with SWCNT densities being proportional to exposure time [9]. Following deposition of the SWCNTs, the second set of electrodes was patterned using the same procedure as the alignment electrodes. These interdigitated electrodes (connection electrodes) were aligned perpendicular to the alignment electrodes with a separation of 500 nm and were used to make an electrical connection across the aligned nanotubes.

Sensor construction on the flexible, plastic substrates was accomplished by coating a cleaned oxidized silicon wafer with a thin film (~12 m) of PI2525 by spin coating at 2000 rpm for 30 seconds. After spinning, the film was cured in air by heating to $200°C$ at $4°C$/min, holding at $200°C$ for 30 minutes, heating to $300°C$ at $2.5°C$/min, and holding at $300°C$ for 60 minutes, followed by a gradual cooling to room temperature. All patterning and SWCNT deposition were performed using the procedures described above for the silicon substrates. After deposition of the connection electrodes, the polyimide film was removed from the silicon surface by dissolving the oxide layer using a 1% HF solution. To prevent the nanotubes from being dislodged during the removal process, the polyimide coated silicon surface was held by forceps so that only an edge of the surface is exposed to the HF. As the HF dissolves the oxide layer, capillary forces draw the HF up along the polyimide-silicon oxide interface, resulting in removal of the polyimide film while protecting the nanotubes from the HF solution. After removal of the film, the polyimide substrate was gently rinsed with de-ionized water to remove any residual HF.

3 RESULTS AND DISCUSSION

The lithographic patterns employed for these devices were designed to maximize the concentration and alignment of SWCNTs in a relatively small area. An optical micrograph of the larger gold wires and alignment marks for electron-beam lithography deposited using photolithography is shown in Figure 1A, and an AFM topography image showing the alignment electrodes is shown in Figure 1B. Finite element simulation of the potential and electric field near an alignment electrode is show in Figure 1C, and indicates that this design is ideal for the concentration and alignment of nanotubes in a relatively small area. The potential on the surface decreases rapidly a short distance away from the electrode, while the electric field is strongest at the tip and oriented toward the second electrode. The addition of the connection electrodes *after* the SWCNT deposition step helped to ensure adequate electrical contact for probing the nanotubes, as the electrodes were formed on top of the nanotubes on the surface. They also provide an anchor for the nanotubes, thus minimizing slippage on the surface. An AFM topography image of deposited SWCNT bundles and connection electrodes is shown in Figure 1D. The nanotube bundles can be clearly seen to span between the two alignment electrodes, and only where the electric field is strongest (i.e. between the two tips of the alignment electrodes). It can also be seen that despite the low concentration of nanotubes used for deposition, some spurious bundles were still deposited during this process. However, they were typically away from the electrodes and caused no interference with the measurements.

Electrical connection between the connection electrodes and the immobilized SWCNTs was investigated using a conventional four-point measurement technique. The measured current as a function of applied voltage is shown in Figure 2 and exhibited the expected linear behavior with a resistance of 10.5 k . The relatively small resistance indicated that the immobilized bundle might have contained several individual nanotubes, thus lowering the resistance.

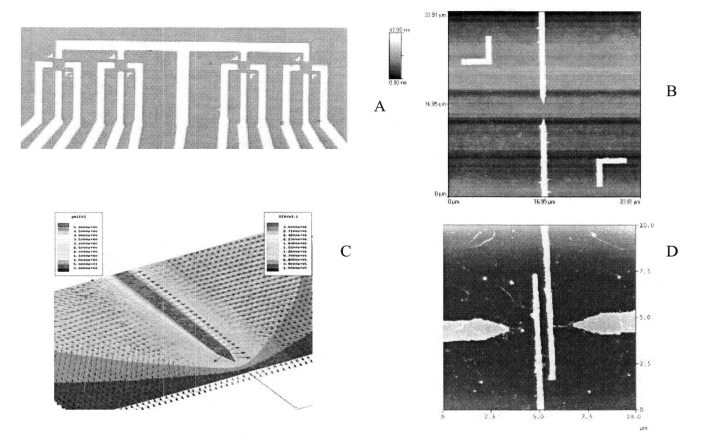

Figure 1. Procedure for designing a SWCNT-based structural health monitor. A) Optical micrograph showing large gold wires and alignment marks deposited using photolithography. B) AFM topography image showing alignment electrodes deposited using electron beam lithography. C) Finite element simulation of the potential and the electric field near an alignment electrode. D) AFM topography image showing a deposited and aligned SWCNT bundle and the deposited connection electrode.

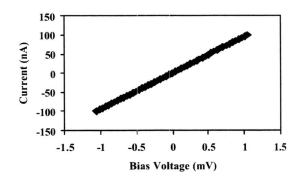

Figure 2. Typical I-V curves measured through a pair of connection electrodes with deposited and aligned SWCNT bundles bridging them. The measurements were obtained using a conventional four point measurement technique, and the calculated resistance of the bundles is ~10.5 k .

With careful control of the deposition conditions (concentration, applied voltage, alignment electrode separation, etc.), smaller bundles and even individual nanotubes can be deposited, resulting in measured resistances of several hundred k .

To determine the efficacy of this approach for fabricating sensors, an early prototype was mounted in an environmental chamber, and the resistance of the nanotubes was measured using a Manipulated Nanoprobe System. The resistance was monitored as the temperature was increased from 100 K to 350 K. The results are shown in Figure 3 and indicate a transition of some of the nanotubes in the immobilized bundles from semiconducting at low temperature to metallic behavior at higher temperatures. To investigate this design for measuring strain, the sensor will be mounted on an aluminum beam and placed in a load frame. To determine the actual strain applied and correlate the response of the sensor, conventional strain sensors will also be applied in the same area. Recent experimental work has demonstrated a dependence on the conductance on the strain of SWCNTs spanning the electrodes. In this work, a scanning probe

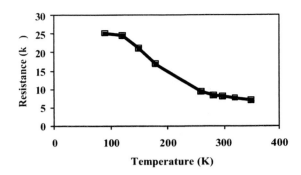

Figure 3. Measured resistance through aligned SWCNT bundles as a function of temperature.

microscope was used to produce in-plane strains to SWCNT ropes deposited across the junction [11].

4   CONCLUSIONS

A prototype design for using SWCNTs as sensors has been presented.   The sensor was designed using conventional electron-beam lithography and photolithography techniques to pattern alignment and connection electrodes. SWCNTs were deposited and aligned from a suspension using a dielectrophoretic technique. This design was chosen so that the concentration and alignment of the nanotube bundles occurred in specific, controllable locations on the substrate, and the prototype was demonstrated to be sensitive for temperature. Tests are currently underway to determine its sensitivity to applied strain. Finally, these sensors are small and robust, and can be easily manufactured on flexible, plastic-based substrates. These advantages will make the implementation of a surface-mountable and multifunctional (large arrays capable of measuring many different parameters) SWCNT-based sensor a near-term achievable goal.

REFERENCES

[1] S. Iijima, Nature, 354, 56-58, 1991.
[2] M. Dresselhaus, G. Dresselhaus and P. Avouris, Eds., "Carbon Nanotubes: Synthesis, Structure, Properties and Applications," Springer-Verlag: Berlin, 2001.
[3] S.J. trans, A.R.M. Verschueren and C. Dekker, Nature, 393, 49-52, 1998.
[4] R. Martel, T. Schmidt, H.R. Shea, T. Hertel and P. Avouris, Appl. Phys. Lett., 73(17), 2447-2449, 1998.
[5] J. Kong, N.R. Franklin, C. Zhou, M.G. Chapline, S. Peng, K. Cho and H. Dai, Science, 287, 622-625, 2000.
[6] P.G. Collins, K. Bradley, M. Ishigami and A. Zettl, Science, 287, 1801-1804, 2000.
[7] X.Q. Chen, T. Saito, H. Yamada and K. Matsuhige, Appl. Phys. Lett., 78(23), 3714-3716, 2001.
[8] L.A. Nagahara, I. Amlani, J. Lewenstein and R.K. Tsui, Appl. Phys. Lett., 80(20), 3826-3828, 2002.
[9] J. Smits, B. Wincheski, J. Ingram, N. Watkins and J. Jordan, "Controlled Deposition and Applied Field Alignment of Single Walled Carbon Nanotubes for CNT Device Fabrication," presented at the Materials Research Society Conference, December 2-6, 2002 and published in the proceedings.
[10] A. Maita, Nature Materials, 2, 440-442, 2003.
[11] J. Smits, B. Wincheski, P. Williams, J. Ingram, N. Watkins and J. Jordan, "In-Plane Deformation of Single Walled Carbon Nanotubes and its Effect on Electron Transport," presented at the Materials Research Society Conference, December 1-5, 2003, and published in the proceedings.
[12] P. Nikolaev, M.J. Bronikowski, R.K. Bradley, F. Rohmund, D.T. Colbert, K.A. Smith and R.E. Smalley, Chem. Phys. Lett., 313(1-2), 91-97, 1999.

Scanning Probe Lithography on InAs Substrate

Lionel F. Houlet[*], Hiroshi Yamaguchi[**] and Yoshiro Hirayama[***]

[*]NTT Basic Research Laboratories, NTT Corporation, Atsugi, Kanagawa 243-0198, Japan,
lhoulet@will.brl.ntt.co.jp
[**]NTT Basic Research Laboratories, NTT Corporation, Atsugi, Kanagawa 243-0198, Japan,
hiroshi@will.brl.ntt.co.jp
[***]NTT Basic Research Laboratories, NTT Corporation, Atsugi, Kanagawa 243-0198, Japan. SORST-
JST, Kawaguchi, Saitama 332-0012, Japan, hirayama@will.brl.ntt.co.jp

ABSTRACT

In this study, we focus our interest on patterning a conventional electron beam resist by electron field emission exposure using Atomic Force Microscope (AFM). We have fabricated 50-140 nm deep structures in InAs with the resolution less than 100 nm through 20 nm thick PMMA resist that was exposed. According to our knowledge, electron field emission exposure of resist has already been performed on gold, silicon but never on InAs substrate.

Compared with other semiconductors, InAs has the Fermi level pinned in the conduction band leading to the superior electric property even for nanometer scale structures. Using electron field emission exposure technique, structures deeper than 100 nm can be performed that is enough to release suspended free structures after chemical etching, being ideal for nanowire and NanoElectroMechanical Systems (NEMS).

In the following, we will describe the experimental set-up and results of fabrication.

Keywords: nanolithography, atomic force microscope, scanning probe lithography, nanowire, InAs.

1 INTRODUCTION

AFM is nowadays a versatile tool for nanotechnology and has been used to perform many processes like nanomachining [1-2], electron field emission exposure [3], field evaporation [4], anodic oxidation [5], thermomechanical writing [6], polymer crazing [7] and electrostatic nanolithography [8].

In this study we focus our interest on the electron field emission exposure process also called Scanning Probe Lithography (SPL). SPL has many advantages for nanolithography with respect to the closest competitive technologies: Electron Beam Lithography and Scanning Tunneling Lithography that use an electron beam apparatus or a scanning tunneling microscope respectively. SPL can be performed in air and the feedback control of the gap between the tip and the exposed surface is decoupled from the exposure mechanism allowing exposure on insulating and conductive substrate. Compared with other semiconductor systems, InAs is an attractive material for

nanotechnology, because the Fermi level is pinned in the conduction band leading to the superior electric property even for nanometer scale structures (about 15 nm [9]). That enables the fabrication of submicroscopic conductive structures with confined 2D electron gas in the near surface region [10-12].

2 EXPERIMENTAL SET-UP

Fig. 1 describes the set-up for our SPL experiment. The sample made of InAs substrate is coated by PMMA resist with a thickness of about 20 nm. Such thickness is obtained by dilution of PMMA in xylene that is a low vapor pressure solvent. During spin coating, such solvent has a slow evaporation rate resulting in very thin resist thickness [13].

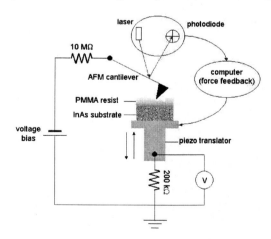

Figure 1: Schema of the scanning probe experiment.

A short length AFM cantilever, with a spring constant of about 40 N/m, was used in order to decrease the oscillations of the tip due to electrostatic forces between the resist and the cantilever. The tip of a radius of 20 nm was coated with a 30 nm titanium layer that has good mechanical and electrical properties. The tip is first approached to the surface of the resist. The force feedback of the AFM is set in order to maintain a repulsive contact force between the tip and the resist of about 1 nN. A DC voltage of about 20 Volts is then applied between the AFM tip and the backside of the substrate. High electrical resistance is added in the circuit in order to avoid electrical short cut if the tip falls inside a hole of the resist layer. The tip is negatively biased with respect to the substrate and a current of a few tenth of nano Ampere is measured. The metallic like behavior of InAs allows applying similar voltage as for gold surface [3].

For this voltage, PMMA resist behaves like a negative resist so that the exposed area corresponds to the final pattern after developing in acetone. The electrons are field emitted from the AFM tip and tunnel in air or a water meniscus before tunneling inside the resist. The energy from the electrons induces crosslinkings of the molecules in the exposed resist. After exposure, the patterns were transferred into the substrate by dry etching using a BCl_3 ECR-RIE system.

3   RESULTS OF FABRICATION

We have fabricated submicroscopic 50-140 nm deep structures in InAs through 20 nm thick PMMA resist using a commercial AFM. 3D structures like grating, square mound and wire were shaped (see Fig. 2 and 3).

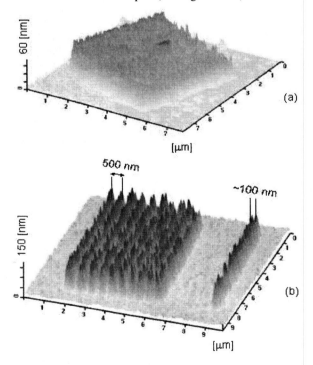

Figure 2: AFM topography of fabricated InAs structures: 5x5 μm^2 square mound (a) and 8 μm length grating (b).

The grating has a pitch of 500 nm and the width of the lines is about 100 nm that corresponds to the diameter of the AFM tip. The roughness observed on the structures in Fig. 2 may come from partial etching of the top of the

structures where the very thin resist mask might have been completely etched away.

According to our knowledge, SPL has already been performed on gold [3], silicon [14] but never on InAs substrate. The conductive property of InAs provides a large advantage in future technological applications, especially for nanowires [15] and NEMS [10]. The advantage of SPL with respect to Electron Beam Lithography [11] is that patterning the wire can be done without alignment marks and precisely positioned with scanning prior to exposure.

Fig. 3 shows an AFM image of a fabricated wire between two Ti/Au pads and a close view taken by SEM of the center part of the wire.

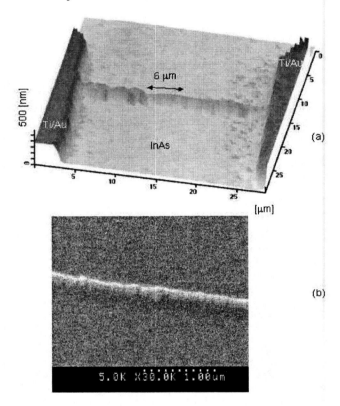

Figure 3: AFM topographic image of an InAs wire (a) and 40° tilted SEM view of the center part of the wire (b).

The Ti/Au pads were first fabricated and the resist mask was then prepared by SPL for patterning the wire. After developing in acetone, only the location of the wire is protected by the resist mask so that the pads and the unexposed area of InAs substrate are dry etched. We are currently working on obtaining good contact between the wire and the pads.

A cross section view of the center part of the wire shows that a minimum width of about 100 nm is obtained (Fig. 4). Due to the finite size of the tip, the foot of the wire seems to be larger than the actual size. Smaller width can be obtained if the applied voltage between the AFM tip and the substrate is decreased and a current feedback is used to control the variation of the field emission current with respect to the roughness of the exposed surface.

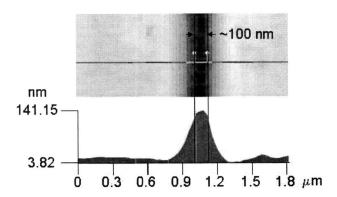

Figure 4: AFM Height profile of the center part of the wire.

AFM anodic oxidation was already performed to produce relatively shallow InAs-based structures [12]. 70 nm deep trenches in GaSb/InAs heterostructures were also obtained by mechanical cutting of 10 nm deep trenches with an AFM tip, followed by wet etching [2]. However, structures deeper than 100 nm can be performed using SPL that is enough to release suspended free structures after chemical etching, being ideal for nanowires and NEMS device fabrications.

4 CONCLUSIONS

In this study, the use of the Scanning Probe Lithography is reported on InAs substrate for the first time.

Submicroscopic 50-140 nm deep structures have been fabricated in InAs through 20 nm thick PMMA resist using a commercial AFM. 3D structures like grating, square mound and wire between Ti/Au pads were shaped. A wire width of less than 100 nm is achieved.

We are currently working on improving that structure to get a wire isolated from the substrate in order to study the electro-magnetic properties of InAs wires.

This study is partly supported by NEDO International Joint Research Program, "Nano-elasticity".

REFERENCES

[1] Y. Kim, C.M. Lieber, Science, 257, 375, 1992.
[2] R. Magno, B.R. Bennett, Appl. Phys. Lett., 70 (14), 1855, 1997.
[3] A. Majumdar, P.I. Oden, J.P. Carrejo, L.A. Nagahara, J.J. Graham, J. Alexander, Appl. Phys. Lett., 61 (19), 2293, 1992.
[4] S. Hosaka, H. Koyanagi, A. Kikukawa, Jpn. J. Appl. Phys., 32, L464, 1993.
[5] H.C. Day, D.R. Allee, Appl. Phys. Lett., 62(21), 2691, 1993.
[6] H.J. Mamin, D. Rugar, Appl. Phys. Lett., 61(8), 1003, 1992.
[7] A.S. Körbes, L.J. Balk, J.W. Schultze, Mater. Phys. Mech., 5, 56, 2002.
[8] S.F. Lyuksyutov, R.A. Vaia, P.B. Paramonov, S. Juhl, L. Waterhouse, R.M. Ralich, G. Sigalov, E. Sancaktar, Nat. Mat., 2, 468, 2003.
[9] H. Yamaguchi and Y. Hirayama, Jpn. J. Appl. Phys. 37, 1599, 1998.
[10] H. Yamaguchi, Y. Hirayama, Appl. Phys. Lett., 80 (23), 4428, 2002.
[11] M. Inoue, K. Yoh, A. Nishida, Semicon. Sci. Technol., 9, 966, 1994.
[12] S. Sasa, T. Ikeda, K. Anjiki, M. Inoue, Jpn. J. Appl. Phys., 38, 480, 1999.
[13] D.E. Haas, J.N. Quijada, S.J. Picone, D.P. Birnie, SPIE Proc. 3943, 280, 2000.
[14] S.W. Park, H.T. Soh, C.F. Quate, S.-I. Park, Appl. Phys. Lett., 67 (16), 2415, 1995.
[15] C. Thelander, T. Martensson, M.T. Björk, B.J. Ohlsson, M.W. Larsson, L.R. Wallenberg, L. Samuelson, Appl. Phys. Lett., 83 (10), 2052, 2003.

A New Wet Chemical Approach for the Selective Synthesis of Silver Nanowires

Emil Avier Hernandez*, Bertina Posada*, Roberto Irizarry** and Miguel E. Castro*

*Chemical Imaging Center, The University of Puerto Rico at Mayaguez, Mayaguez, Puerto Rico 00680, mcastro@uprm.edu
**Dupont Microelectronics, Manati, Puerto Rico
Roberto.irizarry@pri.dupont.com

ABSTRACT

A new wet chemical approach is presented for the selective synthesis of silver nanowires based on the assembly of silver nanoparticles covered with a bi-functional thiol. The particles are formed spontaneously upon the addition of trace amounts of the thiol to silver nitrate. Dilution of the particle with water leads to the formation of a clear dispersion of white flakes. The silver nanowires are formed while drying the flakes in air. The silver nanowires are characterized with scanning electron microscopy (SEM) and have aspect ratios of about 46. The results are discussed in the context of an assembly mechanism driven by intermolecular interactions among thiol molecules.

Keywords: silver nanowires, silver nanoparticles, assembly of particles into wires, nanowires SEM

1 INTRODUCTION

There is tremendous interest in the synthesis of silver nanowires due to the important role they are to play in the manufacture of electronic devices. Long aspect ratio silver nanowires may find applications in nanoscaled electrical circuits requiring long electrical connections. Wires with short aspect ratios find applications relevant to the wireless telecommunication industries. Here we report on a simple, wet chemical approach to the selective synthesis of silver nanowires of varied aspect ratios. The method is based in the assembly of silver nanoparticles using a bi-functional alkyl thiol, as illustrated in scheme 1.

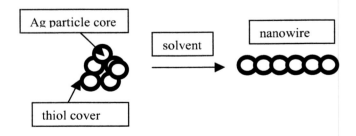

Scheme 1: assembly of silver nanoparticles into nanowires.

2 EXPERIMENTAL

The silver nanoparticles are prepared in a glass reactor. The reactor has a cylindrical shape and is 5 cm long and 1 cm in diameter. Scanning electron microscopy measurements were performed in a JEOL 6460 HV/LV instrument. Fourier transform infrared (FTIR) measurements reported here were performed in a Matheson FTIR instrument coupled to an Olympus microscope. All FTIR microscopy measurements reported here were performed in the surface reflection mode with an aluminum coated objective and with a times thirty (30 x) magnification. UV absorption measurements were performed with a deuterium lamp with an Ocean Optics 2000 spectrograph.

3 RESULTS
3.1 Synthesis of thiol-silver material

A yellow material is spontaneously formed upon the addition of a few μL of thiol to 2 mg of silver nitrate. The yellow material is then diluted with 10 mL of water and the reactor is closed. A bright yellow solution is immediately

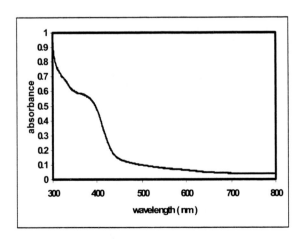

Figure 1: UV-visible absorption spectrum of a thiol-silver material diluted with 10 mL of solvent.

formed. The yellow solution turns to a bright green color a few seconds after the mixture is prepared. A clear dispersion of white flakes forms a few minutes after the formation of the green solution. The dispersion was stored in the dark to avoid the breakdown of the thiol and release of silver nanoparticles into the aqueous solution.

A typical UV-visible absorption measurement of the dispersion, prepared as described above, is displayed on figure 1. A band centered at 388 nm is the only peak observed in UV-visible absorption measurements. This band is typically found in silver nanoparticles a few nanometers in size. The lack of bands due to the transverse and longitudinal plasmon modes indicates that no other silver nanostructures, like nanowires or nanorods, are formed at this point in the synthesis process.

3.2 Synthesis of silver nanowires

A representative SEM image on a deposit from the solutions described above is displayed on figures 2a and 2 b. The deposit was allowed to dry in the dark for over 24 hours prior to the insertion in the SEM chamber. Silver

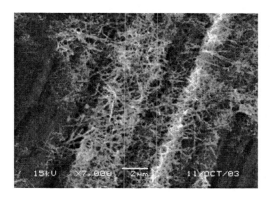

Figure 2: SEM images of two different regions of deposits of thiol-silver materials diluted with 10.0 mL of solvent.

nanowires are the only nanostructures observed. Evidence for the formation of silver nanospheres or rods was searched for, but not found. The wires displayed on figure 2b are (12 + 3) mm in diameter and (260 ± 40) nm in diameter and have an average aspect ratio of 46. Significantly, not a single particle is observed in SEM images of these deposits, indicating that the method is very selective for the synthesis of nanowires.

3.3 Mechanism of silver nanowires synthesis

A representative SEM image on deposits of the thiol-silver deposit following the addition of a trace amount of solvent is displayed on figure 3. The SEM image is characterized by a large number of agglomerates of particles. The agglomerates range from 10 to 1 micrometer in size. The particles forming the agglomerates are about (300 ± 100) nm in diameter.

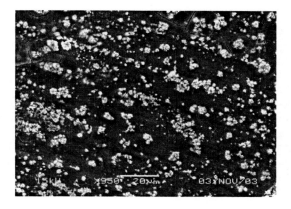

Figure 3: SEM image of deposit of silver-thiol material wetted with a trace amount of solvent.

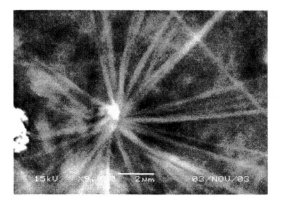

Figure 4: SEM image of a silver-thiol material following the addition of 500 μL of solvent.

Silver nanowires emerge from these agglomerates upon the addition of larger amounts of solvent, as illustrated in the SEM image displayed in figure 4. The wires are about 200 nm in diameter and do not have any specific orientation. Taken together, the SEM images presented in figures 3 and 4, are consistent with the aseembly of the silver nanoparticles into wires assisted by the solvent.

4 DISCUSSION

Silver nanoparticles are formed from the reduction of silver cations with mercapto acetic acid. This conclusion is supported by the observation of the 390 nm band in the UV-visible absorption measurements of dispersions prepared by diluting the pastes with water. The observation of particles in the SEM measurements performed on fresh silver-thiol materials is consistent with this conclusion. Furthermore, the SEM measurements on deposits prepared from the addition of trace amounts of water to these materials provide further support to this conclusion. Particle and particle agglomerates are formed upon the addition of 100 µL of water to the paste. Silver nanowires are formed when pastes are mixed with larger amounts of water. The lack of absorption bands due to the longitudinal and transverse plasmon modes in UV-absorption measurements of silver-thiol materials diluted with water indicates that the wires are formed while drying the deposits. The observations listed above lead us to conclude that the nanowires form from the assembly of silver nanoparticles, as illustrated in scheme 1. The assembly of particles into wires is mediated by the thiol, as no wires are observed in a typical silver nitrate reduction. Water plays a central role in the synthesis of the silver nanowires described here. Water partially removes the thiol from the particles, similar to its role in the preparation of photo resists and masks. The partial removal of thiol and sulfur oxides from the particle surface facilitates the combination of particles and the formation of wires.

5 ACKNOWLEDGEMENT

Financial support from the National Science Foundation, Nanotechnology Undergraduate Education Program (award number 0304348) is gratefully acknowledged. EAH acknowledges financial support from the Puerto Rico AMP program.

6 REFERENCES

[1] S. Bhattacharrya, S.K. Saha, and D. Chakravorty, Applied Physics Letters, 76(26), (2000), page 3896.

[2] S. Bhattacharrya, S.K. Saha, and D. Chakravorty, Applied Physics Letters, 77 (23), (2000), page 3770.

[3] G. Sauer, G. Brehm, S, Schneider, K. Nielsch, R.B. Wehrspohn, J. Choi, H. Hofmeister and U. Gosele, Journal of Applied Physics, 91(5), (2002), page 3243.

[4] Miaden Barbic, Jack J. Mock, D.R. Smith and S. Shultz, Journal of Applied Physics, 91(11), (2002), page 9341.

[5] Jia-Yaw Chang, Jia-Jiu Chang, Bertrand Lo, Shin_hwa Tzing, Yong-Chien Ling, Chem. Phys. Letters, 379, (2003), pages 261-267.

[6] Yiwei Tan, Lei Jiang, Yongfang Li, and Daoben Zhu, J. Phys. Chem. B. 106, (2002), pages 3131-3138.

[7] Andrea Tao, Franklin Kim, Christian Hess, Joshua Goldberg, Rongrui He, Yugang Sun, Younan Xia and Peidong Yang, Nanoletters, 3(9), (2003), pages 1229-1233.

Thermal Noise Response Based Static Non-Contact Atomic Force Microscopy

Anil Gannepalli*, Abu Sebastian*, Murti V. Salapaka* and Jason P. Cleveland**

* Department of Electrical and Computer Engineering, Iowa State University
Ames, Iowa 50011, USA, murti@iastate.edu
** Asylum Research, 341 Bollay Drive
Santa Barbara, CA 93117, USA, jason@AsylumResearch.com

ABSTRACT

Micro-cantilever based devices have revolutionized imaging and they are the primary tools for investigation and control of matter at the nanoscale. In this paper a novel approach based on the thermal noise response of the cantilever is developed that makes non-contact AFM possible in static mode. This technique exploits the dependence of cantilever's resonant frequency on the tip-sample separation to maintain a small tip-sample separation by regulating the equivalent resonant frequency. The resonant frequency is estimated from cantilever's response to the thermal noise. The experiments performed in ambient room conditions have achieved tip-sample separations as small as 4 nm for time periods in excess of 20 min. Based on this control technique a new static non-contact mode operation of AFM has been demonstrated. This method has given rise to an extremely powerful non-contact imaging technique capable of detecting sub-angstorm features at a bandwidth of 200 Hz with a force sensitivity of a few pN.

Keywords: thermal noise response, cantilever resonance, non-contact force microscopy, static mode operation, tip-sample separation control

1 INTRODUCTION

In many putative studies a micro-cantilever based investigation of extremely small forces evolving over large time scales is of considerable interest. One such application is the study of conformational changes of proteins that is fundamental to investigating the function of proteins. Another application is the detection of single electron spin that is a key requirement for quantum computing. In these applications in order to detect the highly localized forces it is essential to maintain a tip-sample separation in the order of a few Å over extended periods of time.

In such applications the cantilever tip is too obtrusive to the observation if it encounters the repulsive region of the tip-sample potential. This necessitates maintaining the tip in the attractive regime of the tip-sample potential. In many samples this also achieves the objective of maintaining sub-nanometer tip-sample separation. A primary hurdle in achieving this goal is the drift of the system that becomes particularly detrimental due to the large time scales involved. These drift effects are due to uncertain factors like changes in the deflection detector [3], [6], thermal bending [5], [7] and drift in the piezo based sample positioner. Deflection based force detection cannot differentiate between attractive and repulsive interactions thereby making it unsuitable for maintaining the tip in the attractive regime. The dynamic modes of operation, viz., amplitude modulation (AM) [2] and frequency modulation (FM) [1], are not applicable due to the large amplitude oscillations of the cantilever tip that are needed in these methods.

In this letter we present an approach based on the thermal noise response of the cantilever that promises to meet the demands of maintaining sub-nanometer separations over large time periods and consequently enable a static non-contact mode operation of the AFM.

2 MODEL

The microcantilever modeled as a single spring-mass-damper system, as shown in Figure 1, is described by

$$m\ddot{p}(t) + c\dot{p}(t) + kp(t) = \eta(t) + F(t) \quad (1)$$

where $p(t)$ is the cantilever deflection, m is the mass of the cantilever, c ($= \frac{m\omega_0}{Q}$) is the damping coefficient , k is the spring constant and $\eta(t)$ is the Langevin thermal noise forcing term and $F(t)$ describes other external forces acting on the cantilever. The noise power spectral density of a cantilever in thermal equilibrium in the absence of external forcing is given by

$$S_{pp}(\omega) = \frac{2k_B T}{m} \frac{\gamma}{(\omega_0^2 - \omega^2)^2 + \gamma^2 \omega^2} \quad (2)$$

where k_B is the Boltzmann's constant, T is the temperature, $\omega_0 = \sqrt{k_0/m}$ is the resonant frequency of the cantilever and $\gamma = c/m$ is the damping constant.

When the tip interacts with the sample, the tip-sample forces alter the effective spring constant thereby changing the resonant frequency. For small tip-sample forces the resonant frequency shift $\Delta\omega$ can be approximated by the relation

$$\frac{\Delta\omega}{\omega_0} = \frac{k_s}{2k}, \quad k_s = \frac{\partial F_s}{\partial l}. \quad (3)$$

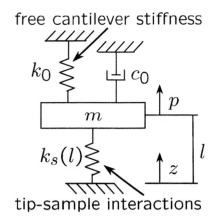

free cantilever stiffness

k_0 c_0

m p

$k_s(l)$ z l

tip-sample interactions

Figure 1: The cantilever is modeled by the spring-mass-damper system. The effect of the tip-sample interactions is modeled by a nonlinear spring whose stiffness depends on the tip-sample separation.

where k_s is the gradient of the tip-sample interaction force F_s with respect to the tip-sample separation l. The equivalent resonant frequency ω_{eq} ($= \omega_0 + \Delta\omega$) decreases (increases) when the tip-sample interaction force is attractive (repulsive). Thus, by observing the equivalent resonant frequency the attractive and repulsive regimes of the interaction potential can be differentiated. The information about ω_{eq} is available in the power spectral density of the thermal noise response as a shift in the peak position of the power spectrum. The main contribution of this letter is to utilize this fact to estimate the equivalent resonant frequency from the power spectrum to control the tip-sample separation. One fundamental requirement is that the cantilever spring constant be large enough to overcome any jump-to-contact instabilities.

3   STATIC NON-CONTACT MODE OPERATION

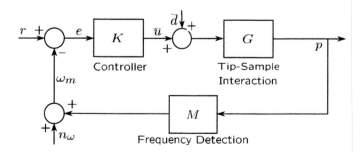

Figure 2: A schematic block diagram of the closed loop

As described above maintaining a constant tip-sample separation translates into a problem of regulating the resonant frequency of the cantilever at a desired value as illustrated in the proposed control scheme in Figure 2.

The estimation of frequency from the cantilever's thermal noise response is performed by Pisarenko harmonic decomposition [4]. The deflection signal is assumed to be a single sinusoid buried in noise. The frequency of this sinusoid corresponds to the equivalent resonant frequency of the micro-cantilever.

The controller (K) should be designed such that it is capable of compensating for the disturbances afflicting the system due to drift. Any other disturbance that has a bandwidth greater than the closed-loop bandwidth of the system, will go unchecked by the controller and will potentially show up as a variation in the cantilever's resonance. This is the principle behind the use of this technique for imaging to monitor and observe variations in tip-sample interaction forces.

4   EXPERIMENTAL RESULTS

The thermal noise based non-contact mode operation has been demonstrated in a variety of experiments few of which are discussed below. The experiments are performed on a Digital Instruments Multimode AFM in ambient environment. The signal processing for frequency estimation and the controller are implemented on a TMS320C44 digital processing platform. The cantilevers used are of Silicon with a nominal $Q = 300, k = 40$ N/m and $\omega_0 = 350$ kHz and the sample surface was freshly cleaved HOPG.

Figure 3(a) shows the variation in the cantilever eigenfrequency as a function of tip-sample separation during approach and retraction. It is seen that the resonant frequency decreases due to the long range attractive tip-sample interactions. However, a similar effect is not observed in the deflection as the maximum observable deflection (see Figure 3(b)) estimated to be approximately 4 pm is much smaller than the deflection sensitivity of the instrument.

The following experiment demonstrates the feasibility of the proposed method to control the tip-sample separation. In this experiment a step change is given to the reference frequency and the recorded cantilever resonant frequency, control signal and deflections are shown in Figure 4(a), (b) and (c). In the initial stages of the control, the tip is not interacting with the sample and the measured resonance is 397 kHz (see Figure 4(a)), which is the free resonant frequency of the cantilever. The controller, therefore, acts to move the sample towards the tip as seen in Figure 4(b). Once the desired tip-sample separation is achieved, indicated by the resonant frequency being close to the reference, the control action counteracts the drift in the instrument. At approximately 1600 s into the experiment the step change in the reference is introduced and the controller is able to track this change. As the reference is reduced, implying a smaller desired tip-sample separation, the controller moves the sample towards the tip and is seen as

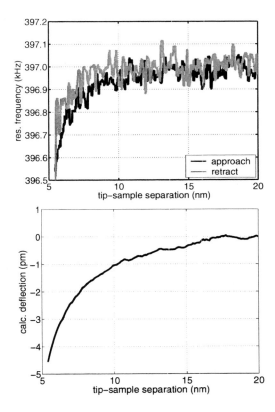

Figure 3: The variation in (a) resonant frequency with tip-sample separation. (b) The calculated deflection $p_{calc} = \frac{2}{\omega_0} \int \Delta\omega dl$ for the frequency observed in (a).

a small "jump" in Figure 4(b) at 1600 s shown by an arrow and magnified in the lower left inset. This control action results in a reduction in tip-sample separation of approximately 0.8 nm as seen in the upper right inset. This correlates well with the change required for a reduction in the resonance from 396.9 kHz to 396.75 kHz (see Fig 3(a)). The new reference is reached in approximately 7 s. As reasoned earlier, the variations in deflection in Figure 4(c) can be attributed to the drift in the deflection sensor as the tip-sample forces are too small to induce any perceivable change in the deflection. Figure 4(d) shows the estimated tip-sample distance is 6 nm in good agreement with the corresponding separation for a resonant frequency of 396.75 kHz in Figure 3(a). From the above discussion and Figure 4 it can be inferred that a tip-sample separation of approximately 7 nm is being maintained for over 15 min (from 600 s to 1600 s) and a separation of around 6 nm for time periods in excess of 5 min (from 1610 s to 1930 s). Similar experiments have yielded a tip-sample separation of under 3 nm for over 20 min.

Figure 5 demonstrates the non-contact imaging capability of this method. In this experiment the force gradient is modulated by moving the sample in a sinusoidal manner while the tip-sample separation in being

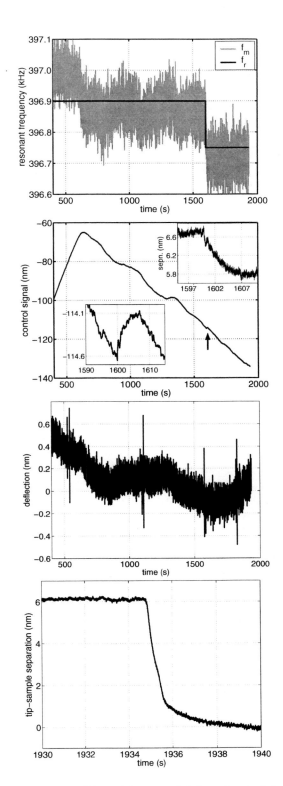

Figure 4: Time history of (a) estimated frequency (b) control effort and (c) deflection while tracking a step change in reference frequency shown in (a). The insets in (b) show the step response in tip-sample separation (upper right) and control effort (lower left). (d) Tip-sample separation just before tip-sample contact.

maintained. The resonant frequency of the cantilever in Figure 5(a) during control is lower than its free resonance, signifying the non-contact mode of operation. The drift compensation seen in Figure 5(b) indicates that the closed-loop bandwidth of 0.5 Hz is larger than the bandwidth of the drift processes in the system. During control the sample position is modulated by a 0.5 Å, 5 Hz sine wave such that the modulation frequency is greater than the closed-loop bandwidth and hence will not be acted upon by the controller. This results in a modulation of the tip-sample separation and consequently the cantilever's resonance modulates at 5 Hz as seen in Figure 5(c). No signature of this modulation is seen in the deflection signal (see Figure 5(d)) indicating an inferior sensitivity to the frequency based gradient detection. The changes in tip-sample forces induced by the modulation of the sample position have been estimated to be a few pN. Recent experiments have resulted in the detection of tip-sample modulations of 0.25 Å up to 200 Hz.

5 CONCLUSIONS

A novel static non-contact mode of operation of AFM based on the thermal noise response of the cantilever has been demonstrated. In this approach cantilever's thermal noise response is used to estimate the changes in its resonant frequency that is fed back for maintaining the tip-sample separation. This method enables an extremely powerful non-contact imaging technique in static mode with bandwidths up to 200 Hz and force sensitivity of a few pN observed in experiments performed in ambient room conditions. Tip-sample separations as small as 4 nm for periods extending over 20 min have been achieved. A better design of instrumentation and controlled experimental conditions promise improved performance of this technology.

REFERENCES

[1] T. R Albrecht, P. Grütter, D. Horne, and D. Rugar. *J. App. Phys.*, 69(2):668, January 1991.

[2] Y. Martin, C. C. Williams, and H. K. Wickramasinghe. *J. App. Phys.*, 61(10):4723, May 1987.

[3] Gerhard Meyer and Nabil M. Amer. *Appl. Phys. Lett.*, 53(24):2400, December 1988.

[4] V. F. Pisarenko. *Geophysics. J. Roy. Astron. Soc.*, 33:347, 1973.

[5] M. Radmacher, J. P. Cleveland, and P. K. Hansma. *Scanning*, 17(2):117, 1995.

[6] D. Rugar, H. J. Mamin, R. Erlandsson, J. E. Stern, and B. D. Terris. *Rev. Sci. Instrum.*, 59(11):2337, November 1988.

[7] M. B. Viani, T. E. Schäffer, and A. Chand. *J. App. Phys.*, 86(14):2258, August 1999.

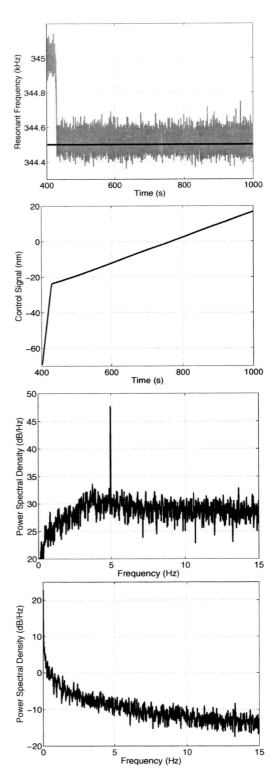

Figure 5: (a) Cantilever's natural resonance is at 345 kHz. A lower resonant frequency (344.5 kHz) during control signifies the tip in attractive region. (b) Drift compensation by the controller. (c) Power spectral density plot of the estimates of the resonant frequency - modulation of the resonant frequency at 5 Hz (d) Power spectral density plot of cantilever deflection. No modulation in the deflection is observed

Metal-Insulator-Metal Ensemble (MIME) Chemical Detectors

R.R. Smardzewski[1,], N.L. Jarvis[2] A.W. Snow[3], H.Wohltjen[4] & R.A. Mackay[2]

[1]Geo-Centers, Inc., Gunpowder Branch PO Box 68
Aberdeen Proving Ground, MD 21010-0068
[2]U.S. Army Edgewood Chemical Biological Center
Aberdeen Proving Ground, MD 21010-5424
[3]Naval Research Laboratory, Washington, DC 20375
[4]Microsensor Systems, Inc., Bowling Green, KY 42103

ABSTRACT

A new class of nanometer-scale, low power, solid-state devices is being investigated for the detection of hazardous vapors. These chemical vapor sensors are comprised of nanometer-sized gold particles encapsulated by monomolecular layers of alkanethiol surfactant deposited as thin films on interdigitated microelectrodes (Fig. 1). These new, alkylthiol-stabilized, gold nanocluster materials are appropriately categorized as metal-insulator-metal ensembles (MIME). When chemical (agent, hazmat) vapors reversibly absorb into these thin MIME films, a large modulation of the electrical conductivity of the film is observed. The measured tunneling current between gold clusters is extremely sensitive to very small amounts of monolayer swelling or dielectric alteration caused by absorption of vapor molecules. For chemical agent simulants, a large dynamic range (5-logs) of sensitivities is observed and extends down to well below sub-ppm vapor concentrations. Tailored selectivities of the sensors are accomplished by incorporation of chemical functionalities at the terminal structure of the alkanethiol surfactant or substitution of the entire alkane structure. Current research efforts are focused on examining the molecular mechanism(s) of conduction and mapping the selectivity and sensitivity of sensor elements. Targeted applications include: low-cost, low-power CB agent sensors, filter residual life indicators and orthogonal detector applications.

1. SENSING PRINCIPLE

Chemical vapors, reversibly absorbing into these thin films, effect large changes in the electrical conductivity of the film (Fig. 1). Tunneling currents between gold clusters are extremely sensitive to very small amounts of monolayer swelling or dielectric alteration caused by absorption of vapor molecules. Response times are extremely fast for monolayers. Selectivity depends on chemical functionalization of the alkanethiol. For simulants, a large dynamic range (5-logs) of sensitivities is observed and extends down to well below sub-ppm vapor concentrations.

2. CONCEPT

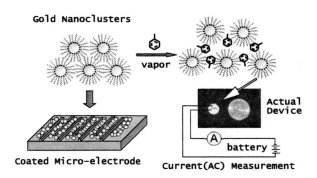

Fig. 1. MIME nanosensor concept.

3. NANOCLUSTER SIZES

Individual nanoclusters (Fig. 2) are prepared by the controlled reduction of gold chloride in the presence of suitable alkylthiols. As colloidal gold particles form, a monomolecular layer of alkythiol molecules absorbs on their surface producing a highly stable dispersion of uniformly sized gold particles. Colloidal dimensions (nanometers) are determined by the molecular ratio of gold to alkylthiol reactants.

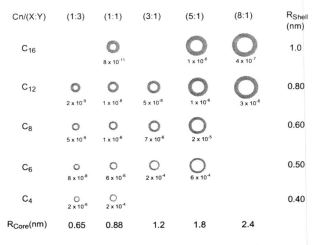

Cn/(X:Y)	(1:3)	(1:1)	(3:1)	(5:1)	(8:1)	R_Shell (nm)
C_{16}		8×10^{-11}		1×10^{-8}	4×10^{-7}	1.0
C_{12}	2×10^{-9}	1×10^{-8}	5×10^{-8}	1×10^{-6}	3×10^{-6}	0.80
C_8	5×10^{-9}	1×10^{-6}	7×10^{-6}	2×10^{-5}		0.60
C_6	8×10^{-8}	6×10^{-6}	2×10^{-4}	6×10^{-4}		0.50
C_4	2×10^{-6}	2×10^{-4}				0.40
R_{Core}(nm)	0.65	0.88	1.2	1.8	2.4	

Fig. 2. Sizes & conductivities of gold nanoclusters.

As might be expected for unfunctionalized thiol ligands, electrical conductivities (ohm-cm^{-1}) of these nanoclusters increase with an increase in core size or a decrease in ligand shell thickness.

4. SENSOR RESPONSE

Tailored selectivities of MIME sensors can be accomplished by incorporation of chemical functionalities at the terminal structure of the alkanethiol surfactant or substitution of the entire alkane structure. Differing responses are observed for the various functionalized MIME coatings. In some cases, in fact, a conductivity *__increase__* is observed upon exposure. This was observed for several phosphorus and nitrogen-containing compounds (viz., Fig. 3).

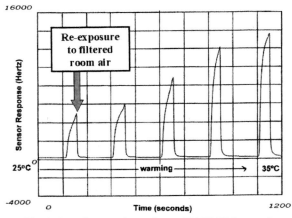

Fig 3. Headspace detection - 5% TNT in sand.

5. ADVANTAGES

Some significant advantages of this sensor class include:
- Low power devices (microwatts)
- Reversible sensing activity
- Fast responding (vis-à-vis polymer films)
- Large array size possible (minimal crosstalk)
- Large dynamic range (4-6 logs)
- Low humidity response

6. POTENTIAL APPLICATIONS

Potential applications (a non-exhaustive list) which are suggested include:
- Handheld chemical vapor monitors
- Drop-off/expendable chemical sensors
- Filter residual life indicators
- Orthogonal detector applications
- Explosives detectors

7. PROTOTYPE

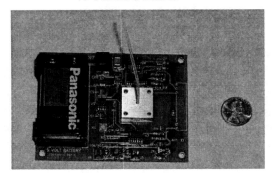

8. ACKNOWLEDGEMENT

Dr. N. Lynn Jarvis is currently working at the Edgewood Chemical & Biological Center under the auspices of a National Research Council Senior Resident Associateship.

9. REFERENCES

Jarvis, N.L., Snow, A.W., Wohltjen, H. and Smardzewski, R.R., "CB Nanosensors", 11th Int. Conf. Surf. & Coll. Sci., 15-19 Sep 2003, Iguassu Falls, Brazil.

Snow, A.W. and Wohltjen, H., "Size-Induced Metal to Semiconductor Transition in a Stabilized Gold Cluster Ensemble", *Chem. Mat.,* 10, No. 4, 947 (1998).

Silicon Nanostructures Patterned on SOI by AFM Lithography

I. Ionica[*], L. Montès[*], S. Ferraton[*], J. Zimmermann[*], V. Bouchiat[**] and L. Saminadayar[**]

[*]Institut de Microélectronique, Electromagnetisme et Photonique (IMEP), UMR CNRS-INPG-UJF 5130
23 Rue des Martyrs, BP257, 38016 Grenoble Cedex1, France
Phone +33 (0) 476 85 60 46   Fax : +33 (0) 476 85 60 70   E-mail : ionica@enserg.fr
[**]Centre de Recherche sur les Très Basses Températures (CRTBT), UPR CNRS 5001
25 Rue des Martyrs, BP166, 38042 Grenoble, France

ABSTRACT

The actual trends in microelectronics are the reduction of the dimensions and the search of new devices standing upon new phenomena as in the case of a Single Electron Transistor (SET) that is based on the Coulomb blockade. The nowadays limitation for device dimensions is that we are reaching the resolution limits of the lithography techniques. To go beyond this problem, we use the Atomic Force Microscope (AFM) nanolithography. To all the advantages brought by this technique we add those of using a silicon-on-insulator (SOI) substrate. In this article we are showing an example of a nanostructure fabricated by this method. Transport measurements and simulations performed on the device are in good agreement.

We demonstrate the potential of using AFM lithography fabricated devices for applications like multi-gate transistors.

Keywords: AFM lithography, SOI, single electron devices

1   INTRODUCTION

Scanning probe microscopy (SPM) is a reliable technique that allows imaging and modification of the surface structure of the materials down to nanometric scales. The main advantage of SPM tools is the use of near-field interactions that allow a precise positioning of the probe (tip) next to the surface of the sample, which implies atomic resolution in the vertical plane. Other advantages of AFM techniques are its compatibility with the actual CMOS technologies and the fact that there are no proximity effects like in the e-beam technique.

The probe can induce different kind of changes in the surface, for example the local oxidation of the surface of the sample by application of a voltage on the tip. The feasibility demonstration was made in 1990 by using another SPM technique, the Scanning Tunneling Microscopy (STM) oxidation on a Silicon surface [1]. Different ways of improving the technique were proposed : oxidation in tapping mode [2], the use of a pulsed tension on the tip [3]. We have chosen to use the oxidation by AFM in contact mode.

Another advantage of our technique is the use of silicon-on-insulator (SOI) substrates. They ensure very thin monocrystalline top Si films with high quality interfaces, that are very important if we want to validate a reproducible process to fabricate the nanostructures.

Using the AFM induced local oxidation on a SOI substrate, we have obtained silicon nanowires with lateral gates.

2   FABRICATION PROCESS

The principle of AFM lithography on SOI was described for the first time by Campbell et al. [4]. Later results by this technique are presented in the references [5], [6].

The first step is the passivation of the Si wafer during a HF (2%) treatment. This allows the saturation of the silicon surface in Si-H. The hydrogen atoms can be locally replaced by oxygen under the influence of a negatively biased tip, as shown in Figure 1. It is similar to an electrochemical anodization. The only precaution is that the tip voltage must be more negative than the threshold voltage (-2.7V) [7], [8].

As the process appears only if there is a thin layer of water at the surface of the sample, the humidity must be kept constant and about 30%-40%. The speed of the tip while drawing the oxide line is $0.1\mu m/s$. With these parameters we have obtained oxide patterns of 0.5-1.5nm high and about 50-70nm wide [9].

In figure 2 we are showing an example of a nanowire with lateral gates obtained by this technique.

The oxide patterns can be used as a mask during the silicon etching process. We are using a wet etching technique, by tetramethyl-amino-hydroxide (TMAH), that ensures a very high selectivity Si/SiO_2 (2000:1) [10].

The substrates used are Unibond ® silicon-on-insulator [11] samples with an ultra-thinned monocristalline Si layer down to 15nm and with a 400nm buried oxide thickness. The samples are doped by Arsenic ion implantation at 8keV, with two different surface doses : $5 \times 10^{11} cm^{-2}$ and $2 \times 10^{13} cm^{-2}$. This means that we can perform electrical measurements on samples of two doping levels, $2 \times 10^{17} cm^{-3}$ and $10^{19} cm^{-3}$.

Conductive pads are already present on the samples. They are doped by Phosphorus and their doping level is about $2.5 \times 10^{20} cm^{-3}$.

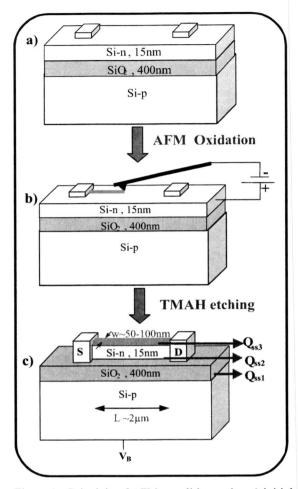

Figure 1 : Principle of AFM nanolithography: a) initial substrate, b) AFM oxidation, c) final structure after TMAH etching. In c), Q_{SS} represents the charge levels at each Si/SiO_2 interface.

Figure 2 : SiO_2 nanowire designed by AFM lithography. In this example the wire is about 2.5μm length, 50nm width and 1nm high.

3 BACKAGATE EFFECT ON I-V CURVES

The first electrical tests were done on a nanowire of 2.5μm length, 15nm height, 100nm width and $2\times10^{17}cm^{-3}$ doping level. The influence of the backgate voltage, V_B, on the I-V characteristics is shown in Figure 3.

We are observing the same kind of behavior as in a MOS transistor, normally on. For a negative bias on the backgate, the wire is in depletion and the resistance is decreasing as a space charge region appears and is increasing. For a positive bias, the wire is in accumulation.

On this wire, measurements were performed for biases down to −50V on the backgate; the wire was still conducting even if it should be completely deserted for V_B smaller then −3V. One hypothesis is that there is a native oxide layer that is formed on the surface and the interface between this oxide and the thin silicon layer does not have a very high quality. In this case, the conduction through the wire seems to be ensured by the charges located at the interface.

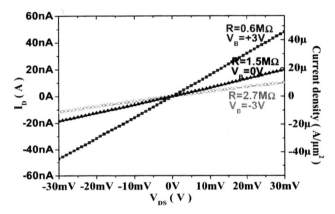

Figure 3 : Influence of the backgate voltage, V_B, on the I-V characteristics. The transistor-like behavior is evidenced.

4 ELECTRICAL SIMULATIONS FOR BACKGATE EFFECT

Simulations were performed using Silvaco simulator [12]. The structure simulated has the same geometrical and physical properties as the one used for electrical measurements (Figure 4). The influence of the backgate bias is the same as in the measurements for positive values. For negative values, the current through the nano-wire is zero. In this example, the interface charges have not been considered.

In Figure 5, we have introduced in our simulation interface charges at all the Si/SiO_2 interfaces (as shown in the Figure 1c). We can observe that the current for negative backgate biases is no longer zero, but the values of the resistance are changing quite a lot from those measured during the electrical characterization.

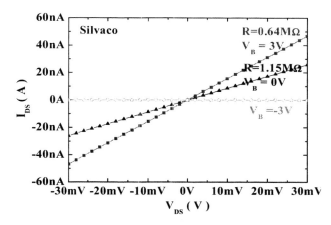

Figure 4 : Silvaco simulations without considering interface charges.

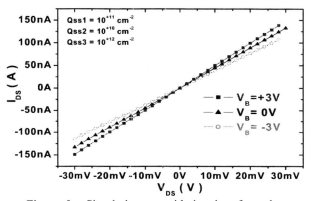

Figure 5 : Simulations considering interface charges. The position of the various Q_{SS} is shown in the Figure 1c.

5 SURFACE TREATMENTS

The fact that the interface charges seem to have such an importance on the conduction in the wire gave us the idea of trying to investigate the effect of surface treatments on SOI. We have tested the effect of a rapid thermal annealing (RTA) on the conduction through a SOI thin layer.

During the cleaved step we create defects in the structure of the thin layer of silicon and this induces an increase of the resistance value. If we add a rapid thermal annealing step, we are able to improve the qualities of this layer, so the resistance decreases (Table1).

Treatment	Resistance
SOI	65Ω
Cleaved SOI	106Ω
RTA SOI	82Ω

Table 1: Evolution of the resistance at different fabrication stages.

6 LATERAL GATES EFFECT ON I-V CURVES

The tests were done on a wire with two lateral gates having the same type of geometry as the first one, but with a doping level of $10^{+19} cm^{-3}$. Figure 6 shows the effect of the lateral gates bias on the conduction through the wire. The measurement was done without applying any voltage on the substrate. The shape of the I-V is the same as for a classical TMOS transistor.

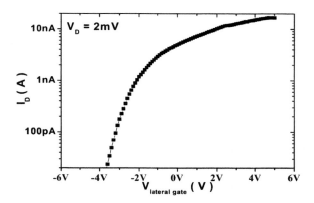

Figure 6 : Influence of the lateral gate voltage on the I-V characteristics

7 CONCLUSION

We have shown the feasibility of fabricating nanostructures on SOI substrate by using an AFM nanolithography technique. This tool is very promising for future applications because it is completely compatible with the actual CMOS technology, since we operate on SOI substrates, and it has the resolution required for the nanotechnology devices. The electrical characterization showed the influence of the backgate bias on the conduction in the wire. The same kind of behavior is observed for the lateral gates bias. So we have demonstrated the functioning of these nanostructures as multi-gates transistors.

Simulations had shown that the interface charges have a large impact on the conduction in the wire. Surface treatments like rapid thermal annealing (RTA) can improve the characteristics of the material.

In perspective, we shall make measurements of the electrical transport, especially at low temperatures, in order to study monoelectronic effects like Coulomb blockade. Comparisons between different architectures and different doping levels are going to be done. The impact of different surface treatments on the conduction will be studied.

The objectives of our work are to design nanostructures that can be used in applications based on monoelectronic devices: Single Electron Memory or Single Electron Transistor...

Acknowledgements

The authors are thanking D. Fraboulet, D. Mariolle and J. Gautier from LETI/CEA Grenoble for providing the SOI thinned down substrates.

REFERENCES

[1] J. A. Dagata, J. Schneir, H. H. Harary, C. J. Evans, M. T. Postek and J. Bennett, "Modification of hydrogen-passivated silicon by a scanning tunneling microscope operating in air", Applied Physics Letters, 56, 2001-3, 1990

[2] J. Servat, P. Gorostiza, F. Sanz, F. Perez Murano, N. Barniol, G. Abadal and X. Aymerich, "Nanometer scale lithography of silicon (100) surfaces using tapping mode atomic force microscopy", Journal of Vacuum Science & Technology A (Vacuum, Surfaces, and Films), 14, 1208-12, 1996

[3] B. Legrand and D. Stievenard, "Nanooxidation of silicon with an atomic force microscope: A pulsed voltage technique", Applied Physics Letters, 74, 4049-51, 1999

[4] P. M. Campbell, E. S. Snow and P. J. McMarr, "Fabrication of nanometer-scale side-gated silicon field effect transistors with an atomic force microscope", Applied Physics Letters, 66, 1388-90, 1995

[5] V. Bouchiat, M. Faucher, T. Fournier, B. Pannetier, C. Thirion, W. Wernsdorfer, N. Clement, D. Tonneau, H. Dallaporta, S. Safarov, J. C. Villegier, D. Fraboulet, D. Mariolle and J. Gautier, "Resistless patterning of quantum nanostructures by local anodization with an atomic force microscope", Microelectronic Engineering, 61, 517-22, 2002

[6] N. Clement, D. Tonneau, H. Dallaporta, V. Bouchiat, D. Fraboulet, D. Mariole, J. Gautier and V. Safarov, "Electronic transport properties of single-crystal silicon nanowires fabricated using an atomic force microscope", Physica E, 13, 999-1002, 2002

[7] F. Marchi, V. Bouchiat, H. Dallaporta, V. Safarov, D. Tonneau and P. Doppelt, "Growth of silicon oxide on hydrogenated silicon during lithography with an atomic force microscope", Journal of Vacuum Science & Technology B Microelectronics and Nanometer Structures. Nov. Dec., 16, 2952-6, 1998

[8] D. Stievenard, P. A. Fontaine and E. Dubois, "Nanooxidation using a scanning probe microscope: an analytical model based on field induced oxidation", Applied Physics Letters, 70, 3272-4, 1997

[9] K. Morimoto, F. Perez-Murano and J. A. Dagata, "Density variations in scanned probe oxidation", Applied Surface Science. May, 158, 205-16, 2000

[10] O. Tabata, R. Asahi, H. Funabashi, K. Shimaoka and S. Sugiyama, "Anisotropic etching of silicon in TMAH solutions", Sensors and Actuators A Physical. July, 1, 51-7, 1992

[11] See the site www.soitec.com for details about the UNIBOND®

[12] www.silvaco.com

On the Design of SET Adders

Mawahib Sulieman and Valeriu Beiu

School of Electrical Engineering and Computer Science, Washington State University
102 Spokane Street (EME), Pullman, WA 99164-2752, USA
Email: {mawahib,vbeiu}@eecs.wsu.edu

ABSTRACT

Single-Electron-Technology (SET) is one of the future technologies distinguished by its small and low power devices. SET also provides simple and elegant solutions for threshold logic gates (TLG's). This paper presents the design of an optimal TLG adder implemented in SET. It provides a detailed procedure for designing capacitive–input SET TLG's for building the adder. The paper also presents design details and characteristics (delay and power dissipation) of a 16-bit Kogge-Stone SET adder.

Keywords: Single-electron technology, threshold logic, adders.

1 INTRODUCTION

Up to this date, integration and scaling provided lower costs and higher performance circuits. However, as devices become smaller, many physical effects retard the advance of micro/nanoelectronics towards higher performance systems. Owing to the difficulty of successfully scaling conventional bulk CMOS technology to meet the increased performance, density and reduced power dissipation required for future technology generations, new technologies are being researched. Single-Electron-Technology (SET) is one of the emerging technologies, and is distinguished by *a very small device size* and *ultra-low power dissipation*. These two properties promise to allow large-density integration without exceeding the power density physical limits [1]. Besides, SET provides simple and elegant solutions for implementing threshold logic gates (TLG's). A TLG is more powerful than a Boolean gate and its principle of operation is different [2]. TLG's implement threshold functions expressed as:

$$f(x_1,...,x_\Delta) = sign\left(\sum_{j=1}^{\Delta} w_j x_j + \theta\right) \quad (1)$$

where w_j, θ, Δ are the weights, the threshold and the *fan-in* respectively. The TLG computes the weighted sum of its inputs, and compares this sum with a threshold value. If the sum is higher than the threshold, the TLG outputs a one, otherwise the output becomes a zero.

One of the SET logic circuits which has been the focus of several recent studies is the classical full adder [3]–[6]. The focus of these reports has been on a single-bit adder, and only a few articles extend it to multi-bit adders [3]. Even more, such extensions used simple adder structures and did not investigate advanced adder architectures.

In this paper we will describe an optimal structure for parallel-prefix TLG adders and its implementation in SET. At the gate level, the capacitive-input SET inverter (C-SET) lends itself well to the design of TLG's, and it has been used to design majority gates. This paper will generalize the design to arbitrary TLG's, and provide a detailed procedure for designing C-SET TLG's. With regard to characterizing the adder, we shall present and discuss both the delay and the power dissipation of our novel adder. In the literature about SET, power dissipation was only investigated for single gates (inverters), and was not reported for larger systems.

This paper is organized as follows. The architecture of the adder is described in section 2. This is followed by a detailed procedure for designing C-SET TLG's in section 3. Section 4 presents the simulation results of the 16-bit adder. Concluding remarks are provided in section 5.

2 ADDER ARCHITECTURE

Addition is among the functions which allow for simpler TLG implementations, *i.e.* shallow depth and polynomial size, while trading the size of weights for *fan-ins*. The sum function which is traditionally expressed as an XOR function, can also be represented in a linearly separable form. This can be done by writing it in terms of the *carry-in* (c_{i-1}) and *carry-out* (c_i):

$$s_i = (a_i \cdot b_i \cdot c_{i-1}) + [c_i^{\cdot} \cdot (a_i + b_i + c_{i-1})] \quad (2)$$

which may be represented as:

$$s_i = sign\ [a_i + b_i + c_{i-1} - 2c_i - 0.5] \quad (3)$$

having weights (1,1,1,2). This solution was detailed in 1969 [7], used later in 1997 [8], and rediscovered in 1999 [9]. All the fastest TLG adder solutions use this (1,1,1,2) TLG to produce the sum. This alleviates the need for an XOR gate that requires two layers of TLG's. The fastest TLG adders also use fast parallel-prefix architectures. The main difference between these ultra fast adders is their way of implementing the *carry-merge* stages. Basically, two different approaches have been proposed for designing the *carry-merge* layers in fast TLG adders.

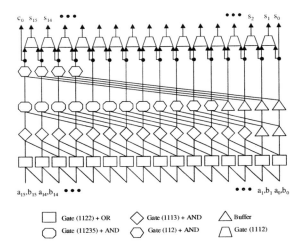

Fig. 1. TLG-optimized 16-bit adder.

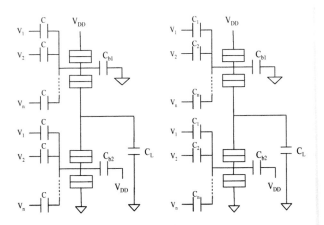

Fig. 2 Majority (left) and general TLG (right) capacitive-input SET gates.

One approach is based on *Fibonacci-weighted* TLG's: 1,1,2,3,... (*FIB*) [10], while the other uses *power-of-two* TLG's: 1,1,2,2,... (*PT*) [11]. The *FIB* approach has certain advantages. First of all, such TLGs can be immediately used in all the well-known architectures for Boolean adders. Secondly, its basic TLG has a small sum-of-weights for practical values of the *fan-in*. The main disadvantage is that it requires a first layer for computing the *propagate (p)* and *generate (g)* bits for each group of addend and augend bits. Obviously, this layer increases the overall depth of a *FIB* adder by one (when compared to *PT* adders). The *PT* approach does not require this first layer but its basic gate has a larger sum-of-weights than the *FIB* TLG. Our adder uses *PT* TLG's in the first layer and interfaces them to upper *FIB* layers being an optimal hybrid solution [12].

The adder architecture is shown in Fig. 1. It is based on the Kogge-Stone [13] adder. For $n = 16$, the adder consists of five layers:

- The first layer is a *PT* layer comprising the (1,1,2,2) TLG which produces 2-bit generate signals (G_i); and 2-input OR gates for the propagate signals (P_i).
- The second layer works as an interface to the upper *FIB* layers. The G_i TLG in this layer has *weights* (1,1,1,3). The P_i TLG's are simple majority gates implementing AND functions.
- The third and fourth layers are *FIB* layers and constitute the *carry-merge* tree. The third layer contains two types of TLG's:
  - A 3-input *FIB* TLG (1,1,2) performs a radix-2 carry merge $G_i = G_i + P_i \cdot G_{i-1}$.
  - A 5-input *FIB* TLG (1,1,2,3,5) performs a radix-3 carry merge $G_i = G_i + G_{i-1} \cdot P_i + G_{i-2} \cdot P_{i-1} \cdot P_i$. It also contains AND gates for the propagate signals (P_i).
- The fourth *FIB* layer contains only one type, namely the 3-input TLG (1,1,2).
- The last layer is the well-known TLG sum layer (1,1,1,2) described previously.

3 SET THRESHOLD LOGIC GATES DESIGN

The TLG's design is based on the capacitive-input SET inverter. This structure was introduced for FET transistors in 1966 [14], and rediscovered in 1992 [15]. Since then, it has been known as the neuron-MOS (or vMOS). In [3] the application of this structure to majority SET gates was presented. In that article the adequacy of this approach to SET circuits was demonstrated and a full adder example was given.

The first step in our design was to augment the majority based design by generalizing it to arbitrary TLG's as shown in Fig. 2. With this modification, the new design reduces the number of components, the delay and power dissipation. Table 1 compares the characteristics of full adders based on majority gates and TLG's. The results were obtained using SIMON which is a Monte Carlo simulator for SET [16].

The TLG design is based on the SET inverter proposed by Tucker [17], and the inverter parameters are chosen such as to produce a step characteristic [3]. Since each TLG is based on an inverter, the output will always be the complement of the desired function. One approach to design a layered structure based on this TLG is to insert inverters between consecutive layers. A better solution is to implement the desired function in one layer and the dual of the function in the next layer. The first layer will produce the output complements, while the next layer will take the complements as inputs, implement the complement of the function and produce the desired function at the inverter output. This alternation continues for all remaining layers.

To design one TLG, we first determine the threshold equation of the particular function and/or the dual of the function. Secondly, the weight ratios are used to calculate the values of the capacitors. As an example we will show here the design of the TLG used in the last layer of the

	Majority	TLG
Number of components	57	28
Adder Delay (ns)	0.26	0.20
Power Consumption (pW)	1..95	0.75

Table 1: Comparison between Majority and TLG full adders.

TLG	C_1	C_2	C_3	C_4	C_5	C_b
(1,1,2,2)	0.4	0.4	0.8	0.8	-	0.6
(1,1,3)*	0.45	0.45	0.45	1.35	-	0.6
(1,1,2,3,5)	0.2	0.2	0.4	0.6	1.0	0.6
(1,1,2)	0.7	0.7	1.4	-	-	0.4
(1,1,2)*	0.6	0.6	1.2	-	-	0.6
(1,1,1,2)	0.6	0.6	0.6	1.2	-	-

Table 2: Capacitor values (in aF) for our 16-bit adder TLG's (* means the dual of the function).

adder (1,1,1,2). The threshold equation was given in equation (3). Eliminating the negative weight it can be written as: $s_i = a_i + b_i + c_{i-1} + 2c_i' - 2.5$. Hence, the dual of this function can be written as: $s_i' = a_i' + b_i' + c_{i-1}' + 2c_i - 2.5$. The fact that s_i and s_i' have equal weights and thresholds translates into identical TLG's.

The second step is to determine the capacitor values. These are calculated according to the weights and the sum of input capacitance. The sum is determined as part of the inverter parameters which produce a step inverter. Using the parameters given in [3], the sum is 3 aF. To calculate the capacitors, define C as the unit capacitance corresponding to a weight of 1. Since the threshold is 2.5, three capacitor units should be larger than 1.5 aF and two units less than this value. Mathematically: $3C > 1.5$ and $2C < 1.5$, hence $0.5 < C < 0.75$. For $C = 0.6$, the input capacitors are: $C_1 = C_2 = C_3 = 0.6$ aF and $C_4 = 1.2$ aF. Some TLG's require a bias capacitor in addition to the input capacitors. For example, consider the function implemented by the 3-input Fibonacci TLG (1,1,2). This function can be described as: $G_i = 2G_i + P_i + G_{i-1} - 1.5$ and its dual as: $G_i' = 2G_i' + P_i' + G_{i-1}' - 2.5$. In the adder described above, we implemented both G_i and G_i'. The latter with a threshold of 2.5 requires $0.5 < C < 0.75$. Using $C_1 = C_2 = 0.6$ and $C_3 = 1.2$ gives a total of 2.4 aF. Hence this TLG requires a bias capacitor of 0.6 aF which should be connected to ground for proper TLG operation. Table 2 shows the capacitor values used for all TLG's that constitute our 16-bit adder. Several AND/OR gates are used in this adder and each one has equal input capacitances and differs only in the values of bias capacitors. The parameters common to all gates are: $C_{b1} = C_{b2} = 9.0$ aF, $C_L = 24$ aF, $V_{dd} = 6.5$ mV.

4 ADDER CHRACTERISTICS

The 16-bit optimal adder was fully constructed, and was simulated using SIMON [16]. Due to the limited user interface, a MATLAB program was written to facilitate building the adder circuit. The program consists of two main modules, one to build the circuit from elementary gates and the other to specify stimuli signals. Simulation results showed that the adder functions properly.

Fig. 3 shows the delay of the 16-bit adder. The LSB of one input has a transition from '0' to '1' at 10 ns, and the output bits follow after some delay. Since the simulator is based on stochastic processes, the delay was calculated as the average of the delays for different random numbers. This 'average' delay is about 2 ns.

The power dissipation of the adder was thoroughly investigated. Fig. 4 shows the power of the 16-bit adder when running at different frequencies. The values obtained agree with reported results for SET inverters [18], taking into consideration the differences in load capacitance, voltage supply and scaling by the number of gates.

The simulation results mentioned above were obtained at helium temperature ($0 \div 4$ K). For getting an insight into the scaling of the power with respect to temperature, we have simulated one inverter at different temperatures. The results are shown in Fig. 5. The total power is obviously increased by temperature. This is due to the increase in static power caused by thermally generated tunneling.

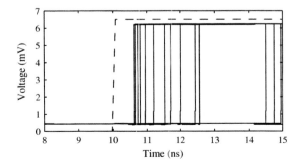

Fig. 3. Adder outputs (solid lines) and LSB of one input (dashed line)

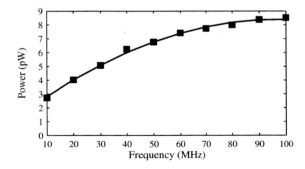

Fig.4. Adder power dissipation (with second order fitting).

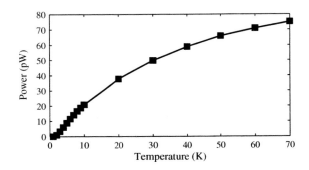

Fig. 5. Inverter power dissipation vs temperature.

With regard to integration, these results show that at liquid helium temperatures, an IC with 10^{11} transistors should dissipate below 1 W. As the temperature increases, a limit is reached where the gate does not function properly. Increasing the voltage supply can restore the functionality but will increase the power dissipation. Nevertheless, even if the supply voltage is increased 10 times (65 mV), an IC with 10^{10} devices should dissipate about 75 W.

5 CONCLUSION

A 16-bit adder was designed using TLGs. Each adder node consists of one or two capacitive-input SET TLG's. This can be compared to relatively complex Boolean gates used in *carry-merge* stages in CMOS adders. Simulation results showed quantitatively the ultra-low power dissipation of SET circuits. The circuit delay is high when compared to CMOS, as was expected for SET devices. The major problems encountered while doing this work were:

- The limited user interface of SIMON, which made it cumbersome to build the circuit.
- The very long simulation run which is typical of Monte Carlo based simulators. This could be alleviated by using SPICE simulations: either using specific models for SET [19], or by using a universal device model [20] on which a specific SET model can be defined.

REFERENCES

[1] V. Zhirnov, R. Cavin, J. Hutchby and G. Bourianoff, "Limits to binary logic switch scaling—A gedanken model," *Proc. IEEE*, vol. 91, Nov. 2003, pp. 1934–1939.

[2] S. Muroga, *Threshold Logic and Its Applications.* NewYork: John Wiley & Sons, 1971.

[3] H. Iwamura, M. Akazawa and Y. Amemiya, "Single-electron majority logic circuits," *IEICE Trans. Electron.*, vol. E81-C, Jan. 1998, pp. 42–48

[4] Y. Ono, H. Inokawa and Y. Takahashi, "Binary adders of multi-gate single-electron transistor: Specific design

using pass-transistor logic," *IEEE Trans. Nanotech.*, vol. 1, Jun. 2002, pp. 93–99.

[5] C. Lageweg, S. Coţofană and S. Vassiliadis, "A full adder implementation using SET based linear threshold gates," *Proc. Intl. Conf. Electronics, Circuits and Systems,* Sep. 2002, pp. 665–668.

[6] T. Oya, T. Asai, T. Fukui and Y. Amemiya, "A majority logic device using an irreversible single-electron box," *IEEE Trans. Nanotech.*, vol. 2, Mar. 2003, pp. 15–22.

[7] R. Betts, "Majority logic binary adder," U.S. Patent 3 440 413, Apr. 22, 1969.

[8] S. Coţofană and S. Vassisliadis, "Low weight and fan-in neural networks for basic arithmetic operations," *Proc. IMACS World Congress Sci. Comp., Modeling and Appl. Maths.*, vol. IV, 1997, pp. 227–232.

[9] J.F. Ramos and A.G. Bohórquez, "Two operand binary adders with threshold logic," *IEEE Trans. Comp.*, vol. 48, Dec. 1999, pp. 1324–1337.

[10] V. Beiu, "Neural addition and Fibonacci numbers," *Proc. Intl. Work-conf. Artif. Neural Networks,* Springer, LNCS 1607, vol. II, 1999, pp. 198–207.

[11] S. Vassiliadis, S. Coţofană, and K. Berteles, "2–1 addition and related arithmetic operations with threshold logic," *IEEE Trans. Comp.*, vol. 45, Sep. 1996, pp. 1062–1067.

[12] M. Sulieman and V. Beiu, "Optimal practical adders using perceptrons," *Intl. Conf. Neural Networks and Signal Proc.* Nanjing, China, Dec. 2003, to appear.

[13] P.M. Kogge, and H.S. Stone, "A parallel algorithm for the efficient solution of a general class of recurrence equations," *IEEE Trans. Comp.*, vol. 22, 1973, pp. 783–791.

[14] J.R. Burns, "Threshold circuit utilizing field effect transistors," U.S. Patent 3 260 863, Jul. 12, 1966.

[15] T. Shibata and T. Ohmi, "Functional MOS transistor featuring gate-level weighted sum and threshold operation," *IEEE Trans. Electron Dev.*, vol. 39, Jun. 1992, pp. 1444–1455.

[16] C. Wasshuber, H. Kosina and S. Selberherr, "SIMON: A simulator for single-electron tunnel devices and circuits," *IEEE Trans. Comp. Aided Design of Integ. Circ. and Sys.*, vol. 16, Sep. 1997, pp. 937–944.

[17] J.R. Tucker, "Complementary digital logic based on the Coulomb blockade," *J. Appl. Phys.*, vol. 72, Nov. 1992, pp. 4399–4413.

[18] Y-H Jeong, "Power consumption considerations of C-SET logics for digital application," *Proc. Intl. Conf. Solid-State and Integrated Circuits*, vol. 2, 2001, pp. 1373–1377.

[19] S.-H. Lee, "A practical SPICE model based on the physics and characteristics of realistic single-electron transistors," *IEEE Trans. Nanotech.*, vol. 1, Dec. 2002, pp. 226–232.

[20] M. Ziegler, G. Rose and M.R. Stan, "A universal device model for nanoelectronic circuit simulation," *Proc. IEEE Conf. Nanotech*, Aug. 2002, pp. 83–88.

In situ Transformations of Gold Contacts Studied by Molecular Dynamics Simulations

A.V.Pokropivny[*], A.Lohmus[**], R.Lohmus[**], D.Erts[***], V.V.Pokropivny[*] and H.Olin[****]

[*]Institute for Problems of Materials Science, National Academy of Sciences of Ukraine, Krzhyzhanovsky str. 3, 03142, Kiev, Ukraine, e-mail: dep40@materials.kiev.ua Institute of Technology,
[**] Institute of Physics, University of Tartu, 142 Riia Str., 51014 Tartu, Estonia
[***] Institute of Chemical Physics, University of Latvia, LV-1586 Riga, Latvia
[****]Physics and Engineering Physics, Chalmers University of Technology, SE-412 96 Gothenburg, Sweden

ABSTRACT

We apply molecular dynamics methods for simulation of in situ processes in new combined TEM/SPM technique. The atomic structure transformations of a gold nanobridge between two contacts are studied in processes of loading-unloading and friction cycles, vertical, lateral, diagonal and zigzag motion. In all cases only a single-atom contact is broken at the final stage of deformations. The deformation process strongly depends on the velocity of fracture and schemes of motion.

Keywords: molecular dynamics simulation, TEM-SPM

1 INTRODUCTION

Contact phenomena between metallic and ceramic nanoparticles play a key role in the processes of adhesion, seizure, friction, indentation, sintering, recrystallization, deformation, etc. Till recently the nanocontact phenomena studies were limited only to observations and analysis of transformations ex situ. Recently developed TEM/AFM technique [1,2] allows us to visualize an atomic structure of contacts under deformation in situ in couple with a simultaneous measurement of driving tip-sample nN-forces. However, mechanism of contact deformations during such experiments is not sufficiently understood.

Modeling and simulation is an effective theoretical method for observation of a cantilever and tip motion in SPM [3-9]. Molecular dynamics (MD) simulations were widely used for calculation of nanowires deformation with picosecond resolution during lateral and normal motion of tips [3-4,6,8]. The purpose of this report is to investigate the mechanisms of atomistic transformations of gold nanocontacts under lateral, normal and zigzag motion. We combined theoretical MD technique and experimental TEM/AFM [1,2] to study contact phenomena at the nanolevel for gold.

2 CALCULATION METHODS

To simulate the dynamical evolution of the contact we used original SIDEM software . MD simulation was performed using original empirical short-range pair potential for Au.

The potential ensures the stability of fcc gold lattice. We modeled the contacts by two different ways.

The first modeling was the loading-unloading and friction cycles calculations of gold-tip with different radii and gold surface. An example of contact of Au-tip of 4 nm in radius with Au (100)-surface is shown in fig.1.a. The loading-unloading cycle was simulated by moving of the tip by a step of 0.01a to the sample (1-200 steps) and then away from it (201-400 steps), were a=0.408 nm - lattice period for Au. The friction cycles were modeled by moving of the tip with the same step size along the surface (1-200 steps) and backwards (201-400 steps) at the different heights.

The second modeling is the nanobridge deformations calculations. The calculated gold nanobridge of 8X8X8 lattice period total contains 2433 atoms. The nanobridge consists of two grains or parts, upper and lower. The all 1132 atoms of upper part at each step undergo the the following displacement: (i) in lateral direction with a step of 0.01a, (ii) in normal direction with the step 0.01a, 0.1a, 0.3a, (iii) in diagonal direction with a step of 0.014a and (iv) zigzag motion with the step 0.01a. Cases (i) and (ii) correspond to shear and strain deformations, respectively. Case (iii) corresponds to more complex deformation with both components. Case (iv) corresponds to motion with lateral vibration. The total numbers of displacement steps were 1000.

For both calculations at every step a sufficient time was given for the tip-surface or nanobridge system relax in a stable state. Dynamical relaxation was carried out after each displacement step. The structures of contact as well as radial distribution function were calculated also after each step of the relaxation. The process of dynamical evolution in all cases is represented as the series of snapshots. The snapshots image of all atoms in the narrow plane dy=(0,a), crossing the centre of tip-surface or nanobridge in plane (x,z). Atoms in the (0, a/2) and (a/2, a) planes are shown by open and black circles, respectivelyallowing the evolution of two central atomic planes may be analyzed. We did not use any thermostat for calculation of temperature dependences. There were not any important differences between the results obtained at 12 K and 290 K in the resent molecular dynamics experiments [6].

3 RESULTS

3.1 Loading-unloading and friction cycles

The results for loading-unloading cycle are presented at fig.1. At the loading ther "adhesion avalanche" instability was oserved .Neighboring tip and surface atoms jumps abruptly to meet each other at a height of 0.46 nm. A number of contact bonds changes from 16 to 905, adhesion force drops from -1.6 nN to -50 nN. The radius of created contact equals to ~1 nm, closely to experimental ones. The contact is compressed (at loading) and stretched (at unloading) so as the interaction type is elastic. Then, at subsequent unloading several instabilities and reorganization of contact appear. Calculated values of yield strength at the points of contact reorganizations (points A, B, C and D in fig.1.c) equal to 3-6 GPa, which is consistent with experimental values of 5 GPa calculated by us in the TEM/AFM experiment [2]. At the final image a reorientation of tip structure from (001) to (111) are observed (fig.1.b).

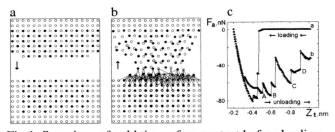

Fig.1. Snapshots of gold tip-surface contact before loading (a) and after unloading (b) and calculated adhesion force variations (c). Open and black circles show atoms in adjacent (001)-planes.

The frictional cycles were modeled for different heights, namely 0.3a, 0.5a, 0.6a, 0.7a, 0.9a. It was found that the critical height was 0.6a. If the height of scanning was below of this critical value the tip was blunting and wearing. In this case the scanning run with the defect formation, such as blocks, twins and cracks formation and its healing. The effect of subsurface atoms seizure by tip apex atoms was observed. This "subsurface seizure" effect consists in formation of ideal crystalline structure of subsurface atoms with apex atoms and cracks in subsurface itself. The contact structure in this case is an atomic "iceberg" submerged in the surface "ocean". These findings allow us to find a more effective mode of the TEM holder motion.

3.2 Nanobridge deformations

Just before breaking of nanobridge in TEM/AFM experiment the forces up to 10 nN was observed, see [2]. Junging from this breaking force the number of breaking atoms could have correspond to ten. Beside this the creation of the double neck was observed. We tried to explain these results. For this purposes we modeled the elongation of nanobridge in three different directions: [100], [010] and [110] and employed several different schemes of upper grain motion, including vibration.

In all cases strong rearrangements of the nanobridge are observed. The nanobridge thins step by step with defect formation, such as vacancies, twins, grain boundaries, surface steps. At the final stage of the deformation only single atom contact is broken in any cases. But the mechanism of deformation differs for different cases. During strain deformations reorientation from (100) to (111) planes for all atoms of nanobridge take place with formation of zig-zag vacancy cavities like in loading-unloading modelling (fig1.b). During shear deformations the tilt and slips along (110) plane take place with formation of twins. During shear and strain deformations rotation of grains take place with formation of several grain boundaries, vacancies and steps (fig.2).

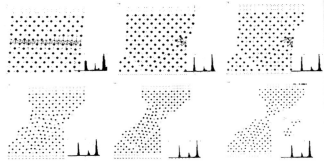

Fig.2. Snapshots of the nanobridge during shear and tensile strains at N-steps DX=0.01a and DY=0.01a.

It should be noted that Marszalek at. al. [9] experimentally confirmed the sliding of crystal planes within the gold nanowires with changing of the local structure from fcc to hcp. Such reconstraction during strength are well known for plastic deformation of fcc metals [10].

The velocity influences on the deformations were studied. For study of speed effects on the deformation process we have studied the tensile strain with different step, namely 0.01a, 0.1a and 0.3a. If the velocity of atomic relaxation more then the velocity of fracture than plastic behaviour takes place. If the velocity of fracture more then the velocity of an atomic relaxation the brittle behaviour takes place.

We modelled the lateral vibration of TEM holder as zigzag motion of upper grain. The motion at every step were modelled as total displacement in the elongation direction, summarized from two diagonal motion. During vibration we observed formation of two nanobridges so as in the real experiment [1,2].

CONCLUSION

We may conclude that the process of nanocontact deformation is more complex in comparison with the well known MD simulations [3-8]. During modelling of nanocontact evolution we must take into account the directions of deformation (not only parallel or perpendicular directions), the velocities of deformation and the velocities of atomic relaxation, the vibration and temperature effects. If velocity of atomic relaxation greater then velocity of fracture than plastic behaviour takes place, including twining, reconstruction, sliding and nanocracks healing. If the velocity of fracture greater then the velocity of an atomic relaxation the brittle behaviour takes place, including formation of vacancies, their clusters, nanockracks and formation of free surface. The schemes of deformation may be represented as "tilt-slip-tilt-slip-thining" for shear strains, "elongation-recrystallisation-elongation-thining" for tensile strains, "tilt-slip-step formation-tilt-slip-thining" for diagonal strain and more complex with rotation of grains for zigzag motion. In all schemes at the final stage of demormations only a single-atom contact is broken.

ACNOWLEDGEMENTS

This work was supported by the Estonian Science Foundation (grant no. 5015) and the European Science Foundation program "Nanotribology"

REFERENCES

[1] Erts D., Olin H., Olsson L., Ryen L., Tholen A. EUREM 12, Brno, Czech Republic (2000) I265.

[2] Erts D., Lohmus A., Lohmus R., Olin H., Pokropivny A.V., Ryen L., Svensson K. Applied Surface Science. 188 (2002) 460.

[3] Landman U., Luedtke W.D., Burnham N.A., Colton R.J. Science. 248 (1990) 454.

[4] Buldum A., Ciraci S., Batra I.P. Phys. Rev. B. 57 (1998) 2468.

[5] Agrait N., Yeyati A.L., Ruitenbeek J.M. Physics Reports. 377 (2003) 81.

[6] Pokropivny V.V., Skorokhod V.V., Pokropivny A.V. Model. Simul. Mater. Sci. Eng. 5 (1997) 579.

[7] Pokropivny A.V. Physics Low-Dim. Struct. 3/4 (2001) 117.

[8] Sorensen M.R., Brandbyge M., Jacobsen K.W. Phys. Rev. B. 57 (1998) 3283.

[9] Marszalek P.E., Greenleaf W.J., Li. H., Oberhauser A.F., Fernandez J.M. PNAS. 97 (2000) 6282.

[10] Smith M.C. Principles of physical metallurgy. Happer and Brothers, NY, 1956.

Development of the Nano Aligner for CNT-tip fabrication

Gyungsoo Kang[*], Junsok Lee[**], Jaiseong Choi[***], Yoonkeun Kwak[****] and Soohyun Kim[#]
[*]Mechanical Engineering, KAIST, Republic of Korea, gskang@kaist.ac.kr
[**]JamesLee@kaist.ac.kr, [***]sant@kaist.ac.kr, [****]ykkwak@kaist.ac.kr
[#]Mechanical Engineering, KAIST, Republic of Korea, soohyun@kaist.ac.kr

ABSTRACT

AFM tip has been used for surface profiling with a fine resolution, but there is a barrier to improve its performance because of the low aspect ratio. Many researchers have solved this problem with attaching carbon nanotube (CNT) to Si-tip. In this paper, we proposed the aligner system that composed of dual type stage system, and these stages could attach a carbon nanotube to tungsten-tip in vacuum condition. We used tungsten tip instead of Si-tip because of its conductivity. The aligner system proposed in this paper has 10 degree-of-freedom that 3 in the first stage and 7 in the second stage. With picomotors and piezotube, the first stage has the resolution about several tens of nm and the second stage has a resolution about 1 nm. With this aligner system, we could make nanotip and with 2 nanotips, we can make nano tweezer.

Keywords: aligner system, dual type stage system, carbon nanotube, picomotor, piezotube

1 INTRODUCTION

Because of its high aspect ratio, CNT-tip can be useful for measurement of surface profile like high-dense semi-conductor which has narrow, deep valley structure [1]. CNT also can be used for nano tweezer composed of two CNT-tip, that can grip and operate nano meter sized materials [2].

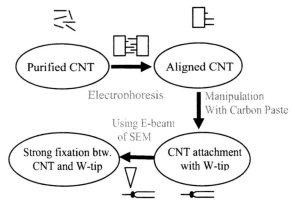

Figure 1 : Schematic diagram to make CNT-tip

As shown in Figure 1, fabrication process of CNT-tip needs SEM because we deal CNT with several tens of nm diameter, and precise manipulation system for aligning CNT and W-tip.

The aligner system must have efficient degree-of-freedom to place carbon nanotube and tungsten-tip to the operating area in SEM.

2 REQUIREMENTS FOR THE NANO ALIGNER AND ACTUATORS

2.1 Requirements for the Nano Aligner

For manipulation of attachment of W-tip and CNT, following conditions are needed.

1) Aligner system has dual-type two-stage for placement W-tip and CNT separately.
2) Stage1 and stage2 can be moved along X, Y axis for placing to the SEM gun spot region.
3) Both stages should be moved along Z axis for focusing.
4) After attachment process, we should check the axial-attachment condition with the rotating movement.
5) Since the diameter of CNT is several tens of nm, nanometer-order resolution is needed in stage system for fine motion.
6) Since SEM chamber has limited space size of height and diameter, the size of aligner system should be as small as possible.

For satisfaction of these requirements Zyvex company made dual-type stage which the stage1 has linear motion along X, Y axis with picomotors, stage2 can move linearly along Z axis and rotate with X axis. And with a piezotube, they experimented the attachment of CNT on the STM probe, also measured the mechanical and electrical properties of CNT [3]. KRISS in Korea made CNT-tip with an AFM and CNTs, using two stages which can move linearly along X, Y, Z axis [4].

But there are some problems for experiment with the Zyvex`s proposed system. Cause both 2 stages can`t move along X, Y, Z axis, we must use the stage in SEM chamber and that is uncomfortable, also need a good skill. And, it is impossible for fast experiment with variable tips, because attachment regions are different with where the tips are used.

As mentioned in 5), a fine motion actuator like piezotube is needed. Piezotube has a resolution of about 1 nm and this makes the improvement of resolution in whole system.

2.2 Principle of Picomotor's movement

For the purpose of giving linear movement along X, Y, Z axis to the stage, linear picomotors are adopted which can go back and forth with screw motion. And rotating picomotor is used for rotating motion.

Linear picomotor has a screw and if the screw rotate in CW direction then picomotor's ahead ball go forward, in CCW direction then picomotor's ahead ball go backward. The PZT exists in picomotor and the screw moves with the slow force by high static friction, and does not move with the fast force by low dynamic friction [5].

Namely, if picomotors are in the condition of long rising time and short falling time, the screw rotates with CW direction; but the other case of short rising time and long falling time, the screw rotates with CCW direction. The principles of movement for rotating picomotor and general motors are similar.

2.3 Principle of Piezotube's movement

But the resolution of linear picomotor is about 30 nm, we obtained more precise motion using 4-quadrant piezotube which made more fine motion possible. We used simple cylindrical shaped piezotube, and this is similar to a hollow bamboo. Electrodes are coated on inner and outer surface of piezotube, and outer plane is divided to 4 parts. We experimented with 4 divided type, calling 4-quadrant piezotube, and outer electrodes are called +X, +Y, -X, -Y domain. +X is placed opposed to –X, and +Y, -Y are also. If we commit high voltaged opposite alternating current to the +X and –X electrodes, then a piezotube swing along X-axis. Meanwhile giving same voltage to 4-outter electrodes and ground to inner electrode makes it move along axial direction.

Scanning range of piezotube depends on its length, inner diameter, wall thickness, strain coefficient of piezotube, and operating voltage. The piezotube can scan from several tens of Å to about 100 μm in lateral direction, and below Å to about 10 μm in axial direction. The equation of range in axial direction is shown in Eq. (1) [6].

$$\Delta L \approx d_{31} \cdot L \cdot \frac{U}{d} \qquad (1)$$

where, $\triangle$L: scan range in Z [m]
d_{31}: strain coefficient [m/V]
L : length of the PZT ceramic tube [m]
U : operating voltage [V]
d : wall thickness [m]

We used the piezotube which has specification like Table 1.

	PT 130.14
Length (mm)	30
Outside dia. (mm)	6.5
Inside dia. (mm)	5.35
Max. operating voltage (V)	500
Strain coefficient (m/V)	2.1×10^{-10}

Table 1 : Specification of the piezotube

Using the specification in Table 1 and Eq. (1), we can obtain that the range of axial motion is 2.3 μm. And it is considered that the clamping length is is 5 mm.

In similar way, it can contract in same displacement, so the final range of axial direction is ± 2.3 μm. In addition to axial direction, the range of lateral direction can get from following equation.

$$\Delta X \approx \frac{2\sqrt{2} \cdot d_{31} \cdot L^2 \cdot U}{\pi \cdot ID \cdot d} \qquad (2)$$

where,
ΔX : scan range in X and Y [m]

Using the Eq. (2), the range of deflection in X and Y can be calculated to 9.6 μm. And the final range of lateral direction is ± 9.6 μm.

3   DESIGN OF NANO ALIGNER

3.1   System Structure

From above condition, we made up with the aligner like Figure 2, stage1 can move along X, Y, Z axis and aligned CNTs are adopted here. The stage2 also moves along X, Y, Z axis, and it has a rotating picomotor. At the end of the axle, 3 DOF piezotube with sleeve is put on. With the other sleeve, W-tip is inserted at the other platform of piezotube. So overall aligner system has 10 DOF with 3 DOF in stage1 and 7 DOF in stage2.

3.2   Principle of Operation

The used stage body is Sigma-Koki TSDS-255 SL which is composed of X, Y Stage and bracket type Z stage. X, Y, Z stage has dimension of 25×25×60 mm, that is quite small enough to place in SEM chamber. Picomotors are attached to the stage removed microhead. The operating range of the X, Y, Z axis is ±5 mm. The rotating picomotor is put to the Z-bracket of stage2 with holder.

The total size of the aligner system is about 170×135×65 mm³, and it can be put in the SEM chamber.

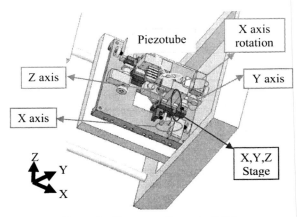

Figure 2 : The nano aligner in SEM

3.3 System Composition

The schematic diagram of the system is shown Figure 3. Stage1 is composed of 3 linear picomotors, and stage2 is composed of 3 linear picomotors, a rotating picomotor and a piezotube. 5 input channels are needed to operate piezotube, and we used 2 piezotube drivers (PI, E-463.00) which has 3 output channels. And for picomotors, we used 9 channel picomotor driver (New Focus, Model 8769) to operate 7 picomotors.

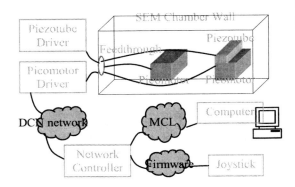

Figure 3 : Schematic Diagram of the System

4 FABRICATION OF THE NANO ALIGNER

We make some of holders, mountings and stages to make the nano aligner. These components have the unique shape to satisfy the requirements of the nano aligner. As mentioned previously, some of sleeves are used to connect piezotube and insulated from the high input voltages on piezotube.

The whole shape of system is shown in Figure 4.

Figure 4 : The Nano Aligner

For the purpose of operating the nano aligner in SEM chamber, the electrical feedthrough is fabricated using the hole in chamber wall. The vacuum condition is maintained using the elliptical O-ring in the feedthrough.

The connectors are the Amphenol`s product and it has 12 pins in one connector. Since 19 pins are needed to operate the nano aligner, two connectors are used. The input signals are 5 for piezotube and 14 for picomotors. In Figure 5, the manufactured feedthrough is shown.

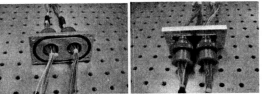

Figure 5 : Electrical feedthrough (left : inside view, right: outside view)

Among the total system, the movement of picomotors can be controlled to the desired displacement at the uniform speed or stayed at the wished speed by a supplied DCN (Distributed Control Network) with the picomotor driver. The motion program of picomotor is done by Labview and Visual C++. And, the motion of piezotube can be controlled by the voltage input of piezotube driver (PI, E-463.00).

5 CONCLUSION

In this paper, we designed and fabricated the nano aligner system which has 10 DOF and small size. The system has the picomotors which has a resolution of several tens of nm and a piezotube which has a resolution of about 1 nm.

This system could make CNT-tip that is composed of carbon nanotube and W-tip with a precise manipulation in SEM vacuum chamber. With two CNT-tips, the system can make a nano tweezer.

6 ACKNOWLEDGEMENT

This research was supported by a grant (M102KN010001-03K1401-01120) from Center for Nanoscale Mechatronics & Manufacturing of 21st Century Frontier Research Program.

REFERENCES

[1] Hongjie Dai and Jason H. Hafner, et al., "Nanotubes as nanoprobes in scanning probe microscopy," Nature, Vol.384, No.6605, pp.147-150, November 1996.

[2] Philip Kim and Charles M.Lieber, "Nanotube Nanotweezers," Science, Vol.286, No.5447, pp.2148-2150, December 1999.

[3] MinFeng Yu and Mark J Dyer, et al., "Three-dimensional manipulation of carbon nanotubes under a scanning electron microscope," Nanotechnology, Vol.10, No.3, pp.244-252, 1999.

[4] Byoung Chun Park, et al., "Development of CNT tip and its application to the high resolution scanning probe microscopy," Development of Advanced Measurement Technology, pp.41-65, 2001.

[5] New Focus, User`s guide.

[6] Physik Instrumente, Catalogue for users.

Nanogripper using Carbon Nanotube

Junsok Lee[*], Soongeun Kwon[**], Jaiseong Choi[***], Gyungsoo Kang[****], Yoonkeun Kwak[*****] and Soohyun Kim[#]

[*]Mechanical Engineering, KAIST, Republic of Korea, jameslee@kaist.ac.kr
[**]kluivert@kaist.ac.kr, [***]sant@kaist.ac.kr, [****]gskang@kaist.ac.kr, [*****]ykkwak@kaist.ac.kr
[#]Mechanical Engineering, KAIST, Republic of Korea, soohyun@kaist.ac.kr

ABSTRACT

Nanogripper was made from two nanotips and the nanotip is composed of tungsten tip and carbon nanotube. The tungsten tip was made by the electrochemical etching with KOH solution. The initial and final diameter of tungsten tip were 500 micrometer and several micrometer respectively. The carbon nanotube is multi-wall carbon nanotube grown on the Al_2O_3 by chemical vapor deposition. The average diameter of carbon nanotube is 130 nanometer. The procedure to make nano-tip is monitored by optical microscope of two thousand magnifications. The length of carbon nanotube in nanotip was controlled by the electrochemical etching to satisfy in special application. The nanogripper was made by the fixing stage that has three degree of freedom.

Keywords: nanogripper, nanotip, electrochemical etching, carbon nanotube

1 INTRODUCTION

Sumio Iijima firstly discovered carbon nanotube at 1991[1]. Many research results about the properties of carbon nanotube have been reported in many fields. And, Hongjie Dai proposed the first report about nanotip in 1996[2]. In that report, carbon nanotube was attached on the conventional AFM tip with acrylic adhesives. The lateral resolution of nanotip was compared with the conventional silicon tip. In 1999, Philip Kim proposed the nanogripper firstly[3] and he made nanogripper using glass electrode under optical microscope. In 2001, Seiji Akita reported the nanogripper that was made from conventional AFM tip under scanning electron microscope[4]. The proposed nanogrippers in previous researches are made from one substrate and attached two carbon nanotube. But, in this paper, we proposed a new method to make nanogripper using nanotip and it is possible to make various types of nanogripper.

2 MANUFACTUREING NANOTIP

2.1 Tungsten Tip

The proposed nanogripper was made from nanotips and nanotip was composed with tungsten tip and carbon nanotube. In this section, the method to make tungsten tip is described.

The electrochemical etching method is used to make tungsten tip. The initial and final diameter of tungsten tip are 500 micrometer and several micrometer respectively. In Figure 1, the system to make tungsten tip and manufactured tungsten tip are shown.

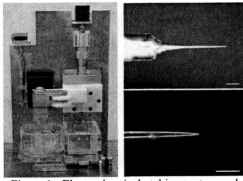

Figure 1 : Electrochemical etching system and manufactured tungsten tip (scale bar : 200 μm(upper), 10 μm (lower))

As shown in Figure 1, the diameter of the end region of tungsten tip is several micrometer. The used electrolyte is 5 mole of potassium hydroxide and the immersion depth of tungsten tip is controlled by the step motor and several optical stages.

2.2 Preparation of Carbon Nanotube

In this paper, we used multi-wall carbon nanotube(MWCNT) which was grown on the Al_2O_3 with chemical vapor deposition(CVD). The average diameter of carbon nanotube is 130 nanometer and the range of length is from 5 to 50 micrometer.

It is essential to purify the MWCNT to use in the process to make nanotip. We obtained some of purified MWCNT from 600W sonicator in the isopropyl alcohol(IPA) for 2 hours. The purified MWCNT in IPA is poured on the slide glass and the IPA is evaporated. Then, some of purified carbon nanotube can be obtained. In Figure 2, the original MWCNT, which was used in this paper captured by SEM in commercial company, and the purified MWCNT on the slide glass captured by optical microscope are shown.

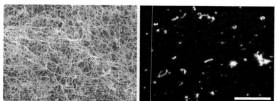

Figure 2 : The original MWCNT entangled and purified MWCNT on the slide glass

2.3 Manufacturing Nanotip

We used two tungsten tips and carbon tape to make nanotip. The method to make nanotip is shown in Figure 3.

Tungsten Tip 1

Tungsten Tip 2

Figure 3 : How to make nanotip

As shown in Figure 3, the carbon nanotube was attached on 1st tungsten tip with scratching on the slide glass covered with purified carbon nanotube. In 2nd tungsten tip, some of carbon tape is painted in the end region under optical microscope. The process to make nanotip is shown in Figure 4 and the all process was monitored by the optical microscope.

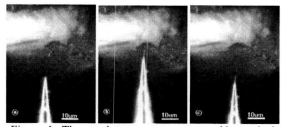

Figure 4 : The attachment process captured by optical microsocpe

As shown in Figure 4, the attachment process is composed of 3 steps. The first step is to find out the appropriate carbon nanotube in 1st tungsten tip. The second step is to approach the 2nd tungsten tip to the found carbon nanotube using optical and piezo stages and dwells several seconds for firm attachment. The final step is the separation of the 2nd tungsten tip from the 1st tungsten tip. Then the carbon nanotube is dragged out to the 2nd tungsten tip due to the acrylic adhesive force by carbon tape.

2.4 Length control of Nanotip

Because the nanotip made by this process has arbitrary length and improper end shape for nanogripper, the length of carbon nanotube must be controlled for its special application. The schematic diagram of cutting procedure of carbon nanotube is shown in Figure 5.

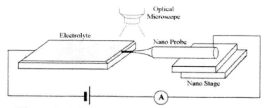

Figure 5 : Schematic diagram of cutting system

The cutting mechanism is to use the electrochemical etching process using KOH solution and DC power supply, and the all procedures are monitored by optical microscope. The nanotip is approached to the electrolyte by nano-stage such as pizo actuator under optical microscope. Since the immersed depth of carbon nanotube in the electrolyte can be controlled by nano-stage, the length can be controlled with nano scale. The start and end of the process was checked with the value of current. The overall procedure is displayed in Figure 6 that was captured by optical microscope.

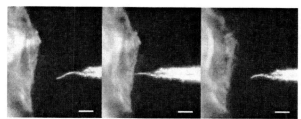

Figure 6 : The process to cut carbon nanotube in nanotip (scale bar : 5 μm)

In Figure 6, the left is the electrolyte and the right is nanotip. When the nanotip contacts to the electrolyte with DC power supply, the electrochemical etching and polishing process will be produced with the current flow through closed electrical loop.

The detail views of nanotip are shown in Figure 7 which were captured by SEM. As shown in Figure 7, the cut length of nanotip is about 2 micrometer.

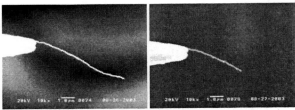

Figure 7 : The SEM images of nanotip prior(left) and post(right) cutting process

Since the process is monitored by optical microscope, it is possible to cut repeatedly. In Figure 8, the example of repeated cutting and the results of the cutting are shown.

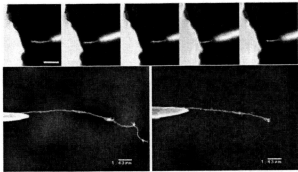

Figure 8 : The repeated cutting(upper, scale bar:5 μm) and the results of cutting captured at prior and post process

Through the cutting process, we can obtain the increase of the contact force between tungsten tip and carbon nanotube.

3 MANUFACTURING NANOGRIPPER

Because we made nanogripper using two nanotips, the fixing stage, which is used to approach two nanotips each other, is required. Using this method, we have some advantages as followings.

- It is possible various type of nanogripper.
- It is possible large grip motion using fixing stage.
- It is possible to use conducting substrate.
- It is not necessary to deposit additional electrode.

In Figure 9, the nanogripper is shown that was made from the fixing stage and two nanotips.

As shown in Figure 9, we made nanogripper using fixing stage. The fixing stage has three degree of freedom and is operated by the bolt with ball and spring. After manufacturing the nanogripper, the fixing stage is become the body of nanogripper. So, the fixing stage must be small and not expensive.

Figure 9 : Manufactured nanogripper (scale bar in right image : 5 μm)

The movement of nanogripper and gripping of nano particle are doing now. From the simulation of nanogripper, we found that the desired nanogripper size is 7 μm of length, 1 μm of initial gap between the end of nanogripper and 50 nm of diameter of carbon nanotube.

4 CONCLUSION

In this paper, we proposed a new method to make nanogripper using two nanotips. The tungsten tip was made by electrochemical etching and the purification of carbon nanotube was done by sonication. The nanotip was made by two tungsten tips and mechanical attachment with carbon tape. The length of carbon nanotube in nanotip was controlled by the electrochemical etching and the carbon nanotube was dissolved in electrolyte. Although the type of nanogripper in previous research is confined with one type, it is possible to make various type of nanogripper with proposed method such as a tripod.

5 ACKNOWLEDGEMENT

This research was supported by a grant (M102KN010001-03K1401-01120) from Center for Nanoscale Mechatronics & Manufacturing of 21st Century Frontier Research Program.

REFERENCES

[1] Sumio Iijima, "Helical microtubules of graphitic carbon", Nature, Vol.354, pp.56-58, 1991
[2] Hongjie Dai, Jason H.Hafner, Andrew G.Rinzler, Daniel T.Colbert, Richard E.Smalley, "Nanotubes as nanoprobes in scanning probe microscopy", Nature, Vol.384, pp.147-150, 1996
[3] Philip Kim, Charles M.Lieber, "Nanotube nanotweezers", Science, Vol.286, pp.2148-2150, 1999
[4] Seiji Akita, Yoshikazu Nakayama, Syotaro Mizooka, Yuichi takano, Takashi Okawa, Yu Miyatake, Sijenori Yamanaka, Masashi Tsuji, Toshikazu Nosaka, "Nanotweezers consisting of carbon nanotubes operating in an atomic force microscope", Applied Physics Letters, Vol.79, No.11, pp.1691-1693, 2001

Displacement Detection Using Quantum Mechanical Electron Tunneling in Micro and Nano-electro-mechanical Systems

D. Taylor[*], and K.L. Ekinci[**]

[*]Physics Dept., Boston University, Boston MA 02215, USA, drtaylor@bu.edu
[**]Aerospace and Mechanical Eng. Dept., Boston University, Boston MA, 02215, ekinci@bu.edu

ABSTRACT

Here, we investigate the applicability of quantum mechanical electron tunneling to displacement sensing in micro and nano-electro-mechanical systems (MEMS and NEMS). Our experiments were performed using an ultrahigh vacuum (UHV) scanning tunneling microscope (STM). A micro-scale silicon nitride membrane with clamped boundary conditions was fabricated using standard lithographic techniques. The STM tip was then brought into tunneling range over the center of the membrane. Dynamic motion of the device was detected in the tunnel current and measurements of the electromechanical response of the device were taken. In a second set of measurements, the tunnel bias between the membrane surface and the STM tip was modulated at a selected frequency to produce a signal at the "mixed down" or difference frequency — thus avoiding attenuation due to the amplifier. Finally, the sensitivity limits of this technique are discussed for displacement detection in NEMS.

Keywords: microelectromechanical systems (MEMS) nanoelectromechanical systems (NEMS), scanning tunneling microscopy, displacement detection

1 INTRODUCTION

Micro-electro-mechanical systems (MEMS) and, more recently, nano-electromechanical systems[1] (NEMS) are emerging as candidates for a number of important technological applications — such as ultra-fast actuators, sensors, and high frequency signal processing components. Finding precise methods for characterizing the electromechanical behavior of MEMS and in particular NEMS, however, remains a challenge.

Quantum mechanical tunneling of electrons between two metallic surfaces — with its strong dependence on the tunnel gap — has been shown to be a suitable method for studying displacements at this length scale. Displacement detection using a scanning tunneling microscope (STM) has been demonstrated in a number of micro-systems such as accelerometers[2],[3] and magnetometers[4]. Our experiment utilizes electron tunneling to measure the resonance frequency, $\omega_0 / 2\pi$, and the quality (Q) factors of a MEMS device. In the first set of measurements, the STM was used to detect the motion of this device *directly* in the vicinity of $\omega_0 / 2\pi$. In the second set of measurements, we made use of the inherent non-linearity in the tunneling current across the tunnel junction. The tunnel current signal was "mixed down" by modulating the potential difference between the STM tip and the device. Measurements made at the difference frequency exhibited less attenuation from the amplifier.

These results provide information about the specifics of electromechanical resonances in MEMS and NEMS and form a basis for future investigations of the mechanical properties of micro-scale and nanoscale devices using electron tunneling.

2 DEVICE FABRICATION

Now, we turn to a detailed description of the experimental procedure. We first illustrate the method for fabrication of a thin single crystal silicon nitride structure.

The fabrication procedure is illustrated in Figure 1 (a-g). The starting material for device fabrication is a 125-μm-thick undoped silicon (100) wafer, coated on both sides with an LPCVD grown silicon nitride (Si_3N_4). The Si_3N_4 thickness is 300 nm. Fabrication begins by defining an area approximately 120 μm^2 using electron beam lithography. Reactive ion etching is then performed with a plasma of He and SF_6 at a pressure of 300 mTorr with respective flow rates of 21 sccm and 13 sccm and a microwave power of 100 W. The etch rate under these conditions is ~5 nm/s. The vertical etch removes the Si_3N_4 layer exposing the Si. A subsequent wet etch in potassium hydroxide (KOH) dissolves the Si preferentially along the (100) plane, at a 57° angle to the (111) plane, with high selectivity. The wet etch terminates at the Si_3N_4 device layer to form a 300 nm thick membrane with clamped boundary conditions. Additional lithography is then performed on the device layer to prepare the sample for analysis in the STM. Metal contacts of layered Cr and Au are deposited with respective thicknesses of 7 nm and 25 nm.

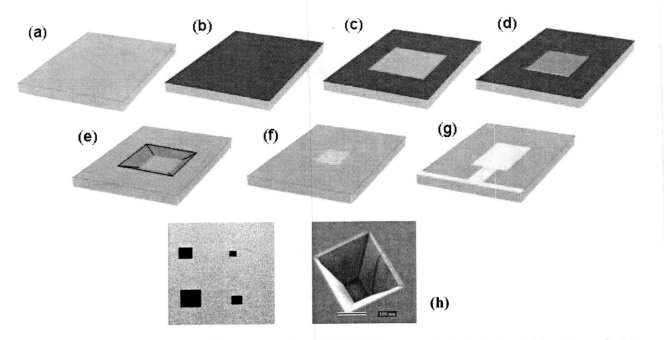

Figure 1: (a) the Si₃N₄-Si- Si₃N₄ wafer pre-processing (b) PMMA is spin-coated onto the backside of the wafer (c) electron beam lithography is used to pattern the backside of the wafer (d) reactive ion etching opens the SiN layer (e) A KOH wet etch dissolves the Si layer along the (100) plane. (f) shows the front side of the wafer with a 300 nm thick clamped SiN membrane. (g) additional lithography is used to pattern a metallic layer for actuation and STM analysis (h) an SEM micrograph, shown from the front, of four membranes with an SEM photograph, shown from the back, of a single membrane.

3 MEASUREMENTS

A custom built RHK Technology, Inc.[5] UHV STM was used to probe the small mechanical displacements of the fabricated devices. First, electrical connections were made to the device layer as well as to a gold layer beneath the membrane for electrostatic actuation. The sample was then loaded into the UHV chamber containing the STM. The tip was brought into tunneling (approximately a few angstroms gap distance) over the central region of the membrane.

The device was driven electrostatically as shown in Figure 2 (a). Capacitive coupling between the actuation and device layers induced motion in the membrane when both AC and DC signals were applied to the actuating layer. A network analyzer was used to sweep the AC signal over a range of frequencies while simultaneously monitoring the response in the tunneling current at the STM tunnel current pre-amplifier output.

The maximum response was observed in the tunnel current at the natural resonance frequency of the device, measured at 12.248 kHz with a Q factor of 50 as shown in Figure 3. These values agreed well with the previously measured ones from an optical interferometry setup described elsewhere[6]. The amplitude of this response was seen to be dependent upon the amplitudes of both applied signals as well as the value of the tunneling current

set. Additionally, significant attenuation of the signal was observed due to the fact that the value of the natural resonance exceeded the 3dB point of the current pre-amplifier, measured to be at ~3 kHz. The attenuation around 12 kHz due to this effect was estimated to be ~20dB.

A more complex measurement scheme is needed to get rid of the aforementioned attenuation. Such a scheme will be valuable for characterizing the behavior of smaller, high frequency devices.

In an effort to extend the applicability of STM measurements to higher frequency devices, we experimented with a second technique that makes use of the inherent non-linearity of the IV characteristics of a tunnel junction[7]. The tunnel current, I_T, across the tunnel gap is given by, $I_T \propto e^{-kz}$ where z is the gap separation and k is the electron tunneling decay constant proportional to the work function of the metal. The gap distance is given by a fiducial separation defined by the feedback setting added to the small amplitude of the membrane center's oscillatory motion at frequency ω_1.

To measure higher frequency signals, a bias modulation signal of frequency ω_2 is applied to the tip itself. The nonlinearity in the IV and Iz characteristics of the tunnel gap then gives rise to a signal proportional to the tunnel gap at the sum and difference frequencies, $i.e.$

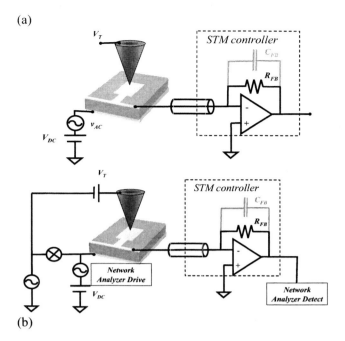

(a)

(b)

Figure 2: (a) The measurement scheme used to find the resonance and Q directly through the tunnel current. (b) The signal at the difference frequency is measured by modulating the tip-sample bias and sweeping the AC signal on the actuator using a network analyzer. The tunnel current is then fed back into the network analyzer.

$\omega = \omega_1 - \omega_2$ and $\omega = \omega_1 + \omega_2$. Thus, by carefully selecting ω_1 and ω_2, one can "mix down" the device response to low frequencies.

Initially, a constant potential difference of 15 V was established between the sample's actuation and device layers. Then, a fixed 11.250 kHz (100 mV) signal was applied simultaneously to both the bias and the actuation layer. A network analyzer signal sweeping 0.5 to 1.5 kHz (500 mV) at a 10 Hz resolution bandwidth was conventionally mixed with the AC signal on the actuator.

The resulting sum frequency signal swept 11.750 kHz to 12.750 kHz, across the natural resonance of the membrane at 12.248 kHz. The peak, shown in Figure 3, was observed at approximately 1 kHz possessing the same width as the previously observed peak at 12.248 kHz. As expected, this position in the sweep corresponds to where the summed frequency signal crosses 12.248 kHz. Subsequent measurements conducted using different values for the fixed AC signal verify that the peak at 1 kHz corresponds to the "mixed down" signal of the natural resonance of the device.

4 NOISE ANALYSIS

Ultimately, the displacement measurement sensitivity of the technique is limited by the amount of current noise present in the measurement system. Given a small change in the gap separation, there will be a corresponding shift in the tunnel current. If the change is below the current noise threshold, then it becomes impossible to distinguish the response from the noise.

The tunnel current may be written in terms of the work function of the metal. Expressed in units of nanoamperes and angstroms,

$$I \propto e^{-1.025\sqrt{\phi}z}. \tag{1}$$

Inserting the theoretical work function for gold[8] of 5.4 eV gives

$$I \propto e^{-2.38z} \tag{2}$$

Thus, the infinitesimal change in tunnel current for a small displacement is

$$\frac{\partial i}{\partial z} = -2.38I \tag{3}$$

indicating that the so-called "responsivity" changes for

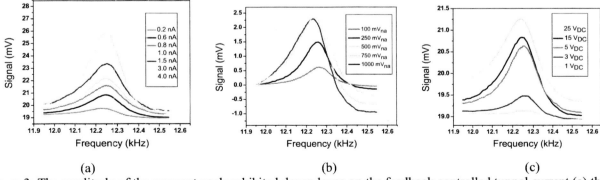

(a)   (b)   (c)

Figure 3: The amplitude of the resonant peak exhibited dependence on the feedback controlled tunnel current (a) the amplitude of the AC swept signal (b) and the amplitude of the DC voltage applied to the actuator.

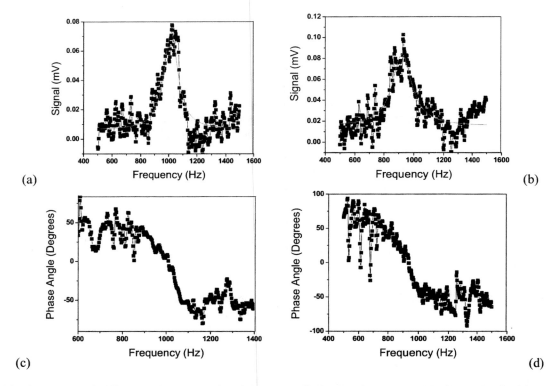

(a)

(b)

(c)

(d)

Figure 4: (a) The measured difference frequency signal for an applied AC voltage at 11.250 kHz mixed with a swept signal from 0.5-1.0 kHz. The summed frequency signal sweeps over the natural resonance of the device simultaneously with the difference frequency signal sweeping over 1 kHz (b) The same measurement performed with an applied AC voltage of 11.350 kHz (c)-(d) the respective phase plots.

different gap heights. Again, using a typical value for the feedback set tunnel current of 1 nA (corresponding to a gap of about 0.39 on gold) gives $\partial i / \partial z = -2.38$.

For our experiment, we used an RHK IVP 200 pre-amplifier. Previous measurements[9] quote average current fluctuations of 1.2 pA (parasitic capacitance) for a 1.5 kHz bandwidth. This corresponds to a current noise spectral density, $\sqrt{S_i} \sim 8 \times 10^{-4}$ pA/$\sqrt{\text{Hz}}$. To estimate the displacement sensitivity, we convert the current noise spectral density into displacement noise spectral density as

$$S_x = \frac{S_i}{\left(\partial i / \partial z\right)^2} .$$ (5)

Our displacement sensitivity is then $\sqrt{S_x} \sim 1.4 \times 10^{-2}$ fm/$\sqrt{\text{Hz}}$.

5 CONCLUSIONS

The use of quantum mechanical electron tunneling appears to be applicable to displacement measurements of high frequency devices. Using a scheme for "mixing down" the tunnel current signals, displacement detection at high frequencies can be realized. The sensitivity limits of

the current measurements are set by amplifier noise processes.

We gratefully acknowledge support from the NSF under grant no. 324416.

REFERENCES

[1] A. Cleland and M. Roukes, Appl. Phys. Lett. 69, 2653 1996.
[2] T.W. Kenny et al., J. Vac. Sci. Technol. A 10 (4) 2114, 1992.
[3] S.B. Waltman and W.J. Kaiser, Sensors and Actuators 19, 201 1989.
[4] A.A. Baski, T.R. Albrecht, C.F. Quate, J. Microscopy 152, 73 1988.
[5] RHK Technology, Inc. 1050 East Maple Road, Troy MI, 48083.
[6] T. Kouh, private communications.
[7] J. Yang, P. Voigt, and R. Koch, APL 82, 1866 2003.
[8] Stroscio, J. "Scanning Tunneling Microscopy" Academic Press, Inc. 1993.
[9] Technical brief published at http://www.rhk-tech.com.

A Nanoscale Electromechanical Contact Switch for Logic Applications

C. C. Huang [*] and K. L. Ekinci [**]

[*] Electrical and Computer Engineering, Boston University, Boston MA 02215, jchuang@bu.edu
[**] Aerospace and Mechanical Engineering, Boston University, Boston MA 02215, ekinci@bu.edu

ABSTRACT

Here, we describe the operation of a nano-electro-mechanical contact switch. The switch design relies on a sharp contact tip integrated to a compliant nanomechanical beam. When the beam is actuated electrostatically, the tip establishes intimate electrical contact between two electrodes. A preliminary contact switch was fabricated on a Si_3N_4 membrane using optical lithography, electron beam lithography and reactive ion etching. Its quasi-static operation was subsequently demonstrated.

Keywords: nano-electro-mechanical systems, nano-electro-mechanical switch, mechanical logic

1 INTRODUCTION

Recent efforts to scale micro-electro-mechanical systems (MEMS) [1] down into the nanometer scales[i] have created *integiable* nanomechanical structures with *extremely high resonance frequencies* — approaching the pace of semiconductor electronic devices. These nano-electro-mechanical systems (NEMS) come with extremely high resonance frequencies, diminished active masses, tolerable force constants and high quality (Q) factors of resonance. These attributes collectively make NEMS suitable for a multitude of technological and scientific applications.

Here, we demonstrate the operation of a NEMS contact switch — intended for use in electromechanical logic applications. The switch offers such prospects as ultra-fast, ultra-low power operation. In this domain of reduced length scales, however, there are unique physical phenomena that challenge the operation of the switch. For this work, we have considered various physical processes that would effect the switch operation and designed an operable structure. Our design is based on a nanoscale contact between a suspended high frequency resonator and a cathode electrode. The device was fabricated on a Si_3N_4 membrane and was tested in the quasi-static (low-frequency) operation mode. Our preliminary results indicate that contact resistances in the "*on*" state greatly are higher than our estimations.

2 DEVICE FABRICATION

Now, we turn to a detailed description of the device fabrication procedure.

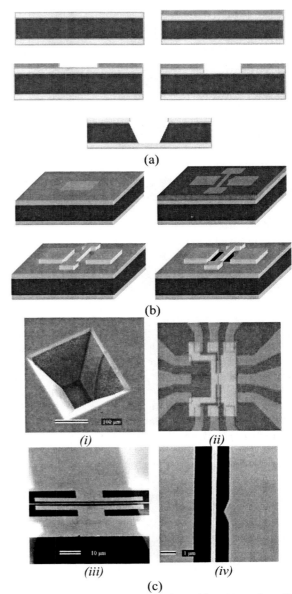

Fig. 1 The device is fabricated by a "top-down" fabrication approach. (a) First, a suspended membrane of Si_3N_4 is fabricated. (b) A contact NEMS switch is nanomachined atop the membrane. (c) *(i)* The back side of the completed membrane, *(ii)* top view of the completed device, *(iii)* and *(iv)* scanning electron micrograph of the switch and contact tip.

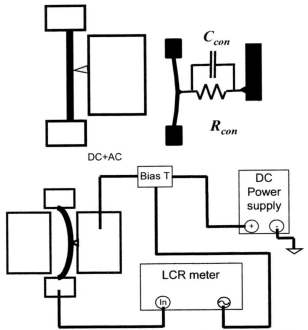

Fig. 2: (a) A schematic of the NEM switch. (b) An equivalent circuit for the switch on contact. (c) The setup employed to measure the contact resistance and capacitance.

The fabrication process has two main steps: manufacturing the suspended silicon nitride (Si₃N₄) membrane, and nanomachining the switch structure atop the membrane.

The membrane fabrication procedure is illustrated in Figure 1 (a). Initially, a silicon (100) wafer of thickness 125 micron is coated on both sides with 125-nm thick silicon nitride using an LPCVD process. Photolithography is then used to pattern a square into the backside of the wafer. Next, reactive ion etching (RIE) with He and SF₆ gases is used to open the nitride layer, exposing the silicon sacrificial layer. A potassium hydroxide (KOH) wet etch is used to dissolve the silicon preferentially in the (100) direction at ~57° angle to the (111) plane. The etch terminates at the Si₃N₄ device layer. The final result of this primary process is a 70 μm x 125 μm Si₃N₄ membrane, clamped at its boundaries.

The device fabrication procedure is illustrated in Figure 1 (b). Fabrication begins on the membrane by defining large area contact pads by optical lithography. A 60 nm-thick layer of Cr is then evaporated and, subsequently, standard liftoff is carried out with acetone. Samples are then coated with a bi-layer PMMA (polymethyl methacrylate) resist prior to patterning by electron beam lithography. After resist exposure and development, 5 nm of Cr and 70 nm of Au are evaporated on the samples, followed by lift-off in acetone. After lift-off, the membrane is removed by using RIE, again. Finally, we use plasma of He and SF₆ at a pressure of 250 mTorr with respective flow rates of 21 sccm and 13 sccm, and a microwave power of 100 W — to suspend the structure. The completed switch structure is shown in Fig. 1(c).

3 DEVICE CONCEPT

A NEMS contact switch thus fabricated is shown in Figure 1 (c). The switch has two electrical terminals: the first terminal is connected to the central nanomechanical beam structure; the second, side electrode is used for actuation as well as detection as shown in Figure 2 (a). The center of the beam displaces in-plane and accomplishes mechanical and electrical contact with the side electrode as shown in Figure 2 (b).

The actuation is realized by using electrostatic forces, i.e. by applying a voltage, V_{DC}, to the side electrode. In this scheme, the switch body is kept at electrical ground; the conductance and capacitance through the contact between the nanomechanical beam and the side electrode is measured using a lock-in technique as shown schematically in Figure 2 (c).

To clearly characterize the operation principle of the switch, one needs to compare several competing forces: the *electrostatic* actuation force, the *elastic* restoring force and the *stiction* forces arising from mechanical contact. When the actuation voltage is applied, the structure experiences an attractive electrostatic force towards the side electrode given by

$$F_{es} = \frac{1}{2} \frac{\partial C}{\partial x} V_{DC}^2 \approx \frac{1}{2} \frac{C}{(D-x)} V_{DC}^2 \qquad (1)$$

where C is the capacitance between the beam and the side electrode, D is the initial separation between the beam and the side electrode and x is the displacement of the center of the beam. When the beam bends towards the side electrode, there an elastic restoring force acts upon it given by

$$F_r = -k_{eff} x . \qquad (2)$$

Here, $k_{eff} = 32Etw^3 / l^3$ is the effective stiffness of the doubly-clamped beam with Young's modulus, E, and dimensions $l \times t \times w$. The actuation force will move the beam until it is equaled by the elastic restoring force of the flexing beam. Using this concept, one can obtain an estimate for the maximum displacement, x_{max}, of the center of the beam.

Upon release of the switch voltage, the elastic restoring force, $F_r \approx -\kappa_{eff} x_{max}$ must bring the switch back to its original position to create an open circuit. Given that the beam and the side electrode are in physical contact, one must consider the attractive (stiction) forces between the two surfaces. Van der Waals interactions between the atoms (molecules) of the two surfaces can give rise to significant attractive forces, F_{vdW}, at small separations.

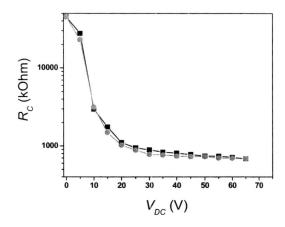

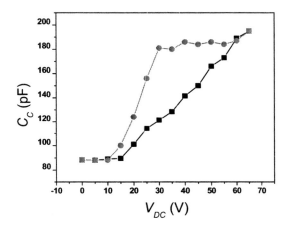

Figure 3: (a) The contact resistance as a function of the actuation voltage. (b) The capacitance as a function of the actuation voltage. The (■) are for increasing voltage values.

The contact resistance, R_C, of the switch as a function of the actuation voltage, V_{DC}, is presented in Figure 3 (a). The measurements are taken for increasing and decreasing V_{DC} values. R_C appears to drop precipitously for $V_{DC} < 10$ V. With increasing V_{DC}, R_C saturates around 800 kΩ. R_C displays hysteretic behavior with increasing and decreasing V_{DC} values.

The capacitance of the contact as a function of V_{DC}, is presented in Figure 3 (b). The capacitance, C_C, of the junction appears to increase as expected from the simple expression

$$C_C = \frac{\varepsilon_r \varepsilon_0 A}{(D-x)} \tag{5}$$

C_C also exhibits hysteretic behavior.

We believe that the hysterisis in R_C and C_C, are caused by the stiction phenomenon in the junction.

5 CONCLUSION

In conclusion, a nanomechanical contact switch was fabricated and tested in this work. We have characterized the quasi-static behavior of the switch. The switch displays higher contact resistance than expected. In future work, we will characterize the dynamic behavior of the switch.

REFERENCES

[1] Roukes, M. L. "Nanoelectromechanical systems face the future." *Physics World* **14**, 25-31 (2001).

[2] A. Erbe, R. H. Blick, A. Tilke, A. Kriele, and J. P. Kotthaus, "A mechanically flexible tunneling contact operating at radio frequencies" Appl. Phys. Lett. 73, 3751 (1998)

[3] P. G. Datskos, T. Thundat, "Nanocantiliever Signal Transduction by Electron Transfer" J. Nanosci. Nanotech. 2002, Vol.2, 369 (2002)

[4] W. H. The, J. K. Luo, M. R. Graham, A. Pavlov, C. G. Smith, "Switching characteristics of electrostatically actuated miniaturized micromechanical metallic cantilevers" J. Vac. Sci. Technol. B 21(6), (2003)

[5] E. Buks, M. L. Roukes, "Stiction, adhesion energy, and Casimir effect in micromechanical systems" Phy. Rev. B, Vol.63, 033402 (2001)

[6] J. Israelachvili, "Intermolecular and Surface Forces," *Academic Press* (1992).

[7] Dawn Bonnell, "Scanning Probe Microscopy and Spectroscopy : Theory, Techniques, and Applications," *John Wiley & Sons* (2000).

Especially in small systems — where surface to volume ratios are large — such short range forces can dominate the system behavior [6, 7].

We have designed our switch with these considerations in mind. The switch structure shown in Figure 1 (c), for instance, has $k_{eff} \approx 0.008$ N/m. The separation between the side electrode and the beam is 500 nm, giving sufficient actuation force at $V_{DC} \sim 50$ V. The contact radius can be estimated as $r \sim 100$ nm, giving rise to negligible stiction forces.

4 RESULTS AND ANALYSIS

We now turn to a discussion of our measurements. In these measurements, the contact impedances are measured at a frequency of 1 kHz using a rms amplitude of 50mV.

Defects and Fault Characterization in Quantum Cellular Automata

Mehdi Baradaran Tahoori, Mariam Momenzadeh, Jing Huang, Fabrizio Lombardi

Department of Electrical and Computer Engineering,
Northeastern University, Boston, MA, 02115
{mtahoori ,mmomenza ,hjing, lombardi}@ece.neu.edu

ABSTRACT

In this paper, a detailed simulation-based characterization of QCA defects and study of their effects at logic-level are presented. Different defect mechanisms for QCA active devices and interconnects are considered and the appropriate fault models at logic-level are investigated. Different implementations of QCA logic devices and interconnects are compared in term of defect tolerance and testability.

Keywords: Quantum Computing, Nano systems and devices, Defect Characterization

1 INTRODUCTION

There has been extensive research in recent years at nano scale to supersede CMOS technology. It is anticipated that these technologies can achieve a density of 10^{12} devices/cm^2 and operate at THZ frequencies.

Among these new devices, *quantum dot cellular automata* (QCA) not only gives a solution at nano scale, but also it offers a new method of computation and information transformation [5]. In terms of feature size, it is projected that a QCA cell of few nanometer size can be fabricated through molecular implementation by a self-assembly process. The unique feature of QCA based designs is that logic states are not stored in voltage levels as in conventional electronics, but they are represented by the position of individual electrons.

For QCA the cells must be aligned precisely at nano scales to provide correct functionality, so proper testing of these devices for manufacturing defects and misalignment plays a major role for quality of these circuits.

In this paper, the defect characterization of these devices has been extensively studied; the effects of defects at logic-level are investigated. Also, testing of QCA is compared with testing of conventional CMOS implementations of these logic devices.

The approach used in this work is based on simulating different manufacturing misalignments, investigating their effects at logic level and identifying the test vectors for fault detection. Different fabrication schemes of the majority voter at cell level are studied and compared in terms of defect tolerance and testability.

The rest of this paper is organized as follows. In Sec. 2, the defect characterization for QCA is presented. In Sec. 3, test sets and defect and fault coverage are discussed. Finally, Sec. 4 concludes the paper.

2 DEFECT CHARACTERIZATION

In this section, the robustness of the QCA Majority Voters and Binary Wires is investigated. Various configurations of QCA devices are studied using the QCADesigner[1] v1.20 simulation tool. For accuracy, the bistable model is used, i.e. a quantum mechanical engine using Jacobi algorithm to calculate the eigenvalues/vectors of the Hamiltonian.

2.1 Defect and Failure Modes

To perform a defect characterization of QCA devices and circuits and study their effects at logic-level, appropriate defect mechanisms and models must be considered which 1) can be simulated using available simulation methods and 2) are realistic to model manufacturing and fabrication defects.

A *cell displacement* is a defect in which the defective cell is misplaced within its original direction. Several cell displacement defects are shown in Fig. 1. In a *cell misalignment* defect, the direction of the defective cell is misplaced. Some examples of cell misalignments are shown in Fig.2. In this paper, the following defects are simulated for QCA devices: all possible combinations of cell displacement with respect to the central cell under different distances, cell misalignment in different directions, and rotation.

2.2 Majority Voter Defect Analysis

The basic logic gate in QCA is the *majority voter* (MV). The MV with logic function MV(A, B, C) = AB + AC + BC, can be realized by only 5 QCA cells [5]. Logic AND and logic OR functions can be implemented from a majority voter by setting one input permanently to 0 and 1, respectively. In [5], [4] the design of a set of universal logic gates based on the majority voter and inverter gate is presented.

There has been a study of the fault tolerant properties of the MV under some manufacturing misalignments [1]–[3]. Based on this simulation-based study, a

---

[1]QCADesigner is the product of an ongoing collaboration between the ATIPS laboratory and University of Notre Dame.

fault tolerant MV block has been proposed. It has been shown that MV is more vulnerable to misalignment in the vertical direction than in the horizontal direction.

In this work, different defects in the MV (cell displacement and misalignment) are considered and simulated. The faulty results for cell displacement and misalignment are shown in Table 1 and 2, respectively. Only faulty entries are shown in the tables. The considered defect free majority gate has a $5nm$ dot and a $20nm \times 20nm$ cell size, with a $5nm$ cell distance.

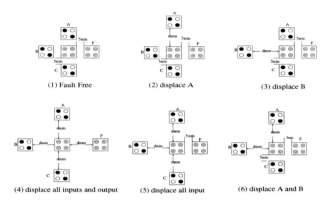

(1) Fault Free   (2) displace A   (3) displace B

(4) displace all inputs and output   (5) displace all input   (6) displace A and B

Figure 1: Displacement in Majority Voter

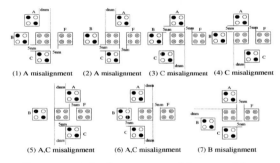

(1) A misalignment   (2) A misalignment   (3) C misalignment   (4) C misalignment

(5) A,C misalignment   (6) A,C misalignment   (7) B misalignment

Figure 2: Misalignment in Majority Voter

The data shows that in most cases the horizontal input cell (i.e. cell B) is the dominant cell. For misalignment, any single cell misalignment greater or equal to half a cell causes malfunction (fault at logic-level). In some cases the error margin is even smaller.

2.3 Defect Analysis of Rotated Majority Voter

The simulation results show that MV is robust with respect to rotation of all input and output cells around the center cell, i.e. the logic-level behavior of the rotated MV is the same as the original device. Based on this observation, some simulations are performed to investigate the robustness of the rotated MV (RMV). The basic functionality of MVs is based on the Coulombic interaction among four neighboring QCA cells, which strongly depends on the precision and geometry of its implemen-

displace cell A: fig 1(2)	
$d \leq 15nm$ Normal Operation	$d \geq 20nm$, F=B

displace cell B: fig 1(3)		
$d \leq 40nm$ Normal Operation	$d \geq 45nm$	
	A B C	F
	001	Z (no polarization)
	011	Z(no polarization)
	100	Z (no polarization)
	110	Z (no polarization)

displace all input/output cells: fig 1(4)		
$d \leq 10$ or $30 \leq d \leq 40nm$ Normal Operation	$15 \leq d \leq 25nm$	
	A B C	F
$d \geq 45nm$	010	0/1
F=Z (no polarization)	101	1/0

displace all input cells: fig 1(5)			
$d \leq 15$ or $d = 40nm$ Normal Operation	$d \geq 45nm$ F=Z (no polarization)		
$20 \leq d \leq 25$ or $d = 35nm$		$d = 30nm$	
A B C	F	A B C	F
010	0/1	000	0/1
101	1/0	010	0/1
		101	1/0
		111	1/0

displace cells A and B: fig 1(6)	
$d \leq 5nm$ Normal Operation	$d \geq 10nm$, F=C

Table 1: Results for displacement in Majority Voter

tation. We focus on validating different configurations of MV in the 45° rotation, as shown in Fig. 3.

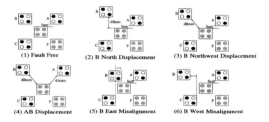

(1) Fault Free   (2) B North Displacement   (3) B Northwest Displacement

(4) AB Displacement   (5) B East Misalignment   (6) B West Misalignment

Figure 3: Rotated Majority Voter (Fault-free, with displacement or misalignment)

The simulation results show that the RMV functions normally, except when moving:

- A input north, with dA$\geq 10nm$ for ABC = 001, 110 (the output follows the C input). A similar output appears when moving A to northeast with dB$\geq 10\sqrt{2}nm$.

- B input north, with dB$\geq 40nm$. The output is unknown (unpolarized) for ABC = 001, 011, 100, 110. A similar output appears when moving B to the northwest with dB$\geq 30\sqrt{2}nm$.

- C input south, with dC$\geq 15nm$ for ABC = 011, 100 (the output follows the A input). A similar output appears when moving C to the southwest with dC$\geq 10\sqrt{2}nm$.

- A, B, C or A, B, C, Z away for d$\geq 30\sqrt{2}nm$. The output is undefined for all input combinations.

- A and B inputs away with d$\geq 10\sqrt{2}nm$ for ABC = 001, 110 (the output follows the C input).

- A and C inputs away with d$\geq 10\sqrt{2}nm$ for ABC = 010, 101 (the output follows the B input).

- B and C inputs away with d$\geq 10\sqrt{2}nm$ for ABC = 011, 100 (the output follows the A input).

move A toward west: fig 2(1)	
$d \leq 5nm$ Normal Operation	$d \geq 10nm$, F=B

move A toward east: fig 2(2)		
$5 \leq d \leq 15nm$		$d = 20$ or $d = 30nm$ Normal Operation
A B C	F	
001	0/1	
010	0/1	$d = 25nm$ F=A
101	1/0	
110	1/0	

move C toward west: fig 2(3)	
$d \leq 5nm$ Normal Operation	$d \geq 10nm$ F=B

move C toward east: fig 2(4)		
$5 \leq d \leq 15nm$		$d = 20$ or $d = 30nm$ Normal Operation
A B C	F	
010	0/1	
011	1/0	$d = 25nm$ F=C
100	0/1	
101	1/0	

move A,C toward west: fig 2(5)
$d \geq 5nm$ F=B

move A,C toward east: fig 2(6)		
$d = 5, 20, d \geq 30nm$ F=B	$10nm \leq d \leq 15nm$	
$d = 25nm$ Normal Operation	A B C	F
	000	0/1
	010	0/1
	101	1/0
	111	1/0

move B toward south/north: fig 2(7)		
$d \leq 5nm$ Normal Operation	$d \geq 45nm$	
	A B C	F
	001	0/1
	011	1/0
	100	0/1
	110	1/0

Table 2: Results for misalignment in Majority Voter

Cell misalignment defects for RMV are also considered (e.g., Fig. 3(5,6)). The following shows the results:

- Shifting the input A west (half/full cell size), leads the output F to follow input A, while shifting A east effects the output such that it follows input C.

- RMV functions normally when input B is shifted west for a half or full cell size. However, the output is undefined for inputs ABC = 001, 011, 100, 110 when dB$\geq 40nm$.

- The output follows the input B, when B is shifted east for a half or full cell size.

- The output follows the input A when C is shifted west, and follows C when C is shifted east.

2.3.1 OMV and RMV Comparison

The results for different configurations of the Original Majority Voter (OMV) and the Rotated Majority Voter (RMV) are illustrated in Table 3. MV is completely robust with respect to rotation of all inputs and output cells around the central cell. This gives a significant degree of freedom for synthesizing designs based on QCA, as RMV can be used as the Original Majority Voter block. However, the original block is more dependent on the middle input (B) than the other inputs (A and C), in terms of displacement and misalignment. In the rotated version, this dependency can be completely changed based on the degree of rotation. An overall comparison in the table confirms that RMV is more fault-tolerant than the Original MV. Note that

only half and full misalignments are considered.

Config.	Faults	OMV	RMV
A move	distance	$d \geq 20nm$	$d \geq 10(N)$ or $10\sqrt{2}nm$ (NE)
	# of faults	2	2
B move	distance	$d \geq 45nm$	$d \geq 40(W)$ or $30\sqrt{2}nm$ (NW)
	# of faults	4	4
C move	distance	$d \geq 20nm$	$d \geq 10(S)$ or $10\sqrt{2}nm$ (SW)
	# of faults	2	2
ABC move	distance	$20 \leq d \leq 35$ or $d \geq 45nm$	$d \geq 30\sqrt{2}nm$
	# of faults	2/4/8	8
ABCF move	distance	$15 < d < 25$ or $d \geq 45nm$	$d \geq 30\sqrt{2}nm$
	# of faults	2/8	8
AB move	distance	$d \geq 7.5nm$	$d \geq 10\sqrt{2}nm$
	# of faults	2	2
AC move	distance	$d \geq 7.5nm$	$d \geq 10\sqrt{2}nm$
	# of faults	2	2
Z move	distance	$d \geq 45nm$	$d \geq 30\sqrt{2}nm$
	# of faults	8	8
AC misalignment	# of faults	4	4
B misalign. West	# of faults	4	0
B misalign. East	# of faults	4	2

Table 3: Original Majority Voter vs. Rotated Majority Voter

2.4 Binary Wires and Inverter Chains

The effect of cell displacement defects on two parallel binary wires as well as two parallel inverter chains are investigated.

2.4.1 Double Binary Wires

Two defect-free binary wires are shown in Fig. 4(a); the wires are denoted as the upper wire ($i1$ to $o1$) and the lower wire ($i2$ to $o2$). The cells have a size of $20nm \times 20nm$, and the dot diameter is $5nm$. In the defect-free case, the cells in the same wire are separated by $15nm$ and the wire distance is $60nm$.

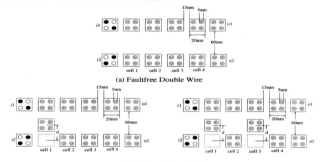

(a) Faultfree Double Wire

(b) Defects in Double Wire

Figure 4: Displacement in Binary Double Wires

The displacement defects are simulated by moving one or two cells in the lower wire toward the upper wire (by displacement d) as shown in Fig. 4(b).

The simulation results are shown in Table 4. The results show that the upper wire is dominant in most cases: $o1$ and $o2$ are either equal to $i1$ or $i1'$, depending on which cell(s) are displaced and the value of the displacement, d. In most cases, the upper wire functions normally, i.e. $i1 = o1$. However, in some cases the upper wire behaves as an inverter. Clearly, unlike CMOS designs, the coupling defects at QCA device-level do not behave as the *wired bridging fault* model. However, these defects manifest themselves as the dominant model (at logic level) in which the output of a wire is determined by the value of the coupled wire.

move cell1 OR cell2		
$d \leq 40nm$ Normal	$d = 45 - 50nm$ $o1 = i1, o2 = i1$	$d \geq 55nm$ $o1 = i1, o2 = Z$
move cell3 OR cell4		
$d \leq 35nm$ Normal	$d = 40 - 50nm$ $o1 = i1, o2 = i1'$	$d \geq 55nm$ $o1 = i1, o2 = Z$
move cell1 AND cell2		
$d \leq 35nm$ Normal	$d = 40 - 50nm$ $o1 = i1, o2 = i1$	$d \geq 55nm$ $o1 = i1, o2 = Z$
move cell1 AND cell4; OR move cell 2 AND cell 3; OR move cell3 AND cell4		
$d \leq 35nm$ Normal	$d = 40 - 50nm$ $o1 = i1, o2 = i1'$	$d \geq 55nm$ $o1 = i1, o2 = Z$

move cell1 AND cell3			
$d \leq 35nm$ Normal	$d = 40 - 50nm$ $o1 = i1, o2 = i1$	$d = 45nm$ $o1 = i1, o2 = i1'$	$d \geq 55nm$ $o1 = i1, o2 = Z$

move cell2 AND cell4				
Normal	$d=20\text{-}25nm$ $d=40\text{-}45nm$ $o1 = i1$	$d=30\text{-}35nm$ $o1 = i1$ $o2 = i1$	$d = 50nm$ $o1 = i1'$ $o2 = i1$	$d \geq 55nm$ $o1 = i1$ $o2 = Z$

Table 4: Results for Double Binary Wires

2.4.2 Double Inverter Chains

The double inverter chain is shown in Figure 5(a). The simulation results for moving one cell in the bottom wire toward the upper wire, with displacement d, (Figure 5(b)) are presented in Table 5. The displacement defects behave as the *dominating bridging fault* model at logic level. Moreover, a comparison with the binary wires shows that binary wires are more defect tolerant than inverter chains in case of displacement coupling defects.

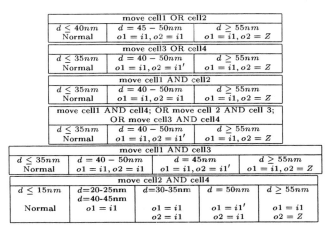

(a) Fault Free Inverter Chain   (b) Single Cell Displacement Defect

Figure 5: Displacement in Double Inverter Chains

Fault Free: $o1 = i1'$; $o2 = i2'$		
move cell1 OR cell2 OR cell3		
$d \leq 35nm$ Normal	$d = 40nm - 50nm$ $o1 = i1', o2 = i1'$	$d \geq 55nm$ $o1 = i1', o2 = Z$
move cell4		
$d \leq 30nm$ Normal	$d = 35nm - 50nm$ $o1 = i1', o2 = i1'$	$d \geq 55nm$ $o1 = i1', o2 = Z$

Table 5: Results for Double Inverter Chains

3   TEST SETS, DEFECT COVERAGE AND FAULT MODEL

The effectiveness of different stuck-at test sets is evaluated for the simulated defects on a single majority voter. The main results are as follow:

- In all simulations, *super exhaustive* input patterns (i.e. all possible input transitions) are used. The

results show that there is no sequence dependent behavior at logic level; i.e. none of the manufacturing misalignments introduces a state dependency.

- Except for a single case (i.e. the displacement of all inputs and output cells) faults are detected using a subset of some 100% stuck-at fault test sets. Note that not all of these 100% stuck-at test sets are the same.

- A particular 100% 2-detect stuck-at test set (each fault is detected by two vectors) can detect all manufacturing defects, except for the simultaneous displacement of the top and left inputs.

- Moreover, a particular 100% single stuck-at test set (001,010,011,101) can detect all simulated defects.

4   CONCLUSIONS

Quantum dots cellular automata (QCA) are novel devices which are promising in the era of nano scale computing. In this paper, a detailed defect characterization for QCA logic devices and some representative circuits has been presented. As shown in this paper, the coupling mechanisms and behavior of defects at logic-level (i.e. the faults) are not similar to those for conventional CMOS. The effectiveness of different stuck-at test sets in detecting QCA defects has been studied. Our results show that to achieve a high defect coverage, the specific QCA implementation of each function must be considered for test generation.

REFERENCES

[1] C.D. Armstrong, W.M. Humphreys, A. Fijany, "The Design of Fault Tolerant Quantum Dot Cellular Automata Based Logic", 11th NASA Symposium on VLSI Design, 2003.

[2] C.D. Armstrong, W.M. Humphreys, "The Development of Design Tools for Fault Tolerant Quantum Dot Cellular Automata Based Logic", 2nd International Workshop on Quantum Dots for Quantum Computing and Classical Size Effect Circuits, 2003.

[3] A. Fijany and B. N. Toomarian, "New design for quantum dots cellular automata to obtain fault tolerant logic gates", Journal of Nanoparticle Research, vol. 3, pp. 27-37, 2001.

[4] C.S. Lent and P.D. Tougaw, "A Device Architecture for Computing with Quantum Dots", Proceedings of IEEE , vol. 85(4), pp. 541-557, 1997.

[5] P.D. Tougaw and C.S. Lent, "Logical Devices Implemented Using Quantum Cellular Automata", Journal of Applied Physics, vol. 75(3), pp. 1818-1825, 1994.

Investigation of Robust Fully-Silicided NMOSFETs for Sub-100 nm ESD Protection Circuits Design

Jam-Wem Lee[1], Howard Tang[2] and Yiming Li[1,3,*]

[1]Department of Nano Device Technology, National Nano Device Laboratories, Hsinchu 300, TAIWAN
[2]United Microelectronics Corporation, Hsinchu 300, TAIWAN
[2]Mircoelectronics and Information Systems Research Center, National Chiao Tung University, Hsinchu 300, TAIWAN
[*]P.O. Box 25-178, Hsinchu 300, TAIWAN, ymli@faculty.nctu.edu.tw

ABSTRACT

This paper demonstrates a fully-silicided ESD protection device design in sub-100nm integrated circuits. No drain ballast resistor required is the most significant feature that makes the new device differ form the conventional ones. Accordingly, a simplified manufacturing process and a reduced device area could be obtained simultaneously. It is believed that the achievements are caused from the floating charge effects during the ESD stressed. On the other hand, in avoiding the device function affected by the floating charges, a switch is incorporated at the body electrode. The newly designed device structure is very attractive in novel ESD design for the consideration of its low cost, small size and high efficiency.

Keywords: Fully-silicided, ESD protection, floating body design, thin gate-oxide, turn on voltage, turn on resistance, efficiency

1 INTRODUCTION

Electrostatic discharge (ESD) has long played an important role in integration circuit design; in roughly estimating, ESD responses for several percents of the chip failure in modern VLSI manufacturing [1-12]. Accordingly, how to design a robust ESD protection circuit with a high efficiency is an important issue for IC designs [1-2]. Traditionally, in having a strong ESD protection circuit, the drain ballast resistor made from the non-silicided diffusion region has been placed between the drain contacts and polysilicon gates [3-6]. However, it requires an additional photo-mask, enlarges the devices area, and an extra device models in VLSI circuit design. The consequence will make a higher process cost, a lower devices density, and a poorer circuit performance. Therefore, a fully-silicded MOSFET device is a good choice in ESD protecting circuits. Unfortunately, it suffers a low ESD robustness [7-10], especially for the sub-100nm era.

In this paper, we propose a novel device structure to improve ESD robustness of fully-silicided MOSFETs. This alternative lets the body electrode in floating state under ESD stress. The experimental results show that the floating body devices reduce the turn voltage, improve the ESD performance, and have a better robustness for sub-100nm device design. The test devices are fabricated with standard CMOS process with a 90 nm lengthen gate and a 1.2 nm thicken gate-oxide. The ESD characteristics are verified with the Barth model 4200 TLP systems with a pulse width of 100 ns. It can be found that the body floating one has a much better ESD robustness than the body grounded one. The improvement could be caused from the impact ionization holes that accumulate and built up an electrostatic potential at the substrate; consequently, decrease the turn on voltage of parasitic BJT. A decrement of turn on voltages will suppress the damage of gate oxide and reduces the drain electric fields. We can also indicate that there is a wider current path existed in the body floating devices, as a result, the device sustains a better ESD strength than the grounded body one does.

We have presented a constructive alternative for designing a fully-silicided ESD protection device. It exhibited an excellent efficiency on both protection and chip area, and is attractive to sub-100 nm CMOS circuit era.

2 DEVICE FABRICATION AND MEASUREMENT

The test devices were fabricated by using a standard twin well CMOS process. The minimum gate length of this process is 90nm and the thickness of gate-oxide is 1.2nm. Cobalt silicide has long been used in deep sub-micrometer devices for its high thermal stability and low resistance in narrow width silicide line. Thus, in our experiments, a 20 nm thicken cobalt silicide is formed by using two steps rapid thermal process in exploring the fully-silicded ESD protection devices.

Interconnections were achieved by copper line and low dielectric constant (low K) insulator. The ESD characteristics were verified by using Barth model 4200 TLP system shown in figure 1, which generates a pulse wave with a width of 100 nano-second. Comparing with experimental results, the measured result is reflected to the human body model (HBM) value.

Transmission line

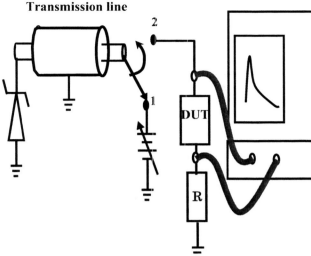

Figure 1: An illustration of the TLP measurement configuration in the study.

3 RESULTS AND DISCUSSION

Drain ballast resistor is the most used design methodology for ESD protection devices. The main purpose of the design is altering the current flow path at the drain junction. Make it more clearly that, with adding the resistor, the currents will prefer flowing through the deep junctions instead of drain extensions. The ESD current flowing through the deep junctions will take many benefits:

1. Extensions are much shallower than the deep junctions that confining current flowing in a very narrow path. Thus, current flowing through the deep junctions will result in a much wider path, cause a lower current density and sustains a better ESD robustness; and

2. The current paths at the deep junctions will reduce the electric field and the hot carrier at the channel region of the MOSFET; consequently, the gate-oxide damage from the hot carriers will be greatly reduced.

Conventionally, the ballast resistors are achieved by using a photo-mask that removes all the silicide on this region. The process will produce a non-silcided region that provides a ballast resistor at drain side. The ballast resistor, however, causes two main problems in designing of ESD protection devices. First, the non-silicided region will occupy a large area, for the most designs, the resistor dominates more than one half of the protection device. This result could be easily obtained form the comparison between Fig. 3 and Fig. 5. Second, the resistor will increase the turn on resistance of the protection devices; therefore, make the core circuits have a higher risk from ESD damages. According to previously discussion, a fully silicided ESD protection device is greatly preferred by all the circuit designers. Unfortunately, it is well known that the structure suffers a weak ESD robustness. The result, in

our theoretical analysis is only true for normally used body grounded MOSFETs but not for the floating ones. The layouts of those two structures are presented in the Fig. 3 and Fig. 4 respectively; moreover, the schematics are also shown in the Fig. 2 (a) and 2 (b) below.

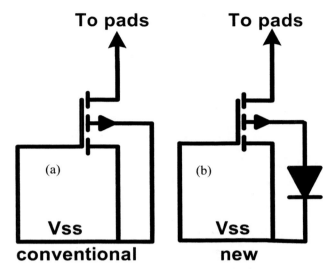

Figure 2: The schematics of the fully-silicided NMOSFET with (a) body grounded (b) Floating body.

Figure 6 presents the IV characteristics measured from the TLP. A clear result could be found that the floating body one has a much better ESD robustness than the grounded one. The results are highly matched with our theoretical predicted results.

The improvement could be due to an accumulation of impact ionization holes at substrate, which decrease turn on voltage of parasitic BJT. Owing to a decrement of turn on voltage, the damage of gate-oxide could be suppressed and the drain electric field could be also lowered. Moreover, for the body floating devices, a wider distribution of current flowing paths will keep away from the channel interface that could also enhance the ESD strength. Thus, the floating body one not only has a lower temperature distribution, but also prevents gate-oxide from hot carrier damage. On the other hand, the grounded body one has a much narrower distribution of current paths that located much closer to the gate-oxide interface. The current distribution not only produces a higher temperature distribution but could also have a risk of gate-oxide damage.

The novel ESD protection structure proposed in this experiment has been demonstrated having a superior ESD robustness. In comparing with the TLP IV characteristics of the normally used partially silicided and our proposed fully-silicided ESD protection devices in the Fig. 7, it could be found that the fully-silicided one has a much lower turn on resistance and a lower turn on voltage. Similar results could be made from the observation of Fig. 8 that an increment of the non-silicided width will result in a higher turn on resistance. Owing to those two characteristics, the voltage dropping through the gate-oxide of the input stage will be

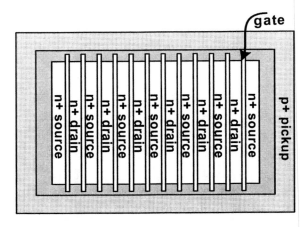

Fig. 3. Top view of the conventional fully-silicide multi-fingers device.

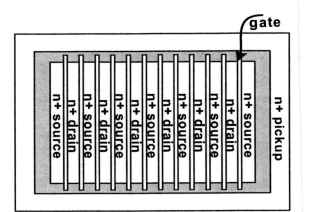

Fig. 4. Top view of the new fully-silicide multi-fingers device.

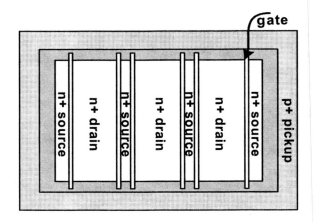

Fig. 5. Top view of the conventional silicide blocked multi-fingers device.

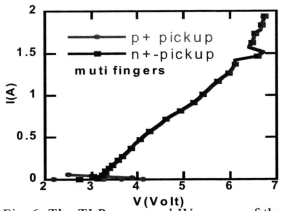

Fig. 6. The TLP measured IV curves of the two fully-silicided devices (W=320um).

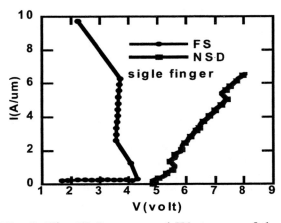

Fig. 7. The TLP measured IV curves of the new fully-silicided and the conventional silicide blocked devices (W=20um).

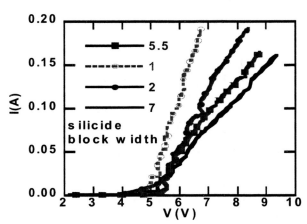

Fig. 8. The TLP measured IV curves of the conventional silicide blocked devices with different block width.

largely reduced. Those features are greatly helpful for keeping the core circuit away form ESD damages. Furthermore, in our experiences, the drain ballast resistances occupied over one half area of the device; thus, in comparing with the normally used ESD protection device, the fully silicided devices save over fifty percents of the area.

The fully silicided structure has been also tested in the whole chip circuit design. It is found that the benefit of the floating charge effects would not be eliminated due to the common substrate. This result is mainly caused from the fact the common substrate is too inefficient to pick the hole out form the well region.

It could be further addressed that the earlier turn on characteristics is actually a very wonderful property in protecting the ultrathin gate-oxide devices. Owing to the drain ballast resistor cannot lower the turn on voltage of the protection device; the device could have a gate-oxide breakdown earlier than the parasitic breakdown. When this situation occurred, the drain ballast resistor is not functional. Here, we can make a brief summary that the floating body fully silicided structure has a very high efficiency in saving the chip area and protecting circuit from ESD damage. It is especially useful in the ultrathin gate-oxide devices.

4 CONCLUSIONS

We have presented a constructive alternative for designing a fully-silicided ESD protection device. It exhibited an excellent efficiency on both protection and chip area, and is attractive to sub-100 nm CMOS circuit era. Floating body fully silicided MOSFET devices have been explored and demonstrated a good ESD robustness in this paper. Those devices have a lower turn on voltage, a lower turn on resistance, a good ESD robustness, and a high efficiency in both the area usage and protection. It is very attractive in novel ESD protection circuit design, especially suitable for ultrathin gate-oxide devices design. For example, the sub-100nm CMOS integration circuit designs.

5 ACKNOWLEDGEMENTS

This work is supported in part by the TAIWAN NSC grants: NSC-92-2112-M-429-001 and NSC-92-2815-C-492-001-E. It is also supported in part by the grant of the Ministry of Economic Affairs, Taiwan under contract No. 91-EC-17-A-07-S1-0011.

REFERENCES

[1] W. Fichtner, K. Esmark, and W. Stadler, "TCAD software for ESD on-chip protection design," Technical Digest of International Electron Devices Meeting (IEDM) 311, 2001.

[2] A. Salman, R. Gauthier, W. Stadler, K. Esmark, M. Muhammad, C. Putnam, and D. Ioannou, "Characterization and investigation of the interaction between hot electron and electrostatic discharge stresses using NMOS devices in 0.13 μm CMOS technology," Proceedings of The International Reliability Physics Symposium (IRPS) 219, 2001.

[3] T.-Y. Chen and M.-D. Ker, "Investigation of the gate-driven effect and substrate-triggered effect on ESD robustness of CMOS devices," IEEE Transactions on Device and Materials Reliability 1, 190, 2001.

[4] T.-Y. Chen and M.-D. Ker, "Design on ESD protection circuit with very low and constant input capacitance" Proceedings of International Symposium on Quality Electronic Design 247, 2001.

[5] M.-D. Ker and H.-C. Jiang, "Whole-chip ESD protection strategy for CMOS integrated circuits in nanotechnology" Proceedings of the 1st IEEE Conference on Nanotechnology (IEEE-NANO) 325, 2001.

[6] K.-H. Oh, C. Duvvury, K. Banerjee, and R. W. Dutton, "Investigation of gate to contact spacing effect on ESD robustness of salicided deep submicron single finger NMOS transistors" Proceedings of The International Reliability Physics Symposium (IRPS) 148, 2002.

[7] J.-W. Lee, Y. Li, and Howard Tang, "Silicide Optimization for Electrostatic Discharge Protection Devices in Sub-100 nm CMOS Circuit Design," Proceedings of The 2003 International Conference on VLSI (VLSI'03), CSREA Press 251, 2003.

[8] M.-D. Ker, C.-H. Chuang, and W.-Y. Lo, "Layout design on multi-finger MOSFET for on-chip ESD protection circuits in a 0.18-μm salicided CMOS process," Proceedings of The 8th IEEE International Conference on Electronics, Circuits and Systems (ICECS) Vol. 1, 361, 2001.

[9] C.-S. Kim, H.-B. Park, B.-G. Kim, D.-G. Kang, M.-G. Lee, S.-W. Lee, C.-H. Jeon, W.-G. Kim, Y.-J. Yoo, and H.-S. Yoon, "A novel NMOS transistor for high performance ESD protection devices in 0.18 μm CMOS technology utilizing salicide process," Proceedings of Electrical Overstress/Electrostatic Discharge Symposium 407, 2000.

[10] J.C. Smith, "An anti-snapback circuit technique for inhibiting parasitic bipolar conduction during EOS/ESD events," Proceedings of Electrical Overstress/Electrostatic Discharge Symposium 62, 1999.

[11] J.-W. Lee, Alice Chao, Y. Li, and H. Tang, "A Study of ESD Protection under Pad Design for Copper-Low K VLSI Circuits," Extended Abstracts of the 2003 International Conference on Solid State Devices and Materials (SSDM 2003) 672, 2003.

[12] Y. Li, J.-W. Lee, and S. M. Sze, "Optimization of the Anti-punch-through Implant for ESD Protection Circuit Design," Japanese Journal of Applied Physics 42, 2152, 2003.

Self-Assembly of Tetraazamacrocyclic Compounds on Single-Walled Carbon Nanotubes: Experimental and Theoretical Studies

E.V. Basiuk[*], V.A. Basiuk[**], E.V. Rybak-Akimova[***],
D. Acosta-Najarro[****], I. Puente-Lee[*****], and J.M. Saniger[*]

[*]Centro de Ciencias Aplicadas y Desarrollo Tecnológico, Universidad Nacional Autónoma de México (UNAM), 04510 México D.F., Mexico, elenagd@servidor.unam.mx
[**]Instituto de Ciencias Nucleares, UNAM, 04510 México D.F., Mexico, basiuk@nuclecu.unam.mx
[***]Department of Chemistry, Tufts University, Medford, Massachusetts 02155, USA, erybakak@granite.tufts.edu
[****]Instituto de Física, UNAM, 04510 México D.F., Mexico
[*****]Facultad de Química, UNAM, 04510 México D.F., Mexico

ABSTRACT

To explore the possibility of reversible modification of carbon nanotube (CNT) sidewalls with aromatic ligands and their metal complexes employing aromatic stacking phenomena, we studied (experimentally and theoretically) interaction of a series of tetraazamacrocyclic compounds tetraazaannulene, *meso*-tetraphenylporphine, Ni(II) and Cu(II) complexes of tetramethyltetraazaannulene with single-walled carbon nanotubes. All of them are conjugated systems with a high degree of aromaticity. Molecular mechanics modeling was used to estimate what kind of arrangements can form. In some cases HRTEM results were obtained consistent with theoretical results. Adsorption of aromatic molecules, having a bent shape and/or hydrophobic substituents, allows to reversibly attach modifying chemical species. The latter stick strongly to the nanotube walls due to $\pi-\pi$ and hydrophobic interactions and remain stable in aqueous solutions, but can be removed by washing with non-polar solvents.

Keywords: carbon nanotubes, tetraazamacrocyclic, porphines, metal complexes, molecular mechanics

1 INTRODUCTION

All the existing methods of chemical modification of carbon nanotubes (CNTs) can be divided in two groups, depending on whether functionalizing moieties are introduced onto the nanotube tips or sidewalls. The latter approach [1-3] offers wider opportunities to change CNT properties, since it allows high coverages with modifying groups. It can rely upon either covalent bond formation, or simple adsorption on CNTs *via* non-covalent interactions (hydrophobic, π-stacking, etc.).

To explore the possibility of reversible modification of CNT sidewalls with aromatic ligands and their metal complexes employing similar stacking phenomena, we studied (experimentally and theoretically) interaction of a series of tetraazamacrocyclic compounds tetraazaannulene (H_2TAA), *meso*-tetraphenylporphine (H_2TPP), Ni(II) and Cu(II) complexes of tetramethyltetraazaannulene (H_2TMTAA) (see structures in Figure 1) with single-walled carbon nanotubes (SWNTs). All of them are conjugated systems with a high degree of aromaticity.

2 EXPERIMENTAL

We used open-end SWNTs commercially available from Iljin Nanotech Co., Ltd., Korea (by arc-discharge process, 95%+ purified by thermal oxidation and chemical treatment). H_2TPP ligand and the macrocyclic complexes NiTMTAA and CuTMTAA were synthesized according to the procedures described elsewhere [4-7].

Typically, H_2TPP, NiTMTAA or CuTMTAA was dissolved in benzene (for H_2TPP) or ethanol (for the complexes), then SWNTs were added to the resulting solution and left overnight. After centrifugation of the product, it was additionally washed with a small amount of the corresponding solvent, then centrifuged again and dried under ambient conditions. Upon washing, the solvent acquired only a pale green (for NiTMTAA) or red-brownish (for CuTMTAA and H_2TPP) coloration, indicating a relatively strong adsorption of the tetraazamacrocyclic compounds on SWNTs.

The NiTMTAA—SWNT adduct had a noticeable dark-greenish color and metallic luster, common for NiTMTAA; the CuTMTAA—SWNT product was more alike the starting nanotube material in appearance. The amount of adsorbed tetraazamacrocyclic compound was calculated by measuring UV-Vis absorption of the solution before and after the adsorption. It was about 80% in both cases. This percentage corresponds to a SWNT:tetraazamacrocyclic complex mass ratio of about 5:4. The H_2TPP—SWNT adduct also had a strong metallic luster. The amount adsorbed calculated from the UV-Vis absorption differences, was about 50%, which is lower as compared to

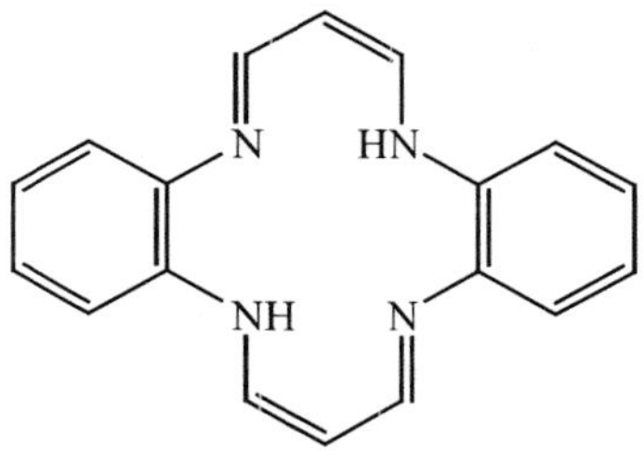

H$_2$TAA

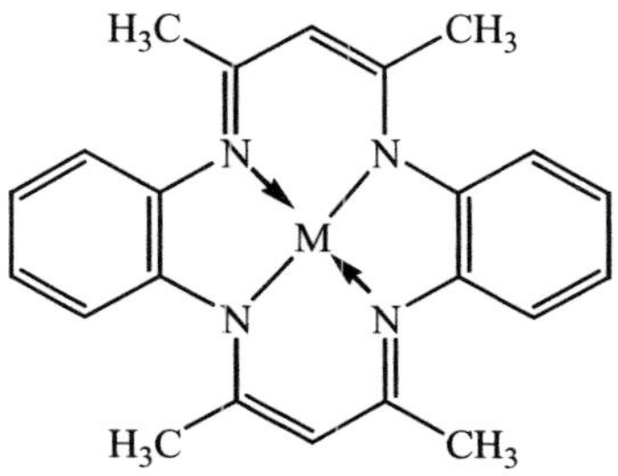

NiTMTAA (M=Ni^{2+})

CuTMTAA (M=Cu^{2+})

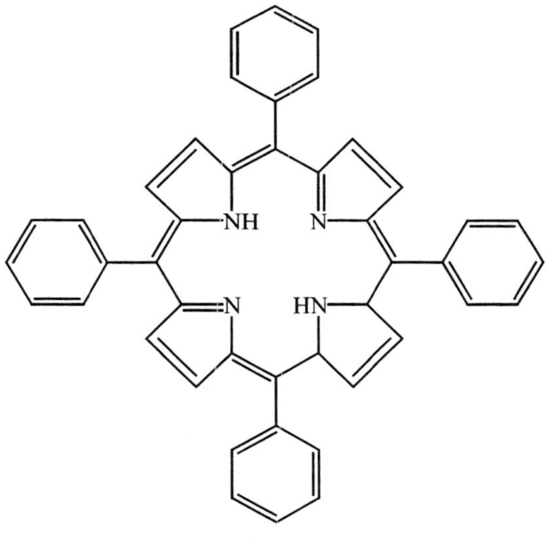

H$_2$TPP

Figure 1. Chemical structures of the tetrazaamacrocyclic ligands tetraazaannulene (H$_2$TAA), metal complexes of tetramethyltetraazaannulene (H$_2$TMTAA) with Ni(II) and Cu(II), and *meso*-tetraphenylporphine (H$_2$TPP).

the previous case. This corresponds to a SWNT:H$_2$TPP mass ratio of ca. 10:3.

Infrared spectra were recorded on a Nicolet 5SX FTIR spectrometer (in KBr); UV-Vis spectra, on a Shimadzu UV-160U spectrophotometer. Microscopic measurements were performed on a JEOL 100CX transmission electron microscope (TEM) and a JEOL 4000EX high-resolution transmission electron microscope (HRTEM).

In our theoretical simulations, we used universal force fields UFF and MM+, included into Gaussian 98W [8] and HyperChem version 5.1 (by HyperCube Inc., Canada) packages, respectively. For geometry optimizations with Gaussian 98W, the default convergence criteria were applied. When HyperChem 5.1 was used, the optimizations were performed with the Polak-Ribiere conjugate gradient algorithm and a root mean square gradient of 0.001 kcal Å^{-1} mol^{-1}. Both methods produce essentially similar results.

3 RESULTS AND DISCUSSION

In the case of TMTAA complexes (for a preliminary report, see [9]), their geometry is distorted from the plane due to the presence of four methyl substituents interfering with the benzene rings. As a result, the molecules of NiTMTAA and CuTMTAA adopt a saddle shape conformation, with the CH$_3$ groups and benzene rings turned to opposite sides of the MN$_4$ coordination plane. This geometry was especially attractive for our purpose, since it matches the curvature of SWNT sidewalls. One can expect that preferable mutual orientation of the metal complexes and the nanotube is 'saddle-on-horseback', increasing the contact area and thus adsorption strength.

Using molecular mechanics (MM) modeling, we tried to estimate what kind of arrangement can correspond to SWNT:complex mass ratios close to the one of 5:4 found experimentally. According to our results, given a uniform distribution of the macrocyclic molecules, an approximately monolayer coverage forms. A saddle-shaped conformation of the macrocyclic molecules helps their alignment along SWNT axis and better accommodation on the outer nanotube walls (Figure 2a), where the annulene complexes are arranged in a chess-like order (Figure 3a). No manifestations of the above molecular assembly can be distinguished by TEM. However HRTEM observations [9] revealed on SWNT bundles periodic transversal formations (period close to 1 nm), consistent with the MM results.

H$_2$TAA ligand molecules, due to the absence of methyl substituents in the macrocyclic ring, are flat. At the same time, their two benzene rings remain to be possible sites for π–π and hydrophobic interactions with SWNT walls, in addition to the annulene system. MM modeling showed that adsorption on the nanotube walls is more energetically preferable when H$_2$TAA molecules are oriented along SWNT axis (Figure 2b). Nevertheless, the difference for the longitudinal and transversal orientation is small, and in the case of many H$_2$TAA molecules adsorbed, they interfere the position of each other. As a result, no highly ordered molecular assembly was found (Figure 3b). Related experimental data have not been obtained so far.

In the case of H$_2$TPP, the orientational effect can be expected due to the existence of four C$_6$H$_5$ groups. Steric hindrance between the latter and the neighboring pyrrole

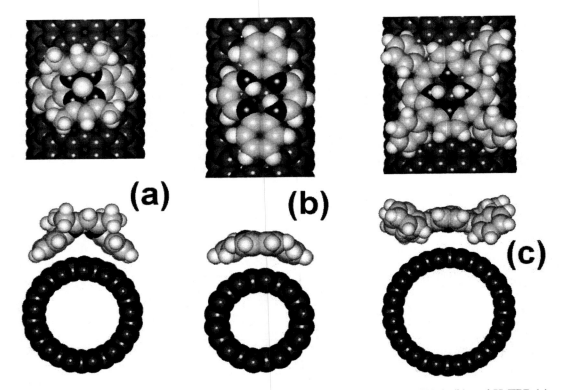

Figure 2. Energetically preferable orientation of adsorbed NiTMTAA (a), H_2TAA (b) and H_2TPP (c) molecules on SWNT walls, according to molecular mechanics simulation. Upper row, side views; lower row, cross sections.

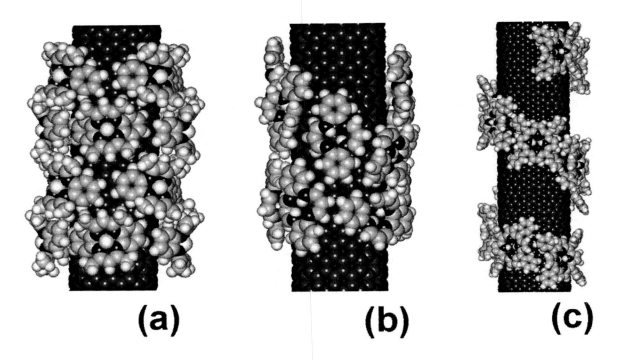

Figure 3. Assembly of the tetraazamacrocyclic compounds on SWNT walls, according to molecular mechanics simulation: (a) NiTMTAA, (b) H_2TAA, and (c) H_2TPP.

rings makes the C_6H_5 moieties turn with respect to the macrocyclic plane (Figure 2c). This can produce ordered molecular assemblies like 'staircase' or 'DNA strand' patterns (Figure 3c; according to MM), although TEM techniques were unable to detect such formations.

As a conclusion and further prospect, adsorption of aromatic molecules, having a bent shape and/or hydrophobic (in particular aromatic) substituents, can be useful in the modification of CNT materials. It allows to reversibly attach modifying chemical species. The latter stick strongly to the nanotube walls due to $\pi-\pi$ and hydrophobic interactions and remain stable in aqueous solutions (and probably in some other polar media), but can be removed by washing with non-polar solvents.

ACKNOWLEDGEMENTS

The authors acknowledge financial support from the National Council of Science and Technology of Mexico (grant CONACYT-40399-Y) and from the National Autonomous University of Mexico (grants DGAPA-IN100402-3 and -100303). This work was supported in part by Tufts University Faculty Research Award (to ERA) and by the NSF (CHE 0111202). EPR facility at Tufts was supported in part by the NSF (CHE 9816557). The authors are grateful to Prof. Maria Stephanopoulos and Ms. Qi Fu for access to and help with SDR UV-Vis spectrophotometry.

REFERENCES

[1] J.L. Bahr and J.M. Tour, Chem. Mater. 13, 3823, 2001.

[2] J.L. Bahr and J.M. Tour, J. Mater. Chem. 12, 1952, 2002.

[3] A. Hirsh, Angew. Chem., Int. Ed. 41, 1853, 2002.

[4] A.D. Adler, F.R. Longo, J.D. Finarelli, J.Goldmacher, J. Assour and L. Korsakoff, J. Org. Chem. 32, 476, 1967.

[5] D.A. Place, D.P. Ferrara, J.J. Harland and J.C. Dabrowiak, J. Heterocycl. Chem. 17, 439, 1980.

[6] J.R. Chipperfield and S. Woodward, J. Chem. Ed. 71, 75, 1994.

[7] K.H. Reddy, M.R. Reddy and K.M. Raju, Polyhedron 16, 2673, 1997.

[8] M.J. Frisch et al., Gaussian 98W, Revision A.9, Gaussian, Inc., Pittsburgh, PA, 1998.

[9] E.V. Basiuk (Golovataya-Dzhymbeeva), E.V. Rybak-Akimova, V.A. Basiuk, D. Acosta-Najarro, and J.M. Saniger. Nano Lett. 2, 1249, 2002.

Computing Metallofullerenes as Agents of Nanoscience: Gibbs Energy Treatment of Ca@C$_{72}$, Ca@C$_{82}$, and La@C$_{82}$

Zdeněk Slanina,*,** Kaoru Kobayashi* and Shigeru Nagase*

*Department of Theoretical Studies
Institute for Molecular Science
Myodaiji, Okazaki 444-8585, Japan
**Institute of Chemistry, Academia Sinica
128 Yen-Chiu-Yuan Rd., Sec. 2
Nankang, Taipei 11529, Taiwan - R.O.C.

ABSTRACT

Three endohedral fullerene systems are simulated computationally, combining the treatments of quantum chemistry and statistical mechanics. Relative concentrations of four isomers of Ca@C$_{72}$, nine isomers of Ca@C$_{82}$, and four isomers of La@C$_{82}$ are evaluated using the Gibbs energy. The results illustrate the enthalpy-entropy interplay in the systems produced under very high temperatures.

Keywords: Carbon-based nanotechnology; molecular electronics; metallofullerenes; optimized syntheses; Gibbs-energy evaluations.

1 INTRODUCTION

Various endohedral cage compounds have been suggested as possible candidate species for molecular memories. One approach is built on endohedral species with two possible location sites of the encapsulated atom [1,2] while another concept of quantum computing aims at a usage of spin states of N@C$_{60}$ [3]. In this work, three systems related to the first approach are simulated computationally, combining the treatments of quantum chemistry and statistical mechanics. Relative concentrations of four isomers of Ca@C$_{72}$, nine isomers of Ca@C$_{82}$, and four isomers of La@C$_{82}$ are computed using the Gibbs energy.

2 COMPUTATIONS

The computations started from the structures [4-6] optimized at the Hartree-Fock (HF) level in a combined basis set: 3-21G basis for C atoms and a dz basis set [7] with the effective core potential on Ca (for the sake of simplicity, denoted here HF/3-21G~dz). Now, the structures are reoptimized using DFT, namely Becke's three parameter functional [8] with the non-local Lee-Yang-Parr correlation functional [9] (B3LYP) with the above basis set (B3LYP/3-21G~dz). The analytical energy gradient was used in the geometry optimizations. All the reported computations are carried out with the Gaussian 98 program package [10].

In the optimized B3LYP/3-21G~dz geometries the harmonic vibrational analysis was carried out with the analytical force-constant matrix. In the same B3LYP/3-21G~dz optimized geometries a higher-level single-point energy calculation was also performed, using the standard 6-31G* basis set for all atoms (if possible). The electronic excitation energies were evaluated by means of the ZINDO method [11,12], known also as the ZINDO/S method, a semiempirical SCF method combined with the configuration interaction technique and specifically parametrized for calculation of electron excited states. Moreover, in some cases the electronic transitions were also calculated with time-dependent (TD) DFT response theory [13] at the B3LYP/3-21G~dz level. Singlet and triplet excited states were evaluated as they both are relevant for the electronic partition function of a singlet species under the conditions of thermodynamic equilibrium.

Relative concentrations (mole fractions) x_i of m isomers can be expressed [14] through their partition functions q_i and the enthalpies at the absolute zero temperature or ground-state energies $\Delta H_{0,i}^o$ (i.e., the relative potential energies corrected for the vibrational zero-point energies) by a compact formula:

$$x_i = \frac{q_i exp[-\Delta H_{0,i}^o/(RT)]}{\sum_{j=1}^{m} q_j exp[-\Delta H_{0,j}^o/(RT)]}, \quad (1)$$

where R is the gas constant and T the absolute temperature. Eq. (1) is an exact formula that can be directly derived [14] from the standard Gibbs energies of the isomers, supposing the conditions of the inter-isomeric thermodynamic equilibrium. Rotational-vibrational partition functions were constructed from the calculated structural and vibrational data using the rigid rotator and harmonic oscillator approximation. No frequency scaling is applied as it is not significant [15] for the x_i values at high temperatures. The geometrical symmetries of the optimized cages were determined not only by the Gaussian 98 built-in procedure [10] but also by a procedure [16] which considers precision of the computed coordinates. The electronic partition function was constructed by directed summation from the ZINDO or TD electronic excitation energies. In fact, just a few first electronic excited states matter for the partition function. Finally, the chirality contribution [17] was included accordingly (for an enantiomeric pair its partition function q_i is doubled).

3 RESULTS AND DISCUSSION

Ca@C$_{72}$, Ca@C$_{82}$, and La@C$_{82}$ are among the first metallofullerenes to which the combined stability computations have been applied. Ca@C$_{72}$ was isolated [18] though its observed structure is not yet available. It follows from its very first computations [4,19] that there are four isomers especially low in potential energy. In fact, C_{72} has only one [20] isolated-pentagon-rule (IPR) structure. The endohedral $Ca@C_{72}$ species created by putting Ca inside the sole IPR cage has been labeled [4] by (a). The other three $Ca@C_{72}$ isomers considered in ref. [4] are related to two non-IPR C_{72} cages (b) and (c), and to a C_{72} structure with one heptagon (d).

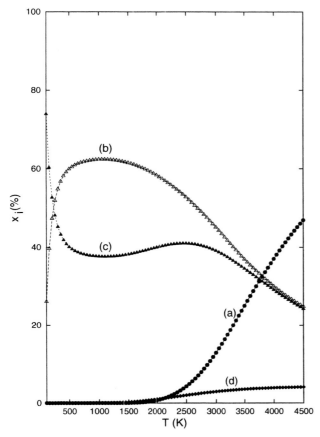

Fig. 1. Relative concentrations of the $Ca@C_{72}$ isomers based on the B3LYP/6-31G* energetics and the B3LYP/3-21G~dz entropy.

The extended computations [21] started from the four optimized structures [4] derived using *ab initio* HF treatment with the combined 3-21G~dz basis set. The structures were reoptimized at the B3LYP/3-21G~dz level. In the optimized B3LYP/3-21G~dz geometries the harmonic vibrational analysis was carried out with the analytical force-constant matrix. In the same geometries single-point energy calculations were also performed at the B3LYP/6-31G* level. The electronic excitation energies were evaluated by means of TD DFT response theory at the B3LYP/3-21G~dz level.

Fig. 1 presents the temperature development of the relative concentrations of the four $Ca@C_{72}$ isomers in a high temperature region. At very low temperatures (not shown in Fig. 1) the structure lowest in the $\Delta H^o_{0,i}$ scale must be prevailing. However, already at a temperature of 226 K (that has no practical meaning) the relative concentrations of the (c) and (b) structures are

interchanged and beyond the point the (b) structure is always somewhat more populated. Even more interesting is the behavior of the IPR-satisfying (a) structure. As the structure is the highest in the potential energy, it must be the least populated species at low temperatures. However, later on the entropy contributions (low symmetry, some lower vibrational frequencies and some lower electronic excitation energies) elevate the (a) isomer into the status of a minor isomer that could also be observed. On the other hand, the (d) isomer has the least chances to be detected. Interestingly enough, the concentration order at high temperatures for $Ca@C_{72}$ is quite similar to that previously computed [22] for $Mg@C_{72}$.

The second illustrative system, $Ca@C_{82}$, exhibits the richest isomerism among the Ca endohedrals [23-28]. Shinohara *et al.* [25] isolated four isomers of $Ca@C_{82}$ and labeled the isomers by (I), (II), (III), and (IV). Dennis and Shinohara concluded [29] from the ^{13}C NMR spectra of $Ca@C_{82}$(III) its symmetry as C_2. The ultraviolet photoelectron spectra measured by Hino *et al.* [30] support the finding; a similarity with $Tm@C_{82}$(II) was also noted [31]. Very recently, Achiba *et al.* [28] measured the ^{13}C NMR spectra of the all four isomers and assigned the symmetry of isomers (I), (II), (III), and (IV) as C_s, C_{3v}, C_2, and C_{2v}, respectively.

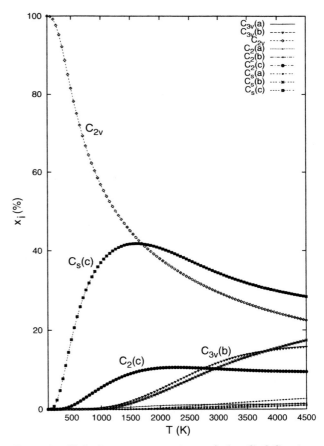

Fig. 2. Relative concentrations of the $Ca@C_{82}$ isomers based on the B3LYP/6-31G* energetics and the B3LYP/3-21G~dz entropy.

The Ca@C_{82} structure-energetics relationships were also computed [5] and a qualitative agreement with the experiment found [2]. The computations were performed at the HF and DFT levels and in both cases the C_{2v} structure was the lowest isomer in the potential energy. There were still three other low energy species - C_s, C_2, and C_{3v}. The combined stability computations are now also available for the full set of nine isomers of Ca@C_{82} considered in Ref. [5].

The nine C_{82} IPR structures [20] produce nine Ca@C_{82} cages with the following symmetries recognized [5] at the HF level: $C_{3v}(a)$, $C_{3v}(b)$, C_{2v}, $C_2(a)$, $C_2(b)$, $C_2(c)$, $C_s(a)$, $C_s(b)$, and $C_s(c)$. It has turned out for the structures reoptimized at the B3LYP/3-21G~dz level that in five cases the original HF structures after the DFT reoptimizations within the same symmetry lead to saddle points with imaginary vibrational frequencies, not to the required local energy minima. When the five saddle points are relaxed and reoptimized, the following local minima are obtained: $C_{3v}(b) \rightarrow C_s$, $C_{2v} \rightarrow C_s$, $C_2(a) \rightarrow C_1$, $C_2(b) \rightarrow C_1$, $C_s(b) \rightarrow C_1$.

Fig. 2 presents the temperature development of the relative concentrations of the nine Ca@C_{82} isomers in a wide temperature region. At very low temperatures the structure lowest in the $\Delta H_{0,i}^o$ scale must be prevailing. However, at a temperature of 1700 K the relative concentrations of the $C_{2v} \rightarrow C_s$ and $C_s(c)$ structures are interchanged and beyond the point the $C_s(c)$ structure is always somewhat more populated. The $C_s(c)$ isomer and also $C_2(c)$ exhibit a temperature maximum. Then, there are still two other structures with significant populations at high temperatures: $C_{3v}(b) \rightarrow C_s$ and $C_s(b) \rightarrow C_1$. Although the former species is a bit more populated, their concentrations are rather close. Fig. 2 is in a reasonable agreement with the qualitative population information [25,28] in a relatively wide temperature interval though the fifth isomer has not been observed yet.

The third illustrative case deals with La@C_{82}, i.e., an electronic open-shell system. The La@C_{82} metallofullerene is one of the very first endohedrals that was macroscopically produced [32] and solvent extracted. La@C_{82} has attracted attention of both experiment [33-42] and computations [43-49]. Recently structures of two its isomers were clarified [41,42] using ^{13}C NMR spectra of their monoanions generated electrochemically. The major isomer [41] was thus assigned C_{2v} symmetry and the minor species [42] C_s. The C_{2v} structure was moreover confirmed by an X-ray powder diffraction study [40]. Two isomers could also be extracted [34,36,50] for Sc@C_{82} and Y@C_{82}. The findings stand in a contrast to Ca@C_{82} with four known isomers. Computations at ab initio HF and DFT levels pointed out [6] just three IPR cages with a sufficiently low energy after La atom encapsulation: C_{2v}, $C_{3v}(b)$, and $C_s(c)$. The fourth lowest La endohedral species, $C_2(a)$, is actually already rather high in energy to be significant in experiment.

An agreement with experiment can be reached (Fig. 3) for temperatures roughly from 1000 to 1300 K when the C_{2v} species is the major isomer followed by an isomer that undergoes C_{3v}/C_s symmetry reduction while the genuine C_s species comes as a still less populated third product. It is possible that the C_{3v} isomer is suppressed in the condensed phase by higher reactivity, however, some additional data are needed.

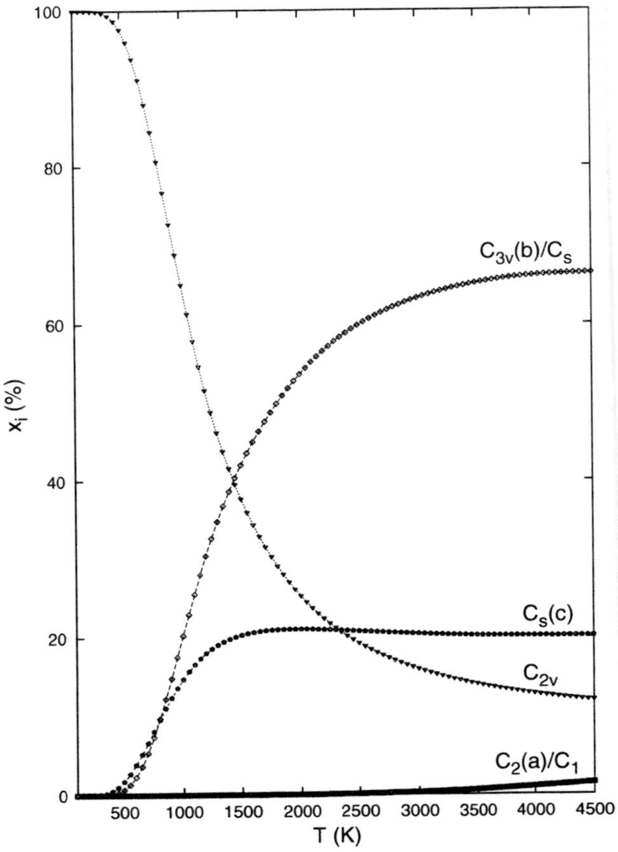

Fig. 3. Relative concentrations of the La@C_{82} isomers based on the B3LYP/6-31G*~dz energetics and the B3LYP/3-21G~dz entropy.

ACKNOWLEDGMENTS

The reported research has been supported by a Grant-in-aid for NAREGI Nanoscience Project, Scientific Research on Priority Area (A), and Scientific Research (B) from the Ministry of Education, Culture, Sports, Science and Technology of Japan, and also by the Japan Society for the Promotion of Science.

REFERENCES

[1] J. K. Gimzewski, in The Chemical Physics of Fullerenes 10 (and 5) Years Later, Ed. W. Andreoni, Kluwer, Dordrecht, 1996, p. 117.

[2] K. Kobayashi and S. Nagase, in Endofullerenes - A New Family of Carbon Clusters, Eds. T. Akasaka and S. Nagase, Kluwer Academic Publishers, Dordrecht, 2002, p. 155.

[3] W. Harneit, M. Waiblinger, C. Meyer, K. Lips and A. Weidinger, in Recent Advances in the Chemistry and Physics of Fullerenes and Related Materials, Vol. 11, Fullerenes for the New Millennium, Eds. K. M. Kadish, P. V. Kamat and D. Guldi, Electrochemical Society, Pennington, 2001, p. 358.

[4] K. Kobayashi, S. Nagase, M. Yoshida and E. Ōsawa, J. Am. Chem. Soc. 119, 12693, 1997.

[5] K. Kobayashi and S. Nagase, Chem. Phys. Lett. 274, 226, 1997.

[6] K. Kobayashi and S. Nagase, Chem. Phys. Lett. 282, 325, 1998.

[7] P. J. Hay and W. R. Wadt, J. Chem. Phys. 82, 299, 1985.

[8] A. D. Becke, J. Chem. Phys. 98, 5648, 1993.

[9] C. Lee, W. Yang and R. G. Parr, Phys. Rev. B 37, 785, 1988.

[10] M. J. Frisch, G. W. Trucks, H. B. Schlegel, G. E. Scuseria, M. A. Robb, J. R. Cheeseman, V. G. Zakrzewski, J. A. Montgomery, Jr., R. E. Stratmann, J. C. Burant, S. Dapprich, J. M. Millam, A. D. Daniels, K. N. Kudin, M. C. Strain, O. Farkas, J. Tomasi, V. Barone, M. Cossi, R. Cammi, B. Mennucci, C. Pomelli, C. Adamo, S. Clifford, J. Ochterski, G. A. Petersson, P. Y. Ayala, Q. Cui, K. Morokuma, D. K. Malick, A. D. Rabuck, K. Raghavachari, J. B. Foresman, J. Cioslowski, J. V. Ortiz, B. B. Stefanov, G. Liu, A. Liashenko, P. Piskorz, I. Komaromi, R. Gomperts, R. L. Martin, D. J. Fox, T. Keith, M. A. Al-Laham, C. Y. Peng, A. Nanayakkara, C. Gonzalez, M. Challacombe, P. M. W. Gill, B. Johnson, W. Chen, M. W. Wong, J. L. Andres, C. Gonzalez, M. Head-Gordon, E. S. Replogle and J. A. Pople, *Gaussian 98, Revision A.9*, Gaussian, Inc., Pittsburgh, PA, 1998.

[11] D. R. Kanis, M. A. Ratner, T. J. Marks and M. C. Zerner, Chem. Mater. 3, 19, 1991.

[12] R. D. Bendale and M. C. Zerner, J. Phys. Chem. 99, 13830, 1995.

[13] M. E. Casida, C. Jamorski, K. C. Casida and D. R. Salahub, J. Chem. Phys. 108, 4439, 1998.

[14] Z. Slanina, Int. Rev. Phys. Chem. 6, 251, 1987.

[15] Z. Slanina, F. Uhlík and M. C. Zerner, Rev. Roum. Chim. 36, 965 1991.

[16] M.-L. Sun, Z. Slanina, S.-L. Lee, F. Uhlík and L. Adamowicz, Chem. Phys. Lett. 246, 66, (1995).

[17] Z. Slanina and L. Adamowicz, Thermochim. Acta. 205, 299, 1992.

[18] T. S. M. Wan, H.-W. Zhang, T. Nakane, Z. Xu, M. Inakuma, H. Shinohara, K. Kobayashi and S. Nagase, J. Am. Chem. Soc. 120, 6806, 1998.

[19] S. Nagase, K. Kobayashi and T. Akasaka, J. Mol. Struct. (Theochem) 462, 97, 1999.

[20] P. W. Fowler and D. E. Manolopoulos, *An Atlas of Fullerenes*, Clarendon Press, Oxford, 1995.

[21] Z. Slanina, K. Kobayashi and S. Nagase, Chem. Phys. Lett. 372, 810, 2003.

[22] Z. Slanina, X. Zhao, X. Grabuleda, M. Ozawa, F. Uhlík, P. M. Ivanov, K. Kobayashi and S. Nagase, J. Mol. Graphics Mod. 19, 252, 2001.

[23] L. S. Wang, J. M. Alford, Y. Chai, M. Diener, J. Zhang, S. M. McClure, T. Guo, G. E. Scuseria and R. E. Smalley, *Chem. Phys. Let.* 207, 354, 1993.

[24] Y. Kubozono, T. Ohta, T. Hayashibara, H. Maeda, H. Ishida, S. Kashino, K. Oshima, H. Yamazaki, S. Ukita and T. Sogabe, Chem. Lett. 457, 1995.

[25] Z. D. Xu, T. Nakane and H. Shinohara, J. Am. Chem. Soc. 118, 11309, 1996.

[26] F. G. Hopwood, K. J. Fisher, P. Greenhill, G. D. Willett and R. Zhang, J. Phys. Chem. B 101, 10704, 1997.

[27] T. Kimura, T. Sugai and H. Shinohara, Int. J. Mass Spectrom. 188, 225, 1999.

[28] T. Kodama, R. Fujii, Y. Miyake, K. Sakaguchi, H. Nishikawa, I. Ikemoto, K. Kikuchi and Y. Achiba, Chem. Phys. Lett. 377, 197, 2003.

[29] T. J. S. Dennis and H. Shinohara, Appl. Phys. A 66, 243, 1998.

[30] S. Hino, K. Umishita, K. Iwasaki, M. Aoki, K. Kobayashi, S. Nagase, T. J. S. Dennis, T. Nakane and H. Shinohara, Chem. Phys. Lett. 337, 65, 2001.

[31] T. Kodama, N. Ozawa, Y. Miyake, K. Sakaguchi, H. Nishikawa, I. Ikemoto, K. Kikuchi and Y. Achiba, J. Am. Chem. Soc. 124, 1452, 2002.

[32] Y. Chai, T. Guo, C. Jin, R. E. Haufler, L. P. F. Chibante, J. Fure, L. Wang, J. M. Alford, R. E. Smalley, J. Phys. Chem. 95 (1991) 7564.

[33] R. D. Johnson, M. S. de Vries, J. Salem, D. S. Bethune, C. S. Yannoni, Nature 355 (1992) 239.

[34] S. Suzuki, S. Kawata, H. Shiromaru, K. Yamauchi, K. Kikuchi, T. Kato, Y. Achiba, J. Phys. Chem. 96 (1992) 7159.

[35] S. Bandow, H. Kitagawa, T. Mitani, H. Inokuchi, Y. Saito, H. Yamaguchi, N. Hayashi, H. Sato, H. Shinohara, J. Phys. Chem. 96 (1992) 9609.

[36] M. Hoinkis, C. S. Yannoni, D. S. Bethune, J. R. Salem, R. D. Johnson, M. S. Crowder, M. S. De Vries, Chem. Phys. Lett. 198 (1992) 461.

[37] K. Kikuchi, S. Suzuki, Y. Nakao, N. Nakahara, T. Wakabayashi, H. Shiromaru, K. Saito, I. Ikemoto, Y. Achiba, Chem. Phys. Lett. 216 (1993) 67.

[38] C. S. Yannoni, H. R. Wendt, M. S. de Vries, R. L. Siemens, J. R. Salem, J. Lyerla, R. D. Johnson, M. Hoinkis, M. S. Crowder, C. A. Brown, D. S. Bethune, L. Taylor, D. Nguyen, P. Jedrzejewski, H. C. Dorn, Synth. Met. 59 (1993) 279

[39] K. Yamamoto, H. Funasaka, T. Takahasi, T. Akasaka, T. Suzuki, Y. Maruyama, J. Phys. Chem. 98 (1994) 12831.

[40] E. Nishibori, M. Takata, M. Sakata, H. Tanaka, M. Hasegawa, H. Shinohara, Chem. Phys. Lett. 330 (2000) 497.

[41] T. Akasaka, T. Wakahara, S. Nagase, K. Kobayashi, M. Waelchli, K. Yamamoto, M. Kondo, S. Shirakura, S. Okubo, Y. Maeda, T. Kato, M. Kako, Y. Nakadaira, R. Nagahata, X. Gao, E. van Caemelbecke, K. M. Kadish, J. Am. Chem. Soc. 122 (2000) 9316.

[42] T. Akasaka, T. Wakahara, S. Nagase, K. Kobayashi, M. Waelchli, K. Yamamoto, M. Kondo, S. Shirakura, Y. Maeda, T. Kato, M. Kako, Y. Nakadaira, X. Gao, E. van Caemelbecke, K. M. Kadish, J. Phys. Chem. B 105 (2001) 2971.

[43] S. Nagase, K. Kobayashi, T. Kato, Y. Achiba, Chem. Phys. Lett. 201 (1993) 475.

[44] S. Nagase, K. Kobayashi, Chem. Phys. Lett. 228 (1994) 106.

[45] W. Andreoni, A. Curioni, Phys. Rev. Lett. 77 (1996) 834.

[46] K. Kobayashi, S. Nagase, Chem. Phys. Lett. 282 (1998) 325.

[47] J. Lu, X. W. Zhang, X. G. Zhao, S. Nagase, K. Kobayashi, Chem. Phys. Lett. 332 (2000) 219.

[48] K. Kobayashi and S. Nagase, in Endofullerenes - A New Family of Carbon Clusters, eds. T. Akasaka, S. Nagase, Kluwer Academic Publishers, Dordrecht, 2002, p. 155.

[49] K. Kobayashi, S. Nagase, Mol. Phys. 101 (2003) 249

[50] E. Nishibori, M. Takata, M. Sakata, M. Inakuma, H. Shinohara, Chem. Phys. Lett. 298 (1998) 79.

Electrical property of fullerene films polymerized by argon plasma treatment

Ryuichiro Maruyama

Material Laboratories, Sony Corporation, 2-1-1 Shinsakuragaoka, Hodogaya-ku, Yokohama 240-0036, Japan
Ryuichiro.Maruyama　p.sony.com

Abstract

Photoemission spectral analysis was carried out on plasma polymerized fullerene (C60) films prepared by sublimating C60 under an argon plasma. In the photoemission analysis, the overall distribution of the shallow valence band states of the plasma-treated C60 was compared with pristine C60. These changes suggest that the pi-conjugation of the shallow valence band states of two C60 molecules is induced by polymerization during the plasma treatment. To investigate the changes in the onset of the shallow valence band states brought about by polymerization, the orbital energy levels in C60 dimers were examined using the semi-empirical AM-1 method. Calculations showed that the highest occupied molecular orbital (HOMO) levels of the dimers were shallower than that of C60.

Introduction

Fullerene (C60) is a promising material for nano-carbon device research. Control of the electrical property of C60 is necessary for applications. C60 polymer has great promise because its unique structure allows its electrical and optical properties to be controlled. We reported that argon glow discharge technique using a high-frequency plasma was applied to the polymer synthesis.[1] In this method, because a non-equilibrium gas phase in the glow discharge is induced by the non-equilibrium plasma, C60 is polymerized without the need for a high temperature. The electrical property of the polymerized C60 films were investigated using the photoemission and the conductivity measurement.

Experimental

For thin-film deposition, 50-100 mg of 99.95%-purified C60 powder was loaded into a molybdenum boat, as shown in Fig. 1. The substrate was located on a steel stage on the anode

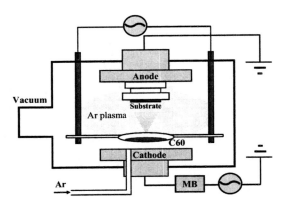

Fig. 1. Block diagram of plasma polymerization system.

and was 10 cm from the boat, which was positioned between the anode and cathode. A quartz-crystal microbalance was used to monitor the rate of deposition and the film thickness. The details of the film-deposition procedure have been reported elsewhere. A steel substrate and a silica-glass substrate that had been deposited with gold electrodes were used to measure the plasma-power dependence of the density of the shallow valence band states and the temperature dependence of the conductivity, respectively. The polymer samples were first deposited in a plasma polymerization chamber, then transferred to an inter-connected chamber to measure the photoemission spectra. They were then transferred to a final inter-connected chamber to measure the conductivity, since oxygen has proved to be one of the most important intercalants that can affect the electrical properties of C60 films.

The electronic structure of the C60 dimers was calculated using AM-1 parameterization as implemented in the program system MOPAC. Figure 2 shows the structure of the [2+2] C60 dimer as a reference, and 4 possible forms of the peanut-type C120, all of which were optimized

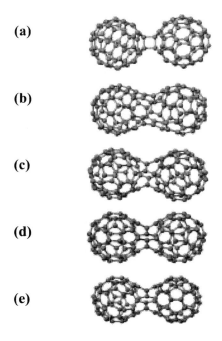

(a)

(b)

(c)

(d)

(e)

Fig. 2. The structure of the [2+2] C60 dimer and possible forms of the peanut-type C120.

using the MNDO Hamiltonian with the AM-1 parameterization, as implemented in the program system MOPAC. The onset of the shallow valence band states of the polymerized C60 was considered using the highest occupied molecular orbitals (HOMO) for each calculated dimer model.

The valence structures of the prepared C60 polymer and C60 films were measured using photoemission spectra. 50 nm thick films were deposited on the conductive steel stage and measured. Excitation of photoemissions was achieved using the HeI resonance line (21.2 eV) from a discharge lamp attached to the photoelectron chamber and the spectra were acquired at an analyzer resolution of 10 meV. The energy axis of all the photoemission spectra presented in this paper was referenced to the Fermi level of the Au film deposited in situ. Based on this, we obtain a work function of 4.9 eV for Au.

To investigate the electrical transport properties of the films, we fabricated a C60 polymer/gold structure for electrical characterization. Gold was deposited on a silica-glass substrate in the form of coplanar electrodes with a 0.5 mm interelectrode spacing and a 100 mm meander length. A 200 nm thick

film of the C60 polymer was deposited at 0.1 nm/sec on the silica-glass substrate coated by the gold electrodes. During the C60 polymer deposition, 50 W of plasma power was supplied in an argon atmosphere of 3.9 Pa. The temperature dependence of the conductivity of the C60 polymer film was measured in vacuum using the dc conductivity measurement system of a Hewlett-Packard HP4515A.

Results and Discussion

Figure 3 shows the calculated molecular orbital energy of C60 and C120 near the Fermi level. As a consequence of the nature of the bonds for the orbitals around the Fermi energy, these orbitals are strongly affected by variations in the dimer structure. The HOMO level of the [2+2] and the peanut-type becomes shallow, because the two cages structure lowers the molecular symmetry and induces a split between orbitals of the same level and tends to make the HOMO shallow by dimerization. The lowest unoccupied molecular orbital (LUMO) also decreases. The HOMO levels of C120 with D2v and C2h symmetries increase above the HOMO level of the [2+2]. The shrinkage of the HOMO-LUMO gap of the peanut-type is more than that of the [2+2], because there is an expansion of the pi-electron system over the two cages in the peanut-type. There are no sigma bonds in the peanut-type structure. The structure shows a step in the change from the [2+2] into the final C120 tube, which has a narrow band gap.[2],[3] The HOMO states of C60 are split by the polymerization, and the states in the [2+2] and peanut-type are distributed from -9 to -10 eV. The center of the distribution is equal

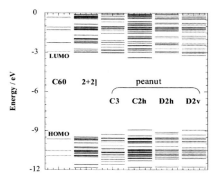

Fig. 3. The molecular orbital energy levels in C60 and C120s calculated by AM-1

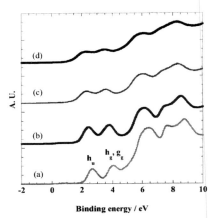

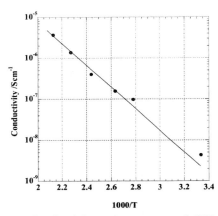

Fig. 4. Photoemission (hv =21.2eV) results for plasma polymerized C60 films (a) 0 W, (b) 20 W, (c) 40 W, and (d) 50 W. The spectra are plotted in arbitrary units

Fig. 5. Conductivity measurement of C60 film polymerized at 50 W. The measured conductivity (log scale) is plotted vs 1000/T on an Arrhenius plot

to the HOMO level of C60. This indicates that the polymerization induces the broadening of the onset of the shallow valence band states. Additional evidence supporting this proposition is provided by photoemission measurements of the shallow valence band states of plasma-polymerized C60 (Fig. 4). Figure 4 provides an overview of the distribution of states for C60 and plasma polymerized C60 showing the onset of the shallow valence band states after annealing at 500 K. The plasma polymerized film contains dimers, trimers, polymer chains and planes.[1], [4] The spectra are plotted in arbitrary units. As shown in Fig. 4, the valence electronic structures of the C60 polymer are broadened and shifted by the plasma treatment. The peaks between 0 and 7 eV essentially correspond to the pi-electron states.[5] The valence-band peaks broaden with increased plasma power. Because the [2+2] bond between the two C60 molecules was eliminated by annealing, the broadening of the states with sigma and pi-parentage originate from coalesced fullerene including a peanuts-like structure. This broadening suggests that there is a split of the degenerate states of the C60 molecules brought out by the molecular elongation introduced during plasma treatment.

Given the exponential dependence of the conductivity on the Fermi level in a semiconductor, conductivity measurements can provide an extremely sensitive way of detecting the electronic state in the band gap or near the mobility edge of the polymerized C60. Figure 5

shows the relationship between the logarithm of the conductivity and the inverse of the temperature for the polymer film deposited in a 50 W argon plasma. The conductivity increased exponentially in the temperature range from 300 to 500 K. The exponential behavior in this temperature range indicates that semiconductor-like properties do dominate and no graphitic character is revealed within the film. In the high-temperature region, the conductivity of the film shows Arrhenius type behavior, thus the conductivity can be described by the Arrhenius equation

$$\sigma = \sigma_0 \exp(-\Delta E_a / kT).$$

Here σ is the conductivity, σ_0 is the pre-exponential factor, ΔE_a is the activation energy and T is the absolute temperature. The value of ΔE_a estimated from the slope of this Arrhenius plot is 0.42 eV, which is different from the 0.84 eV value for pure C60 reported by Pevzner et al.[20] ΔEa for the electron transfer of a semiconductor is typically equal to $E_c - E_f$ (E_c is the mobility edge and E_f the Fermi level). Therefore, the shallow E_f originates from the narrow mobility gap and the pi-conjugated bonds between the C60 cages brought on by polymerization.

Conclusion

The molecular orbitals of the C60 dimer, which was the most abundant product after the polymerization process were calculated using the

AM-1 method. This result suggests that the polymerization process causes a narrowing of HOMO-LUMO gap and a broadening of the mobility edge into the gap. With increased plasma power, the overall distribution of onset of the shallow valence band states was detected using photoemission spectra. The [2+2] cycloaddition between two C60 molecules is eliminated by 500 K annealing. Therefore, the structure responsible for the change in the onset of the shallow valence band states is the coalesced structure containing peanut-type material. The electrical properties of the C60 polymer suggest that E_f is shallow due to the shrinkage of the mobility gap caused by polymerization and the pi-conjugated bonds between the C60 cages.

References

[1] M. Ata, N. Takahashi and K. Nojima, J. Phys. Chem., 98, 40, 9960 (1994)

[2] Y. H. Kim, I. H. Lee, K. J. Chang and S. Lee, Phys. Rev. Lett., 90 (6), 655011 (2003)

[3] Yufeng Zhao, B. I. Yakobson and R. E. Smalley, Phys. Rev. Lett., 88 (18), 1855011 (2002)

[4] M. Ata, K. Kurihara and N. Takahashi, J. Phys. Chem. B, 101, 5 (1997)

[5] N. Troullier and J. L. Martins, Phys. Rev. B, 46 (3),1754 (1992)

Nano 6H Diamond Polytype Polycrystalline Powder

P. Darrell Ownby

Ceramic Engineering Department, University of Missouri-Rolla, Rolla, MO 65401, USA,
ownby@umr.edu

ABSTRACT

In 1967, "Hexagonal diamond - a new form of carbon" was reported to be found in meteorites. This hexagonal form, 2H, was named Lonsdaleite, after the noted crystallographer, Kathleen Lonsdale. Since the corrected structures and x-ray diffraction data for all six possible non-cubic polytypes was published in 1992, the considerably increased interest in CVD diamond has resulted in finding others of the predicted polytypes, including 6H, 8H and 15R. The rapid crystallization and cooling characteristics of CVD diamond, and that produced by rapid implosion in meteorites and from man-made explosions, tend to produce non-cubic polytypes because of the rapid and short-lived transformation to the more dense crystal structure. Powder from the latter source has been characterized by infrared spectroscopy, and Rietveld x-ray and neutron diffraction. For particle size fractions in the range of 10 to 1000 nanometers, the non-cubic, 6H hexagonal polytype phase is present in amounts $\geq$ 50% and the cubic diamond phase is present at the $\leq$ 50% level. In all of these size ranges each particle is polycrystalline containing crystallites of nanometer and sub-nanometer dimensions.

Keywords: diamond, polytypes, nanopowder

1 INTRODUCTION

Cubic, 3C diamond is the most abundant crystalline form of diamond found in nature. It is also the form produced by high-pressure hot-pressing graphite in the diamond stable range. In 1967, "Hexagonal diamond – a new form of carbon" was reported to be found in meteorites.[1,2] This hexagonal form, 2H, was named Lonsdaleite, after the noted crystallographer, Kathleen Lonsdale. Other non-cubic polytypes were proposed in 1973[3] and 1990[4] but only the 2H and 6H polytypes had actually been found in CVD produced diamond[5,6] when the corrected structures and x-ray diffraction data for all six non-cubic polytypes was published in 1992.[7] Since that time the considerably increased interest in CVD diamond has resulted in finding several of the other predicted polytypes, including, 8H[8], 15R[9], and other forms such as "X - Diamond."[10] The rapid crystallization and cooling characteristics of CVD tends to produce some non-cubic polytypes.

In the mid 1980's it was theorized that diamond produced by rapid implosion from man- made explosions, like natural meteorite impacts on the earth's surface, would be more likely to contain non-cubic diamond polytypes because of the rapid and short-lived transformation to the more dense crystal structure. Furthermore, these polytype powders might be used to toughen polycrystalline diamond or ceramic matrices because they could reverse transform more easily to higher volume graphite by a displacive transformation as opposed to the diffusive transformation required in the cubic diamond case. Consequently, a U.S. Patent was issued in 1990[11] and a paper was published in 1991[12] showing an example of toughening of carbides by this means which were hot-pressed at 5.5 GPa and 1500°C. In the latter it was shown that the toughening effect didn't show up until the diamond polytype additive particle size became very small, in the nanosize range. This paper also showed the comparison between one micron size GE diamond, which was 100% 3C and the implosively produced one micron diamond containing 12% hexagonal polytype thought to be 2H. It was thought, but never proven, that the smallest size ranges were effective in their role as toughening agents because they contained an even higher fraction of non-cubic polytypes. These implosively formed diamonds were produced by duPont using their widely used plastic wrap explosives to implode graphite particles entrained in a copper matrix. They were marketed as Mypolex suspensions of diamond powder in aqueous ammonium hydroxide or as dry powders. Although they were sold in various size fractions, duPont did not have a good handle on very small particle size measurement so their smaller size classifications were more than an order-of-magnitude in error.

In 1999, the Swiss firm, Rudolf Spring, AG, purchased duPont's Mypolex polycrystalline diamond powder business. Rudolf Spring has been micronizing diamond powders since the 1950's and 1960's and at the time of the Mypolex acquisition they had a very highly developed facility for micronizing and precisely classifying diamond powders. Since acquiring the Mypolex product they now manufacture three size classes smaller than the smallest ever available from duPont.

2 STRUCTURAL ANALYSES

The diamond polytype phases contained in these powders were determined qualitatively and quantitatively by three different analytical techniques. Infrared spectroscopy determined the existence of higher order polytypes. The Rietveld whole profile powder X-ray and neutron diffraction technique is able to resolve overlapping peaks of multiple phases. This technique determines the percentage of each phase present. These powders are unique in that in spite of their nano-size, each particle is polycrystalline and therefore made up of even smaller crystallites. The technique also determines the crystallite size of each phase. The analyses were performed on five different particle size range classifications. They were designated by their mean particle size in nanometers.

2.1 Infrared Spectroscopy

The diamond powder samples were dispersed in KBr powder and pressed into pellets. The pellets were placed in a Perkin Elmer 1760-x Infrared Fourier Transform spectrometer and the data was analyzed using Spectrum v. 2.00, Perkin Elmer Ltd. 1988 software.

The difference between the diamond polytypes is the stacking sequence between lattice planes. The cubic, 3C stacking sequence is ABC/ABC/... The hexagonal, 2H sequence is AaBb/AaBb/... where the lower case letters represent mirror images of their upper case counterparts. 3C stacking is "pure" cubic and 2H is "pure" hexagonal. They are the simplest and represent the two extremes or end members of the polytypes. The stacking sequences for the other polytypes are more complex because they are "mixtures" of the hexagonal and cubic stacking. The end member polytypes, 3C and 2H are not infrared active whereas all of the other higher order polytypes are infrared active in their first order spectra[4]. These powders displayed discrete IR peaks indicating that diamond polytypes other than the end member 3C and 2H were present. Therefore the powders must contain one or more of the higher order polytypes, 4H, 6H, 8H, 15R, or 21R.

2.2 Rietveld X-ray Diffraction

Five samples with different particle size ranges were analyzed. Their mean particle size ranged from 25 to 175 nanometers. All samples were run in a Siemens D5000 diffractometer with Cu Kα radiation. Prior to running any experiments, the diffractometer was aligned and characterized with LaB$_6$ and Al$_2$O$_3$ standards from NIST. Three preliminary tests were run to characterize the diffractometer. These were a check of peak positions, peak half-widths, and peak intensities. Data from these scans were used by the Rietveld refinement to adjust for systematic errors and to confirm that the machine was well within the allowable alignment errors. A thin coating of a powdered sample was deposited onto a quartz zero background holder and then placed into the diffractometer. Tests were run with a divergence slit of 1 mm, an anti scatter slit of 0.1 mm, and a detector slit of 0.2 mm. The step size was 0.05° with a counting time between 30 and 120 seconds/step. The scans were run over the angular range from 10° to 100°. The Riqas software was used to analyze the data. A pseudo-Voigt fitting program was chosen to model the crystalline peaks and a 5th order polynomial was used to fit the background. All the single crystal data was taken from the ICSD database. The very small amount of graphite phase revealed only the single strong peak. Therefore the graphite polytype, 2H or 3R could not be determined. The 2H form was chosen and the refinement constrained making the information on the graphite phase incomplete. It was sufficient, however to discern the particle size and amount. The superposition of many of the 3C and 6H peaks required that the 3C structure be fit and refined first and then constrained. Next the 6H was refined and constrained. This process was repeated a second time and then the refinements were allowed to continue unconstrained.

All of the sample patterns could be fit with the three phases, 3C and 6H diamond and 2H graphite. Several other combinations of diamond polytypes were tried but none were as satisfactory. The quality of the refinement as indicated by the weighted residual, Rwp, was in the 2.4% to 4.7% range as shown in Table 1. Figures 1, 2 and 3 show the structure of these three crystalline phases.

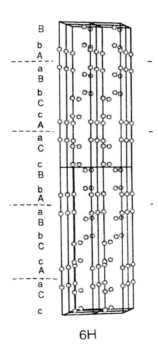

6H

Figure 1 The 6H Diamond Polytype Unit Cell

Figure 4 shows a typical Rietveld fit of the XRD data. The expected scans of each of the three crystalline phases from the ICSD database are shown below the experimental scan which is fit by iteratively varying the amounts of each phase in the sample

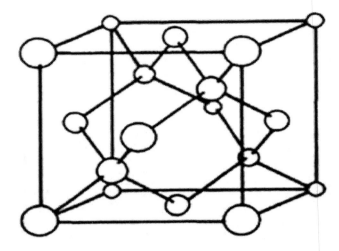

Figure 2 The Cubic Diamond, 3C, Unit Cell

Figure 3 The Graphite Crystal Structure

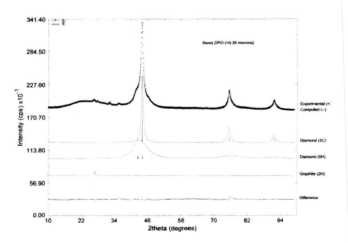

Figure 4 Rietveld Fit of Typical X-ray Diffraction Pattern

2.3   Rietveld Neutron Diffraction

Neutron data were collected using the high-resolution diffractometer[13] at the University of Missouri Research Reactor (MURR). This instrument uses a bent, perfect Si crystal monochromator and position sensitive detector, to take advantage of scattering space focusing of the neutron beam, achieving high throughput for small samples. The detector spans 20° (2θ) and the full spectrum is collected by moving the detector in five 20° steps to span the range from 5° to 105°. The incident neutron wavelength is 1.487Å. Data collection with this instrument can be as short as 1 hour for well-crystallized materials, but in this case where weak second phase peaks are hard to detect and peak broadening effects are prominent, roughly 12 hours of data collection was used for each sample.

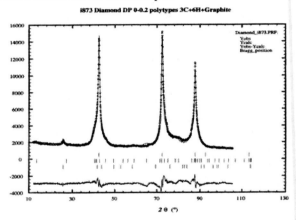

Figure 5 Rietveld Fit of Typical Neutron Diffraction Pattern

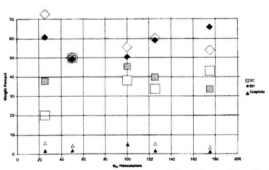

Figure 6   Phase % vs. Mean Particle Size   (Symbols from XRD are solid and from Neutron diffraction are open)

The phase distribution difference between the XRD data and the ND may at least partially be accounted for by the ~ 2 yr space between obtaining the XRD and ND samples. When the XRD samples were obtained, the supplier is reported to have been mixing powder from another non-US source with that from the Pennsylvania plant.

Each tiny powder particle is not a single crystal, but polycrystalline, containing many smaller crystallites with different orientations & structures.

- From X-ray Diffraction

d50	3C	6H
10	128	11
25	116	16
50	102	9
75	98	9

- From Neutron Diffraction

		8
		8

Table 1: Crystallite size in nanometers

- From X-ray Diffraction
  Rwp   2.4% to 4.7%

- From Neutron Diffraction
  Rwp   5.8% to 6.1%

Table 2 Rietveld Degree of Confidence

3 CONCLUSIONS

Diamond powder manufactured by explosive shock wave has a different structural character than static, high temperature and high pressure formed diamond powder. Statically formed diamond powder is all in the 3C cubic diamond crystal structure. Moreover, each particle is typically a single crystal. Crystalline diamond powder particles in the size range of 10 to 1000 nanometers formed by explosive shock wave have been shown to contain $\leq$ 50% 3C, cubic diamond structure and $\geq$ 50% higher order non-cubic polytype. The single higher order polytype that fits both the X-ray and Neutron diffraction data remarkably well is the 6H polytype. These individual nanosize powder particles are not single crystals but contain crystallites of nanometer and sub-nanometer dimensions. The cubic, 3C crystallites appear to be concentrated in the higher size range of each size fraction, while the hexagonal, 6H polytypes are concentrated in the smaller size side of each range. This work is first in characterizing a commercial source of the 6H diamond polytype powder.

REFERENCES

[1] F. P. Bundy and J. S. Kasper, "Hexagonal Diamond – A new Form of Carbon," J. Chem. Phys. 46, 3437-46, 1967.

[2] R. E Hanneman, H. M. Strong, and F. P. Bundy, "Hexagonal Diamond in Meteorites: Implications," Science 155, 995-97, 1967.

[3] C. E. Holcombe, "Calculated X-ray Diffraction Data for Polymorphic Forms of Carbon," Report No. Y-1887, Oak Ridge Y-12 Plant, Oak Ridge, TN, 1973.

[4] K. E Spear, A. W. Phelps, and W. B. White, "Diamond Polytypes and Their Vibrational Spectra," J. Mater. Res. 5, 2272-85, 1990.

[5] M. Frenklach, R. Kematick, D. Huang, W. Howard, K. E. Spear, R. Koba, and A. W. Phelps, "Homogeneous Nucleation of Diamond Powder in the Gas Phase," J. Appl. Phys. 66, 395-99, 1989.

[6] W. Howard, D. Huang, J. Yuan, M. Frenklach, K. E. Spear, R. Koba, and A. W. Phelps, "Synthesis of Diamond Powder in Acetylene Oxygen Plasma," J. Appl. Phys. 68, 1247-51, 1990.

[7] P. D. Ownby, Xi Yang and J. Liu, "Calculated X-ray Diffraction Data for Diamond Polytypes," J. Am. Ceram. Soc. 75, 1876-83, 1992.

[8] R. Kapil, B.R. Mehta, and V.D. Vankar, "Growth of 8H Polytype of Diamond using Growth Etch Cyclic Oxy-Acetylene Flame Setup, Thin Solid Films 312, 106-110, 1998.

[9] R. Kapil, B.R. Mehta, and V.D. Vankar, "Synthesis of 15R Polytype of Diamond in Oxy-Acetylene Flame Grown Diamond," Thin-Films, Appl. Phys. Lett. 68, 2520-2522, 1996.

[10]. M. Rossi, G. Vitali, M. L. Terranova and V. Sessa, "Experimental Evidence of Different Crystalline Forms in Chemical Vapor Deposited Diamond Films," Appl. Phys. Lett. 63, 2765-2767, 1993.

[11].P.D. Ownby, "Fracture Toughening of High Pressure Sintered Diamond and Carbide Ceramics", U.S. Patent No. 4,968,647, Issued Nov. 6, 1990.

[12]P.D Ownby, and J. Liu, "Nano Diamond Enhanced Silicon Carbide Matrix Composites" Ceram. Eng. Sci. Proc. 12, 1345-1355, 1991.

[13] Yelon, W.B., Berliner R., and Popovici M., "A Perfect Match for High Resolution Neutron Powder Diffraction: Position Sensitive Detection and Focusing Monochromators", Physica B241-3, 237 1997.

ACKNOWLEDGEMENTS

The author expresses his appreciation and thanks to the following that performed the analytical work:
  X-ray diffraction - William E. Mayo and Eric W. Bohannan
  Neutron diffraction – William B. Yelon, Q. Cai and William J. James
  Infrared and Raman spectroscopy – Francisco P. Delacruz
Gratitude is also expressed to Microdiamant, AG for donating the diamond powder samples for this work.

Structural and mechanical properties of double wall carbon nanotubes

R. R. Bacsa[*], Ch. Laurent[*], A. Peigney[*], P. Puech[**], H. Hubel[***], D. Dunstan[***] and W. S. Bacsa[**]

[*]CIRIMAT Université Paul Sabatier, UMR-CNRS
118 route de Narbonne, Toulouse 31062, France, bacsa@chimie.ups-tlse.fr
[**]Laboratoire de Physique des Solides, Université Paul Sabatier, UMR-CNRS
118 route de Narbonne, Toulouse 31062, France, bacsa@lpst.ups-tlse.fr
[***]Department of Physics, Queen Mary, University of London
Mile End Road, London, United Kingdom, d.dunstan@qmul.ac.uk

ABSTRACT

Narrow diameter double wall carbon nanotubes can be grown with high yield by catalytic chemical vapor deposition. Double walled carbon nanotubes are the molecular analogs of coaxial cables. The external tube protects the internal tube from its environment and isolates the electric conductor of sub-nanometer dimension over micrometer long distances. While it is still particularly difficult to grow tubes of controlled helicity, inelastic light scattering is a powerful tool to find the helicity of narrow diameter internal tubes and explore their mechanical properties. We have examined the inelastic light scattering spectrum of narrow double walled carbon nanotubes to specify their helicity and to study their mechanical stability under hydrostatic pressure.

Keywords: nanotubes, carbon, microscopy, light scattering, phonons

1 INTRODUCTION

Carbon nanotubes (CNTs) are model systems in nanoscience. They combine molecular extension in two dimensions and macroscopic extension in one dimension. The electronic properties of carbon nanotubes depend on the orientation of the two dimensional honey-comb lattice of graphite with respect to the tube axis. This is described by two indices (m,n) which define the circumference in the honey comb lattice and determine the helicity. For each helicity there are two chiralities (left and right). Depending on the helicity the boundary condition of the electronic wave functions is different and accordingly the electronic properties depend on helicity. The combination of a metallic internal tube with an insulating external tube forms an interesting example of a nanowire in its most simple form. In the case of two metallic tubes, they represent the molecular analog to a coaxial cable. Deviations from sp² hybridized carbon are larger for narrow diameter CNTs due to their larger curvature. The weak interaction between the walls makes it possible to slide the two tubes with respect to each other. The double wall CNT's (DWCNTs) we report here have external diameters comparable to single wall CNTs (SWCNTs). We find that DWCNTs have a higher mechanical stability towards deformations than SWCNTs. The internal tube is furthermore well protected from its environment.

Double walled carbon nanotubes were first observed by Dai et al (1) who reported the presence of a small amount of DWCNTs along with SWCNTs during the disproportionation of CO on alumina supported Mo particles at 1200°C. The presence or absence of double walled tubes was related to the nature of the catalyst by Hafner et al (2). Double walled tubes were also observed by Laurent et al (3) during the decomposition of methane on iron catalysts generated *insitu* during the reduction of a solid solution of iron and aluminum oxides. In this process, a mixture of single and double walled tubes was obtained. Similar results were obtained for Co catalyst. A high proportion of double walled tubes could be obtained also by the selective removal of single walled tubes from a mixture of the two. The interesting feature in the last two results is that the diameter of the internal tube is often smaller than that of the smallest observed diameter for single walled tubes in the sample. Double walled tubes have also be obtained by filling C₆₀ molecules in single shell carbon nanotubes followed by annealing (4) or by changing the catalyst in the sublimation of graphite in an electric arc (5).

Inelastic light scattering provides information about the dynamics of the atomic lattice and mechanical properties and can also provide information on low energy electronic excitations. The technique is extremely sensitive to small structural changes of the atomic lattice. The inelastic light spectrum depends on the energy, momentum and polarization of the incident photon. At low Raman shifts (50-450cm^{-1}) a radial symmetric breathing mode is observed for nanotubes whose energy is inversely proportional to the diameter ($\propto 1/d$) (6). This means that the sensitivity of this mode to the diameter increases with smaller tube diameter. The calibration factor depends, however, on the environment of the tube and ranges from *224-248cm^{-1}/nm* (6). Some authors have reported that the interaction with the support surface shifts the mode energy by 10-20cm^{-1} (7). The fact that helicity and diameter are related to each other and the number of helicities for a given energy range of the breathing mode decreases with smaller diameter gives the possibility to attempt to assign the helicity of the internal tubes (8). We report here the assignment of the radial breathing mode of DWCNT's an the corresponding optical phonon bands, we show the influence of tube-tube interaction using spectroscopic imaging and we investigate the mechanical properties under hydrostatic pressure to learn about the pressure gradient in DWCNTs.

2 EXPERIMENTAL RESULTS

We use a micro Raman setup (Dior XY, Renishaw) with an argon or a krypton ion laser.

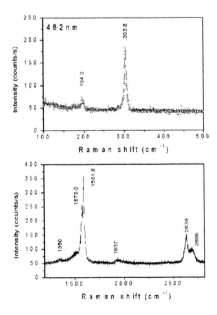

Figure 1

Inelastic light scattering spectrum of a double wall carbon nanotube in the low (top) and high (down) energy range.

Figure 1 shows an inelastic light spectrum of double wall carbon nanotubes with the incident beam set at 482nm. The tubes have been diluted in ethanol and sonicated. A droplet of the nanotube solution has been dried on glass slide before the spectroscopic inspection. Figure 1 shows a spectrum of a highly diluted region were we believe that we detect the signal of only a few DWCNTs. The top figure shows the low energy part with the radial breathing modes and at the bottom we show the corresponding high energy part of the spectrum. We can deduce the diameter of (0.75nm/ 0.83nm, error 0.01nm) from the two peaks at 294.2cm^{-1} and 303.8cm^{-1}. The uncertainty in the calibration factor translates into a uncertainty in the assignment of the helicity which ranges (m,n) for the peak at 194.2cm^{-1} between (9,3) and (10,0) and for the peak at 303.8cm^{-1} between (6,6) and (9,1). When we compare the so determined tube diameters and observed energy of the radial breathing modes with the calculated values for DWCNTs using the valence force model (9) we find that the peak at 303.8 agrees well with determined diameter using a calibration factor of 223.75cm^{-1}. The peak at 194.2cm^{-1} falls between the values calculated for internal and external tubes. Using the same calibration factor, the calculated values are in agreement with the observed spectra for a external tube if we assume that the interaction with the substrate shifts the radial mode by 20cm^{-1}. The two corresponding diameters are too close to come from the same DWCNT. The fact that we do not see corresponding radial mode of the internal or external tube indicates that the observed intensity is strongly diameter dependent. The one dimensional tubes give rise to particularly strong singularities in the electronic density of states which has the effect that the observed scattering intensity is expected to depend on tube diameter.

At high energy we observe a narrow doublet (1572cm^{-1}/1592cm^{-1}, G-band) of the tangential optical modes, a shoulder at 1500cm^{-1} attributed to the optical modes of metallic tubes and a small band at 1350cm^{-1} (D-Band) attributed to disorder induced double resonant scattering (10). Second order spectral bands are observed at higher energy shifts (G' band at 2600cm^{-1}). The doublet at 1580cm^{-1} falls in the same energy range of the optical phonon bands found for SWCNTs but differs by the fact that the lower energy peak is more intense.

Figure 2 shows a collection of spectra recorded in the energy range of the radial breathing mode along a single tube bundle in steps of 0.2μm. Each line corresponds to a radial breathing mode of a single tube. They are well defined and change significantly within a distance of 0.2μm, the step size. For tubes with a diameter below 1 nanometer the low energy breathing modes are very sensitive to the diameter and the peak positions are for neighbouring helicities are displaced by more than the spectral width (2-5cm^{-1}) of the peak. The number of spectroscopic lines or radial breathing modes corresponding to a particular tube changes along the tube bundle (vertical axis in figure 2). One of the most intense peaks at 255cm⁻¹ can be traced along the entire image. The number of peaks changes in particular between 1-2 μm in figure 2 which we attribute to a tube intersection. The phonon peak at 255cm⁻¹ in this same region (1-2μm, figure 2) shifts slightly and shows that the intersection with the tubes influences the local strain in the CNTs.

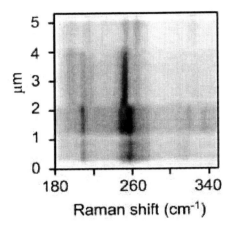

Figure 2

Raman image of a double wall carbon nanotube along a single tube bundle in steps of 0.2μm.

The mechanical properties of DWCNT's can be examined by applying a hydrostatic pressure to a liquid pressure medium where the tubes are suspended, and observe the changes of the phonon bands as a function of the applied pressure. For DWCNT we expect that the external tube protects the internal tube and the pressure is reduced for the internal tube. Fig. 3 shows the spectra at low and high pressure. With increasing applied pressure we observe that the optical phonon band splits into two bands. For SWCNTs and graphite only one band is observed which shifts linearly with applied pressure. The two bands observed for DWCNT's can be associated to the optical phonon band of the internal and external tube. The linear pressure coefficient for the two bands is 3.11 cm⁻¹/GPa for the internal tube and 5.59 cm⁻¹/GPa for the external tube. The pressure coefficient for the internal tube is the lowest observed for three folded carbon and directly indicates that the pressure reduction due to the external tube. It is interesting to note that the boundary conditions for the internal and external tubes are different. While the external tube experiences a radial stress from the pressure medium from outside and the internal tube from within, the internal

tube experiences only a radial stress from the external tube. While the radial stress is continuous, the weak inter-tube interaction leads to a discontinuous tangential stress component. This explains why the optical phonon band, sensitive to the tangential strain, splits into two bands for DWCNT's corresponding to the optical phonon band of the internal and external tubes. If we can apply the elastic continuous shell model with a continuous radial and discontinuous tangential stress component and we find perfect agreement with our experimental observation (11).

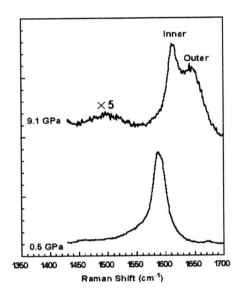

Figure 3

Inelastic light scattering spectra (633nm) of double wall carbon nanotubes under hydrostatic pressure.

In figure 3 we also see that the band due to the external tube which shifts more strongly has a large line width while the band from the internal tube has a constant width. We find that due to the different boundary conditions that the pressure coefficient for the external tube depends on the diameter while for the internal tubes the pressure coefficient is constant. The external tube is stabilised by the internal tube. Our sample contains a distribution of DWCNTs, centred at 1.1 nm and ranging from 0.6 to 2.4nm. The diameter distribution has therefore the effect to broaden the optical phonon line of the external tubes. Each external diameter will shift differently with applied hydrostatic pressure. But for the internal tubes of different diameter, all the internal tubes will shift the same amount with applied hydrostatic pressure.

3 Conclusion

We have assigned the helicity of radial breathing modes of internal and external tubes of DWCNT's and compared the experimental results to valence force calculations. This comparison shows that the smaller value of the calibration factor for the radial breathing mode is in better agreement and this indicates that the interaction with the substrate shifts the energy of the external tube by 10-20cm^{-1} as has been suggested by earlier model calculations (7). Systematic spectroscopic inspection along a line shows that the radial breathing mode is influenced by the presence of intersecting tubes. When applying a hydrostatic pressure on DWCNTs we observe a discontinuity of the tangential stress. The optical phonon due to the in-plane mode is very sensitive to tangential stress and splits with applied pressure. This observation can be explained by the elastic continuum shell model. The same model can also explain the different line broadening observed for internal and external tubes. The tangential stress depends for external tubes on the diameter whereas it stays constant for the internal tubes. The different mechanical behaviour of DWCNTs are expected to find interesting applications in sensors and activators in the near future [12].

REFERENCES

[1] H. Dai, A. Rinzler, P. Nikolaev, A. Thess, D. Colbert, R. Smalley, Chem.Phys.Lett. 260 (1996) 471.

[2] J. Hafner, M. Bronikowski, B. Azamian, P. Nikolaev, A. Rinzler, D. Colbert, K. Smith, R. Smalley, Chem.Phys.Letters, 296 (1998) 195-202.

[3] A. Peigney, Ch. Laurent, F. Dobigeon , A. Rousset J. Mater. Res. 12 (1997) 613. R. Bacsa, Ch. Laurent, A. Peigney, T. Vaugien, E. Flahaut, W. Bacsa, A. Rousset, Journal of American Ceramic Society Vol. 85 (2002) 2666

[4] J. Wei, L. Ci, B. Jiang, Y. Li, X. Zhang, H. Zhu, C. Xu and D. Wu, J. Mater. Chem., 13 (2003) 1340. S. Bandow, T. Takizawa, K. Hirahara, M. Yudasaka, S. Iijima, Chem. Phys. Lett. 337 (2001) 48.

[5] S. Lyu, T. Lee, C.Yang , C. Lee, Chemical Communications (2003) 1404

[6] S. Bandow, S. Asaka, Y. Saito, A. Rao, L. Grigorian, E. Richter, P. Eklund, Phys. Rev. Lett. 3779 (1998) 80; S. Bandow, G. Chen, G. Sumanasekera, R. Gupta, M. Yudasaka, S. Iijima, P.C. Ecklund, Phys. Rev. B 66, (2002) 075416, A. Jorio, R. Saito, J. H. Hafner, C. M. Lieber, M. Hunter, T. McClure, G. Dresselhaus, and M. S. Dresselhaus, Phys. Rev. Lett. 86, 1118-1121 (2001). V. W. Brar, Ge. G. Samsonidze, M. S. Dresselhaus, G. Dresselhaus, R. Saito, A. K. Swan, M. S. Ünlü, B. B. Goldberg, A. G. Souza Filho, and A. Jorio Phys. Rev. B 66, 155418 (2002), A. G. Souza Filho, A. Jorio, G. Dresselhaus, M. S. Dresselhaus, R. Saito, A. K. Swan, M. S. Ünlü, B. B. Goldberg, J. H. Hafner, C. M. Lieber, and M. A. Pimenta, Phys. Rev. B 65, 035404 (2002)

[7] J. Benoit, J.P. Buisson, O. Chauvet, C. Godon, S. Lefrant, Phys. Rev. B 66, (2002) 073417-1

[8] R. Bacsa, A. Peigney, Ch. Laurent, P. Puech, W. Bacsa Phys. Rev. B, Vol. 65 (2002), R161404. J. Maultzsch, S. Reich, P. Ordejon, R. Bacsa, W. Bacsa, E. Dobardzic, M. Damnjanovic and C. Thomsen, Physics and Chemistry of Fullerenes and Derivatives (World Scientific), proceedings of the international winterschool on electronic properties of novel materials, (2003) Vol. 685, 324-327.

[9] V.N. Popov, L. Hennard, Phys. Rev. B. 65 (2002) 235415-1

[10] S. Reich and C. Thomsen, Phys. Rev. B 62, 4273 (2000); J. Maultzsch, S. Reich, C. Thomsen, Phys. Rev. B 65, 233402 (2002)

[11] P. Puech, H. Hubel, D. Dunstan, R.R. Bacsa, C. Laurent, W. Bacsa (submitted to Physical Review)

[12] R. Saito, G. Dresselhaus, M. Dresselhaus, Physical properties of Carbon Nanotubes (Imperial College Press, London, 1998)

Interaction of laser light and electrons with nanotubes

H. R. Sadeghpour*, Brian E. Granger**, and Petr Král***

* ITAMP, Harvard-Smithsonian Center for Astrophysics, Cambridge, Massachusetts 02138
** Dept. of Physics, Santa Clara University Santa Clara, CA 95053
** Department of Chemical Physics, Weizmann Institute of Science, 76100 Rehovot, Israel

Abstract

Designer electronic states in confined geometries are attracting considerable attention. In this work, a novel set of electronic states around nanotubes with long lifetime and exotic properties is discussed.

Keywords: Carbon nanotube, image states, ultrafast rotation, Rydberg-like states, long lifetimes

1 Introduction

Technological leaps of the last two decades have opened windows of opportunity for designing and controlling electronic states in novel systems. The fabrication of quantum dots in semiconductor devices has led to the invention of single-electron transistors [1] and controllable single photon emitters [2]. Such designer atoms are ideal settings for the control and manipulation of electronic states. Carbon nanotubes (CNT), first synthesized in 1991 by Iijima [3] as graphitic carbon needles, have remarkable electrical and mechanical properties. Carbon nanotubes are ideal for investigation at the interface of atomic and nanoscopic physics. Our aim has been to bring atomic physics techniques to the study of interaction of light and particles with nanotubes. Below, we will discuss a number of applications of atomic physics in the nanoscopic scale.

2 A brief primer on nanotubes

A carbon nanotube (CNT) is constructed by rolling a graphene sheet, defined by its primitive lattice vectors, $\mathbf{a}_1$ and $\mathbf{a}_2$, along a vector $\mathbf{C}_h$, the so-called chiral vector, around an axial or translation vector $\mathbf{T}$ (see Fig. 1). The chiral vector,

$$\mathbf{C}_h = n\mathbf{a}_1 + m\mathbf{a}_2 \equiv (n, m),$$

and translation vector,

$$\mathbf{T} = [\frac{(2m+n)}{d_R}\mathbf{a}_1 - \frac{(2n+m)}{d_R}\mathbf{a}_2],$$

are determined by two lattice indices, n and m. Here,

$$d_R = \left(\begin{array}{c} 3d \text{ if } n - m = 3qd \\ d \text{ if } n - m \neq 3qd \end{array} \right)$$

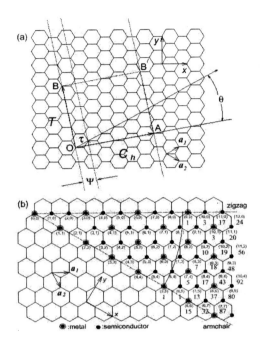

Figure 1: A graphene honeycomb lattice showing the primitive vectors and the chiral and translation vectors. From Ref. 4.

where q is an integer and d_R is the greatest common divisor of n and m.

The diameter of an (n, m) CNT is $d = C_h/\pi$, where $C_h = \sqrt{3}a_{c-c}[m^2 + n^2 - mn]^{1/2}$ is the magnitude of the chiral vector, and $a_{c-c} = 1.42\text{Å}$ is the carbon-carbon bond length. The chirality of a nanotube is characterized by the chiral angle,

$$\theta = \tan^{-1}\sqrt{3}m/(m + 2n)],$$

which is the angle between the vectors $\mathbf{C}_h$ and $\mathbf{a}_1$.

Nanotubes having either (but not both) $m = 0$ or $n = 0$ have $\theta = 0°$ and are called zigzag nanotubes due to the zigzag nature of the bonds at the tube ends. Likewise, tubes with $n = m$ have $\theta = 30°$ and are called armchair tubes[4], [5]. All other tubes with $n \neq m$ are called chiral nanotubes. One of the most remarkable properties of carbon nanotubes is that depending on their chirality, conducting, semiconducting or insulating nanotubes are possible[6].

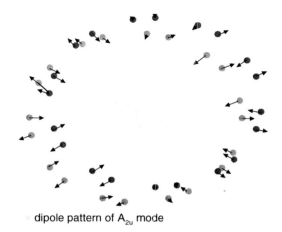

dipole pattern of A_{2u} mode

Figure 2: Ir-active A_{2u} phonon mode in (10,10) carbon nanotube at $\omega_{A_{2u}} = 870$ cm^{-1}.

3 Photoabsorption in carbon nanotubes: mechanical rotation

In one such application, we calcualted the absorption of circularly-polarized infrared photons by the optically-active resonant phonon modes in CNT[7]. The vibrational modes of carbon nanotubes have been investigated by several workers and the phonon dispersion energy curves have been calculated for a large number of CNTs using the zone-folding method [5], [8]. Of particular interest are two optically-active infrared (IR) phonon modes at $\omega_{A_{2u}} = 870$ cm^{-1} and $\omega_{E_{1u}} = 1580$ cm^{-1}. Both of the IR-active modes absorb photons and carry angular momentum.

Fig. 2, gives the dipolar pattern of the carbon atom displacements [9] in the doubly-degenerate A_{2u} phonon mode for a (10,10) CNT. We found that through the transfer of photon angular momentum, at a rate of $\dot{L}_{A_{2u}} = \hbar \dot{n}^+_{A_{2u}} \approx 2.5 \times 10^{-29}$ Nm, to the phonon modes and their eventuall decay into two acoustical phonons- the observed decay lifetime of the A_{2u} mode is roughly 2.2 ps- CNTs would rotate with a frequency of $\omega_{rot} \approx 28$ GHz. The steady-state angular momentum of $L_{A_{2u}} = \hbar \Delta n^+_{A_{2u}} = \dot{L}_{A_{2u}} \tau_{A_{2u}} \approx 5.2 \times 10^{-41}$ Js. This angular momentum is transferred to the tube following the two-acoustical phonon decay of the A_{2u} mode near the $k = 0$ branch.

This proof-of-concept study demonstrated that nanotubes with their particular mechanical and electrical properties could form parts of nanoscopic motors, centrifuges or stabilizers.

4 Rydberg-like electronic states near nanotubes

Rydberon states are ubiquitous in atomic and molecular physics [10]. Collisions between ultracold neutral and Rydberg atoms in magneto-optical traps have been predicted to form exotic classes of Rydberg molecules

dubbed "trilobite" and "butterfly" molecules [11].

Rydberg electronic states have also been observed above conducting surfaces. This is possible due to the attractive interaction of an electron with its image charge. The resulting interaction potential for an electron at a distance z above a flat surface having a dielectric constant ϵ, is: $V(z) = -\frac{e^2}{4z}\left(\frac{\epsilon-1}{\epsilon+1}\right)$. Because of the Coulomb interaction, this potential supports an infinite number of "image states" that form the familiar Rydberg-series, $E_n = -\frac{13.60}{16n^2}\left(\frac{\epsilon-1}{\epsilon+1}\right)^2$ eV, where n is the principal quantum number. Femtosecond two-photon pump-probe spectroscopy of electronic image states near Cu(100) surfaces has revealed quantum beats due to the interference of different eigencomponents in the resulting electronic wavepackets above the surface [12]. The collapse of these states into the surface is the primary reason for their short lifetimes.

In a recent work, we showed that conducting CNTs do indeed have Rydberg-like excitations[13]. However, because an electron can have angular momentum about the nanotube axis, a centrifugal barrier forms, dramatically increasing the lifetimes of the states compared to their counterparts above flat surfaces, see Fig. 3. We take the charge (the electron) to be located outside the tube at the position $(\rho_0, 0, 0)$ and $(I_m(x), K_m(x))$ are the regular and irregular modified Bessel functions. The electrostatic force between the electron and the conducting cylinder is calculated by differentiating the induced scalar potential Φ_{ind} with respect to ρ:

$$
\begin{aligned}
F(\rho_0) &= -q \, \partial \Phi_{ind}/\partial \rho|_{(\rho_0,0,0)} \\
&= \frac{2q^2}{\pi a^2} \int_0^\infty dx \left[A_0(x) + 2\sum_{m=1}^\infty A_m(x) \right], \\
A_m(x) &= \frac{I_m(x)}{K_m(x)} K_m(x\rho_0/a) \, x \, K'_m(x\rho_0/a). \quad (1)
\end{aligned}
$$

The potential energy $V(\rho_0) = -\int^{\rho_0} F(\rho) \, d\rho$ can be calculated numerically from these expressions. Alternatively, physical intuition can be gained by an asymptotic analysis of the potential. The result,

$$
V(\rho_0) \sim \frac{q^2}{a} \text{li}\left(\frac{a}{\rho_0}\right) \approx -\frac{q^2}{a} \frac{1}{(\rho_0/a)\ln(\rho_0/a)}, \quad (2)
$$

is dominated by the $m = 0$ term in Eq. 1 and is given in terms of the logarithmic integral $\text{li}(x) \equiv \int_0^x dt/\ln(t)$.

The total effective potential that an electron would "see" in front of a conducting nanotube is

$$
V_{eff}(\rho) = V(\rho) + \frac{(l^2 - \frac{1}{4})}{2m\rho^2}, \quad (3)
$$

where m is the mass of the electron. This potential for different values of the angular momentum, is shown in

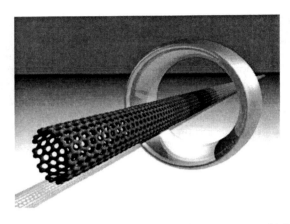

Figure 3: Visualization of an electron in a tubular image state around a $(10, 10)$ metallic carbon nanotube. From Ref. 13

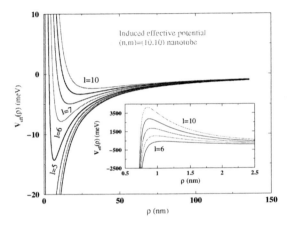

Figure 4: The effective potential between an electron and a conducting nanotube as a function of the angular momentum, l. From Ref. 13.

Fig. (4) for a (10,10) carbon nanotube of radius $a = 0.68$ nm. For moderate angular momenta ($l \geq 6$), the effective potential possesses extremely long-range wells that support bound states. The inset of Fig. 4 shows that high ($1 - 2$ eV) potential barriers separate these wells from the tube surface. Indeed, for angular momentum values of $l < 6$, no long-range wells exist so that low angular momentum states will exist close to the surface and have short lifetimes. However, because the radial overlaps between the high angular momentum and lower angular momentum states are small (they are localized at different radial distances) the decay of the high angular momentum states to ones with lower angular momentum through spontaneous radiative decay will be greatly suppressed.

The lifetimes due to tunneling increase exponentially with l and are substantially longer than those of image states above planar surfaces. Interaction with holes in the nanotube and the coupling of these states to the phonon modes in the tube can reduce their lifetimes.

These decay mechanisms can be suppressed by operating at temperatures ($T < 10$ K) lower than the transition energy between different image states. Spontaneous radiative transitions between different states also limit their lifetimes. The calculated lifetimes between $l_f = l_i \pm 1$ are found to be 5-10 ms and stimulated transitions due to blackbody radiation cut these lifetimes to probably hundreds of microseconds.

Two likely mechanisms for forming these states are inverse photoemission and charge exchange with ultracold Rydberg atoms. The former process favors low-energy recombination with rate coefficients of $\alpha^{rr} \sim 10^{-13} - 10^{-12}$ cm^3 s^{-1}. In the latter approach, the cross section to capture electrons from Rydberg atoms scales geometrically as n^4, where n is the Rydberg atom's principle quantum number.

Recent experimental realization of suspended CNT network[14] provides the necessary ground for observing the proposed states. We are extending our treatment to a periodic array of carbon nanotubes. The preliminary calculations show rich electronic structures with band gaps for the propagation inside the array.

This work was supported by the US National Science Foundation.

REFERENCES

[1] J. A. Folk *et al.*, *Science* **299**, 679 (2003); D. Goldhaber-Gordon et al., *Nature* **391**, 156 (1998).

[2] M. Pelton et al., *Phys. Rev. Lett.* **89**, 233602 (2002).

[3] S. Iijima, *Nature* **354**, 56 (1991).

[4] M. S. Dresselhaus, G. Dresselhaus, and P. C. Eklund, *Science of Fullerenes and Carbon Nanotubes* (Academic Press Inc., San Diego, 1996).

[5] R. Saito, G. Dresslhaus, and M. S. Dresselhaus, *Physical Properties of Carbon Nanotubes, Imperial College Press, London, 1998).

[6] J. W. Mintmire, B. I. Dunlap, and C. T. White, *Phys. Rev. Lett.* **68**, 631 (1992).

[7] P. Kral and H. R. Sadeghpour, *Phys. Rev. B* **65**, R-161401 (2002).

[8] M. S. Dresselhaus and P. C. Eklund, *Adv. Phys.* **49**, 705(2000).

[9] R. Saito (private communication).

[10] T. Gallagher, *Rydberg Atoms*, Cambridge University Press, New York (1994).

[11] C. H. Greene, A. S. Dickinson, and H. R. Sadeghpour, *Phys. Rev. Lett.* **85**, 2458 (2000); E. L. Hamilton, C. H. Greene, and H. R. Sadeghpour, *J. Phys. B:* **35**, L199 (2002).

[12] U. Höfer et al., *Science* **277**, 1480 (1997).

[13] B. E. Granger, P. Kral, H. R. Sadeghpour, and M. Shapiro, *Phys. Rev. Lett.* **89**, 135506-1 (2002).

[14] Y. Homma, Y. Kobayashi, and T. Ogino, *Appl. Phys. Lett.* **81**, 2261(2002).

Synthesis of Carbon Nanotubes by Decomposition of Hydrocarbon Using an Arc-Jet Plasma

Shinil Choi[*], Junseok Nam, Sookseok Choi, Chanmin Lee, Jongin Kim, and Sang Hee Hong[**]

Department of Nuclear Engineering, Seoul National University
Seoul 151-742, Korea
[*]hopefor@fusma.snu.ac.kr, [**]hongsh@snu.ac.kr

ABSTRACT

We present a method to produce carbon nanotubes by decomposition of hydrocarbons using an arc-jet plasma of high temperature (5000–20000 K). Carbon nanotubes are continuously produced in a floating condition by introducing a hydrocarbon gas and a metal precursor into the arc-jet plasma. In this experimental work, argon and hydrogen are used as a plasma gas, nickel powder as a metal precursor, and methane as a carbon source. For the investigation of morphology, crystallization degree and purity of the carbon nanotubes in the soot produced under various processing conditions, we have performed the material analyses by using SEM, Raman spectroscopy and TGA. Since the suggested arc-jet plasma process is continuous and easily scalable, it is expected to be a promising technique for a large-scale commercial production of carbon nanotubes.

Keywords: carbon nanotubes, arc-jet plasma, decomposition of hydrocarbon, synthesis

1 INTRODUCTION

Because of the unique properties of carbon nanotubes (CNTs), they have a great potential of practical applications to field emission display, reinforcing component, hydrogen storage medium, and so on. However, in order to realize their industrial applications, a continuous process for large-scale production of CNTs is essential. To date, several methods such as arc discharge, laser ablation, chemical vapor deposition (CVD) and catalytic pyrolysis usually have used for CNTs production.

As one of catalytic pyrolysis methods, we suggest a promising method for continuous large-scale production of carbon nanotubes by decomposition of hydrocarbon using an arc-jet plasma. The arc-jet plasma is generated by blowing a cold gas into current path of a dc arc discharge between two electrodes. Since the arc-jet plasma has thermal plasma properties of high temperature (5000-20000 K) and high velocity (100-1000 m/s) [1], it has been widely used for plasma spray coating [2], fine powder synthesis [3], and hazardous waste treatment [4]. When the arc-jet plasma technique is applied for the CNTs synthesis, this approach is analogous to the conventional catalytic pyrolysis in respect that CNTs grow in a floating condition by

introducing carbon and catalyst sources simultaneously to a reaction zone. But, instead of a conventional electric furnace in the catalytic pyrolysis method, the arc-jet plasma is used as a heat source for decomposition of hydrocarbon and catalyst. Therefore, we can obtain a synthesis ambience of very high temperature, which is not achievable by the conventional electric furnace. Thus the arc-jet plasma method is somewhat similar to the conventional arc discharge method in respect of synthesis ambience.

In the synthesis of CNTs in the arc-jet plasma, the expected process characteristics are as follows: i) It is possible to achieve high rates of hydrocarbon decomposition and CNTs growth resulted from high temperature and high enthalpy of the thermal plasma; ii) Due to high temperature and enthalpy sufficient for evaporating the metal powder of large size in the arc-jet plasma, this technique itself is regarded as a process for the metal nano-particle synthesis [3], and therefore does not use nano-sized metal powders or organic metals as a catalyst source for the CNTs synthesis like the conventional catalytic pyrolysis. The use of coarse powder is economically benefit for a catalyst supply; iii) This process can be easily scale-up through the plasma power-up. Moreover, the plasma power-up can be achieved in a small volume without expansion of a hot plasma zone; iv) We can easily collect synthesized materials because of one directional flow of the jet, which permits a continuous process of the CNTs synthesis with a continuous supply of carbon and catalyst sources.

In this communication, we have performed the material analyses of the soot obtained from the arc-jet plasma synthesis by using SEM, Raman spectroscopy and TGA to investigate morphology, crystallization degree and purity of the carbon nanotubes in the soot. And, we present effects of hydrogen as a plasma gas and flow rate of methane as a carbon source on synthesis of carbon nanotubes using this arc-jet plasma method.

2 EXPERIMENAL

A schematic diagram of the synthesis system is shown in Fig. 1. The arc-jet plasma is generated by a non-transferred plasma torch. The torch consists of a tungsten cathode of conical shape and a copper anode of nozzle shape. While a plasma gas is introduced into a channel between the anode and the cathode, a current path forms between the electrodes by a dc arc discharge, and the resultant arc-jet plasma is ejected from the exit of the anode. Both electrodes are cooled by water in order to prevent them from erosion caused by enormous heat flux from the arc.

In this experiment, a cathode diameter of 12 mm and a conical angle of $60°$, and an inner anode diameter of 9 mm were taken. Argon and hydrogen were used as a plasma gas. Argon flow rate was fixed at 25 slpm and hydrogen flow rate was varied in the range of 0-3 slpm. Methane was used as a hydrocarbon, and nickel powder of about 3 μm was used as a metal precursor. Methane of 1-4 slpm and nickel powder of 0.44 g/min were introduced, through stainless tube of 1.5 mm inner diameter, into the arc-jet plasma at the exit of anode. The nickel powder has been commonly used by some researchers for the conventional arc discharge method. The arc-jet plasma was generated with a dc current of 350 A. The reactor was maintained at atmosphere pressure.

The CNT samples produced in the experiment were collected by a stainless steel collection tube for material analyses and tests by using field emission scanning electron microscopy (FE-SEM, JEOL, JSM-6330F), thermal analysis system (TA Instruments, TGA2050) and Raman spectrometer (Jobin-Yvon/Spectra-Physics, T64000). The FE-SEM was operated at 10 kV. Approximately 8 mg of the produced soot was used for TGA, in which measurement conditions were 10 °C/min ramp rate from room temperature to 800 °C at an air flow rate of 100 sccm.

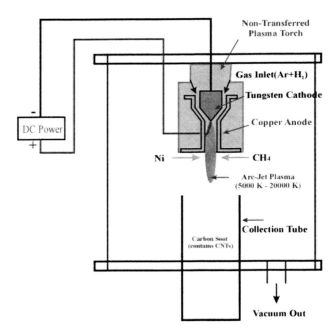

Figure 1. Schematic of an arc-jet plasma reactor used for carbon nanotubes synthesis by decomposition of hydrocarbon.

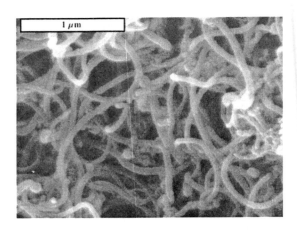

Figure 2. Typical SEM image of raw soot produced by decomposition of methane in the arc-jet plasma (H_2: 1 slpm, CH_4: 2 slpm)

3 RESULTS AND DISCUSSION

Some raw soot was obtained from the collection tube after the synthesis process, and carbon nanotubes were eventually found in the soot through its material analyses, which were conducted without a purification process.

Figure 2 shows a typical image of the soot produced by arc-jet plasma method for the case of 1-slpm hydrogen and 2-slpm methane. In this figure, high density of carbon nanotubes can be seen along with other particles. The carbon nanotubes are estimated to have diameters of 30-60 nm and lengths of longer than 1.5 μm. The exact length cannot be measured because no carbon nanotubes were observed with both ends of carbon nanotubes appeared at the same time.

Figure 3 presents a typical TGA graph of the raw-soot produced in the condition of 1-slpm hydrogen and 1-slpm methane. In this figure, the bold, dash and dot lines correspond to TG, DTG and Lorentzian fit of DTG curve, respectively. This graph indicates that a residual weight of the soot is about 0.85, which is contributed mainly from nickel particles. This implies that the metal precursor was excessively introduced in this experiment.

From this figure, it is also observed that an initial burning temperature of the raw-soot is about 300 °C and that there are two stepwise weight losses which can be assigned to amorphous carbon and muti-wall carbon nanotubes. The peak temperature of Lorentzian fit for amorphous carbon is about 430 °C, which is somewhat higher than that of other methods for a unclear reason. It is well known that DTG is a common method for quantitative analysis, and the content of each species can be determined from its peak area in DTG. Thus the yield of carbon nanotubes is estimated as about 27 % in the soot produced by this arc-jet plasma method.

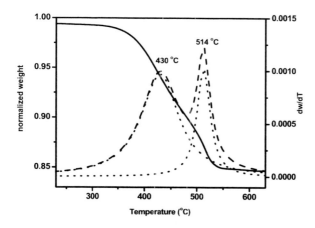

Figure 3. Typical TGA curve of the raw-soot produced by decomposition of methane in the arc-jet plasma (H₂: 1 slpm, CH₄: 1 slpm)

Figure 4 is the typical Raman spectra of the raw-soot produced under the condition of 1-slpm hydrogen and 1-slpm methane. This spectral curve was recorded at room temperature, using an excitation wavelength of 514.5 nm. In the high-frequency range from 1200 to 1700 cm^{-1}, two peaks are observed at 1354 and 1583 cm^{-1}. A peak at 1354 cm^{-1} corresponds to a disorder-induced phonon mode (D-band), and a peak at 1583 cm^{-1} can be assigned to a Raman-allowed phonon mode (G-band). It is known that the intensity ratio, G/D, indicates the crystallization degree of grown carbon nanotubes. Figure 4 shows that the intensity of G band is a bit higher than that of D band. Generally, it is known that the ratio is very high in the arc discharge method, while low in the catalytic pyrolysis method. According to Fig. 4 for the present arc-jet plasma synthesis, the crystallization degree of carbon nanotubes is very low compared with that by the arc discharge method, but a little higher than that by the catalytic pyrolysis.

Figure 5 shows SEM images of the raw-soot produced with 2-slpm methane at the different flow rates of hydrogen; (a) 0 slpm, (b) 1 slpm, and (c) 3 slpm. As seen in Fig. 5(a), carbon nanotubes are hardly found in the case that hydrogen gas is not mixed as a plasma gas. However, in Fig. 5(b), a number of carbon nanotubes were produced when 1-slpm hydrogen gas was mixed as a plasma gas. In the latter case, it was measured that the arc voltage increased by 1.5 times compared with the former case at the same arc current of 350 A, which means that the input electric power into the arc-jet plasma accordingly increases. And other studies on arc-jet plasma have suggested that mixing of hydrogen gas makes the plasma temperature and enthalpy higher. When hydrogen is not mixed with argon gas, nickel particles of large size were observed in the SEM image. But, when 1-slpm hydrogen was mixed with argon gas, they were not found in the image. It is concluded from this observation that in the pure argon plasma, the catalytic nickel powder does not evaporate sufficiently for forming

nano-sized metal particles essential for growth of carbon nanotubes, and that a high enthalpy plasma by addition of hydrogen gas is required for production of carbon nanotubes.

But, in Fig. 5(c) for 3-slpm hydrogen, the fraction of carbon nanotubes in the soot has reduced again when an excessive hydrogen gas was mixed. In this case, the arc-jet plasma was unstable and the growth of carbon nanotubes appeared to be disturbed by plasma instability caused by the excessive hydrogen mixing.

Figure 6 shows SEM images of the sample produced with hydrogen of 1 slpm and different flow rates of methane; (a) 1 slpm and (b) 4 slpm. The experimental result for 2-slpm methane was given previously in Fig 2. In the present experiment, methane gas was used as a carbon source. Therefore, the flow rate of methane is expected to determine the production rate of solid carbon material when the introduced methane is converted to solid carbon. However, it is shown in Figs. 5 and 6 that the flow rate of methane also has an influence on purity of carbon nanotubes in the soot. When the flow rate of methane was so low as 1-2 slpm, a considerable amount of carbon nanotubes were produced as seen in Fig. 5(b) and Fig. 6(a). However, for a high flow rate of 4-slpm methane, carbon nanotubes were hardly found as observed in Fig. 6(b), where nickel particles of large size were also observed. Since an excessive flow rate of methane lowers the plasma temperature and enthalpy, the nickel powder appears not to be evaporated sufficiently.

From the above experimental results, it is important for the carbon nanotubes synthesis by decomposition of methane using the arc-jet plasma to provide the sufficiently high temperature and enthalpy of the plasma during the synthesis process.

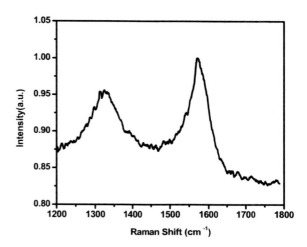

Figure 4. Typical Raman spectra of the raw-soot produced by decomposition of methane in the arc-jet plasma (H₂: 1 slpm, CH₄: 1 slpm)

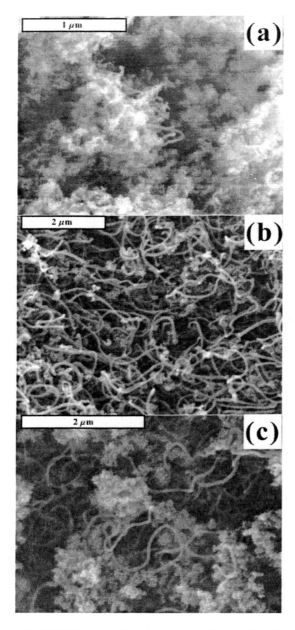

Figure 5. SEM images of carbon nanotubes produced with the different flow rates of H_2: (a) 0 slpm, (b) 1 slpm, and (c) 3 slpm

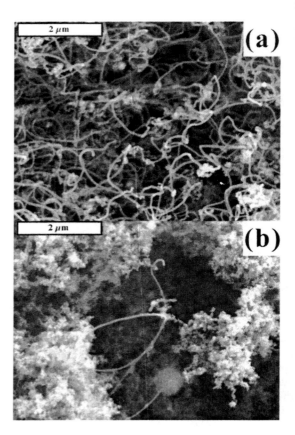

Figure 6. SEM images of carbon nanotubes produced with the different flow rates of CH_4: (a) 1 slpm and (b) 4 slpm

method, but a little higher than that of the catalytic pyrolysis. From the experimental study performed under the various operating conditions, it was found that the fraction of carbon nanotubes in the soot considerably increases by mixing of hydrogen properly with argon as a plasma gas, and that the excessive flow rate of methane reduces the purity of carbon nanotubes in the soot.

The synthesis process by this arc-jet plasma method has not been optimized yet and further studies on production of carbon nanotubes are required. But, we feel that this method will be a promising technique for large-scale commercial production of carbon nanotubes because of its advantages of continuous process, easy scale-up of plasma power, and simple collection of product.

4   CONCULUSION

We have demonstrated a synthesis method of carbon nanotubes by decomposition of hydrocarbon using the arc-jet plasma in this experimental work. It has been observed from the SEM images that high density of carbon nanotubes was typically produced with diameters of 30-60 nm and lengths of longer than 1.5 μm. And the weight purity of the carbon nanotubes in the soot was found about 27% and the crystallization degree of carbon nanotubes produced by this method was quite lower than that of the arc discharge

REFERENCES

[1] Maher I. Boulos, IEEE T. Plasma Sci. 19, 1078, 1991
[2] Pierre Fauchais and Armelle Vardelle, IEEE T. Plasma Sci. 25, 1258, 1997
[3] N. Rao, S. Girshick, J. Heberlein, P. McMurry, S. Jones, D. Hansen and B. Micheel, Plasma Chem. Plasma Process. 15, 581, 1995
[4] H.R. Snyder, C.B. Fleddermann and J.M. Gahl, Waste Manage. 16, 289, 1996

Lattice Directional Growth of Single-Walled Carbon Nanotubes Arrays

H. Lin[*], C. Harnett[*] and D. M. Tanenbaum[**]

[*] Department of Applied and Engineering Physics, Cornell University, Ithaca, NY, USA
lh77@cornell.edu, ch33@cornell.edu
[**] Department of Physics and Astronomy, Pomona College, Claremont, CA, USA
dtanenbaum@pomona.edu

ABSTRACT

In this report, 2D arrays of single-walled carbon nanotubes (SWNTs) have been grown on Silicon substrate of both Si(100) surface and Si(111) surface by using chemical vapor deposition (CVD). The two-fold growth symmetry on Si(100) and three-fold growth symmetry on Si(111) show preferred growing direction of carbon nanotubes along the lattice orientation of the underneath substrate. This confirms the earlier simulation results [1, 2].

Keywords: lattice-directional-growth, single-walled carbon nanotubes, arrays, two-fold symmetry, three-fold symmetry

1 INTRODUCTION

The research on single-walled carbon nanotubes (SWNTs) has long been an interesting research topic due to their unique mechanical, electronic and thermal properties as well as chemical stability [3,4]. Utilizations of these properties have shown wide range applications including nanoelectronic devices [5], sensors [6], scanning probes [7] and field emission display [8]. These applications, especially the one of making carbon nanotube transistors [9], make it very desirable for the integration of carbon nanotubes (CNTs) with long established Si technology. However, there is one long-standing issue that needs to be tackled in order to accomplish large integration: growing ordered nanotube architecture with high yield. In the past, obtaining ordered SWNT architecture in aligned orientations has been challenging. Many efforts have been put into making well-defined 2D networks of SWNTs and nanowires (NWs) [10-11]. But the alignment achieved is limited. Here we propose a simple but effective way of building SWNT arrays with perfect alignment to lattice orientation through the lattice directional growth mechanism

The three-fold symmetry of the discrete orientations between CNTs and hexagonal graphene surfaces has shown that carbon nanotubes (CNTs) have preferred orientations on the highly oriented pyrolytic graphite (HOPG) substrate where they have sharp potential energy minima [12]. This was demonstrated by the manipulation of CNTs with AFM on HOPG substrate showing the motion changing from sliding/rotating in-plane to stick-roll associated with dramatic increase of lateral manipulating force when CNTs are in commensurate contact with HOPG. And this was additionally justified by the simulation results using molecular statistics and dynamics method [1]. Later, on silicon surfaces, direct synthesis of SWNTs with relevant simulation [2] using the same method as in [1] has revealed that the orientations of grown SWNTs can be determined by the lattice of the Si substrate, though the experimental results obtained are not distinct and the yield of SWNTs is low. In our approach [13], arrays of SWNTs with discrete orientations have been synthesized through the same mechanism on different silicon substrates by chemical vapor deposition using methane. The yield has been largely improved.

2 SYNTHESIS

Our synthesis begins with the patterning of catalytic strips on silicon substrate using micro contact printing method [13]. Firstly, liquid phase catalyst are prepared by dissolving $AlCl_3 \cdot 6H_2O$ (2.4g), $FeCl_3 \cdot 6H_2O$ (90mg), MoO_2Cl_2 (4mg) and P123 copolymer (1g) in a mixed ethanol and butanol solution using previous developed recipe [14, 15] for growing SWNTs. A conditioning catalyst is also adopted to enhance the yield [10]. Secondly, a 1cm 1cm PDMS stamp with strip pattern on the surface is inked with catalyst and used to print catalyst strip pattern onto the silicon surface [13]. On the surface of silicon substrate, there is naturally grown native oxide. The strips are spaced at 2-micron pitch. Thirdly, the patterned substrate is transferred into 1-inch quartz tube furnace to proceed to CVD growth at 900°C with methane flowing rate at 1000mL/min. The flowing of methane is maintained for 10 to 20min and is preceded and followed by the flowing of argon.

3 RESULTS

We have conducted the experiments on substrate of both Si (100) surface and Si (111) surface. We characterized our samples with field emission electron microscope (SEM) and atomic force microscope (AFM).

3.1 Si (100) surface

In Fig. 1(a), the feature of the two-fold growth symmetry is clearly seen from the arrays of white lines.

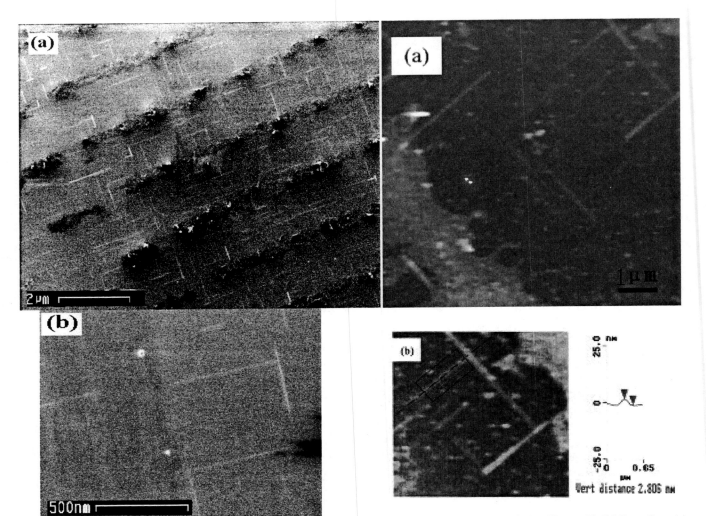

Figure 1: SEM images of SWNTs on single crystalline Si (100) surface substrates. (a) Arrays of SWNTs are grown between patterned catalyst strips. (Strip pattern are broken into disconnected islands.) (b) SWNTs are grown across each other with 90°.

Figure 2: AFM images of SWNTs on Si (100) surface (a) and a cross section analysis of one tube.

The two preferred orientations of these SWNTs as shown in Fig. 1(b) characteristically reflect the lattice structure underneath. With further cross section characterization by AFM (Fig2), we can measure their diameters. The range of their diameters is from 1-5nm with a means of 1-2nm and lengths up to 6 micrometers [13].

3.2 Si (111) surface

We repeated the same experiment on substrate of Si(111) surface. Three-fold growth symmetry is clearly shown in Fig 3(a) and 3(b). The orientations of these nanotubes being separated by multiples of 60°, additionally confirm a perfect matching between SWNTs and the lattice orientation of the Si (111) surface. Dark area of strip patterns is catalyst printed from PDMS stamp.

Contrary to earlier result [2], in our experiments under the same condition, SWNTs grown on Si(111) aren't any shorter than those grown on Si (100), though from the previous simulation, energy barriers for sliding the nanotubes on Si(111) is 2 times higher than that on Si (100).

One very interesting observation that we have in the above images is that it is rare to find any two carbon nanotubes grown to cross each other while they can easily be grown across the catalyst. This proposes a method of controlling the length of carbon nanotubes and forming a 2-D network architecture. Especially in figure 3c, the branching feature is observed. Whether they are inherently grown to connect to each other is yet to be determined.

In this work, we have confirmed in further steps the preferred growing direction of SWNTs along Si substrate lattice orientation. We observed that other tubes lying across to them limit the lengths of these SWNTs when they are grown into arrays. This, combined with our improved yield of growing SWNTs, emphasized the great potential of

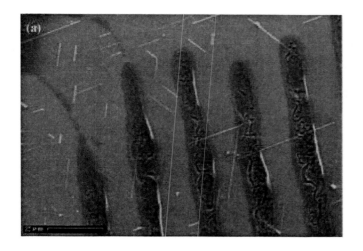

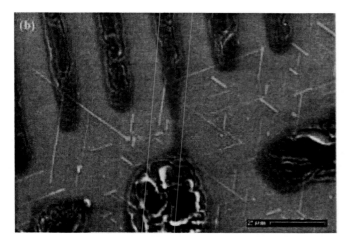

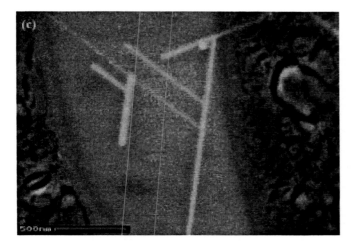

Figure 3. (a)-(c) SEM images of SWNTs on single crystalline Si(111) surface substrate. Dark area is the pattern of catalyst. SWNTs are aligned to each other with an angle of multiples of 60°. No tubes are observed to cross each other. Branching feature is clearly seen in (c).

the technique to the integration of making nanotube devices with conventional silicon technology.

Acknowledgement:
This work was completed with support of the NSF through both the Cornell Center for Materials Research and the Cornell Nanofabrication Facility. The authors wish to acknowledge both stimulating conversations with members from Professor Paul McEuen's group and the experimental resources provided by Professor Harold Craighead and Professor Michael Isaacson over the course of the research

REFERENCES

[1] Buldum, A.; Lu, J. P. *Phys. Rev. Lett*, 83(24), 5050-5053. 1999

[2] Su, M.; Li, Y.; Maynor, B.; Buldum, A.; Lu, J. P.; Liu, J. *J. Phys. Chem. B.* 104(28), 6505-6508, 2000.

[3] Dresselhaus, M. S.; Dresselhaus, G.; Eklund, P. C. *Science of Fullerences and Carbon Nanotubes* Academic Press: San Diego, CA 1996

[4] Dekker, C. *Phys. Today* 52, 22-28, 1999.

[5] McEuen, P. L. *Phys. World* June,31-36, 2000.

[6] Kong, J.; Franklin, N.; Chou, C.; Pan, S.; Cho, K. J.; Dai, H. *Science* 287, 622, 2000.

[7] Hafner, J.; Cheung, C.; Lieber, C. *Nature* 398, 761-762, 1999.

[8] Fan, S.; Chapline, M.; Franklin, N.; Tombler, T.; Cassell, A.; Dai, H. *Science* 283, 512-514, 1999

[9] Martel, R.; Schmidt, T.; Shea, H. R.; Hertel, T.; Avouris Ph. *Appl. Phys. Lett.* 73(17) 2447-2449, 1998.

[10] Franklin, N. R.; Dai, H. *Adv. Mater.* 12, 890-893, 2000.

[11] Huang, Y.; Duan, X.; Wei, Q.; Lieber, C. M. *Science* 291, 630-633, 2001.

[12] Falvo, M. R.; Steele, J.; Taylor II, R. M.; Superfine R. *Phys. Rev. B.* 62(16) R10665-R10667, 2000.

[13] Lin, H. Master thesis, Cornell University, 2001.

[14] Cassell, A. M.; Raymakers, J. A.; Kong, J.; Dai, H.; *J. Phys. Chem. B.* 103, 6484-6492, 1999.

[15] Dai, H.; Kong, J.; Zhou, C.; Franklin, N.; Tombler, T.; Cassell, A.; Fan, S.; Chapline, M. *J. Phys. Chem. B* 103, 11246-11255, 1999.

Continuous Mass Production of Carbon Nanotubes by 3-Phase AC Plasma Processing

F. Fabry[*], T. M. Gruenberger[**], J. Gonzalez Aguilar[**], H. Okuno[****], E. Grivei[*], N. Probst[*], L. Fulcheri[**], G. Flamant[***] and J.-C. Charlier[****]

[*]Timcal Belgium S.A., 534, av. Louise, B-1050 Brussels, Belgium, f.fabry@be.timcal.com
[**]Ecole des Mines de Paris, ENSMP, rue Claude Daunesse B.P. 207, F-06904 Sophia Antipolis, France, laurent.fulcheri@ensmp.fr
[***]IMP-CNRS, B.P. 5 Odeillo, F-66125 Font Romeu, France, flamant@imp.cnrs.fr
[****]University of Louvain, Unit of Physico-Chemistry and Physics of Materials, LLN, Belgium, charlier@pcpm.ucl.ac.be

ABSTRACT

For the synthesis of carbon nanotubes (CNTs), the plasma process is an original new approach. Hereby, the carbon mass flow is no longer limited by a physical ablation rate (which is the limiting step in the production rate of the classical nanotube processes), but is freely adjustable. Moreover, the process is operated at atmospheric pressure and the NT rich soot is extracted continuously. The feasibility for producing a large variety of different nanostructures at high selectivity including Carbon MWNTs, SWNTs, nanofibers and necklaces nanostructures is well established. From the characteristics observed, it is concluded that the AC plasma technology shows a significant potential for the continuous production of bulk quantities of carbon-based nanotubes of controlled properties and novel nanostructures.

In this paper, the plasma process for continuous nanotube synthesis, typical process conditions prevailing and products are presented.

Keywords: Mass production, thermal plasma, carbon nanotubes, carbon nanostructures

1   INTRODUCTION

Since many years, nanotechnologies are considered as very promising, capable of revolutionising entire sectors of industry. However, a real breakthrough of nanotechnologies has not been achieved yet, most probably because of the failure to transfer these novel technologies to existing industries. The synthesis of pure nanotubes with selected properties is now achievable, but only in very small quantities. The unrestricted availability of nanotubes at a reasonable price will generate the growth of a range of technological areas.

The current state-of-the-art shows that constant progress is made towards the production of nanomaterials at low cost. These materials concern mainly the production of nanopowders of different species (e.g. ceramics and metals) and carbon nanotubes. CNTs have been identified to have many potential applications in a large field of different areas like biology, material and surface science, energy storage, gas storage, environmental technology, electronics, etc. However, their market is still limited because their methods of production are not yet efficient enough to produce them with controlled properties at a relatively low cost.

Present production methods for novel nanostructured materials applied in research or for commercial production can only provide very limited quantities of the order of few grams per hour. Accordingly, current market prices are prohibitive, which is, therefore, limiting research activity. Currently known processes for carbon nanotube synthesis use either of the following methods: Arc [1, 2, 3], laser [4], solar [5, 6], plasma [7, 8] or catalytic method [10, 11]. The first group, comprising the first four processes corresponds to the high temperature techniques in which CNTs formation process is based on the sublimation and recondensation of the carbon precursor. High temperature processes are used for the production of CNTs with a high degree of graphitisation (defect free), however, product yields necessitate a purification of the products to remove amorphous carbon. Within this group, arc, laser and solar processes are having many features in common, being all of them based on the ablation of a solid carbon target, working in batch mode at reduced pressure. These three processes having reached their limits in production rate. An original approach is taken with the plasma process. Hereby, the carbon mass flow is no longer limited by a physical ablation rate, but is freely adjustable. Moreover, the process is operated at atmospheric pressure and the NT rich soot is extracted continuously. Over the last ten years an industrial consortium composed of Timcal Belgium, Ecole Nationale Supérieure des Mines de Paris (ENSMP) and Centre National de la Recherche Scientifique (CNRS) has developed a new process based on a 3-phase AC plasma technology for the continuous mass production of carbon nanostructures.

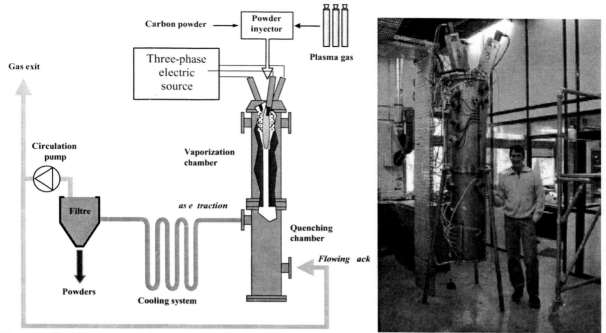

Figure 1: (Left) Scheme of the new configuration of plasma facility [9]. (Right) Plasma reactor in its present form.

This 3-phase AC plasma system; initially developed and optimised for the synthesis of novel grades of carbon black [12, 13] and later modified for the continuous synthesis of fullerenes [14, 15], has been adapted for the synthesis of carbon nanotubes and tube-like structures [7, 8, 9].

2 PLASMA PROCESS

These adaptations lead to the process scheme shown in Figure 1. The process can be briefly described as follows:

The plasma is generated by an arc discharge between three graphite electrodes placed in the upper part of the reactor. Simultaneously, an inert gas flows across the arc and the carbonaceous precursor and catalysts are injected into the high temperature region in or after the arc by a special powder injection system. Due to the high density of enthalpy obtained, the carbon and the metal catalysts are vaporised completely while passing through the graphite nozzle.

The quenching/sampling system collects the high temperature gas at a predetermined position in the reactor and cools it rapidly (quenching). The gas is filtered and a part of it is re-injected into the reactor. The fraction that corresponds to the flow rate initially entering the reactor as plasma gas is exhausted. This set-up (with recirculation) allows the extraction of a gas volume superior to the initial flow of plasma gas and therefore disconnects the dependency of these two parameters, which leads to an additional degree of freedom in relation to process operation.

Moreover, the reactor set-up is permanently being improved and industrial production of carbon nanotubes is envisaged.

3 RESULTS AND DISCUSSION

A large number of process parameters was investigated. A brief overview is given in Table 1.

Nature of precursor	Carbon Black + Ni, Co and Y at different concentrations. Ethylene + Ni
Nature of catalysts	Coating on carbon black Metal powder
Nature of plasma gas	Helium Nitrogen
Flow rates	Plasma gas Precursor
Quenching and cooling conditions	Location of product extraction

Table 1: Main process parameters investigated for nanotube production.

First efforts trying to correlate operating conditions to obtained product, with the final aim of understanding carbon nanotube formation inside the plasma system, show very promising characteristics. High temperature measurements inside the reactor revealed that an initial cooling to a plateau temperature above metal catalyst solidification and a following abrupt quenching are realized and can be freely adjusted and controlled, which presents one of the major advantages of the plasma process over conventional high temperature synthesis processes. Full details on these first correlations will be published by T. M. Gruenberger et al. [8].

All samples have been analysed using electron microscopy techniques (SEM/TEM/HRTEM) whereby multi-wall nanotubes (MWNT) have been found in many samples (Bamboo-like structures - Figure 2).

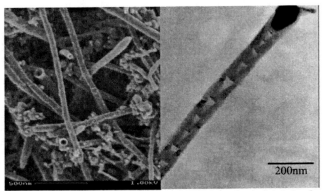

Figure 2: SEM image (left) and TEM analysis (right) of carbon bamboo-like structures (MWNT) [16]

The presence of single-wall nanotubes (SWNT) bundles and isolated SWNTs has been observed and has even been confirmed for samples collected from the product filter. Furthermore, the formation of carbon nanotubes in the gas phase by means of plasma processing seems to be confirmed also. The presence of bamboo-like, urchin-like, necklace, and others structures is frequently observed, which seems to be an indication that this process is capable of producing any kind of nanostructures. Finally, the reproducibility of the obtained products is confirmed, even for novel structures obtained under very special operating conditions.

The carbon samples originating from the internal wall of the reactor in the high temperature zone (wall sample) and from the bag filter (filter sample) have been analysed in detail. Electron microscopy analyses have allowed identifying three families of original carbon nanostructures, which can be associated to operating conditions.

3.1 Single Wall Nanotubes

SWNTs (Figure 3) have been produced for different operating conditions but the best results can be associated to a specific operating condition: Extraction of the product from the nozzle entrance zone. In this configuration, the process shows better general time-temperature characteristics required for nanotube formation than other configurations. The temperature just before the extraction zone is about 3000 – 3500 K. A very abrupt quenching is realised just after this zone ($\sim 10^6$ K/s) and can be freely adjusted and controlled by the flow rate of the recirculation

gas, which presents another major advantage of the plasma process over conventional high temperature synthesis processes.

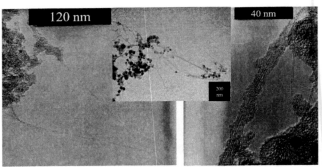

Figure 3: HRTEM Pictures of long SWNTs and SWNT bundles [16]

3.2 Carbon Fibres

The second family of nanostructures, are the carbon fibres (Figure 4). These nanostructures are produced with ethylene as carbon feedstock. The diameter of these carbon fibres is in the range of 20 – 50 nm with a length of few µm. The structure of these fibres is composed of an axial stacking of carbon layers with a low degree of organisation. The growth of these fibres is the result of the thermal decomposition of a hydrocarbon in presence of catalysts. This is a structure typically obtained at low temperature (< 1500 K) in presence of hydrogen.

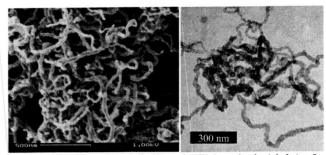

Figure 4: SEM image (left) and TEM analysis (right) of carbon nanofibers [16].

3.3 Carbon Necklaces

The third and most original group of nanostructures, are the carbon necklaces (Figure 5). These nanostructures are produced with Nitrogen as plasma gas. Carbon nano-necklaces usually curve smoothly and entangle together. The bamboo-like appearance of successfully joined segments is clearly seen. The segments in the carbon nano-necklaces are actually short variable-diameter compartments with one end sealed and the other one open. Some of these segments are completely filled with the Ni-Co catalyst used during the synthesis.

TEM and HRTEM pictures show a high degree of graphitic organisation in the thickness of the segments. The best well-organised structures have been identified on the wall samples. High temperature measurements inside the reactor revealed that with Nitrogen plasma gas, the axial and radial temperature distributions in the plasma reactor are much higher than with Helium (temperature along the graphite wall can reach 2500 K) and seems to be a fundamental parameter in the production of carbon necklaces [16].

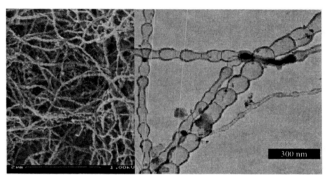

Figure 5: SEM image (left) and TEM analysis (right) of the carbon necklace nanostructures [16].

4 CONCLUSIONS AND PERSPECTIVES

The feasibility for producing a large variety of different nanostructures at high selectivity including carbon MWNTs, SWNTs, nanofibers and necklace nanostructures is well established. From the characteristics observed, it is concluded that the 3-phase AC plasma technology shows a significant potential for the continuous production of bulk quantities of carbon-based nanotubes of controlled properties and novel nanostructures.

As microscopic analyses give only qualitative information on the sample composition and do not allow the quantification of the nanotube content, purification (oxidation and acidic attack) of the most promising samples will be carried out to quantify the NT yield. However, rough estimations derived from the micrographs let assume that yield for CNTs is in the range of 10 to 50 % in as-produced soot with a production rate of 250 g/h.

Although, quantitative evaluations of the products can not be correlated to operating conditions yet, but due to its flexibility and controllability in terms of residence times and quenching rates, the plasma process shows a great potential for further development. First efforts trying to correlate operating conditions to obtained products, with the final aim of understanding CNT formation inside the plasma system, show very promising characteristics and seem to be an indication that this process is capable of producing any kind of nanostructures. Future efforts will need to focus on the product selectivity of the production process.

The plasma process is addressing products currently obtained by the so-called arc method and novel products. The other high temperature processes having reached their limits in production rate, the 3-phase AC plasma process can be considered as an improved highly flexible version of the same family with an enormous potential for further up-scaling to an industrial size at commercially viable cost.

ACKNOWLEDGEMENT

This specific research and technological development programme has been supported by the European Commission under the Competitive and Sustainable Growth Programme, contract PLASMACARB, G5RD-CT-1999-00173 and under the TMR Programme, contract NANOCOMP, HPRN-CT-2000-00037

REFERENCES

[1] M. S. Dresselhaus, "Down the Straight and Narrow," Nature, Vol. 358, pp. 195-196, (16. Jul. 1992).

[2] T. W. Ebbesen and P. M. Ajayan, "Large Scale Synthesis of Carbon Nanotubes." Nature Vol. 358, pp. 220-222 (1992).

[3] C. Journet, et al., Nature 388, 756 (1997).

[4] A. Thess, et al., Science 273, 483 (1996).

[5] D. Laplaze, et al., Carbon 1998, 36, 685.

[6] T. Guillard, et al., "Towards the large scale production of fullerenes and nanotubes by solar energy". Solar Forum 2001, Washington, DC, April 21-25, 2001.

[7] Patent DE 10312494.2 "Carbon nanostructures and process for the production of carbon-based nanotubes, nanofibers and nanostructures".

[8] T. M. Gruenberger, et al., "Production of Carbon Nanotubes and Other Nanostructures Via Continuous 3-Phase AC Plasma Processing", Fullerenes, Nanotubes and Carbon Nanostructures, in press.

[9] L. Fulcheri, et al., "Production of carbon nanostructures ranging from carbon black over fullerenes to nanotubes by thermal plasma", 16 th International Symposium on Plasma Chemistry (ISPC 16), Taormina (ITALY) (2003).

[10] V. Ivanov, et al., Chem. Phys. Lett 223, 329 (1994).

[11] A. Li, et al., Science 274, 1701 (1996).

[12] PCT/EP94/00321, AC Plasma Technology for Carbon Black and Carbon Nanoparticles, 1993.

[13] L. Fulcheri, N. Probst, G. Flamant, F. Fabry, E. Grivei, and X. Bourrat, Carbon 40, 169, 2002.

[14] PCT/EP98/03399, Fullerene Production in AC Plasma, 1997.

[15] L. Fulcheri, Y. Schwob, F. Fabry, G. Flamant, L. F. P. Chibante, and D. Laplaze, Carbon 38, 797, 2000.

[16] H. Okuno, et al., "Synthesis of carbon nanotubes and nano-necklaces by thermal plasma process", Carbon, in press.

Sol-Gel Prepared Single Wall Carbon Nanotube SnO$_2$ Thin Film for Micromachined Gas Sensor

Jianwei Gong, Quanfang Chen*

Department of Mechanical, Materials & Aerospace Engineering, University of Central Florida, Orlando, FL 32816-2450, USA, * qchen@mail.ucf.edu

ABSTRACT

We report here a novel micromachined single wall carbon nanotube (SWNT)/tin dioxide (SnO$_2$) nano film gas sensor. A polymeric sol-gel process in combination with purified SWNT has been successfully developed in fabricating SWNT/SnO$_2$ thin film nanocomposite sensor on an alumina substrate. The test of the novel nanocomposite gas sensors has been conducted for hydrogen detection, in comparison with sol gel derived SnO$_2$ thin film. The test results show that SWNT has greatly improved the nano-SnO$_2$ thin film's gas sensing property, in terms of lower working temperature, greater sensitivity (a factor of 3), faster response time and recovery time, and less drift of resistance. The improved capabilities are credited to the large surface to volume ratio of gas sensing thin film with nano passes created by SWNT, and the distance between adjacent passages being less than electron depletion layer thickness is defined by SWNT. The fabrication process is compatible with IC industry thus it is cost effective for batch production.

Keywords: MEMS, SMO, Nanocrystalline SnO$_2$, SWNT Sol Gel

1. INTRODUCTION

Semiconductor Metal Oxide (SMO) gas sensors have been studied for decades owning to their advantages of more robust, less sensitive to moisture and temperature, simple interface electronics, faster response time and recovery time [1]. SMO gas sensors based on pellets or thick film technology are normally not chosen for the fabrication of intelligent Microsystems. This is not only due to processing incompatibility but also due to the high power consumption. Therefore, SMO thin film gas sensor has been regarded as the next generation gas sensing elements. In recent years, nanotechnology has emerged for developing novel gas sensing materials with unusual properties. Nanotechnology has proven very promising in SMO gas-sensing materials. Since SMO's gas detection principle is based on variations of the depletion layer at the grain boundaries, when either reducing or oxidation gases are present. As the results, the adsorbed gas leads to a variation in the height of the energy barriers for free charge carriers (e.g. electrons in SnO$_2$) [2]. When particle size decreases, the ratio of the depleted volume to the whole bulk grain volume increase greatly (related to the surface to volume ratio which is reverse proportional to the particle size). It is recognized that SnO2 semiconductor thin film can have maximum gas sensitivity only if the nanocrystallite size within the film is comparable with its space-charge layer thickness (3nm for SnO$_2$, [3]). Currently most SnO2 films are fabricated with physical vapor deposition and sol gel process. The grain size is normally in a range of 10nm to 100nm, which is larger than the depletion thickness (3nm).

Carbon nanotubes (CNTs) are hollow nano pipes that could be used to provide nano passes in sensing materials. The CNTs are also proven gas sensors such as oxygen [4] and carbon dioxide [4], nitrogen oxide (NO$_2$) methane [5], and ammonia (NH$_3$)[6]. Single walled carbon nanotubes (SWNT) have limitations in detecting some gases such as CO [6], while SnO$_2$ does have good response to a wide range of chemical gases.

The objective of this research article is to take advantage of both SnO$_2$ and SWNT and to improve SnO$_2$ gas sensing property in terms of sensitivity, response time and recovery time by using SWNT reinforcement. The function of SWNT can be summarized into two-fold. The first one is using SWNT to achieve permanent nano passes for gas sensing after sintering SnO_2. The second one is using SWNT to define the distance between gas accessing boundaries to be less than two time of depletion zone of SnO2 (6nm).

MEMS (Micro-Electro-Mechanical-System) has been relative more mature than nano technology. By combing nanotechnologis with MEMS techniques, novel gas sensor has been developed and it is described in this article. This approach provides low cost minimized device that lower power consumption and compatible to IC fabrication will be realized.

2. EXPERIMENTAL

2.1 Gas Sensor Fabrication

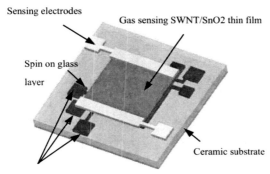

Fig. 1 Micromachined gas sensor developed

The MEMS based gas sensor cell is sketched in Figure 1 and the microfabrication process is shown in Figure 2. Aluminum wafer was used as substrate for better thermal and electrical isolation. Thin film Pt is used as micro heater and temperature sensor and it was fabricated by thermal deposition and lift off process. High dielectric spin-on glass was spin coated on as the electric isolation layer. SnO2/SWNT gas sensing thin film was fabricated by spin coating prepared Sol. The sensing electrode was deposited on top by thermal evaporation and patterned by liftoff after sintering the SnO_2 film.

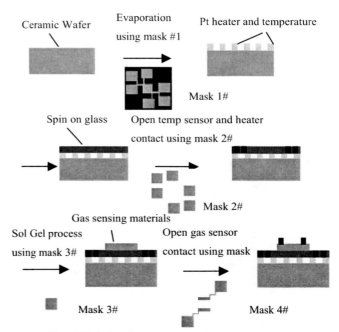

Fig. 2 Fabrication Process of gas sensor

2.2 Preparation of Sol

Porous nanocrystalline tin oxide layer was developed by polymeric sol gel technique. A precursor of tin isopropoxide ($Sn(O^iPr)_4$, 10%, Alfa Aesar) 25ml was firstly dissolved in 50ml isopropyl alcohol followed by magnetic stirring for half an hour. Then a complexing agent, acetylacetone (AcAc) 3ml was added to stabilize the hydrolysis of tin isopropoxide.

2.3 SnO_2 thin film by Sol Gel process

Nanostructured SnO_2 film was fabricated by sol gel spin coating followed by sintering. During spin coating, the rotating speed was gradually increased to speed up to 3,000rpm for 30 seconds. Then the gel film was dried in air for 10 min. The process was repeated for 3 times until desired thickness was achieved. After spin coating, the gel film was dried in an oven at 100°C for 30 min. Finally, sintering was done in a furnace with a heating rate (2°C/min) to 500°C and kept for 2 hours). Fig. 3 shows an SEM micrograph of nano SnO_2 film (grain size is about 15nm).

Acquired SWNT was mixed into a SnO_2 sol, after a specific purification process developed. After 24 hours of magnetic stirring and another 10 min of ultrasonic cell disrupting, the light dark transparent sol solution containing SWNT suspension is obtained. The SWNT's diameter is approximately 1.4 nm. The viscosity of the sol was adjusted with addition of PVA (poly vinyl alcohol). After sintering at 500°C within an argon environment for 2 hours, nano-sized SnO_2 containing uniformly distributed SWNT is indicated in Figure 4, in compared to pure SnO_2 particles (Figure 3).

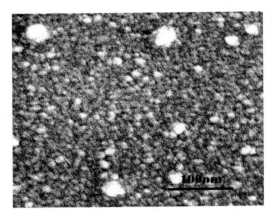

Fig.3 SnO_2 grains (15nm) after sol gel process

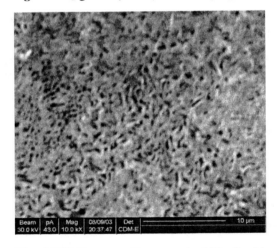

Fig. 4 SWNT wrapped by nano sized Tin dioxide

2.4 Gas Sensor Test System

The testing system includes H_2, N_2 and pure air gas cylinders with regulators (Airgas), 2cc/min mass flow controller (Omega), flow meter, gas chamber oscilloscope (TDS 224) and NI data acquisition system. The gas sensitivity was determined by:

$$Sensitivity = (R_1-R_0)/R_0 \times 100$$

Where R_1 is resistance values of gas sensing thin film in air, and R_0 represents resistance values in gas environment.

3. RESULT AND DISCUSSION

Both pure nano SnO_2 and SWNT enhanced nano SnO_2 sensors were tested in 1500ppm hydrogen under different temperatures for comparison. Typical tested results are presented in Figure 5 and the comparison is presented in Table 1. It is obvious that SWNT/SnO_2 composite sensor has much better gas-sensing properties than the pure SnO_2 sensor. The sensitivity is 3 times higher (Table 1) at various temperatures tested. It has a good sensing at low temperatures (150°C) at which the pure SnO_2 doesn't. The undoped SWNT/SnO_2 sensor has much better sensitivity than doped with catalysts [7].

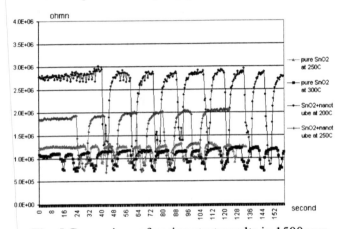

Fig. 5 Comparison of various test results in 1500ppm

Table 1: Comparison of Pure SnO_2 sensor and SWNT/SnO2 sensor at different working temperature

	Sensitivity	Response time	Recovery time
Pure SnO2 at 150°C	N/A	N/A	N/A
SnO2/SWNT at 150°C	~50%	5~7s	15~18s
Pure SnO2 at 200°C	~39%	5~7s	25~30s
SnO2/SWNT at 200°C	~115%	3~5s	3~5s
Pure SnO2 at 250°C	~40%	3~5s	6~8s
SnO2/SWNT at 250°C	~147%	2~3s	4~5s
Pure SnO2 at 300°C	~56%	3~5s	5~7s

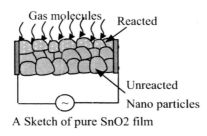

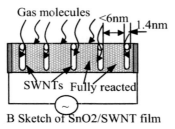

Fig. 6 Sketch of gas sensing mechanism between pure SnO2 film and SnO2/SWNT film

The enhanced sensitivity of nano SnO$_2$/SWNT thin film attributes to two aspects as sketched in Figure 6. First, SWNT provide permanent nano passes for gas sensing after SnO$_2$ sintering. As a result, the surface to volume ratio increases largely, which leads to better sensitivity. Secondly, SWNT with 1.4 nm diameters was used to define the distance between SnO$_2$ crystalline grains and thus confine gas-accessing boundaries to be less than 6nm. According to the model brought up by Xu et al. [8], depending on the nanocrystalline size (D) relative to its space charge layer thickness (L), the transducer function is operated by a mechanism of grain-boundary control (D>>2L), neck control (D>2L) or grain control (D≤2L), respectively. The gas sensitivity of nanocrystalline SnO$_2$ thin film is enhanced only when the film resistance is controlled by the latter two mechanisms, especially by the grain control mechanism. For the SnO$_2$ thin film, the space-charge layer thickness has been calculated to be ~3 nm at 250 °C [8]. As a result, the sensitivity increase steeply as *D* decreases to be comparable with or less than 2*L* (≈ 6 nm).

4. CONCLUSION

MEMS based sensor with nanocrystalline SnO$_2$/SWNT sensing films was developed using polymeric sol-gel process in combination with purified SWNT. Compared to pure SnO$_2$ type, this sensor produce improved sensitivity (by a factor of 3), faster response time and recovery time to 1500ppm hydrogen at even relatively low working temperature. Since this gas sensor is fabricated using micro fabrication process, the power consumption is limited to low level, MEMS based device also make low cost per device possible.

5. ACKNOWLEDGEMENTS

Financial support for this study is provided by UCF in house research funding.

The authors would like to thank UCF-Cirent Materials Characterizaton Facility at UCF for materials characterization.

REFERENCES

1. Won JaeMoon, Ji Haeng Yu and Gyeong Man Choi, 2002, Sensors and Actuators B 87 464-470
2. B.K.Miremadi, K.Colbow: Sensors and Actuators B46 (1998) 30-3
3. S. Seal, S. Shukla, Nanocrystalline SnO gas sensors in view of surface reactions and modifications, JOM 54 (9) (2002) 35-38, 60.
3. Simon, I. et al., 2001, Sensors and Actuators B 73 1-26
4. K.G. Ong, K. Zeng, C.A. Grimes, IEEE Sensor J. 2 (2002) 82-88.
5. Jing Kong, et al, 28 JANUARY 2000 VOL 287 SCIENCE
6. M. Bienfait, B. Asmussen, M. Johnson, P. Zeppenfeld, Surf. Sci. 1-3 (2000) 243-248.
7. Jianwei Gong, et al, IEEE Sensors 2003 Toronto, Canada, October 22-24, 2003.
8. C. Xu, J. Tamaki, N. Miura, and N. Yamazoe, Sensors and Actuators B 3, 147 (1991).

Gated Carbon Nanotube Field Emission Enhancement and Regeneration by Hydrogen

David S.Y. Hsu and Jonathan L. Shaw

US Naval Research Laboratory, Washington, DC, USA
dhsu@ccs.nrl.navy.mil; jon.shaw@nrl.navy.mil

ABSTRACT

We report large increases in field emission current when operating carbon nanotubes in substantial pressures of hydrogen, especially when the nanotubes were contaminated. We have previously demonstrated two different configurations of integrally gated carbon nanotube field emitter arrays (cNTFEAs), CNTs grown inside microfabricated gate apertures with and without silicon posts [1-4]. Salient features of these in-situ grown microgated cNTFEAs include the absence of electrical arcing, low operating voltage, and enhancing effect of some residual ambient gases. Operating both configurations of cNTFEAs without special pre-cleaning in greater than 10^{-5} Torr hydrogen produced orders of magnitude enhancement in emission. For a cNTFEA intentionally degraded by oxygen, the operation in hydrogen resulted in a 340-fold recovery in emission. The results suggested a dependence on atomic hydrogen produced from the interaction between emission electrons and molecular hydrogen. The observed emission enhancement could be due to removal of oxygen-containing surface species, a surface dipole formation, or hydrogen doping.

Keywords: carbon nanotubes, field emission, field emitter arrays, gate apertures, hydrogen

1 INTRODUCTION

Carbon nanotubes have become premier candidates for use as field emitters because of their large geometric field enhancement/low voltage operation, lack of electrical arcing (due to the lack of a surface oxide [5]), and robustness with certain ambient gases (due to the relative chemical inertness and high work function of carbon). These combined qualities overcome many of the shortcomings of conventional metal and silicon tip FEAs. Potential applications include flat panel displays, high frequency amplifiers, spacecraft electric propulsion systems, high voltage and high temperature electronics, miniature mass spectrometers and x-ray sources, multi-beam electron beam lithography, etc.

Most of the published nanotube field emission work involved a diode configuration in which the cNTs, either grown or placed as dense mats on substrates, were placed at a known separation (usually many tens of microns) from an anode. Although the nanotubes produce emission at very low electric fields, the operating voltages are too high for most applications (usually hundreds of volts). In order to reduce the gate voltage, we have grown multi-walled cNTs inside microfabricated gates. We have demonstrated two different configurations of gated cNT field emitters; one consists of cNTs grown on top of gated silicon posts [1] and the second cNTs grown inside open gated apertures [2]. Five patents have been awarded to the Navy on these cNTFEAs [3]. Turn on-voltages below 20 volts and current densities up to 1mA at 40 volts from a 33,000-cell array with 0.5 mm^2 area were measured [1,4]. In addition, we observed a high degree of robustness such as a lack of arcing, emission unaffected by xenon and high temperature, and enhancements by water vapor and hydrogen [4]. We previously observed a 60% increase in emission in 1.5 x 10^{-5} Torr hydrogen, in which case the cNTFEA had been carefully degassed and cleaned [4]. In the same experiment, about a 20% emission enhancement was observed at 1 x 10^{-6} Torr hydrogen. This is to be contrasted with the lack of any effect observed by Dean et. al. [6] at 1 x 10^{-6} Torr H_2 from their ungated single walled carbon nanotube emitter. Wadhawan et. al. [7] observed no effect due to 1 x 10^{-7} Torr hydrogen on their ungated nanotubes.

Studies by Dean et. al. [6] and Wadhawan et. al. [7] have shown that nanotube emission can be adversely and sensitively affected by oxygen contamination. Since the surface of as-grown nanotubes can be in various stages of contamination, including oxygen-containing groups, our motivation is to "clean up" the nanotube surface by operating the emitters in hydrogen. One expects the emission enhancement to depend on the degree of initial contamination. Our present investigation demonstrates that operation in hydrogen can indeed recover emission from even severely contaminated cNTFEAs, resulting in very large enhancement factors. These findings have implications in regard to emitter lifetime and cost-savings.

2 EXPERIMENTAL

CNTFEAs in both the cNT-on-Si post and the cNT-in-open aperture configurations were used in the present investigation. With the exception of some modifications to the former, the details of the fabrication were the same as those published in Refs. [1] and [2], respectively, for the 2 designs.

2.1 Modified Fabrication of cNT-on-Si Post

The structure and fabrication of the gated device were slightly different from those described in our earlier

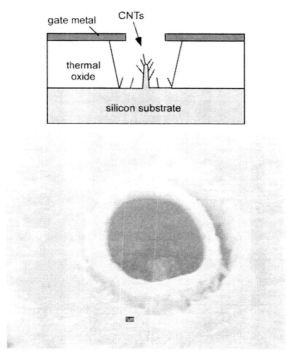

Fig.1. Gated carbon nanotube-on-silicon post field emitter cell at 30° tilt angle. Top: schematic drawing of cell; bottom: SEM of cell with nanotubes on a silicon post centered in a 2.5 micron diameter Pt gate aperture.

publication [1]. Isotropic etching reduced the height and the diameter of the silicon post to about 1 micron and 0.25 micron, respectively. The gate material was platinum instead of chromium. A thin layer of Ti was sputter deposited before sputter-deposition of the Ni catalyst (~200A). Instead of a HF dip to lift off catalyst from the oxide regions, glancing-angle sputtering at 15° from the substrate was used to remove the catalyst from the top surfaces of the substrate. All other growth parameters were the same as previously [1] (same hot-filament assisted cold wall CVD reactor, same temperatures and gas (ethylene and ammonia) flow rates). The resulting cell structure consisted of multi-walled nanotubes protruding from the top of the Si post in a generally random direction and is shown in the scanning electron micrograph in Fig. 1. Only a very small fraction of the cells contained nanotubes on the Si posts in this array of 3840 cells.

2.2 Fabrication of cNT-in-Open Aperture

The cNTFEA with the open aperture design was from the same sample as the one published in Ref. [2]. Open apertures were first reactive-ion-etched through chrome/silicon gates and silicon dioxide insulator on a silicon substrate. A sidewall silicon dioxide spacer was formed by conformal silicon dioxide layer deposition by CVD, followed by etch back. Fe catalyst was sputter-deposited onto the sample

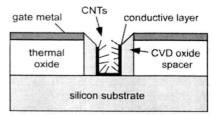

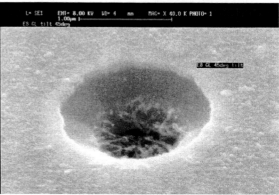

Fig.2. Gated carbon nanotube-in-open aperture emitter cell at 45° tilt angle. Top: schematic drawing of cell; bottom: SEM of cell with nanotubes grown on the sidewall oxide spacer (and cell bottom). The gate diameter and the oxide spacer thickness are 1.7 and 0.35 microns, respectively.

consisting a small arrays of 10 to 40 cells, followed by 15° glancing angle sputter-removal of the Fe from the top surface. Hot-filament assisted CVD described in Ref. [2] was used to grow the nanotubes inside the apertures, including on the vertical sidewall spacer. Figure 2 shows a scanning electron micrograph of such a cell.

2.3 Emission Measurements

Current-voltage emission characterization for both configurations of emitters was carried out in an UHV chamber (base pressure 10^{-10} Torr) equipped with a load lock, sample stage heater, and computerized data collection. Tungsten probes contacted the cathode (substrate) and the gate and the emission was collected on an anode probe biased at 200 V and placed about 1 mm from the sample. Hydrogen was admitted through a leak valve and dynamically pumped using an oil-free turbo-molecular pump. For anode current-time measurements (Fig. 5), a separate turbo-pumped chamber (base pressure 10^{-8} Torr) was used. The gate pads of arrays of the cNT-on-Si post configuration were contacted with gold wire bonding and an anode made of a Pt mesh at 200 V bias was placed at about 2 mm from the sample. Purified hydrogen from a Pd diffusion cell was used in all the experiments.

3. RESULTS AND DISCUSSION

3.1 CNT-on-Si Post Emitters

Fig. 3 shows the anode current vs. gate voltage characteristics obtained first under UHV and then at 10^{-4} Torr of pure hydrogen from a 3840-cell array of the cNT-on Si post design (corresponding to Fig.1). The array was operated in an ion pumped UHV chamber for many hours before the UHV data were taken. Exposure to hydrogen increased the emission current by orders of magnitude and reduced the apparent "turn-on" voltage by 30%.

A separate array with the same number of cells was run overnight in a turbo-pumped chamber under a continuous flow of 1×10^{-7} Torr oxygen at a constant gate voltage of 50 V until the emission degraded to about 44 nA. The effect of the addition of a continuous flow of hydrogen at 9×10^{-5} Torr is shown in the anode current-time plot in Fig.4. A sharp increase in emission is followed by a gradual increase until stabilizing at 15 μA after about 2.8 hr, with an overall recovery factor of 340.

These results suggest that operation in oxygen did not (significantly) consume the nanotubes through reaction with oxygen to form CO or CO_2 gas. Instead, the emission degradation was likely due to surface contamination with oxygen, which was removed by reaction with hydrogen (atoms). We have observed that exposure of the emitters to molecular hydrogen or oxygen when the arrays were not emitting had no effect on the emission produced once the gases are removed. The fact that the emission characteristics do no change when exposed to gases unless field emission is taking place suggests that the nanotubes are inert to the molecular forms of hydrogen and oxygen and that the atomic forms, which are created by electron dissociation, react with surface groups either in removal or attachment processes.

3.2 CNT-in-Open Aperture Emitters

As reported previously [2], this configuration has achieved the lowest gate current to anode current ratio (2.5%) of any nanotube emitters to date. The results from a 40-cell array taken under UHV conditions are reproduced in Fig. 5.

Figure 6 compares the emission anode current from an array of 20 cells obtained under hydrogen pressures of 1×10^{-8} and 1×10^{-4} Torr in the UHV chamber. A large emission increase at the higher pressure was observed (i.e. a factor of 10 at 45 volts). The saturation behavior at higher voltages could be due to faster hydrogen desorption at the higher currents.

We did not observe significant changes in the emission current for hydrogen pressures below 1×10^{-5} torr. The effect increased with pressure up to about 10^{-4} torr, and stayed the same at higher pressures. The emission began to decrease as soon as the hydrogen was removed but some effect remained for several hours after the hydrogen was removed.

The requirement for relatively high pressures ($\geq 10^{-6}$ Torr) of hydrogen again suggests that atomic hydrogen is responsible for the large enhancement and regeneration effects and that atomic hydrogen is created by electron impact from the operating emitters. The production rate of atomic hydrogen is apparently too low at lower pressures.

The effect of the atomic hydrogen may be any or all of the following mechanisms a) chemical removal of oxygen-containing surface species (which may act as p-type dopants and/or increase the work function), b) formation of a surface dipole (reducing the work function), and c) n-type doping by atomic hydrogen.

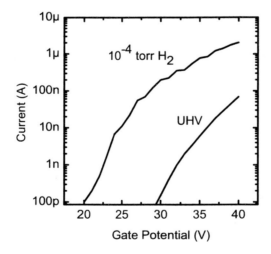

Fig. 3. Anode current- voltage characteristics of array of cNT-on Si post emitters obtained under UHV conditions and under 10^{-4} Torr hydrogen.

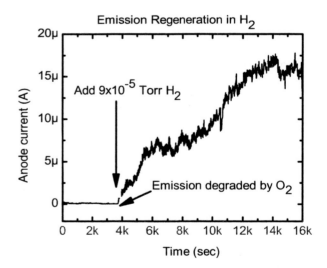

Fig.4. Anode current-time evolution showing regenerative effect of hydrogen on oxygen-degraded cNT-on-Si post emitters.

The results suggest that these beneficial hydrogen-nanotube interaction processes could also be accomplished and speeded up by exposing the emitters to an external source of hydrogen atoms. The inclusion of hydrogen at appropriate pressures (so not to affect electron mean free-path) in devices which use cNT emitters can enhance emitter lifetime and result in large cost-savings.

REFERENCES

[1] David S.Y. Hsu and Jonathan Shaw, Appl. Phys. Lett. **80**, 118 (2002).

[2] David S.Y. Hsu, Appl. Phys. Lett. **80**, 2988 (2002).

[3] D.S.Y. Hsu and H.F. Gray, US Pat. No. 6,333,598 (25 Dec 2001); D.S.Y. Hsu, US Pat. No. 6,440,763 (27 Aug 2002); US Pat. No. 6,448,701 (10 Sep 2002); US Pat. No. 6,568,979 (27 May 2003); US Pat. No. 6,590,322 (8 Jul 2003).

[4] D.S.Y. Hsu and J.L. Shaw, Cold Cathodes II, Electrochem. Soc. Proc. (Electrochem. Soc., Pennington, NJ, 2002), pp13-32.

[5] J. Shaw, J. Vac. Sci/ Technol B **18**, 1817 (2000).

[6] Kenneth A. Dean and Babu R. Chalamala, Appl. Phys. Lett **75**, 3017 (1999).

[7] A. Wadhawan, R.E. Stallcup II, K.F. Stephens II, and J.M. Perez, Appl. Phys. Lett.**79**, 1867 (2001).

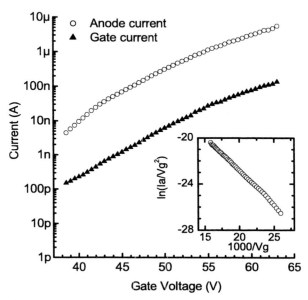

Fig. 5. Emission current-voltage characteristics from an array of 40 cells corresponding to Fig. 2. Inset shows a Fowler-Nordheim plot of the anode current, the linearity of which indicates well-behaved field emission.

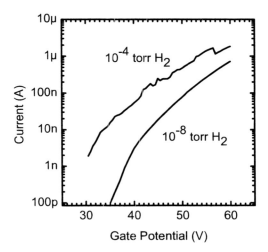

Fig. 6. Emission anode current - voltage characteristics from an array of 20 cells of the cNT-in-open aperture emitters obtained under 1 x 10^{-8} (lower) and 1 x 10^{-4} Torr (upper) hydrogen.

Noncovalent Engineering of Carbon Nanotube Surfaces

Jian Chen[*], Rajagopal Ramasubramaniam[**], Haiying Liu[***]

[*]Zyvex Corporation, 1321 North Plano Road, Richardson, Texas 75081, USA, jchen@zyvex.com
[**]Zyvex Corporation, 1321 North Plano Road, Richardson, Texas 75081, USA, rrajagop@zyvex.com
[***]Department of Chemistry, Michigan Technological University, Houghton, Michigan, USA

ABSTRACT

We report here that the rich functional chemistry of poly(aryleneethynylene)s (such as PPE) allows us to prepare various single-walled carbon nanotubes (SWNTs) with tunable functionalities and solubilities. In contrast to pristine SWNTs, soluble SWNTs can be homogeneously dispersed in commercial polymers such as polycarbonate, polystyrene etc. These composites showed dramatic improvements in the electrical conductivity and mechanical properties.

Keywords: carbon nanotubes, chemical functionalization, dissolution, composites

1 INTRODUCTION

Single-walled carbon nanotubes (SWNTs), due to their novel structural, thermal, electrical, mechanical and optical properties, are expected to find applications in many fields. In order to take advantage of the full potential of SWNTs, it's necessary to address the fundamental issues (cutting, solubilization, chemical functionalization, purification, manipulation, and assembly) in molecular engineering of carbon nanotubes.[1-14] Pristine SWNTs are generally insoluble in common solvents, and difficult to functionalize controllably. We recently reported a non-wrapping approach to noncovalent engineering of carbon nanotube surfaces by short, rigid functional conjugated polymers, poly(aryleneethynylene)s (PPE). This method enables the superior control of the relative placement of functionalities on the nanotube surface while still preserving nearly all of the nanotube's intrinsic properties.[15]

Short, rigid conjugated polymers, poly(aryleneethynylene)s (such as PPE),[16] are used to solubilize SWNTs. In contrast to previous work,[17-19] the rigid backbone of PPE cannot wrap around the SWNTs. The major interaction between polymer backbone and nanotube surface is most likely π-stacking (Figure 1).[20] This approach allows control over the distance between functional groups on the carbon nanotube surface, through variation of the polymer backbone and side chains. This approach represents the first example of carbon nanotube solubilization via π-stacking without polymer wrapping and enables the introduction of various neutral and ionic functional groups onto the carbon nanotube surface.

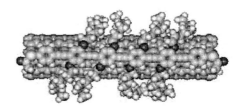

Figure 1. A molecular model of PPE-SWNT complex.[15]

2 RESULTS AND DISCUSSION

We report here that the rich functional chemistry of PPE allows us to prepare various functional SWNTs with tunable functionalities. The PPE-functionalized SWNTs can be solubilized in various organic solvents and water, and the solubilities are as high as 10 mg/ml. The soluble SWNTs with specific functionalities can be used as novel gelators that can gelate various organic solvents such as chloroform, 1-methyl-2-pyrrolidinone *etc.*[21,22]

Two common approaches have been used previously to disperse the SWNTs in a host polymer: 1) Dispersing the SWNTs in a polymer solution by lengthy sonication (up to 48 h).[23] The lengthy sonication, however, can damage/cut the SWNTs, which is undesirable for many applications. 2) *In situ* polymerization in the presence of SWNTs.[24,25] The efficiency of this approach, however, is highly dependent on the specific polymer. For example, it works better for polyimide[24] than polystyrene[25].

Our noncovalent chemistry can provide nanotube solubilization in organic solvents that allows homogeneous dispersion of nanotubes in the host polymer matrix. This approach avoids the lengthy sonication and is applicable to many organic soluble polymers such as polycarbonate, polystyrene and poly(methyl methacrylate) etc.[26]

Homogeneous nanotube polymer composites can be fabricated using noncovalently functionalized, soluble SWNTs (Figure 2.), and these composites show dramatic improvements in electrical conductivity with low percolation threshold (0.05-0.1 wt% of SWNT loading).[26] By significantly improving the dispersion of SWNTs in commercial polymers, we show that only a very small amount of SWNTs are needed to achieve conductivity levels required for different electrical applications without compromising the host polymer's other desired physical properties and processability (Figure 3 and 4). The electrically conductive CNT polymer composites will find various applications such as electrostatic dissipation, EMI shielding, printable circuit wiring, and transparent conductive coatings.

The SEM images of the surface and cross-section of PPE-SWNTs (5 wt% of SWNTs)/polystyrene composite film show the excellent dispersion of PPE-functionalized SWNTs in host polymer matrix. SWNTs are randomly distributed not only along the surface (Figure. 2a), but also through the cross section (Figure. 2b and 2c), indicating the formation of an isotropic, three-dimensional nanotube network in host polymer matrix. This is essential for obtaining composites with isotropic electrical conductivity.

Figure 3a shows the measured volume conductivity (using the standard four-point probe method) of PPE-SWNTs/Polystyrene composites as a function of the SWNT loading. The conductivity of pure polystyrene is about 10^{-14} S/m.

Figure 3b indicates an extremely low percolation threshold at *0.045 wt% of SWNT loading*. The very low percolation threshold is a signature of excellent dispersion of high aspect ratio soluble SWNTs.

Apart from the very low percolation threshold, we also observe that the conductivity reaches *6.89 S/m* at 7 wt% of SWNT loading, which is 14 orders of magnitude higher than that (10^{-14} S/m) of pure polystyrene, and 5 orders of magnitude higher than that (*1.34×10^{-5} S/m*) of SWNTs (8.5 wt %)/polystyrene composite that was prepared by *in situ* polymerization.[25]

Figure 4a shows the measured volume conductivity of PPE-SWNTs/Polycarbonate composites as a function of the SWNT loading. The conductivity of pure polycarbonate is about 10^{-13} S/m. We find that the conductivity of PPE-SWNTs/polycarbonate is generally higher that that of PPE-SWNTs/polystyrene at the same SWNT loading. As shown in Figure 4b, we also observe very low percolation threshold at *0.11 wt% of SWNT loading*.

The resulting polymer composites also show significant enhancement in mechanical strength. The mechanical measurement showed that 2 wt % of soluble SWNT filling resulted in a 79% increase in the tensile strength of polycarbonate.[22] We also observed a stress-induced SWNT alignment in polycarbonate at room temperature, which is impossible to achieve with insoluble SWNTs. The SEM study of the fracture surface indicates excellent interfacial interaction/load transfer between SWNTs and polycarbonate, possibly due to the considerably increased

roughness of carbon nanotube surfaces by PPE non-covalent functionalization.

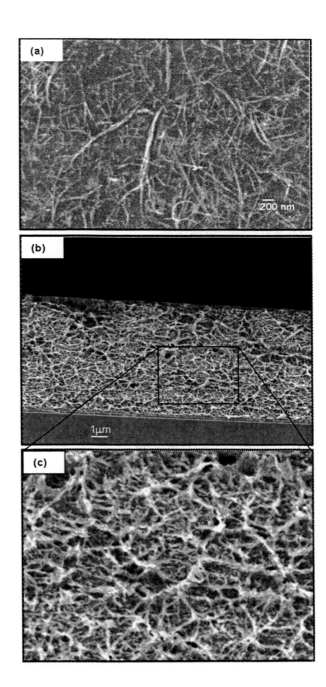

Figure 2. (a) Surface and (b) cross sectional SEM images of PPE-SWNTs (5 wt% of SWNTs)/polystyrene composite film. (c) The zoom-in view of (b).[26]

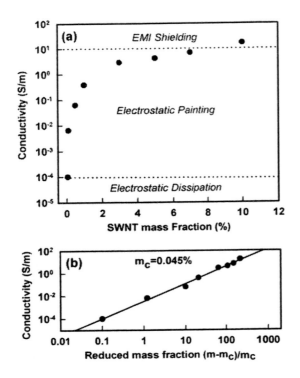

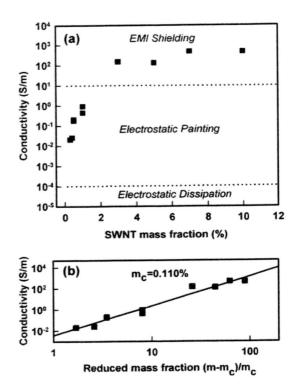

Figure 3. (a) Room temperature electrical conductivity of PPE-SWNTs/polystyrene composite versus the SWNT weight loading. Dashed lines represent approximate conductivity lower bound required for several electrical applications. (b) Room temperature conductivity of PPE-SWNTs/polystyrene composite as a function of reduced mass fraction of SWNTs.[26]

Figure 4. (a) Room temperature electrical conductivity of PPE-SWNTs/polycarbonate composite versus the SWNT weight loading. Dashed lines represent approximate conductivity lower bound required for several electrical applications. (b) Room temperature conductivity of PPE-SWNTs/polycarbonate composite as a function of reduced mass fraction of SWNTs.[26]

REFERENCES

(1) Liu, J. *et al. Science* **1998**, *280*, 1253-1255.
(2) Chen, J. *et al. J. Am. Chem. Soc.* **2001**, *123*, 6201-6202.
(3) Chen, J. *et al. Science* **1998**, *282*, 95-98.
(4) Chen, J. *et al. J. Phys. Chem. B* **2001**, *105*, 2525-2528.
(5) Sun, Y. –P. *et al. Chem. Mater.* **2001**, *13*, 2864-2869.
(6) Bahr, J. L. *et al. Chem. Commun.* **2001**, 193-194.
(7) Ausman, K. D. *et al. J. Phys. Chem. B* **2000**, *104*, 8911-8915.
(8) Wong, S. S. *et al. J. Am. Chem. Soc.* **1998**, *120*, 8557-8558.
(9) Bahr, J. L. *et al. J. Am. Chem. Soc.* **2001**, *123*, 6536-6542.

(10) Holzinger, M. *et al. Angew. Chem. Int. Ed.* **2001**, *40*, 4002-4005.

(11) Georgakilas, V. *et al. J. Am. Chem. Soc.* **2002**, *124*, 760-761.

(12) Zhao, B. *et al. J. Am. Chem. Soc.* **2001**, *123*, 11673-11677.

(13) Chen, J. *et al. J. Am. Chem. Soc.* **2002**, *124*, 758-759.

(14) Diehl, M. R. *et al. Angew. Chem. Int. Ed.* **2002**, *41*, 353-356.

(15) Chen, J. *et al. J. Am. Chem. Soc.* **2002**, *124*, 9034-9035.

(16) Bunz, U. H. F. *Chem. Rev.* **2000**, *100*, 1605-1644.

(17) Dalton, A. B. *et al. J. Phys. Chem. B* **2000**, *104*, 10012-10016.

(18) Star, A. *et al. Angew. Chem., Int. Ed.* **2001**, *40*, 1721-1725.

(19) O'Connell, M. J. *et al. Chem. Phys. Lett.* **2001**, *342*, 265-271.

(20) Chen, R. J. *et al. J. Am. Chem. Soc.* **2001**, *123*, 3838-3839.

(21) Chen, J., Ramasubramaniama, R., Liu, H. presentation at the *Nanotech 2004*, Boston, 2004.

(22) Chen, J. *et al.* unpublished results.

(23) Biercuk, M. J. *et al. Appl. Phys. Lett.* **2002**, *80*, 2767.

(24) Park, C. *et al. Chem. Phys. Lett.* **2002**, *364*, 303.

(25) Barraza, H. J. *et al. Nano Lett.* **2002**, *2*, 797.

(26) Chen, J., Ramasubramaniama, R., Liu, H. *Appl. Phys. Lett.* **2003**, *83*, 2928.

Carbon Nanotube-based Sensor Devices for IC Performance Evaluation

R. Glenn Wright*, Larry V. Kirkland**, Marek Zgol*, David Adebimpe*, Robert Mulligan*

* GMA Industries, Inc., 20 Ridgely Avenue, Suite 301,
Annapolis, MD, USA, {glenn,marek,david,robert}@gmai.com
** US Air Force, OO-ALC, 7278 4th Street, Bldg. 100,
Hill AFB, Utah, USA, larry.kirkland@hill.af.mil

ABSTRACT

A novel approach to the problem of identifying failed integrated circuits on electronic circuit boards has been sought that would allow for drastically reducing the costs of the test equipment development and maintenance. This article describes our ongoing efforts toward developing the original concept of Molecular Test Equipment®. Our focus is on our progress in creating carbon nanotube-based sensors carrying out the integrated circuit performance monitoring functions in an integrated circuit substrate level, as well as on the problem of interfacing the molecular devices with the existing metallic circuitry of the integrated circuits. The two main areas of interest explored recently by our team in order to create the basis for such sensor fabrication were: (i) producing solutions of homogenously dispersed carbon nanotubes in organic solvents, and (ii) developing a method for controlling the process of nanotube deposition at specific location on the silicon substrate.

Keywords: carbon nanotube electronics, nanoscale sensors, IC failure detection

1 INTRODUCTION

Integrated circuit (IC) density is continuing to increase through advances being made in semiconductor physics, IC development tools, and other technologies that support the conception of smaller and more efficient devices. In addition, advances in electron-beam and other lithography techniques enable these increasingly smaller devices to be created on the substrate of an IC. Today's best and most precise lithography techniques can create devices with dimensions on the sub-micron scale. For example, in complementary metal-oxide-semiconductor (CMOS) devices copper/silicon dioxide (Cu/SiO2) interconnect structures exhibit a line width of 0.9 um (900 nm), with line spacing of 0.5 um (500 nm). The ratio of line width to line spacing in this case is approximately two to one.[1] Dimensions for aluminum/silicon dioxide (Al/SiO2) interconnect structures exceed those of copper. Hu and Harper have documented the superior performance of Cu over Al through improved conductivity and reduced electromigration properties.[2] Based in part upon these trends, we anticipate that copper interconnects are likely to replace aluminum in future IC designs. Utilizing the newest developments in lithographic techniques, dimensions are expected to continue to decrease to the 100 nm range (close to the predicted limitation of conventional IC manufacturing capabilities) in the 2005 time frame.[3]

With nanotube device dimensions ranging as small as five to 30 nanometers, we are attempting to take advantage of the "empty space" that exists between and among conventional IC designs to provide internal IC testing capability without interfering with existing chip functionality.

Our approach involves the use of nanotube-based sensors as part of an overall Molecular Test Equipment (MTE) architecture to monitor the internal chemical composition of ICs and to measure electrical signal activity to determine their operational status. Traditional test methodologies have typically employed a "black box" approach whereby the IC is poked and prodded from the outside (stimulus is applied) to see its reaction (response signals are analyzed). If the reaction is generally what we expected to see, we then assume the IC is operational. This approach only approximates a complete test, and cannot tell whether problems internal to the IC exist that may be indicative of imminent failure or degraded performance now or in the future. The present process we have described requires a large investment in test equipment, interface devices, and software.

Figure 1: MTE-augmented Integrated Circuit

The presence of nanotube-based sensors within the IC provides us with access to structures where we can measure chemical and electrical signal activity from the inside – where the actual circuit operations are taking place. This gives us the added advantage of measuring test points directly where failures are most likely to exist, and requires no investment in expensive external hardware and software to determine the outcome of testing. For manufacturing purposes, we anticipate the carbon nanotube-based sensors will be formed in a catalytically- and electrically-controlled self-assembly process; therefore, the cost for incorporating the MTE components within the ICs themselves is expected to be negligible comparing to the overall cost of the IC chip production. It is our intention that the cost increase will range from fractions of a penny to a few cents per chip, depending on quantity.

Sensor signals are communicated to a surface-mounted display device on the IC through the use of a combination of conventional and molecular wires, as well as conductive nanotubes in highly "crowded" locations. We describe our efforts for growing nanotubes using electrode material consisting of existing IC signal paths as a catalyst and address the specification of the conditions at which the properties of the created nanotubes meet the test requirements. These properties include the length of the nanotubes as well as their electrical properties (i.e., semiconducting or metallic).

2 MTE ARCHITECTURE

The nanotube-based chemical and electrical sensors comprise a fundamental portion of the overall MTE architecture. Additional components of this architecture include a surface-mounted failure indicator and one or more communication conduits through which sensor signals are sent to the failure indicator.

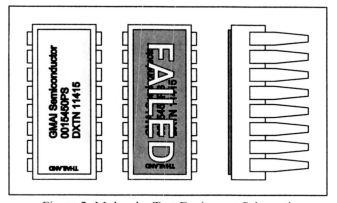

Figure 2: Molecular Test Equipment Schematic

A carbon nanotube, with an appropriate chiral structure allowing it to act as a metal wire, will conduct an electrical signal from the nanotube-based sensor to the surface mounted failure indicator. The input to the nanotube is a signal from the sensor. We are currently attempting to perform a comprehensive study to determine the range of voltages that a single nanotube may conduct. We are also considering the use of conventional micro-wires as a supplement to carbon nanotubes as a communication conduit.

The failure indicator provides a visual indication to a technician as to whether the IC is good or faulty. This indicator is designed for mounting upon the surface of the chip to ensure adequate visibility. A key requirement of this indicator is that there be minimal impact to the form, fit, and function of the chip. For example, we anticipate low power requirements with a small footprint on the surface of the chip with little increase in chip height. Fault indicator fabrication is being accomplished to produce a simple TN-LCD device with the undemanding requirement that it has a low driving voltage. We anticipate a maximum driving voltage limit of 2.5 V, and a current requirement in the microampere regime for a 0.5 cm x 0.5 cm LCD device.

Translation of these architectural features into practice where ICs that can fully test themselves and notify a technician of a failure through a visual indicator is pictured in the Figure 1 showing a photograph of our MTE demonstration model. In the event of failure, an electrical signal is transmitted to the fault indicator. The indicator is expected to be extremely thin and not require any significant increase in IC height what is illustrated schematically in Figure 2.

3 RESEARCH PROGRESS

Figure 3: CNT-based metal migration sensor

One of the first types of sensors being targeted for the demonstration of our failure analysis system are metal migration sensors intended to detect IC failures coming from electromigration, hillock growth on metal conductors or electronic components, as well as the melting of some low melting-point metals. Carbon nanotubes have shown many specific properties, one of them is the high conductivity. Based upon their small size, one-dimensional molecular wires can be fabricated. When this carbon nanotube wire is placed on the empty area between signal paths within ICs, if electromigration of signal path materials or melting of signal paths occurs, a change in conductivity of the nanotube will occur and the event can be detected. This configuration is illustrated in Figure 3.

Essential to using carbon nanotubes as functional elements in electronic test equipment their processability allowing for achieving our desired configuration, and also their ability to perform electrical tests on the engendered nanoscale structures. We currently are assessing:

- techniques that will result in the assembly of molecular wires and carbon nanotubes into desired architectures on desired substrates, in readiness for their connection to metal electrodes,
- techniques that will result in the creation of robust contacts and connections between the molecular wires, carbon nanotubes, and the metal electrodes that they eventually will be connected to, and
- the possibility of using a combination of scanning tunneling microscopy (STM) and conductive probe – atomic force microscopy (CP-AFM) to evaluate the nature of the nanostructured electronic ensembles required for the development of our nanosensor-pin testing paradigm.

Our efforts to date have primarily targeted the areas related to the development of suitable solvent systems able to produce homogenously dispersed suspensions of carbon nanotubes and the issue of controlled deposition of carbon nanotubes on desired substrates. Our findings related to this matter are presented in the paragraphs that follow.

3.1 Homogenous Dispersion Of Carbon Nanotubes

Well-dispersed carbon nanotube (CNT) solutions are a basis for not only the correctly performed deposition process but for a whole spectrum of procedures, including carbon nanotube purification, CNT functionalization, creating CNT-based composites, and many others. Developing a methodology for untangling the nanotube-ropes and dispersing CTNs into a uniform solution would create a major impact in the entire field of nanotechnology. We tested our original idea of surfactant-assisted dispersion of CNTs in organic and inorganic solvents. Although the procedure we used still requires optimization, we managed to prepare nanotube solutions with dispersion rates far better than those obtained in the commonly used processes as shown in Figure 4.

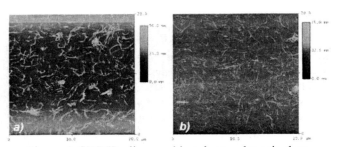

Figure 4: SWNTs dispersed in toluene, deposited on Si/SiO2/APTS wafer.

By using surfactant-based solvent systems, we have successfully untied a great portion of the single walled carbon nanotube (SWNT) ropes and evenly distributed them within the solution. We have also presented a method of controlling the deposition of SWNTs on the silicon substrate by tuning the concentration of the SWNT solution. Non-polar solvents, specifically toluene and N-Methyl-2-pyrrolidinone (NMP), presented the best results out of the group tested. At this point it is unclear what was the exact mechanism of the improved dispersion caused by the surfactant addition, however, it is expected that it was one of the following two processes: (1) the long aliphatic chain of the surfactant was wrapping around the nanotubes "pulling" them out of the bundle, or (2) the interactions between the aligned surfactant's aromatic rings and the nanotubes form structures, where the aliphatic chains are left sticking out preventing the nanotubes from re-tangling together.

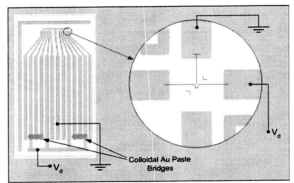

Figure 5: SWNT deposition on a test bed.
Deposition potential V_d = 10V peak-peak AC (5MHz)

3.2 Nanotubes Deposition and CNT-wire Contacts Formation

Collaborative activities between GMA Industries, Inc. and NASA Langley Research Center have also resulted in well-developed competences to deposit, align, and subsequently connect carbon nanotubes with metal and metallized electrodes, through a process known as AC electrophoresis [4],[5].

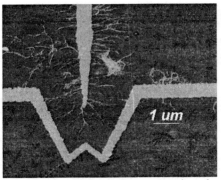

Figure 6: AFM image of the carbon nanotubes deposited at the central area of our W-shaped gold test bed circuit on Si/SiO2 wafer.

In our experimental method, nanotubes homogeneously dispersed in our developed solvent system were deposited onto a test bed fabricated for this purpose. One electrode of the bed was thereafter grounded and a potential was applied to the other electrode to create an electric field between the electrodes (as depicted in Figure 5). Upon the application of a potential, robust connection between the nanotubes and the metal electrodes was realized. AFM micrographs revealed that these nanotubes were well aligned and in concert with applied electric field lines (Figure 6). Measurements showed that the resistance through the test areas was 35 kΩ, which proved the robust connection formation; these connections maintained their rigidity even after being blown with nitrogen. It is clearly evident that the localization of the nanotubes can be controlled to great extent using the electric field alignment method. This concept will assist efforts to fabricate nanotube-based circuitry and specifically the MTE sensors including the metal migration sensors.

4 CONCLUSIONS

Annual non-recurring costs for the development of automatic test equipment and test program sets and the recurring costs for their operation and maintenance runs into the billions of dollars. Our approach has a potential to significantly reduce costs associated with maintenance of circuit boards by embedding MTE embedded within ICs to enable them to continuously test themselves during normal operation, and to provide a visual indication that they have failed. This approach brings failure detection down to the atomic level, where we will be able to recognize changes in the structure of the material itself.

Future work will build upon this success through the development of prototype demonstration technology that will form the basis for use in IC production. It is likely that future nanoscale ICs will be extremely fast, small and energy efficient, and low cost devices that can be embedded within new IC designs as they evolve. Further, this technology can be adapted to provide a wide range of additional information regarding device operation, such as hours of operation, device temperature, etc. Our MTE concept is anticipated to be capable of being retrofitted into existing IC designs with minimal impact on IC functionality or change to current fabrication methods. As IC technology continues to shrink into the nanometer realm, MTE will be fully capable of being integrated directly into these new IC designs.

ACKNOWLEDGMENTS

We gratefully acknowledge the participation of Buzz Wincheski and the Nondestructive Evaluation (NDE) Research Group at NASA LaRC in the experimental portion of the research presented in this paper.

REFERENCES

[1] http://www.zdefects.com
[2] http://www.radio-electronics.com
[3] Advanced Electronic Packaging: With Emphasis On Multi-Chip Modules, Editor: W. D. Brown, Chapter 12, Testing and Qualification, S. Kolluru and D. Berleant, Wiley-IEEE Press, 2001
[4] K. Yamamoto, S. Akita, and Y. Nakayama, J. Phys. D: Appl. Phys., 31 (8), L34 -36 (1998).
[5] P. A. Williams, B. Wincheski, J. Smits, D. Adebimpe and M. Zgol, J. Mater. Res (in press)

Modeling and Simulation of Multi-walled Carbon Nanotubes using Molecular Dynamics Simulation

C. H. Wong[*,***], K. M. Liew[*,***], X. Q. He[*], M. J. Tan[**] and S. A. Meguid[***]

[*]Nanyang Centre for Supercomputing and Visualisation, Nanyang Technological University, Nanyang Avenue, Singapore 639798
[**]School of Mechanical and Production Engineering, Nanyang Technological University, Nanyang Avenue, Singapore 639798, mkmliew@ntu.edu.sg
[**]Engineering Mechanics and Design Laboratory, Department of Mechanical and Industrial Engineering, University of Toronto, 5 King's College Road, Toronto, Ontario, M5S 3G8, Canada

ABSTRACT

Molecular dynamics simulation is performed on the buckling behavior of single and multi-walled carbon nanotubes under axial compression. Brenner's 'second generation' empirical potential is used to describe the many-body short range interatomic interactions for single-walled carbon nanotubes, while the Lennard Jones 12-6 model for van der Waals potential is added for multi-walled carbon nanotubes to describe the interlayer interactions. Results indicate that there exists an optimum diameter for single-walled nanotubes at which the buckling load P_{cr} reaches its maximum value. The buckling load P_{cr} for single-walled nanotube increases rapidly with the increase of the diameter d up to the optimum diameter. However, any further increase in the diameter d after the optimum diameter will result in a slow decline in buckling load P_{cr} until a steady value is reached. The buckling behavior of multi-walled nanotubes is also presented The effects of layers on the buckling load of multi-walled nanotubes are examined.

Keywords: carbon nanotube, molecular dynamics, buckling, nanomechanics, simulation

1 INTRODUCTION

The discovery of carbon nanotubes in the early 1990s by Iijima [1] has sparked a revolution in chemical physics and materials science in recent years. Since then, much research has been done on these new forms of carbon due to its exceptional mechanical properties. Among these research, computer simulations using empirical pair potentials, such as Brenner's 'second generation' empirical potential [2], are effective methods for the analysis of structural and mechanical properties carbon nanotubes.

In this paper, classical molecular dynamic simulations employing Brenner's 'second-generation' empirical potential function have been carried out for the buckling behavior of perfectly structured single and multi-walled carbon nanotubes under compressive deformation. Spontaneous plastic collapse of the nanotubes are described, which is in qualitative agreement with the experimental observations of Lourie et al. [3] and the simulated observations of Srivastava et al. [4] using the quantum generalized tight binding molecular scheme. From these buckling analyses, which are detailed in the following sections, the mechanical properties such as the buckling loads of the nanotubes can be determined.

2 NUMERICAL SIMULATION AND DISCUSSION

Using the Brenner's 'second generation' reactive empirical many-body bond order potential energy expression [2], which is computed as

$$E_{POT} \sum_{i} \sum_{j i} V^{R} r_{ij} \quad \bar{b}_{ij} V^{A} r_{ij} \,, \tag{1}$$

the buckling behavior of single-walled and multi-walled carbon nanotubes are simulated by solving the equations of motions using the Gear's predictor-corrector algorithm.[22] The distance between atoms i and j are denoted by r_{ij}, the pair additive repulsive and attractive interactions are represented by V^{R} and V^{A} respectively, and the reactive bond order is expressed by $\bar{b}_{ij}$. The axial compression of perfectly structured single and multi-walled carbon nanotube is achieved by applying a rate of 20 m/s and 10 m/s respectively at both ends. At the same time, the atoms at both ends of the nanotube are kept transparent to the interatomic forces. The end atoms are then moved inwardly along the axis by small steps, followed by a conjugate gradient minimization method while keeping the end atoms fixed.

2.1 Single-walled Carbon Nanotubes

To characterize the buckling behavior of single-walled carbon nanotubes, a (8,0) single-walled carbon nanotube is used in the simulation. The nanotube has a length l 43Å and diameter d 6.3Å. Each time step used in this simulation is equivalent to 1 fs and the simulations were allowed to run for 20,000 time steps.

Figure 1 shows the plot of strain energy per atom, which is determined as the difference in total energy per atom of the strained and unstrained carbon nanotube, against strain, which is defined as the ratio of elongation/original length for the (8,0) carbon nanotubes after it is compressed axially using the Brenner's 'second generation' empirical potential function [2].

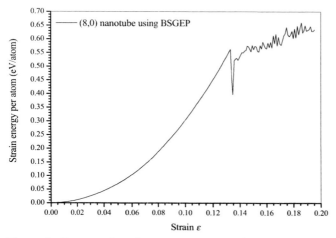

Figure 1: Computed strain energy per atom for (8,0) single-walled carbon nanotubes using Brenner's 'second generation' empirical potential (BSGEP) [2].

The (8,0) single-walled carbon nanotube undergoes elastic deformation before collapsing catastrophically at a certain critical strain of ε 0.13, resulting in a 30% spontaneous drop of strain energy per atom to 0.4 eV/atom. The nature of the spontaneous plastic collapse of single-walled nanotubes reported in this paper is in qualitative agreement with the experimental observation of Lourie et al. [3] and the simulated observation of Srivastava et al. [4] using the GTBMD scheme. The calculated critical stress σ_{cr} 149 GPa for the (8,0) single-walled carbon nanotube in this work is also close to the simulation work of Srivastava et al. [4] who obtained σ_{cr} 153 GPa. The (8,0) carbon nanotube is subjected to acute morphological changes, hence, higher strains are found in the (8,0) carbon nanotube, particularly around the kinks.

A length/diameter (l/d) ratio of 7.7:1 is used to illustrate the relationship between the diameters d and the buckling loads P_{cr} of selected zig-zag single-walled carbon nanotubes. From Fig. 2, it shows that as the diameter d of the zig-zag nanotube increases, there is a rapid increase in its buckling load P_{cr}. However, as the diameter d reaches an optimum value d 11.76 Å, any further increment beyond that will cause the buckling load P_{cr} to decrease slowly to a steady value of about 1.10×10^{-7} N. It is evident that the length l of the carbon nanotubes will affect the critical strain ε_{cr}.

It is worthy to note that the classical shell theory can be used for local buckling analysis of nanotubes. For a layer of

cylindrical shell with length l, radius r, thickness t, the Young's modulus E, the Poisson's ratio v, and m and $2n$ longitudinal and circumferential wave numbers, the critical stress for the buckling of the cylindrical shell is obtained as[23]

$$\sigma_{cr} \quad \frac{Et^3}{12\,1\;v^2}\quad \frac{n^2\quad\left(\dfrac{m^2\pi^2 r^2}{l^2}\right)^2}{r^2\quad\dfrac{m^2\pi^2 r^2}{l^2}}\quad \frac{E\;\dfrac{m^2\pi^2 r^2}{l^2}}{n^2\quad\left(\dfrac{m^2\pi^2 r^2}{l^2}\right)^2} \quad (2)$$

In predicting the buckling load of nanotubes by using cylindrical shell model, almost all previous literatures adopted the interlayer separation of graphite, i.e. $t = 3.4$ Å as the representative thickness of single-walled nanotubes. However, the buckling loads obtained from Eq. (2) are larger than our molecular dynamics simulated results if $t = 3.4$ Å is used. Here, we take the diameter of carbon atom (1.54 Å) as the thickness of the nanotube. The Young's modulus $E = 1.28$ TPa is directly extracted from the experiment result by Wong et al. [5] for the calculation of buckling loads using Eq. (2). As the buckling load is not sensitive to the value of Poisson's ratio, it is taken as $v = 0.25$. Using the cylindrical shell formula in Eq. (2) and molecular dynamics, buckling loads are computed for various zig-zig single-walled nanotubes, as shown in Fig. 2.

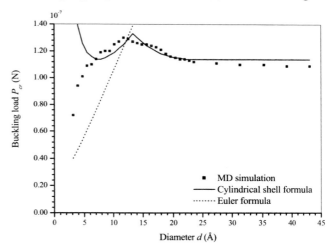

Figure 2: Comparison between cylindrical shell formula and molecular dynamics simulation for buckling loads P_{cr} of various zig-zig single-walled nanotubes with different diameters

It is observed from Fig. 2 that the two sets of results are in good agreement with diameter d ranging from 5.49 Å to 43.1 Å or larger. In spite of this, the results obtained from the cylindrical shell formula [Eq. (2)] for diameters d smaller than 4.7 Å do not agree well with that of the molecular dynamics simulation. This is because the nanotube behaves more like a rod than a cylindrical shell with thickness/radius (t/r) ratio larger than 0.66. In view of this, the buckling loads of zig-zag nanotubes with fixed ends are also calculated using Euler's formula [6] and are presented in Fig. 2,

$$P_{cr} \quad \frac{4\pi^2 EI}{l^2}, \tag{3}$$

where I is the moment of inertia. For the fixed length/diameter (l/d) ratio, the buckling loads obtained from Euler's formula for nanotubes with fixed ends are proportional to the diameter d of nanotubes, as shown in Fig. 2, and is obvious from the trend that the Euler's formula is more reasonable than the cylindrical shell model in the estimation of the buckling loads of nanotubes with (t/r) ratio larger than 0.66.

2.2 Multi-walled Carbon Nanotubes

Comparisons are also made between various multi-walled carbon nanotubes to determine the effect of the number of layers on the properties of the multi-walled nanotubes. Three configurations of multi-walled nanotubes are considered in the simulation: the first is a two-walled (5,5) and (10,10) nanotube; the second is a three-walled (5,5), (10,10), and (15,15) nanotube; and the last is a four-walled (5,5), (10,10), (15,15), and (20,20) nanotube. The lengths of all three multi-walled nanotubes are $l \quad 60\,\text{Å}$. And their diameters d are 13.52 Å, 20.36 Å, and 27.15 Å for the two, three and four layers multi-walled nanotubes respectively. In these simulations, the long-range van der Waals potential is added into the short-range covalent potential for the interlayer interaction using Lennard-Jones 12-6 potential [7]

$$V_{ij} \quad r_{ij} \quad 4\frac{\xi}{\sigma} \quad \frac{\sigma}{r_{ij}}^{12} \quad \frac{\sigma}{r_{ij}}^{6}. \tag{4}$$

where coefficients of well-depth energy ξ and the equilibrium distance σ are $4.2038 \quad 10^{-3}$ eV and 3.4 Å, respectively [8].

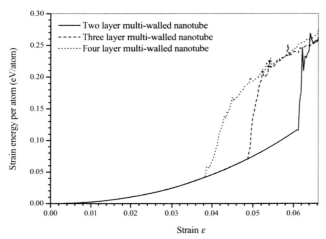

Figure 3: Strain energy per atom for two-walled, three-walled, and four-walled carbon nanotubes.

The van der Waals potential is nonzero only after the covalent potential is zero, such that there is no artificial reaction barrier formed by the steep repulsive wall of the Lennard-Jones 12-6 potential to prevent non-bonded atoms from chemical reaction.

The three multi-walled nanotubes are compressed axially using the same method as the single-walled nanotubes. Each time step is equivalent to 1 fs and there are a total of 20,000 time steps, which corresponds approximately to $\varepsilon \quad 0.066$.

Figure 3 depicts that the two-walled carbon nanotube manages to keep its elasticity for the longest with the largest critical strain at $\varepsilon \quad 0.06$, while the four-walled carbon nanotube has the lowest critical strain at $\varepsilon \quad 0.038$. Instead of a spontaneous decrease of strain energy upon buckling as evidenced in single-walled carbon nanotubes, the strain energy per atom increases until all the layers in the multi-walled carbon nanotubes are fully buckled.

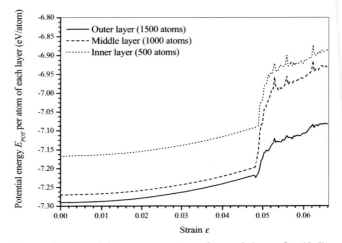

Figure 4: Potential energy per atom for each layer for (5,5), (10,10) and (15,15) three-walled carbon nanotube with diameter 20.36 Å.

The sudden increase in strain energy per atom for each layer is due to the spontaneous increase in potential energy of each atom for each layer as shown in Fig. 4. From the plot, it is observed that the middle layer has the largest increment of potential energy of each atom for each layer followed by the inner layer and then the outer layer. Being sandwiched between the outer and inner layer, when the three-walled carbon nanotube starts to buckle, the atoms at the middle layer will form more nearest-neighbors than those of the outer and inner layer. This causes the formation of sp^2 to sp^3 configuration, which leads to a larger increase in the empirical bond order function and consequently the potential energy.

Figure 5 shows the three-dimensional and cross-sectional three-walled nanotube at different strains. It is evident from Fig. 5(a) that the outer layer starts to deform first into a ring pattern at $\varepsilon \quad 0.0484$, while the middle and inner layers are undeformed and their strain energies remain low. However, further compression at ε 0.0486 causes the middle layer to deform too, resulting in a sudden increase in the strain energy per atom because of the increase in potential energy due to the atoms in the middle layer. Similarly at this time, the inner layer maintains the elasticity of the nanotube temporarily as shortly after at ε 0.0493, the inner layer also starts to buckle, therefore increasing the overall potential energy and hence increasing the strain energy per atom further. After ε 0.052, all the layers are deformed, hence there is no further abrupt increase in potential energy due to sudden addition of nearest-neighbors.

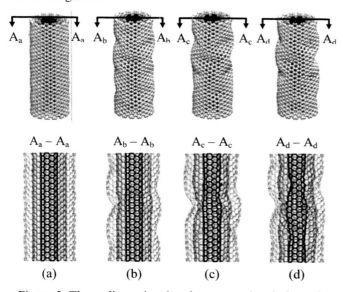

Figure 5: Three-dimensional and cross-sectional view of morphological changes for three-walled (5,5), (10,10), and (15,15) carbon nanotube with diameter 20.4 Å.

The buckling loads P_{cr} for two and three-walled carbon nanotubes are tabulated in Table 1. It is observed from the table that generally multi-walled carbon nanotubes have larger buckling loads P_{cr}. For instance, the (10,10), (15,15) two-walled carbon nanotube has a higher buckling load P_{cr} 2.02 10^7 N compared to the (5,5), (10,10) two-walled carbon nanotube that has a buckling load P_{cr} 1.72 10^7 N (see Table 1).

Two-walled		Three-walled	
$(n_1,m_1)(n_2,m_2)$	P_{cr} (10^7 N)	$(n_1,m_1)(n_2,m_2)(n_3,m_3)$	P_{cr} (10^7 N)
(5,5)(10,10)	1.72	(5,5)(10,10)(15,15)	2.46
(10,10)(15,15)	2.02	(10,10)(15,15)(20,20)	2.73
(15,15)(20,20)	2.09	(15,15)(20,20)(25,25)	2.81
(20,20)(25,25)	2.10	(20,20)(25,25)(30,30)	3.05

Table 1: Buckling loads for selected two and three-walled carbon nanotubes.

Moreover, Table 1 also verifies that as the number of layers in a carbon nanotube increases, its buckling load P_{cr} increases too. According to Table I, the buckling load P_{cr} for a (5,5) single-walled carbon nanotube is 0.88 10^7 N, however if an extra outer layer is added to form a (5,5), (10,10) two-walled carbon nanotube, its buckling load P_{cr} will increase to 1.72 10^7 N as seen in Table II. If another layer is further added to form a (5,5), (10,10), and (15,15) three-walled carbon nanotube, the buckling load P_{cr} will rise to 2.46 10^7 N.

3 CONCLUSIONS

In conclusion, molecular dynamics simulation was used to analyze the structural properties of single and multi-walled nanotubes. The buckling loads for many single and multi-walled carbon nanotubes are determined in the study. The calculation shows that as the diameter of single-walled nanotubes increases, the buckling load P_{cr} increases rapidly up to the optimum buckling load P_{cr}. Any further increase beyond the optimum diameter d, however, will result in a slow decline in buckling load P_{cr} up to a steady value. Therefore, each single-walled carbon nanotubes has an optimum diameter d that yields the highest buckling load P_{cr}. In addition, the number of layers in a multi-walled nanotube will also affect its structural properties. When a single-walled carbon nanotube buckles, there is a sudden decrease in strain energy; however, when a multi-walled carbon nanotube buckles, there is a spontaneous increase in strain energy. This is due to the growth of potential energy as more chemical bonds change from sp^2 to sp^3 configuration.

REFERENCES

[1] Iijima, Nature (London) 354, 56, 1991.
[2] Brenner, Shenderova, Harrison, et al, J. of Phys: Cond Matt 14, 783, 2002.
[3] Lourie, Cox, and Wagner, Phys. Rev. Lett. 81, 1638, 1998.
[4] Srivastava, Menon, and Cho, Phys. Rev. Lett. 83, 2973, 1999.
[5] Wong, Sheehan, and Lieber, Science 277, 1971, 1997.
[6] Timoshenko and Gere, Van Nostrand Reinhold Company, USA, Chapt. 10, 1973.
[7] Lennard-Jones, Proc. Roy. Soc., London 106A, 441, 1924.
[8] Moller, Tildesley, Kim, and Quirke, J. Chem. Phys. 94, 8390, 1991.

Electrochemical Storage of Hydrogen in Carbon Nanotubes

S.-R. Chung, K.-W. Wang and T.-P. Perng*

Department of Materials Science and Engineering
National Tsing Hua University, Hsinchu, Taiwan, *tpp@mse.nthu.edu.tw

ABSTRACT

The electrochemical hydrogenation property of carbon nanotubes (CNTs) was investigated. The CNTs were prepared by different methods, including chemical vapor deposition (CVD) using different catalysts (Co, Ni, etc.) or laser ablation (LA) using Co or Ni catalyst. All of the CNTs contained various amounts of residual metal catalyst. A high discharge capacity of 197 mAh/g of the CVD-Co electrode was obtained, followed by CVD-Ni (48 mAh/g) and LA (36 mAh/g) at room temperature. If the CNTs were purified by HNO_3 and similar test was conducted, it was found that the discharge capacities were dramatically reduced. When the purified CNTs were deposited with a controlled amount of metal on the surface, the discharge capacity was improved significantly. It is postulated that the electrochemical hydrogen storage by carbon nanotubes might be influenced by the catalytic effect.

Keywords: electrochemical properties, carbon nanotube, discharge capacity, catalytic effect

1. INTRODUCTION

Carbon nanotubes (CNTs) have unique mechanical, electric, and magnetic properties. The applications of CNTs include field emission display, composite, SPM tip, hydrogen storage, catalyst support for fuel cells, Li-ion battery, and son on. To study the hydrogen storage property, there are two ways to measure the amount of absorbed hydrogen, i.e., gas-phase or electrochemical hydrogenation. Dillon et al. [1] first measured the hydrogen adsorption capacity of an as-prepared carbon soot containing only about 0.1 to 0.2wt% SWNT at 133K. Based on this result, they predicted that hydrogen adsorption capacity of pure SWNTs was between 5 to 10wt%. Generally, for physical sorption of hydrogen, a pressure of 100 bars and low temperature 77K is needed. Hydrogen adsorption by CNT is due to its large surface area [2]. Opening the caps on the end of the nanotubes [3] or increasing the spacing between the graphene sheets [4] and heat treated at temperatures higher than 1800°C [5, 6] can all increase the hydrogen storage capacity of CNT. It has been pointed that hydrogen can be stored in SWNT at room temperature [7, 8]. Metal oxide catalyst or residual catalyst may participate in hydrogen adsorption onto

carbon nanotubes [9, 10]. Tibbetts et al. studied the sorption of hydrogen with nine different carbon materials at pressures up to 11MPa and temperature from -80 to 500°C [11]. Unfortunately, very little hydrogen can be adsorbed by those carbon materials.

Until now, electrochemical method for hydrogen storage has been less extensively employed compared to physical methods. However, there have been more papers focusing on the electrochemical property of CNT, because the results can be directly used in rechargeable battery. For example, a capacity of 110mAh/g under very low discharge current density was observed for an SWNT electrode [12]. The discharge capacity of CNT can be improved if the cap is removed by acid treatment [13]. Metallic impurities, such as Co and Ni, could increase the capacity of CNT [14]. To date, the mechanism of reversible charge and discharge of hydrogen in carbon materials is not clear. It may be affected by specific surface area [15], residual metal catalyst, diameter of CNT [16], the functional group produced after acid treatment [17], charge/discharge current density, crystallinity of CNT [18], and so on. Qin et al. pointed out that for the electrode formed with MWNT and Ni powder a high discharge capacity of 200mAh/g was obtained [19]. When carbon nanotubes were decorated with metallic particles, the discharge capacity of CNT can be significantly improved [20, 21]. The electrochemical discharge capacity increased from 101 to 149mAh/g by increasing the Ni-P content on the outer surface of CNTs owing to the synergistic effect between the metal and carbon in the electrochemical reaction [22].

The purpose of this study is to offer a new idea about catalytic effect of deposited metal particles on electrochemical hydrogenation of CNTs.

2. EXPERIMENTAL

2.1 Preparation of CNTs

The carbon nanotubes were prepared by catalytic chemical vapor deposition (CVD) or laser ablation (LA) using Co and/or Ni as the catalyst. In the CVD process, a quartz tube reactor was used and C_2H_2 gas was introduced to grow carbon nanotubes on various catalysts (Co, Ni, etc.) at 750°C. For the LA process, Co or Ni was used as catalyst. All of the CNTs contained various amounts of residual metal catalyst. Nitric acid was used to remove the residual metallic particles for all samples, and then the CNTs were soaked with $Co(NO_3)_2$ at different atomic ratios, followed by reducing in H_2 at 300°C for 30min. The salt was reduced to metallic Co or oxidized to Co_3O_4. In addition to the above samples, some CNTs with a specified purity of 95% were also tested.

The phase structure of the samples was identified by X-ray diffraction (XRD). The diameter and surface morphology of CNTs were examined by field-emission scanning electron microscopy (FESEM).

2.2 Battery Assembly

The charge and discharge curves were measured in a test cell, which contained two pieces of positive electrode, one piece of carbon nanotube (0.02g) as the negative electrode, and polypropylene as the separator. The electrolyte was 6M KOH + 1wt% LiOH. The positive electrode material consisted of nickel hydroxide, 5wt% Co, and 5wt% CoO. Each of the positive and negative electrode material was mixed with 3 wt% PTFE to form a paste, and coated on a piece of Ni-foam. The electrode plate was cold pressed at a pressure of 100 kgf/cm^2 for 30s. The charge/discharge current and temperature were set at 10 mA/10 mA and 25°C, respectively, and the cut-off voltage was 900 mV.

3. RESULTS AND DISCUSSION

3.1 Characterization of CNTs

Fig. 1 shows the XRD patterns of as-prepared and HNO_3 treated Ni-CNT and Co-CNT prepared by CVD. After the purification treatment, although most of the metallic particles have been removed by HNO_3, there still exists some residual catalyst. The SEM micrographs of the as-prepared CNTs show that the samples contain carbon nanotubes, amorphous carbon, metallic particles, and some nanocapsules.

3.2 Electrochemical Test

Fig. 2 shows the discharge capacities of the as-prepared and HNO_3 treated CNTs (Co-, and Ni-), and of Co-CNT decorated with 5at% Co. The maximum discharge capacity of the as-prepared Co-CNT is 200 mAh/g, equal to an H/C value of 0.74wt%. After the acid treatment, however, the hydrogen storage capacity of Co-CNT is dropped to nearly 0. After decoration with Co, the discharge capacity increases very significantly, and the cyclic stability is improved compared with the as-prepared sample. The discharge capacity for as-prepared Ni-CNT is quite low compared with that of Co-CNT. However, they have the same behavior, i.e., the discharge capacity is

decreased after removing the metal catalyst. Although Ni is a good catalyst for hydrogen dissociation in alkaline solution, the catalytic effect of Ni seems not as good as Co for the CNT electrode.

When the commercial CNTs with a purity of 95% were tested, the discharge capacity was also very low, as shown in Fig. 3. After decoration with Co, the discharge capacity was increased markedly. The maximum discharge capacity reaches 231mAh/g when 10at% Co has been deposited to the CNTs. This phenomenon can also be observed in CNTs prepared by different methods. Fig. 4 shows the discharge capacities of an SWNT electrode prepared by laser ablation with a purity of 70%. The maximum discharge capacity of as-prepared sample is only 25mAh/g. It increases dramatically to 150mAh/g after decoration with 10at% Co. The discharge capacity of the

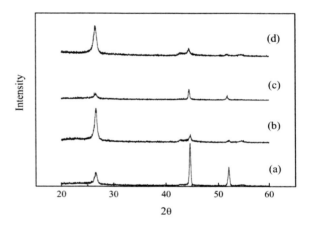

Fig. 1   XRD patterns of Ni- and Co-CNT prepared by CVD. (a) as-prepared Ni-CNT, (b) acid treated Ni-CNT, (c) as-prepared Co-CNT, and (d) acid treated Co-CNT.

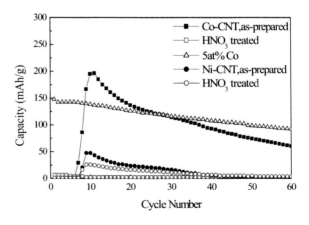

Fig. 2   Discharge capacities of Co- and Ni-CNT electrodes.

modified electrode increases with the Co content.

The results of the above electrochemical tests are summarized in Table 1. It is concluded that CNTs of higher purity or after acid treatment have lower discharge capacity.

4. CONCLUSION

Several kinds of CNT were used as the negative electrode material in alkaline solution, and their hydrogen storage capacities were measured by electrochemical method. The CNTs of high purity or after acid treatment exhibit lower hydrogen storage capacity. Decoration with Co on CNTs is effective to enhance the electrochemical hydrogen storage capacity.

5. ACKNOWLEDGMENTS

This work was supported by Ministry of Education of ROC under contract A-91-E-FA04-1-4 and Ministry of Economic Affairs under contract 91-EC-17-A-08-S1-0003.

REFERENCES

[1]   A. C. Dillon, K. M. Jones, T. A. Bekkedahl, C. H. Kiang, D. S. Bethune, and M. J. Heben, *Nature* 386(1997) 377.

[2]   A. Züttel, P. Sudan, Ph. Mauron, T. Kiyobayashi, Ch. Emmenegger, and L. Schlapbach, *Int. J. Hydrogen Energy 27 (2002) 203.*

[3]   L. F. Sun, J. M. Mao, Z. W. Pan, B. H. Chang, W. Y. Zhou, G. Wang, L. X. Qian, and S. S. Xie, *Appl. Phy. Lett. 74 (1999) 644.*

[4]   P. Chen, X. Wu, J. Lin, and K. L. Tan, *Science 285 (1999) 91.*

[5]   X. S. Li, H. Zhu, L. J. Ci, C. L. Xu, Z. Q. Mao, B. Q. Wei, J. Liang, and D. H. Wu, *Carbon 39 (2001) 2077.*

[6]   L. J. Ci, H. W. Zhu, B. Q. Wei, C. L. Xu, and D. H. Wu, *Appl. Surf. Sci. 205 (2003) 39.*

[7]   C. Liu, Y. Y. Fan, M. Liu, H. T. Cong, H. M. Cheng, and M. S. Dresselhaus, *Science 286 (1999) 1127.*

[8]   R. Andrews, D. Jacques, D. Qian, and E. C. Dickey, *Carbon 39 (2001) 1681.*

[9]   A. Lueking and R. T. Yang, *J. Catal. 206 (2002) 165.*

[10] F. H. Yang and R. T. Yang, *Carbon 40 (2002) 437.*

[11] G. G. Tibbetts, G. P. Meisner, and C. H. Olk, *Carbon 39 (2001) 2291.*

[12] C. Nützenadel, A. Züttel, D. Chartouni, and L. Schlapbach, *Electeochem. Solid-State Lett. 2 (1999) 30.*

[13] N. Rajalakshmi, K. S. Dhathathreyan, A. Govindaraj, and B. C. Satishkumar, *Electrochem. Acta 45 (2000) 4515.*

[14] A. Züttel, C. Nützenadel, P. Sudan, P. Mauron, C. Emmenegger, S. Rentsch, L. Schlapbach, A. Weidenkaff, and T. Kiyobayashi, *J. Alloys Comp. 330 (2002) 676.*

[15] A. Züttel, P. Sudan, P. Mauron, T. Kiyobayashi, C. Emmenegger, and L. Schlapbach, *Int. J. Hydrogen Energy 27 (2002) 203.*

[16] G. P. Dai, C. Liu, M. Liu, M. Z. Wang, and H. M. Cheng, *Nano Lett. 2 (2002) 503.*

[17] X. S. Li, H. W. Zhu, L. J. Ci, C. L. Xu, Z. Q. Mao, B. Q. Wei, J. Liang, and D. H. Wu, *Chinese Science Bulletin 46 (2001) 1358.*

[18] J. M. Skowronski, P. Scharff, N. Pf nder, and S. Cui, *Adv. Mater. 15 (2003) 55.*

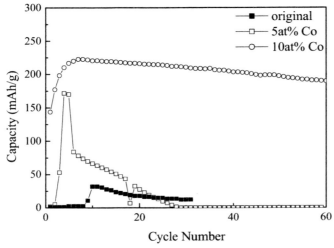

Fig. 3   Discharge capacities of commercial CNTs with a purity of 95%.

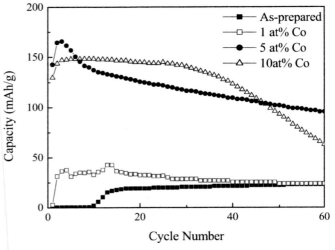

Fig 4.   Discharge capacities of CNTs prepared by LA with a purity of 70%.

Table 1   Discharge capacities (mAh/g) of the CNT samples.

Sample	As-prepared	Decorated with Co			HNO$_3$ Treated
		1at%	5at%	10at%	
Co-CNT	197	-	143	-	7
Ni-CNT	48	-	-	-	26
Commercial	-	-	172	231	32
LA-CNT	36	43	166	148	-

Note: 100mAh/g = 0.37 wt% H.

19]   X. Qin, X. P. Gao, H. Liu, H. T. Yuan, D. Y. Yan, W. L. Gong, and D. Y. Song, *Electeochem. Solid-State Lett. 3 (2000) 532.*

[20]   X. P. Gao, Y. Lan, G. L. Pan, F. Wu, J. Q. Qu, D. Y. Song, and P. W. Shen, *Electeochem. Solid-State Lett. 4 (2001) A173.*

[21]   Gautam Gundiah, A. Govindaraj, N. Rajalakshmi, K. S. Dhathathreyan, and C. N. R. Rao, *J. Mater. Chem. 13 (2003) 209.*

[22]   X. Q. Yan, X. P. Gao, Y. Li, Z. Q. Liu, F. Wu, Y. T. Shen, and D. Y. Song, *Chem. Phys. Lett. 372 (2003) 336.*

Theoretical studies on relative stabilities of C_{70} fullerene dimers

Xiang Zhao [a, *], Zdenek Slanina [b], and Hitoshi Goto [a]

a. Laboratory of Theoretical/Computational Chemistry, Department of Knowledge-Based Information Engineering, Toyohashi University of Technology, Toyohashi 441-8580, Japan
b. Department of Theoretical Studies, Institute for Molecular Science, Okazaki 444-8585, Japan

ABSTRACT

Five stable fullerene dimers $(C_{70})_2$ with [2+2] bridges between hexagon-hexagon bonds, which are experimentally characterized very recently, have been investigated by several semi-empirical MO approaches and *ab initio* Hartree-Fock self-consistent field calculation and hybrid density functional theory treatment. Energy difference among the five isomers is predicted to be quite small, where two of them seem to be isoenergetic. The computed B3LYP/6-31G temperature-dependent relative concentrations indicate the two lower energy isomers to be the most thermodynamically populated one with a ratio of 1:1 over a wide temperature area. This finding agrees reasonably well with the recently reported experimental observations.

Keywords: fullerene dimer, ab initio method, density-functional theory, entropy-enthalpy interplay

INTRODUCTION

A great deal of considerable interest has been paid to fullerene dimers [1,2] since the discovery of C_{60} [3]. Recently, several experimental and theoretical investigations on $(C_{60})_2$, $(C_{70})_2$ and even $(C_{59}N)_2$ have been reported [4-7], which seem to be important to elucidate the unique physical properties of fullerene polymeric materials. Very recently, Shinohara *et al.* [8] reported the first experimental synthesis and characterizations of five [2+2] structural isomers of fullerene dimers $(C_{70})_2$. They also reported the five separated isomers in a relative production ratio of 0.8/1.0/0.5/0.5/0.2. Although the relative isomeric mixtures in experiment are not yet clear before the individual structures are assigned, such an experimental effort obviously leads to the corresponding theoretical study. In the present report this fullerene dimer system is addressed with quantum-chemical calculations.

COMPUTATIONS

According to the previous calculations by Fowler *et al.* [6], all 15 lower energy $(C_{70})_2$ dimer structures were considered and pre-optimized with several semi-empirical MO methods (i.e. PM3, AM1, and MNDO). It is found, in agreement with previous results, that five energetically reasonable isomers indeed exist dominantly over the rest of isomers. The five low-energy structures are labeled same as previous code combined with its symmetry – **cis**:C_{2v}, **trans**:C_{2h}, **abc**:C_1, **c1**:C_{2h}, and **c2**:C_{2v}. In order to promote the energetic accuracy, the five isomers are further subjected to the fully geometry optimizations at the HF/6-31G and B3LYP/6-31G levels of theory, with the Gaussian 98 program package [9]. The harmonic vibrational analyses are carried out on the five isomers at both AM1 and PM3 levels in order to evaluate their relative concentrations at high temperatures using both enthalpy and entropy terms.

Relative concentrations (mole fractions) x_i of m isomers can be expressed through the partition functions q_i and the enthalpies at the absolute zero temperature or the ground-state energies $\Delta H^{o}_{0,i}$ by a compact formula [10]:

$$x_i = \frac{q_i \exp[-\Delta H^{o}_{0,i}/(RT)]}{\sum_{j=1}^{m} q_j \exp[-\Delta H^{o}_{0,j}/(RT)]}$$

where R is the gas constant and T the absolute temperature. Since PM3 energetic results show some contradictions to other applied methods, the rotational-vibrational partition functions are constructed from the AM1 calculated structural and vibrational data under the rigid rotator and harmonic oscillator approximation. Chirality's contribution, frequently ignored, is included accordingly. Finally, temperature-dependent relative stabilities under thermodynamic equilibrium can be evaluated as a key output (mole fractions in the isomeric mixture) for comparisons with the available experimental observations.

Table 1: The relative energies (kJ/mol) of five $(C_{70})_2$ dimers

isomer:sym.	B3LYP	HF	AM1	PM3	MNDO
abc:C_1	9.34	5.69	1.37	-0.27	4.48
c1:C_{2h}	20.16	11.37	2.81	-0.47	9.06
c2:C_{2v}	20.88	12.02	2.95	-0.33	9.23
cis:C_{2v}	0.28	0.21	0.04	0.04	0.04
trans:C_{2h}	0.0	0.0	0.0	0.0	0.0

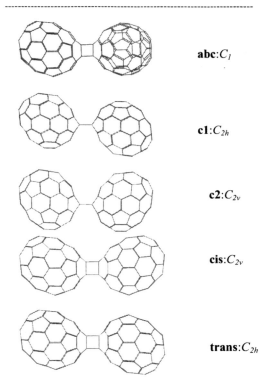

abc:C_1

c1:C_{2h}

c2:C_{2v}

cis:C_{2v}

trans:C_{2h}

Fig. 1: Five stable isomers of $(C_{70})_2$ dimers

RESULTS AND DISCUSSION

Table 1 presents the computed energetics for the five stable fullerene dimers $(C_{70})_2$ with [2+2] bridges between hexagon-hexagon bonds (see Figure 1). The structure trans:C_{2h} is predicted as the lowest energy species in the dimeric system only with exception of PM method. Based on the AM1, MNDO, HF/6-31G, and B3LYP/6-31G energetic results, it is clear that the energy difference among the five isomers is rather small and two dimeric structures (**trans**:C_{2h} and **cis**:C_{2v}) are almost isoenergetic (only 0.28 kJ/mol at B3LYP level).

The relative energetics themselves cannot always predict relative stabilities in an isomeric system at high temperatures as stability interchanges induced by a significant enthalpy-entropy interplay. In our treatment the Gibbs free energy contributions are taken into account to predict the general relative stabilities within higher temperatures.

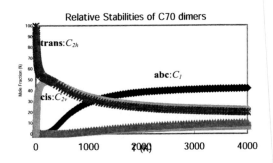

Fig. 2: Relative stabilities of C_{70} dimers

Figure 2 presents the development of the relative concentrations x_i in the five lower energy dimeric system for a wide temperature interval based on the B3LYP/6-31G energetics and the AM1 rotational/vibrational data. At very low temperatures, of course, the **trans**:C_{2h} structure has to be prevailing. When temperatures increase, the population of the **cis**:C_{2v} structure jumps very rapidly and the first relative stability interchange

in this dimeric system should happen around the room temperature (300K) between the **trans**:C_{2h} and the **cis**:C_{2v} structures. As temperatures further increase, the third dimer **abc**:C_l becomes much more important and finally the most thermodynamically stable species over the temperature of 1250K. Two other dimers (**c1**:C_{2h}, and **c2**:C_{2v}) show some relatively small but non-negligible populations at higher temperatures.

Interestingly enough, at a selected temperature of 2000K for instance, the computed B3LYP/6-31G concentration ratio of the five dimers (**cis**:C_{2v} / **trans**:C_{2h} / **abc**:C_l / **c2**:C_{2v} / **c1**:C_{2h}) shows about 0.7/0.7/1.0/0.2/0.2 of the relative abundances. This result agrees reasonably well with the experimental observation of Shinohara et al. [8], where the production ratio of the five dimers is given with 0.8/1.0/0.5/0.5/0.2.

The achieved theory-experiment correspondence is certainly encouraging, though still more advanced computations on energetics and vibrations are needed to clarify possible sources of some disagreement. Further efforts and improvement [11] are still ongoing.

ACKNOWLEDGEMENTS

The reported work has been supported by a Grant-in-aid Scientific Research Project from the Ministry of Education, Culture, Sports, Science and Technology of Japan, and also by the Japan Society for the Promotion of Science (JSPS).

REFERENCES

[1] G. W. Wang, K. Komatsu, Y. Murata, M. Shiro, *Nature*, **387**(1997), 583

[2] J. L. Segura, N. Martin, *Chem. Soc. Rev.*, **29**(2000), 13

[3] W. Kratschmer, L. D. Lamb, K. Fostiropoulos, D. R. Huffman, *Nature*, **347**(1990), 354

[4] M. Fujitsuka, C. Luo, O. Ito, Y. Murata, K. Komatsu, *J. Phys. Chem. A.* **103**(1999), 7155

[5] S. Lebedkin, W. E. Hull, A. Soldatov, B. Renker, M. M. Kappes, *J. Phys. Chem. B.* **104**(2000), 4101

[6] T. Heine, F. Zerbetto, G. Seifert, P. W. Fowler, *J. Phys. Chem. A.* **105**(2001), 1140

[7] J. C. Hummelen, N. Knight, J. Pavlovichm, R. Gonzalez, F. Wudl, *Science* **269**(1995), 1554

[8] G. S. Forman, N. Tagmatarchis, H. Shinohara, *J. Am. Chem. Soc.* **124**(2002), 178

[9] M. J. Frisch, *et al.*, *aussian 98, Rev. A.11*; Gaussian, Inc., Pittsburgh PA, **2001**.

[10] Z. Slanina, X. Zhao, and E. Osawa, in *Adv. Strained Interesting Org. Mol.*, **7**, (JAI Press Inc., Connecticut, 1999), p. 185.

[11] X. Zhao, Z. Slanina, H. Goto, manuscript in preparation.

----

* Corresponding Author. Fax: +81-532-485588.
Email: *zhao cochem2.tutkie.tut.ac.jp*

Electrocatalysis at a Thin Multi-Wall Carbon Nanotube Film Modified Glassy Carbon Electrode

J. Fei, Y. Sun, S. Hu and Z. Gao[1*]

Department of Chemistry, Wuhan University, Wuhan, 430072, China.
[1]Institute of Bioengineering and Nanotechnology, 51 Science Park Road, Singapore 117586
Republic of Singapore, *Corresponding author: zqgao@ibn.a-star.edu.sg

ABSTRACT

A sensitive and selective amperometric procedure, based on a glassy carbon electrode (GCE) modified with a thin multi-wall carbon nanotube (MWNT) film, for the determination of ascorbic acid (AA) was developed. The modified electrode showed excellent electrocatalytic activity towards the oxidation of AA. The oxidation potential of AA shifted negatively by as much as 460 mV to -18 mV (vs. SCE) at the modified electrode. Experimental variables were investigated and optimized. Amperometric tests of AA were performed at -10 mV. A linear response was obtained in the range of 5.0×10^{-6} to 4.0×10^{-3} M with a detection limit of 2.0×10^{-6} M. Interference studies indicated that the modified electrode is highly selective to AA in a mixture of AA, uric acid and dopamine. The proposed procedure was applied to the determination of AA in pharmaceutical and food samples with little or no sample pretreatment.

Keywords: Carbon nanotube, voltammetry, amperometry, electrode, ascorbic acid

INTRODUCTION

There has been much interest in the research of carbon nanotubes since their discovery.[1] Carbon nanotubes are molecular-scale wires with high electrical conductivity, good chemical stability and excellent mechanical strength and modulus.[2] Utilization of these properties has found applications of carbon nanotubes in scanning probes, nanoelectronic devices, batteries, potential hydrogen storage material and chemical sensors.[2] Generally, carbon nanotubes consist two categories: single-wall carbon nanotubes (SWNTs) and multi-wall carbon nanotubes (MWNTs). The subtle electronic behavior of carbon nanotubes reveal that they have the ability to promote electron-transfer when used as electrode material in electrochemical reactions.[3,4,5] Since they are insoluble in most solvents, attaching carbon nanotubes to an electric circuit in a controllable manner is still a technical challenge. The MWNTs have been used to fabricate carbon nanotube modified electrodes by mixing them with binders such as bromoform, mineral oil or liquid paraffin. The resulting electrodes have been used in the electrochemical oxidation of dopamine,[3] electrocatalysis of O_2[4] and electrochemistry of protein.[5] SWNTs have been dispersed into DMF and cast on carbon electrodes to form carbon nanotube films. These films exhibit stable electrochemical behavior and are used to catalyze electrochemical oxidations of some biomolecules, such as dopamine, epinephrine, ascorbic acid (AA), and cytochrome *c*.[6] The evaporation of DMF is very slow, and therefore the formation of carbon nanotubes films requires a long waiting time or heating of the electrode surface. Moreover the resulting films are brittle.

It is well known that AA exists extensively in fruits and is involved in many biological reactions. Recently, there has been a considerable effort in the development of voltammetric procedures for the determination of AA selectively and sensitively in biological systems and foodstuff. It is generally believed that direct electrooxidation of AA at bare electrodes is irreversible and requires high overpotentials. Furthermore, direct oxidations of uric acid (UA) and dopamine (DA) take place at potentials similar to that of AA. Electrochemical determination of AA therefore often suffers from interferences from these species. To differentiate voltammetric responses of these species and accurately determinate each of them in a mixture, Gao *et al.* fabricated an ultrathin polypyrrole-teradecyl sulfate film modified gold electrode for simultaneous voltammetric determinations of UA, DA and AA.[7] It was observed that all three species were resolved and determined voltammetrically at the modified electrode.

In this work the voltammetric response of AA at the MWNT film modified GCE was investigated. MWNTs were easily dispersed into water in the presence of an anionic surfactant, dihexadecyl hydrogen phosphate (DHP). The stable suspension of MWNTs was cast on GCE to form a carbon nanotube film after the solvent was evaporated. The electrochemical behavior of AA at SWNT modified electrodes has been previously reported.[6] The oxidation peak potential of AA was observed at about 0.16 V (*vs.* SCE), which overlaps with that of DA (0.18 V). Compared to the SWNT film modified electrode, the MWNT film modified electrode not only exhibited stronger catalytic activity toward the oxidation of AA, but also resolved the overlapping voltammetric responses of UA, DA, and AA into three independent voltammetric peaks. Thus, a sensitive and selective procedure based on the MWNT film modified electrode for the amperometric detection of AA was developed.

EXPERIMENTAL SECTION

The multi-wall carbon nanotubes (obtained from the Institute of Nanomaterials, Central China Normal University, China) were synthesized by a catalytic pyrolysis method and purified with hot concentrated HNO_3.[8]

2.5 mg of the purified MWNTs and 5.0 mg of DHP were mixed with a 5.0 ml of doubly distilled water. A homogeneous and stable suspension of MWNT-DHP was achieved with the aid of ultrasonic agitation. The GCE was coated by casting a 5.0 l aliquot of the MWNT dispersion and dried under ambient conditions. The amount of MWNT on the GCE surface was about 0.036 mg cm^{-2}. The freshly prepared modified electrode was activated in a phosphate buffered saline (PBS) solution by repetitive potential cycling between -0.40 and 0.80 V.

Voltammetric measurements were performed with a model CHI 660A electrochemical workstation (CH Instruments, Austin, USA) at room temperature. All solutions were deoxygenated with high-purity nitrogen for at least 5 min prior to each experiment. Voltammograms of AA on the modified electrode were obtained by scanning the potential at a rate of 50 m s^{-1}. Amperometric tests were carried out at a poised potential of –10 mV.

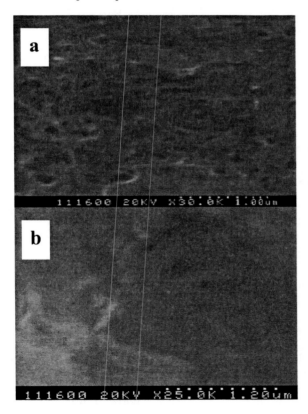

Fig. 1. (a) SEM image of a MWNT-DHP film and (b) a pure DHP film on GCE.

RESULTS AND DISCUSSION

The MWNT-DHP suspension was cast on the GCE and a SEM image of the film formed is shown in Fig. 1a. It can be seen that the GCE is completely and homogeneously coated by a thin MWNT-DHP film, with diameters of the MWNT bundles in the film spreading with a range from 20 - 40 nm. It can also be seen from this image that the MWNT-DHP film contained very small portion of amorphous carbon impurities. The SEM of a pure DHP film on a GCE is shown in Fig. 1b. As can be seen, a uniformed and compact DHP film readily forms on the GCE.

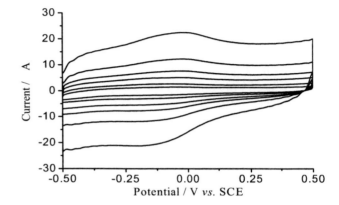

Fig. 2. Cyclic voltammograms of the MWNT film modified electrode. Scan rates from the innermost to the outermost: 50, 100, 200, 300, 500 and 1000 mV s^{-1}.

As shown in Fig. 2, when the modified electrode was immersed in PBS, a pair of reduction and oxidation waves were observed and the reduction and oxidation peak potentials were centered at 0.12 and –0.049 V, respectively. Similar to the case of MWNT microelectrodes,[3-5] the reduction and oxidation waves of the MWNT film were very broad. Meanwhile, a large charging current of the modified electrode was observed, which might be due to the ultrahigh surface area of the MWNT film. Voltammetric experiments also demonstrated that the electrochemical response of the MWNT film remained very stable after numerous successive potential scans and after exposing the MWNT film to air under ambient conditions for several days. The effect of potential scan rate on the voltammetric response of the MWNT film was studied by varying the scan rate from 50 to 1000 mV s^{-1}. With increasing scan rate, both the reduction and oxidation peak currents increased linearly and the difference between the reduction and oxidation peak potentials remained unchanged even at a scan rate of 1000 mV s^{-1}, suggesting a very fast charge transfer process of the MWNT film. Voltammetric experiments at various pH showed that both the reduction and oxidation peak potentials shifted negatively with increasing pH, implying that protons are probably involved in the redox reaction of the MWNT film.

Fig. 3 shows cyclic voltammetric responses of AA at different electrodes. At a bare GCE, a broad oxidation peak was observed at 0.45 V with $E_p - E_{p/2} = 0.25$ V (Fig. 3 trace a). Compared to the bare GCE, the oxidation peak current of AA increased significantly at the modified electrode and the oxidation peak potential shifted negatively to -0.18 mV with $E_p - E_{p/2} = 0.045$ V (Fig. 3 trace c). The obvious increase in the oxidation current and the decrease in overpotential of about 468 mV for AA demonstrated unambiguously the existence of an efficient catalytic reaction between the MWNT film and AA in solution. The shift in the overpotential is due to kinetic effect, or in other words, the increase in charge transfer rate of AA, which results in the improvement of the reversibility of the electron transfer processes and the increase in oxidation current.[9] The electrochemical behavior of AA at the pure DHP film coated electrode was also examined. As shown in Fig. 3 trace b, the oxidation of AA disappeared completely at the DHP film modified electrode. Since pK_{a1} of AA is 4.20, AA is anionic in PBS. Considering that DHP is anionic and forms a compact film on the GCE, the disappearance of the oxidation peak of AA is obviously due to the impermeable nature of the anionic DHP film to anionic AA.

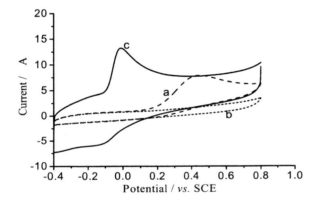

Fig. 3. Voltammograms of 5.0 10^{-4} M AA in PBS at (a) a bare GCE, (b) the DHP modified and (c) the MWNT film modified GCE. Potential scan rate: 50 mVs⁻¹.

The relationship between the amount of MWNT on GCE and the oxidation peak current of AA was studied. It was found that the oxidation peak current increases gradually with increasing the amount of MWNT and starts to level off at 0.036 mg cm⁻². However, the charging current increased semi-linearly with increasing the amount of the MWNT on the electrode surface. Therefore large amounts of MWNT prevent from determining AA at low concentrations. In this report, the amount of MWNT on the GCE surface was controlled at 0.036 mg cm⁻².

The pH of the supporting electrolyte had significant influence on the oxidation of AA at the modified electrode, altering both the oxidation peak current and potential. For examples, the oxidation peak current of AA increased with increasing pH from 2 to 6.3, reached maximum over the pH range from 6.3 to 7.2 and decreased quickly with further increase in pH beyond 7.2

The study of the effect of pH on the oxidation peak potential (E_{pa}) of AA at the modified electrode showed a negative shift in E_{pa} with increasing pH (Fig. 4). An estimated [Epa / pH] of –59.4 mV pH⁻¹ was obtained between pH 2.0 and 4.2, indicating that the total number of electrons and protons involved in the electrochemical oxidation of AA is the same in this pH range, whereas at higher pH the [Epa / pH] decreased to 30.8 mV pH⁻¹, suggesting a 2e⁻ /1H⁺ oxidation process.

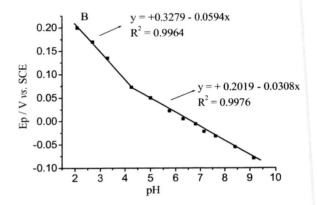

Fig. 4. Dependence of E_{pa} on pH. 5.0x10⁻⁴ M AA in PBS, potential scan rate: 50 mVs⁻¹.

The influence of potential scan rate on AA oxidation at the modified electrode was examined by cyclic voltammetry from 10 to 300 mVs⁻¹. It was found that the peak current of AA is proportional to the square root of the scan rate indicating that AA is undergoing a diffusion-controlled process. The oxidation peak potential of AA depended linearly on the logarithm of the scan rate according to the following equation $E_p(V) = - 0.0777 + 0.0363 \log v$, $R^2 = 0.997$. From the slope of the plot, 0.0363, the value of an, where α is transfer coefficient of the oxidation of AA and n is the number of electrons transferred during the oxidation of AA, was estimated to be 0.82. As the oxidation of AA is generally known to occur via a two-electron transfer process, thus α is 0.41.

Usually, poor reproducibility (fouling) is encountered when unmodified electrodes are employed in the determination of biomolecules. Our experimental results showed that no fouling effect was observed for the determination of AA and this might be due to that AA cannot penetrate into the highly compact and conductive MWNT film and cannot adsorb on the anionic MWNT film neither. Therefore, good reproducibility is expected. For example, for a series of 10 repetitive determinations of AA, carried out in a 2.0 10^{-4} M AA solution in PBS, a RSD of 2.2% was obtained. The modified electrode also showed high stability. The voltammetric response decreased by only 3.5 % after 4 h of continuous tests.

As discussed earlier, the electrochemical determination of AA, particularly amperometric, is often affected by the oxidations of DA and UA. It is necessary to solve the problem of resolving the electrochemical oxidation processes of these species. Since the overpotential of AA oxidation was significantly reduced at the modified electrode, it is highly possible that selective determination of AA can be achieved through a careful selection of poise potential. The overall facility of the modified electrode for selective determination of AA was demonstrated in Fig. 5. As shown in Fig. 5, the presence of the MWNT film at the electrode surface resolved the mixed voltammetric response of UA, DA and AA into three well-defined voltammetric peaks at potentials of 0.29, 0.15 and −0.018 V, corresponding to the oxidations of UA, DA and AA, respectively. It was found that an applied potential of −10 mV is sufficient for amperometric determinations of AA, which is significantly lower than those using other modified electrodes. This is a considerable advantage since the number of interferences in the amperometric determination increases drastically with increasing poise potential and most interferences are expected not to be oxidized at −10 mV. Our experiments showed that additions of 5.0 10^{-4} M DA and 1.0 10^{-3} M UA have no obvious effect on the amperometric response of 1.0 10^{-5} M AA.

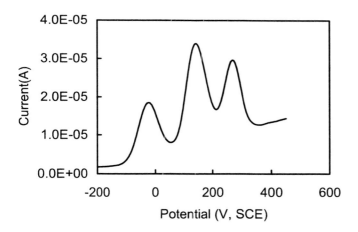

Fig. 5. Differential pulse voltammogram of 5.0 10^{-4} M UA, DA and AA the modified electrode. Potential scan rate: 10 mV s^{-1}, pulse amplitude: 50 mV.

Amperometric determinations of AA were carried out in PBS with the poise potential of -10 mV. Under optimal conditions, the amperometric response was linearly related to the AA concentration over the range of 5.0 10^{-6} to 4.0 10^{-3} M with a sensitivity of 14.2 $nA/\mu M$, an intercept of 67.3 nA and a correlation coefficient, r, of 0.998. The limit of detection, defined as a signal-to-noise ratio of 3:1, was found to be 2.0 10^{-6} M.

To confirm the applicability of the proposed method, we have employed it to determine AA in tomato juice and multi-vitamin tablets without any sample pretreatment For the multi-vitamin tablet analysis, three tablets of 100 mg

each were finely powered, dissolved in double distilled water and then made up to 100 ml. An aliquot of the sample solution was subsequently added to PBS and analyzed amperometrically. The results agreed well with the indicated vitamin C contents and the recoveries were found to be 97.5 ~ 101.9 %. Since the proposed amperometric method is very sensitive, a small aliquot of the tomato juice was adequate for the determination of AA and the average AA concentration in the tomato juice was found to be 9.85 mg ml^{-1} and the recoveries obtained were found to be 96.3 ~ 103.7%.

CONCLUSIONS

An MWNT film modified electrode was fabricated and used to investigate the electrochemical oxidation of AA in detail. In PBS, the modified electrode reduced the overpotential of AA oxidation and increased the oxidation current significantly, which clearly demonstrated the excellent electrocatalytic activity of the MWNT film toward the oxidation of AA. Due to a much less positive poise potential in the amperometric determination of AA, most interfering species, including DA and UA, cannot be oxidized at this potential. Thus a sensitive and highly selective amperometric procedure was developed for the determination of AA in real world samples with little sample pretreatment. Its sound results showed that the method is quite valuable and seems to be of great utility for further development.

REFERENCES

[1] S. Iilima *Nature* 354, 56-58, 1991.
[2] P. M. Ajayan *Chem Rev.* 99, 1787 – 1799, 1999.
[3] P. J. Britto, K. S. V. Santhanam and P. M. Ajayan *Bioelectrochem. Bioenerg.* 41, 121-125, 1996.
[4] P. J. Britto, K. S.V. Santhanam, V. Alonso, A. Rubio and P. M. Ajayan *Adv Mate.* 11, 154-157, 1999.
[5] J. J. Davis, R. J. Coles and H. A.O Hill *J Electroanal Chem.* 440, 279-282, 1997,
[6] J. Wang, M. Li, Z. Shi, N. Li, and Z. Gu *Anal Chem.* 74, 1993-1997, 2002,
[7] Z. Q. Gao, and H. Huang, *Chem. Comm.* 2107-2108, 1998.
[8] S. C. Tsang, Y. K. Chen, P. J. F. Harris and M. L. H. Green *Nature* 372, 159-162, 1994.
[9] J. Osteryoung and J. J. O'Dea, in *Electroanalytical Chemistry,* ed. A. J. Bard, Marcel Dekker, New York, vol. 14, 86-118, 1988.

Nanotube Stochastic Resonance: Noise–enhanced Detection of Subthreshold Signals

Ian Y. Lee, Xiaolei Liu, Bart Kosko*, and Chongwu Zhou**

Department of Electrical Engineering
University of Southern California
Los Angeles, CA 90089, USA
*kosko@sipi.usc.edu

ABSTRACT

Noise can help signal detection at the nano–level. Experiments on a single–walled carbon nanotube transistor confirm that a threshold exhibits stochastic resonance: a judicious amount of noise can help a threshold–like nanotube transistor detect subthreshold signals while large amounts of noise overwhelm the signals. The nanotube produced this stochastic–resonance effect using three types of *synchronized* discrete–time white noise and two performance measures: mutual information and input–output correlation. Experiments in a cryostatic vacuum chamber added Gaussian, uniform, and impulsive (Cauchy) electrical noise. The electrical noise corrupted a random digital (Bernoulli) voltage sequence that acted as the subthreshold input for the nanotube transistor. The noisy signal stimulated the transistor's gate and produced a sequence of random output (Bernoulli) current in the nanotube. Shannon's mutual information and simple correlation measured the nanotube system's performance gain by comparing the input and output sequences. Neither measure assumed any special nanotube structure. The observed nanotube SR effect was robust: it persisted even when infinite–variance Cauchy noise corrupted the signal stream. Such noise–enhanced signal processing at the nano–level promises applications to signal detection in wideband communication systems and biological and artificial neural networks.

Keywords: stochastic resonance, nanotube transistor, single-walled carbon nanotube, threshold system

Electrical noise can help carbon nanotube transistors detect subthreshold electrical signals by increasing the transistor's input–output mutual information or cross–correlation [1]. The threshold stochastic resonance (SR) occurs for various types of threshold units or neurons [2-7]. Nanotube experiments confirmed the specific SR prediction that simple memoryless threshold neurons exhibit SR for almost all finite–variance and infinite–variance noise types [8]. The experiments used three types of discrete–time additive white noise: Gaussian, uniform, and infinite–variance Cauchy electrical noise. White Gaussian noise gave the nonmonotonic signature of SR in figure 1. The modes of the mutual–information and cross–correlation curves occurred for nonzero noise strength with a standard deviation of at least 0.01.

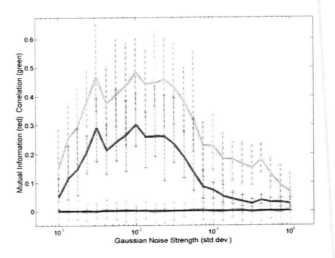

Figure 1: Stochastic resonance with additive white Gaussian noise (see reference [1]). The nanotube detector's mutual information (dark red curve) and zero-lag correlation (light green curve) increase for small amounts of noise and then decrease for larger amounts. The control experiments gave the flat non–SR mutual information (dark red line) and correlation (light green line) when no nanotube bridged the source and drain electrodes. The SR mode or optimal noise level was the same standard deviation value of 0.01 for both performance measures.

The experiments applied a nanotube transistor [9-14] whose threshold–like nonlinearity (figure 2) approximated a threshold detector. A semiconductor carbon nanotube acted as the nanometer–scale conduction channel of the transistor. A noise–corrupted voltage signal S or V stimulated the transistor's gate and produced a nano–amp scale output current Y or I for suprathreshold input voltages. The transconductance G related the drain–to–source current I to the gate voltage V and the threshold voltage V_T in a memoryless function: $I = G(V - V_T)$ if $V < V_T$ and zero otherwise. The nanotube transistor was undoped and exhibited p–type current–voltage characteristics so its transconductance G was negative.

The experiments produced the SR effect for the Shannon mutual information and the input–output correlation. Both measures compared the random input and output sequences. The detector performance computed the mutual information by subtracting the nanotube system's

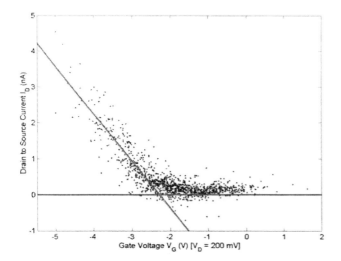

Figure 2 The nanotube transistor's threshold–like gate effect (see reference [1]). Each point shows the detector's response to a random input symbol. The experimental data showed that the nanotube detector behaved as a threshold in response to the noisy input signal stream. The gate effect showed little hysteresis.

output conditional entropy $H(Y|S)$ from its unconditional entropy $H(Y)$: $I(S, Y) = H(Y) - H(Y \mid S)$ where the input signal S was a random binary (Bernoulli) voltage and the output Y was a random transistor current. The detector performance computed the correlation measure [15] by removing the mean from the input and output sequences and normalizing the scalar zero–lag value of their cross–correlation sequence:

$$r_{SY}(l) = \sum_{k=1}^{N} s(k) y(k-l) \tag{1}$$

Each of the nanotube experiments applied 32 independent trials of 1000–symbol input sequences for 24 noise levels per type and over a range of gate voltages. The 24 sampled noise levels ranged from 0.001 to 1 standard deviation (dispersion for infinite–variance Cauchy) linearly in logarithmic scale. The noisy input was a *synchronized* Bernoulli sequence of independent random (subthreshold) ON/OFF values and additive white noise of three types. The discrete-time noise was white because the noise samples were uncorrelated in time. So the discrete-time Fourier transform was 2π–periodic and produced a flat noise power spectrum over the interval $[0, 2\pi]$ [16, 17]. Synchronization allows the nanotube systems to implement a variety of algorithms from signal processing and communications.

The ON/OFF values in figure 1 were ON = –1.6 volts and OFF = –1.4 volts. The input updated the symbols about once every 10 milliseconds. A 200–millivolt drain-source voltage biased the nanotube at room temperature in vacuum. The experiment measured and averaged 10

samples of the detector output at 100 kilo–symbols per second near the end of each symbol interval to estimate the output sequence.

The experiments made no assumptions about the nanotube structure. Nor did they impose a threshold scheme to interpret the detection. A histogram of the output sequences gave the discrete probability density function (pdf) $P(Y = Y_i) = p_i$ that computed the unconditional Shannon entropy:

$$H(Y) = -\sum_{i=1}^{N} p_i \ln p_i \tag{2}$$

for mutual information. A histogram of the sorted output sequences (based on the input symbols) gave the conditional output pdf $P_{Y|S} (Y = Y_i \mid S = S_j) = p_{ji} / p_j$ conditioned on the input symbols that computed the conditional entropy:

$$H(Y \mid S) = -\sum_{i=1}^{N} \sum_{j=1}^{N} p_{ji} \ln\left(\frac{p_{ji}}{p_j}\right) \tag{3}$$

The Shannon mutual information was the difference between the unconditional and the conditional entropies: $I(S, Y) = H(Y) - H(Y \mid S)$. A discrete correlation computed the cross–correlation sequence from the input and output symbol sequences. The zero–lag value of the cross–correlation sequence:

$$r_{SY}(0) = \sum_{k=1}^{N} s(k) y(k) \tag{4}$$

gave a scalar representation for the correlation measure. We converted the ON/OFF voltages to the bipolar values +1 and –1. This converted the input sequence into a nearly zero mean sequence. Subtracting the sample mean from the output sequence improved the match between similar input and output sequences. The normalized correlation measure divided the zero–lag cross correlation $r_{SY}(0)$ by the square root of the energy of the input and the output sequences:

$$C(S,Y) = \frac{\sum_{k=1}^{N} s(k) y(k)}{\sqrt{\sum_{k=1}^{N} s(k) s(k)} \sqrt{\sum_{k=1}^{N} y(k) y(k)}} \tag{5}$$

where the energy of a sequence is the same as the zero–lag value of its autocorrelation:

$$|x| = \sum_{k=1}^{N} x^2(k) = \sum_{k=1}^{N} x(k) x(k-l) \Big|_{l=0} = r_{xx}(0) \tag{6}$$

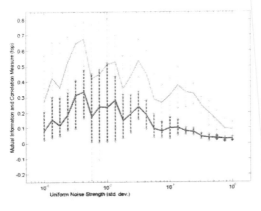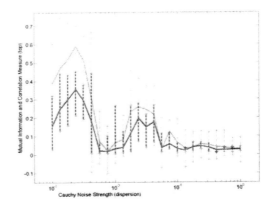

Figure 3: (a) Stochastic resonance with additive white uniform noise (see reference [1]). The noisy subthreshold input produced a clear SR response in both mutual information (dark red curve) and input-output correlation (light green curve) just as with additive white Gaussian noise. The figure shows the SR effect for the subthreshold signal ON = –1.8 V and OFF = –1.6 V. The SR mode is at 0.04 standard deviations. (b) Robust stochastic resonance with additive white Cauchy noise [1]. This highly impulsive noise has infinite variance and infinite higher-order moments. The figure shows an approximate SR effect for the subthreshold signal ON = –2 V and OFF = –1.8 V. The SR mode lies at about the 0.003 dispersion value. Several SR researchers have found multiple modes in the plot of system performance against noise strength [50-52].

The experiment found the SR effect for mutual information and correlation for Gaussian and uniform noise and for four combinations of binary symbols (–2.0, –1.8), (–1.8, –1.6), (–1.6, –1.4), and (–1.4, –1.2) volts. Figure 1 shows the SR effect for additive white Gaussian noise and the subthreshold signal pair ON = –1.6 V and OFF = –1.4 V. The SR mode of the mutual–information curve is six times the value at minimal noise. The SR mode of the correlation curve is three times the value at minimal noise. Figure 3a shows the SR effect for additive white uniform noise and the signal pair ON = –1.8 V and OFF = –1.6 V.

We also passed impulsive or infinite–variance white noise through the nanotube detector to test whether it was robust to occasional large noise spikes. We chose the highly impulsive Cauchy noise [2] for this task. This infinite–variance noise probability density function had the form:

$$p(n) = \left(\frac{1}{\pi}\right)\frac{\gamma}{n^2 + \gamma^2} \qquad (7)$$

for zero location and finite dispersion γ. Figure 3b shows that a diminished SR effect still persists for Cauchy noise with subthreshold signal pair ON = –2.0 volts and OFF = –1.8 volts. Not all Cauchy experiments produced a measurable SR effect.

We note that the SR results occurred in a nanotube transistor that shows time–varying hysteresis [18-20] and sensitivity to adsorbed molecules [21-23].

These SR results suggest that nanotubes can exploit noise in other signal processing tasks if advances in nanotube device technology can overcome the problems of hysteresis and parasitic capacitance that affect logic circuits [24] and high-frequency signals [25]. The nanotube signal

detectors might apply to broadband [26, 27] or optical communication systems [28] that use sub-micro amp currents and attenuated signals in noise because our nanotube detectors used nano-amp current and could distinguish between subthreshold binary symbols. The detectors might apply to parallel signal processing [29] at the nano level because they could have small minimum feature size [30] in vast parallel arrays of nanotubes. The parallel detectors could apply to spread spectrum communications: each nanotube can act as an antenna [31] that matches a separate frequency channel [32] in frequency hopping and perhaps in other types of spread spectrum communications [33]. A nanotube's length can code for a given frequency [34] while chemical adsorption can tune a nanotube's threshold [22, 23]. The detectors might apply to chemical detection and parallel field programming by tuning the threshold chemically. The nanotube detectors can also operate in a biological environment such as saline solution [35]. The nanotube detectors could interface with biological systems because an electrolyte can act as their gate [35, 36]. The nanotube detectors might also help implement pulse-train neural networks and exploit noise in biological [37-48] or robotic systems because the detectors are threshold devices similar to spiking neurons [49].

REFERENCES

[1] Lee, I. Y.; Liu, X.; Kosko, B.; Zhou, C.; *Nano Letters* **3** (12), 1683-1686, 2003.

[2] Kosko, B.; Mitaim, S. *Neural Networks* **16**, 755, 2003.

[3] Kosko, B.; Mitaim, S. *IJCNN'02 IEEE Proceedings of the International Joint Conference on Neural Networks* **2**, 1980, 2002.

[4] Moss, F.; Pierson, D.; O'Gorman, D. *International Journal of Bifurcation and Chaos* **6**, 1383-1397, 1994.

[5] Gingl, Z.; Kiss L.; Moss, F. *Europhysics Letters* **29**, 191, 1995.

[6] Inchiosa, M. E.; Robinson, J. W. C.; Bulsara, A. R. *Physical Review Letters* **85**, 3369-3372, 2000.

[7] Gammaitoni, L. & Bulsara, A. R. *Physical Review Letters* **88**, 230601, 2002.

[8] Kosko, B.; Mitaim, S. *Physical Review E* **64**, 051110, 2001.

[9] Tans, S. J.; Verschueren, R. M.; Dekker, C. *Nature*, **393**, 49, 1998.

[10] Zhou, C.; Kong, J.; Dai, H. *Applied Physics Letters*, **76**, 1597, 2000.

[11] Martel, R.; Schmidt, T.; Shea, H.R.; Hertel, T.; Avouris Ph. *Applied Physics Letters*, **73**, 2447, 1998.

[12] Cheung, C. L.; Kurtz, A.; Park, H.; Lieber, C. M. *Journal of Physical Chemistry B*, **106**, 2429, 2002.

[13] Liu, J. et al., *Science*, **280**, 1253, 1998.

[14] Saito, R.; Fujita, M.; Dresselhaus, G.; Dresselhaus, M. S. *Applied Physics Letters*, **60**, 2204, 1992.

[15] Collins, J. J.; Chow, C. C.; Capela, A. C.; Imhoff, T. T. *Physical Review E*, **54**, 5575, 1996.

[16] Oppenheim, A.V.; Schafer, R.W.; Buck, J.R. *Discrete-Time Signal Processing, Second Edition*; Prentice-Hall, Inc.: Upper Saddle River, NJ, 1999.

[17] Ingle, V. K.; Manolakis, D. G.; Kogon, S. *Statistical and Adaptive Signal Processing: Spectral Estimation, Signal Modeling, Adaptive Filtering and Array Processing*; McGraw-Hill Co.: New York, NY, 1999

[18] Kim, W. et al. *Nano Letters*, **3**, 193, 2003.

[19] Fuhrer, M. S.; Kim, B. M.; Durkop, T.; Bringlinger, T. *Nano Letters*, **2**, 757, 2002.

[20] Radosavljevic, M.; Freitag, M.; Thadani, K. V.; Johnson, A. T. *Nano Letters*, **2**, 761, 2002.

[21] Ong, K. G.; Zeng, K.; Grimes, C. A. *IEEE Sensors Journal*, **2**, 82, 2002.

[22] Collins, P. G.; Bradley, K.; Ishigami, M.; Zettl, A. *Science*, **287**, 1801, 2000.

[23] Kong, J.; Franklin, N. R.; Zhou, C.; Chapline, M. G.; Peng, S.; Cho, K.; Dai, H.; *Science*, **287**, 622, 2000.

[24] Bachtold, A.; Hadley, P.; Nakanishi, T.; Dekker, C. *Science*, **294**, 1317, 2001.

[25] Harris, D.; Horowitz, M. A. *IEEE Journal of Solid-state Circuits*, **32**, 1702-1711, 1997.

[26] Bulsara, A. R.; Zador, A. *Physical Review E*, **54**, R2185, 1996.

[27] Barbay, S.; Giacomelli, G.; Marin, F. *Physical Review E*, **63**, 051110-051118, 2000.

[28] Chen, Y.-C et al., *Applied Physics Letters*, **81**, 975, 2002.

[29] Stocks, N. G. *Physical Review E*, **63**, 041114-041122, 2001.

[30] Rochefort, A.; Ventra, M. D.; Avouris, Ph. *Applied Physics Letters*, **78**, 2521, 2001.

[31] Liang, W.; Bockrath, M.; Bozovic, D.; Hafner, J. H.; Tinkham, M.; Park, H. *Nature*, **411**, 665 – 669, 2001.

[32] Blake, L. V. *Antennas second edition*; Artech House Inc.: Dedham, MA, 1984.

[33] Irmer, R.; Fettweis, G.; *The 5th International Symposium on Wireless Personal Multimedia Communications*, **2** (27-30), 412- 416, 2002.

[34] Huang, S.; Cai, X.; Liu, J. *Journal of American Chemical Society,* **125** (19), 5636 –5637, 2003.

[35] McEuen, P. L.; Fuhrer, M. S.; Park, H. *IEEE Transactions on Nanotechnology*, **1**, 78, 2002.

[36] Kruger, M.; Buitelaar, M. R.; Nussbaumer, T.; Schonenberger, C.; Forro, L. *Applied Physics Letters*, **78**, 1291, 2001.

[37] Russel, D. F.; Wilkens, Lon. A.; Moss, F. *Nature*, **402**, 291, 1999.

[38] Jaramillo, F.; Wiesenfeld, K. *Nature Neuroscience*, **1**, 384, 1998.

[39] Bulsara, A. R.; Boss, R. D.; Jacobs, E. W. *Biological Cybernetics*, **61**, 211–222, 1989.

[40] Pantazelou, E.; Dames, C.; Moss, F.; Douglass, J.; Wilkens, L. *International Journal of Bifurcation and Chaos*, **5**, 101–108, 1995.

[41] Pei, X.; Wilkens, L.; Moss, F. *Journal of Neurophysiology*, **76**, 3002–3011, 1996.

[42] Douglass, J. K.; Wilkens, L.; Pantazelou, E.; Moss, F., *Nature*, **365**, 337–340, 1993.

[43] Levin, J. E.; Miller, J. P. *Nature*, **380**, 165–168, 1996.

[44] Miller, J. P.; Jacobs, G. A.; Theunissen, F. E. *Journal of Neurophysiology*, **66**, 1680–1689, 1991.

[45] Miller, J. P.; Jacobs, G. A.; Theunissen, F. E. *Journal of Neurophysiology*, **66**, 1690–1703, 1991.

[46] Braun, H. A.; Wissing, H.; Sch¨afer, K.; Hirsch, M. C. *Nature*, **367**, 270–273, 1994.

[47] Gluckman, B. J.; Netoff, T. I.; Neel, E. J.; Ditto, W. L.; Spano, M. L.; Schiff, S. J.; *Physical Review Letters*, **77**, 4098–4101, 1996.

[48] Collins, J. J.; Imhoff, T. T.; Grigg, P. *Journal of Neurophysiology*, **76**, 642–645, 1996.

[49] Rieke, F.; Warland, D.; Steveninck, RdRv. *Spikes. Exploring the neural code*; MIT press: Cambridge, Massachusetts, London, England 1999.

[50] Vilar, J. M. G.; Rubí, J. M. *Physical Review Letters*, **78**, 2882-2885, 1997.

[51] Lindner, J. F.; Breen, B. J.; Wills, M. E.; Bulsara, A. R.; Ditto, W. L. *Physical Review E*, **63**, 051107, 2001.

[52] Matyjaskiewicz, S.; Krawiecki, A.; Holyst, J. A.; Kacperski, K.; Ebeling, W. *Physical Review E*, **63**, 026215, 2001.

Synthesis and Characterization of HighTemperature Stable Nanocomposite Catalysts

M. Kirchhoff[*], U. Specht[*] and G. Veser[*,**]

[*] Max-Planck-Institute for Coal Research, 45470 Mülheim an der Ruhr, Germany
[**] Chemical Engineering Department, University of Pittsburgh, Pittsburgh, PA, USA, gveser@pitt.edu

ABSTRACT

Unusually active and sinter-resistant nanocomposite materials were synthesized by combining the high reactivity of nanosized noble metal particles with the excellent high-temperature stability of hexa-aluminates through a simple one-step microemulsion-templated sol-gel synthesis. The novel nanocomposite catalysts were characterized by TEM, XRD, DTA and BET measurements after calcination at temperatures between 600°C and 1300°C, and were tested in the high-temperature partial oxidation of methane. The materials showed very high activity and selectivity along with excellent stability with respect to surface area, noble metal dispersion and catalytic activity over prolonged operation at reaction temperatures between 900 and 1100°C. To our knowledge, this is the first time that metal nanoparticles have been stabilized to such extreme high temperature conditions. We see these novel materials therefore not only as very promising candidates for high-temperature catalytic applications, but generally view this as a possible route to expand the parameter range for nanoparticle applications.

Keywords: nanocomposite materials, high-temperature stability, heterogeneous catalysis, hexaaluminates, methane oxidation

1 INTRODUCTION

The large surface areas of nanoscale particles as well as reports on novel chemical reactivity of particles with nanometer dimensions makes these materials highly interesting for heterogeneous catalysis. A particularly simple and well-controlled way to produce such materials is through micelle-templated syntheses, since the self-assembling properties of surfactants offer unique possibilities for the engineering of nanostructured materials [1-3]. Beyond allowing the template-directed synthesis of pore morphologies for nano- and mesoporous materials, the nanometer-sized droplets in reverse (water-in-oil) microemulsions also constitute a controlled environment for chemical reactions, such as sol-gel reactions, in which the micelles act as individual 'nano-reactors' [4-6].

However, the use of such nanoscale materials is to-date restricted to low-to-moderate temperature conditions (T<500°C), since the thermal stability of particles decreases rapidly with decreasing diameter. This imposes severe limitations for the application of these materials, particularly as catalysts for chemical processes. Here, the conversion of hydrocarbons to hydrogen, synthesis gas and olefins at high-temperature, millisecond contact-time conditions has found much interest in recent years as a novel way for more efficient utilization of fossil and biomass-derived hydrocarbon resources [7-12]. While the high reactivity of noble metal nanoparticles would make such materials in principle ideally suited for millisecond contact-time catalysis, the extreme reaction temperatures (T = 800–1200°C) currently preclude such use.

In this contribution we report on the successful combination of the high reactivity of nanosized metal particles with the high-temperature stability of ceramics, resulting in unusually active and highly sinter-resistant nanocomposite materials for high-temperature catalytic applications.

2 SYNTHESIS

The synthesis of the nanocomposite materials is based on the simultaneous synthesis of a ceramic hexa-aluminate matrix and metal nanoparticles, using a microemulsion-templated sol-gel route. Figure 1 shows a schematic overview of the sequential synthesis steps: An aqueous noble metal salt solution, isooctane and one out of a number of different non-ionic surfactants are combined to yield a reverse (water-in-oil) microemulsion. Next, a solution of aluminum- and barium-isopropoxide in iso-propanol is added to the microemulsion, and the reaction mixture is aged for up to 72 hours under constant stirring at room temperature.

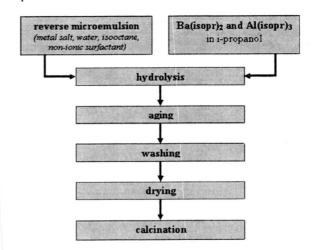

Fig. 1: Steps in the synthesis of noble metal barium-hexaaluminate catalysts.

Following these steps, the microemulsion is cooled to -20°C to achieve phase separation via thermally induced phase separation (TIPS). The oil phase is decanted and the product phase is washed several times to remove the remaining surfactant. Finally, the residue is dried in high vacuum and calcined in air at various temperatures between 600°C and 1300°C.

This procedure yields a fine powder which was used as synthesized for characterization as well as reactive tests.

3 CHARACTERIZATION

Pure barium hexaaluminate (BHA) samples as well as Pt-BHA nanocomposite materials which were synthesized as described above were characterized through XRD, DTA, BET and TEM measurements as described in the following.

3.1 Barium Hexaaluminate (BHA)

XRD measurements of a pure BHA sample synthesized according to the above describe route (without the addition of metal salt solutions to the microemulsion) reveal a largely amorphous ceramic structure at the lowest calcination temperatures of 600°C which persists even at temperatures as high as 1100°C (see figure 2). Only at calcination temperatures in excess of 1200°C the ordered structure of the high-temperature stable barium-hexaaluminate (BHA) structure becomes detectable.

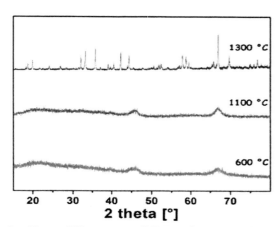

Fig. 2: X-ray diffractogram (XRD) of BHA synthesized in a reverse microemulsion and calcined at temperatures of 600°C, 1100°C and 1300°C.

This transition agrees well with DTA measurements of BHA samples shown in figure 3, which indicate a crystallization point of around 1200°C. As can be expected, the re-arrangement of atoms occurring upon crystallization of the hexaaluminate phase leads to a collapse of the large surface area.

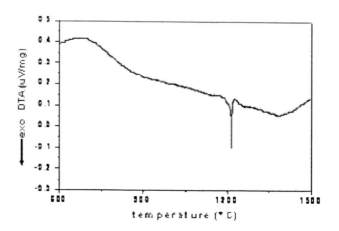

Fig. 3: DTA scan of a pure BHA sample.

The structural stability of the ceramic matrix is furthermore reflected in the change of BET surface area of these materials with increasing temperature (see fig. 4): the materials as synthesized show very large surface areas in excess of 300 m²/g even after calcinations at 600°C. This large surface area is preserved fairly well up to calcinations temperatures as high as 1100°C, where surface areas of around 200 m²/g were measured. However, upon crossing the crystallization point for the ceramic host material (T>1200°C), the BET surface area collapses to values of 20 m²/g and less.

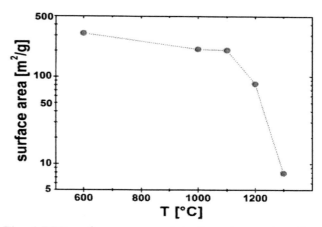

Fig. 4 BET surface area vs calcinations temperature for a BHA sample after calcination for 5 hours in air at increasing temperatures.

3.2 Pt Nanoparticles

The above described characteristics of the hexaaluminate matrix remained unchanged upon the addition of noble metal components to the synthesis. TEM investigations of the nanocomposite materials show that the samples consist of loose agglomerates of nanoparticles with diameters of about 5 – 15 nm after calcination at 600°C (figure 5). Noble metal and hexaaluminate form separate but finely intermixed phases, i.e. the materials consist of metallic nanoparticles which are homogeneously dispersed in the matrix of the ceramic host material.

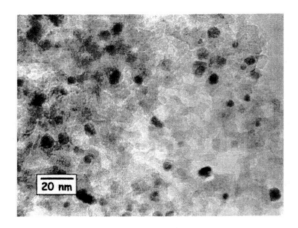

Fig. 5: TEM image of a Pt-BHA sample showing Pt nanoparticles (black) embedded in a matrix of BHA nanoparticles (grey). The sample shown had been calcined in air at 600°C for 5 h.

Both the particle distribution as well as the size of the Pt nanoparticles show very little change up to very high temperatures, indicating that through the reverse-microemulsion synthesis route, the metal nanoparticles appear to "inherit" some of the high-temperature stability of the ceramic host materials. TEM measurements show that the metal nanoparticles are stable up to temperatures around 1300°C (figure 6). However, both XRD and TEM also indicate that a small number of metal nanoparticles show an early sintering at lower temperatures which then dominates the XRD spectrum (T> 700°C, figure 7).

Interestingly, the unusual stability of the noble metal nanoparticles is a highly sensitive function of the exact synthesis conditions. Several tests with post-impregnation

of pure BHA powders and even apparently minor modifications of the synthesis steps showed that only the above described simultaneous synthesis of metal and ceramic nanoparticles yields the reported high-temperature stable metal nanoparticle sizes and distribution.

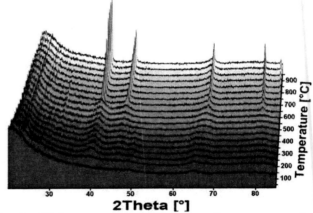

Fig. 6: High-temperature XRD scans of a nanocomposite Pt-BHA catalyst showing the onset of some sintering of Pt nanoparticles at about 700°C..

4 CATALYTIC TESTS

The catalytic activity of the nanocomposite materials was tested in a simple tubular quartz-glass reactor using high-temperature catalytic partial oxidation of methane to synthesis gas as test reaction. We have previously reported on the drastic improvements in syngas selectivities as well as methane conversion over a Pt-BHA composite catalyst derived from a reverse microemulsion synthesis compared to a conventional Pt-coated alumina foam monolith [13].

Here, we highlight the extremely high reactivity of the obtained nanocomposite materials by showing results from an investigation in which we studied the influence of the platinum loading incorporated in the Pt-BHA catalyst. A series of syntheses was conducted in which the amount of platinum salt added during the material synthesis was

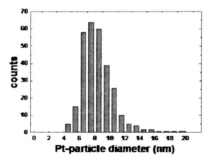

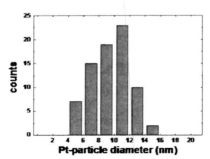

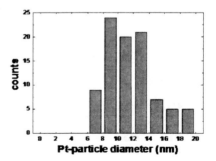

Figure 7: Particle statistics from TEM images of Pt-BHA catalysts after calcination at T= 600°C (left), 1100°C (middle) and 1300°C (right). Very little sintering of the Pt particles is occurring with mean diameters increasing from 8.4 to 9.2 and 12.3 nm, respectively.

varied, yielding materials with a Pt weight loading between 0 and 30 wt%. TEM investigations showed that this increase in Pt-content in the material results in a higher platinum particle density while leaving particle sizes unchanged at about 5 to 10 nm. Also, the homogeneity of the distribution of the Pt-particles in the ceramic matrix remained unaffected.

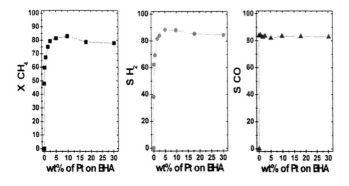

Fig. 7: Methane conversion, hydrogen selectivity and CO selectivity (from left to right) versus Pt weight loading of the Pt-BHA catalysts. Experimental conditions: $CH_4:O_2:N_2$ = 2:1:4, total inlet gas flow at standard conditions: 0.24 m^3/s, adiabatic reactor operation.

The influence of the Pt amount in the Pt-BHA catalyst on the catalytic activity of the material is shown in figure 7. While pure BHA does not show any significant activity towards partial oxidation at any temperature studied in this investigation, even very small amounts of platinum already result in a noticeable catalytic activity and selectivity: A Pt-BHA catalyst with as little as 0.06 wt% Pt gave a methane conversion of 48% and a hydrogen selectivity of 38%. It seems remarkable that both oxygen conversion (not shown) and CO selectivity jump immediately upon Pt addition to their final values of 100% and ~82%, respectively.

Increasing the platinum content further, methane conversion and hydrogen selectivity increase very quickly to 60 % and 62%, respectively, at 0.24 wt% Pt, and reach a maximum of about 82% and 88%, respectively, at about 5 wt% Pt. Higher Pt-loadings do not lead to any significant changes in selectivity or conversion any more.

These results appear truly remarkable, since at the lowest Pt loadings and our typical reactor catalyst loading of 50 mg, the catalyst sample contains about 30 *micro*grams of platinum with significant conversion and yield at very high gas flow rates! This metal loading is several orders of magnitude below the loading that is necessary in conventional supported Pt catalysts and indicates the highly unusual catalytic activity of these nanocomposite materials.

Finally, it should be noted that even with the low weight loadings we did not see any catalyst deactivation over the course of our experiments (at least several hours of continuous operation), which confirms our previous findings that the high-temperature stability of these catalysts is in fact maintained at reaction conditions [13].

5 SUMMARY

A novel type of catalyst material for high-temperature applications was synthesized via a reverse-microemulsion templated sol-gel synthesis in which noble metal and ceramic nanoparticles are synthesized simultaneously. The resulting material shows large, high-temperature stable surface areas and a very homogeneous distribution of noble metal nanoparticles dispersed in a ceramic host matrix.

The ceramic phase is largely amorphous up to about 1200°C, while the metal nanoparticles show a rather narrow size distribution and very little sintering up to the same temperature range.

The catalysts were tested in the autothermal partial oxidation of methane to synthesis gas. They show excellent and selectivity at extremely low noble metal loadings (below 1 wt%). Most importantly, the materials appear high-temperature stable even after prolonged operation at reactive conditions

To our knowledge, this is the first time that metal nanoparticles were successfully stabilized to temperatures in excess of 800°C. We see these materials therefore not only as very promising candidates for high-temperature catalytic applications – such as hydrogen production by catalytic partial oxidation of hydrocarbon fuels - but generally view this as a possible route to expand the parameter range for nanoparticle applications.

REFERENCES

1. Schmidt-Winkel, P., et al., Chem. Mat., 2000. 12(3): 686-696.
2. Pileni, M.P., Langmuir, 1997. 13(13): 3266-3276.
3. Pileni, M.P., Supramolecular Science, 1998. 5(3-4): 321-329.
4. Pileni, M.P., et al., Langmuir, 1998. 14(26): 7359-7363.
5. Zarur, A.J., et al., Langmuir, 2000. 16(24): 9168-9176.
6. Zarur, A.J., H.H. Hwu, and J.Y. Ying, Langmuir, 2000. 16(7): 3042-3049.
7. Schmidt, L.D., M. Huff, and S. Bharadwaj, Chem. Eng. Sci., 1995. 49: 3981-3994.
8. Goetsch, D. and L.D. Schmidt, Science, 1996. 271: 1560-1562.
9. Bodke, A.S., et al., Science, 1999. 285(5428): 712-715.
10. Bizzi, A., et al., Chem. Eng. J., 2002. 90(1-2): 97-106.
11. Beretta, A., E. Ranzi, and P. Forzatti, Chem. Eng. Sci., 2001. 56(3): 779-787.
12. Veser, G. and J. Frauhammer, Chem. Eng. Sci., 2000. 55(12): 2271-2286.
13. Schicks, J., et al., Catal. Today, 2003. 83: 287-296.

Nanostructure of a High-Permeability, Hydrogen-Selective Inorganic Membrane

S. T. Oyama, D. Lee, P. Hacarlioglu, Y. Gu, R. F. Saraf

Virginia Tech, Dept. of Chemical Engineering, Blacksburg, VA, USA, oyama@vt.edu

ABSTRACT

This paper describes the preparation and properties of an inorganic membrane with permeability for H_2 comparable to palladium and with over 99.9% selectivity over larger species like CO, CO_2 and CH_4. The membrane is a composite formed by the deposition of a thin, 20 nm SiO_2 layer on an alumina support. The alumina support is obtained by the deposition of a boehmite sol on top of a porous substrate, so as to create a uniform structure with small pore sizes. The permeation of the small gas species, H_2, He, and Ne through the silica layer is analyzed in detail in order to obtain insight about the transport mechanism and the structure of the silica. The order of permeance through the silica layer is highly unusual, He > H_2 > Ne, following neither molecular weight nor size. The order of permeation is quantitatively explained using a statistical mechanics approach, which takes into consideration the density of solubility sites for the various species and the vibrational frequency of the species within the sites. An extension of the Masaryk-Fulrath treatment for glasses combined with the Percus-Yevick model is used to estimate the vibrational frequency (7.0×10^{12} s^{-1}), solubility site density (3.0×10^{26} m^{-3} for H_2) and the average distance between sites (0.84 nm). This is the first time an inorganic membrane has been described in detail at the nanometer level.

Keywords: silica/alumina membrane, hydrogen permeation, statistical mechanics theory, solubility sites

1 INTRODUCTION

Silica-based membranes have been studied extensively because of their excellent properties in selective hydrogen permeation. The first membranes were obtained [1,2,3] by placing silica inside the pores of Vycor glass substrates using reactive chemical vapor deposition (CVD). In a subsequent development it was found that better results were obtained by placing the silica on the outer surface of the substrate to form a thin compact layer [4]. Until recently the mechanism of hydrogen permeation through silica membranes has not been studied in detail. In this paper we develop the theory of gas permeation through the silica layer building upon an existing description for vitreous silica glasses [5]. A key aspect of the existing theory is the presence of solubility sites for the permeating species. In this work we calculate the distance between solubility sites using the Percus-Yevick treatment for a dense liquid, which assumes it is composed of random, non-interacting spheres.

The order of permeance through the silica layer is highly unusual, He > H_2 > Ne, following neither molecular weight nor size. This is the same order as observed in vitreous silica glass, but occurs with lower activation energies in the silica layer. The order of permeation is explained for the first time using a statistical approach, which takes into consideration the density of solubility sites for the various species and the vibrational frequency of the species within the sites.

The membranes are generally prepared by chemical vapor deposition of a silica precursor at low temperatures using H_2O, O_2 or O_3 as co-reactants. This work describes the preparation of a special highly hydrogen permeable silica membrane, referred to as Nanosil, obtained by an adaptation of the method. Instead of using low temperatures, the SiO_2 layer is deposited at high temperatures by thermal decomposition. This gives rise to a composite membrane with excellent selectivity (~10^3) for the small gas molecules (He, Ne and H_2) over other larger gas molecules (CO_2, CO, and CH_4).

2 EXPERIMENTAL

The silica/alumina membrane was prepared by depositing a thin silica layer on a porous γ-alumina support by the thermal decomposition of tetraethylorthosilicate (TEOS) at 873 K in an argon stream. The membrane support used in this study was purchased from US Filter (Part No. S700-0011), and had a tubular geometry with an outside diameter of 10 mm and a thickness of 1.5 mm. This membrane support had a multi-layered structure consisting of a coarse α-Al_2O_3 tube coated with finer layers of α-Al_2O_3 and an inner top layer of γ-Al_2O_3 of average 5 nm pore size. A 4 cm section of the alumina membrane support was connected at both ends to two pieces of dense alumina tubing using a high temperature glass glaze (Duncan, IN, Part No. 1001). Gas tight connections between the membrane support and the dense tubing were obtained after 0.5 h of thermal treatment at 1150 K.

After the joint connection, an additional γ-Al_2O_3 layer was introduced on top of the existing γ-Al_2O_3 layer of the alumina support to reduce defects or pinholes that give rise to a low hydrogen selectivity in the silica layer. A 0.05 M dispersion of boehmite (γ-AlOOH) sol was prepared following the method reported by Uhlhorn [6]. Aluminum tri-sec-butoxide (Aldrich, 97%) was added to boiling water with vigorous stirring, and 0.07 mole HNO_3 per mole butoxide was added. This colloidal solution was boiled until most of the butanol was evaporated, then was refluxed for 20 h. Polyvinylalcohol (PVA, Fluka, M.W. 72000) solution was prepared separately by adding 3.5 g of PVA to 100 cm^3 of boiling water followed by 5 cm^3 of 1 M HNO_3. The PVA solution was then refluxed for 4 h. A final 0.05 M boehmite sol was prepared after adding 660 cm^3 of the PVA solution per mol of boehmite followed by stirring for 3 h at 353 K. The alumina support tubing was dip coated with the 0.05 M boehmite sol for 10 s, dried for 24 h at room temperature, and then calcined at 873 K for 24 h (heating rate 1 K min^{-1}).

For the CVD of the silica layer, the alumina support substrate was installed concentrically inside another piece of glass tubing of 14 mm inside diameter using machined Swagelok fittings with Teflon ferrules. After placing the assembly in an electric furnace, argon gas flows were introduced on the outer shell side (19 $\mu mol\ s^{-1}$) and inner tube side (15 $\mu mol\ s^{-1}$) of the reactor (flow rates in $\mu mol\ s^{-1}$ may be converted to $cm^3\ min^{-1}$ (NTP) by multiplying by 1.5), and the temperature was raised to 873 K. A flow of tetraethylorthosilicate (TEOS, Aldrich, 98%) was introduced on the inside of the porous alumina substrate using a bubbler (at 298 K) with argon (4 $\mu mol\ s^{-1}$) as a carrier gas. This stream was mixed with the tube stream of argon before introducing it to the tube side to produce a stream with a TEOS concentration of 0.02 mol m^{-3} (0.045 mol %). The synthesis of the silica membrane was studied by varying the silica deposition time. The CVD process was interrupted at various times and the permeance of H_2, CH_4, CO, and CO_2 were measured at different deposition times at 873 K

General gas permeation measurements were conducted in the temperature range of 373 – 873 K by flowing 40 $\mu mol\ s^{-1}$ of a pure gas at 160 kPa through the inner tube. The permeation rate of each gas exiting from the shell side of the reactor assembly was measured with a sensitive bubble flow meter at atmospheric pressure. The permeance of gas was obtained from the expression $Q_i = F_i / A\ \Delta P_i$, where Q_i is the permeance (mol $m^{-2}\ s^{-1}\ Pa^{-1}$) of species i, F_i is the gas flow rate on the shell side (mol s^{-1}), A is the surface area (m^2) of the membrane section, and ΔP_i is the pressure difference (Pa) between the shell and tube side. For higher sensitivity the permeance of the gases was also measured with a gas chromatograph (GC) equipped with a thermal conductivity detector (SRI, Model 8610B). The tube side gas flow rates and pressure conditions were the same as above, however, on the shell side an argon flow was introduced as a sweep gas for the permeated gas. The shell side outlet gas flow rate was measured using a bubble flow meter, and the flow was injected into the GC to obtain the concentration of the permeated gas. The permeance was then calculated using the outlet gas flow rate and the concentration of the permeated gas on the shell side.

3. RESULTS and DISCUSSION

The evolution of the gas permeance on the silica membrane was measured at 873 K as a function of the silica deposition time to monitor the formation of the silica layer on the membrane support. The results are shown in Fig. 1.

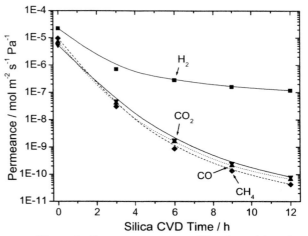

Figure 1. Gas permeance vs. silica deposition time.

Initially, the fresh alumina support showed very high permeance for all of the gases ($H_2 = 2.2 \times 10^{-5}$, $CH_4 = 9.7 \times 10^{-6}$, $CO = 7.1 \times 10^{-6}$, $CO_2 = 5.4 \times 10^{-6}$ mol $m^{-2}\ s^{-1}\ Pa^{-1}$). The permeance of all the gases decreased with silica deposition time. For hydrogen, the permeance decreased rapidly after 3 h of silica deposition to the order of 10^{-7} mol $m^{-2}\ s^{-1}\ Pa^{-1}$, and then decreased slowly with further silica deposition. In contrast, the permeance of CH_4, CO, and CO_2 showed a continuous and rapid drop with silica deposition. After 12 h of deposition the permeance of hydrogen remained at 1.2×10^{-7} mol $m^{-2}\ s^{-1}\ Pa^{-1}$, whereas the permeance of the other gases dropped off significantly ($CH_4 = 4.3 \times 10^{-11}$, $CO = 6.7 \times 10^{-11}$, $CO_2 = 8.0 \times 10^{-11}$ mol $m^{-2}\ s^{-1}\ Pa^{-1}$). The results indicate that a complete silica layer was formed on the alumina support after 12 h of deposition, and that the layer was selective for H_2 transport while significantly excluding passage of CH_4, CO, and CO_2 through the membrane. Before the silica deposition, the permeance order of the gases through the alumina support was $H_2 > CH_4 > CO > CO_2$ showing an inverse dependence on the molecular weight of the gases, in agreement with the Knudsen diffusion mechanism. However, after 3 h of silica deposition the permeance order changed to $H_2 > CO_2 > CO > CH_4$, which followed molecular size ($H_2 = 0.289$ nm, $CO_2 = 0.33$ nm, $CO = 0.376$ nm, $CH_4 = 0.38$ nm [7]), and this order was retained with further silica deposition. This gives evidence that the mechanism of molecular differentiation by the silica layer is through size selectivity.

The hydrogen selectivity over other gases is shown as a function of hydrogen permeance in Fig. 2. The selectivity for hydrogen on the fresh alumina support was low for all the gases. The selectivity was characteristic of Knudsen diffusion. After 3 h of silica deposition the selectivity for hydrogen showed a slight increase accompanied by a large decrease in H_2 permeance from the order of 10^{-5} to 10^{-7} mol m^{-2} s^{-1} Pa^{-1}. The H_2 selectivity then increased rapidly with further silica deposition, with only a small drop in hydrogen permeance. After 12 h of silica deposition the hydrogen selectivity of the membrane increased to over 1000 for all the gases (CH_4: 2800, CO: 1800, CO_2: 1500). This is a purity above 99.9 %.

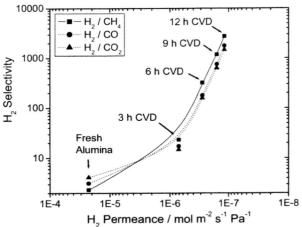

Figure 2. Selectivity of hydrogen vs. H_2 permeance.

The temperature dependence of small gas transport through the silica membrane obtained after 12 h of silica deposition was investigated by measuring the permeance of He, H_2 and Ne at various temperatures (373 – 873 K). It was found that the permeance of these gases through the silica membrane was activated, and increased with temperature. This differed from the permeance in the fresh alumina membrane support where the permeance decreased with temperature in accordance with the Knudsen transport mechanism.

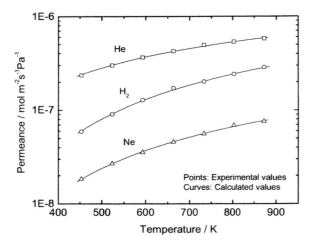

Figure 3. Theoretical and Experimental Permeance Curves.

The permeance order of these gases was unusual

$$He > H_2 > Ne$$

as it did not follow the size (He = 0.26 nm, H_2 = 0.289 nm, Ne = 0.275 nm [7]) nor the mass of the species (He = 4.0 au, H_2 = 2.01 au, Ne = 20.1 au).

The results can be explained [8] using an equation originally derived to describe permeability of monatomic gases in vitreous glass that was based on a mechanism involving jumps between solubility sites [5]. The equation used a classical statistical mechanics approach [9] and assumed equilibrium sorption in the sites [10], random motion, and a transition state with two degrees of vibrational freedom and one degree of translational freedom. For the case of hydrogen this equation needs to be adapted to account for the partial loss of rotational degrees of freedom as the molecule passes through the doorways.

$$Q = \frac{1}{6L}\left(\frac{d^2}{h}\right)\left(\frac{h^2}{2\pi m k T}\right)^{\frac{3}{2}}\left(\frac{\sigma h^2}{8\pi^2 I k T}\right)^{0.2}\frac{(N_S/N_A)}{(e^{hv'/2kT} - e^{-hv'/2kT})^2}e^{-\Delta E_K/RT} \quad (1)$$

In this equation Q is the permeance of a gas, L is the thickness of the membrane, d is the jump distance, m is the mass of the species, h is Planck's constant, k is Boltzmann's constant, v^* is the vibrational frequency of the species in the doorways between the sorption sites, T is temperature, N_s is the number of solubility sites available per m^3 of glass volume, N_A is Avogadro's number, R is the gas constant, and ΔE_K is the activation energy for hopping between sorption sites. For non-monatomic gases σ is the symmetry number of the permeating species, with $\sigma = 2$ in the case of hydrogen, and I is the moment of inertia. The results are shown in by the curves in Fig. 3. As can be seen, very good fits are obtained to the experimental points.

A concern about the application of the permeability equation (1) is the use of an arbitrary jump distance. The jump distance should not be an independent parameter, but should be the average distance, d, between the sites. If the sites do not interact and are non-overlapping (i.e., the sites are subject to an excluded volume interaction), then d should only be a function of number density, N_s and diameter of the sites, σ.

A collection of randomly distributed sites embedded in a solid may be treated as an assembly of hard spheres that are mutually impenetrable and have no interactions. This is a situation that can be addressed by the Percus-Yevick treatment in the theory of liquids, where the spheres represent the atoms of the liquid. The representation of an amorphous solid as a liquid is appropriate as both have only short range order [11].

Considering that the sites are non-interacting and may not overlap, only the short-range correlations will be important. This is called the Percus-Yevick (PY) approximation. For the case of hard spheres, the direct correlation function C(k), can be calculated analytically using the PY approximation [12] given in the next page.

$$C(k) = 4\pi \int_0^\sigma \left[\frac{(1+2\eta)^2}{(1-\eta)^4} + \frac{6\eta(1+\frac{1}{2}\eta)^2}{(1-\eta)^4}\frac{r}{\sigma} - \frac{\eta(1+2\eta)^2}{2(1-\eta)^4}\left(\frac{r}{\sigma}\right)^3 \right] \frac{Sin(kr)}{kr} r^2 dr \quad (2)$$

where, the σ is the diameter of the site and $\eta = \pi N_s \rho \sigma^3/6$ is the volume fraction occupied by the sites. Details will be given in a future publication [13].

Although analytical solutions for the radial distribution function and its derivatives are available [14], these still involve infinite sums and are cumbersome to use. In this work the radial distribution function is calculated numerically

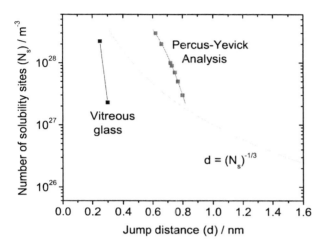

Figure 4. Relationship between number of solubility sites and jump distance

The figure compares values used in the vitreous glass literature, the P-Y calculation results, and the prediction for a random array of points $(N_S)^{-1/3}$. It can be seen that the jump distances used in the description of vitreous glass are the smallest, even smaller than the predictions from the array of points of zero volume, which is impossible. These previously used jump distances are consequently in error. The jump distances calculated by the P-Y treatment are larger than those predicted by the array of points, as they should be for a collection of sites of finite diameter. In this case the size of the solubility sites was taken to be 0.3 nm, which is what is expected for both the glass and the silica membranes.

Using these results and fitting the experimental results (Fig. 3) to equation (1) gives the following values of the parameters. The numbers are physically realistic.

Table 1. Parameters in equation (1).

Species	He	H$_2$	Ne
ν (s^{-1})	5.3×10^{12}	7.0×10^{12}	2.7×10^{12}
N_S (m^{-3})	4.7×10^{26}	3.0×10^{26}	4.0×10^{26}
ΔE_K (kJ mol^{-1})	8.0	13.8	16.6

REFERENCES

[1] W. G. Perkins, D. R. Begeal, "Diffusion and permeation of He, Ne, Ar, Kr, and D$_2$ through silicon oxide thin films", J. Chem. Phys. 54, 1683, 1971.

[2] T. Okubo, H. Inoue, "Introduction of specific gas selectivity to porous glass membranes by treatment with tetraethoxysilane", J. Membr. Sci. 42, 109, 1989.

[3] G.R. Gavalas, C.E. Megiris, S.W. Nam, "Deposition of H2-permselective SiO$_2$ films", Chem. Eng. Sci. 44, 1829, 1989.

[4] A.K. Prabhu, S.T. Oyama, "Highly hydrogen selective ceramic membranes: application to the transformation of greenhouse gases", J. Membr. Sci. 176, 233, 2000.

[5] J.S. Masaryk, R.M. Fulrath, "Diffusivity of helium in fused silica", J. Chem. Phys. 59, 1198, 1973.

[6] R.J.R. Uhlhorn, M.H.B.J. Huis In't Veld, K. Keizer, A. J. Burggraaf, "Synthesis of ceramic membranes, Part I. Synthesis of non-supported and supported γ-alumina membranes without defects", J. Mater. Sci. 27, 527, 1992.

[7] D.W. Breck, "Zeolite Molecular Sieves: Structure, Chemistry and Use", Wiley, New York, 636, 1974.

[8] D. Lee, S. T. Oyama, "Gas Permeation Characteristics of a Hydrogen Selective Supported Silica Membrane", J. Membr. Sci. 210, 291, 2002.

[9] R. M. Barrer, D. E. W. Vaughan, "Solution and Diffusion of Helium and Neon in Tridymite and Cristobalite", Trans. Faraday Soc. 63, 2275, 1967.

[10] P.L. Studt, J. F. Shackelford, R. M. Fulrath, "Solubility of gases in glass-a monatomic model", J. Appl. Phys. 41, 2777, 1970.

[11] J. M. Ziman, "Models of Disorder", Cambridge Univ. Press, Cambridge, 87, 1979.

[12] C.A. Croxton, "Introduction to Liquid state Physics", John Wiley & sons, New York, 84, 1975.

[13] S. T. Oyama, D. Lee, P. Hacarlioglu, R. F. Saraf, In preparation for J. Membr. Sci.

[14] J. Largo, J.R Solana, "A simplified analytical expression for the first shell of the hard-sphere fluid radial distribution function", Fluid Phase Equil. 167, 21, 2000.

The authors acknowledge the Director, National Science Foundation, Division of Chemical and Thermal Systems for award of Grant No. CTS-0321979 in support of this work. The authors also thank the ConocoPhillips Company for their sponsorship of the research program.

Characterisation of a novel self-association of an alternating copolymer into nanotubes in solution.

Cecile Malardier-Jugroot[1,2], T.G.M. van de Ven[1,2], M.A. Whitehead[1].

[1]Chemistry Department, McGill University, Montreal, Canada.
[2]Pulp and Paper Research Centre, McGill University, Montreal, Canada.

ABSTRACT

The characterisation of the association of an alternating copolymer was performed using theoretical methods (quantum mechanics and molecular mechanics) and experimental methods (cryo-Transmission Electron Microscopy, Neutron Reflectivity and Neutron Scattering). The most stable conformation obtained for the self-association at pH 7 using theoretical methods is a tubular structure in which eight SMA molecules make one twist of a helix. The tubes can grow in length by continued regular stacking of benzene rings. The nanotubes have inner and outer diameters of about 28 and 41 Å respectively. The hydrophobic groups are mainly located inside the tube and the hydrophilic groups are mainly on the exterior surface of the tube. They can also associate with themselves creating planes of aligned tubes, which can stack upon each other. The association of alternated copolymer into nanotubes has not been recognized before to the best of our knowledge[1].

The association of SMA octamers into a tubular structure at pH7 was confirmed experimentally by cryo-TEM and the nanotubes observed were several micrometers long. The shape as well as the inner and outer diameter of the nanotubes were also characterised by neutron scattering and the conformation at the air-water interface by neutron reflectivity.

Keywords: alternating copolymer, self-assembly, quantum mechanics, cryo-Transmission Electron Microscopy, Neutron Scattering.

INTRODUCTION

Self-assembly is a very efficient method to obtain well defined and often defect-free nano-architectures. The association and variety of shapes and properties of block copolymers have been extensively studied. Applications of those associations range from foam stability to drug delivery. In contrast, the associations of alternating copolymers have been rarely studied. The present paper investigates the association of a specific alternating copolymer, poly(styrene-alt-maleic anhydride) (SMA), both numerically and experimentally.

Poly(styrene-alt maleic anhydride) is an alternating copolymer used in the pulp and paper industry as a surface sizing agent to enhance the printing. SMA is composed of alternating hydrophobic and hydrophilic group and its chemical structure is pH dependent; the different ionization states are shown in figure 1. The association of the SMA chains has previously been observed at neutral pH by Dynamic Light Scattering[2] over a wide range of molecular weights, with hydrodynamic radius in the order of microns. We chose a theoretical approach to characterize precisely the association of SMA in solution and to explain the pH dependence of SMA conformations.

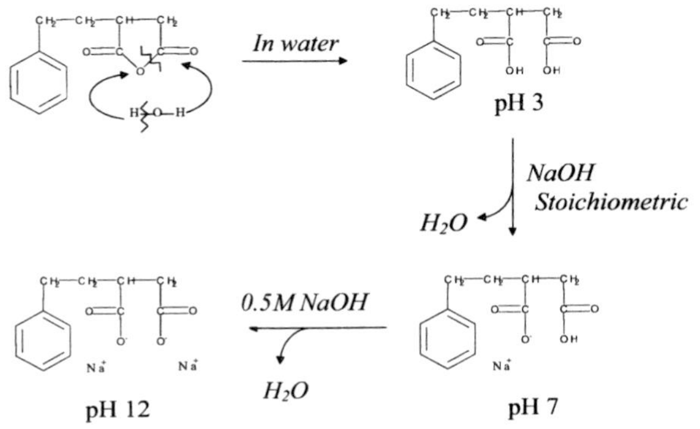

Figure 1: pH dependence of the molecular structure of poly(styrene-alt-maleic anhydride).

1 MONOMER CONFORMATIONS

Two methods of optimisation have been combined to obtain the ground state configuration of monomers at different pH values: the Tree Branch method[3] and a series of scans in energy[4] using the molecular modelling softwares Gaussian 98W[5] and Hyperchem[6]. The calculations performed in this study are gas phase calculations and the difference in pH will be modelled using the different degrees of ionization of the polymer in solution (Fig. 1) The configurations obtained for the monomer at three different pH values are shown in figure 2. The main difference between the configurations occurs at pH 7 where the presence of a hydrogen bond stiffens the conformation of the molecule and as a result, the orientations of the binding sites for a second monomer are very different for pH 7 from those at pH 3 or 12. This hydrogen bond is very strong (1.68 Å), compared to a typical hydrogen bond (1.80 Å for water) and therefore is expected to be present in aqueous solution.

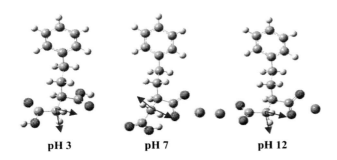

pH 3   pH 7   pH 12

Figure 2: Optimised conformation of styrene maleic anhydride at 3 different pH values at the RHF/6-31G** level of theory. The orientations of the binding sites (———►) at pH 7 are very different from the orientations of the binding sites at pH 3 or 12.

2 DIMER CONFORMATIONS

The dimer structure of SMA possesses two chiral centres and therefore all the structures need to be investigated. For the dimer conformation, four structures will be optimised: the RR, the RS, the SR and the SS conformations. The structures were optimised at the PM3 level of theory using series of scans in energy. These structures were reoptimised at the RHF/6-31G**.

The structures of the dimers obtained at pH 3 and pH 12 show a 90 degrees angle between the first monomer and the second (Fig. 3). The structures obtained at pH7 are very linear compared to pH 3 and 12. In addition the hydrogen bond observed in the monomer conformation is still present in the dimer structures (Fig. 3).

	pH 3	pH 7	pH 12
RS	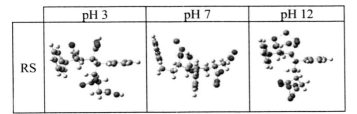		

Figure 3: Optimised structure of dimers of SMA of RS chirality at pH 3, 7 and 12.

3 QUADRIMER CONFORMATIONS

The study of the quadrimer conformations at different pH values shows that the polymer at pH 7 is linear, unlike at pH 3 and at pH13 (Fig. 4). The linearity arises from the stiffening of the molecule by the internal H-bond observed in the monomer and dimer conformation. This conformation does not depend on the chirality of the chain[7].

In addition the orientation of the benzene groups at pH7 are very similar for all the different possible chiralities of the chain. The linearity of the conformation and the similar orientation of the benzene groups therefore allow strong association between the polymer chains.

At low or high pH no association was observed because the conformation is an irregular helix. The irregularity of the helix at pH 3 and 13 is due to the chirality dependence of the conformation at low and high pH[7]. The study of the quadrimers, pentamers and hexamers showed that the conformation of SMA at pH7 is repetitive from the trimer[8].

pH	Chirality	Optimised structure	Backbone
pH 3	SR RR SR	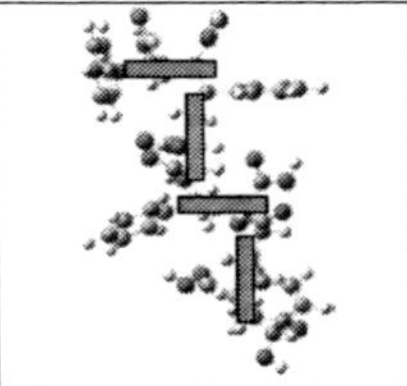	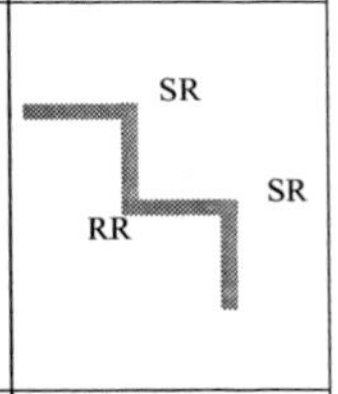
	SR SR SR	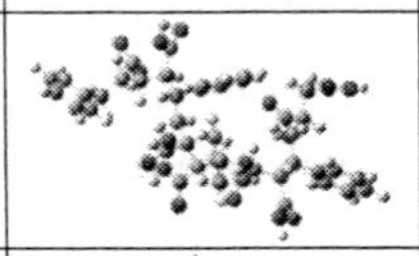	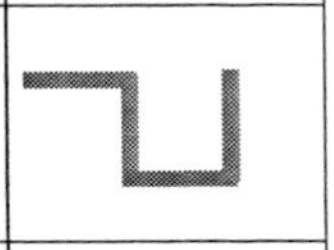
pH 7	SR RR SR	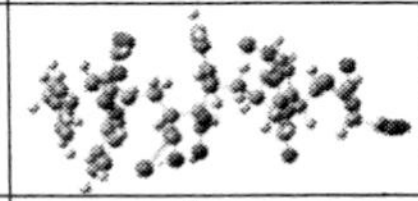	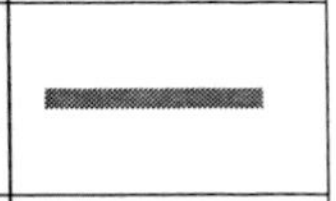
	SR SR SR		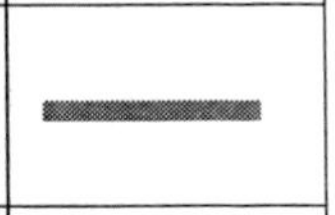
pH 12	SR RR SR	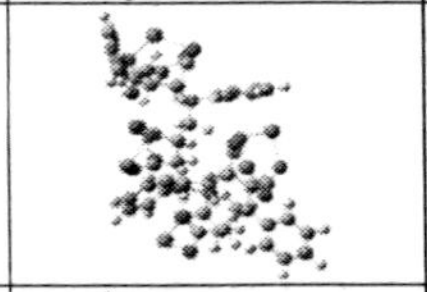	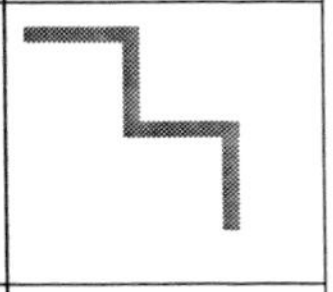
	SR SR SR	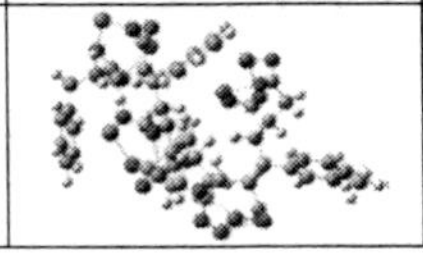	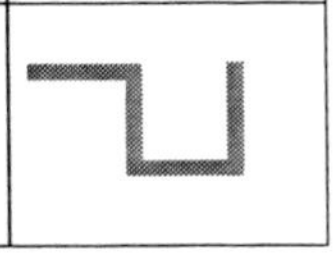

Figure 4: Optimised structure and schematic backbone representation of quadrimers of SMA of different chirality at pH 3, 7 and 12. At pH 7, the structures are linear, independent of the chirality of the chains.

4 ASSOCIATION AT PH 7

When two molecular chains at pH 7 associate, the angle between them is 90°, therefore by associating eight SMA molecules a tube can be formed. The front view of this tube is a square due to the 90° angle between two chains. As an example, the association of dodecamers was studied using molecular modelling and the most stable conformation was found to be a tubular conformation for which the inside of the tube is mainly hydrophobic, whereas the exterior is mainly hydrophilic (Fig. 5). The cavity of the tube was determined to be about 28 Å and the external diameter about 41 Å. The tubular structure is maintained by stacking interactions between the benzene groups of poly(styrene-maleic anhydride) and therefore this interaction, studied in the gas phase, is likely to be stable in water. This way of self-assembly in nanotubes has not been recognized before:

a recent review on organic nanotubes mentions four possible strategies for the molecular assembly of tubular materials[1], but the self-assembly of rigid alternating copolymers is not considered. The interactions occurring between the chains are hydrophobic interactions, protecting the benzene groups from contact with water.

Figure 5: Front view and side view of the configuration of the tubular association of SMA dodecamers at pH 7 at the molecular mechanical level.

This association can grow in two directions: at the tube ends to increase the length (longitudinal growth) and between the tubes to increase the width (radial growth) (Fig. 6). In addition the association does not depend on the molecular weight nor on the polydispersity of the SMA chains; indeed small chains can easily associate with long chains (Fig. 6). The tubes can also associate with themselves creating planes of aligned tubes (Fig.6). The planes can then associate with an angle of 45° between the orientations of the tubes in the planes due to the orientation of the benzene groups with respect to the tube growth.

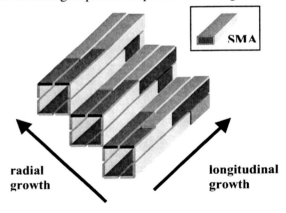

Figure 6: Schematic representation of the growth of the SMA association at pH 7 in the radial and longitudinal directions.

5 CRYO-TEM CHARACTERISATION

We performed experiments to observe the tubular structures in water by cryo-TEM. In these experiments, a drop of an aqueous solution of octamers of SMA (of molecular weight 1600) at a given pH was refrigerated in liquid nitrogen. The frozen sample was cut, and a replica of the top surface was obtained by first spraying a platinum-carbon mixture at 45° with respect to the surface and then by spraying carbon normal to the surface. In preparing samples for cryo-TEM, the surface often fractures at locations were discontinuities in structure occur. The results obtained with this method for pH 7 show very long lines with a spacing of about 50 Å, which closely corresponds to the diameter of the nanotubes calculated theoretically. These lines are organized in sheets, one on top of the other, with the lines of one sheet making an angle of 45° with the other sheet (Fig. 7). This angle is due to the stacking interaction between benzene groups, which make a 45° angle with the main axis of the tube. No regular structure was observed by cryo-TEM at pH 12.

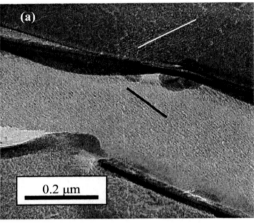

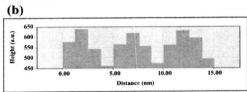

Figure 7: (a) Cryo-TEM picture of the association of SMA in solution 1% wt at pH 7. The lines represent the direction of the tubular association of SMA.

(b) Height profile of three lines of the cryo-TEM image, showing that the diameter of one tube is about 50 Å.

6 NEUTRON SCATTERING CHARACTERISATION

In order to fully characterise the association at the molecular level and the macromolecular properties influencing the structure of SMA at pH 7, a neutron scattering study has been performed at ILL (Grenoble, France).

The profiles obtained were fitted with a model using nanotubes interacting with a spherical potential. The fitted result for a 2% wt SMA solution at pH 7 (molecular weight 40,000) is shown in figure 8. The fitted inner and outer radii of the nanotube are 14 Å and 19 Å respectively, which closely correspond to the predicted radii. The profile at pH 7 can be compared to the profile at pH 12 where no structure factor was observed (Fig. 9).

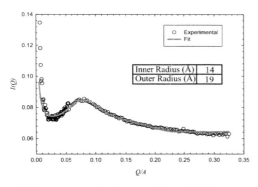

Figure 8: Neutron scattering profile and fitting curve of 2% wt SMA solution at pH 7.

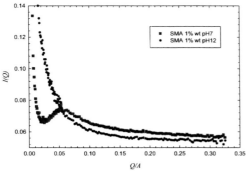

Figure 9: Neutron scattering profile of 1% wt SMA solution at pH 7 and pH 12.

7 NEUTRON REFECTIVITY

To understand the mechanism of association at pH 7 and the influence of the conformation of SMA at the air-water interface, neutron reflectivity experiments[9,10] were performed at NIST (Washington, USA). These experiments were performed with 1%wt solutions of the deuterated ethyl esters of poly(styrene-maleic anhydride), used in order to label the hydrophilic part of SMA.

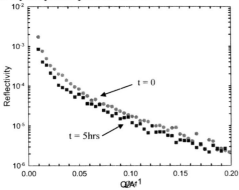

Figure 10: Neutron reflectivity profile of the ethyl ester of SMA, residing at an air-water interface at pH 7.

The results obtained show that a complete monolayer of SMA is formed at the air-water interface at pH 3 and 7, with a coverage of about 1.2 mg/m². At pH 13, the polymer is more soluble and forms an incomplete monolayer (coverage 0.9 mg/m²). SMA adsorbs in a flat configuration, with the benzene groups sticking out in the air. The

association does influence the conformation at the air-water interface at pH 7, as a 5 hr delay is observed in obtaining a stable signal (Fig. 10). (Such a delay was not observed at low or high pH). This delay is likely due to the slow establishment of an equilibrium between associated SMA molecules in solution and single SMA molecules at the air-water interface.

CONCLUSION

The association of an alternating copolymer observed by DLS was characterized by theoretical and experimental methods. The association consists of eight SMA chains associated into a nanotube with an inner diameter of 28 Å and an outer diameter of 41 Å. The nanotubes can then associate to form a 2 dimensional structure, the planes obtained by self-association of the nanotubes can stack upon each other by hydrophobic interaction and form a 3 dimensional architecture. The association of alternated copolymer into nanotubes has not been recognized before.

The structure predicted by theoretical calculations was confirmed and observed by cryo-TEM and neutron scattering. In addition, an equilibrium was observed between associated SMA molecules in solution and single SMA molecules at the air-water interface by neutron reflectivity.

ACKNOWLEDGMENTS

This work was supported by the NSERC (Canada) and by the "Wood Pulp" Network of Centers of Excellence. The NIST (Washington, USA) and NCNR are thanked for support and neutron reflectometer facilities. The ILL (Grenoble, France) is thanked for support and neutron scattering facilities. We gratefully acknowledge Dr. T. Cosgrove and Dr. R. Richardson (University of Bristol, U.K.) for help with the neutron reflectivity and scattering experiments and Dr. H. Vali and Dr. S.K. Sears (McGill University, Canada) for the cryo-TEM experiments.

REFERENCES

[1] Bong T.B., et al., Angew. Chem. Int. Ed. 2001, 40, 988.

[2] Garnier G., et al., Langmuir, 16, 3757, 2000.

[3] Villamagna F., Whitehead, M.A.; J. Chem. Soc., Faraday Trans., 90(1), 47, 1994.

[4] Malardier-Jugroot C., et al., J. Mol. Struc. (THEOCHEM), 623, 263, 2003.

[5] Gaussian 98W (Revision A.5), Frisch, M. J.; et al., Gaussian, Inc., Pittsburgh PA, 1998.

[6] HyperChem release 5.11 (1999) Hypercube Inc.

[7] Malardier-Jugroot C., van de Ven, T.G.M.; Whitehead, M.A.; submitted to Nature (December 2003).

[8] Malardier-Jugroot C., van de Ven, T.G.M.; Whitehead, M.A.; (to be submitted to J. Mol. Struc. (THEOCHEM)).

[9] Penfold J., et al., J. Chem. Soc., Faraday Trans., 93(22), 3899, 1997.

[10] Richardson R., et al., Macromolecules, 33, 6269, 2000.

Carbon Nanotube- and Nanofibre-reinforced Polymer Fibres

M.S.P. Shaffer[1], J.K.W. Sandler[2], S. Pegel[2,3], A.H. Windle[2], F. Gojny[3], K. Schulte[3], M. Cadek[4], W.J. Blau[4], J. Lohmar[5], M. van Es[6]

[1] Department of Chemistry, Imperial College London, London SW7 2AY, UK
[2] Department of Materials Science and Metallurgy, University of Cambridge, Cambridge CB2 3QZ, UK
[3] Polymer Composites, Technical University Hamburg-Harburg, D-21073 Hamburg, Ger
[4] Materials Ireland Polymer Research Centre, Department of Physics, University of Dublin - Trinity College, Dublin 2, Ireland
[5] Degussa AG, D-45764 Marl, Germany
[6] DSM Research, 6160 MD Geleen, The Netherlands

ABSTRACT

A range of multi-wall carbon nanotubes and carbon nanofibres were mixed with a polyamide-12 matrix using a twin-screw microextruder, and the resulting blends used to produce a series of reinforced polymer fibres. The aim was to compare the dispersion and mechanical properties achieved for nanofillers produced by different techniques. A high quality of dispersion was achieved for all the catalytically-grown materials and the greatest improvements in stiffness were observed using aligned, substrate-grown, carbon nanotubes. The use of entangled multi-wall carbon nanotubes led to the most pronounced increase in yield stress. The degrees of polymer and nanofiller alignment and the morphology of the polymer matrix were assessed using X-ray diffraction and calorimetry.

Keywords: nanotubes, nanocomposites, polymer fibres, mechanical properties

1 INTRODUCTION

Although carbon nanotubes were observed at least as early as 1976, it was only more recently that their importance was recognised. Since then, enormous attention has been paid to their fundamental properties and related applications [1], including considerable efforts to exploit the remarkable mechanical properties of individual nanotubes in macroscopic composites [2]. Whilst some encouraging results have been obtained, significant improvements over conventional fillers have generally proved elusive, for a number of reasons. Therefore, it is particularly interesting to consider the use of nanotubes to reinforce structures in which conventional fillers cannot physically be accommodated, such as within micro-components, or fibre composite matrices. As an example, this study focuses on the fabrication and properties of nanofibre and nanotube reinforced polymer fibres. This approach necessitates the use of only small quantities of materials, readily enabling comparisons of different types of nanotube. The development of such nanocomposite fibres will involve the detailed study of dispersion, alignment, and matrix interactions, that will be fundamental for exploiting the full potential of bulk nanotube composites, whilst, at the same time, generating property enhancements that may be applied immediately.

Earlier work, focusing on carbon nanofibre (CNF) reinforced PEEK fibres [3], showed promise, demonstrating that high quality fibres with excellent surface finish and filler dispersion could be obtained using simple thermoplastic processing. The resulting fibres had significant improvements in yield strength and stiffness that could not otherwise be readily achieved. In addition, the presence of nanotubes is not damaging to the key properties of the PEEK matrix, namely, high temperature, chemical, and abrasion resistance; indeed, there is evidence that the wear performance of the blend is substantially improved [4]. A final benefit is that, because PEEK is a high performance material, the nanofibres are no more expensive than the matrix. However, the system has some complications; the degree of crystallinity is affected by the presence of the nanofiller and must be taken into account when assessing the composite properties. This paper describes a series of experiments comparing the CNF used in the PEEK study with three types of carbon nanotube, in an attempt to establish the potential for improvement. In order to simplify processing, and to study a matrix already used commercially for nanotube composites, we decided to focus on a polyamide-12 (PA12).

2 EXPERIMENTAL

The arc-grown nanotubes (AGNT) [5], aligned catalytically-grown nanotubes (aCGNT) [6], and entangled catalytically-grown nanotubes (eCGNT) [7] were produced using previously reported techniques. The catalytically-grown nanofibres (CNF) were purchased from Applied Sciences Inc., USA, grade PR-19-PS. The average outer diameters of these multi-walled nanomaterials were 15, 43, 10, and 155 nm respectively. All of the catalytically-grown materials were essentially pure except for the presence of the catalytic transition metal. The AGNT sample contained significant graphitic and nanoparticulate impurities, with a total nanotube weight content around 40 % [5]. Pellets of PA-12 (VESTAMID L1700, from Degussa) and as-produced carbon powders were weighed, dried, and blended in a DSM twin-screw microextruder, operating at 220 °C at

80 rpm. The extrudate was roughly chopped and fed into a Rheometrics Scientific capillary rheometer at 220 °C and a single strand was spun from a 1 mm diameter die to produce fibre diameters in the range 100 - 200 μm. Nanocomposites with loading fractions up to 15 wt% were produced, depending on the availability of the nanofillers. Composite samples were fractured under liquid nitrogen and examined in a JEOL 6430F FEG SEM, operating at 10 kV after chromium coating. 2D X-ray fibre diffraction patterns were collected using Ni-filtered CuKα radiation on a Photonics CCD system, calibrated using silicon powder. Individual fibre tensile testing was performed on a TA Instruments 2980 Dynamic Mechanical Analyzer applying a constant force ramp of 0.1 N/min, at 30 °C, with a gauge length of 5.5 mm. At least 3 fibres were tested for each sample. Differential scanning calorimetry (DSC) was performed using a TA Instruments DSC 2920 operating at 10 °C/min.

3   RESULTS

Macroscopically, all of the nanocomposite fibres appeared to be of generally good quality, with a reasonable surface finish and uniform diameter, regardless of catalytic nanotube filler type and loading fraction. The entangled multi-wall carbon nanotubes led to a slightly rougher fibre surface compared to the carbon nanofibres and aligned multi-wall carbon nanotubes. Only the PA12 nanocomposite fibres containing the arc-grown multi-wall carbon nanotubes showed non-uniformity in diameter which was independent of filler weight content. All nanocomposite fibres were coloured uniformly black. Typical SEM images of the composite fibres are shown in Figure 1; similarly high levels of dispersion were observed for all of the catalytically-grown materials. However, the unpurified AGNTs showed a poor overall level of dispersion in all cases, with aggregates leading to the formation of voids within the fibres. Maximum pull-out lengths appeared to be around 400 - 500 nm for both CNF and eCGNT but reached several microns for the straighter, more crystalline aCGNT samples.

An example of the two-dimensional X-ray fibre diffraction patterns of the extruded fibres is presented in Figure 2. The main features of the 2D X-ray patterns are an amorphous polymer halo and relatively low intensity crystalline reflections, typical of the γ-phase. The results were confirmed by detailed equatorial diffraction studies. In neither case is there evidence of a change in crystal structure on the addition of nanotubes, as has been observed previously in polypropylene and polyvinylidene fluoride composites [8]. Nor is there the variation in the degree of crystallinity observed with the PEEK samples [3]. The graphitic (002) reflections of the aCNGT and CNF indicate a considerable degree of alignment parallel to the fibre axis due to the shear flow during processing. The AGNT samples showed a random distribution, consistent with the presence of equiaxed nanoparticles and aggregated

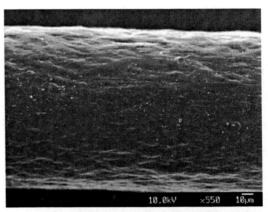
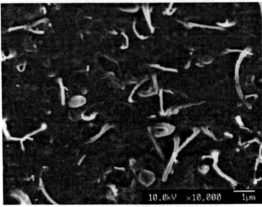

Figure 1. SEM images of the nanocomposite fibres, containing 5 wt% aCGNT, showing the fibre surface as produced (top) and a fracture surface (bottom).

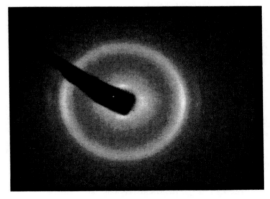

Figure 2. 2D X-ray fibre diffraction pattern for sample containing 5 wt% aCGNT. The pattern shows a strong amorphous halo overlayed by a crystalline peak, and, at larger angles, the (002) graphitic spacing of the nanotubes.

nanotubes. Equatorial X-ray diffraction patterns of the nanocomposite fibres, show similar diffraction features. By comparing the areas under the peaks, it can be established that the overall degree of crystallinity across all of the as-processed PA12 samples was constant, estimated at around 10 %; this figure is a slight overestimate due to the modest orientation of the polymer crystallites parallel to the axis.

Figure 3 shows DSC thermograms obtained on heating as-spun nanocomposite fibres; as a function of aCGNT content; similar curves were obtained for all the nanocomposites. There is a clear glass transition at around 50 °C, followed by an endothermic melting peak above 170 °C. No significant variations were observed in either shape or onset of the glass transition and melting features. Nevertheless, there appeared to be a change in the heat capacity of all as-spun fibres between 100 °C and the onset to melting at around 170 °C, indicated by small deviations in the gradient. Detailed DSC studies, under different heating rates and cycled conditions, provided further evidence for a broad, underlying exotherm. Such a broad crystallisation above T_g during a DSC experiment is common for polyamides and reflects the relaxation of processing stresses [9]. In an attempt to establish a comparative degree of crystallinity of all nanocomposite fibres the total melting peak area was evaluated, normalised to the fraction of polymer present. This approach led to a roughly constant degree of crystallinity for all the samples between 23 to 27 % (based on an enthalpy of melting of 209.34 J/g for 100% crystalline material [10]). These results are an overestimation of the true, initial, degree of crystallinity of the as-spun fibres due to the recrystallisation process; hence, the X-ray based estimate is more reliable. However, we can conclude that the nanofillers do not affect the maximum crystallinity that can be obtained in this system, although more detailed analysis does indicate a nucleation effect and an associated change in crystal size [11].

A typical set of tensile stress-strain curves of nanocomposite fibres is shown in Figure 4. The pure polymer and many of the nanocomposites revealed a two-stage yielding behaviour between about 3 and 30 % of strain, followed by a stress plateau up to about 190 % strain corresponding to drawing. This particular yielding behaviour is due to the experimental set-up (load control) and the occurrence of strain-induced crystallisation after initial fibre necking. The occurrence of the characteristic two-stage yielding behaviour could be seen for all nanocomposites up to a filler concentration of about 5 wt% (about 10 wt% in case of the carbon nanofibres). At higher filler loading fractions the tensile behaviour changed to a simple yielding, followed by a steady stress increase.

As shown in Figure 5, the nanocomposite fibre stiffness and yield strength increase linearly with filler content in all cases. As can be seen, the aCGNTs show the steepest increase in composite stiffness, closely followed by the eCGNTs. These results are in agreement with the trend in nanotube crystallinity observed in Raman data. The AGNTs present the worst increase in nanocomposite stiffness; a result that can be attributed to their poor dispersion and alignment. The yield stress data showed similar trends with both the eCGNTs and aCGNTs revealing the most prominent increases as a function of filler weight content, although the eCGNT come out top most likely as result of increased constraint of the polymer matrix due to their

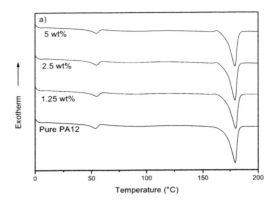

Figure 3: DSC data obtained for aCGNT nanocomposite fibres, with increasing filler loading fraction, at +10 °C/min.

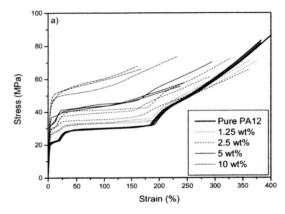

Figure 4: Nominal stress-strain curves with increasing filler weight content obtained under stress control at 30 °C for nanocomposite fibres containing aCGNTs.

relatively high surface area. The results are qualitatively consistent with earlier studies on single filler types (in some case single loading fractions) but a detailed comparison is beyond the scope of these proceedings.

4   CONCLUSIONS

Nanofillers provide interesting opportunities to reinforce polymer fibres. Carbon nanofibres have already been shown usefully to improve the strength, stiffness, and wear properties of PEEK fibres, as well as a range of other thermoplastic matrices. This study has successfully explored the potential of various multi-wall carbon nanotubes and nanofibres as mechanical reinforcements in polyamide-12 composite fibres. Linear improvements in yield strength and stiffness were observed that were greater for well-dispersed nanotubes than for nanofibres. Further improvements might be achieved using nanotubes of greater intrinsic quality and straightness. Many questions remain as to the characteristics of the optimal nanotube reinforcement that will only be answered by continued comparative studies.

5 ACKNOWLEDGEMENTS

The authors would like to thank the National Centre for Biomedical Sciences in Ireland and DSM for the use of their equipment and Prof J-B Nagy (Namur) for the supply of nanotube material. Financial support from the EC Thematic Network "CNT-NET" [G5RT-CT-2001-05026], the EPSRC, and the Cambridge European Trust is gratefully acknowledged.

REFERENCES

1. R.H. Baughman, A.A. Zakhidov, and W.A. de Heer. Science, **297,** 5582, 787-792 (2002)
2. E.T. Thostenson, Z.F. Ren, and T.W. Chou. Composites Science and Technology, **61,** 13, 1899-1912 (2001)
3. J.K.W. Sandler, M.S.P. Shaffer, P. Werner, D. V., A. V., and W. A.H. Journal Of Materials Science, **38,** 10, 2135-2141 (2003)
4. P. Werner, O. Jacobs, R. Jaskulka, V. Altstädt, J.K.W. Sandler, M.S.P. Shaffer, and A.H. Windle. Wear, **Submitted,** (2003)
5. M. Cadek, R. Murphy, B. McCarthy, A. Drury, B. Lahr, R.C. Barklie, M. Panhuis, J.N. Coleman, and W.J. Blau. Carbon, **40,** 6, 923-928 (2002)
6. C. Singh, M. Shaffer, I. Kinloch, and A. Windle. Physica B-Condensed Matter, **323,** 1-4, 339-340 (2002)
7. K. Hernadi, A. Fonseca, P. Piedigrosso, M. Delvaux, J.B. Nagy, D. Bernaerts, and J. Riga. Catalysis Letters, **48,** 3-4, 229-238 (1997)
8. B.P. Grady, F. Pompeo, R.L. Shambaugh, and D.E. Resasco. Journal of Physical Chemistry B, **106,** 23, 5852-5858 (2002)
9. Y.P. Khanna and W.P. Kuhn. Journal of Polymer Science Part B-Polymer Physics, **35,** 14, 2219-2231 (1997)
10. S. Gogolewski, K. Czerniawska, and M. Gasiorek. Colloid & Polym. Sci., **258,** 10, 1130-1136 (1980)
11. J.K.W. Sandler, S. Pegel, M. Cadek, F. Gojny, M.v. Es4, J. Lohmar, W.J. Blau, K. Schulte, A.H. Windle, and M.S.P. Shaffer. Polymer, **Sumbitted,** (2003)

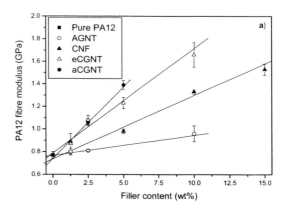

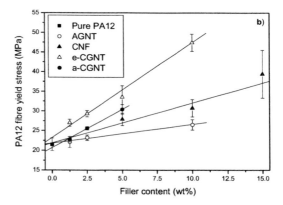

Figure 5: Plots summarising the relationship between nanoscale filler weight content and (a) tensile modulus and (b) yield stress.

Thermal and Morphological Study of a New Type of Nanocomposites:
Functionalized PP / PA / Clay Nanocomposites

Hideko T. Oyama[*], Yuko Oono[**], and Kazuo Nakayama[***]

Research Center of Macromolecular Technology, National Institute of Advanced Industrial Science and Technology (AIST), 2-41-6 Aomi, Kohtoh-ku, Tokyo, 135-0064, Japan, *hideko-oyama@aist.go.jp, ** y-oono@aist.go.jp, *** kazuo-nakayama@aist.go.j

ABSTRACT

In the present work we explore a new type of nanocomposite, in which clay is distributed in a multi-phase polymer system. Our nanocomposites consist of an immiscible polymer pair, polypropylene grafted with maleic anhydride (PP-MAH) and polyamide 6 (PA), mixed with organo-modified synthetic mica. It was shown that (20/80/4) PP-MAH/PA/mica systems indicated that the mica was predominantly localized in the PA matrix in a composite prepared from PP-MAH with the lower MAH concentration. By contrast, in a composite prepared from PP-MAH with higher MAH concentration, the clay was located in both the PA and PP phases as well as at their interface. In this case, the PP dispersed phase, which was thermally unstable, was protected from thermal decomposition both by the wall of clay located at the interface and the dispersed clay in the PP phase. It was also found that copolymers formed *in situ* at the interface during melt-mixing also contributed to the thermal stability in a temperature range above 400 °C.

Keywords: nanocomposite, morphology, polypropylene, polyamide, clay

1 INTRODUCTION

The finding that nanoscale distribution of clay significantly improves polymer properties and extends their utility has stimulated extensive work in the field of polymer/clay nanocomposites in the last decade [1]. However, most studies thus far were carried out with simple single-phase polymer/clay systems. In the present study we explore new polymer/clay nanocomposites containing a multi-phase polymeric system.

2 EXPERIMENTAL

1.1 Materials

Table 1 lists the characteristics of PA and PP samples used in this study. Three types of PA with different molecular weights were obtained from Unitika, Japan, and two types of PP functionalized with different concentration of maleic anhydride (PP-MAH) were purchased from Aldrich Chemical Co. The organo-clay used in this study was a synthetic mica treated by dipolyoxyethylene alkyl(coco)methylammonium cation purchased from CO-OP Chemical, Japan, and its CEC was reported to be 120 meq/100g.

1.2 Melt-mixing

Materials were first dried under vacuum at 80 °C for 48 h prior to experiments. Melt-mixing was carried out in a HAAKE Minilab Rheomex with conical twin screws, which was operated at 240/260 °C at 50/100 rpm using a heating bath. The mixing time was fixed to be 10 min.

2.3 Analysis

<u>Structure</u>   The obtained composites were cryotomed and

their morphology was examined by transmission electron microscopy (TEM). Two types of transmission electron microscopes were used for characterizing the samples. One was a Hitachi H-7000 TEM operated at an accelerating voltage of 75 kV for specimens stained by the vapor of ruthenium tetraoxide and the other was a LEO 922 energy-filtering TEM (EFTEM) operated at 200 kV for unstained specimens. Furthermore, light scattering measurements (LS) were carried out with an Otsuka Electronics Dyna 3000 in order to measure the mean particle size of the dispersed phase [2]. The composites were also characterized by wide-angle X-ray diffraction (WAXD), which was conducted using Cu K_α radiation (40kV, 300 mA) with an X-ray diffractometer (Rint 2500 VH/PC, Rigaku Co. Ltd.).

Properties A differential scanning calorimeter (Perkin Elmer DSC-7) was operated with a heating rate of 10 °C/min under a N_2 atmosphere. Dynamic thermal mechanical analysis (DMA) was carried out with Rheovibron DDV-25FP from Orientec Co. Ltd. at frequencies of 1, 3, 10, and 100 Hz at a heating rate of 2 °C/min between 150-250 °C. Moreover, thermogravimetric analysis (TG) was performed using a Perkin Elmer Pyris 1 TGA at a heating rate of 10 °C/min under air atmosphere from room temperature to 800 °C.

3 RESULTS AND DISCUSSION

Figure 1 shows the WAXD profiles of organo-treated mica and two kinds of composites composed of (20/80/4) PP-MAH/M-PA/Mica with different MAH concentrations. The (001) peak originally observed for the mica disappears in both composites, which indicates that the clay does not have regular gallery space anymore in the composites.

Figure 2 and Figure 3 are EF-TEM micrographs of the composite prepared from PP-*l* MAH and from PP-*h* MAH, respectively. Figure 2 shows that the clay is predominantly localized in the PA matrix, whereas Figure 3 shows that it is located in both the PA and PP phases as well as at their interface. The LS data showed that the mean particle size of the dispersed phase decreases by addition of the clay, probably due to a contribution of enhanced melt viscosity of the PA matrix.

Next, the thermal stability of the composites in air was examined by TG. The results of a non-reactive blend of (20/80) i-PP/M-PA and the component polymers are shown in Figure 4. The comparison with the results of a reactive blend of (20/80) PP-*l* MAH/M-PA given in Figure 5 demonstrates that the reactive blend exceeds the non-reactive blend in thermal stability in a temperature range above 400 °C, which would be due to a contribution of copolymers formed *in situ* at the interface during melt-mixing. It is also shown that incorporation of clay significantly suppresses thermal decomposition of the PP dispersed phase in a temperature range between 300 and 400 °C. Figure 6 shows a difference in thermal stability between the composites prepared from PP-MAH with different MAH concentrations. It is clearly demonstrated that in a temperature range, in which the PP phase is decomposed, the composite prepared from PP-*h* MAH is much more stable compared to that prepared from PP-*l* MAH. This difference is probably caused by the difference in the clay distribution shown in Figures 2 and 3. The higher polarity of the PP-*h* MAH phase than the PP-*l* MAH phase would enable the clay to be located in both PA and PP phases as well as at the interface, which results in the effective protection of the PP dispersed phase from thermal decomposition.

Furthermore, DMA was measured in both composites, with the results given in Figure 7. A difference in dynamic storage modulus, *E*', in the glassy state (*e.g.* –50 °C) and in the rubbery state (*e.g.* 120 °C) significantly decreases in the composites compared to neat PA. It is worth mentioning that the heat distortion temperature (HDT), which is located between T_g and T_m, would significantly increase in the composites, especially in the composite prepared from PP-*h* MAH. The decrease in E' was negligible in this composite at T_g of PP, although it was not in the composite prepared from PP-*l* MAH. To shed further light on the situation DSC was undertaken in order to estimate the crystallinity, X_c, of the PP phase in both systems. The heat of fusion of PP was assumed to be 165 J/g [3], which resulted in X_c = 4.9 % in PP-*h* MAH/M-PA/Mica and 7.2 % in PP-*l* MAH/M-PA/Mica. This means that restriction of PP chain movements at T_g of PP in the former system is not due to crystallinity, but probably due to interactions with clay or crosslinkages formed in the dispersed phase. In conclusion, nanocomposites composed of multi-phase polymer system were observed to have unique stability properties in the present study.

REFERENCES

[1] A. Usuki, M. Kawasumi, Y. Kojima, Y. Fukushima, A. Okada, T. Kurauchi, and O. Kamigaito; *J. Mater. Res.,* **8**, 1179 (1993)

[2] K. Yamanaka, T. Inoue; Polymer 30, 662 (1989)

[3] J. E. Mark, Ed.; *Polymer Data Hand Book*, Oxford University Press: New York, Oxford (1999)

Table 1: Characteristics of materials.

	M_n (g/mol)[*1]	Functionality[*1]		T_m[*2] (T_g[*3])
		(wt%)	(#/chain)	(°C)
L-PA	14,000	0.44	1 COOH, 1 NH$_2$	219.5
M-PA	17,500	0.35	1 COOH, 1 NH$_2$	220.0 (62.7)
H-PA	21,500	0.28	1 COOH, 1 NH$_2$	219.0
PP-*l* MAH	10,000	0.6	0.6 MAH	163.2 (5.0)
PP-*h* MAH	3,900 M_w= 9,100	8~10	3.2~3.9 MAH	153.7 (--)

*1 reported values not analyzed values   *2 DSC data   *3 E'' peak at 1 Hz measured by DMA

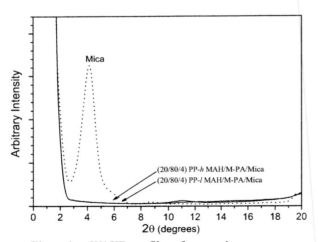

Figure 1:   WAXD profiles of composites.

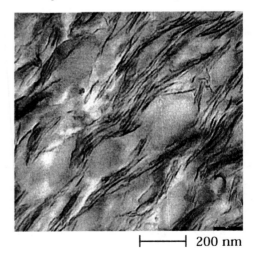

Figure 2: EFTEM micrograph of (20/80/4) PP-*l* MAH/M-PA/Mica.

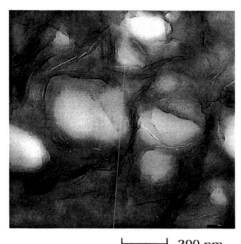

├──────┤ 200 nm

Figure 3: EFTEM micrograph of (20/80/4)
PP-*h* MAH/M-PA/Mica.

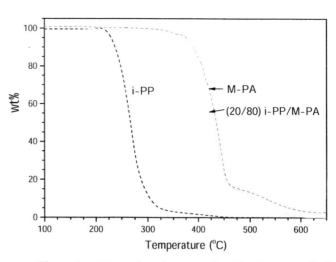

Figure 4: TG results of non-reactive blend composed of
(20/80) i-PP/M-PA and the component polymers in air.

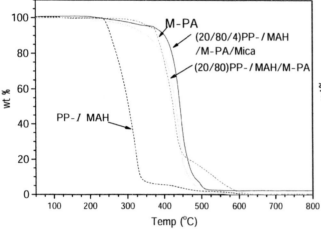

Figure 5: Effects of clay and copolymers formed *in situ*
at the interface on thermal stability of composites in air.

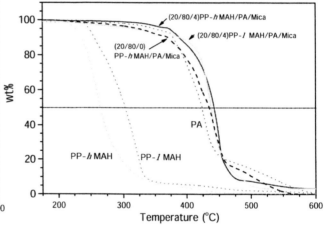

Figure 6: Effects of the MAH concentration
incorporated to PP on thermal stability of
composites in air.

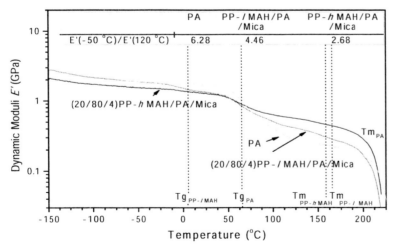

Figure 7: Dynamic storage
modulus, *E'*, of the composites
[DMA results measured at 1 Hz].

Corrosion Protection by Multifunctional Stratified Coatings

Sarah L. Westcott[*], Nicholas A. Kotov[**], John W. Ostrander[***],
Arif A. Mamedov[*], Dennis K. Reust[*] and Joel P. Roark[*]

[*] Nomadics, Inc., 1024 South Innovation Way, Stillwater, OK, USA, swestcott@nomadics.com,
amamedov@nomadics.com, dreust@nomadics.com, jroark@nomadics.com
[**] University of Michigan, Ann Arbor, MI, USA, kotov@umich.edu
[***] Oklahoma State University, Stillwater, OK, USA, john.ostrander@lycos.com

ABSTRACT

We have demonstrated a hybrid coating consisting of both a stratified layer that incorporates corrosion inhibiting ions and a silicate barrier layer. The stratified layer is applied to the aluminum alloy using the layer-by-layer assembly (or electrostatic self-assembly) method and comprises alternating layers of charged polyelectrolytes and nanometer-thickness clay platelets. These nanocomposite coatings can be doped with corrosion inhibiting ions and allow lateral diffusion of those ions through the coating to have high 'throwing power'. Over the LBL coating is applied an organic-inorganic hybrid coating of organically modified silicates (ORMOSILs) to provide additional barrier protection and bind to primers and paints. These types of hybrid coatings have electrochemically shown superior corrosion protection compared to hexavalent chromium coatings.

Keywords: corrosion, layer-by-layer, polyelectrolyte, silicate, montmorillonite

1 INTRODUCTION

Because the hexavalent chromium conversion coatings traditionally used to protect aluminum alloy aircraft are environmentally hazardous, replacements are sought. In the search for a corrosion protection scheme, there are two general approaches: barrier layers and active inhibitors. Barrier layers prevent water and ions from reaching the metal surface and can provide excellent protection, but if mechanical damage results in a path for water through the barrier layer, corrosion is rapid. The most effective active inhibitor is hexavalent chromium; no other material on its own has provided such good protection to multiple metals and alloys in a variety of environments. However, chromium is such an environmental and occupational hazard that phasing out its use is mandated. Combining non-chromium active inhibitors with barrier layers might overcome these weaknesses. Unfortunately, while active inhibitors can be trapped and incorporated into barrier layers, the barrier layer may be so dense and non-porous that the active inhibitor cannot travel to the corrosion site and has no throwing power. Thus we are considering a stratified approach to a corrosion protection coating. The layers closest to the substrate can incorporate inhibitor ions while permitting their transport parallel to the substrate and further from the substrate, a barrier layer provides protection.

2 STRATIFIED COATINGS

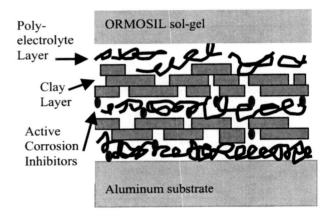

Figure 1: Layers of clay platelets, polyelectrolytes, and active corrosion inhibitors are assembled on the aluminum substrate and coated with an ORMOSIL sol-gel.

The general coating is shown in Figure 1. Near the substrate, alternating layers of polyelectrolytes (high molecular weight molecules or polymers with bound charges) and clay platelets are prepared using layer-by-layer (LBL) deposition. LBL is a technique that involves the adsorption of charged species onto an oppositely charged substrate. The charged species can be polyelectrolytes, clay platelets [1], nanoparticles [2], carbon nanotubes [3], and even biological molecules. Most often, the substrate is immersed in a solution of the charged species. However, as this is not practical for application to aircraft, we are also investigating spraying [4]. Our aluminum substrate (which has negatively charged surface groups) is sprayed with a solution of positively-charged poly(diallyldimethyldiammonium chloride), PDDA. Approximately a monolayer of the PDDA will remain attached to the substrate, making its apparent charge positive. Further layers are built up with negatively-charged poly(acrylic acid), PAA, followed by another PDDA layer, followed by a layer of montmorillonite clay. In solution, this clay exfoliates into platelets, 1-4 nm thick and approximately 200 nm in diameter. When deposited on a substrate by the LBL method, the platelets tend to align themselves parallel to the substrate. This deposition sequence of 4 layers can be repeated to build a thicker film. Other species could be incorporated into these LBL films by substituting for the solutions of appropriate charge. In the LBL coatings, the charged species can attract ionic active corrosion inhibitors, providing storage, while the layered structure allows for lateral diffusion of these inhibitors to corrosion sites.

To provide a barrier function, hybrid organic-inorganic coatings derived from organically modified silicates (ORMOSIL) were used because of the dense and chemically inert characteristics that can be achieved. A further benefit of the organic groups is that they are compatible with the paints used as topcoats. The corrosion resistance properties of these ORMOSIL coatings made from different combinations of organic and inorganic groups have been studied [5]. In parallel tests, they meet the performance characteristics of chromated conversion coatings, except if there is a mechanical breach of the coating.

The high density of the ORMOSIL coating does not permit active inhibitor ions to have throwing power if directly incorporated in the coating, but they can be used in combination with LBL coatings made by immersion [6].

3 EXPERIMENTAL PROCEDURE

PDDA with MW 450,000-500,000 and PAA, with MW ~450,000 were purchased from Aldrich and used as 1% solutions in water. The pH of those solutions was typically between 3.2-3.7 and 2.9-3.5, respectively.

The clay was SWy-2 sodium montmorillonite from the University of Missouri – Columbia. Solutions were prepared by mixing 5 g of clay in 200 mL of deionized, ultrapure water. The clay solution was sonicated to mix well. The solution was then allowed to settle for at least 3 days. After the initial settling period, the remaining suspension is stable and can be used for weeks or months. That solution had a typical pH of 8.

To remove dust and grease from aluminum alloy 2024-T3 samples, they were first cleaned with an anionic detergent, then sonicated in hexanes or acetone for 10 to 40 minutes, and then desmutted using approximately 50 g of Oakite aluminum cleaner 164®. The samples were stirred in the cleaning solution for 30 minutes at 55°C before being rinsed and allowed to dry.

In addition to 2024-T3 aluminum alloy samples, we also used glass microscope slides (for absorbance measurements which required transparent substrates) and freshly cleaved mica surfaces (extremely flat, smooth surfaces for atomic force microscopy). An airbrush was used for spraying the solutions onto these substrates. For each layer, a reservoir containing the appropriate polyelectrolyte or clay solution was attached to the airbrush, which was sprayed with an effort to keep the spraying time and distance from the samples relatively constant. Typically 3-5 seconds of spraying from the airbrush is sufficient to wet the substrate surface. Note that it was important to spray some water through the airbrush to clean out one solution before spraying the next. The charge attraction between the polyelectrolytes and the clay resulted in aggregates clogging the nozzle otherwise.

Ultraviolet-visible (UV-vis) absorbance measurements were carried out in a Hewlett Packard 8453A spectrophotometer. Atomic force microscopy (AFM) images were acquired on a Nanoscope III (Digital Instruments, Inc.) using standard silicon nitride tips in air at room temperature. Samples for scanning electron microscope (SEM) were coated with a thin conductive layer of gold using a Denton Desktop II sputterer (Denton Vacuum, Inc.). The images were taken with a JSM 6400 microscope.

Electrochemical measurements were taken using a Gamry PCI400 potentiostat and a three-electrode cell with a platinum counter electrode and an Ag/AgCl/Cl⁻ (3M KCl) reference electrode. The measurements were conducted in dilute Harrison's solution (0.35 wt% $(NH_4)_2SO_4$ and 0.05 wt% NaCl) at room temperature. The treated 2024-T3 aluminum alloy panels were used as the working electrode. After all electrodes were placed in the electrolyte solution, the open circuit potential was monitored for at least 15 minutes to allow time to reach steady state potential. Electrochemical Impedance Spectroscopy (EIS) was carried out over the range from 10 kHz to 30 mHz with 10 mV rms applied potential relative to the open circuit potential (E_{oc}). Potentiodynamic scans (PDS) were carried out from –0.05 (vs. E_{oc}) to 0.5 V (vs. E_{oc}). The corrosion potential (E_{corr}) and corrosion resistance (R_{corr}) were determined as the intercept and slope, respectively, of the potential vs. current/area in the region from –0.05 to 0.05 V (vs E_{oc}). The pitting potential was determined using the criterion described by Kelly *et al* wherein pitting would have occurred by the time the anodic current density reached 3 x 10^{-5} A/cm^2 [7].

4 RESULTS

The first concern with the spraying process was to verify the presence of the polymer and clay layers. The sprayed films can be seen by eye after about 5 layers.

An AFM image of 1 bilayer of PDDA/clay on a mica substrate is shown in Figure 2. In the phase contrast image on the right, the shape of the clay platelets lying parallel to the substrate can be seen. The height image shows only a modest variation in height.

In Figure 3, SEM images are shown of the cleaned aluminum substrate and the LBL coatings. The LBL coating with 3 layers is noticeably rougher than the 10 layer coating. This may be because the first few layers show the effects of the roughness of the aluminum substrate, but by 10 layers, the LBL deposition has compensated for that roughness. Some larger aggregates are visible for 20 layers; this is interesting when related to the electrochemical measurements shown later.

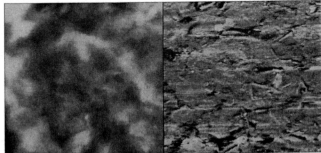

Figure 2: Atomic force microscopy image of film sprayed with 1 layer of PDDA and 1 layer of clay on a mica sheet. Both images are of the same region 3 micrometers square.

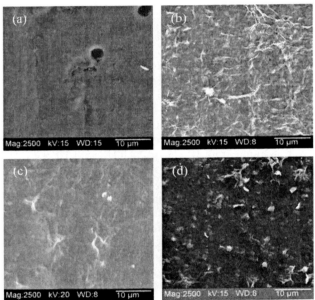

Figure 3: SEM images of (a) cleaned aluminum substrate; (b) 3 layers of (PDDA/PAA/PDDA/clay); (c) 10 layers; and (d) 20 layers.

While AFM and SEM are useful for visualizing the deposited layers, UV-vis absorbance measurements are often used to determine whether an equal amount of material is deposited in each (PDDA/PAA/PDDA/clay) layer. This does require a transparent substrate and so glass microscope slides were used. In Figure 4, the absorbance for layers of (PDDA/PAA/PDDA/clay) is shown to increase linearly.

Electrochemical measurements of the films on aluminum substrates were used to test their corrosion resistance properties. Potentiodynamic scans (PDS) for the cleaned aluminum substrate, with has a native aluminum

oxide on the surface, and for a substrate with 8 layers are shown in Figure 5.

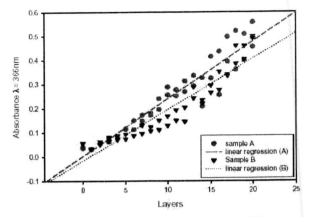

Figure 4: UV-visible absorbance measured at 366 nm increases linearly with the number of (PDDA/PAA/PDDA/clay) layers.

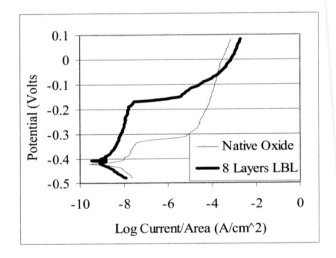

Figure 5: PDS scans of the bare substrate and with 8 layers of (PDDA/PAA/PDDA/clay).

The corrosion potential, E_{corr}, is the potential at the "zero" current. The pitting potential, Epit, is the potential at which the current reaches 3×10^{-5} A/cm^2 which has been empirically found to be the current when corrosion pits are forming on the substrate. For both of these potentials, a more positive value means the aluminum alloy is less likely to corrode. The corrosion resistance, R_{corr}, is determined by the slope near zero current. A higher value of R_{corr} indicates that the sample is less likely to corrode.

For each type of sample, at least three specimens were prepared and measured to determine average values, shown in Table 1. The corrosion potential and pitting potential increased only slightly with the LBL coating, but the corrosion resistance increased, with the best values between 8 and 15 layers. A 20 layer film had poorer resistance. The values determined by PDS and EIS measurements were similar.

Number of Layers	E_{corr} PDS (V)	E_{pit} PDS (V)	R_{corr} PDS (kΩ cm^2)	Phase Angle EIS	R_{tot} EIS (kΩ cm^2)
0	-0.46	-0.16	580	-41	520
3	-0.53	-0.10	640	-44	430
5	-0.52	-0.08	740	-51	870
8	-0.45	-0.17	1620	-65	1420
10	-0.47	-0.13	830	-66	1040
13	-0.47	-0.10	1130	-69	1810
16	-0.45	-0.10	1460	-66	1660
20	-0.69	-0.27	190	-39	480

Table 1: Corrosion characteristics determined by PDS and EIS measurements.

Electrochemical Impedance Spectroscopy (EIS) measurements were also made on these samples. A typical EIS scan is shown in Figure 6. The impedance response as a function of frequency can be fitted to an electrical circuit model [6] and values of film resistance and capacitance can be determined. Even without fitting, the phase angle measured at the minimum frequency of 10 mHz, provides an indication of the film quality. A value close to –90 degrees indicates a film which behaves as a capacitor even at low frequencies and is therefore a good protective film. This phase angle and the resistance are shown in Table 1.

5 CONCLUSIONS

We have investigated spraying polyelectrolyte and clay solutions to form LBL assembled films for corrosion protection of aluminum alloy substrates. The films show increased corrosion resistance. Incorporation of active inhibitor ions and an ORMOSIL barrier layer are expected to improve the protection.

6 ACKNOWLEDGEMENTS

The authors acknowledge the assistance of Edward Knobbe and Olga Kachurina of Sciperio, Inc. in Stillwater, OK.

This material is based upon work supported by the United States Air Force under Contract No. F49620-03-C0034. Any opinions, findings, and conclusions or recommendations expressed in this material are those of the authors and do not necessarily reflect the views of the United States Air Force.

REFERENCES

[1] Z. Tang, N. A. Kotov, S. Magonov, B. Ozturk, Nature Mater. 2, 413-418, 2003.
[2] N. A. Kotov, MRS Bulletin 26, 992-997, 2001.
[3] A. A. Mamedov, N. A. Kotov, M. Prato, D. M. Guldi, J. P. Wicksted, A. Hirsch, Nature Mater. 1, 190-194, 2002.
[4] J. B. Schlenoff, S. T. Dubas, T. Farhat, Langmuir 16, 9968-9969, 2000.
[5] T. L. Metroke, R. L. Parkhill, E. T. Knobbe, Prog. Inorg. Coatings 41, 233-238, 2001.
[6] O. Kachurina, T. L. Metroke, J. W. Ostrander, E. Knobbe, N. A. Kotov, International Journal of Nanotechnology , accepted.
[7] G. O. Ilevbare, J. R. Scully, J. Yuan, R. G. Kelly, Corrosion 56, 227-242, 2000.

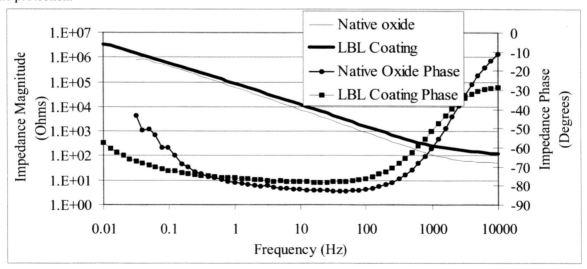

Figure 6: EIS data for the native oxide and the 13 layer LBL coating. An increase in impedance magnitude and a decrease in phase angle at low frequencies indicate the LBL film is providing protection.

On the Development of a Nanocrystalline Yttria Stabilised Zirconia Ceramic Capable of Superplastic Deformation at Relatively Low Temperatures

S.S. Bhattacharya[*], U. Betz[**] and H. Hahn[**]

[*]Materials Testing Facility, Department of Metallurgical & Materials Engineering
Indian Institute of Technology Madras, Chennai – 600036, India, ssb@iitm.ac.in
[**]Thin Films Division, Institute of Materials Science, Darmstadt University of Technology,
Petersenstrasse 23, D-64287 Darmstadt, Germany, hhahn@nano.tu-darmstadt.de

ABSTRACT

Superplastic behaviour is well established in fine-grained zirconia ceramics at high temperatures (well above $0.5T_m$, where T_m is the absolute melting point). In this study, an attempt was made to develop a nanocrystalline 5 mol.% yttria partially stabilised zirconia ceramic (5Y-PSZ) capable of exhibiting superplastic flow at relatively lower temperatures. A physical vapour processing route was used to synthesise the powders which was consolidated and subjected to a pressureless sintering route. Dense specimens with grain sizes in the nanometer range were subjected to tensile as well as compressive tests in the temperature range of 1283-1523K at different stress or strain rate levels in order to determine the superplastic deformation behaviour. The stress – strain – strain rate response of the material was analysed by a model for grain boundary sliding controlled superplastic flow. It was demonstrated that the strain rates predicted by the model are in close agreement with the experimentally observed ones.

Keywords: nanocrystalline ceramics, 5Y-PSZ, superplastic deformation, grain boundary sliding, activation energy

1 INTRODUCTION

Superplasticity is the ability of a polycrystalline material to exhibit very large deformations prior to failure. Wakai et al. [1,2] first established superplastic deformation behaviour in zirconia ceramics by stabilising the tetragonal or the cubic phase with a small quantity of yttria or other oxides. Since then several workers [3-11] have reported superplastic deformation in these ceramics at fairly high temperatures. The typical temperature range where superplastic behaviour has been observed lies well over $0.5T_m$ (where T_m is the absolute melting point) and requires special heating arrangements during testing. However, only recently results on the superplastic deformation in these ceramics at lower temperatures have been reported [12-15]. This is of interest not only from a scientific standpoint, but also from a practical and commercial point of view. In this study, the superplastic deformation behaviour of a nanocrystalline 5 mol. % yttria partially stabilised (5Y-PSZ) zirconia was examined at relatively lower temperatures.

2 EXPERIMENTAL

Nanocrystalline zirconia and yttria were separately synthesised using an Inert Gas Condensation (IGC) process [13, 15]. The crystallite sizes were determined from the peak broadening of the X-Ray Diffractogram (XRD). A crystallite size of 8 nm in case of zirconia and 14 nm for yttria was computed using the Scherrer formula. Specific surface areas determined for both ceramic powders by nitrogen adsorption measurements using a 5 point BET method [16] revealed particulate sizes of 8 and 15 nm respectively for zirconia and yttria, thus demonstrating little or no agglomeration in the powders.

The powders were then mixed in the correct proportion of 5 mol.% yttria in zirconia in isopropanol, given a stirring and ultrasonic agitation treatment to ensure uniform dispersion and finally, the organic medium was evaporated and the powder dried and collected. The dried powder was then uniaxially compacted followed by a cold isostatic compaction step. The sintering behaviour of the green bodies was first investigated. A pressureless sintering route was selected which essentially consisted of the following steps: (a) heating at a fixed rate in an oxygen flux to an intermediate temperature, (b) heating further under vacuum to the target sintering temperature, (c) holding at the target temperature for a predetermined amount of time with the oxygen flux restarted about midway through this period, and (d) cooling at a predetermined rate to room temperature under an oxygen flux. Specimens were sintered at different target temperatures and the resulting grain size (from line broadening of the XRD peaks as well as SEM micrographs) and the relative density (by Archimedes' principle) measured.

The green bodies for the tensile tests were prepared by uniaxial pressing in a 40 mm die at a pressure of 200 MPa followed by isostatic compaction at 400 MPa. The green bodies for the compression tests were prepared by uniaxial pressing in a 5.8 mm die at a pressure of 400 MPa and an isostatic compaction step at 520 MPa. For all the specimens, a sintering temperature of 1423 K with a hold time of 120 mins was selected as this resulted in a optimal combination of a relatively high density and a small grain size, well below 100 nm.

The tensile tests were carried out in the constant crosshead speed mode at temperatures of 1283 and 1403 K at initial strain rates in the range of 10^{-6} to 10^{-4} s^{-1} till failure. Reference specimens were also placed alongside the deforming specimens in order to monitor the static behaviour at these temperatures. The compression tests were carried out on a constant load type creep testing machine at temperatures of 1363, 1463 and 1523 K at different stress levels in the range of 12-90 MPa. The compression tests were not carried out till failure as (a) even after substantial time and deformation there was no failure seen and (b) density, deformation etc., measurements are simpler to make on a non-damaged specimen. From the tests, the strain – time data (obtained through a data acquisition set-up) at a given temperature and stress level were analysed for the deformation behaviour. The final density (from Archimedes' principle), grain size (from XRD) etc., were measured for all the specimens tested. To determine the strain rate sensitivity and the temperature dependence under tensile conditions, a stress jump test at a constant temperature of 1383 K and a temperature jump test at a constant stress of 20 MPa respectively, were carried out.

3 RESULTS AND DISCUSSION

The variation of the grain size and the relative density as a function of the sintering temperature is depicted in Figure 1. It is clear from the plot that a maximum relative

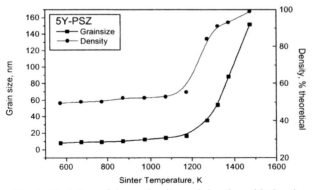

Fig. 1: Variation of the grain size and density with the sinter temperature

density of over 95% can be obtained in a specimen with a grain size well below 100 nm by sintering at a temperature of about 1400K. A sintering temperature of 1423 K for the test specimens resulted in a density of about 98% with a grain size of about 65 nm. Analysis of the sinter trajectory revealed that during the sintering process the rate controlling mechanism for grain growth is pore drag whereas the pore growth mechanism is rate controlled by a pore coalescence process. Both processes require an apparent activation energy of about 450 kJ mol^{-1}. The detailed analyses of grain growth and pore size distribution during sintering are reported elsewhere [15].

The results of the tensile tests, i.e., the stress – strain responses at different initial strain rates, in the case of a test temperature of 1333 K are depicted in Figure 2. From the

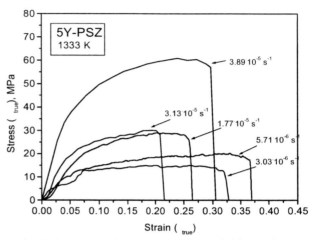

Fig. 2: Stress – strain curves at different initial strain rate levels obtained from the tensile tests at 1283 K

tensile tests carried out at 1403 K, a maximum elongation of 70 % at an initial strain rate of 3.13×10^{-5} s^{-1} was obtained clearly demonstrating that the ceramic is superplastic at this temperature. The results of the compression tests in the case of a temperature of 1523 K at various initial stress levels are shown in Figure 3. From the compression tests a

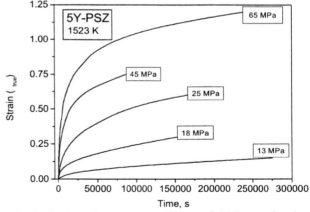

Fig. 3: Strain – time curves at different initial stress levels obtained from the compression tests at 1523 K

true compressive strain of more than 1.4 was established at 1463 K and 50 MPa initial stress level (again indicating the superplastic deformation capability of the ceramic). From the stress jump test an average strain rate sensitivity index of 0.33 was calculated [15]. The data obtained from the temperature jump test carried out at a constant stress level of 20 MPa, shown in Figure 4, was analysed to give an apparent activation energy, Q_{app} ($\ln \dot{\varepsilon} / 1/T$, where $\dot{\varepsilon}$ is the instantaneous strain rate and T is the absolute temperature) of 530 kJmol^{-1}. The term apparent is used

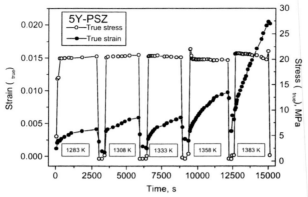

Fig. 4: Strain – time response of 5Y-PSZ during the temperature jump test carried out at a constant stress level of 20 MPa

because the computed activation energy would change with the stress level at which the experiment is carried out and must be distinguished from the true activation energy for the rate controlling process.

The mechanical deformation (stress - strain rate) data, taken at a constant material condition, were analysed by a model for grain/interface boundary sliding controlled flow [17], which has recently been shown to be the most suited in explaining superplastic deformation. The constitutive equation arising from this model is given as follows:

$$\dot{\varepsilon} = C_1 \left[\sigma_o \left\{ 1 + \ln \frac{\sigma_o}{\sigma_m} \right\} - \sigma \left\{ 1 + \sigma_m \exp \left(\frac{\sigma}{\sigma} \right) \right\} \right] \quad (1)$$

where C_1 is a material and temperature dependent constant, σ_o the threshold stress for plane interface formation, $\beta = 1/(\sigma^{0.5})$ with σ being the standard deviation and σ_m the mean stress of the internal stress distribution arising from the sliding process, $\dot{\varepsilon}$ and σ being the strain rate and the stress respectively (see ref. 17 for details). The true activation energy for the rate controlling sliding process can be computed from the following equation:

$$C_1 = C \exp \left(-\frac{Q}{RT} \right) \quad (2)$$

where C is a material constant inversely proportional to the grain size, Q is the true activation energy, R is the universal gas constant and T the absolute temperature.

As eqn. (1) is transcendental in nature, a step-scan regression procedure using a standard mathematical analysis software was used to obtain the best fit values of the constants, C_1, σ_o and σ_m for a given temperature and material condition. From the values of C_1 at different temperatures, the activation energy for the rate controlling process was computed [18]. Table 1 lists the activation

energy computed for constant true strain (material) conditions with the thermal vibration frequency factor, $\nu = kT/h$ (when considered as in diffusion [19]) and $\nu =$ constant ($10^{13} s^{-1}$ when it is considered to be constant for a solid [20]). Using the constants C, σ_o, and σ_m and the

True strain level	Activation Energy, kJ mol^{-1}	
	$\nu = kT/h$	$\nu = 10^{13} s^{-1}$
0.15	456	468
0.20	476	488
0.25	479	491

Table 1: Computed values of the activation energy at different material conditions for the 5Y-PSZ ceramic

activation energy term, the strain rates were predicted for the different stress levels and compared with the experimentally observed values. The predicted strain rates were found to be in reasonable agreement in most of the cases. Figure 5 shows a comparison between the experimentally observed and predicted strain rates obtained by the model at a true strain level of 0.25.

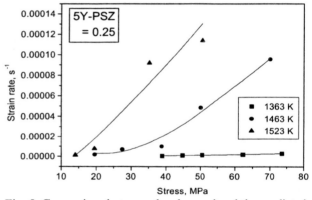

Fig. 5: Comparison between the observed and the predicted strain rates in case of the 5Y-PSZ ceramic at a true strain level of 0.25

Using eqn. (1), the apparent activation energy ($Q_{app} = \partial (\ln \dot{\varepsilon}) / \partial (1/T) |_\sigma$) at a constant true stress level, σ, could be predicted. At a true stress level of 20 MPa, an apparent activation energy of 543 kJ mol^{-1} was predicted, which compares very well with the experimentally obtained value.

4   CONCLUDING REMARKS

Mechanical deformation studies of a nearly fully dense, nanocrystalline 5 mol. % yttria stabilised zirconia ceramic at elevated temperatures demonstrated clearly that this material is capable of exhibiting superplastic deformations at relatively lower temperatures under tensile and compressive stresses. The stress – strain rate response was analysed by a model for grain/interface boundary sliding controlled superplastic flow. The true activation energy for the rate controlling process as well as the different deformation parameters that govern the constitutive relation were computed and it was shown that the strain rates predicted by the model are in reasonable agreement with the experimentally observed ones. The apparent activation energy at a constant stress level, predicted by the model, compares well with the experimentally observed result.

ACKNOWLEDGEMENTS

One of the authors (SSB) would like to thank the Alexander von Humboldt Foundation, Bonn, Germany for providing financial support with a fellowship for this research programme. The financial support by the Deutsche Forschungsgemeinschaft (HA 1344/2-3) is gratefully acknowledged. The authors would also like to thank Prof. K.A. Padmanabhan, currently Vice Chancellor, South Asian International Institute, Hyderabad, India for many fruitful discussions and for permitting the use of his tensile testing facilities at IIT Kanpur, India under a DAAD programme.

REFERENCES

[1] F. Wakai, S. Sakaguchi and Y. Matsuno, "Superplasticity of yttria-stabilised tetragonal ZrO_2 polycrystals," Adv. Ceram. Mater., 1, 259-263, 1986.

[2] F. Wakai, S. Sakaguchi, K. Kanayama and H. Onishi in Ceramic Materials and Components for Engines, W. Bunk and H. Hausner (eds.), Deutsche Keramische Gesellschaft, Saarbrücken, 315, 1986.

[3] C. Carry and A Mocellin in Deformation of Ceramics II, R.E. Tressler and R.C. Bradt (eds.), Plenum, New York, 391, 1984.

[4] W.R. Cannon in Structure and Properties of MgO and Al_2O_3 Ceramics, W.D. Kingery (ed.), The American Ceramic Society, Colombus, OH, 741, 1985.

[5] C. Carry and A Mocellin, "Structural superplasticity in single phase crystalline ceramics," Ceram. Intl., 13, 89-98, 1987.

[6] Yasuhiro Maehara and Terence G. Langdon, "Review – Superplasticity in ceramics," J. Mater. Sci., 25, 2275-2286, 1990.

[7] T.G. Nieh and J. Wadsworth, "Superplastic behaviour of fine-grained yttria-stabilised zirconia polycrystals (Y-TZP)," Acta Metall. Mater., 38, 1121-1133, 1990.

[8] F. Wakai, Y. Kodama, S. Sakaguchi N. Murayama, K. Izaky and K. Niihara, "A superplastic covalent crystal composite," Nature, 34, 421-423, 1990.

[9] I-W. Chen and S.-L. Hwang, "Shear thickening creep in superplastic silicon nitride," J. Am. Ceram. Soc., 75, 1073-1079, 1992.

[10] M. Jimenez-Melendo, A. Dominguez-Rodriguez and A. Bravo-Leon, "Superplastic flow of fine-grained yttria stabilised zirconia polycrystals: Constitutive equations and deformation mechanisms," J. Am. Ceram. Soc., 81, 2761-2775, 1998.

[11] E. Sato, H. Morioka, K. Kuribayashi and D. Sundararaman, "Effect of small amount of alumina doping on superplastic behaviour of tetragonal zirconia," J. Mater. Sci., 34, 4511-4518, 1999.

[12] U. Betz, G. Scipione, E. Bonetti and H. Hahn, "Low temperature deformation behaviour of nanocrystalline 5 mol.% yttria stabilised zirconia under tensile stresses," Nanostructured Mater., 7, 845-853, 1997.

[13] Ulrich Betz, Subramshu S. Bhattacharya and H. Hahn, "On the development of a low temperature, superplastic nanocrystalline 5 mol.% yttria stabilised zirconia: Synthesis, processing and mechanical characterisation," J. Metastable Nanocrystalline Mater., 8, 539-544, 2000

[14] U. Betz, K.A. Padmanabhan and H. Hahn, "Superplastic flow in nanocrystalline and sub-microcrystalline yttria-stabilized tetragonal zirconia," J. Mater. Sci., 36(24), 5811-5821, 2001.

[15] U. Betz, in Synthese und superplastizitaet von nanokristallinen zirkondioxid-kermiken, Ph.D. Dissertation, Institute for Material Science, Darmstadt University of Technology, Darmstadt, Germany, June 2000.

[16] S. Brunauer, P. Emmet and E. Teller, "Adsorption of gases in multimolecular layers," J. Am. Chem. Soc., 60, 309-319, 1983.

[17] K.A. Padmanabhan and J. Schlipf, "A model for grain boundary sliding and its relevance to optimal structural superplasticity – I. Theory," Mater. Sci. Technol., 12, 391-399, 1996.

[18] S.S. Bhattacharya and H. Hahn, "On the low temperature superplastic deformation behaviour of a nanocrystalline yttria stabilised zirconia ceramic," under preparation.

[19] S. Glasstone, K.J. Laider and H. Eyring in The Theory of Rate Processes, McGraw Hill, New York, 516, 1946.

[20] P. Shewmon in Diffusion in Solids, 2nd Edn., The Minerals, Metals and Materials Soc., Warrendale, Pa, USA, 74, 1989.

Synthesis of Spinel Ferrite Particle/Organic Hybrid from Metal-Organics

Toshinobu Yogo, Satoshi Nakamura and Wataru Sakamoto

Center for Integrated Research in Science and Engineering,
Nagoya University, Furo-cho, Chikusa, Nagoya 464-8603, Japan,
yogo@cirse.nagoya-u.ac.jp

ABSTRACT

Nanocrystalline spinel particle/organic hybrid was successfully synthesized from iron(III) 3-allylacetylacetonate (IAA) and nickel acetylacetonate (NA) or nickel diammine 3-propenylacetylacetonate (NPA) by *in situ* hydrolysis. Spinel ferrite particle/organic hybrid was synthesized by hydrolysis of the IAA-NA by addition of methylhydrazine. Crystalline particles of below 10 nm in size were found to be dispersed in an organic matrix. Nanocrystalline particles were identified to be $NiFe_2O_4$ by X-ray diffraction analysis, electron diffraction and energy dispersive X-ray analysis. The saturation magnetization of the hybrid was dependent upon hydrolysis conditions. Nickel ferrite particles in the hybrid from IAA-NPA had a higher crystallintiy and saturation magnetization than those from IAA-NA. Nanometer-sized nickel ferrite particle/organic hybrid showed a typical superparamagnetic behavior.

Keywords: spinel ferrite, metal-organics, hydrolysis, magnetic properties, hybrid

1 INTRODUCTION

Inorganic/organic hybrid materials attract great attention as newly emerging composite materials. Inorganic/organic hybrid material consists of a composite structure at nanometer level between the inorganic phase and organic phase, such as molecules, particles, layers and networks [1]. Nanocrystalline magnetic particles show unique phenomena of superparamagnetism [2] and quantum size effect [3]. The magnetic properties of the fine particles depend upon the crystalline phases, crystallinity and particle sizes. Magnetic particle/organic hybrid has various potential applications in magnetic and medical uses, such as magnetic recording [4], magnetic fluid [2], magnetic ink, magnetic resonance imaging [5] and thermomagnetic surgery [6]. Magnetic particle/organic composites have been synthesized using various methods, such as mixing of magnetic particles and organic phases [7], ion-exchange gels from ferrofluid materials [8] and particle formation in a polymer matrix [9]. Blending of small particles into organics is difficult because of aggregation due to van der Waals force. Especially, magnetic moment results in the aggregation of magnetic particles. Therefore, *in situ* formation of magnetic particles in an organic matrix is one of the most favorable methods for the synthesis of such composites.

Iron allylacetylacetonate (IAA) has carbon-carbon double bonds for polymerization and chelated Fe-O bonds. The chelated Fe-O bond was reported to be hydrolyzed with ammonia water affording oxide particles [10]. The authors synthesized superparamagnetic α-Fe_2O_3/oligomer hybrid from the controlled hydrolysis of iron (III) allylacetylacetonate (IAA) and IAA oligomer [11]. Magnetic particle/oligomer hybrid was synthesized from the IAA oligomer in the presence of hydrazine [12]. Transparent magnetic particle/organic film was successfully synthesized from IAA under controlled polymerization and hydrolysis conditions, and revealed the quantum confinement effect [13].

This paper describes the synthesis of spinel oxide particles/oligomer hybrid material from metal acetylacetonates. IAA, nickel acetylacetonate (NA) and nickel (II) diammine 3-propenylacetylacetonate (NPA) were used as starting metal-organics. The reaction conditions for the formation of $NiFe_2O_4$ were investigated. Magnetic properties of spinel ferrite particle/organic hybrid were also evaluated. Nanocrystalline spinel ferrite particle/organic hybrid was successfully synthesized from IAA-NA or IAA-NPA under controlled hydrolysis conditions

2 EXPERIMENTAL

2.1 Synthesis of Spinel Particle/Organic Hybrid

Iron (III) tris(3-allylacetylacetonate) (IAA) and nickel (II) diammine 3-propenylacetylacetonate (NPA) were prepared by the method described in the literature [14]. Ethanol was dried over magnesium ethoxide and then distilled before use. Commercial methylhydrazine (CH_3NHNH_2) was used as received. Commercial nickel (II) acetylacetonate dihydrate was used after drying at 130°C and 10 Pa for 2h.

The experimental procedure is shown in Fig. 1. IAA and nickel acetylacetonate ($Ni(acac)_2$, NA) were weighed with a molar ratio of 2, and dissolved in ethanol yielding a homogeneous solution. The solution was hydrolyzed with a mixture solution of methylhydrazine and water. The reaction mixture was then heated below 80°C from 0.5 to 24h. The solid product was obtained after removal of

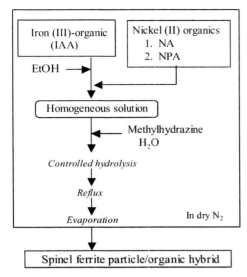

Fig.1 Experimental procedure for synthesis of spinel ferrite particle/organic hybrid

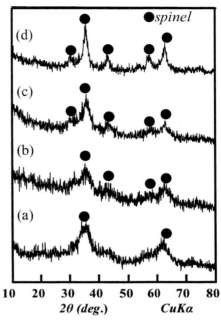

Fig.2 XRD of hybrid, (a) ammonia water, (b) metal/ methylhydrazine/H_2O =1/6/10 (molar ratio), (c)1/6/20, (d)1/6/40.

solvent followed by drying under vacuum at room temperature.

2.2 Characterization

The amount of organics of the hydrolysis products was analyzed by DTA-TG (Rigaku, TAS-100). The infrared absorptions of starting compounds and product were measured with FT-IR spectroscopy (Nicolet, Nexus-470). The phase of the product was identified by X-ray diffraction (XRD) using CuKα radiation with a monochromator (Rigaku, RINT-2500). The hybrid powder was analyzed by XPS (Jeol, JPS-9000MC). The magnetic particles in an organic matrix were observed by transmission electron microscopy (TEM, Hitachi, H-800) with an energy dispersive X-ray (EDX) analyzer. The magnetic properties were measured at room temperature using a vibrating sample magnetometer (VSM, Toei Industry, VSM-5).

3 RESULTS AND DISCUSSION

3.1 Synthesis of Spinel Particle/Organic Hybrid from IAA-NA by *in situ* Hydrolysis

In order to synthesize a homogeneous precursor solutions, $Ni(acac)_2 \cdot 2H_2O$ was dehydrated at 130°C and 10 Pa for 2h. After the treatment, DTA-TG analysis revealed no weight loss due to removal of water at around 130°C. Thus, anhydrous $Ni(acac)_2$ (NA) was obtained under the conditions. IAA and NA was mixed with a molar ratio of 2 yielding a brown solution.

The solution was hydrolyzed under various conditions. First, the solution was hydrolyzed ammonia water at 80°C for 24h. The XRD of the product revealed two broad diffractions shown in Fig. 2(a). Then, the starting solution was hydrolyzed with methylhydrazine. Methylhydrazine (MH) was selected based upon the previous results by the authors [12,13]. The XRD patterns of the products are shown from Fig. 2(b) to 2(d). When the solution was hydrolyzed with (IAA-NA)/methylhydrazine(MH)/H_2O=1/6/10, several broad diffractions are observed as shown in Fig. 2(b). The pattern was almost similar to that obtained with ammonia. With increasing water in amount from 20 to 40, the diffractions increased in intensity as shown in Fig. 2(c) and 2(d). The diffraction pattern shown in Fig. 2(d) is in good agreement with that of spinel. Water promotes hydrolysis of metal-organics and crystallization of spinel particles.

Both $NiFe_2O_4$ and Fe_3O_4 have a cubic symmetry of crystal structure, and have a quite similar lattice constant to each other. Therefore, XRD analysis is not sufficient for phase identification. The hybrid shown in Fig. 2(d) was analyzed by XPS. The XPS spectrum exhibited not only Fe $2p_{3/2}$ peak at 711 eV but also Ni $2p_{3/2}$ peak at 855 eV. The ratio of peak area of Ni to Fe was 1/2.

Microstructures of spinel particle/organic hybrid was observed by TEM. The hybrid synthesized at (IAA-NA)/MH/H_2O=1/6/10 at 80°C for 24h consisted of black particles below 10 nm and an organic matrix. The selected area diffraction (SAD) pattern of the particles revealed spots on the rings, which supported the particles to be crystalline. The d values of the SAD pattern are in good agreement with those of nickel ferrite. The EDX analysis revealed that nickel and iron existed at a 1/2 ratio.

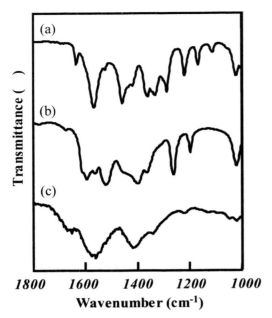

Fig. 3 FT-IR spectra of (a) IAA, (b) NA (c) $NiFe_2O_4$ particle/organic hybrid.

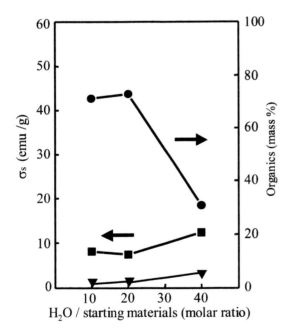

Fig. 4 Changes of saturation magnetization and organics of hybrids with amount of hydrolysis water, ▲ hybrid, ■ corrected σs value.

Figure 3 shows IR spectra of starting IAA, NA and hydrolyzed product. The spectrum of IAA shown in Fig. 3(a) include a C=C absorption band at 1640 cm^{-1} and chelated carbonyl absorptions at 1570 and 1460 cm^{-1}. Similarly, the absorptions of chelated carbonyl groups are observed at 1590 and 1520 cm^{-1} in the spectrum of NA (Fig. 3(b)). After the reaction of (IAA-NA) with methylhydrazine and water at (IAA-NA)/MH/H_2O=1/6/10, the C=C absorption of IAA at 1640 cm^{-1} disappears as shown in Fig. 3(c). The disappearance of C=C suggests the polymerization of allylacetylacetonate ligand during heating at 80°C for 24h. The absorptions of chelated carbonyl groups at 1565 and 1415 cm^{-1} increase in bandwidth. The broadening of absorption indicates the coordination of the chelated ligand to nano particles.

The changes in saturation magnetization and amount of organics in the hybrids with water amount are summarized in Fig. 4. The hybrid was synthesized at (IAA-NA)/MH=1/6 at 80°C for 24h. The corrected values based upon the amount of $NiFe_2O_4$ and organics are also shown in Fig. 4. The amount of organics was measured by DTA-TG analysis. The saturation magnetization increases with increasing amount of added water for hydrolysis. The increase of saturation magnetization is attributable to the increase in size and crystalline regularity of spinel particles precipitated out in the organic matrix. The corrected saturation magnetization of the hybrid is about 20 emu/g, when the precursor was hydrolyzed at (IAA+NA)/MH/H_2O=1/6/40. The value was about 36 % of bulk $NiFe_2O_4$ (56 emu/g).

3.2 Synthesis of Spinel Particle/Organic Hybrid from IAA-NPA by *in situ* Hydrolysis

A precursor consisting of IAA and nickel (II) diammine 3-propenylacetylacetonate (NPA) was treated with a mixture solution of methylhydrazine and water. Figure 5 shows the XRD profiles of spinel particle/organic hybrid from IAA-NA and IAA-NPA with (Fe+Ni)/MH/H_2O=1/6/10 at 80°C for 24h. Spinel particles

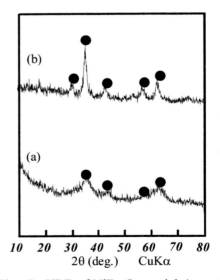

Fig. 5 XRD of $NiFe_2O_4$ particle/organic hybrid from (a) IAA-NA, (b) IAA-NPA at metal/MH/H_2O=1/6/10.

were also synthesized from IAA-NPA as shown in Fig. 5(b). The diffractions of $NiFe_2O_4$ particles in the organic matrix (Fig. 5(b)) is much more distinct than that from IAA-NA. The higher crystallinity of $NiFe_2O_4$ of the former than that of the latter is considered to derive from the reactivity of metal-organic bond of precursors. The reactivity of nickel-organic bond of IAA-NPA is more susceptible to hydrolysis and crystallization than that of IAA-NA.

Figure 6 shows the room-temperature magnetizations versus applied fields for specimens shown in Fig. 5. The magnetic induction increases with increasing applied field, although no hysteresis is observed. The magnetizations shown in Fig. 6(a) saturates rapidly at a low magnetic field. The curve shows neither residual magnetization nor coercive field. These results imply the formation of superparamagnetic particles. The corrected saturation magnetization of the hybrid shown in Fig. 6(a) was 43.5 emu/g, which was 78 % that of bulk $NiFe_2O_4$. The curve shown in Fig 6(b) exhibits a similar behavior. Even at the high applied field of 10 kOe, the hybrid show no decrease in magnetic induction as shown in Fig. 6(a). This is considered to result from the high dispersion of magnetic particles in the organic matrix through chemical bonding.

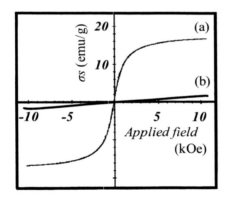

Fig. 6 BH curves of hybrids synthesized from (a) IAA-NPA, (b) IAA-NA at metal/MH/H_2O=1/6/10.

4   CONCLUSIONS

Nanocrystalline spinel ferrite particle/organic hybrids were successfully synthesized through controlled hydrolysis of iron- and nickel-organics below 100˚C. The results were summarized as follows:

1.   The hybrid consisted of uniformly dispersed nano-sized $NiFe_2O_4$ particle and organic matrix.
2.   The crystallinty of $NiFe_2O_4$ particles was dependent upon the amount of hydrolysis water in iron (III) tris(3-allylacetylacetonate) (IAA) and nickel (II) acetylacetonate system (NA).

3. $NiFe_2O_4$ particles from IAA-nickel (II) diammine 3-propenylacetylacetonate (NPA) had higher crystallinity than those from IAA-NA.
4.   The hybrid revealed superparamagnetic behavior. The saturation magnetization of the hybrid was 43.5 emu/g at room temperature.

REFERENCES

[1] "Organic/inorganic Hybrid Materials-2002", edited by C.Sanchez, R.M.Laine, S.Yang and C.J.Brinker, Mater. Res. Soc. Symp. Proc., 726, Mater. Res. Soc., Warrendale, 2002.
[2] S.W.Charles and J.Popplewell, "*Ferromagnetic Materials*", Vol. 2, 509, edited by E.P.Wohlfarth, North-Holland, Amsterdam, 1980.
[3] L.Zhang, G.C.Papaefthymiou and J.Y.Ying, *J. Appl. Phys.*, **81**, 6892, 1997.
[4] L.Gunther, *Physics World* **3**, 28, 1990.
[5] M.B.E.Olsson, B.R.B.Persson, L.G.Salford, U.Schröder, *Mag. Reson. Imaging*, **4**, 437, 1986.
[6] R.W.Rand, H.D.Snow, D.G.Elliott and M.Snyder, *Appl. Biochem. Biotech.*, **6**, 265, 1981.
[7] L.Nixon, C.A.Koval, D.A.Noble and G.S.Staff, *Chem. Mater.*, **4**, 117, 1992.
[8] J.Ugelstad, A.Berge, T.Ellingsen, O.Aune, L.Kilass, T.N.Nilsen, R.Schmidt, P.Stenstad, S.Funderud, G.Kvalheim, K.Nustd, T. Lea, F. Vartdal and H.Danielsen, *Makromol. Chem. Macromol. Symp.*, **17**, 177, 1988.
[9] R.F.Ziolo, E.P.Giannelis, B.A.Weinstein, M.P. O'Horo, B.N.Ganguly, V.Mehrotra, M.W.Russell and D.R.Huffman, *Science*, **257**, 219, 1992.
[10] K.Higuchi, S.Naka and S.Hirano, *Adv. Ceram. Mater.*, **1**, 104, 1986.
[11] T.Yogo, T.Nakamura, K.Kikuta, W.Sakamoto and S.Hirano, *J. Mater. Res.*, **11**, 475, 1996.
[12] T.Yogo, T.Nakamura, W.Sakamoto and S.Hirano, *J. Mater. Res.*, **14**, 2855, 1999.
[13] T.Yogo, T.Nakamura, W.Sakamoto and S.Hirano, *J. Mater. Res.*, **15**, 2114, 2000.
[14] H.A.Tayim and M.Sabri, *Inorg. Nucl. Chem. Lett.*, **9**, 753, 1973.

Assisted Alignment of Carbon Nanotubes

Yousef Haik[*1], Jhunu Chatterjee[1] and Ching Jen Chen[1]

[1]Center for Nanomagnetics and Biotechnology, Tallahassee, Florida 32310

*United Arab Emirates University, Department of Mechanical Engineering, Al Ain, UAE

ABSTRACT

Extensive research has been focused on the processing of carbon nanotubes polymeric composites. The ability to impose a preferred alignment of carbon nanotubes in a composite will increase the effectiveness of utilizing nanotubes in composite applications. The alignment of nanotubes will maximize the interfacial bonding across the nanotube matrix interface. Nanomagnetic particles made of magnetite are used to impose preferred orientation on carbon nanotube in a composite matrix. The nanomagnetic particles are attached by surface adsorption to the carbon nanotubes surface. The external applied magnetic field is utilized to attract the nanomagnetic particles toward the higher field. The carbon nanotubes are stretched along the field line because of the nanoparticles movement toward the higher field. Industrial applications of carbon nanotube composites become more feasible with this technology.

Keywords Nanotubes, Alignment, Nanomagnetics

1. INTRODUCTION

Carbon nanotubes have unique physical properties with a very high length to diameter ratios which has made its use as one of the strongest fiber known. Different techniques were introduced to produce uniform polymer nanotubes composites [1-4].

Various methods of alignment of nanotubes have been reported, such as carbon arc discharge [5], clipping of epoxy resins [6], rubbing of films [7], Chemical vapor deposition [8-10] and mechanical stretching of nanotubes in polymer matrix [11]. Magnetic orientation based on difference in magnetic susceptibility between the carbon nanotubes and the polymer has also been reported [12-13]. However, the magnetic susceptibility of both the polymer and the carbon nanotubes is very weak (in the order of 10^{-6}) [14], and require high magnetic fields (~15-25 tesla) to induce the orientation. The technique becomes very expensive for industrial applications.

In this study, magnetic nanoparticles are used to induce preferred orientation of magnetic nanotubes in a polymer composite. The magnetic nanoparticles are attached to the carbon nanotube surface. Because of the Van der Waals forces on the carbon nanotubes, they will have a strong physical adsorption capacity. The strong Van der Waals force exerted by the surface of the nanotubes can attach almost its weight equivalent of iron oxide onto its surface.

Because of its graphene sheet structure, the large surface area is responsible for this kind of attachment. The physical attachment of nanomagnetic particles with the carbon nanotubes caused much better alignment of the nanotubes in a weak magnetic field. Nanotubes with attached magnetic particles are observed using ESEM, STEM and AFM. The alignment in the presence of the magnetic fields has been confirmed by STEM, ESEM and by Raman laser microscopy

2. IRON OXIDE AND SWNT COMPOSITE

Iron oxide is prepared by the conventional coprecipitation procedure. The precipitate was sonicated. 0.10 mg of SWNT (from Carbon Nanotechnologies Inc. Texas) was dispersed in methanol by ultrasonication. The black precipitate of iron oxide (magnetite) in alkaline medium was added to it and the mixture was thoroughly stirred. SWNT-magnetite composite was formed almost immediately. The whole mass was being attracted by 0.5 T magnets.

SWNT-Magnetite Polymer Composite

SWNT-magnetite was mixed with commercially available epoxy resin (PR 2032) and further crosslinked with a hardener (PH 3660) at room temperature. The crosslinked SWNT magnetic composite were formed by keeping the above mixture at room temperature. 12 ml of this resin mixture and 1ml of SWNT magnetite dispersion was mixed thoroughly and put in a 4 ml standard plastic cuvette with a lid. The resin and SWNT mixture was allowed to set for 24 hours. After 24 hrs, a solid block of the resin was formed and was released by breaking the plastic cuvette. To align the SWNT-iron oxide composite, another

Fig 1 Exp Setup

cuvette with above composition of resin and SWNT was placed in between two 0.5tesla (square shaped) permanent magnets (Fig 1). 100nm and 50nm films were cut from these blocks by microtoming.

Fig 2 shows the STEM micrograph for iron oxide coupled with SWNT aligned in epoxy with the help of the external magnetic field. The STEM was done in 100nm thick microtomed samples obtained from the epoxy SWNT composite block. The rope-like bundles of nanotubes containing a cluster of iron oxide nanoparticles attached sporadically on the tubes were observed. ESEM micrograph (fig 3) obtained with 100 m thick film shows the presence of unaligned SWNT and aligned SWNT in magnetic field at different magnifications very clearly. Care was taken to avoid while cutting the samples. It has been reported that the tubes in a polymer resin composite can be aligned by preferential cutting process [6]. Alignment of carbon nantubes was confirmed by use of near-infrared Raman spectra.

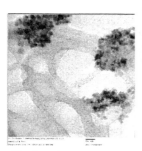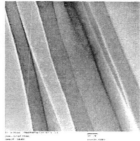

Fig 2 a&b Magnetic particle adsorbed to nanotube and aligned nanotubes

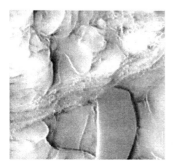

Fig 3 a&b unaligned nanotubes

3. CONCLUSION

SWNT were aligned by assistance of magnetic particles. The magnetic nanoparticles adsorb to the carbon nanotubes surface. An external weak applied field can be used to induce preferred alignment of magnetic nanotubes in the polymer composite. The alignment of nanotubes in the polymer was confirmed by Raman spectra. Utilization of magnetic nanoparticles to induce alignment of nanotubes in composites is relatively inexpensive and may facilitate industrial applications of carbon nanotube reinforced composites.

References:

1. R. Andrews, D. Jacques D, A.M. Rao, T. Rantell, and F. Derbyshire, Appl. Phys. Letters 75, (9), (1999) 1329-31
2. S. Kumar, H. Doshi, M. Srinivasarao, J.O. Park, Polymer 43, (2002), 1701-03.
3. J.H. Chen, Z.P. Huang, D.Z. Wang, S.X. Yang et al., Appl. Phys. A, 73, (2001) 129-31.
4. E. Bermejo, T. l. Mercier, M. Quarton, J of the Amer. Ceramic Soc. 78, 365-368 (1995).
5. A. Holloway, R. W. McCallum, S. R. Arrasmith, J. of Mat. Res. 8, Apr (1993).
6. X. K. Wang, X. W. Lin, V. P. Dravid, J. B. Ketterson, R. P. H. Chang, Appl. Phys. Lett. 62, (1993), 1881.
7. P. M. Ajayan, O. Stephan, C. Colliex, D. Trauth, Science, 265 (1994), 1212.
8. W. A. de Heer, W. S. Bacsa, A. Chatelain, T. Gerfin, R. Humphery-Baker, L. Forro, D. Ugarte, Scince, 268, (1995), 845.
9. W. Z. Li, S. S> Xie, L. X. Qian, B. H. Chang, B. S. Zou, W. Y. Zhou, G. Wang, Science, 274, (1996), 1701.
10. T. Kyotani, L. Tsai, A. Tomita, Chem. Mater., 8, (1996), 2109.
11. S. Fan, M. G. Chapline, N. R. Franklin, T. W. Tombler, A. M. Cassell, H. Dai, Science, 283 (1999), 512
12. L. Jin, C. Bower, O. Zhou *Appl. Phys. Lett.* 73 (1998) 1997
13. B. W. Smith, Z. Benes, D. E. Luzzi, and J. E. Fischer, Appl. Phys. Lett., 77(5), (2000) 663-5.
14. M. Fujiwara, E. Oki, Y. Tanimoto, I. Mukouda, Y. Shimomura, J. Phys. Chem., 105(18), (2001) 4383-86.
15. . Chauvet, L. Forro, W. Bacasa, Phys. Rev. B, 52 (10), (1995) 6963-6.
16. C Velasco-Santos, A. L. Martinez-Hernandez, M. Lozada-Cassou, A. Alvarez-Castillo,V.M.Castano *Nanotechnology* 13 (2002) 495
17. M. Knez, M. Sumser, A. M. Bittner, C. Wege, H. Jeske, S. Kooi, M. Burghard, K. Kern. *J.Electroana.Chem.*522 (2002) 70
18. T. Hertel, R. E. Walkup, P. Avouris *Phys. Rev. B* 58 (1998) 13870

Synthesis & Mechanical Characterization of Carbon/Epoxy Composites Reinforced with SiC Nano Particles

Nathaniel Chisholm, Hassan Mahfuz[1], Vijaya Rangari, Reneé Rodgers, and Shaik Jeelani

Tuskegee Center for Advanced Materials (T-CAM)

Tuskegee University, Tuskegee, Alabama 36088. U.S.A

ABSTRACT

Room temperature cure epoxy resin has been doped with SiC nanoparticles through an ultrasonic cavitation process. The modified resin was then utilized in a Vacuum Assisted Resin Transfer Molding (VARTM) process with satin weave carbon fibers to manufacture nanophased composite panels. The nanoparticles were spherical in shape and about 29 nm in diameter. The loading of nanoparticles into the resin ranged from 1.5 to 3.0 percent by weight. The mixing of nanoparticles with Part-A of SC-15 resin was carried out in a *Sonics Vibra Cell* ultrasonic liquid processor at 55% of the amplitude for about 30 minutes. At this time, the dispersion of nanoparticles seemed uniform through visual observation. In order to avoid rise in the temperature during sonication, cooling was employed by submerging the mixing beaker in a mixture of ice and water. Part-B (hardener) was then added with the mixture at a ratio of 3:10 and the mixing was carried out mechanically for about 10 minutes using a high speed mechanical stirrer. In the next step, the mixture was used in a VARTM set up with satin weave carbon preforms to fabricate nanophased composite panels. Once cured, test coupons were extracted and subjected to both quasi-static and dynamic loading. Under quasi-static loading tensile and flexural response were of particular interests. Dynamic tests were also carried out at low velocity impact and at high rates of strain. Details of manufacturing, analyses of mechanical tests and enhancements in properties due to nanoparticle infusion are discussed in the paper.

Keywords: nanocomposites, carbon/epoxy, SiC nano scale fillers, reinforced composites, nanotechnology

1.0 INTRODUCTION

Mineral fillers, metals, and fibers have been added to polymers for decades to form composite materials [1-5]. Compared to neat resins, these composites have a number of improved properties including strength, heat distortion temperature, and modulus. As a result, for structural applications composite materials have become an essential entity. With the advent of the new millennium, and with our relentless quest for lighter and stronger materials, the demand for materials has taken a new dimension. No longer are the traditional composite materials capable of satisfying our stringent requirements, nor can they be engineered to control properties at the atomic or molecular levels. The essence of such control in properties has derived from the fact that the aggregate properties of materials under external excitations such as force, pressure or temperature, are largely dictated by their molecular level orientation. Eventually, materials which are fashioned at the molecular level will be highly sophisticated to be engineered according to specific requirements. It is surprising to know that the newly developed material system known as *"nanocomposite materials"* offers the ability to build at the molecular level to create large structures with fundamentally new molecular organizations.

The ability to reorganize materials at the molecular level is what makes nanocomposites an attractive tool for fabricating materials to meet our stringent requirements for materials with enhance mechanical and thermal properties. What makes nano scale building blocks attractive in polymers, is the extremely high surface area which is created by the nanoparticles when interaction takes place with the polymer chain. This interaction creates large interfaces in a composite therefore enhances adhesion energy which translates into increase bonding. This increase in chemical bonding improves the polymer cross-linking and enhances mechanical and thermal properties. An interphase of 1 nm thick represents roughly 0.3% of the total volume of polymer in case of micro particle filled composites, whereas it can reach 30% of the total volume in case of nanocomposites [6]. A negligible contribution made by the interphase provides diverse possibilities of performance tailoring and is able to influence the properties of the matrices to a much greater extent under rather low nano-filler loading. Significant improvement in the tensile properties of polypropylene composites has also been reported in terms of stiffening, strengthening and toughening with a low filled content of about 0.5% [7]. Other studies

---

1. Corresponding author, telephone number (334)-727-8985, and e-mail address: ememah@tuskegee.edu

have shown that by using silica clay platelet nano-fillers in polymers, thermal properties can be improved significantly when compared to conventional filled polymers [8].

While these studies revealed significant findings, in a variety of areas such as tensile modulus, strength, and thermal stability in polymers, the development of nanocomposites which can meet the growing demand for structural applications have been slow. With the understanding of how nano fillers interact with polymers, it is feasible to develop a structural composite fashioned from a nano-phased polymer. In retrospect, the use of nanocrystalline materials provides an excellent means for the development of polymer fibrous composites with both enhanced mechanical and thermal performance. In the present study, epoxy resin has been doped with SiC nanoparticles through an ultrasonic cavitation process. The modified (nano-phased) resin was then utilized in a Vacuum Assisted Resin Transfer Molding (VARTM) process with satin weave carbon fibers to manufacture composite panels. Test coupons were extracted and subjected to various loading conditions. Details of manufacturing and analyses of mechanical tests due to nanoparticle infusion are discussed in the following sections.

1.1 Manufacturing of Nanocomposites

The fabrication of nanophased carbon/epoxy composites was carried out in three steps. In the first step, spherical SiC nanoparticles of about 29 nm in diameter (manufacturer: MTI Corporation Inc. 2700 Rydin Road, Unit D, Richmond, CA. 94804, USA) were ultrasonically mixed with part-A (mixture of: Diglycidylether of Bisphenl A, 60 to 70%, Aliphatic Diglycidylether, 10 to 20% and epoxy toughner 10 to 20%) of SC-15 epoxy resin (manufacturer: Applied Poleramic, Inc 6166 Egret Court, Benicia, CA. 94510, USA). SC-15 is a two phase toughened epoxy resin system, it cures at room temperature and is extensively used in Vacuum Assisted Resin Transfer Molding (VARTM) processes. The loading of nanoparticles ranged from 1.5 to 3.0 percent by weight of the resin. The mixing was carried out in a *Sonics Vibra Cell* ultrasonic liquid processor (Ti-horn, frequency=20 kHz, intensity=100W/cm^2) as shown in Fig. 1. The mixing was carried out at 55% of the amplitude for about 30 minutes. At this time, the dispersion of nanoparticles seemed uniform through visual observation. In order to avoid rise in temperature during sonication, cooling was employed by submerging the mixing beaker in a mixture of ice and water. In the next step, Part-B (hardener, cycloaliphatic amine 70 to 90% and polyoxylalkylamine 10 to 30%) was added with

Fig 1. Vibra-cell ultrasonic processing

the mixture at a ratio of 3:10 and the mixing was carried out mechanically for about 10 minutes using a high speed mechanical stirrer. In the final step, the reaction mixture was used in a VARTM set up [9-11] with satin weave carbon fiber preforms to fabricate Carbon/Epoxy nanocomposite panels as shown in Fig. 2. Test coupons were extracted from each category of panels to conduct various mechanical tests.

Fig 2. A Typical VARTM Setup

2.0 RESULTS AND DISCUSSION

2.1 Quasi-Static Tests.

Two types of mechanical tests, namely flexure and tensile, were performed to evaluate the bulk stiffness and strength of each of the material systems on an MTS 8010 tensile testing machine.

A typical stress strain behavior from the flexural test is shown in Fig. 3. It is observed that the system with 1.5% SiC infusion has the highest strength and stiffness among the three systems indicated in Fig. 3. Gain in strength and stiffness

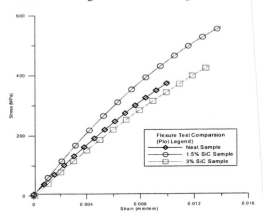

Fig. 3. Flexural testing of Carbon/Epoxy nanocomposites

of the 1.5% system is around 30% and 12% respectively over the neat as shown in Table-1. Enhancement in strength by 30% during flexure was somewhat surprising due the fact that previous studies with nanoclays [12, 13] did not show such improvement. It is also shown in Table-1 that if the particle infusion is increased to 3% SiC there is no proportional improvement in properties. Rather, there is a very nominal increase in strength, and a significant reduction in stiffness with the 3% SiC wt system. Similar reflection in properties with somewhat different ratios is observed during tensile tests as shown in Fig.4 and

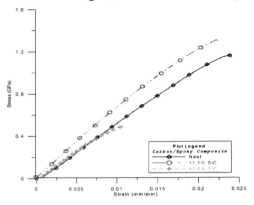

Fig 4. Tensile testing of Carbon/Epoxy nanocomposites

Table-2. It is noted that the enhancement in strength and stiffness is consistent with those shown in Fig. 3. The strain to failure of 1.5% SiC system is seen to be around 2.25% which is almost identical with that of the neat indicated in Fig. 4. Possible reasons for such behavior could be that since the reinforcement is in cloth (satin woven) form it did not contribute much to the elongation, and that the failure was mostly controlled by matrix and delamination related failure modes.

2.2 Dynamic Tests.
Low velocity impact analysis and high strain rate testing were preformed to determine the dynamic mechanical performance of the nano-phased carbon/epoxy systems compared to the neat composite. For low velocity impact testing, the analysis was performed using a Dynatup 8210. Specimens were cut to a 10.2 cm square with a thickness of 3 mm and tested with a pointed head alloy steel impactor weighing 6.33 kg. An electronic sensor was used to record the displacement of the impactor and the data was relayed to the Dynatup/GRC software for analysis. The potential energy (P.E.) was calculated from a height of 37 cm. The maximum calculated P.E. was 23.0 J for all three material systems. The absorption energy versus time graph is shown in Fig. 5. It is observed that the system with 1.5% SiC

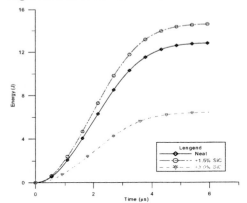

Fig 5. Low Velocity Impact testing at 37 cm height.

infusion has the highest energy absorption among the three material systems indicated in Fig. 5. Gain in impact energy of the 1.5% SiC system is around 14.7% over the neat as shown in the graph. It is also seen in Fig. 5 that if the particle infusion is increased to 3% SiC a degrading of properties are observed. When comparing the neat and 1.5% SiC systems to that of the 3.0%, the behavior of the 3.0% SiC system profile shows a rapid decline in the absorption energy rate. It is observed from then

Fig. 5 that slope of the 3.0% SiC system is much lower by comparison with the neat and 1.5% SiC material systems. This gives a sharp indication to a material system with lower dynamic modulus. The lower energy absorption rate indicates a weaken matrix which does not transfer load from matrix to fiber effectively.

Split Hokinson's pressure bar (SHPB) was used to evaluate dynamic compression and strain rate. For SHPB testing, three specimens were tested for neat, 1.5%, and 3.0% SiC systems with the dimension of 1.6 cm square and a thickness of 1.27 cm. The corresponding inlet pressure for each material system was 124.1 KPa. The specimens were tested in the transverse direction. The incident pulse e_i, reflected pulse e_r (input) and transmitted pulse e_t (output) was recorded using CEA-13-240UZ-120 measurements group strain gages. The data was acquired using gagescope software. Both voltage versus time and strain versus time graphs were plotted. VP3 software was used to merge both plots and stress versus strain response was obtained. A typical dynamic compression versus compressive strain is shown in Fig. 6. It is observed that the system with 1.5% SiC

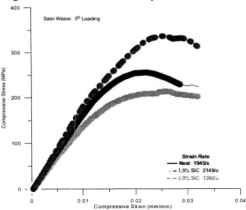

Fig 6. Split Hokinson's Pressure Bar testing at 121.1 KPa.

infusion has the highest dynamic compression strength as well as modulus among the three systems indicated in Fig. 6. Improvement in compressive strength is shown to be approximately 40% for the 1.5% SiC system. Compressive modulus is around 10.7% over the neat as shown in Table-3. When comparing the 3.0% SiC system with the neat, once again, a drop is noted for the compressive strength as well as compressive modulus. This was expected because of early indications from low velocity impact testing of the three material systems. By inspection of the 3.0%

SiC material system, the damaged specimens showed multiple areas of delimination within the specimen when compared to the 1.5% SiC and neat material systems. The multiple delimination within the specimen strongly indicates a weakened bond between matrix and fibers. Enhancements also could be seen in the strain rate from the 1.5% SiC when compared to the neat.

3.0 SUMMARY AND CONCLUSION

The following is a summary of the above investigation.
- A low cost but reliable manufacturing procedure has been introduced to fabricate large scale laminated structural composites with a nanophased matrix.
- Nanoparticles loading into the base matrix are found to be optimal around 1.5% by weight to derive maximum gain in the mechanical and thermal properties of the structural composites.
- With 1.5% loading, an average of 20-30% increase in mechanical properties has been observed both in tension and flexure.
- Impact properties have been seen to improve by 10-14% with 1.5% SiC loading.
- Compressive strength and modulus showed significant improvements with the 1.5% SiC loading.
- Nanoparticles loading at higher loads tend to degrade material properties.

4.0 ACKNOWLEDGEMENT

The authors would like to thank the Office of Naval Research (Grant No. 00014-90-J-11995) and the National Science Foundation, (Grant No HRD-976871), for supporting this research.

5.0 Tables

Table-1 Flexural test data for carbon/epoxy

Material	Flexural Strength [MPa]		Gain/Loss Strength in [%]	Flexural Modulus [GPa]		Gain/Loss Modulus in [%]
Neat	334.0		—	45.0		—
	390.0			46.0		
	395.0	381.8		39.0	45.60	
	390.0			44.0		
	400.0			54.0		
+1.5% SiC	550.5		+39.2	50.3		+12.4
	505.0			51.0		
	525.0	531.3		47.0	51.26	
	556.0			59.0		
	520.0			49.0		
+3.0% SiC	424.1		+4.7	37.5		-16.4
	390.0			36.0		
	395.0	399.8		39.0	38.10	
	390.0			41.0		
	400.0			37.0		

Table- 2 Tensile test data for carbon/epoxy

Material	Tensile Strength [GPa]		Gain/Loss Strength in [%]	Tensile Modulus [GPa]		Gain/Loss Modulus in [%]
Neat	1.196		—	50.1		—
	1.192			51.5		
	1.193	1.19		39.0	50.1	
	1.194			53.9		
	1.194			55.9		
+1.5% SiC	1.330		+11.6	63.2		+23.5
	1.430			59.5		
	1.320	1.33		61.9	61.8	
	1.290			62.8		
	1.290			61.8		
+3.0% SiC	0.500		-53.9	53.1		+6.3
	0.500			51.3		
	0.450	0.48		61.0	54.0	
	0.380			50.4		
	0.550			54.0		

Table- 3 SHPB Response @ 121.1 KPa

Mat. Type	Strain Rate (/s)	Peak Stress (MPa)	Strain @ Peak Stress (mm/mm)	Modulus (GPa)
Neat	3762	259	0.023	13.9
	3502	273	0.019	14.8
	3837	232	0.020	12.6
Avg.	**3700**	**255**	**0.020**	**13.8**
+1.5 SiC	2530	330	0.022	20.5
	2634	342	0.024	16.3
	2555	365	0.026	18.4
Avg.	**2573**	**335**	**0.025**	**18.4**
+3.0 SiC	3890	224	0.022	11.9
	3901	237	0.020	12.3
	3906	213	0.023	9.50
Avg.	**3899**	**225**	**0.022**	**11.2**

REFERENCES

[1]. Brain P. Rice, Chenggang Chen and Larry Cloos, "Carbon Fiber Composites: Organoclay Aerospace Epoxy Nano-composites, Part I" SAMPE Journal, Vol. 37, No. 5, September/October 2001.

[2]. Michael Alexandre, Philippe Dubois, "Polymer-layered silicate nanocomposites: preparation, properties and uses of a new class of materials" Reports: A Review Journal, No. 28, March 2000.

[3]. Ole Becker, Russell Varley, George Simon, "Morphology, thermal relaxations and mechanical properties of layered silicate nanocomposites based upon high-functionality epoxy resins" Polymer 2002;43: 4365-73.

[4]. T. Agag, T. Koga, T. Takeichi, "Studies on thermal and mechanical properties of polyimide-clay nanocomposites" Polymer 2001, 42: 3399-3408.

[5]. Shiner, C., Timmerman, J., Ebonee, P.M., Williams, and Seferis, J., "Thermal and Mechanical Characteristics of Nano Modified Fiber-reinforced Composites," 48th International SAMPE Symposium, May 11-15, 2003, pp. 2539.

[6]. Kojima, Y., Fukimori, K., Usuki, A, Okada A., and Karauchi, T., Journal of Materials Science Letters, 12, 889 (1993).

[7]. LeBaron, P.C., Wang, Z., and Pinnavaia, T.J., Applied Clay Science, 15, 11 (1999).

[8]. Haque, A., F. Hossain, D. Dean and M. Shamsuzzoha, "S2-glass Fiber Reinforced Polymer Nanocomposites: Manufacturing, Structures, Thermal and Mechanical Properties" *Journal of Composite Materials*, 2002.

[9]. Mahfuz, H., Zaman K., Hisham, M., Foy, Costee, Haque, A. and Jeelani, "Fatigue Life Prediction of Thick-Section S2-Glass/Vinyl-Ester Composites under Flexural Loading," Transaction of ASME, Journal of Engineering Materials and Technology, Vol. 122, October 2000, pp. 402-408.

[10]. Mahfuz, H., Mamun, W. and Jeelani, S., "High Strain Rate Response of Sandwich Composites; Effect of Core Density and Core-Skin Debond," Journal of Advanced Materials, Vol. 34, No. 1, January 2002, pp. 22-26.

[11]. U.K. Vaidya, M.V. Kamath, M.V. Hosur, Mahfuz, H. and S. Jeelani, "Manufacturing and Low Velocity Impact Response of Sandwich Composites With Hollow And Foam Filled Z-Pin Reinforced Core," Journal of Composites Technology and Research, JCTRER, Vol. 21, No.2, April 1999, pp. 84-97.

[12]. Shiner, C., Timmerman, J., Ebonee, P.M., Williams, and Seferis, J., "Thermal and Mechanical Characteristics of Nano Modified Fiber-reinforced Composites," 48th International SAMPE Symposium, May 11-15, 2003, pp. 2539.

[13]. Fukushima, Y, and Inagaki, S., Journal of Inclusion Phenomena, 5, 473 (1987).

Carbon Nanofiber Reinforced Carbon/Polymer Composite

M. Joshi* and A. Bhattacharyya

Department of Textile Technology, Indian Institute of Technology
Hauz Khas, New Delhi – 110016, India
* mangala@netearth.iitd.ac.in

ABSTRACT

Conventional carbon fiber reinforced phenolic resins are widely used to prepare carbon/carbon composites due to their high carbon yield. Property enhancement of carbon fabric/phenolic composites is possible through carbon nanomaterial dispersion in the matrix. The effect of carbon nanofiber dispersion in phenolic resins/carbon fabric composites was investigated. The dispersion efficiency in different systems and conditions was studied and the effect of nanofiber concentration on the properties determined. Carbon nanomaterial incorporation influences the mechanical properties of the composites and gives better thermal stability. Nanofiber dispersion shows better adhesion of phenolic resin to the fabric under fracture surface study.

Keywords: nanocomposites, carbon nanotubes, carbon nanofiber, phenolic resins, composite

1. INTRODUCTION

Phenolic resin is extensively used for fabric reinforced carbon-carbon composite due to its highest yield of carbon after carbonizing. Carbon fiber finds use in both filament and fabric form as reinforcement in polymeric matrices for high performance structural applications [1]. Textile structured composites plays an important role in preparation of carbon-carbon composites [2] and exhibit in plane balanced properties. Consequently they find uses in many high-end applications where their special properties can be taken advantage of e.g. spacecraft re–entry frames. Carbon nanotubes have been successfully used to improve the wear properties of carbon-carbon composites [3]. The use of carbon nanotubes in phenoxy matrix also results in significant rise in dynamic mechanical properties [4]. In the present study, carbon nanofiber is dispersed into phenolic resin before using it as matrix for carbon fabric composites. Although literature mentions studies on carbon nanotube in phenolic resins [3], no study using nanofiber-reinforced phenolic resins as matrices for carbon/polymer composites has been reported.

2. EXPERIMENTAL

2. 1. Raw material used

1. Carbon fabric – 3k x 3k, Plain weave, 195 gm/m^2 (Surabhi International, Kanpur, India.)
2. Carbon nanofiber – Pyrograf III (PR24 AGLD) (dia.: 60 – 150nm).
3. Non-ionic surfactant – Polyoxyethylene 8 lauryl ether (30% w/v), (Loba Chemie)
4. Methanol – Solvent for resin, AR grade
5. Phenolic resin –Prepared in house.

2. 2. Processing

2. 2. 1. Resin preparation

The resin was prepared by the condensation of phenol and formaldehyde and refluxing for 1.5 hrs. at 70°C.

2. 2. 2. Dispersion of carbon nanofiber

1.0% carbon nanofiber (on weight of resin preform) was dispersed into the resin through three different processes:
a) Sonication at 20 KHz for 2 hrs.
b) Mechanical stirring with 0.2% non-ionic surfactant (on volume basis) for 2 hrs.
c) Mechanical stirring with 0.2% non-ionic surfactant (on volume basis) for 1 hr. and subsequently sonication treatment of 20 KHz for 1 hr.

Thin films were prepared and observed using optical microscope. The best method of dispersion was chosen for composite preparation.

2. 2. 3. Preparation of the composite

The fabric was impregnated by dipping in the resin preform and drying overnight. Then it was taken for compression-molding using conditions:
a) Curing temperature: 150°C
b) Curing time: 2 hrs.
c) Force: 3 Metric ton

The average resin fraction on weight basis after curing = 45%

3. CHARACTERIZATION

Nanofiber dispersion was observed with a LEICA optical microscope in transmission mode. The tensile (ASTM D3039-76) and flexural tests (ASTM D790) were carried out on a Zwick Z010 tensile tester. Three- layered fabric composites were made for flexural test purpose. All samples were tested warp way. Dynamic mechanical analysis was carried out on a Perkin-Elmer DMA7 at 1 Hz, 200 mN static load and 160 mN dynamic load from −30°C to 200° C. Thermal conductivities were measured on an ALEMBETA instrument. The fracture surface of the composite was studied with a LEICA microscope.

4. RESULT & DISCUSSION

4. 1. Dispersion of Carbon Nanofibers

The nanofibers were dispersed into the resin and the cured films were observed through transmission mode of LEICA. Among the three processes the use of both surfactant and Sonication process was found most suitable (Figure: 1).

a)

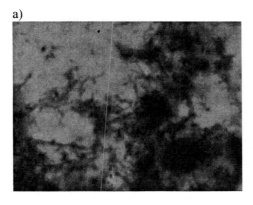

b)

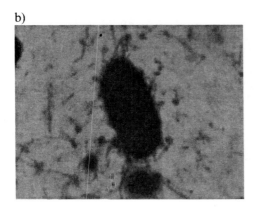

c)

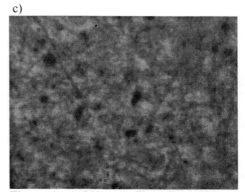

Figure 1: 1. 0% nanofiber dispersion by three different methods (full scale 100 μm x 100μm); a) Sonication for 2 Hrs, b) Surfactant (0.2%), stirring for 2 Hrs, c) Stirring with Surfactant (0.2%) for 1 Hr & Sonication for 1Hr.

Care was taken to minimize the amount of surfactant in this process, as surfactant acts as an impurity in the system. It has been observed that the use of 0.1% surfactant is sufficient to have satisfactory level of dispersion of the nanofiber when the sonication time is 1.5 hrs.

4. 2. Tensile Testing

The tensile properties of the produced composites were estimated and were given in Table 1 below.

Table1. Tensile properties of prepared samples

Samples	Modulus (GPa)	Force at break (MPa)	Strain at break (%)
Carbon/Phenolic resin composite	7.98	285.9	3.98
With 0.5% nanofiber	8.31	290.0	3.26
With 1.0% nanofiber	8.48	309.8	3.29
With 1.5% nanofiber	8.66	320.4	3.16

The addition of nanofiber improves the tensile properties. 1.5% nanofiber addition into the matrix results in 8% increase in modulus and 12% increase in breaking strength of final composite. Nanofiber imparts brittleness into the matrix, which results in reduction of strain at break. The strength improvement is significant considering the fact that the modulus and strength of nanofiber is only slightly higher than the carbon fiber. The improvement is more due to the change in interfacial behavior than the conventional mixture rule.

4. 3. Flexural testing

The flexural properties of the produced composites were estimated and were given in Table 2 below.

Table2. Flexural properties of prepared samples

Samples	Modulus (GPa)	Force at break (MPa)	Strain at break (%)
Carbon/Phenolic resin composite	51.4	654.4	2.01
with 0.5% nanofiber	53.8	688.6	1.77
with 1.0% nanofiber	55.5	648.2	1.60
with 1.5% nanofiber	57.5	695.9	1.27

Result shows the improvement in modulus and breaking strength by 11.5% and 9%, respectively. Decrease in strain at break shows the brittleness due to higher interaction.

4. 4. DMA

Three point rectangular bending test on single-layer fabric composites were carried out from – 30°C to 200°C. Figure 2 reveals that the storage modulus remains more stable over the temperature range. The storage modulus increases from 8.8 GPa to 11.2 Gpa with the incorporation of 1.5% nanofiber into the matrix at -30°C. Heating up to 200°C does not cause sharp fall in modulus when carbon nanofibers are dispersed into the matrix. Similar effect is observed for loss modulus

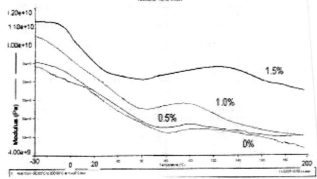

Figure 2: Storage modulus of composite samples with different nanofiber fraction in matrix

4. 5. Thermal conductivity

Table 3 shows that thermal conductivity values increased with the increase in nanofiber content. The thermal conductivity of carbon nanofiber (20 $Wm^{-1}K^{-1}$) is higher than the phenolic resin (0.8 $Wm^{-1}K^{-1}$); hence nanofiber dispersed into the matrix enhances the conduction of the composite.

Table 3. Thermal conductivity of composites

Parameters	Nanofibre content			
	0%	0.5%	1.0%	1.5%
Thermal conductivity (λ) $Wm^{-1}K^{-1}$	52.2 x 10^{-3}	56.7 x 10^{-3}	67.3 x 10^{-3}	70.9 x 10^{-3}

4. 6. Fracture surface study

The reflection mode pictures in LEICA microscope indicate that nanofibers increase the adhesion between the matrix and the carbon fiber. Figure 3 indicates that the carbon fibers of the reinforcing fabric are not fully exposed in the fracture surface when carbon nanofiber is present in the matrix, which indicates better adhesion between fiber and matrix.

a) Full scale 1mm x 1mm

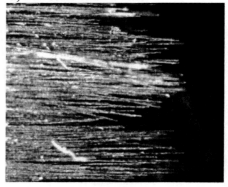

b) Full scale 1mm x 1mm

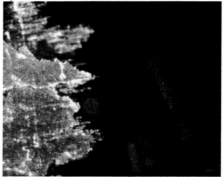

Figure 3: Fracture surface of composite with a) 0% nanofiber, b) 0.5% nanofiber

In transmission mode at higher magnification, it was evident that the nanofiber dispersion into the matrix is satisfactory at the fracture surface (Figure 4a). The Phenolic resin carbon fiber bond strength is not very good as is evident from Figure 3a. But after incorporation of the carbon nanofiber into matrix some fringing from the original fiber occurs (Figure 4b) which results in better load transfer of the resin to the carbon fiber. The carbon nanofiber may act as the coupling between the carbon fiber and the resin. The poor bond strength of phenolic resin is somewhat overcome with carbon nanofiber into the matrix.

a) Full scale 100µm x 100µm

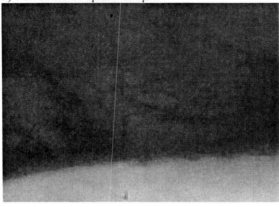

b) Full scale 100µm x 100µm

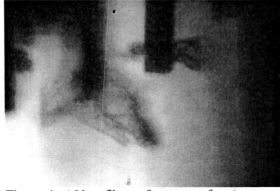

Figure 4: a) Nanofiber at fracture surface (proper dispersion), b) Adhesion of resin with help of nanofiber on fiber

5. CONCLUSION

The use of surfactant is helpful for proper dispersion of the nanoparticle into the matrix. The use of non-ionic surfactant is reported for dispersion of carbon nanotube in epoxy matrix [5]. It has been also found true for carbon nanofiber in phenolic resin as investigated in the present study. Sonication can minimize the amount of surfactant to some extent. The combined process of surfactant-assisted sonication has been found most

suitable for dispersion of carbon nanofiber. Incorporation of carbon nanofiber into phenolic resin matrix of carbon composite enhances its mechanical and thermal properties significantly. It has been observed that nanofiber alters the interface behavior of the carbon fiber and phenolic resin matrix. Thus may give better load transfer into the reinforcement and results in a synergistic effect of nanofiber into the property alteration beyond common mixture rules.

The dynamic mechanical analysis shows that the introduction of the nanofiber into the matrix enhances its stability under wide range of thermo-mechanical condition.

Carbon nanomaterial reinforced polymeric matrix composites with carbon fibre reinforcement are examples of three phase nanocomposite systems – something that has not been studied in great detail before. The interfacial interactions and the stress transfer mechanisms would also be very relevant in the presence of the second reinforcing phase- the fabric. The result of the excellent interphase properties is obtained in a carbon–carbon composite because of its uniphase nature. So the current study has to be extended to carbonization of this composite to find the change in properties because of presence of carbon nanofibers in the phenolic resin.

6. REFERENCES

1. N. K. Naik, S. I. Tiwari and R. S. Kumar: An analytical model for compressive strength of plain weave fabric composites; Composites Science and Technology **63** 5 (2003), pp. 609-625.
2. T.W. Chou and F.K. Ko, Editors: Textile structural composites; Elsevier, Amsterdam (1989), pp. 209–263.
3. Dae-Soon Lim, Jeong-Wook An and Hwack Joo Lee: Effect of carbon nanotube addition on the tribological behavior of carbon/carbon composites; Wear **252** 5-6 (2002), pp. 512-517.
4. H. W. Goh, S. H. Goh, G. Q. Xu, K. P. Pramoda and W. D. Zhang: Dynamic mechanical behavior of in situ functionalized multi-walled carbon nanotube/phenoxy resin composite; Chemical Physics Letters **373** 3-4 (2003), pp. 277-283
5. S. Cui, R. Canet, A. Derre, M. Couzi and P. Delhaes: Characterization of multiwall carbon nanotubes and influence of surfactant in the nanocomposite processing; Carbon **41** 4 (2003), pp. 797-809

Nanoparticle Composites for Coating Applications

Roger H. Cayton

Nanophase Technologies Corporation, Romeoville, IL, USA, rcayton@nanophase.com

ABSTRACT

Nanoparticles can be incorporated into polymeric coatings to enable significant improvement in targeted properties, i.e., scratch resistance, UV resistance, conductivity, etc. Commercial processing methods have been developed to surface treat nanoparticles to render compatibility between the nanoparticle and polymer matrix to control the average particle size and particle size distribution of the dispersed nanoparticles. In addition, commercial processes to disperse the surface treated nanoparticles have been developed. These integrated technologies allow transparent coatings containing nanoparticles to be formed in a wide range of resin formulations. This paper will address the preparation and analytical characterization of nanoparticle dispersions, as well as their performance attributes.

Keywords: nanoparticles, dispersion, coatings, transparency, scratch-resistant

1. INTRODUCTION

During the past several years, advances in nanomaterials have allowed them to be formulated into numerous applications. The majority of these applications sought performance improvements that were previously unobtainable. Examples of such applications containing nanomaterials that have been commercialized include scratch/abrasion resistant transparent coatings, sunscreen lotions to provide visible transparent UV protection, polishing slurries to provide pristine surfaces for optics, and environmental catalysts to reduce pollution.

The quest for improved scratch/abrasion resistant coatings is a goal for many coating formulators. Thousands of scratch resistant coating applications are present in our everyday lives. Examples of these applications include coatings for wood floors, safety glasses, electronic displays, automotive finishes, and polycarbonate panels. Improving the mar, scratch and/or abrasion in these transparent coating applications is a major challenge, particularly with regard to not affecting the other performance attributes of the coating.

Incorporation of inorganic fillers into coatings to improve mechanical properties is well known. Drawbacks associated with this approach can include loss of transparency, reduced coating flexibility, loss of impact resistance, increase in coating viscosity, and appearance of defects. To overcome these defects a filler material should impart improved scratch resistance without causing the aforementioned detriments. Nanomaterials have the potential to overcome many of these detriments because of their inherent small size and particle morphology.

Maintaining transparency in a coating containing inorganic filler particles is a challenge. Four properties dictate the degree of transparency in a composite material: Film thickness, filler concentration, filler particle size, and the difference in refractive index between the bulk coating and the filler particle. Mie theory describes the relationship between particle size, concentration, refractive index, and light scattering for spherical particles dispersed in a bulk phase as shown in equation 1.

$$I_s = (Nd^6/\lambda^4) \{[(n_p/n_c)^2-1]/[(n_p/n_c)^2+2]\}(I_i) \qquad (1)$$

I_s = Intensity of scattered light
N = Number of particles
d = Particle diameter
λ = Wavelength
n_p = Particle refractive index
n_c = Coating refractive index
I_i = Intensity of incident light

As is evident in equation 1, the magnitude of light scattering in a particle/coating composite is strongly influenced by the particle size. In addition, the greater the difference between the refractive indexes of the particle and that of the bulk coating, the greater the degree of light scattering.

Silica particles, colloidal or fumed, and clays are among the most widely studied inorganic fillers for improving the scratch/abrasion resistance of transparent coatings. These fillers are attractive from the standpoint that they do not adversely impact the transparency of coatings due to the fact that the refractive indices of these particles (fumed silica = 1.46, bentonite clay = 1.54) closely match those of most resin-based coatings. The drawback to silica-based fillers is that high concentrations of the particles are generally required to show a significant improvement in the scratch/abrasion resistance of a coating, and these high loadings can lead to various other formulation problems associated with viscosity, thixotropy, and film formation.

The use of alumina particles in transparent coatings is much more limited even though alumina is significantly harder than silica-based materials, and as a scratch and abrasion-resistant filler, higher performance at lower loadings is often observed. For alumina particle sizes greater than 100-nm, the high refractive index (1.72) results in significant light scattering and a hazy appearance in most clear coatings. Currently, only high refractive index coatings, such as the melamine-formaldehyde resins used in laminate production, can use submicron alumina for scratch resistance and maintain transparency.

2 NANOPARTICLE PRODUCTION

To use alumina as scratch-resistant filler in transparent coatings, the particle size must be sufficiently small to overcome its refractive index mismatch. Nanophase Technologies Corporation (NTC) has developed the Physical Vapor Synthesis (PVS) process that is capable of producing metal oxide nanoparticles via a bottoms-up method starting from metallic feed. This process allows production of nonporous crystalline metal oxides having primary particle sizes less than 100 nm at economically viable rates with essentially no byproducts or waste streams.

NTC produces two grades of aluminum oxide using the PVS process: NanoTek™ and NanoDur™ alumina. Both grades feature a mixture of γ and δ crystal phases and are spherical in shape, but the grades differ in terms of primary particle size. NanoTek™ alumina has a surface area of 35 m^2/g corresponding to a mean particle size of 48 nm, whereas NanoDur™ alumina has a surface area of 45 m^2/g with a mean particle size of 37 nm. A TEM image of NanoTek™ alumina is shown in Figure 1.

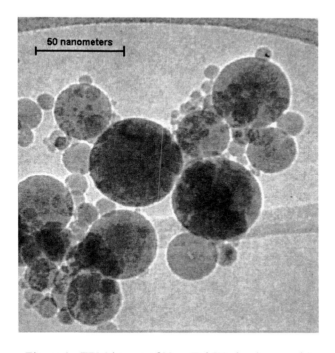

Figure 1: TEM image of NanoTek™ aluminum oxide

3. NANOPARTICLE DISPERSION

For nanoparticles to be of use in transparent coatings, it is critical that aggregates present in the powder be dispersible to their primary particle size in the coating formulation to avoid rapid settling and excessive light scattering. In addition, it is critical that the dispersed primary particles avoid re-aggregation during the coating curing process.

NTC has developed a proprietary particle dispersion-stabilization process that involves specific surface treatments designed to yield nanoparticles that are compatible with a variety of different coating formulations. For example, stable dispersions of metal oxide nanoparticles can be prepared in solvents such as water, alcohols, polar and nonpolar hydrocarbons, plasticizers, and even directly in acrylate monomers with the appropriate surface treatment process. These surface treatments allow solids levels of up to 60 wt% to be dispersed and yet maintain a sufficiently low viscosity for ease of blending.

The use of highly concentrated, non-aggregated, nano-particle dispersions allows incorporation of the nanoparticles into a coating formulation without substantial dilution of the formulation with the dispersion liquid. This feature is particularly important in 100% solids coating formulations wherein the nanoparticle is dispersed in one of the reactive monomers.

4. NANOPARTICLE COMPOSITES

Using concentrated dispersions of aluminum oxide nanoparticles in various solvents and reactive monomers, composites were prepared for a variety of transparent coating formulations including emulsion-based polyurethanes and polyacrylates, solvent-based two-component polyurethanes and melamine-polyols, and 100% solids UV-curable coatings. Importantly, the alumina nanoparticles also featured surface-treatments designed specifically for the formulation chemistry of the coating type in which it was dispersed.

Nanoparticle coating composites were prepared using both the NanoTek™ and NanoDur™ grades of aluminum oxide in a solvent-based transparent melamine-polyol coating to compare the haze of the two different particle sizes. The haze results are shown in Figure 2.

Three aspects of the haze data in Figure 2 are noteworthy. (1) As expected, the larger particle size NanoTek™ alumina (48 nm, mean) results in higher haze in the melamine-polyol nano-composite compared to those made with NanoDur™ alumina (37 nm, mean), at a given alumina loading level. (2) If transparency in a clear coating is defined as < 1% haze, it is apparent that the particle size of alumina used in the composite be no greater than 50 nm in order to maintain transparency. (3) Even with an alumina nanoparticle size grade of < 50 nm, improvement in scratch/abrasion resistance of the coating composite must be attainable with relatively low alumina loadings to stay within the transparency limit.

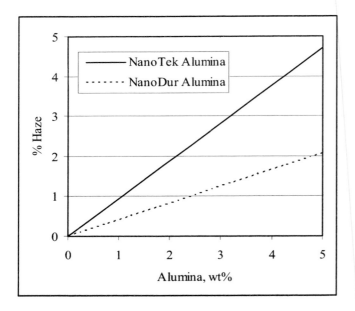

Figure 2: % Haze of alumina/melamine-polyol nano-composite coatings thermally cured to 1 mil thickness

Extension of the alumina particle size haze study to a variety of coating formulations with different refractive indices allowed the derivation of the empirical formula shown in equation 2 to predict the haze resulting from alumina incorporation in any nanocomposite coating.

$$\% \text{ Haze} = (8.8 \times 10^{-6} d^3 - 1.4154 n_c + 2.02)(\text{wt \%})(\text{mil}) \quad (2)$$

d = Alumina particle diameter, nm
nc = Coating refractive index
wt % = Alumina loading level, wt % of total solids
mil = Cured coating thickness

5. SCRATCH-RESISTANT NANOCOMPOSITES

To evaluate the performance of alumina nanoparticles as a scratch-resistant filler in a transparent coating, a nano-composite was prepared with NanoDur™ alumina dispersed in a UV-curable coating formulation. The alumina nanoparticles were dispersed in 1,6-hexanedioldiacrylate, a reactive monomer, at 30 wt%, and this dispersion was blended with a UV-curable formulation to provide composite coatings with variable levels of alumina particles between 0.2 and 2.0 wt%. These composites were subjected to a scratch test involving 200 double rubs with a 0000 grade steel wool pad and the level of scratching quantified by measuring the increase in haze due to the scratches. The

performance of the alumina-containing composite coatings was compared with the neat coating without alumina particles. The results of this scratch study are shown in Figure 3.

The performance of the alumina nanoparticles in Figure 3 is expressed as X times improvement in scratch resistance compared with the neat coating. It is evident the alumina nanoparticles significantly improve the performance of the UV-curable coating, up to a 9-fold improvement, even with very low levels of alumina incorporated in the composite.

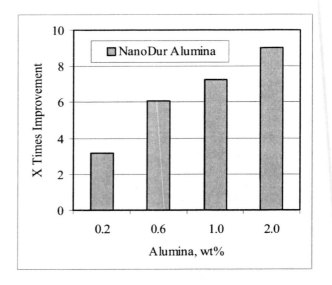

Figure 3: Scratch-resistance performance of NanoDur™ alumina particles in a UV-curable transparent coating

The scratch resistance properties of alumina nano-particles were also compared with silica particles at equivalent loading levels. The silica used in the study was the commercially available Nanocryl silica from Hanse Chemie. The silica particles are surface-treated and dispersed in 1,6-hexanedioldiacrylate and were blended into the same UV-curable coating formulation as the alumina particles. The comparative performance is shown in Figure 4.

As is evident in Figure 4, the alumina particles provide much better scratch resistance protection for the UV-curable coating compared to the silica particles at equivalent particle loadings. The much harder alumina particles are superior at preventing steel wool scratching compared to the softer silica particles.

The scratch resistance performance of alumina nano-particles incorporated into a variety of other transparent coating compositions was also evaluated. A level of 1 wt% alumina particles was used in all cases, and the alumina particles were introduced in the coating formulations by dispersing the alumina at high concentration into the appropriate solvent used in the coating formulation, then blending into the formulation at a level to yield 1 wt% particles with respect to the total solids in the cured coating.

The scratch resistance performance was measured using the same steel wool scratch test as was applied to the UV-curable coatings. The results of this study are summarized in Figure 5.

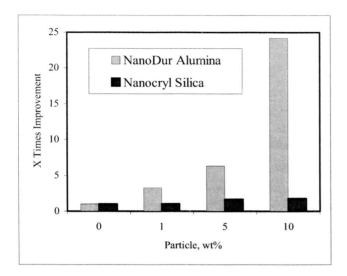

Figure 4: Comparison of the scratch-resistance performance of NanoDur™ alumina particles and Nanocryl™ silica particles in a UV-curable transparent coating

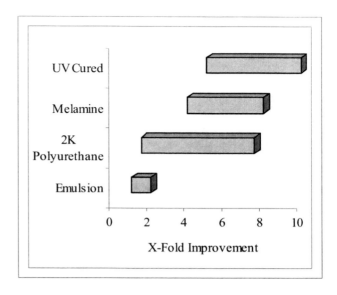

Figure 5: Scratch-resistance performance of NanoDur™ alumina particles at 1 wt% loading in various transparent coating formulations

Several features can be noted by comparing the relative alumina performance in different transparent coating formulations in Figure 5. First, there is a range of scratch improvement dependent upon the particular formulation within the coating class. For example, the scratch resistance UV-curable coatings can be improved anywhere from 5 to 10 fold with the incorporation of 1 wt% alumina, depending on the type and concentration of reactive acrylate components used in the formulation. Performance ranges for the melamine-based coatings, 2K polyurethane coatings, and emulsion-based coatings evaluated were also observed.

Within a given coating class, those formulations that resulted in harder/stiffer coatings tended to show greater improvement with alumina incorporation than did those formulations that lead to softer/more thermoformable coatings.

In addition, those transparent coating formulations which exhibit cross-linking upon curing, such as UV-curable, 2K polyurethane, and melamine-based coatings, showed greater improvement in their scratch-resistance upon alumina nanoparticle incorporation compared to transparent coatings which do not cross-link but rather coalesce, such as emulsion-based coatings.

Sustainable Nanocomposite Materials from Cellulosic Plastics

M. Misra[*], Hwanman. Park[*], A. K. Mohanty[**], L. T. Drzal[*]

[*] Composite Materials and Structures Center,
Department of Chemical Engineering and Material Science,
2100 Engineering Building, Michigan State University,
East Lansing, MI 48824, USA, misraman@egr.msu.edu
[**] The School of Packaging, 130 Packaging Building
Michigan State University, East Lansing, MI 48824, USA

ABSTRACT

Injection molded nanocomposites were successfully fabricated from triethyl citrate (TEC) plasticized cellulose acetate (CA) and organically modified clay. Maleic anhydride grafted cellulose acetate butyrate (MA-g-CAB) was used as a compatibilizer. The effect of compatibilizer contents on the performance of these nanocomposites were evaluated. The mechanical properties of these nanocomposites were correlated with the XRD and TEM observations. Cellulosic plastic-based nanocomposites with 3 wt.% compatibilizer showed better exfoliated structure than the counterpart having 0 or 7.5 wt.% compatibilizer contents. The use of compatibizer improved the tensile strength, modulus of elasticity, and thermal stability of these nanocomposites as compared to the counterpart nanocomposite in absence of the said specific compatibilizer.

Keywords: Biopolymers, clay, nanocomposites, cellulose ester, compatibilization,

1 Introduction

Renewable resource-based biodegradable polymers including cellulosic plastic (plastic made from wood), polylactic acid (PLA; corn-derived plastic) and polyhydroxyalkanoate (PHA; bacterial polyesters) are potential biopolymers, which by effective reinforcement with nanoclay can generate the so-called 'green' nanocomposites [1]. Cellulose from trees is attracting interest as a substitute for petroleum feedstock in making plastic (cellulosic plastic - cellulose esters) in the commercial market [2]. Cellulosic derived plastics such as cellulose acetate (CA), cellulose acetate propionate (CAP), and cellulose acetate butyrate (CAB) are thermoplastic materials produced through esterification of cellulose. However, the main drawback of cellulose acetate plastic is that its melt processing temperature exceeds its decomposition temperature. In order to overcome this problem, these plastics should be plasticized. The phthalate plasticizer, used in commercial cellulose ester plastic, is now under environmental scrutiny and perhaps poses a health threat raising concerns about their long-term use. One main objectives of this research is to find a viable replacement for phthalate plasticizer with eco-friendly plasticizers such as citrate [3], blends of citrate, and derivatized vegetable oil [4]. Melt processing through extrusion-injection molding was adopted in fabricating the nanocomposites. By adding organically modified montmorillonite clay into plasticized CA matrix during melt extrusion with high shear force, we expect to get exfoliation and/or intercalation of clay inside the continuous matrix. Optimization of processing conditions and effect of different compatibilizer contents are investigated in this paper in order to obtain better exfoliate clay hybrid. Morphological (X-ray diffraction, XRD, Transmission Electron Microscope, TEM), mechanical and thermal properties were evaluated.

2 Experimental Details

Materials: Cellulose acetate, CA (CA-398-30) without additives in powder form and triethyl citrate (TEC, Citroflex 2) were supplied by Eastman Chemical Co., Kingsport, TN and Morflex, Inc, North Carolina, respectively. The degree of substitution of cellulose acetate (CA) is 2.45. Organically modified montmorillonite (organoclay) Cloisite 30B was purchased from Southern Clay Co. The ammonium cations present in Cloisite 30B are methyl tallow bis-2-hydroxyethyl quaternary ammonium. As compatibilizer, maleic anhydride grafted cellulose acetate butyrate (MA-g-CAB) was synthesized [5] and characterized [5, 6].

Melt compounding: Prior to processing, CA, MA-g-CAB and organoclay were dried in a vacuum oven at 80° C for 24 hrs. The CA powder and TEC plasticizer (CA:TEC = 75:25 by wt. %) were mixed mechanically with a high speed mixer for 5 min and this mixture was stored in a zip-lock bag for 75 minutes. The pre-plasticized mixtures were then mixed with 5wt% organoclays and compatibilizer MA-g-CAB 0 to7.5wt% followed by mixing with the high-speed mixer. Then such mixtures various formulations (pre-plasticized CA + organoclay + compatibilizer) were melt compounded and plasticized simultaneously at 200-210° C for 6 minutes at 100 rpm with a micro-compounding molding equipment, DSM Micro 15 cc compounder, DSM research, Netherlands [5].

1

Table 1. Comparison of tensile, flexural, Notched Izod impact, HDT, and coefficient of thermal expansion (CTE) properties of plasticized CA /Cloisite 30B/compatibilizer (MA-g-CAB) hybrids

Sample No	CA/TEC Matrix (wt%)	Organo-clay contents (wt%)	Compa-tibilizer (wt%)	Tensile Strength (MPa)	Tensile Modulus (GPa)	Tensile Elonga-tion (%)	Flexural Strength (MPa)	Flexural Modulus (GPa)	Impact Strength (J/m)	HDT (°C)	CTE (um/m °C)
a	100	0	0	70.0 ± 5.1	2.20± 0.1	8.8±1.2	65.4 ± 2.0	2.4±0.2	55 ± 6	85± 3	125 ± 7
b	92.5	0	7. 5	74.0± 6.3	3.0 ± 0.9	7.5±0.8	67.3± 0.9	2.4 ± 0.2	35 ± 13	96 ±2	110 ± 9
c	95	5	0	81.8± 5.9	3.6 ± 0.6	10.0±0.2	74.7± 0.9	2.7 ± 0.5	25 ± 4	90 ±3	102 ± 3
d	92	5	3	84.7± 0.9	3.7 ± 0.6	9.5±1.2	81.7± 0.9	3.1 ± 0.6	29 ± 1	98 ±3	94 ± 10
e	87.5	5	7. 5	84.2± 1.9	3.7 ± 0.4	8.6±0.9	77.4± 1.2	3.0 ± 0.4	13 ± 1	94 ±4	92 ± 6

Characterization of nanocomposites: The samples made by injection molding were used for different characterizations. XRD studies of the samples were carried out using a Rigaku 200B X-ray diffractometer (45 kV, 100 mA) equipped with CuK radiation (= 0.1516 nm). A transmission electron microscope (TEM) (Jeol 100CX) was used to analyze the morphology of nanocomposites at an acceleration voltage of 100 kV. Microtomed ultra thin film specimen with thickness of 70 nm were used for TEM observation. A dynamic mechanical analyzer (2980 DMA, TA instruments, USA) was used to measure the heat deflection temperature (HDT) of nanocomposites with a load of 66 psi according to ASTM D648, and dynamic storage modulus and tan d was measured. Tensile properties and flexural properties of injection mold specimens were measured with a United Testing System SFM -20 according to ASTM D638 and ASTM D790 respectively. Notched Izod impact strength was measured with a Testing Machines Inc. 43-02-01 Monitor/Impact machine according to ASTM D256 with a 1 ft-lb pendulum.

3. Results and Discussion

3.1 Microstructure of hybrids

In order to obtain better exfoliated nanocomposites, compatibilizer synthesized by us was added to pre-plasticized CA /Cloisite 30B composition. Figure 1 shows the XRD patterns of pure Cloisite 30B clay and plasticized CA/ Cloisite 30B nanocomposites with different compatibilizer MA-g-CAB contents. The XRD peak shifted from 5.0° for pure Cloisite 30B to 2.36° for plasticized CA /organoclay (95/5 wt%) nanocomposite without compatibilizer (Figure 1a). This indicates significant intercalation and slight exfoliation in the hybrid structure.

For plasticized CA /Cloisite 30B/compatibilizer nanocomposites with 3wt% and 7.5wt% of compatibilizer, no clear peak was observed at 2.36° (Figure 1b and 1c), suggesting complete exfoliation of organoclays in the CA/TEC matrix. The XRD curves (figure 1) show that about 3-wt%

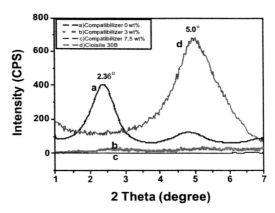

Figure 1. XRD patterns of the plasticized CA / compatibilizer /Cloisite 30B 5 wt% hybrids with different compatibilizer MA-g-CAB: a) compatibilizer 0 wt%, b) compatibilizer 3 wt%, c) compatibilizer 7.5wt%, d) Neat clay (Cloisite 30B).

compatibilizer is needed for almost completely exfoliated clay nanocomposite from the plasticized CA /organoclay system. In case of plasticized CA /organoclay hybrid nanocomposites, the maleated CAB will not only enter into the clay gallery but also will react with the free OH groups of the CA structure; thus improving the overall comptibilization of entire system.

Figure 2 shows TEM, which explains the morphology of the composites. The TEM images show that the plasticized CA / organoclay/ compatibilizer hybrid (Figure 2b and 2c) have better exfoliation than the counterparts without compatibilizer (Figure 2a).

2

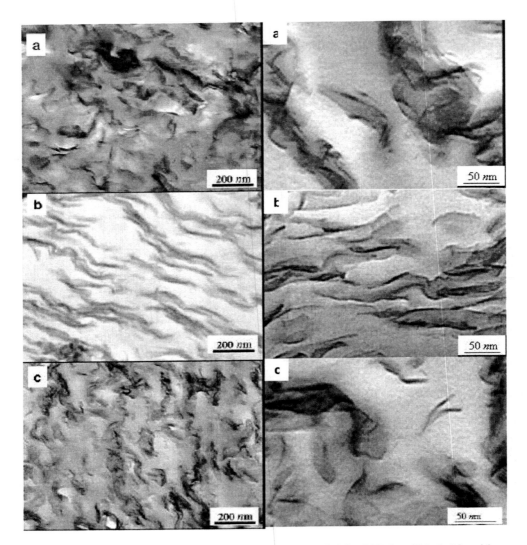

Figure 2 .TEM of the plasticized CA / compatibilizer /Cloisite 30B 5 wt% hybrids with different wt % of compatibilizer MA- g -CAB: a) 0 wt% compatibilizer b) 3 wt% compatibilizer, c) 7.5wt%compatibilizer.

From figure 2a, it can be seen that some the intercalation and aggregation of clay remain in the matrix. It can also be observed from the XRD and TEM results that for exfoliation of clay in CA polymer, the optimum loading of compatibilizer MA-g-CAB is important. The TEM observations of nanocomposites are correlated to validate XRD results.

3.2 Mechanical Properties

Table 1 shows the tensile, flexural, and notched Izod impact, heat deflection temperature (HDT), and coefficient of thermal expansion (CTE) properties of the plasticized CA /Cloisite 30B/MA-g-CAB hybrids with different compatibilizer MA-g-CAB contents. The strength and modulus (flexural and tensile) of the hybrid sharply increased with increasing compatibilizer contents. On the other hand, the notched Izod impact strength and the tensile

elongation at break of the hybrids decreased with increasing compatibilizer contents. This indicates that compatibilizer is effective in increasing the adhesion between organoclay and CA molecular chains. The tensile strength and modulus of hybrids with 3-wt% compatibilizer increased by 20%, and 68% respectively compared to hybrid with no compatibilizer. Also flexural strength and modulus were increased about 20 and 25% respectively.

HDT was increased 10% and CTE was decreased 24%, at 3-wt% MA-g-CAB loaded hybrid making the nanocomposites more stable in comparison to the pure matrix. But Izod impact strength was decreased about 50%. HDT property is closely related to the T_g of thermo plastics, therefore, HDT behavior of CA/TEC matrices and hybrids were similar to T_g behavior from DMA curve (Figure 4)

The above results indicate that the better exfoliation and good dispersion of clay in the CA/TEC matrix give good

3

mechanical properties. 3-wt% compatibilizer loading is optimum for HDT, CTE, and mechanical improvement.

3.3 Dynamic Mechanical Properties

Figures 3 and 4 show temperature dependent of storage modulus, tan d, and glass transition temperature (T_g) for the pristine plasticized CA matrix and the hybrid nanocomposites respectively. Storage modulus & T_g of plasticized CA (figures 3) and plasticized CA /clay hybrid composites (figures 4) increased with increasing compatibilizer contents. A possible explanation for improvement of modulus with reinforcement of clay might be attributed to the creation of a three-dimensional network interconnecting long silicate layers, thereby strengthening the material through mechanical percolation [7]. In figure 3 Tg was shifted from $112°C$ (for 0% compatibilizer, matrix) to $134\ °C$ (for 7.5% compatibilizer, matrix). Increase in compatibilizer content as well as addition of clay decreases the segmental motions in the CA backbone due to cross-linking and thereby increasing the T_g.

4 Conclusions

We have successfully developed nanocomposite formulations from cellulose acetate powder, triethyl citrate plasticizer, compatibilizer, and organically modified clay. The tensile strength, modulus and thermal stability of cellulosic plastic reinforced with organoclay was improved by increasing compatibilizer contents where as the impact strength decreased. Nanocomposites with 3 wt.% compatibilizer contents showed better exfoliated structure than the counterpart having 0 or 7.5 wt.% compatibilizer contents.

Acknowledgements

Authors are thankful NSF-NER 2002 Award # 0210681 and NSF 2002 award # DMR-0216865, under "Instrumentation for Materials Research (IMR) Program for fnancial support. The collaboration with FORD MOTOR COMPANY and EASTMAN CHEMICAL Co. is gratefully acknowledged. The authors also thankful to Eastman Chemical Company, Kingsport, TN for the cellulose ester samples as well as to Mr Arief Wibowo, CMSC, MSU for preparing the maleated CAB for this research.

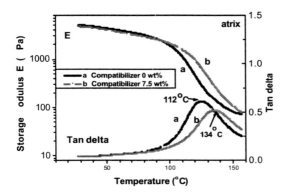

Figure 3. DMA curves of the plasticized CA /compatibilizer matrix with different wt % of compatibilizer (MA-g-CAB) contents.

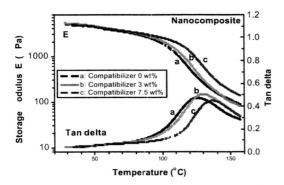

Figure 4. DMA curves of the plasticized CA /compatibilizer / Cloisite 30B hybrid with different wt% compatibilizer (MA-g-CAB) contents.

References

1. A. K. Mohanty, L. T. Drzal and M. Misra, Polymeric Materials Science & Engineering, 88, 60-61, 2003.
2. S. L. Wilkinson, Chem. Eng. News, January 22, 61, 2001.
3. A. K. Mohanty, A. Wibowo, M.Misra and L. T. Drzal, Polymer Engineering & Science, 43 (5), 1151-1161, 2003.
4. P. F. Bruins, In Plastic Technology; New York Reinhold Publishing Co. Chapman & Hall, Ltd.,:London, 1, 1-7,193-199, 1965.
5. A. Wibowo, A. K. Mohanty, M.Misra, and, L. T. Drzal, Polymer preprint, Spring 2004 (accepted), 227th ACS National Meeting, Anaheim, CA, March 25, 2004.
6. D. Carlson, L.Nie, R. Narayan and P. Dubois, J. Appl. Polym. Sci., 72, 477, 1999.
7. E.P. Giannelis and P.B. Messersmith, US Patent 5554670, Sep. 10, 1996.

4

A novel synthesis of Co Au nanoparticles and characterization

Zhihui Ban, C. J. O'Connor

Advanced Materials Research Institute
University of New Orleans, New Orleans, LA, USA, zban2@uno.edu

ABSTRACT

A homogeneous non-aqueous solution reactions method has been developed to prepare gold-coated cobalt (Co@Au) nanoparticles. After the samp le was washed with 8% HCl, XRD (X-Ray Diffraction), TEM (transmission electron microscopy), and magnetic measurements SQUID (Superconducting Quantum Interference Device) are utilized to characterize the nanocomposites. XRD shows the pattern of sample, which is only responding to gold. TEM results show that the average size of Co@Au nanoparticles is about 10 nm and we can find core-shell structure of the sample. SQUID results show that the particles are ferromagnetic materials at 300K. So the gold-coated cobalt nanoparticles (Co@Au) can be successfully prepared by the homogeneous non-aqueous approach. This kind of core-shell materials is stable in acid condition, which would give many opportunities for bio- application.

Keywords: core-shell structure, cobalt, gold, non-aqueous method, magnetic materials

1 INTRODUCTION

A combination of nanotechnologies and biosciences will be one of the leading areas of research and development in the 21st century; magnetic nanoparticles will certainly play an extremely important role. However, as the particle size decreases, the reactivity of the particle increases, and the magnetic properties are influenced more by surface effects[1]. So far the passivation of magnetic nanoparticles (? - Fe_2O_3, Co, Fe, et. al.) by another inert layer (SiO_2, gold, polymer et. al.) has been developed[2-7]. However, the nonmetallic layer potentially reduces the magnetic properties of the magnetic nanoparticles. Gold has become a favored coating material because of a simple synthetic procedure and its chemical functionality[8]. The presence of the diamagnetic coating passivates the magnetic core thus protecting the magnetic properties without having a pronounced effect on the magnetic properties. The presence of the diamagnetic coating allows for a surface that can be derivatized to allow for greater versatility while not reducing the magnetic properties. It is expected that iron nanoparticles can avoid being oxidized and maintain their magnetic properties (such as coercivity or blocking temperature) by gold coating. The reverse micelles method for synthesis of Fe@Au nanoparticles had been developed by our group(9). Here, in this paper, we will present another novel method for synthesis the gold-coated cobalt

nanoparticles, which can be stable in acid condition. The idea of the method is using reducing agents (Na, K, Li) to reduce MLn (M = Fe, Co, Ni; L = Cl, Br; n = 2, 3) to form metallic core nanoparticles, and then using the surface of metallic core particles to reduce Au^{3+} to form gold coated mettallic nanoparticles, as shown in scheme 1.

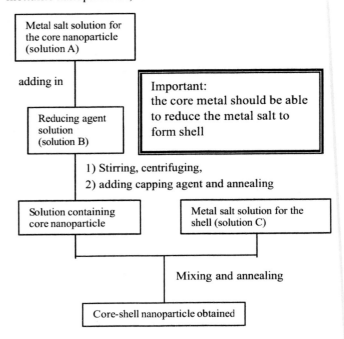

scheme 1. the reaction route to synthesize core-shell nanoparticles.

2 EXPERIMENT

The reduction reactions were performed in solutions of polar aprotic solvents. 2 mmol of $CoCl_2$ were dissolved in NMPO (1-Methyl-2-Pyrrolidinone) to form blue solution (Designated as solution A). 4 mmols of Na were dissolved in NMPO with 4 mmols of naphthalene to form a dark green solution (Designated as solution B). 0.6 mmol of dehydrated $HAuCl_4$ was dissolved in NMPO (Designated as solution C). The solution A was added into solution B quickly with intensive stirring at room temperature. The color of reaction system changed to dark brown immediately. The mixture was further stirred for two hours and then centrifuged to remove the sodium chloride. A small amount of capping agent, 4-benzylpyridine, was then put into the solution. Heated and refluxed at 165 °C for 3 hours. After cooled down to 60 °C, the obtained black mixture was added with solution C, then stirring and heat to 125°C for 4 hours and get light blue solution. After that the reaction solution was centrifuged to get black paste. All the above operations are done in glove box. The sample was then washed with EtOH, 8% HCl solution, water and EtOH for three times, respectively. After dried in vacuum, the product was characterized with SQUID, XRD, and redispersed into hexane for TEM observation.

3 RESULTS AND DISCUSSIONS

3.1 TEM RESULTS

Figure 1 shows the morphology of gold coated cobalt nanoparticles. The magnification is 400K and the scale bar is 20 nm. The core-shell structure (Co@Au) can be observed in the image. Mean core size of the nanoparticles is ca. 6 nm with shells of ca. 2 nm thick (as shown in the two small figures in figure 1. From the TEM results, we can see that the size distribution is not very well, some small size particles should be gold particles.

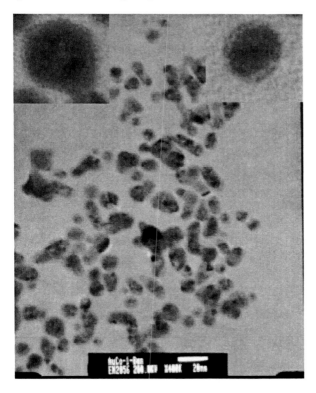

Figure 1 TEM results for Co@Au sample

3.2 XRD RESULTS

The XRD pattern of Co@Au nanoparticles is shown in Figure 2. In the result, there are only the diffraction peaks

which correspond to PDF card of Au (4-784). The diffractions attributed to cobalt cannot be observed probably due to the overlap with gold or amorphous of cobalt. But the magnetic property of the product as shown in Figure 3 shows the existence of pure cobalt. And in the XRD result, Cobalt oxides cannot be found, which would suggest that the cobalt was coated with gold very well.

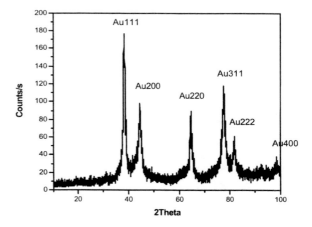

Figure 2 XRD pattern for the Co@Au sample

3.3 MAGNETIC PROPERTIES

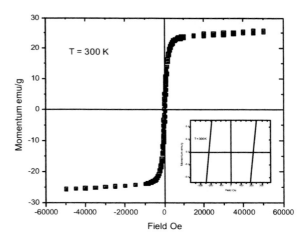

Figure 3 Hysterisis cure for Co@Au sample at 300 K

The SQUID results of the product shows that it is a ferromagnetic material with coercivity of 220 Oe (as shown in figure 3). The magnetic saturation is about 25 emu/g. If we consider the composition of the product obtained from EDS (including gold) and weight loss of TGA (which is organic materials), the magnetic saturation can be converted to over 140 emu/g. This indicates nanoparticles of Co coated completely with gold shell had been synthesized successfully.

4 CONCLUSIONS

Gold-coated cobalt nanoparticles (Co@Au) can be successfully prepared by the homogeneous non-aqueous approach. The Co@Au nanoparticles can be stable in both water and acid conditions, which would provide a wide range of opportunities for biosensors and biosciences. The method presented in this paper also could be used in making core-shell nanoparticles with other compositions.

REFERENCES

1. F. Bodker, S. Morup, and S. Linderoth, Phys. Rev. ett. 72, 282 (1994).
2. C. Pathmamnoharan and A. P. Philipse, J. Colloid Interface Sci. 205, 340 (1998).
3. D. V. Szabo and D. Vollath, Adv. Mater. 11, 1313 (1999).
4. E. E. Carpenter, C. T. Seip, and C. J. O'Connor, J. Appl. Phys. 8, 5184 (1999).
5. Kinoshita, T.; Seino, S.; Okitsu, K.; Nakayama, T.; Nakagawa, T.; Yamamoto, T. A., J Alloys and Compounds 359(1-2), 46(2003).
6. Carpenter, E. E.; Calvin, S.; Stroud, R. M.; Harris, V. G., Chem of Mater, 15(17), 3245 (2003).
7. Teng, Xiaowei; Black, Donald; Watkins, Neil J.; Gao, Yongli; Yang, Hong., Nano Letters, 3(2), 261(2003).
8. M. Brust, D. Bethell, D. J. Schi!rin, and C. J. Kiely, Adv. Mater. 7, 795 (1995).
9. O'Connor, Charles J.; Carpenter, Everett E.; Sims, Jessica Ann. (USA). U.S. Pat. 2002068187

A MODELING TOOL FOR CONTACT PROBE TESTS ON NANO-PARTICLE FILLED MATERIALS

Kevin K. Tseng

Vanderbilt University, Department of Civil and Environmental Engineering
VU Station B 351831, Nashville, TN 37235, USA
E-mail: Kevin.Tseng@Vanderbilt.Edu

ABSTRACT

With the recent advances in nano-scale science and engineering, materials containing reinforcement with superior mechanical properties can be found in many advanced products. The accurate prediction of the mechanical properties of this class of composite materials is important to ensure the reliability of the products. Characterization methods based contact probe such as nano-indentation and scratch tests have been developed in recent years to measure the mechanical properties of the new class of nano materials. This paper presents a constitutive modeling framework for predicting the mechanical properties of nano-particle reinforced composite materials. The formulation directly considers the effects of inter-nano-particle interaction and performs a statistical averaging to the solution of the problem of two-nano-particle interaction. Final constitutive equations are obtained in analytical closed form with no additional material parameters. The predictions from the proposed constitutive model are compared with experimental measurement from nanoindentation tests. This constitutive model for nano-paticle reinforced composites can be used to determine the volume concentration of the reinforcing nano-particles in nano-indentation test.

Keywords: nanocomposite, constitutive modeling, contact probe test, nanoindentation, nanoparticle

1 INTRODUCTION

Composite materials containing reinforcement with superior strength manufactured with nanotechnology have emerged in many applications. For example, carbon nanotubes have been utilized in field emission display as indicated in Qian et al [1]. Thin coatings with nano-particle-filled sol-gel have also become popular in many advanced products including the protective topcoat of recording films in optical storage disks (Malzbender et al. [2]). For the reliability of these advanced products, accurate predictions of the mechanical properties of this class of composite materials are important.

In many existing constitutive models for composite materials, the effects of interaction among reinforcing elements have been neglected mostly due to the following two reasons. First, the difference of the mechanical strength between the reinforcement and the original matrix material may not be too large. Secondly, the volume concentration of the reinforcement may not be high. With the advance in modern nanotechnology, these two assumptions are no longer valid because some materials produced with nano-scale material engineering can have superior mechanical strength than traditional materials. In addition, the size of the reinforcement material can be made very small with nano-scale material engineering and thus greatly increase the homogeneity of the final composite material. This allows the volume concentration of the reinforcement to be increased to achieve much better overall mechanical properties.

This paper presents a material modeling framework for composites with nano-particles. The major advantage of the proposed model is that the effects of inter-nano-particle interaction are captured into the constitutive model through statistical averaging on the solution of a two-particle interaction problem. Analytical closed-form formulations can be derived to predict the mechanical properties of nano-particle reinforced composites. The constitutive equations are expressed in terms of the volume concentration of the nano-particles and the material properties of the original matrix material and the reinforcing material. No additional material parameter needs to be introduced.

2 AVERAGE FIELDS THEORY

The determination of the exact internal local stress or strain field in a composite system is in general formidable due to the high degree of complexity of the arbitrary geometry and concentration of the reinforcing material. In many applications, it is sufficient to have the average of the field concerned. A method based on the averages of stress and strain fields was introduced by Hill [3] to describe the effective properties of composite materials.

In the average field theory, a concentration factor is introduced to represent the relationship between the local field and the average of the global field. For example, the stress at any local point for a specific material phase is related to the average stress for the global composite system via the stress concentration factor. If only the average of the local stress field is required, upon averaging over the local material phase, we have the following relationship

$$\overline{\sigma}_a = \mathbf{P}_a : \overline{\sigma} \tag{1}$$

where σ represents the stress tensor, the fourth rank tensor $\mathbf{P}_\alpha$ is the volume averaged stress concentration factor tensor for phase α, an over-bar represents the volume average of the corresponding quantity, and the subscript α denotes the material phase. Similar definition is made for the strain field

$$\overline{\varepsilon}_a = \mathbf{Q}_a : \overline{\varepsilon} \tag{2}$$

in which ε represents the strain tensor and $\mathbf{Q}_\alpha$ is the volume averaged strain concentration factor tensor for phase α.

From elasticity theory, the elastic stiffness and compliance tensors, $\mathbf{C}_\alpha$ and $\mathbf{D}_\alpha$, respectively, for material phase α relate the local average stress and strain fields according to the following two equations

$$\overline{\sigma}_a = \mathbf{C}_a : \overline{\varepsilon}_a \tag{3}$$

$$\overline{\varepsilon}_a = \mathbf{D}_a : \overline{\sigma}_a \tag{4}$$

Similarly, the macroscopic elastic properties can be expressed by the following equations through the global elastic moduli

$$\overline{\sigma} = \mathbf{C}_* : \overline{\varepsilon} \tag{5}$$

$$\overline{\varepsilon} = \mathbf{D}_* : \overline{\sigma} \tag{6}$$

Subscripts 0, 1, and * denotes the matrix, nano-particle, and overall composite material, respectively. Based on the definitions in Equations (1)-(6), the global effective elastic moduli are expressed in terms of the volume fractions, elastic moduli of the constituent phases, and the concentration factor tensors as shown in the following two equations

$$\mathbf{C}_* = \mathbf{C}_\alpha + \phi_\beta \left(\mathbf{C}_\beta - \mathbf{C}_\alpha \right) \cdot \mathbf{Q}_\beta \tag{7}$$

$$\mathbf{D}_* = \mathbf{D}_\alpha + \phi_\beta \left(\mathbf{D}_\beta - \mathbf{D}_\alpha \right) \cdot \mathbf{P}_\beta \tag{8}$$

From Equations (7) and (8), the global effective elastic moduli for a two-phase composite system can be calculated provided that any one of stress or strain concentration factor tensors is available.

3 INTERACTION OF TWO NANO-PARTICLES

For the simplicity of presentation and mathematical operation, we assume that the material properties for both the matrix phase and the particle phase are isotropic and the loading at any local material point remains within the elastic limit. However, the framework that is presented in this paper is valid for the general composite system with any arbitrary material property for the constituent phase. It is further assumed that the particles do not intersect each other and the material properties of both phases remain unchanged for the loading considered.

Extending from Mura [4], when applying the Eshelby's Equivalence Principle to the inclusion problem without considering the effects of inter-particle interaction, the equation for determining the unknown eigenstrain, which has been proved to be constant throughout the entire spherical region, can be written as

$$-\mathbf{A} : \varepsilon^{*0} = \varepsilon^0 + \mathbf{S} : \varepsilon^{*0} \tag{9}$$

where

$$\mathbf{A} = (\mathbf{C}_1 - \mathbf{C}_0)^{-1} \cdot \mathbf{C}_0 \tag{10}$$

in which $\mathbf{C}_0$ and $\mathbf{C}_1$ are the stiffness tensor for the matrix and inclusion phase, respectively. Tensor product is denoted by $\mathbf{A} \cdot \mathbf{B}$. In Equation (9), $\mathbf{S}$ is the Eshelby's tensor for a spherical inclusion and is defined as

$$\mathbf{S} = \int_\Omega \mathbf{G}(\mathbf{x} - \mathbf{x}') d\mathbf{x}', \qquad \mathbf{x} \in \Omega \tag{11}$$

where the elasticity Green's function tensor $\mathbf{G}(\mathbf{x} - \mathbf{x}')$ is defined by the following equation

$$\varepsilon(\mathbf{x}) = \int \mathbf{G}(\mathbf{x} - \mathbf{x}') : \varepsilon^*(\mathbf{x}') d\mathbf{x}' \tag{12}$$

in which $\varepsilon(\mathbf{x})$ denotes the strain tensor at location $\mathbf{x}$, $\varepsilon^*(\mathbf{x})$ is the tensor of eigenstrain, and $\varepsilon^{*0}(\mathbf{x})$ represents the eigenstrain tensor for the non-interacting particles. The explicit form for the tensor components of $\mathbf{S}$ can be found in Mura [4] for the spherical inclusion considered in the present study. Taking into account the effects of inter-particle interaction, the integral equation governing the distributed eigenstrain can be expressed as

$$-\mathbf{A} : \varepsilon^*(\mathbf{x}) = \varepsilon^0 + \int_{\Omega_i} \mathbf{G}(\mathbf{x} - \mathbf{x}') : \varepsilon(\mathbf{x}') d\mathbf{x}' + \int_{\Omega_j} \mathbf{G}(\mathbf{x} - \mathbf{x}') : \varepsilon(\mathbf{x}') d\mathbf{x}' \tag{13}$$

In the case that we are considering, many equal-sized spherical particles are assumed to distribute randomly among an elastic solid. Based on the solution for Equation (13), which represents the effect of pair-wise interaction, and assuming that the distribution of the particles is uniform and no particle overlaps with each other, ensemble-volume averaged eigenstrain perturbation in a particle can be written as

$$\langle \bar{\varepsilon}^* \rangle = \Gamma : \varepsilon^{*0} \tag{14}$$

where the components for the isotropic *interaction tensor* Γ are defined as

$$\Gamma_{ijkl} = \gamma_1 \delta_{ij} \delta_{kl} + \gamma_2 (\delta_{ik} \delta_{jl} + \delta_{il} \delta_{jk}) \tag{15}$$

in which δ_{ij} is the Kronecker delta,

$$\gamma_1 = \frac{5\phi}{4\beta^2} \left\{ -2(1-v_0) - 5v_0^2 - \frac{4\alpha}{3\alpha + 2\beta}(1+v_0)(1-2v_0) \right\} \tag{16}$$

and

$$\gamma_2 = \frac{1}{2} + \frac{5\phi}{8\beta^2} \left\{ 11(1-v_0) + 5v_0^2 - \frac{3\alpha}{3\alpha + 2\beta}(1+v_0)(1-2v_0) \right\} \tag{17}$$

where

$$\alpha = 2(5v_0 - 1) + 10(1-v_0) \left(\frac{\kappa_0}{\kappa_1 - \kappa_0} \right) \left(\frac{\mu_0}{\mu_1 - \mu_0} \right) \tag{18}$$

and

$$\beta = 2(4 - 5v_0) + 15(1-v_0) \frac{\mu_0}{\mu_1 - \mu_0} \tag{19}$$

In Equations (16) and (17), v, κ, and μ represent the Poisson ratio, bulk modulus, and shear modulus, respectively, for the corresponding material phase which is denoted via the corresponding subscript and ϕ denotes the volume fraction of the particles in the composite material under consideration.

4 CONCENTRATION FACTORS

With an additional averaging over the volume of the representative volume element, the equation relating the average strain $\bar{\varepsilon}$, the uniform remote strain ε^0, and the average eigenstrain $\bar{\varepsilon}^*$ can be expressed as:

$$\bar{\varepsilon} = \varepsilon^0 + \phi S : \bar{\varepsilon}^* \tag{21}$$

With Equations (8), (9) and (14), we get

$$\bar{\varepsilon}^* = B : \bar{\varepsilon} \tag{22}$$

where

$$B = \Gamma \cdot \left[-A - S + \phi S \cdot \Gamma \right]^{-1} \tag{23}$$

Averaging the fundamental equation for the Eshelby's equivalence principle:

$$C_1 : \varepsilon(x) = C_0 : \left[\varepsilon(x) - \varepsilon^*(x) \right] \tag{24}$$

the relationship between the local strain average and the eigenstrain average can be written as

$$C_1 : \bar{\varepsilon}_1 = C_0 : \left[\bar{\varepsilon}_1 - \bar{\varepsilon}^* \right] \tag{25}$$

Further utilizing Equation (10), we arrive at

$$\bar{\varepsilon}_1 = A : \bar{\varepsilon}^* \tag{26}$$

then, with Equation (8),

$$\bar{\varepsilon}_1 = -(A \cdot B) : \bar{\varepsilon} \tag{27}$$

Hence, upon comparing Equation (27) with Equation (2), the strain concentration factor tensor considering the effect of inter-particle interaction can be written as

$$Q_1 = -A \cdot B \tag{28}$$

and the corresponding stress concentration factor tensor can be derived in a similar fashion. The explicit expression for the stress concentration factor tensor takes the following form:

$$P_1 = -C_1 \cdot A \cdot B \cdot \left[I - \phi B \right]^{-1} \cdot C_0^{-1} \tag{29}$$

The tensor components for P_1 can be obtained by carrying out the lengthy tensor operation in Equation (29). The fourth rank tensor P_1 is found to be isotropic and its components are

$$(P_1)_{ijkl} = p_1 \delta_{ij} \delta_{kl} + p_2 (\delta_{ik} \delta_{jl} + \delta_{il} \delta_{jk}) \tag{30}$$

where

$$3p_1 + 2p_2 = \frac{30\kappa_1(1-v_0)(3\gamma_1 + 2\gamma_2)}{(\kappa_0 - \kappa_1)[(3\alpha + 2\beta) + 20(1-2v_0)(3\gamma_1 + 2\gamma_2)]} \tag{31}$$

and

$$p_2 = \frac{15\mu_1(1-v_0)\gamma_2}{(\mu_1 - \mu_0)[\beta + 2(7 - 5v_0)\phi\gamma_2]} \tag{32}$$

5 EFFECTIVE ELASTIC PROPERTIES

As an example, the stress and strain concentration factor tensors derived in the previous section are employed to construct the effective elastic properties for particle-reinforced composites. Through Equations (7) and (29), the effective elastic stiffness tensor incorporating the effect of inter-particle interaction reads

$$\mathbf{C}_* = \mathbf{C}_0 \cdot \left\{ \mathbf{I} - \phi\,\mathbf{\Gamma}\cdot\left(-\mathbf{A} - \mathbf{S} + \phi\,\mathbf{S}\cdot\mathbf{\Gamma}\right)^{-1} \right\} \qquad (43)$$

Since the particles are assumed to distribute uniformly among the matrix material, the composite is isotropic. The effective elastic property can be represented by the effective bulk modulus κ_* and the effective shear modulus μ_* can be explicitly written as

$$\kappa_* = \kappa_0 \left\{ 1 + \frac{30(1-\nu_0)\phi(3\gamma_1 + 2\gamma_2)}{3\alpha + 2\beta - 10(1+\nu_0)\phi(3\gamma_1 + 2\gamma_2)} \right\} \qquad (44)$$

and

$$\mu_* = \mu_0 \left\{ 1 + \frac{30(1-\nu_0)\phi\gamma_2}{\beta - 4(4-5\nu_0)\phi\gamma_2} \right\} \qquad (45)$$

6 COMPARSON WITH EXPERIMENTAL RESULTS

In Malzbender et al. [2], nanoindentation tests were conducted to measure the mechanical properties of nano-particle-filled sol-gel coatings on glass. The material surface is indented with a small indentor loaded with a force. Based on the depth of the indentation, mechanical properties of the thin coating can be calculated.

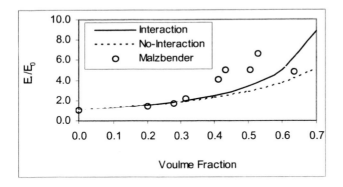

Figure 1: Effects of inter-nano-particle interaction.

Figure 1 compares the experimental results from indentation tests reported in Malzbender et al. [2] with the prediction from the constitutive equations presented in this paper. Nanoindentation has been conducted to measure the mechanical properties of alumina filled coating on glass. The solid curve in Figure 1 represents the mechanical properties predicted by Equations (44) and (45) including the effects of inter-nano-particle interaction. The dashed curve in Figure 1 corresponds to the mechanical properties

predicted by Equations (44) and (45) with neglected effects of inter-nano-particle interaction. The difference in these two curves indicates that as the volume concentration of nano-particles increases, the effects of inter-nano-particle interaction become more important and cannot be neglected. Figure 1 also suggests that the predictions with inter-nano-particle interaction effects are closer to the experimental measurements.

7 CONCLUSIONS

A constitutive modeling framework to predict the mechanical properties of nano-particle reinforced composite materials has been presented. The predictions on mechanical properties from the current model have been compared with experimental data measured by nanoindentation tests. The effects of inter-nano-particle interaction are significant and cannot be neglected especially when the volume concentration of the nano-particles is high.

The constitutive model presented in this paper has the potential of being used for numerical simulation based on finite element analysis to solve practical engineering problems involving composites reinforced with nano-particles up to moderately-high volume concentration. During the constitutive modeling process presented in this paper, no additional material parameter has been introduced. The mechanical properties are related to the volume concentration of the nano-particles. This constitutive model provides a possible tool for the prediction of the volume concentration of nano-particles based on the measurements from nano-indentation tests.

ACKNOWLEGEMENT

This research is supported in part by a nano-initiative grant from Vanderbilt Institute for Nanoscale Science and Engineering.

REFERENCES

[1] D. Qian, G.J. Wagner, W.K. Liu, M.F. Yu, and R.S. Ruoff, Applied Mechanics Review, 55(6), 495, 2002.

[2] J. Malzbender, J.M.J. den Toonder, A.R. Balkenende, and G. de With, Materials Science and Engineering R 36, 47, 2002.

[3] R. Hill, Journal of Mechanics and Physics of Solids, 48, 367, 1963.

[4] T. Mura, Micromechanics of Defects in Solids, Second Revised Edition, Kluwer Academic Publishers, 1987.

Sol-gel synthesis and structural characterization of nanocomposite powder: NiAl₂O₄:SiO₂

P. Muralidharan, I. Prakash, M. Venkateswarlu*, N. Satyanarayana

Raman School of Physics, Pondicherry University, Pondicherry- 605 014, India
Corresponding author: E-mail: nallanis2000@yahoo.com
*Present address: Department of Chemical Engineering, NTUST, Taipei, Taiwan

ABSTRACT

Nanocomposite of NiAl₂O₄/SiO₂ (NAS) was prepared through insitu sol- gel reaction, in which NiAl₂O₄ nanocrystals were dispersed in the silica glassy matrix. The prepared dried NAS gel, at 338 K, was heat treated to different temperature and identified the temperature range (1083 – 1163 K) of the nanocomposite formation through XRD patterns. From XRD patterns, crystallite size of NiAl₂O₄ particles dispersed in the silica matrix was calculated using Scherrer's equation and size of the crystallites is found to be ≈ 8 nm. The structure and thermal behavior of NAS nanocomposite was identified respectively through FTIR and DSC measurements.

Keywords: Sol-gel, NAS, nanocomposite, XRD, FTIR, DSC

1   Introduction

Nanocomposites are important class of materials because of many of their physical and chemical properties show particle size dependence [1-4]. Nano structured metal (Zn, Mg, Ca) aluminate, with spinel structure, dispersed in SiO₂ glassy matrix materials are found to exhibit improved properties such as great thermal stability, hardness, etc. and thus, gained interesting importance in the technological applications like optical, refectories, high alumina cement oxidation catalysts, etc. [5-7]. Nanocomposite materials can be obtained by controlling both the size and polydispersity of the particles in the host matrix through various synthesis methods, under specific conditions, like sol-gel, solid state reaction, etc. [8]. Wide varieties of glass, glass-ceramic monoliths, nano-structured powders, etc. are synthesized through sol-gel technique, since it has many advantages over other methods like low temperature processing, high chemical homogeneity, and purity, etc. Hence, the above mentioned advantages focused our attention to synthesize nanocrystals of spinel structured NiAl₂O₄ dispersed in SiO₂ glassy matrix through insitu sol-gel reaction and characterize by XRD, FTIR and DSC.

2   Experimental

2.1   Sol-gel process

Nanocomposite of NAS gel sample were prepared through sol-gel process, following the composition as 5% NiO-6%Al₂O₃-89%SiO₂, using precursor chemicals of tetraethylorthosilicate (TEOS), Al(NO₃)₃.9H₂O and NiNO₃. Fig. 1 shows flow chart for preparing 5%NiO–6%Al₂O₃-89%SiO₂ (NAS) xerogel through sol-gel process. The precursors were mixed according to their respective molecular weight percentages as mentioned above. The ratio of water to TEOS is maintained as 16:1. The solution was stirred for 2 hrs at 338 K and then allowed to form gel at same temperature.

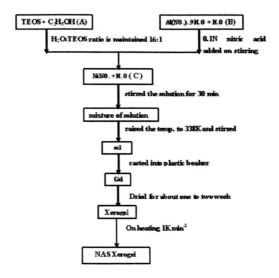

Fig.1. Flow chart for NAS sample preparation by sol-gel process

Transparent green colored monolithic gels were obtained at 338 K and dried at the same temperature for about two weeks. The obtained xerogel was heat treated at 338, 673,

873, 1083, 1123 and 1163 K for two hours at each temperature and is characterized by XRD, FTIR and DSC. The nucleation of $NiAl_2O_4$ crystals appeared at 1073 K and further heated to 1163 K for the growth of the crystals, which was monitored by recording XRD patterns. The synthesis of sol to gel and insitu reaction for the formation of $NiAl_2O_4/SiO_2$ (NAS) composite is described as follows.

$$Si(OC_2H_5)_4 + Al(NO_3)_3 + Ni(NO_3)_2 + H_2O$$

$$\xrightarrow{338\ K} NiO\text{-}Al_2O_3\text{-}SiO_2\ (gel)$$

$$NiO\text{-}Al_2O_3\text{-}SiO_2\ (gel) \xrightarrow{873\ K} NiO\text{-}Al_2O_3\text{-}SiO_2\ (glass)$$

$$NiO\text{-}Al_2O_3\text{-}SiO_2\ (glass)$$

$$\xrightarrow{1083K} NiAl2O4/SiO_2\ (nanocomposite)$$

X-ray diffraction spectra were recorded for NAS sample heat treated at different temperatures using a Rigaku makes miniflex diffractometer with Cu-K$_\alpha$ radiation of wavelength $\lambda = 1.5418$ A$^\circ$ between 70 - 3°, 2θ values, at a scan rate of 2° per minute. All samples were prepared with KBr in the form of pellets at room temperature for recording FTIR spectra. FTIR spectra were recorded for NAS sample to monitor the progress of the dried gel to glass and also the formation of their bond linkages at different temperatures using a Shimadzu FTIR-8300/8700, in the range 4000 - 400 cm^{-1}, 40 nscan. DSC curves were recorded using a Mettler Toledo Stare System, Module DSC 821e/500/575/414183/ 5278 under nitrogen atmosphere at a heating rate of 1K/min for the fine powdered NAS sample placed in an aluminum pan covered with a lid and pressed using the micro pelletizer to form a thin button pellet.

3   Result and discussion

3.1   XRD

Fig. 2 shows the XRD patterns of the NAS sample heat treated in the temperature range 338 – 1163 K. From fig. 2, the observed peak free XRD patterns confirmed that the sample heat treated in the temperature range of 338 – 1073 K remained amorphous in nature. Further, NAS sample heat treated to higher temperatures showed the formation of peaks corresponding to crystalline $NiAl_2O_4$ phase with a broad peak centered at 22°, which is characteristic diffraction pattern of the amorphous phase of SiO_2 glassy matrix [9]. In fig. 2, sample heat treated to 1083 K showed a clear formation of very broad peaks at 35°, 45° and 65° of 2θ value. All the observed peaks were assigned by comparing with the

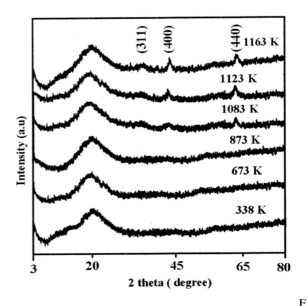

F
ig.2. XRD spectra for NAS sample heat treated at various temperatures

respective JCPDS data and confirmed the formation of crystalline $NiAl_2O_4$ phase. On further heating the NAS sample showed the growth of $NiAl_2O_4$ crystals and the broad peak centered at 22° remained same, which confirmed the dispersion of crystallite $NiAl_2O_4$ in the SiO_2 glassy matrix. The crystallite size was calculated using Scherrer's equation: $D = 0.9\lambda/(\beta\cos\theta)$, where λ X-ray wave length (0.15418 nm), β is full-width half maximum (FWHW) of the peak, and it is found to be ≈ 8 nm. Hence, the XRD patterns confirm the formation of $NiAl_2O_4/SiO_2$ nanocomposite.

3.2   FTIR

Fig. 3 shows the FTIR spectra of the NAS sample, gel to crystalline, heat treated at 338, 673, 1073 and 1123 K. From fig. 3, sample heat treated at 338 K, the observed that the broad bands in the regions of 3500 - 3380cm^{-1} and 1650 - 1635cm^{-1} may be attributed respectively to stretching and bending vibrational modes of O-H of molecular water and the Si-OH stretching of surface silanol hydrogen bond to molecular water [10, 11]. The band at 1376 cm^{-1} and 816 cm^{-1} may correspond to the presence of nitrate groups. The bands at 1234, 1104, 796 and 464 cm^{-1} are attribute to the stretching and bending vibrations of Si-O-Si bonds. The heat treated NAS sample at higher temperatures (673 – 1123 K) showed the disappearance of band positions of 3465, 1640, 1376 and 816 cm^{-1}. The broad band at 1087 - 970 cm^{-1} and 799 cm^{-1} may be due to stretching mode of broken Si-O- bridges [12]. The band at 462 cm^{-1}

corresponds to the deformation mode of Si-O-Si. The appearance of bands at 680 and 580 cm^{-1} for the sample heat treated at 1073 K is attributed to the formation of spinel structure and the intensity of these bands increases with

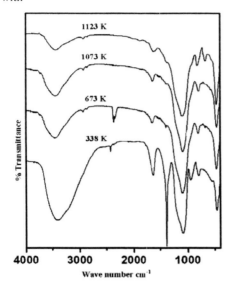

Fig.3. FTIR spectra for NAS sample heat treated at various temperatures

temperature [13]. Hence, the FTIR spectra confirm the formation of NiAl$_2$O$_4$ structure together with SiO$_2$ amorphous matrix to form the nanocomposite, which is in accordance with the XRD results.

3.3 DSC

Fig. 4, shows the DSC curve for dried NAS gel obtained at 338 K. From fig. 4, the DSC curve showed a broad

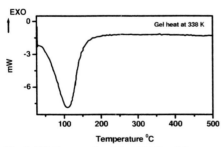

Fig.4. DSC curve for dried NAS gel heat treated at 338 K

endothermic peak between 353 K – 438 K may corresponds to the loss of water molecules present in the dried gels capillaries. From fig. 4, it is important to note that there is no exothermic peak is observed, which may be attributed that there is no formation of crystallization in the sample till

720 K. Thus, DSC results are in accordance with the XRD pattern, where NAS sample showed the amorphous nature till 1050 K.

Conclusions

Green colored transparent NAS gel was prepared at 338 K and on heat treat tthe NiAl$_2$O$_4$/SiO$_2$ nanocomposite could be obtained through new insitu sol-gel process. XRD pattern confirmed that the NAS gel heated to different temperatures is found to be amorphous till 1073 K. Further heat treatment, XRD pattern revealed the formation of crystallite NiAl$_2$O$_4$ particles dispersed in the silica glassy matrix and their crystalline size was found to be $\approx$ 8 nm. Formation of the spinel NiAl$_2$O$_4$ structure together with the SiO$_2$ amorphous matrix was confirmed from the FTIR results.

Acknowledgements

NS thanks the DRDO, CSIR, AICTE and DST for the financial assistance through Research project.

Reference

[1] A.P. Alivisatos, Science 271, 933, 1996
[2] D. Niznansky, N. Viart, J.L. Rehspringer, J. Sol–Gel Sci. Technol. 8, 615, 1997
[3] R. Pool, Science, 248, 1186, 1990
[4] A. Hasselbarth, A. Eychmuller, H. Weller, Chem. Phys. Lett. 203, 271, 1993
[5] T. Ohgushi, S.Umeno, Ull. Chem. Soc.Japan 60 4457 1987
[6] K.T. Jacob, K.P. Jayadevan, R.M. Mallya, Y. Waseda, Adv. Mater. 12, 440 2000
[7] D.A. Fumo, M.R. Morelli, A.M. Segades, Mater. Res. Bull. 31, 1243, 1996
[8] C.J.Brinker and G.W.Scherer, 'Sol-gel science: the physics and chemistry of sol-gel processing' Academic press, New York, 1990
[9] Xiulan Duan, Duorong Yuan, Zhihong Sun, Haiqing Sun, Dong Xu, Mengkai Lv, J. Crystal Growth, 252, 4, 2003
[10] E.A. Hayri and M.Greenblatt, J. Non - Cryst. Solids 94, 387, 1987
[11] E.A.Haryri, M.Greenblatt, T.Tsai and P.P.Tsai , Solid State Ionics 37, 233, 1990
[12] A. Adamczyk, Journal of Mol. Str. 596, 127, 2002
[13] S. Mathur, M. Veith, M. Haas, H. Shen, N. Lecerf, V. Huch, J. Am. Ceram. Soc. 84, 9, 1921, 2001

Optical Thin-Film Structures for Color Analog, and Digital, Long-Term Information Archiving

W.M. Wood and J.E. Nichols

Kuhara Corporation, 1006 Paseo de la Cuma, Santa Fe, NM 87501

monty@kuhara.net, josephnichols@kuhara.net

Abstract

We have demonstrated the use of selective, reproducible, milling of sub-micron feature sizes by focused ion beam of optical thin film structures made from robust materials for color and B&W, analog and digital image and text archiving. This method has also been used to create diffractive structures with controlled, variable diffraction efficiencies. Modeling of the material optical properties, both pre- and post- milling, has enabled accurate prediction of material reflective, transmissive, and diffractive properties, and has effectively shown that ion implantation effects, such as compositional changes and ion deposition, are of second order importance to the optical behaviour of the thin film system.

Keywords: optical thin film, color analog archiving, focused ion beam, digital archiving, diffractive structure

Introduction

It is known that a focused ion beam (FIB) apparatus can be used to alter selected regions on the surface of a medium, either by ion implantation into the medium, or by milling of the medium, and that the regions of implantation and/or milling can be controlled to within a transverse spatial resolution in the nanometer (10^{-9} meters) range[1,2]. Micro-machining of materials can also be achieved by etch sensitization of the material, i.e. by using the fact that a region implanted with ions will etch at a different rate than the un-implanted region when exposed to strong acid or base solutions[3,4,5]. Furthermore, alteration of the medium by ion implantation has direct consequences to the optical properties (reflectivity, transmissivity, absorption,) of the materials[6]. In addition to material changes caused by implantation with ions, the optical properties of thin films has been well understood for many years[7]. Making use of the properties of optical thin films, it is possible to amplify small changes in optical properties (i.e. index of refraction, physical and optical thickness,) of a material through the interference filtration properties of such optical thin films. Results of these changes can be used to create small regions of selectable color, such as pixels within a picture, as well as spatial structures with pre-selectable diffraction efficiencies.

Basis of the effect

The (FIB) apparatus can be used in at least four modes of operation to alter a medium in a predictable and controllable way to record analog (or continuous) and/or digital (or discrete) images or data:

a) implantation of ions within specified surface regions;

b) implantation within and milling of specified surface regions;

c) implantation within and subsequent chemical etching of specified surface regions;

d) implantation within, milling of, and subsequent chemical etching of specified surface regions.

Methods a) and b) were used to create the samples illustrated in this work.

Data is stored on the medium surface based on the predictable and controllable changes in the surface optical properties (reflectivity, transmissivity and absorption) of the medium; changes in optical properties within a single material can be broken down into the following nine general categories:

1) Changes in optical properties due to material amorphization.

2) Changes in optical properties due to material expansion.

3) Changes in optical properties due to material contraction.

4) Changes in optical properties due to formation of color centers within the material.

5) Changes in optical properties due to formation of micro-clusters within the material.

6) Changes in optical properties due to material ordering (crystallization.)

7) Changes in optical properties due to material doping.

8) Changes in optical properties due to material implantation.

9) Changes in optical properties due to material composition changes.

These changes occur within the implantation depth of the ions into the medium.

The actual realization of these optical changes can be enhanced by the use of optical thin film structures. The basic effect can be illustrated with figure 1:

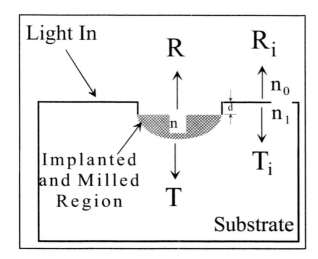

Figure 1: Changes in structure optical properties due to implantation into and milling of a substrate.

$$R \quad r_0\, e^{i2\pi\frac{2d}{\lambda_0}} \qquad T \quad t_0\, e^{i2\pi\frac{d}{\lambda_0}} \qquad (3)$$

The equations governing multiple thin films are easily found by replacing the relevant reflectivity of an interface with the reflectivity of the underlying thin film structure.

These optical changes can be used to create individual regions of color (or black or white,) with resolution of these regions limited only by the longest wavelength of light which will be used to read back the stored information and retain the color information. Because the changes in optical properties are controllable and predictable, the reflection and transmission properties can easily be calculated for any given illumination source. In the case of visible light, the useful minimal size of a region, or "resolution," can be as small as 0.7 microns for standard white light microscopy. For shorter wavelength light, this minimal sized region decreases proportionally to the maximum wavelength of light to be used.

Analog Archiving

Method I: Interference Pixels

Using an optical thin film (ranging in thickness between 1/1000 and 100 times the wavelength of light to be used,) coated on a polished, base substrate or optical thin film structure, it is possible to change the optical depth selectively of the surface layer by implantation and milling with the FIB, as illustrated above. Because optical thin film structures act as interference filters, the local changes to the surface layer (including the change in thickness) serve to change the wavelength tuning of the interference filter structure, resulting in color change on reflection or transmission of white light incident on the region.

Figure 2 shows a diagram of a sample created to illustrate the archiving method described with figure 1. Readback of information can occur in both reflection and transmission modes.

In general, descriptions of the changes in optical properties are described in terms of only the *reflectivity* and the *transmissivity* of a material or material structure, because these are the quantities which are observed directly. (The absorption plays a role in both of these quantities.) Furthermore, the optical changes described above entail changes in both the *amplitude* and the *phase* of the reflected and the transmitted light signals. These changes are described in the following equations.
(The following definitions apply to the equations)

$$r_{ab} \quad \frac{n_a \quad n_b}{n_a \quad n_b} \qquad t_{ab} \quad \frac{2n_a}{n_a \quad n_b} \qquad \delta_a \quad i\frac{2\pi n_a l_a}{\lambda_0} \qquad (1)$$

In the following equations, the subscripted variables r_{ab} represents the amplitude and phase reflectivity (generally a complex number) at the interface between media "a" and "b," for light traveling from medium "a" into medium "b." Similarly, the variables t_{ab} represent the amplitude and phase transmission for light encountering the interface going from "a" to "b." The *energy* or *intensity* reflectivities and transmissions can be found simply by multiplying these numbers by their complex conjugates. These reflectivity and transmissivity values are calculated in terms of the, generally complex, indices of refraction of the media, "n." The index of refraction in medium "a," for example, is written as "n_a." The reflectivity and the transmissivity of the unaltered structures are listed below as R_i and T_i, respectively. The reflectivity and transmissivity for the altered structures are then listed as R and T, respectively. Finally, the subscript on the right hand side of the equations (e.g. n,) represents the properties within the region that has been altered by the focused ion beam.

$$R_i \quad r_{01} \qquad T_i \quad t_{01} \qquad (2)$$

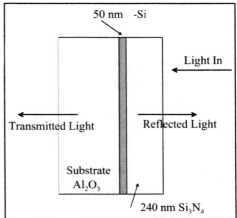

Figure 2: Diagram of archiving substrate.

A substrate of sapphire, Al_2O_3, has been coated with an optical thin film of 50 nanometer thickness -Silicon, and then 240 nanometers of silicon nitride, Si_3N_4. Figure 3 shows the predicted color range under white light illumination for both transmission and reflection perpendicular to the surface for the medium described. Note the color range includes a grey scale that goes from white to black. Figure 4 shows a photograph of a wedge measuring 10 microns on a side, taken through a microscope with light reflected back from the surface. Figure 5 shows an Atomic Force Micrograph (AFM) of the same structure.

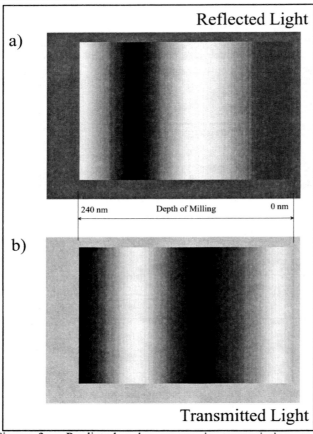

Figure 3: Predicted color ranges in transmission and reflection

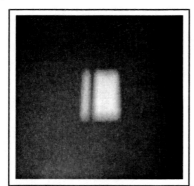

Figure 4: Reflected color from a wedge cut in a sample like that of figure 2.

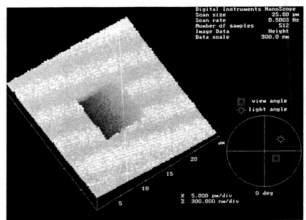

Figure 5: AFM of sample surface.

When used to store color, the smallest pixel must accommodate the longest wavelength in use, e.g. about 0.7 microns for visible light. This limits the pixel size to about 0.5 square microns. When using black and white, or grey scale, images, the reflective or transmissive properties of the surface can be monitored at a single wavelength. Using visible light, the shortest wavelength is about 0.4 microns, leading to effective pixel sizes of 0.16 square microns. Using a UV source of 0.2 microns, the pixel size for a grey scale image is further reduced to 0.04 square microns.

Method II: Diffractive Pixels

Analog storage can also be achieved using diffractive structures. By reducing the silicon nitride layer shown in figure 2 to 65 nanometers, the comparable color range (similar to that shown in figure 3) achieved by milling through the silicon nitride region is relatively insensitive to light color, yielding an effective grayscale. By selectively milling periodic patterns to alter the reflectivity/transmissivity in controllable way, regions that diffract different colors can be created, controllable both in color AND intensity.

A simple illustration of how this can be used for archiving is shown in figure 5. Defining the direction perpendicular to the surface of the storage medium as the "Z" direction, and also as the observation direction, it is possible to create periodic changes (using one of the six methods outlined above) along the X-direction with one period, and periodic changes along the Y-direction with a different period. This "checkerboard" structure can now be illuminated from a fixed set of angles in the X-Z plane (e.g. 40 degrees from perpendicular in either direction) to diffract one spectrum of light due to the periodic structure along the X-direction, and from a fixed set of angles in the Y-Z plane (e.g. 40 degrees from perpendicular in either direction) to diffract another spectrum of light due to the periodic structure along the Y-direction. In this way, a full range of colors, including black and white, can be created in a small region. Furthermore, intensity at each color can also be controlled by adjusting the contrast depth of the periodic variations (whether this is the peak to peak variation of the phase of the reflected light as a function of position, or the amplitude, or both.)

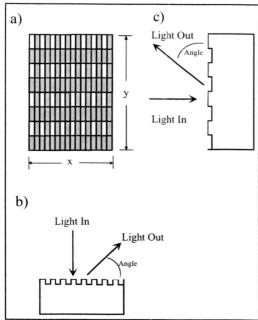

Figure 5: A diffractive pixel for image storage

Pixel sizes of approximately 2.5 microns on a side, or about 6 microns squared, can be achieved with reasonable diffraction efficiencies for use with visible light. Recovery of the image in real color is achieved using a microscope or other, similar, optical viewing apparatus. For black and white storage, visible light could be used monochromatically at 0.4 microns, allowing pixel sizes of 1.2 microns on a side, or 1.5 square microns. For UV illumination, these numbers can be further reduced to a pixel size of 0.7 microns on a side, or 0.5 square microns.

Comparison to State-of-the-Art Analog Image Storage

State-of-the-art analog image archival storage is currently done with photographic film. State-of-the-art resolution is defined by the effective grain size within the film, which defines the minimum "pixel" size for purposes of resolution comparison. For black and white images, the grain size is approximately 3 microns in diameter, giving a pixel area of 10 square microns. For color analog archival storage, this color grain size is on the order of 10 microns, for an effective pixel size of 100 square microns.

Using the two examples outlined above, the relative storage densities for color and black & white archiving are compared in the following table:

Storage Method	Pixel Size (μm^2)	Improve Factor
B&W film	10	1
Color Film	100	1
B&W Interference	0.16	60
B&W UV Interference	0.04	250
Color Interference	0.5	200
B&W Diffraction	2.0	5
B&W UV Diffraction	0.5	20
Color Diffraction	6	15

Clearly, both of the archiving methods outlined in this paper represent an advantage in storage density over microfilm. Furthermore, the materials used for the archiving can be made nearly indestructible.

References

1.   Watkins, REJ et.al., Focused Ion-Beam Milling, *Vacuum*, **v. 36** 961-967. 1986
2. Young, RJ, Micromachining Using a Focused Ion Beam, *Vacuum*, **v. 44** 353-356. 1993
3. Choquette, K.D. et.al., Dry Lithography Using Focused Ion-Beam Implantation and Reactive Ion Etching of SiO$_2$, *Applied Physics Letters*, **62**, 3294-3296, 1993.
4. Matiullah, Durrani, S.A., Measurement of the Etching Characteristics of Proton, Fission Fragment, and Heavy-Ion Tracks in CR-39, *Nuclear Tracks and Radiation Measurements*, **15**, 203-206, 1988.
5. Hosono, K. et.al., Novel Technique for Phase-shifting-mask Repair Using Focused-ion-beam Etch-back Process, *Japanese Journal of Applied Physics part I*, **v. 31**, 4468-4473. 1992
6.   Townsend, P.D. et.al., Optical Effects of Ion Implantation, Cambridge Univ. Press, Cambridge, 1994.
7. Heavens, O.S., Optical Properties of Thin Solid Films, Butterworth's Scientific Publications, London, 1955. (reprinted by Dover Publications, Inc. N.Y. 1991)

Structure Induced Ion Recognition of Self-Assembled Monolayers on Gold

Z. Jiang, S. Liu and Z. Gao[1*]

Department of Chemistry, Wuhan University, Wuhan, 430072, China.
[1]Institute of Bioengineering and Nanotechnology, 51 Science Park Road, Singapore 117586
Republic of Singapore, *Corresponding author: zqgao@ibn.a-star.edu.sg

ABSTRACT

Self-assembled monolayers of anthraquinone-polyethers on gold electrodes were prepared and characterized by various spectroscopic and electrochemical techniques. The monolayers exhibited better thermal stability in wet environments, suggesting that multi-anchoring effect of the molecules improve the integrated adhesion and the binding stability of the monolayers. In contrast to solution electrochemistry, the unique structure and packing of the monolayer lead to selective ion binding to ions with the right size and charge, such as potassium and sodium.

Keywords: monolayer, voltammetry, potassium, sodium, anthraquinones.

INTRODUCTION

Interest in self-assembled monolayers/multilayers (SAMs) arises largely from their excellent microstructural definition, which provides a useful platform to probe the relationship between molecular microstructure on electrode surface and macroscopic electrochemical properties such as electron tunneling, electrocatalysis and sensing.[1] Film durability and stability of these organic monolayers, however, are limited as a result of the relatively easy removal of single organosulfur atoms from solid substrates such as gold.[1] One way to increase the stability of the SAMs is through the use of a multi-anchoring and/or chelating effect.[2-4] For example, extremely stable monolayers of bis-thioctic ester derivatives are easily prepared on gold electrodes.[3,4] No obvious changes are observed for these SAMs after numerous potential scan cycles in the range of 0 - 1000 mV.

Our group has been using self-assembly technique as a means of manipulating molecules with multi-anchoring sites into highly stable monolayers on gold electrodes and exploiting possible ion recognition of these monolayers.[4-6] In two recent reports,[5,6] we described the preparation, self-assembly and electrochemical properties of polymeric self-assembled monolayers. In another paper,[4] crown-ether annelated tetrathiafulvalene derivatives were successfully self-assembled on gold electrodes and the monolayers showed unambiguous ion recognition properties.

In the present work, we report the construction and characteristics of SAMs of the following compounds:

on gold bead electrodes. Surprisingly and in contrast to the behavior of the anthraquinone-polyether compounds in solution, the SAMs showed very high affinity towards alkali metal ions, as evidenced by electrochemical potential shifts of the anthraquinone, even though the molecule contains no polyether macrocycles.

EXPERIMENTAL SECTION

Unless otherwise stated, reagents (analytical-grade or better) were obtained from Sigma-Aldrich (St Louis, MO, USA) and used without further purification. Compounds **1**, **2** and **3** were synthesized via two-step reactions.[7] Gold bead electrodes were prepared from ultrapure gold wire according to the procedure proposed by Schneir *et al.*[8] Anthraquinone-polyether monolayers were formed by immersing freshly prepared gold bead electrodes in a tetrahydrofuran (THF) (or dimethyl sulfoxide) solution containing 0.20 - 1.0 mM of the corresponding compounds. The adsorption time was usually 20 - 24 h. Soaking the gold electrodes for a longer time did not affect the characteristics of the resulting monolayers. After adsorption, the electrodes were copiously rinsed with THF and dried under a nitrogen gas stream. Potentials in this work are referred to a non-aqueous Ag/Ag+ (0.010 M AgNO$_3$ in acetonitrile) reference electrode.

RESULTS AND DISCUSSION

The formation of the monolayers was monitored by various spectroscopic techniques, including ellipsometry and FT-IR. Comparison of the reflection-absorption IR spectra with conventional transmission spectra revealed that weaker absorptions were persistent in all three SAMs for the C=O stretching at 1735 cm^{-1}, -C-O-C- stretching at 1120 cm^{-1} and significant increases in the ratios between the asymmetric and the symmetric C-H stretching in CH$_2$ at about 2926 and 2850 cm^{-1} indicating that the hydrocarbon chains in the monolayer are oriented approximately normal to the gold substrate. The thickness of the SAMs, estimated from ellipsometric measurements, was found to be in the range of 21±2.5 Å. All of the above data indicated that the SAMs are densely packed and well-organized with fully extended polyether chains slightly tilted to the surface normal [9]. The surface coverage of the anthraquinone-polyether SAMs was examined by cyclic voltammetry and electrochemical impedance spectroscopy. As illustrated in Fig. 1, the redox response of ferricyanide/ ferrocyanide was completely blocked by the anthraquinone-polyether monolayer on gold electrode, suggesting the formation of well-packed monolayers with negligible defects.[1] As demonstrated by Rubinstein et al,[10] electrochemical impedance is a more reliable and less perturbing means of evaluating the surface coverage of SAMs. The values of the surface coverage derived from impedance spectra were generally larger than 99.9%, indicating again that densely-packed monolayers were formed on the gold electrodes.

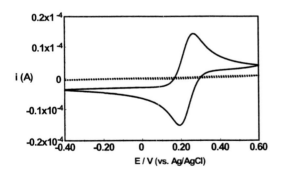

Fig. 1. Voltammograms of 5.0 mM K$_3$Fe(CN)$_6$ at a bare gold (-) and an anthraquinone-polyether monolayer coated gold electrode (- -). 0.10 M KNO$_3$, scan rate 100mV/s.

Since the SAMs contain redox-active anthraquinone moieties, surface-confined electrochemical activities are anticipated. The electrochemical behavior of these compounds in solution is similar to that of other previously reported substituted anthraquinones.[11] For examples, as shown in Fig. 2, the voltammograms of **1** and **2** in THF solution exhibited the usual two quasi-reversible redox waves characteristic of anthraquinones, corresponding to the two one-electron transfer processes leading to the

formation of the dianionic anthraquinone. As can be seen in Fig. 2, the second process, as has been found with other anthraquinone systems, is somewhat chemically and electrochemically irreversible.[11] Addition of various amounts of alkali metal salts to this solution had no observable effect on the voltammetric behavior of these compounds, implying that no significant complexation occurs in solution. This is not surprising in view of the fact that no cyclic polyether macro-ring is present and only three ether oxygen atoms are present on each of the connected chains.

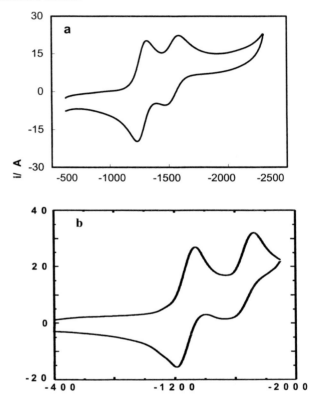

Fig. 2. Voltammograms of 0.20 mM **1** (a) and **2** (b) at 100 mV/s in THF containing 0.10 M TBAPF$_6$ + 1.0 mM KPF$_6$.

Fig. 3 shows cyclic voltammograms of monolayer **1** coated gold electrode in THF containing 0.10 M tetrabutylammonium hexafluorophosphate (TBAPF$_6$). Two pairs of voltammetric peaks appeared at potentials of around -1400 mV and -1600 mV, corresponding to the two one-electron transfer processes of the anthraquinone moieties in the monolayers. Furthermore, unlike the electrochemistry of anthraquinone-polyether derivatives in solution (Fig. 2 and reference 11), both electron transfer processes were clearly resolved and highly reversible. When cyclic voltammetry was conducted at different scan rates, good linearity between the peak currents and the scan rates up to 500 mV/s was obtained, suggesting typical surface-confined voltammetric behavior. Further evidence of the surface-confined voltammetric behavior was obvious from the Gaussian shapes of the voltammetric peaks and the negligibly small peak-to-peak potential separation, generally in the range of 10 - 20 mV. The surface coverage, calculated from the charge under the first reduction peak of the first scan cycle, was 2.37×10^{-10} mol/cm^2. However, the charge under the second peak was always much smaller than that of the first peak, suggesting a partial redox process which may be due to the difficulty in getting bulky TBA$^+$ counter ions into the monolayer, since the anthraquinone-polyether molecules are highly organized and densely packed on the electrode. Upon repetitive potential cycling in the range of 0 to -1800 mV the monolayer showed little change, implying a very robust monolayer due to the multi-anchoring effect of the disulfides. When the potential was scanned to a more negative value than -1800 mV, considerable amounts of the anthraquinone polyether monolayers were removed from the gold electrode, presumably due to the electrochemical reduction of the sulfur-gold bonds. For example, after 20 cycles between 0 and -1900 mV at a scan rate of 100 mV/s, only a small fraction (< 30%, estimated from the loss of charge under the first reduction peak) of the electrode was covered with the anthraquinone-polyether molecules. Similar behaviors were obtained for SAMs of **2** and **3**.

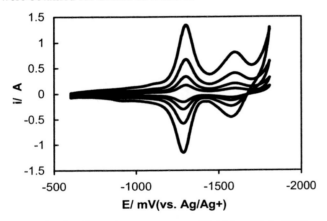

Fig. 3. Cyclic voltammograms of SAMs of **1** in THF solution of 0.10 M TBAPF$_6$. Potential scan rate, from the innermost, 10, 20, 50 and 100 mV/s.

The effect of alkali metal ions on the voltammetric responses of the anthraquinone-polyether monolayers was investigated by cyclic voltammetry in 0.1 M TBAPF$_6$ solution containing 1.0 mM of the hexafluorophosphate or perchlorate salt of the metal ion. Of all of the ions tested (K$^+$, Na$^+$, Li$^+$, Rb$^+$, Cs$^+$, Mg^{2+}, Ca^{2+} and Ba^{2+}), only K$^+$ and Na$^+$ resulted in distinct voltammetric changes. The voltammograms recorded for the SAMs in K$^+$ solution are shown in Fig. 4. For SAMs of **1** (Fig. 4a), both redox waves shifted to less negative potentials, indicating that strong ion binding occurs in solution. [3] Interestingly, for SAMs of **2**, when the potassium salt was added to the electrolyte, a new pair of redox waves appeared at -730 mV and the peak currents increased gradually with the number of potential cycles. Simultaneously, the original peaks at -1375 mV

diminished after a few cycles (Fig. 4b). These changes were observable even at K$^+$ concentration as low as 5.0 M. However, if the potential cycling was conducted between 0 and -1100 mV, i.e. reversing the potential before the first reduction peak, no new peaks were observed after numerous cycles, suggesting that the K$^+$-binding is assisted by the negative charges generated during the reduction of the monolayer.

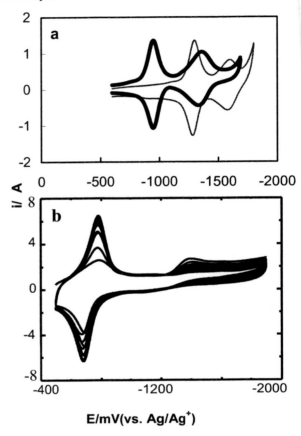

Fig. 4. Cyclic voltammograms of SAMs of **1** (a) and **2** (b) in THF solution containing 0.10 M TBAPF$_6$ and 1.0 mM KPF$_6$. Potential scan rate 100 mV/s.

It appears that the peaks at -730 mV correspond to the reduction of the anthraquinone in close association with K$^+$. In the presence of K$^+$, such anodic shifts of anthraquinone crown ethers due to complexation with M^{n+} in solution are well documented. [11] The anodic shift of the redox potential is closely associated with the cation binding interaction. (For example, an anodic shift of 320 mV corresponds to a cation binding enhancement of 2.6×10^5 times in solution [4,11]). The anodic shift observed here, 645 mV, indicates extremely strong binding of K$^+$. Upon addition of 5.0 mM cryptand [2,2,2] to the solution, the effect of K$^+$ was eliminated and the current peaks shifted back to the original positions. Similar behavior was also observed for Na$^+$, but the anodic shifts of the redox potentials were much smaller. As listed in Table 1, after adding 1.0 mM NaPF$_6$ to the solution, both the first and the second redox processes were anodically shifted by about 195 mV. The effect of Na$^+$ was also reversibly eliminated by adding 5.0 mM cryptand [2,2,2]. Essentially the same voltammetric responses were observed with anthraquinone-polyether monolayers prepared in the presence of alkali metal ions. To verify the ion binding behavior, the polyethers in the anthraquinone derivatives were replaced by hydrocarbons (compound 3). SAMs of 3 showed similar electrochemical characteristics to these of 1 and 2, but alkali ion effect was not observable in solutions containing as high as 5.0 mM of alkali ions.

Table 1. Voltammetric data of monolayers 1 and 2 in THF containing different salts.

Solution	SAM of 1		SAM of 2	
	$E_{1/2}^1$ (mV)	$E_{1/2}^2$ (mV)	$E_{1/2}^1$ (mV)	$E_{1/2}^2$ (mV)
0.10 M TBAPF$_6$	-1300	-1590	-1370	-1690
0.10 M TBAPF$_6$ 1.0 mM KPF$_6$	-950	-1350	-730	------
0.10 M TBAPF$_6$ 1.0 mM NaPF$_6$	-1110	-1420	-1180	-1500
0.10 M TBAPF$_6$ 1.0 mM LiPF$_6$	-1310	-1610	-1375	-1690
0.10 M TBAPF$_6$ 1.0 mM RbClO$_4$	-1305	-1600	-1375	-1690
0.10 M LiPF$_6$ 1.0 mM KPF$_6$	-955	-1355	-730	------

The parameters reflecting the effect of metal ion additions on the voltammetric responses of anthraquinone-polyether monolayers are summarized in Table 1. Note that the K$^+$ induced response was observed even for solutions containing 100-fold excess of other alkali metal ions, indicating excellent selectivity of the anthraquinone-polyether monolayer towards potassium ion binding. As mentioned previously, solution electrochemistry of 1 and 2 with alkali metal ions showed no sign of ion recognition even in the presence of large excesses of alkali metal ions. The ion-recognizing properties of the anthraquinone-polyether monolayer must thus be originated from the self-assembled architecture. The packing of the molecules in the SAMs is so dense (see above) that intermolecular cooperative binding effects are possible, which cannot occur in solution. Analogous behavior was published recently, showing that strong intermolecular hydrogen bonding occurs within densely packed monolayers.[12-13] As schematically illustrated in Fig. 5, it is possible that "pseudo crown ether" structures are formed during self-assembly of the anthraquinone-polyether molecules on the gold electrode, and these show very strong affinity towards K$^+$ and Na$^+$. However, in order to form complexes with such structures in the SAMs, alkali metal ions must be of the appropriate size and charge to be able to penetrate into the monolayer. Cyclic voltammetric experiments indicated that the anthraquinone-polyether molecules are highly organized and densely packed on the gold surface and therefore somewhat impenetrable (Fig. 1). On the other

hand, electrochemical reduction of the monolayer introduces negative charges which attract positively charged species from the solution, especially K$^+$ and Na$^+$, and help them access the binding sites in the monolayer where they form very stable complexes with the monolayers. Removal of K$^+$ or Na$^+$ ions from the monolayers is only possible by introducing a very strong complexing regent such as cryptand [2,2,2].

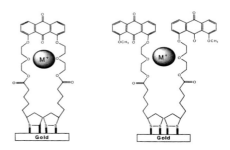

Fig. 5. Illustration of the formation of alkali metal complexes with the monolayers on gold electrodes.

In conclusion, the present work has demonstrated the ability of the anthraquinone-polyether monolayer to bind potassium and sodium ions. The unusually high selectivity is probably the result of the unique structure and packing density of the compound on the surface, which allows cooperative binding to occur but only for ions with the right size and charge to penetrate and fit the monolayer.

REFERENCES

[1]. H.O. Finklea, in Electroanalytical Chemistry, A.J. Bard (Ed) *Vol.* 19, Marcel Dekker, New York, 1996.

[2]. Z. Gao, and H. Chan, *Synth. Met.* 75, 5-10, 1995.

[3]. K. Bandyopadhyay, S. Liu, H. Liu and L. Echegoyen, *Langmuir,* 16, 2706, 2000.

[4]. S Liu, H. Liu K. Bandhyay, Z. Gao and L. Echegoyen, *J. Org. Chem.* **65,** 3292 (2000).

[5]. D. Yang, M. Zi, B. Chen and Z. Gao, *J. Electroanal. Chem.* 470, 114, 1999.

[6]. Z. Gao and K. Siow, *Electrochim. Acta* 42, 315, 1997.

[7]. K. Bandyopadhyay, S. Liu, H.Y. Liu and L. Echegoyen, *Chem-Eur. J.* 6, 4385, 2000.

[8]. J. Schneir, R. Sonnenfeld, Q. Marti, P.K. Hansma, J.E. Demuth and R.J. Hamers, *J. Appl. Phy.* 63, 717, 1988.

[9]. R.G. Nuzzo, F.A. Fusco and D.L. Allara, *J. Am. Chem. Soc.* 109, 2358, 1987.

[10]. E. Sabatani and I. Rubinstein, *J. Phys. Chem,* 91, 6663, 1987.

[11]. R.M. Izatt, K. Pawlak, J.S. Bradshaw and R.L. Bruening, *Chem. Rev.* 95, 2529, 1995.

[12]. O'Hanlon and R.J. Forster, *Langmuir* 16, 702, 2000.

[13]. K. Ariga and T. Kunitake, *Acc. Chem. Res.* 31, 371, 1998.

Synthesis and Characterization of Ferrite Thin Films Obtained by Soft Chemical Methods

G. Caruntu[*], D. Caruntu[*], D. Ganju[**] and Charles J. O'Connor[*]

[*]Advanced Materials Research Institute, Chemistry Department, University of New Orleans, New Orleans, LA, USA, gcaruntu@uno.edu

"Al. I. Cuza" University, Inorganic Chemistry Department, Iasi, Romania, dganju@uaic.ro

ABSTRACT

This paper reports on the synthesis of single phase uniform zinc ferrite thin films prepared the liquid phase deposition (LPD) method. The films with adjustable chemical compositions are identified with a crystal structure as spinel-type and present a spherical/rod-like microstructure. The magnetic films present a superparamagnetic behavior above blocking temperatures which decrease with increasing the Zn content.

KEYWORDS: ferrite thin films, liquid phase deposition, soft solution processing, superparamagnetism

1. INTRODUCTION

With the burst of miniaturization of electronic components, the development of new synthetic strategies for the selective deposition of transition metal ferrites MFe_2O_4 films is critical for the future of microelectronic circuitry. Consequently, a myriad of synthetic strategies which differ in the manufacturability, cost, complexity and environmental hazard have been developed in the last decade [1-5]. The liquid phase deposition (LPD), proposed by Nagayama et al. [6] for the synthesis of dielectric films consists of the direct precipitation of homogenous metal oxide films via the controlled hydrolysis of the corresponding solutions of transition metal-fluoro complexes in presence of a F-scavenger .

$$MF_n^{(n-2m)-} + mH_2O \rightleftharpoons MO_m + nF^- + 2mH^+ \quad (1)$$

Boric acid or aluminum are suitable as fluoride scavengers by virtue of forming stable soluble complexes which cause the shift of the equilibrium reaction with formation of the metal oxide.

$$H_3BO_3 + 4HF \rightleftharpoons BF_4^- + H_3O^+ + 2H_2O \quad (2)$$

However, considerably less is known on using the liquid phase deposition method to synthesize single phase multicomponent oxide films. Gao and coworkers deposited polycrystalline perovskite-type ABO_3 (A=Sr, Ba) thin films with a columnar morphology [7], whereas Deki et al. reported the formation of iron-nickel binary oxide films [8]. The present investigation aims at developing the liquid phase deposition method for the synthesis of zinc ferrite films with different chemical compositions.

2. EXPERIMENTAL

The experiments were performed in open atmosphere using a magnetic hotplate with an external temperature controller. Source chemicals were reagent grade purity and used as received from Alfa Aesar. Prior to deposition, the substrates were degreased by washing repeatedly with acetone and then sonicated in MilliQ water. The parent solution was obtained by dissolving 0.3g of FeO(OH) in a 1M $NH_4F \cdot HF$ aqueous solution. FeOOH was precipitated from an aqueous solution of $Fe(NO_3)_3 \cdot 7H_2O$ upon addition of a diluted solution of ammonia. Then, a separate aqueous solution of Zn^{2+} with a concentration of 2M was prepared by dissolving the corresponding amount of zinc nitrate $ZnNO_3 \cdot 4H_2O$ in distilled water. Zinc ferrite thin films were deposited on non-alkali glass plates (Corning no. 7059) substrates and p-type {111} single crystal Si wafers, respectively. Three separate solutions of iron hydroxides, $Zn(NO_3)_2$ and boric acid (aqueous solution c=0.5M) were mixed in different proportions to obtain a final solution with fixed concentrations of iron and H_3BO_3, whereas the concentration of Zn^{2+} was varied in the range 0.1-0.6M. Substrates were suspended vertically and soaked in the reaction solution at different temperatures ranging between 25°C and 65°C for different periods of time, typically ranging between 3 and 24 hours. To ensure complete crystallization of the zinc ferrite films, samples were subjected to a heat treatment in open air at 600°C followed by a natural cooling to room atmosphere. Surface morphology and microstructure of the films was studied by using a JEOL-JSM 5410 scanning electron microscope, whereas their thicknesses were measured by a surface profile measuring system Dektak-IIA. Identification of the crystalline phases, crystallite size and the phase purity of the films were examined by X-Ray diffraction using a Philips X'Pert System equipped with a curved graphite single-crystal monochromator (Cu K_α radiation).

The metal contents of the deposited zinc ferrite films were determined by inductive coupled plasma (ICP) spectroscopy, using a Varian FT220s flame absorption spectrometer.

Thermal behavior of the as grown films was studied using a TA Instrument TGA 2950. Infrared spectra were collected with a Nicolet Magna 750 FTIR instrument in the range of ν= 4000 to 500 cm^{-1} with a resolution of 5cm^{-1}.

3. RESULTS AND DISCUSSION

A low temperature, single step deposition of solutions of transition metal salts in $NH_4F\cdot HF$ leads to uniform, well adherent zinc ferrite films with thickness which can be easily controlled by varying the deposition time. The slow hydrolysis of transition metal oxy-fluoro-anions leads to supersaturated solutions of oxides/hydroxides which further precipitate onto the substrate to produce a high quality film. For a given deposition time, the increase of the reaction temperature accelerates the chemical deposition which, in turn results in thicker films. The film thickness varies roughly linearly with the deposition time and was found to range typically between 50 and 960nm for a deposition time of 2-24h. In Fig. 1 is represented the variation of the relative amounts of Zn in the deposited ferrite films *vs.* the concentration of Zn^{2+} ions in the reaction solution. Thus, this variation clearly shows that the composition of the film can be strictly controlled by varying the transition metal concentration in the reaction solution. The X-Ray diffraction patterns of the as deposited ferrite films show that they are composed by well crystallized FeO(OH).

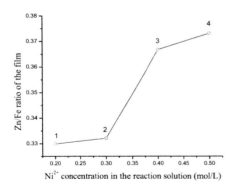

Fig. 1 Variation of the Zn/Fe ratio of the film with the Ni^{2+} concentration of the reaction solution

Although $Zn(OH)_2$ is not detectable by X-Ray diffraction, EDX and TGA experiments revealed that the as prepared films have a Fe/Zn ratio in good agreement with the results obtained from ICP spectroscopy, suggesting that the $Zn(OH)_2$ is amorphous in the as grown films. Fig. 2 displays the TG thermogram of the sample with the composition $Zn_{0.82}Fe_{2.18}O_4$ at a heating rate of 1°C/min under flowing air. The mass loss profile exhibits a very well defined decrease over the temperature range of 25-600°C, which corresponds to the conversion of the intermediate into the final products achieved in three steps. The first inflection point observed at T$\approx$150°C (weight loss =5.11%) is associated with the loss of the hydrated and lattice water, whereas the second one observed at T$\approx$460°C (weight loss of 15.21%) corresponds to the dehydrohylation reaction of the transition metal hydroxides: $2OH^- \rightarrow O^{2-} + H_2O$ and a supplementary loss of water. The third plateau (weight loss of 5.32%) corresponds to the partial reduction of the Fe^{3+} ions to Fe^{2+} with formation of the zinc ferrite spinel structure. Such a large amount of retained water is not surprising, since the films are obtained from aqueous solutions and Zn containing ferrite intermediates are known to retain an unusually amount of associated water [9].

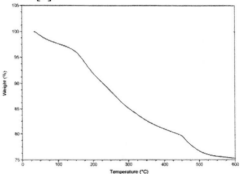

Fig. 2 Thermogravimetric analysis (TG) of hydroxide intermediates with a heating rate of 1°C/min

As shown in Fig. 3, all the experimental diffraction peaks observed in addition to the broad pattern originating from the glass substrate can be assigned to those of the standard polycrystalline $ZnFe_2O_4$ (franklinite, JCPDS 22-1012), which indicates the formation of a spinel-type structure.

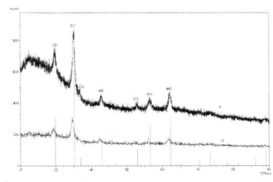

Fig. 3 X-Ray diffraction pattern of the annealed film obtained from the reaction solution with initial Zn^{2+} concentration of 0.2M(a), and 0.3M

The refined value of the cell parameter was a=8.449(3)Å, which is in a good agreement with that of the standard bulk zinc ferrite (a=8.4411Å)[10]. The crystallite size of the film was further determined from the modified Scherrer's formula [11], and found to be close to 20nm. In Fig. 4 are shown typical SEM micrographs the films corresponding to starting solutions whose concentrations of Zn^{2+} ions range from 0 to 0.5M. Highly homogeneous films with a columnar architecture are observed in all cases (Fig. 4b), except the reaction solution whose zinc concentration is 0 and 0.1M where the particles forming the film retain the spherical shape (Fig. 5a) similar to that observed in the case of the α-Fe_2O_3 film [12]. In most of cases films are scratch-resistant and free of cracks. Although the role of Zn^{2+} ions in the treatment solution on the morphology of the ferrite films remains unclear, the experimental data shows that the increase of the Zn^{2+} concentration is accompanied by a change of the morphology of the deposited films from spherical to rod-like type. Additionally, the microstructure of the films doesn't vary noticeably with the thermal treatment.

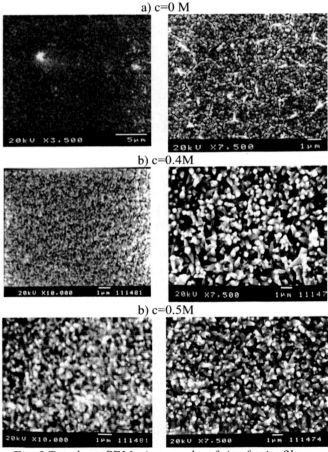

Fig. 5 Top-down SEM micrographs of zinc ferrite films deposited onto a Corning glass substrate from starting solutions with different Zn^{2+} concentrations

Zinc ferrite films were characterized by using standard zero-field-cooling (ZFC) and field-cooling (FC)

procedures. The temperature dependence of the magnetization was measured between 5K and 300K under an external static magnetic field of 100Oe, as shown in Fig. 6. We observe that the ZFC and FC data diverge at low temperature, which is indicative of a superparamagnetic behavior of the zinc ferrite films. The blocking temperature and saturation magnetization roughly decreases with increasing the zinc composition of the sample. Below the critical temperature T_c, the samples exhibit a ferrimagnetic behavior, which is related to the cationic disorder of the two sublattices (denoted by A for tetrahedral and B for octahedral, respectively) of the spinel structure.

The coercivity is found to increase with increasing the Zn content of the films, whereas the values of the magnetization saturation fall within the range reported in literature for zinc ferrites [13].

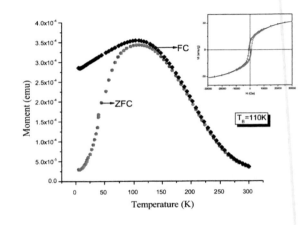

Fig.6 ZFC and FC curves of $Zn_{0.74}Fe_{2.26}O_4$ film. The inset represents the hysteresis loop recorded at 5K

However, the maximum saturation magnetization of 32.11emu/g observed for a Zn/Fe ratio of 0.327 is much lower than is the case of $ZnFe_2O_4$ films obtained by rf sputtering, whose magnetization at 5K was reported to be 90 emu/g for a cation distribution given by the formula $(Zn_{0.4}Fe_{0.6})^{tet}(Zn_{0.6}Fe_{1.4})^{oct}O_4$ [14]. Such a lower value of the saturation magnetization originates from a much smaller fraction of Fe^{2+} ions distributed over the tetrahedral sites which gives rise to a net magnetic moment of the tetrahedral sublattice and, in turn, enhances the A-B interactions at the expense of the B-B ones.

3. CONCLUSIONS

Highly homogenous monophase zinc ferrite thin films can be produced by a simple and flexible soft processing technique, the so-called the liquid phase deposition (LPD) method. Films with a controllable chemical composition present a complex morphology, being constructed by particles with spherical or rod-like shapes.

The dependence with temperature of the magnetic properties of Zn ferrite films with different chemical compositions evidences a superparamagnetic behavior with blocking temperatures ranging between 88 and 108K. The low temperature magnetic properties are closely related to the cationic distribution of Fe^{2+} ions over the two crystallographic sites.

REFERENCES

1. T. Tsurumi, T. Suzuki, M. Yamaze, M. Daimon, *Jpn. J. Appl. Phys.* 1994, **33**, 5192.
2. H. Y. Zhang, B. X. Gu, H. R. Zhai, M. Lu, *Phys. Stat. Sol.*, 1994, **143**, 399.
3. P. C. Dorsey, P. Lubitz, D. B. Chrisey, J. S. Horwitz, *J. Appl. Phys.* 1996, **79**(8), 6338.
4. M. Tachiki, M. Noda, K. Yamada, T. Kobayashi, *J. Appl. Phys.* 1998, **83**, 5351.
5. H. Itoh, T. Uemura, H. Yamaguchi, S. Naka, *J. Mater. Sci,.* 1989, **24**(10), 3549.
6. J. Ino, A. Hishinuma, H. Nagayama, H. Kawahara, *Japanese Patent* 01093443A (Nippon Sheet Glass), June 7, 1988.
7. Y. Gao, Y. Masuda, T. Yinezawa, K. Koumoto, *Chem. Mater.*, 2002, **14**, 5006.
8. S. Deki, Y. Aoi, *J. Mater. Res.*, 1998, **13**(4), 883.
9. E. Auzans, D. Zins, E. Blums, R. Massart, *J. Mater. Sci.*, 1999, **34**, 1253.
10. W. Schiessi, W. Potzel, H. Kartzel, M. Steiner, G. Kalvius, A. Martin, M. Krause, I. Halevy, J. Gal, W. Schafer, G. Will, M. Hillberg, R. Wappling, *Phys. Rev. B.*, 1996, **53**, 9143.
11. H. P. Klug, L. E. Alexander, *X-Ray Diffracton Procedure for Polycrystalline and Amorphous Materials*, Wiley, New York, 2nd Ed., 1974.
12. S. Deki, Y. Aoi, J. Okibe, H. Yanagimoto, A. KAjinami, M. Mizuhata, *J. Mater. Chem.*, 1997, **7**, 1769.
13. J. Smit, H. P. J. Wijn, *Ferrites*, Wiley, New York, 1959.
14. K. Tanaka, S. Nakashima, K. Fujita, K. Hirao, *J. Phys.: Condens. Matter*, 2003, **15**, L469.

Morphology of Nanostructured Films Synthesized via Electrodeposition

R. V. Magan and R. Sureshkumar[*]

Department of Chemical Engineering, Washington University in St. Louis
Campus Box 1198, One Brookings Drive, St. Louis, MO, USA
[*]Email: suresh@poly1.che.wustl.edu

ABSTRACT

In this study we perform Brownian dynamics simulations to investigate the influence of surface reaction rate on the development of size dispersion of interfacial nanostructures that form by electrodeposition of non-interacting ions (particles) onto surfaces with randomly distributed nucleation sites. The surface reaction rate is incorporated into the simulations by using a reaction probability that approaches unity in the case of instantaneous, diffusion-limited deposition. It is found that the size uniformity of the growing particles on the randomly distributed active sites can be improved by decreasing the reaction probability. The simulation results, in qualitative agreement with experiments, show that decreasing the reaction probability is an effective means to weaken the interparticle diffusion coupling by reducing the overlap of diffusion zones surrounding the nanoparticles, and thereby reduce the particle size dispersion.

Keywords: Brownian dynamics, nanostructured films, electrodeposition, morphology, size dispersion

1 INTRODUCTION

In recent times, there has been a growing interest in micro and nanostructured interfaces due to their importance in diverse applications. Examples of interfacial nanostructures include functional coatings and thin films, sessile bacterial colonies on physiological surfaces and, metal semiconductor nanoensembles. There is an interest in carefully designing deposition processes for the manufacture of novel optical and electronic devices with controlled nanostructures. Hence a central issue in the synthesis of interfacial nanostructures is the control of the morphology of the interface.

The size dispersion of growing ensembles is used to characterize the morphology of interfacial structures that are formed by the irreversible deposition of particles or molecules from a bulk solution phase. The arrangement of nucleation sites on the surface have been shown to influence the development of size dispersion during the deposition process. For instance, in the case of non-interacting metal ions that nucleate instantaneously, previous Brownian Dynamics simulation (BDS) for diffusion-controlled growth of metal nanoparticles on planar surfaces [1] has shown that particle size dispersion of metal nano-ensembles develops when active (nucleation) sites are randomly positioned on the electrode surface. The particles become more size-similar with increasing deposition time provided that active sites are regularly arranged. Therefore, size uniformity can be accomplished by distributing the active sites regularly on the electrode surface.

However, in several systems of practical interest, the distribution of active sites is random and the size dispersion develops very quickly with increasing deposition time if the particles nucleate instantaneously and the process is diffusion controlled. Therefore, in order to obtain uniform nanostructures, it is important to seek strategies to control the particle size dispersion for such systems. One of the techniques to reduce size dispersion is to "slow-down" the rate of growth, as demonstrated experimentally in the potentiostatic double-pulse electrodeposition of silver ions. In these experiments, a short initial pulse at a high overpotential is used to initiate the formation of the nuclei, and a longer second pulse at lower overpotential is used to control the growth of the nuclei. Penner [2] demonstrated the benefits of slow growth for double pulse experiments by performing BDS for electrodeposition of metal ions on electrode surfaces. In his simulation, Penner considered an ensemble of nuclei initially grown in the diffusion-controlled regime. He further considered a Nernst layer on the electrode surface, so that changing the overpotential could vary the concentration of ions on the electrode surface, and smaller overpotentials lead to slower growth, thereby reducing size dispersion.

In this paper, we investigate the influence of the surface reaction on the morphology of nanostructures formed by electrodeposition. We present a generic framework to incorporate a finite reaction rate between the free particle and the nucleation site. We also present the results of 3-dimensional BDS for the influence of surface reaction rate on the development of size dispersion of interfacial structures. The balance of the surface reaction rate to the bulk diffusion rate is incorporated into the BDS by using a stochastic reaction probability that approaches unity in the case of diffusion-limited growth [3-5]. For surface-bound electrodeposition, it is anticipated that the reaction probability, can be reduced by decreasing the current density or by changing the electrolyte properties [3].

2 SIMULATION TECHNIQUE

In the BDS we keep track of the coordinates of a set of spherical particles for a dilute system based on the numerical solution of the Langevin equation that represents the force balance for each particle i. The particle trajectory can be computed according to [6]

$$r_i(t + \Delta t) = r_i(t) + \frac{D}{\kappa T} F_E(t)\Delta t + \Delta r^g \qquad (1)$$

where $r(t+\Delta t)$ is the position vector of the center of the particle at time $t+\Delta t$, F_E is the resultant force vector due to all external forces acting on the particle, and Δr^g is the Brownian random displacement due to diffusivity D of the particle and is chosen independently from a Gaussian distribution with zero mean and variance equal to $2D\Delta t$.

The influence of the electric field on the transport of charged species (electromigration) is neglected, thereby assuming that the concenteration of the supporting (inert) electrolyte is high [7]. We further assume quiescient conditions i.e. we neglect forced and natural convection. The particles (ions) are assumed to be diffusing Brownian objects and the particle trajectories are given by

$$r_i(t + \Delta t) = r_i(t) + g_q(\sqrt{2D\Delta t}) \qquad (2)$$

where $\mathbf{g}_q$ is a vector of Gaussian distributed random numbers with zero mean and variance of unity. We define dimensionless space and time variables as

$$\overline{r}_i = r_i / a \ , \ \overline{t} = Dt / a^2 \qquad (3)$$

where a is the particle radius. Then, Eq. 2 is simplified as

$$\overline{r}_i(\overline{t} + \Delta \overline{t}) = \overline{r}_i(\overline{t}) + g_q\sqrt{2\Delta \overline{t}} \qquad (4)$$

The boundary condition at the interface is derived from a balance between diffusion and reaction at the surface. In this study we consider an irreversible first-order surface reaction with lumped surface kinetic rate, k_e. In dimensional form

$$k_e C_{r=R} = D \left. \frac{\partial C}{\partial r} \right|_{r=R} \qquad (5)$$

where, C is the particle concentration and r is the radial distance measured from the center of the nucleus of radius R. Such a mixed (Robin-type) boundary condition on the surface has been used to model electrochemical processes in which ions of the depositing species diffuse to the surface, and are discharged and deposited at the surface [8]. Eq. 5 can be expressed in dimensionless form, as Eq. 6, using the scales introduced in Eq. 3 and $C \dot{=} C_b$, where C_b is the bulk concentration [3,5] as

$$(1 - P_s) \left. \frac{\partial \overline{C}}{\partial \overline{r}} \right|_{\overline{r} = \overline{R}} - P_s \overline{C} \big|_{\overline{r} = \overline{R}} = 0 \qquad (6)$$

where the reaction probability, P_s, represents the relative ratio of the kinetic rate of attachment to the rate of diffusion from bulk to the surface, i.e.,

$$P_s = \frac{Da}{1 + Da} \qquad (7)$$

where Da is a Damkohler number defined as $Da = k_e a/D$. P_s is interpreted as the probability of reaction to take place once a free particle reaches the interface. The limit $P_s \to 1$ denotes growth limited by the diffusion of the particles from the bulk to the surface, and $P_s \ll 1$ indicates that the whole process is controlled by the reaction at the surface.

The reaction probability is dependent on the deposition overpotential, limiting current, exchange current and bulk electrolyte concentration. Voss and Tomkiewicz [9] showed that P_s can be related to the electrochemical parameters as

$$\frac{1}{P_s} = \frac{f_c(\eta) + f_a(\eta)}{f_c(\eta) - f_a(\eta)} + \frac{I_L / I_0}{f_c(\eta) - f_a(\eta)} \qquad (8)$$

where

$$f_c(\eta) = \exp(\alpha_c F \eta / RT); \ f_a(\eta) = \exp(\alpha_a F \eta / RT)$$ and I_0 is the exchange current, I_L is the limiting current , α_c and α_a are the charge transfer coefficients for the cathodic and anodic reactions, η is the deposition overpotential, F is the Faraday number, R is the gas constant and T is the temperature.

The simulations for nanoparticle growth are performed in three dimensional simulation boxes with periodic boundary condition in the x and y directions. At the beginning of the simulation, active (nucleation) sites are randomly distributed on the deposition surface, and the number of these active sites is kept constant during the course of the simulation. The initial system corresponds to an isothermal, monodispersed homogenous system with particle concentration equal to bulk electrolyte concentration. The motion of particles are simulated according to Eq. 4 In the course of the simulation, when the free particles come into contact with any of the active sites on the surface they will attach on it with a finite reaction probability P_s [5]. Once the particle has reacted with the nucleus, the radius of the hemispherical nucleus is increased in such a way that the volume of the hemispherical nucleus is increased by an amount equal to the volume of the free particle. This is done so that the growth of the nuclei is consistent with the Volmer-Weber growth mechanism as observed experimentally for the deposition of silver ions on low surface energy surfaces [10].

3 RESULTS AND DISCUSSIONS

In order to investigate the temporal evolution of the particle size dispersity, the standard deviation of the radii of the growing nanoparticles on the deposition surface, $\sigma(t)$, is used as a quantitative measure. Figure 1 shows the temporal evolution of the particle size dispersion for four different values of reaction probability, ranging from 0.2 to 1.0. Error bars shown in the results indicate one standard deviation of the result that was calculated for the four simulations performed using different seeds on the random number generator. The evolution of the standard deviation with time has similar qualitative trends irrespective of the value of P_s. The size dispersion initially increases rapidly for short times. This is followed by a time period until $t=t_c$ in Figure 1 during which the standard deviation decreases or remains practically unchanged. For $t>t_c$, σ increases with increasing time. However, differences exist among the evolution of σ for different P_s values. Specifically, the long-time evolution of the standard deviation changes from a slow increase for $P_s=0.2$ to a rapid increase for $P_s=1$. As shown in the inset in Figure 1, the slope of the linear best fit of σ vs. t curve at long times, S_σ, increases with reaction probability P_s.

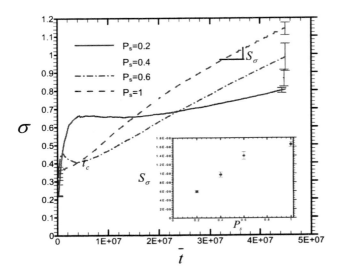

Figure 1: Standard deviation of the particle radius, $\sigma(t)$, as a function of time for different reaction probabilities, P_s. Inset shows slope S_σ of standard deviation versus time plots at different values of P_s.

We further investigated whether the density of active sites could influence the behavior of the σ vs. t curves by performing simulations for lower nucleation densities. These simulation results (not reported here) showed that the influence of P_s on σ is qualitatively independent of nucleation density. However it was observed that for a given P_s, σ is lower for the surface with lower nucleation density. This implies that the size monodispersity of particles can be improved by decreasing the nucleation density on the surface. This result is consistent with experiments in which the relative size monodispersity of particles was found to improve by lowering nucleation density via decreasing the overpotential of the growth pulse [11].

The relative size dispersity (RSD) of the growing nanoparticles (characterized by the ratio of the standard deviation to the mean radius) is shown in Figure 2. At long deposition times, the RSD remains nearly constant at low values of P_s and increases with time for larger P_s. Moreover, as shown in the inset in Figure 2, the slope of the linear best fit of RSD vs. time curve at long times, increases

rapidly as the P_s increases. This observation leads to the conclusion that a strategy to improve the size uniformity of particles growing on a random distribution of nucleating sites on a deposition surface would be to reduce the reaction probability P_s and thereby slow the growth of the nanoensembles.

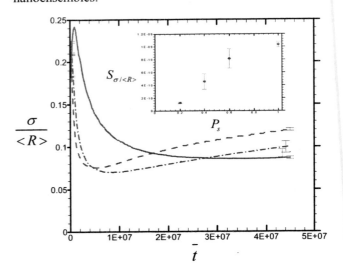

Figure 2: Relative size dispersity of the particle radius, RSD, as a function of time (legend same as in Figure 1). Inset shows slope of relative standard deviation versus time plots at different values of P_s.

The influence of P_s on the size dispersion of the growing nanoparticles can be explained by the dependence of the growth of the nanoparticles on a surface to the distance of the particle to its nearest neighbor, NND from here onwards. A previous study of diffusion-limited growth of nanoparticle ensembles randomly distributed on deposition surface showed an increase in the growth of nearest neighbors as the proximity between the particles increased [1]. This phenomenon of retardation in growth of densely nucleated regions has been referred to as interparticle diffusional coupling [1,2]. In our simulations we investigate the effect of P_s on interparticle diffusional coupling. Towards this end, we compute the increase in the radii of each nucleation site, $\Delta r = r(t)-a$, normalized with respect to the corresponding ensemble averaged value, $<\Delta r>$ for different times. Figure 3 show results from a single simulation of $\Delta r / <\Delta r>$, as a function of NND obtained for $P_s = 0.2$ and $P_s = 1$. It can be seen from these figures that there exists a stronger linear correlation between growth of particles and the NND for $P_s = 1$ as compared to that observed for $P_s = 0.2$. Simulations at other values of P_s support the conclusion that lower values of P_s lead to a reduced correlation between particle radii and NND, and hence result in an improvement in the size uniformity of growing nanoparticles. The simulations at lower nucleation density had a reduced linear correlation between particle radii and NND as compared to the higher nucleation density.

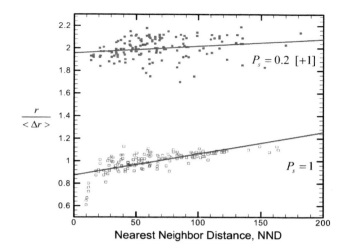

Figure 3: Particle growth, $r / <\Delta r>$, versus the nearest neighbor distance for one simulation performed at $P_s = 0.2$ and $P_s = 1.0$. Results for $P_s = 0.2$ have been incremented by one unit for clarity.

4 SUMMARY

In this paper, the influence of the surface reaction rate on the morphology of interfacial nanostructures is discussed. A reaction probability P_s was introduced into the simulations, to account for the different modes of growth, ranging from diffusion controlled at $P_s = 1$ to surface reaction controlled for $P_s \ll 1$. For the growth of metallic nanoparticles on a surface, it was shown that a strategy for improving the size uniformity of nanoparticles is to reduce the reaction probability, and therefore slow the surface reactions. Lower reaction probability reduce the competition between nearest neighbors, leading to a reduction in particle size dispersion. These results can be used to potentially design deposition processes to create interfacial nanostructures with tailor-made morphology. This can be accomplished by manipulating the process variables that effect P_s by controlling the surface reaction rate (via the applied current density) and/or the bulk diffusivity.

ACKNOWLEDGEMENT

The authors gratefully acknowledge financial support from the Boeing Company.

REFERENCES

[1] J. L. Fransaer and R. M. Penner, J. Phys. Chem. B, 103, 7643, 1999.
[2] R. M Penner, J. Phys. Chem. B., 105, 671,2001.
[3] B. Lin, R. Sureshkumar and J. L. Kardos, Chem. Eng. Sci., 56, 8672, 2001.
[4] B. Lin, R. Sureshkumar and J. L. Kardos, Chem. Eng. Sci., 58, 2445, 2003.
[5] B. Lin, R. Sureshkumar and J. L. Kardos, Ind. Eng. Chem. Res., 41, 1189, 2002.
[6] D. L. Ermak, J. Chem. Phys., 62, 4189, 1975.
[7] I. Rubenstein, "Physical Electrochemistry", Marcel Dekker, 21, 1995.
[8] S. Fletcher, J. Chem. Soc. Faraday Trans., 79, 467, 1983.
[9] R. F. Voss and M. Tomkiewicz, J. Electrochem. Soc., 132, 371, 1985.
[10] J. V. Zoval, R. M. Stiger, P. Biernacki and R. M. Penner, J. Phys. Chem., 100, 837, 1996.
[11] H. Liu, F. Favier, K. Ng, M. P. Zach and R. M Penner, Electrochim. Acta, 47, 671, 2001.

Coating Growth on Nanofibers: Multi-Scale Modeling, Simulations and Experiments

A. Buldum*, C.B. Clemons**, E.A. Evans***, K. Kreider**, G.W. Young**

* Dept. of Physics, Buldum@uakron.edu,
Dept. of Theoretical & Applied Mathematics, * Dept. of Chemical Engineering
The University of Akron, Akron, OH 44325

ABSTRACT

This investigation focuses on the coating of nanotubes and nanofibers with conductive materials using plasma enhanced physical vapor deposition. We examine experimental procedures for coating electrospun polymer nanofibers with metallic materials, then dissolve the inner polymer core to yield a nanotube of the coating material. The interrelationships among processing factors is investigated from a detailed modeling approach that describes the salient physical and chemical phenomena. Solution strategies that couple continuum and atomistic models are used. At the continuum scale we describe the reactor dynamics and deposition of the coatings on the nanofibers. At the atomic level, we use quantum mechanical (QM) and molecular dynamics (MD) simulations to study the deposition mechanisms and migration of atoms in the coating.

Keywords: nanofiber, coating, modeling

1 INTRODUCTION

Nanotubes have attracted great academic and industrial interest in recent years [4]–[6], [13], [14]. Improvement in the ability to synthesize nanotubes of different materials has resulted in the suggestion and development of novel devices based on the properties of the nanotubes [4]. Possible applications for nanotubes in the areas of filtration [3], composites [7], [9], biomedicine [2], [12], and electronics [5] have been suggested. However, several limitations to the widespread synthesis and use of nanotubes can be identified. First, the ability to produce large quantities of nanotubes with controlled electronic and structural properties is still undeveloped. Second, the nanoscale dimensions of these materials often lead to previously unobserved properties that need to be understood and ultimately controlled.

2 EXPERIMENTS

This work addresses some aspects of these issues through a coordinated experimental and modeling program. From the standpoint of nanotube synthesis, we examine physical vapor deposition techniques for applying conductive coatings to electrospun polymer nanofibers. We have successfully coated fibers with carbon, copper, and aluminum films by using a plasma enhanced physical vapor deposition (PEPVD) sputtering process (see Figure 4). The power supply drives a 2 inch diameter electrode which forms the target (or source) material. The nanofibers are placed on a holder that sits 8 cm above the target. A plasma is formed when electrons emitted from the target create ions in the gas phase. Once a plasma is formed, the ions sputter atoms from the target which are then transported to the nanofibers and deposited. The ions also strike the coated nanofibers and tend to make the deposited coating more uniform through a resputtering process. The coating growth rate depends on the rate at which atoms are supplied to the nanofiber surface, the nanofiber temperature, and the ion flux to the nanofiber. The morphology of the coating depends on the mobility of the atoms on the surface and how much time the atoms have to move around before the next atoms hit the surface. The rate at which atoms are supplied to the nanofiber is proportional to the rate at which atoms are sputtered from the target and how far away the nanofiber is from the target. The sputtering rate depends on the ion flux, which is determined by the power applied to the target, the pressure of the system, and the working gas used. The nanofiber temperature is controlled using a heater. The ion flux to the nanofiber is controlled by the potential drop between the plasma and the nanofiber, the working gas used and the pressure.

To determine the effects of these variables on the film growth rate and morphology we analyze the films using TEM. TEM analysis is used to determine the growth rate on the fibers. We compare average thicknesses of the fibers before and after the coating process to determine an average growth rate of the films. To determine coating morphology, TEM images and diffraction patterns are taken. Removing the nanofiber core leaves a polycrystalline nanotube of the coating material.

Figure 1 shows an aluminum-coated fiber. The cylindrical cross-section of a tube is shown in Figure 2, which indicates that the tube did not collapse after the polymer inside had been removed. The smallest inner diameter of the tubes was around 20 nm. The approximate thickness of the wall of the tubes was controlled by the sputtering process. A tube with different wall thickness

Figure 1: TEM images of aluminum-coated fibers.

Figure 2: TEM image of an aluminum nanotube.

is shown in Figure 3.

The above approach can be used to produce tubes of many materials including metals, semiconductors, ceramics and polymers with controlled diameters and a range of nanometer thickness walls.

3 MODELING

The model for the coating of nanofibers is based upon deposition within our traditional PEPVD system. Our objective is to determine the influence of process conditions on the uniformity and morphology of the coating. Our system is characterized by a bulk gas phase dominated by neutral species, and sheath regions that separate the bulk gas phase from the substrate (nanofibers) and the target, as shown in Figure 4. There are several disparate geometrical length scales in the reactor system. The reactor size from target to the top is no more than 20 centimeters in length. The distance from the target to holder is centimeters in length. The sheath thicknesses are on the order of millimeters. The nanofibers are on average 100 nm in diameter. Further, the nanofiber mat is a sparse mesh of the fibers. Hence, we make the assumption that the holder region of nanofibers does not influence the electric field near the target and does not influence the global transport of neutral species within the reactor. Thus, the overall transport of neutral species will be separated into two components. The global or outer component will model the transport away from the target, as shown in Figure 4, without any influence from the sparse mesh of nanofibers and the holder. The local component will

Figure 3: Aluminum nanotube of wall thickness 40 nm.

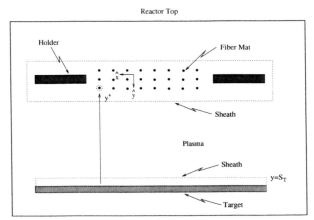

Figure 4: Global schematic of the reactor for neutral species transport within the reactor.

model the neutral species transport in the vicinity of a nanofiber as shown in Figure 5. The global information for neutral species concentration at a nanofiber location, $y = y^*$ from the target, will serve as the far field input to the local model of deposition on the nanofiber.

The mode of transport of the concentration of the deposition material is diffusion for both the global and local systems. Electromagnetic and ion fluid equations govern the transport of ions through sheath regions, and the interaction of the ions with the target and nanofibers. At the local nanofiber scale we examine polar and axisymmetric geometries. Coating deposition equations at the nanofiber include deposition rate parameters and desorption parameters due to ion bombardment. These parameters are functions of the fiber and coating curvature, ion flux to the surface, and ion kinetic energy. These parameters are passed to the continuum equations from molecular dynamics simulations that are described below.

Level set and evolution equation approaches are used to simulate the coating shape. Four basic coating mechanisms are included in these approaches. These are attachment kinetics, curvature effects, etching due to ion bombardment, and solid-state diffusion on the coating surface. These equations are solved numerically and an-

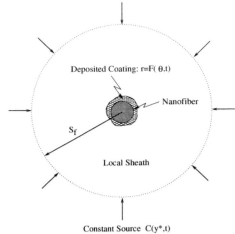

Figure 5: Local nanofiber coordinates
for neutral species transport within the reactor.

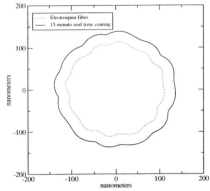

Figure 6: Level set model solution
for nanofiber coating.

alyzed via boundary perturbation techniques. Results from these analyses verify basic experimental observations such as the wavelength and magnitude of the coating roughness is larger in both the axial and azimuthal directions for larger diameter fibers.

At the atomic level we use quantum mechanical (QM) (density functional theory (DFT) [8], [10], SCF-Pseudopotential electronic structure [11], and jellium model calculations) and molecular dynamics (MD) simulations to study the adsorption, reflection and sputtering mechanisms, and migration of atoms in the coating [1]. We input the ion kinetic energy, ion flux, and the thickness of the coating from the continuum models to the QM/MD simulations. This information serves as the initial conditions for the ion bombardment.

Because of the size of the fiber and the computational limitations of the QM/MD approach, it is not possible to simulate the entire circumference of the coating surface. Hence, we examine angular sectors of the nanofiber at various locations around the fiber, and at fixed discrete times, to develop a global picture of the coating growth.

Information from each of the sectors is input to a curve fit to form expressions for the deposition and desorption parameters that are valid around the circumference of the fiber and can be passed to the continuum model. The information obtained from the QM/MD simulations at each time snapshot is also curve fit to obtain expressions that are continuous in time, and for use by the continuum models. This creates a solution methodology that iterates between atomic and continuum length scales.

Putting all of the above pieces together, we present level set simulations of the coating front. The initial polymer nanofiber landscape is taken to be a superposi-

tion of Fourier modes, consistent with models of the electrospinning process. Predictions of the coating thickness are shown in Figure 6. These predictions agree well with our experimental observations.

4 SUMMARY

Polymer nanofibers are produced by electrospinning and then coated using rf magnetron sputtering. The resulting composite structure is then heated to remove the polymer core leaving a nanotube. We develop a comprehensive model integrating across atomic to continuum length scales for simulating the coating process. We investigate global and local models for transport of neutral species and ions within the reactor. We define, solve, and analyze a continuum evolution equation and level set formulation for the coating free surface. Further, we investigate QM/MD simulations of deposition and coating properties. We couple continuum models with atomic simulations. Information is passed between the various length scale models so that the simulations are integrated together. To keep the numerical simulations at a manageable level, asymptotic analyses are used to reduce the complex models to simpler, but still relevant, models.

REFERENCES

[1] M. P. ALLEN AND D. J. TILDESLEY, *Computer Simulation of Liquids*, Oxford University Press, Oxford, 1996.

[2] M. M. BERGSHOEF AND G. J. VANCSO, *Transparent nanocomposites with ultrathin, electrospun nylon-4,6 fiber reinforcement*, Adv. Mater., 11 (1999), p. 1362.

[3] A. BUER, S. C. UGBOLUE, AND S. B. WARNER, *Electrospinning and properties of some nanofibers*, Text. Res. J., 71 (2001), p. 323.

[4] P. G. COLLINS AND P. AVOURIS, *Nanotubes for electronics*, Scientific American, (2000), p. 62.

[5] Y. CUI AND C. M. LIEBER, *Functional nanoscale electronic devices assembled using silicon nanowire building blocks*, Science, 291 (2001), p. 851.

[6] X. DUAN, Y. HUANG, J. WANG, AND C. M. LIEBER, *Indium phosphide nanowires as building blocks for nanoscale electronic and optoelectronic devices*, Nature, 409 (2001), p. 66.

[7] P. W. GIBSON, H. L. SCHREUDER-GIBSON, AND D. RIVIN, *Electrospun fiber mats: Transport properties*, AIChE J., 45 (1999), p. 190.

[8] P. HOHENBERG AND W. KOHN, Physical Review B, 136 (1964), p. 864.

[9] J. S. KIM AND D. H. RENEKER, *Mechanical properties of composites using ultra-fine electrospun fibers*, Polym. Compos., 20 (1999), p. 124.

[10] W. KOHN AND L. J. SHAM, Physical Review A, 140 (1965), p. 1133.

[11] H. MEHREZ, S. CIRACI, A. BULDUM, AND I. P. BATRA, *Conductance through a single atom*, Physical Review B, 55 (1997), p. R1981.

[12] D. SMITH AND D. H. RENEKER. PCT Int. Appl. (2001), Application: WO 2000-2000US27769 20001006. Priority: US 99-158673 19991008; US 2000-571444 20000516.

[13] Y. ZHANG, T. ICHIHASHI, E. LANDREE, F. NI-HEY, AND S. IIJIMA, *Heterostructrures of single-walled carbon nanotubes and carbide nanorods*, Science, 285 (1999), p. 1719.

[14] Y. ZHANG, K. SUENAGA, C. COLLIEX, AND S. IIJIMA, *Coaxial nanocable: Silicon carbide and silicon oxide sheathed with boron nitride and carbon*, Science, 281 (1998), p. 973.

Properties of Self-Assembled Monolayer
as an Anti-adhesion Layer on Metallic Nano Stamper

Sungwoo Choi[*], Namseok Lee[**], Young-Joo Kim[**] and Shinill Kang[*,**]

[*]School of Mechanical Engineering, Yonsei University,
[**]Center for Information Storage Device, Yonsei University,
134 Shinchon-dong, Seodeamun-ku, Seoul, Korea, snlkang@yonsei.ac.kr

ABSTRACT

In the nano replication process, surface quality can be determined by the interfacial phenomena such as the wettability and adhesion force between the metallic stamper and replicated polymeric patterns due to high ratio of surface areas to volume. An experimental method is presented to analyze the temperature dependency on the anti-adhesion property between the stamper and polymer. To analyze the wettability between the stamper and polymer, contact angle of pliable polymer on the metallic stamper was measured at actual molding temperature. To reduce sticking between the stamper and replicated polymeric patterns, SAM (self-assembled monolayer) was applied to the nano replication process as an anti-adhesion layer. Alkanethiol SAM was deposited on nickel surface using solution deposition method. To examine the effectiveness of the SAM deposition on the metallic stamper, contact angle and LFM (Lateral Force Microscopy) were measured.

Keywords: wettability, self-assembled monolayer, anti-adhesion layer, metallic stamper, releasing process

1 INTRODUCTION

With the increasing demand for micro/nano polymeric components, nano-molding using the nano stamper has received much attention. But many technical problems in this molding process have to be overcome for the mass-production of components at low cost. For example, as the features of the nano patterns on the stamper become smaller, surface quality and transcribability are governed by the interfacial phenomena such as the wettability and adhesion force between the stamper and replicated polymeric patterns, due to high ratio of surface areas to volume [1]. High molding temperature and several surface modifications can solve these interfacial problems. Therefore, it is necessary to analyze quantitatively the effects of the mold temperature on the anti-adhesion property between the stamper and polymer in a pliable state especially at temperatures above the glass transition temperature of the polymer. These interfacial problems can also be solved by several surface modifications using an anti-adhesion layer. SAM (self-assembled monolayer) is a candidate material for the anti-adhesion layers deposited on the nano patterns. SAM is stable physicochemically and can

control surface properties of the stamper surface [2,3]. Although nickel substrate is a superb material for the nano stamper, it is not an easy process to deposit SAM on the nickel stamper due to the difficult pretreatment to reduce oxide layer on nickel [4].

To analyze the anti-adhesion property of the stamper precisely, a polymer in a pliable state was used at temperatures above the glass transition temperature. Only a few researchers have measured the contact angle using the polymer in a pliable state, because the high viscosity and the thermal instability of the polymer in a pliable state have made experiments difficult. Grundke et al. [5] and Wulf et al. [6] analyzed the wetting tension and the surface tension of polymer melts as a function of temperature below the melting temperature. However, the anti-adhesion property between the actual stamper and the polymer has not been analyzed. In the analysis of the anti-adhesion property between the stamper and the polymer, the contact angle should be evaluated at the actual molding temperature.

In this study, SAM was applied as an anti-adhesion layer to replication process to reduce the interfacial phenomena between the nickel stamper and replicated polymeric patterns. The effectivness of SAM deposited on nickel stamper was verified though measureing contact angle and lateral friction force. And to analyze the effectivenss of SAM on surface quality of replica in actual nano releasing process, the correlation between the nickel stamper and polymer in actual molding termperature was analyzed as the precedence experiments. To analyze the stability of SAM in actual molding termperature, anti-adhesion properties of SAM were measured for different maximum molding temperature.

2 EXPERIMENTAL PROCEDURES

2.1 Contact angle measurement between the stamper and the polymer at actual molding temperature

Figure 1 shows the schematic diagram of the contact angle measurement system. This system consists of the hot plate and the thermal chamber, which are used to heat the air in the thermal chamber up to the temperature of the actual molding process. The hot plate was connected to the temperature controller and temperature of the stamper surface in the thermal chamber was measured with a thermometer. PC (Polycarbonate) and PMMA (Polymethyl

Methacrylate) were used as the polymer material, and they were used in the measurement of the contact angle on the stamper. The contact angle was measured by using the polymer in a pliable state at a temperature above the glass transition temperature, T_g, of the polymer. The temperature of the hot plate in the contact angle measurement system was maintained at 180°C for about 30 minute. And then, at this temperature the thermal chamber was further heated to the peak mold temperature and maintained at the peak mold temperature for 15 minutes. After the stamper was cooled, the contact angle was measured.

2.2 SAM deposition method and analysis of anti-adhesion properties

The nickel stamper was pretreated by an electrochemical reduction method to remove oxide on the nickel stamper surface. And then using the solution deposition method, alkanethiol SAM as an anti-adhesion layer was deposited on nickel as shown in Figure 2 [4].

To examine the effectiveness of the SAM deposition on the metallic stamper, we analyzed the change in the surface property of SAM deposited nickel stamper. First, to compare the changes in the surface properties between the bare nickel stamper and SAM-deposited nickel stamper, contact angles were measured at room temperature, and the lateral friction force was measured.

However, actual replication processes are performed under various process conditions and environments. To analyze the change in the surface energy of a SAM deposited nickel stamper at these conditions, the contact angle and lateral friction force were measured at the actual processing temperature for the case of compression

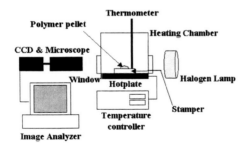

Figure 1: The experimental setup for contact angle measurements

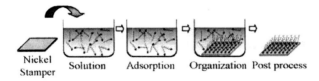

Figure 2: SAM deposition on nickel stamper using solution deposition method

molding and injection molding. Considering conventional molding condition, temperatures were selected from 100℃ to 300℃. In this range, SAM on the nickel substrate was heated for 5 minutes at different maximum molding temperatures. Then, it was cooled at room temperature, and the contact angle and lateral fiction force were measured.

3 RESULT

3.1 Wettability between the stamper and a polymer in a pliable state

Figure 3 shows the dependence of the terminal contact angle at peak mold temperatures of the PMMA and the polycarbonate. As shown in Figure 3, as the peak mold temperature increases, the terminal contact angle values decreases for each polymer material. When peak mold temperatures of PMMA were changed from 210 to 220 °C, the contact angle values decreased markedly. This temperature range includes the melting temperature, T_m, of the PMMA. This behavior indicates that when the fluidity of PMMA at T_m increased, the wettability between the stamper and the PMMA increased.

To evaluate contact angle of the stamper and measure the surface quality and replication quality of the molded substrates at various molding temperatures, we fabricated polymeric replication using the nickel stamper. Molding temperatures were the same as the peak mold temperature in the contact angle measurement in the range of 180 °C to 250 °C. Compression molding with powdered PMMA and PC was used to fabricate a plastic substrates.

Surface roughness of the molded substrate was measured by AFM (Atomic Force Microscopy). Figure 4 shows the effect of mold temperature on the surface roughness of the molded substrate and the contact angle of the stamper. As shown in Figure 4, as the mold temperature increased, the surface roughness of the molded substrates increased while the contact angle value of the stamper decreased. The surface roughness tended to be inversely related to the contact angle value. This result shows that an increase in the wettability of the stamper surface deteriorates anti-

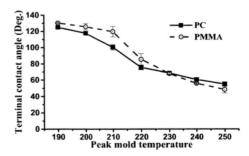

Figure 3: The terminal contact angle by the PMMA and the PC for various peak mold temperature

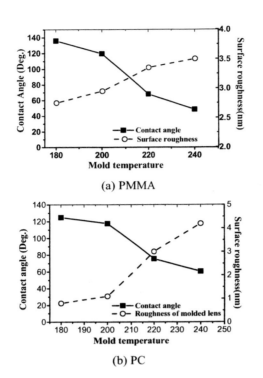

(a) PMMA

(b) PC

Figure 4: Comparison between surface roughness of the molded plastic substrate and contact angle of the stamper at various mold temperature

adhesion properties between the stamper and the polymer. This deterioration of anti-adhesion properties between two surfaces again results in the deterioration of the surface quality of the molded substrates.

3.2 The comparison of the bare nickel stamper and the SAM-deposited nickel stamper

To increase the surface quality of replicated polymeric substrates, SAM was applied to nickel stamper as the anti-adhesion layer. To verify effectiveness of the SAM-deposited nickel stamper, the contact angle and lateral friction force of both the bare nickel stamper and the SAM on the nickel stamper were measured at room temperature. As shown in Figure 5, the contact angle between nickel stampers and D.I water increased from 70.37° to 109.22° after SAM deposition. The comparison of surface properties between the bare nickel stamper and the SAM deposited nickel stamper was summarized in Table 1. The wet energy datas of surfaces were calculated from the contact angle and the surface energy of SAM deposited nickel stamper decreased markedly. Figure 6 shows the lateral friction forces for different normal forces and it is vafied that the higher normal force was, the lower lateral friction force was measured in SAM deposteded nickel

(a)   (b)

Figure 5: Contact angle of (a) the bare nickel stamper and (b) the SAM deposited nickel stamper at room temperature

	Bare nickel stamper	SAM deposited nickel stamper
Contact angle (°)	70.37	109.22
Wet energy (mN/m)	24.46	-23.97
Lateral friction force (eV)	0.0969	0.0605

Table 1 Comparison of surface properties between the bare nickel stamper and the SAM deposited nickel stamper for 10nN normal force of LFM

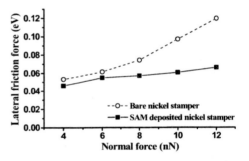

Figure 6: Lateral friction force for different normal forces for the bare nickel stamper and the SAM-deposited nickel stamper.

stamper. The comparison of surface properties between the bare nickel stamper and the SAM-deposited nickel stamper based on these measurement results indicates that surface energy can be reduced by SAM-deposition. Therefore, these results imply the feasibility of SAM being used as an anti-adhesion layer for improving the surface quality of the molded polymeric nano patterns in the nano replication process.

3.3 Change of SAM properties at actual molding conditions

Figure 7 shows the water contact angle and the lateral friction force of SAM-deposited on the nickel stamper for different maximum molding temperatures. The contact

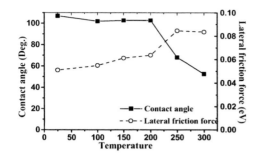

Figure 7: Water contact angle and lateral friction force as function of maximum molding temperature; SAM-deposited nickel stamper was cooled at room temperature after heating at different maximum molding temperatures for 5 minutes

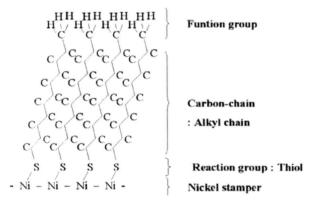

Figure 8: Structure of SAM binding nickel stamper

angle maintained up to the maximum molding temperature of 200℃ and markedly decreased after 200℃, as shown in Figure 7. Also, Figure 7 shows that lateral friction force maintained up to 200℃ similar to contact angle and markedly increased over 200℃.

These tendencies were caused by the damages in carbon-chain structures of SAM, as shown in Figure 8. Since the monolayer is composed by carbon-chains of alkanethiol bound by van der Walls force, the excessive heat energy can break down carbon-chain structure of SAM. The damages, due to heat energy, deteriorated the anti-adhesion properties of SAM and the effectiveness of SAM to metallic stamper may deteriorate. However, it is expected that SAM deposited stamper can be used in the replication process, because the molding temperatures of hot embossing and injection molding are lower than 200℃, the polymer resin in molding is generally filled and cooled in mold cavity for a few seconds.

4 CONCLUSION

We presented the feasibility of SAM as anti-adhesion layer to nickel stamper. To varfy the effectivenss of SAM, it was analyzed that the surface energy of stamper influenced the surface quality of replica. The contact angle was measured at the temperature set for the actual nano-molding process, in which the stamper was heated above the glass transition temperatures of the PMMA and PC. Using these results, we verified that surface quality of replica fabricated using stamper deteriorated as the molding temperature increased. The anti-adhesive property of nickel stamper in replication could be improved by depositing SAM on the nickel stamper. The contact angle and the lateral friction force were measured and anti-adhesion properties of SAM were analyzed. And the stability of SAM in actual molding temperature was analyzed

Acknowledgement
This research was supported by a grant (M102KN 010001-03K1401-00610) from Center for Nanoscale Mechatronics & Manufacturing, one of the 21st Century Frontier Research Programs, which are supported by Ministry of Science and Technology, KOREA.

REFERENCES
[1] K. Seong, S. Moon, and S. Kang, "An optimum design of replication process to improve optical and geometrical properties in DVD-RAM substrate", J. Inf. Stor. Process. Syst. Vol. 3, pp. 169-176, 2001
[2] M. Beck, M. Graczyk, I. Maximov, E. L. Sarwe, T.G.I Ling, M. Keil and L. Montelius, "Improving stamps for 10nm level wafer scale nanoimprint lithography", Microelectric Engineering, Vol. 61-62, pp. 441-448, 2002
[3] Younan Xia, Xiao-Mei Zhao and George M. Whitesides, "Pattern transfer: Self-assembled monolayers as ultrathin resists", Microelectronic Engineering, Vol. 32, pp. 255-268, 1996
[4] Z. Mekhalif et al., "Elaboration of self-assembled monolayer of n-alkanethiols on nickel polycrystalline substrates: time, concentration, and solvent effects", Langmuir, 19, pp. 637-645, 2003
[5] K. Grundke, P. Uhlmann, T. Gietzelt, B. Redlich and H. -J. Jacobasch, "Studies on the wetting behaviour of polymer melts on solid surfaces using the Wilhelmy balance method," Colloids and Surfaces A: Physicochemical and Engineering Aspects Vol. 116, Issues 1-2, pp. 93-104, 1996
[6] M. Wulf, S. Michel, K. Grundke, O. I. del Rio, D. Y. Kwok and A. W. Neuman, "Simultaneous Determination of Surface Tension and Density of Polymer Melts Using Axisymmetric Drop Shape Analysis", Journal of Colloid and Interface Science Vol. 210, Issue 1, pp. 172-181, 1999

Metrology Development for the Nanoelectronics Industry at the National Institute for Standards and Technology

J. A. Dagata, C.A. Richter, R.M. Silver, E.M. Vogel, and J.V. Martinez de Pinillos

National Institute of Standards and Technology,
100 Bureau Drive, Gaithersburg, MD 20899,
john.dagata@nist.gov, curt.richter@nist.gov,
richard.silver@nist.gov, jack.martinez@nist.gov

ABSTRACT

The National Institute of Standards and Technology has provided and continues to provide critical metrology development for the semiconductor manufacturing industry as it moves from the microelectronics era into the nanoelectronics era. This presentation will describe the National Semiconductor Metrology Program, including a detailed discussion of several projects: Nanolithography Using Scanning Probe Oxidation; Atomic-Level Film Characterization; and Nanoelectronic Device Characterization.

Keywords: gate dielectrics, molecular electronics, nanoelectronics, nanolithography, thin films.

1 BACKGROUND

Integrated circuit manufacturing has moved aggressively through the "microelectronics" era into the "nanoelectronics" era. The National Institute of Standards and Technology (NIST) has provided and continues to provide critical metrology for this remarkably productive industry. This presentation will describe broadly the National Semiconductor Metrology Program and discuss in detail several projects in the "nano-regime."

In 1992, the United States Congress, recognizing the critical role of the semiconductor industry for the United States economy, created the National Semiconductor Metrology Program (NSMP) initiative to accelerate semiconductor metrology development at NIST. The NSMP is currently funded at $12.5 million, and leverages approximately an equivalent dollar value of funds from other sources.

The NSMP is administered by the Office of Microelectronics Programs (OMP), which was established to identify projects of high impact to the semiconductor manufacturing industry, to fund those projects, and to monitor the progress. A further function of the OMP is to serve as broker between the industry and NIST, insuring timely transfer of achievements, and to gather critical industry metrology needs to continuously reprioritize NIST projects.

The NSMP projects are grouped into program clusters; (1) **Lithography Metrology**, providing critical optical measurements on materials for next generation lithography (NGL) solutions, calibration of NGL sources, and characterization of NGL resist materials; (2) **Critical Dimension and Overlay Metrology,** delivering measurement techniques and artifacts for length measurements and positioning in the plane of the semiconductor wafer; (3) **Thin Film and Junction Metrology,** developing vertical dimension metrology, film and junction characterization and artifacts; (4) **Interconnect and Packaging Metrology,** reflecting the blurring of back-end-of-wafer interconnect processing and packaging processing, this program explores properties, reliability and mutual compatibilities of the materials used; (5) **Wafer Characterization and Process Metrology,** advancing and refining measurement techniques for the ever tightening requirements for wafer processing; (6) **Test Metrology,** investigating novel techniques for improving high frequency and non-linear circuit testing; and (7) **Manufacturing Support,** contributing to the necessary infrastructure and standards for manufacturing productivity improvement.

2 TECHNICAL MOTIVATION

The CMOS FET (Field Effect Transistor), which is the current basis of ULSI (Ultralarge-Scale Integration) circuits, is beginning to show fundamental limits associated with the laws of quantum mechanics and the limitations of fabrication techniques. By 2005, the Semiconductor Industry Association's (SIA's) 2001 International Technology Roadmap for Semiconductors (ITRS) shows no known solutions for a variety of technological requirements including gate dielectric, gate leakage, and junction depth. Therefore, it is expected that novel fabrication, process monitoring, and device measurement approaches will be needed to continue aggressive CMOS scaling. In addition, it is predicted that entirely new device structures and computational paradigms will be required to augment and/or replace standard planar CMOS devices.

The evolving decrease of the gate dielectric film thickness to an oxide-equivalent value of 1 nm is identified as a critical front-end technology issue in the ITRS [1]. For effective gate dielectric thicknesses below ~ 2.0 nm, SiO_2 is being replaced, initially by oxynitrides or oxide/nitride stacks, and then by either metal-oxides or metal-silicates. Process control tolerance needs for dielectric thickness are projected to be ±4 % (3 σ), which translates to less than 0.1 nm for 2 nm films. Requirements for process control measurements are a factor of ten smaller still. Electrical and reliability characterization methodologies need to be developed and enhanced to address issues associated with both ultra-thin SiO_2 and alternate dielectrics including large leakage currents, quantum effects, and thickness dependent properties. As compared to SiO_2, very little is known about the physical or electrical properties of high dielectric constant gate dielectrics in MOS devices. The use of these films in CMOS technology requires a fundamental understanding of the relationship between the gate dielectric material/interface and device electrical and reliability measurements.

3 ATOMIC-LEVEL FILM CHARACTERIZATION

NIST's work in gate dielectrics ranges from thickness measurements by using spectroscopic ellipsometry to electrical and reliability measurements of MOS devices. The ellipsometry work is directed at determining preferred structural models, spectroscopic index of refraction values, or preferred optical dispersion functions for each of high-k films. Analysis is done with software developed by NIST for spectroscopic ellipsometry; [2] this software allows maximum flexibility for addition of the latest published or custom-developed optical response models as appropriate for each material system investigated. Through collaborations with SEMATECH, IC industry companies, and SRC university staff, as well as with key researchers in other parts of NIST, Project staff are leading and participating in a number of multimethod comparison studies of various ultra-thin gate dielectric films. These multimethod studies utilize techniques such as X-ray and neutron reflectivity, high resolution TEM, EELS, angle-resolved XPS, SIMS, C-V and I-V analysis, as well as spectroscopic ellipsometry and reflectivity. The results of these multimethod studies improve the general understanding of state-of-the-art measurement capability for very thin films, and also allow Project staff to assess the results of various optical models being applied to the analysis of these films with respect to interface layers and structural composition, morphology, and uniformity.

Researchers are also investigating the physics of failure and the reliability testing techniques for ultra-thin SiO_2 and high dielectric constant gate dielectrics [3]. The physical mechanism responsible for "soft" or "quasi" breakdown modes in ultra-thin SiO_2 films and its implications for device reliability are investigated as a function of test conditions and temperature. The understanding generated

in this research is used to continue generating standard measurements through a NIST coordinated collaboration between EIA-JEDEC (Electronic Industries Association Joint Electron Device Engineering Council) and the American Society for Testing and Materials (ASTM). In collaboration with SEMATECH, electrical measurement techniques, procedures and analysis associated with devices having thin oxide and alternate gate dielectrics are investigated. Electrical characterization methodologies are developed to address various issues associated with these films, including large leakage currents, quantum effects, thickness dependent properties, large trap densities, transient (non-steady state) behavior, unknown physical properties, and the lack of physical models. Issues associated with thickness extraction have been addressed [4].

4 ATOM-BASED DIMENSIONAL METROLOGY

At the nanometer scale, the development of nanometer sized structures with accurately measured geometry and position is a primary goal. Samples of this type are not readily available from commercial sources and have to be fabricated in house. In addition, the fabricated features must be measured with tools having high enough imaging and measurement resolution. These samples also have to be dimensionally stable at the nanometer scale and externally accessible to allow transfer to other measurement tools. Chemically-prepared hydrogen-terminated silicon samples are used as substrates and methods for etching nano-scale structures into the silicon have been developed [5]. FIM measured tips can be used to measure features fabricated in-situ (UHV STM and preparation) or prepared externally [6]. These features can be measured accurately with the additional use of a unique picometer-resolution diode laser-based interferometer system [7].

In the atomic regime, we are developing atom-based standards. These are structures of controlled geometry whose dimensions can be measured and traced directly to the intrinsic crystal lattice. This work is based on the preparation and imaging of features with atomic resolution. The atomic spacing is verified with advanced diode laser interferometry. Most current work is also focused on silicon substrates due to our ability to prepare atomically-ordered, hydrogen-terminated samples and their subsequent potential to be etched on the atomic scale. However, other materials for measurement in the atomic regime are being explored for pitch and linewidth artifacts. An essential aspect of this work is the ability to measure features on the atomic scale with a traceable measurementtechnique. We have developed a new implementation of a Michelson interferometer which maintains a fixed number of wavelengths in the measurement path. This method, based on a tunable diode laser, then mixes the tunable laser with a fixed HeNe laser enabling a traceable frequency measurement. An example of this technique is shown below demonstrating the resolution on a graphite lattice.

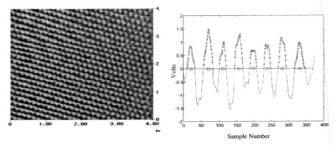

Figure 1. These figures show an STM image of an atomic lattice on the left and a high resolution interferometer measurement on the right. The individual data points in the profile indicate where STM height and interferometer position data were simultaneously acquired.

Figure 2. The figure on the left shows a hydrogen terminated silicon surface. The step and terrace surface is seen with each step being a single atom in height. The triangular structures are a result of complex etching dynamics. The figure on the right shows features written on a silicon surface in a medium vacuum environment. Features in the system have been fabricated below 20 nm in critical dimension.

The fabrication of high resolution features is accomplished in two different environments. The methods used in fabricating these samples have a similar origin to those air ambient methods described in the next section, except that these samples are prepared in UHV or high vacuum complex multi-chamber systems equipped with a STM, FIFEM, and sample preparation facility. The writing process shown in the figure above results in a hard etch mask written directly on a Si (111) substrate. Extensive work has gone into the preparation of atomically ordered

hydrogen terminated surfaces [8]. These surfaces can now be routinely prepared and allow for the systematic development of fabrication processes of sub-20 nm sized features.

5 NANOLITHOGRPHY USING SCANNING PROBE OXIDATION

Scanning probe oxidation is a high-resolution lithography technique pioneered at NIST. It has been in use at research laboratories worldwide for prototyping nanoelectronic, nanoelectromechanical, and nanophotonic devices. NIST projects currently employ this technique for the fabrication and characterization of sub-50-nm linewidth 1-D and 2-D prototype calibration structures for NIST's metrology program and silicon nanowires for novel electronic devices, as illustrated in Figures 1 and 2. Additional collaborations are exploring scanning probe oxidation for local thin-film materials characterization of metallic and nitride Ti, Zr, and Hf films.

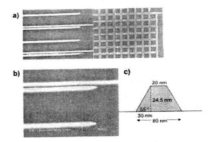

Figure 3. a) 1-D and 2-D prototype structures and nanowires made by scanning probe oxidation and anisotropic etching on a silicon-on-insulator substrate. B) Detail of nanowires. C)Schematic line profile: width along the top of the features is 20 nm and pitch of closely spaced wires is 60 nm.

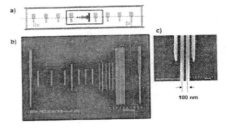

Figure 4. a) 80µm pitch prototype calibration grating with optically and SEM-accessible features made by scanning probe oxidation and anisotropic etching on a silicon substrate. B) Detail of the central SEM accessible pattern. C) Minimum 100 nm pitch obtained on this early prototype.

6 NANOELECTRONIC DEVICES

NIST is developing the metrology needed for nanoelectronic devices. This project is concerned with fundamental research related to possible future devices that will replace or augment standard CMOS technology. The industry for these emerging nanoelectronic devices will require reference data, standards, precision measurement protocols, and standardized test structures and associated measurement protocols to develop into a viable commercial technology.

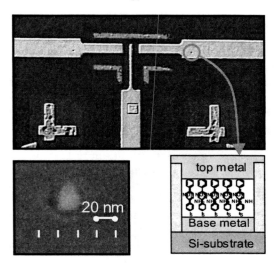

Figure 5: "nanoBucket" molecular electronic test structure. Top: optical micrograph showing three completed nanobuckets of varying areas. Bottom left: SEM micrograph of the base metal of a nominally 30 nm nanoBucket. Bottom right: schematic cross-section of a nanoBucket.

Metrology is under development for two specific areas of nanotechnology: molecular electronics [9] and Si-based quantum electronics [10]. In molecular electronics, we are developing test-structures based upon nanofabrication processing techniques for assessing the electrical properties and reliability of moletronic molecules. Figure 5 shows one example of a molecular electronic, referred to as a nanobucket, successfully fabricated and utilized at NIST. Molecules are incorporated into the test structures via self-assembly to form high-quality SAMs (self-assembled monolayers). In addition to the complexity of the nanofabrication of test structures, the challenges associated with measuring the electrical properties (such as current-voltage and capacitance-voltage as functions of temperature and applied fields) of these small molecular ensembles are daunting. The measured electrical properties are correlated with systematic characterization studies by a variety of advanced analytical probes and the results used in the validation of predictive theoretical models. In Si-based quantum electronics we focus on physical and electrical metrology of the basic building blocks of silicon quantum electronic devices (e.g., quantum layers, wires, and dots of silicon surrounded by silicon dioxide). By working with advanced lithographic techniques such as scanned probe oxidation, device structures are fabricated in order to correlate the physical properties of these silicon building blocks with the ultimate device performance. By identifying and addressing the critical metrology issues associated with these basic building blocks, the basis of metrology for future Si-based ULSI nanotechnology will be defined.

REFERENCES

[1] International Technical Roadmap for Semiconductors, 2003 Edition, Draft, www.public.itrs.net

[2] N.V. Nguyen, "NIST Spectroscopic Ellipsometry Studio," modeling software (2000).

[3] John S. Suehle, "Ultrathin gate Oxide Reliability: Physical Models,Statistics, and Characterization", IEEE TED, vol. 49, no. 6, pp. 958-970, 20

[4] C. A. Richter, A. R. Hefner, and E. M. Vogel, 'A Comparison of Quantum-Mechanical Capacitance-Voltage Simulators,' IEEE Elec. Dev. Lett., vol. 22, p. 35, 2001.

[5] H. Zhou, J. Fu, S. Gonda and R. M. Silver "Effects of Etching Time and Wafer Miscut on the Morphology of Etched Si(111) Surfaces," SPIE Nanotechnology Proc. SPIE Vol. 4608, 2002.

[6] PVM. Rao, C. Jensen, and R. Silver, "An enhanced model for STM tip geometry measured with FIM," To appear in: JVST B, Mar/Apr 2004.

[7] R. M. Silver, S. Gonda, J. Jun, L. Howard, and C. Jensen, "Atomic Resolution Measurements with a New Tunable Diode Laser-Based Interferometer," Optical Eng., Vol. 43, No. 1, Jan. 2004.

[8] J. Fu, H. Zhou, J. Kramar, S. Gonda and R. M. Silver "Dependence of Morphology on miscut angle for Si (111) etched in NH4F," Applied Physics Letters, Vol. 82, Num. 18, 2003.

[9] Tour J.M. Molecular Electronics:Commercial Insights, Chemistry, Devices, Architecture and Programming, (World Scientific Publishing, New Jersey, 2003).

[10] Yasuo Takahashi, Yukinori Ono, Akira Fujiwara and Hiroshi Inokawa, "Silicon single-electron devices," J. Phys.: Condens. Matter 14 R995–R1033, 2002.

Sol-gel Silane Films for DNA Microarray Experiments

K. Saal*/**, T. Tätte*/**, M. Plaado**, I. Kink*, A. Kurg***, R. Lõhmus*, U. Mäeorg**, and A. Lõhmus*

*Institute of Physics, University of Tartu, 142 Riia St., 51014 Tartu, Estonia, saal@fi.tartu.ee
**Institute of Organic and Bioorganic Chemistry, University of Tartu, 2 Jakobi St., 51014 Tartu, Estonia, tanelt@ut.ee
***Institute of Molecular and Cell Biology, Estonian Biocentre, 23 Riia St., 51010 Tartu, Estonia, akurg@ebc.ee

ABSTRACT

Preparation and some characteristic properties of sol-gel derived APTMS-TMOS (APTMS, $H_2NCH_2CH_2CH_2$-$Si(OCH_3)_3$; TMOS, $Si(OCH_3)_4$) hybrid films are discussed on the basis of experimental AFM, FTIR and wettability measurements. The AFM measurements reveal uniform surface of the films that consist of densely packed polysiloxane particles. The films are stable in aqueous environment up to certain relative content of APTMS that extends their applicability in various technologies. As an example, the films were tested as substrates for immobilization of 25-mer oligonucleotide DNA, and results indicated several advantages compared to commercial aminosilanized slides.

Keywords: sol-gel hybrid films, uniform surface, atomic force microscopy, DNA immobilization

1 INTRODUCTION

Silanization of hydroxyl-terminated substrates is an effective and frequently used procedure for modification of chemical and physical properties of the substrates as well as to covalently immobilize a variety of compounds onto them. Therefore, silane coatings serve a number of applications such as protective coatings or adhesion promoters on metal surfaces [e.g. 1,2], adhesives in industrial paints [e.g. 3], selectively binding surfaces for tethering biological molecules in biosensor and DNA chip design [4,5], in scanning probe microscopy (SPM) studies of biomolecules [6], and in chemical force microscopy as probe functionalizing agents [7]. Recently, self-assembling of silane monolayers has received growing attention because it can lead to new technological applications. Focus has mainly been on formation of uniform monolayers of long-chained organosilanes, where alkyltrichlorosilanes, particularly octadecyltrichlorosilane (OTS, CH_3-$(CH_2)_{17}$-$SiCl_3$) on different hydroxylated surfaces such as oxidized silicon or mica [8,9] are among the most studied systems. In contrast, alkyltrialkoxysilanes bearing short tail group have been studied only in a limited number of cases [e.g. 10].

Since introduction [6] as a reliable route for immobilization of DNA for SPM studies the silanization of mica or glass using trialkoxyaminopropylsilanes, particularly 3-aminopropyltriethoxysilane (APTES, $H_2NCH_2CH_2CH_2$-$Si(OC_2H_5)_3$) has become a common procedure in similar research [11,12,13]. However, the formation of uniform trialkoxy- and trichlorosilane monolayers is impeded by self-polymerisation of silane, caused by the trace quantities of water in reaction medium [14]. Considering this, it has been claimed that the surface of APTES layer is heterogeneous and varies from sample to sample [15], and it is not stable in aqueous medium [12].

In our recent work we proposed an alternative silanization technique that substantially improved homogeneity and smoothness of the surfaces [16]. Mica substrate was dip coated with partially pre-polymerized APTMS sol that was thereafter gelled in humid air. Still, instability of the films in water that can be explained with low-rate cross-linkage between individual siloxane molecules because of steric hindrance of aminopropyl groups, needed to be improved. In this work we present our further developments of the approach. APTMS was co-polymerized with TMOS in order to favour gelation of the precursor material, and subsequently gelled on glass substrate. Essential properties of the films were characterized by IR spectroscopy, MALDI TOF mass spectrometry, wettability, and atomic force microscopy (AFM) measurements. The potential of the films as substrates for immobilization of 25-mer oligonucleotide DNA was tested and their advantages compared to commercial analogues (SAL-1 slides, Asper Biotech Ltd., ref. 17) are discussed.

2 EXPERIMENTAL PROCEDURES

In brief, a series of samples were prepared varying molar concentration of APTMS relative to TMOS for characterizing dependence of surface topography on the ratio of silanes using AFM imaging. Thereafter, wettability of the surfaces of the same series was measured and characterized by determining contact angles of the surfaces to correlate relative concentrations of APTMS/TMOS and hydrophobicity of the surfaces. Presence of several chemical groups was verified by IR spectroscopic analysis. One of the possible applications of the new method, DNA spotting was tested by a standard DNA spotting method using 25-mer oligonucleotide DNA (1 part of Cy3 3'labelled 5'-aminomodified 25-mer oligonucleotide DNA mixed with 100 parts of unlabelled 25-mer oligonucleotide DNA).

3 RESULTS

3.1 Spectroscopic measurements

For FTIR measurements, freshly prepared KBr pellets were coated with solutions of pre-polymerized precursor of APTMS-TMOS in methanol. The absorptions corresponding to NH_2, SiOH, SiOSi and CH_2 vibrations were confirmed.

MALDI TOF MS measurements were performed using 1,8,9-trihdroxyanthracene (dithranol) and 2,5-dihydroxy benzoic acid (DHB) as matrixes. The hybrid material had molar masses in range of 800 – 1500 amu. As expected, the spectra had very complicated structure that contained different "families" of oligomers and therefore was not very informative. It was estimated that such mass distribution corresponds to the oligomers containing approximately 7-12 monomers.

3.2 The surface of APTMS-TMOS films

APTMS-TMOS 0:1 (0:1 is relative molar concentration) film exhibited a uniform and smooth surface (average vertical difference 5 nm per 1 μm scan). APTMS-TMOS 1:10 film showed surface consisting of grains with several to a hundred nanometers in diameter and average height distribution of 20 nm/μm (Fig. 1a). The surfaces of APTMS/TMOS 1:5 and 1:3 films were similar to 1:10 film. Starting from APTMS-TMOS 1:1 film the surface profiles ranged between two nanometers, thus showing practically featureless topography in micrometer scale (Fig. 1b).

3.3 Wettability of APTMS-TMOS films

The contact angle measurements indicated that APTMS-TMOS 0:1 and 1:10 films were completely wettable, e.g. no water drops formed on their surfaces. The contact angle of APTMS-TMOS 1:5 film was 14 degrees and abrupt jump to 40 and further on to 60 degrees was observed in the case of APTMS-TMOS 1:3 and 1:1 film, respectively. The contact angles of APTMS-TMOS 3:1, 5:1 and 10:1 films also remained in proximity of 60 degrees, whereas in the case of APTMS-TMOS 1:0 film the contact angle dropped to 50 degrees.

3.4 DNA immobilised to APTMS-TMOS films

Fig. 2 illustrates the binding efficiency of DNA 25-mers to silanized slides. In the case of APTMS-TMOS 0:1 film no binding was detected, which is because of the absence of functional groups in the film. The binding to APTMS-

TMOS 1:10 and 1:5 films was in the order of 10% of the SAL film, indicating to the presence of functional groups on the slide. Further increase in the ratio of APTMS

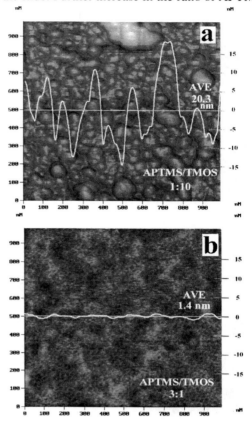

Fig. 1. Semi-contact mode AFM images of APTMS-TMOS hybrid films; scan range 1x1 μm². The scale bar on the right side of the images corresponds to the line profiles drawn in the middle of the scan.

(APTMS-TMOS 1:3 film) gave considerable rise in the amount of DNA immobilised on the surface, but the signal still remained 50% below the level of SAL-glass. The SAL-glass signal was exceeded in the cases of APTMS-TMOS 1:1 and 3:1 films showing substantially higher binding efficiencies (140 % and 135 % of SAL signal, respectively). Further increase in the relative amount of APTMS in the film led to scattered DNA spots and thus, no comparative binding efficiencies could be obtained. The scattering can be explained by dissolution of the films in aqueous environment due to lower rate of cross-linking between aminosilaxane oligomers.

The dimensions of the DNA spots decreased with the increase of the amount of APTMS in the films (Fig. 3), which can be explained by the decrease in wettability, caused by additional amount of hydrophobic aminopropyl groups. The dimensions of the spots on APTMS-TMOS 1:3, 1:1 and 3:1 films were close to the spot sizes on SAL-glass. Starting from APTMS-TMOS 5:1 film, the spots were not clearly outlined due to the dissolving of silane coating in aqueous medium.

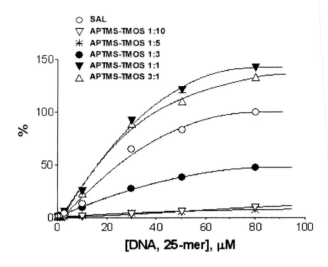

Fig.2. The binding curves of 25-mer oligonucleotide DNA to APTMS-TMOS hybrid films, normalized to the signal of 80 μM DNA spot on the SAL-slide. Each data point corresponds to an average fluorescence intensity of 16 independent spots.

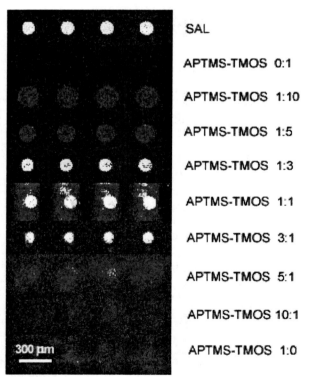

Fig.3. Fluorescence images of 80 μM DNA spots on SAL-slide and on APTMS-TMOS hybrid films.

CONCLUSIONS

It was shown that APTMS-TMOS hybrid films have potential as substrates for immobilisation of biomolecules. The ratio of APTMS/TMOS determines the density of functional groups on the surface. The degree of polymerization can be readily monitored using FTIR spectroscopy and MALDI TOF MS methods, respectively. Such films are stable in aqueous environment up to APTMS content 3:1 in APTMS-TMOS hybrid film. The binding rates of DNA 25-mers to APTMS-TMOS 1:1 and 3:1 films were ≈ 140 % of the binding to commercial SAL-glass, which we believe is due to the uniform distribution of functional groups. Furthermore, the use of APTMS-TMOS hybrid films increases reproducibility of the fraction of immobilised biomolecules, because of the formation of new surface that is virtually independent of the underlying topography of the glass support. Still, the optimisation of the preparation of precursor material and films, as well as the detailed binding characteristics of biomolecules remain to be resolved.

ACNOWLEDGEMENTS

This work was supported by the Estonian Science Foundation grants no. 5015 and 5545. I.K. acknowledges support by EC FW5 "Centres of Excellence" programme (ICA1-1999-70086). The authors acknowledge Asper Biotech Ltd. for helpful cooperation.

REFERENCES

[1] N. Tang, W.J. van Ooij and G. Górecki, *Prog. Org. Coat.* **30** (1997), p. 255.
[2] A. Rattana, J.D.Hermes, M.-L. Abel and J.F.Watts, *Int. J. Adhes. Adhes.* **22** (2002), p. 205.
[3] B. Orel, Z. C. Orel, A. Krainer and M. G. Hutchins, *Sol. Energ. Mat.* **22** (1991), p. 259.
[4] L. Henke, P.A.E. Piunno, A.C. McClure and U.J. Krull, *Anal. Chim. Act.* **344** (1997), p. 201.
[5] C.A. Marquette, I. Lawrence, C. Polychronakos and M.F. Lawrence, *Talanta* **56** (2002), p. 763.
[6] Y.L. Lyubchenko, A.A. Gall, L.S. Shlyakhtenko, R.E. Harrington, B.L. Jacobs, P.I. Oden and S.M. Lindsay, *J. Biomol. Struct. Dyn.* **10** (1992) p. 589.
[7] L.A. Wenzler, G.L. Moyes, L.G. Olson, J.M. Harris and T.P. Beebe, *Anal.Chem.* **69** (1997), p. 2855.
[8] D.A. Styrkas, J.L. Keddie, J.R. Lu, T.J. Su and P.A. Zhdan, *J. Appl. Phys.* **85** (1999), p. 868.
[9] G.A. Carson and S. Granick, *J. Mat. Res.* **5** (1996), p. 1745.
[10] K.C. Popat, R.W. Johnson and T.A. Desai, *Surf. Coat. Tech.* **154** (2002), p. 253.
[11] Y.L. Lyubchenko, R.E. Blankenship, A.A. Gall, S.M. Lindsay, O. Thiemann, L. Simpson and L.S. Shlyakhtenko, *Scanning Microsc. Suppl.* **10** (1996), p. 97.

[12] Y.L. Lyubchenko and L.S. Shlyakhtenko, *Proc. Natl. Acad. Sci. U.S.A.* **94** (1997), p. 496.

[13] Q. Weiping, X. Bin, Y. Danfeng, L. Yihua , W. Lei, W. Chunxiao, Y. Fang, L. Zhuhong and W. Yu, *Mat. Sci. Eng.* **C 8-9** (1999), p. 475.

[14] M. Hu, S. Noda, T. Okubo, Y. Yamaguchi and H. Komiyama, *Appl. Surf. Sci.* **18** (2001), p. 307.

[15] H.X. You and C.R.Lowe, *J. Colloid Interface Sci.* **182** (1996), p. 586.

[16]. T. Tätte, K. Saal, I. Kink, A. Kurg, R. Lõhmus, U. Mäeorg, M. Rahi, A. Rinken and A. Lõhmus, *Surf. Sci.* **532-535** (2003), p. 1085.

[17]. For details see http://www.asperbio.com.

The measurement of ferroelectric thin films using piezo force microscopy

M G Cain[*], S. Dunn[**] and P. Jones[***]

[*]National Physical Laboratory, Teddington, Middlesex, UK, markys.cain@npl.co.uk
[**]Cranfield University, Cranfield, UK, S.C.DUNN@cranfield.ac.uk
[***]Cranfield University, Cranfield, UK, pjones4981@yahoo.com

ABSTRACT

The use of Atomic Force Microscopy to evaluate the properties of ferroelectric thin films is often associated with poor contrast images and quantitative analysis is often not possible. In this paper elements of the metrology associated with this technique are explored, and results comparing different materials types and surface cleanliness are described.

Keywords: piezo force microscopy, ferroelectric, thin films, AFM, characterisation

1 INTRODUCTION

Ferroelectric materials in bulk or thin film form are used in a large variety of applications and with the emergence of Micro-Electro-Mechanical-Systems (MEMS) have driven the reduction of scale for devices that utilise them. The piezoelectric effect (for which ferroelectrics are a sub-set of) is used in such devices to convert motion or stress into electrical signals whilst the converse piezoelectric effect is used to enable motion in structures. More recently, ferroelectric thin films have been used in the development of non-volatile memory devices, such as the Sony Playstation 2. In order to understand and characterise the thin films used for these devices it is necessary to investigate the domain structure and polarisation of the material at the nanometre length scale. At present, one of the best techniques to undertake these investigations is Piezo Force Microscopy (PFM). Using an electrically conductive tip the Atomic Force Microscope (AFM) can detect the piezoresponse induced vibrations of the film when an alternating current (ac) electric field is applied across the film thickness.

PFM is based on a uniquely configured Atomic Force Microscope and makes use of the converse piezoelectric effect in piezo-materials [1][2][3]. The conductive tip of the AFM is used as the top electrode on a ferroelectric thin film sample which has a grounded base electrode. When an ac signal is passed through the tip, with zero dc offset, an alternating electric, E, field is generated between the two electrodes. This E-field causes the generation of strain in the film by way of the converse piezoelectric effect. The value of this strain is $\varepsilon = d_{33}E$, where ε is the strain, d_{33}

is the *out-of-plane* piezoelectric coefficient and E the applied E-field.

As well as the out-of-plane strain there can also be an *in-plane* strain where the material will expand or contract parallel to the surface. The magnitude of the in-plane strains is dependent on $\varepsilon = d_{13}E$ and $\varepsilon = d_{23}E$, where d_{13} is the piezoelectric coefficient in the x-axis and d_{23} the piezoelectric coefficient in the y-axis (In many ferroelectric perovskites, these two tensor components are equivalent because of the symmetry of the crystalline system). The resulting strain causes the film to deform in a periodic manner with respect to the driving frequency, which is detected by the AFM optics. By adding a dc offset to the driving ac voltage, in excess of the materials coercive field, it is possible to pole (electrically orient) individual grains of the ceramic thin film [4]; the poling orientation being controlled by the polarity of the dc offset.

Some of the more commonly used ferroelectric materials used in MEMS are based on lead zirconium titanate (PZT) due in part to their high mechanical strain under an applied E-field [5]. Other materials such as gallium nitride also find use in opto-electronic systems [6]. Most macroscale measurements of these materials ferro/piezoelectric response give information only about its bulk properties. It has been observed that thin films possess a reduced *effective* piezoelectric coefficient compared to their bulk equivalents. In order to understand and characterise ferroelectric thin films for use at the nanometre scale, in devices like memory, piezoelectric micro-actuators and pyroelectric detectors [7], it is necessary to observe the dynamics of domains within the material at the nanometre scale [8]. Piezo-Force Microscopy is ideally suited to the investigation of the individual domains within these materials [6]. Though other techniques have been used to investigate ferroelectric films [9] such as scanning/transmission electron microscopy, they do not have the flexibility that is offered by PFM in both resolution and sensitivity [10].

It has been found that the PFM technique can be used to give greater information about the domain structure in ferroelectric materials, but the actual interactions that take place between the SPM cantilever and sample are not fully understood. A review of the literature reveals that images obtained by PFM are not always repeatable even when the same settings are used for each scan. A thorough investigation into the nature of the interaction between the AFM tip and sample has not yet been carried out. Our research will investigate the nanoscale interactions between tip and sample surface including effects of contamination and environment, and modelling of the E-fields generated around the tip and in the ferroelectric samples. This short paper describes some early results showing the effect of surface contamination on PFM image quality.

1.1 Effects of surface contamination

Surface contamination comes from a variety of sources including initial film formation, absorbed water from the surrounding environment, and even conductive layers deposited by contact with other objects and oxidation. Often it is stated that there is no surface preparation required before PFM is carried out [11], with most experiments being undertaken at atmospheric pressures and humidity.

Desheng et al [12] proposed that one of the reasons why the measured piezo coefficients, using PFM, were so low was because an ultra-thin air gap could exist between the tip and the sample. At nanometre scales this could have a noticeable affect on the E-field. An alternative explanation is that, as most of these experiments are operated in atmospheric conditions, absorbed water fills any space between tip and sample creating a meniscus on the tip, introducing a thin dielectric layer between tip and sample.

PFM is ideally suited to explore these issues because; by changing the tip force interaction (via changing the bias voltage on the AFM cantilever and scanning the tip) we are able to 'wipe' successive layers of contaminant material from the surface of the ferroelectric thin film.

2 EXPERIMENTAL PROCEDURE

All the samples were manufactured using the sol-gel fabrication method and were deposited on two substrate types:

·   ITO/Glass, coated with PZT (30,70) at 210nm thickness, which formed a rosette like structure surrounded by an amorphous matrix on the surface [13].
·   Pt/Ti/SiO2/Si, coated with PZT (30,70) at 200nm thickness and formed a very fine grain structure [14].

On each sample, one corner of the PZT was carefully scraped off using a scalpel and fine wire anchored in place with conductive epoxy resin. This enabled the bottom electrode to be grounded during experimentation. In the experiments described in this paper, no separate top electrodes were deposited onto the ferroelectric thin film.

The AFM was configured for PFM operation, utilising a digital lock-in amplifier and signal generator. Details may be found in references [1][2]. The grounding wire of the sample was connected to the ground of the signal generator output. The output from the signal generator was set to a frequency of 18kHz and amplitude of 4Vpk-pk. For all the samples used, an initial scan of 25μm x 25μm at 0V deflection set-point was undertaken. On completion of the initial scan an area of interest was selected, and a series of scans then followed, using an initial deflection set point of 0V and ending with 6V, in increments of 1V. This had the effect of increasing the force between the Si-AFM tip and the ferroelectric thin film. Two sets of materials were investigated. One set was several years old that had been stored in a normal laboratory environment. The second set was the same material type but had been cleaned using a standardized si-wafer cleaning process. The effect of removal of surface contamination on PFM image contrast was then established.

3 EXPERIMENTAL RESULTS

3.1 Un-cleaned (aged) samples

The first scans at 0V deflection set point (low tip force) resulted in poor PFM image contrast. Increasing the tip force resulted in an improvement in image contrast up to a certain level beyond which the contrast did not improve noticeably. The difference between the contained area and the scrubbed area can be seen clearly in Figure 1. The image on the left of the figure is a topographic AFM image and the image on the right is the PFM image, where bright regions indicate a high degree of piezoelectric induced strain.

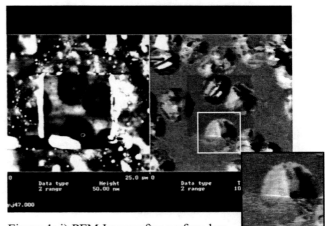

Figure 1: i) PFM Image after surface has been scrubbed clean by tip & inset ii) zoomed in area across the scrubbed un-scrubbed interface.

The magnified rosette in Figure 1ii shows how the demarcation between the scrubbed and un-scrubbed areas affects the contrast. The top half of the rosette is in the scrubbed and the bottom in the un-scrubbed areas.

3.2 Cleaned (aged) sample

The results of the scans on the cleaned sample, Figure 2, differ from those seen when un-cleaned, in that the contrast is very clear from the first low tip force scan to the last. It is interesting to note that the contrast does not alter when different tip forces are used and so the PFM contrast results appear to be tip force independent.

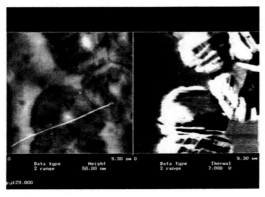

Figure 2: i) 0v Deflection set point

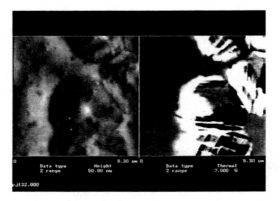

Figure 2: ii) 6v Deflection set

The images on the left of figure 2 are conventional AFM images (height) and the images on the right are PFM data.

The initial scans for the contaminated samples, Figure 1, gave an improved contrast as the tip force increased. However, the improvement with tip force reached a point of saturation beyond which no further improvements were observed. In addition, on reducing the tip force the contrast did not diminish but stayed much the same. The initial improvement in image quality was therefore due to the thick contamination layer being scrubbed from the sample surface by the scanning tip. This either allowed the tips electrical field to make better contact to the ferroelectric material and/or allowed an enhanced mechanical coupling between tip and surface (resulting in a better measurement

of resultant piezo-strain). Both improvements would result in a clearer higher contrast PFM image.

The processing of the ferroelectric thin films is known to leave a residue of a surface contamination layer of lead oxide and lead hydroxy-carbonate on all the samples. When manufacturing the thin films by the sol-gel it is normal to add excess lead to the mixture in order to guarantee that there are enough lead atoms to fill the perovskite structure (loss of volatile lead is a known challenge affecting processing of lead based perovskites). The top surface of the lead oxide film will eventually react with carbon dioxide to form a thin layer of lead carbonate, which, when exposed to water, will also form lead hydroxy-carbonate. Thermodynamic analysis carried out using NPL's MTDATA (x) software shows that only trace amounts (10^{-8}atm) of CO_2 are required for the formation of lead carbonate.

4 DISCUSSION

The lead oxide/carbonate contaminant layers are masking the piezoresponse of the thin films. The masking effect could be either electrical and/or mechanical.

Mechanically, the layer of lead carbonate can be thought of as a hard crust on top of a softer layer of lead oxide. This layer would act as a buffer allowing only a certain amount of coupling between the PFM tip and the piezoresponse of the film, reducing the effective contrast of the images obtained. The initial improvement in scan quality would be due to the PFM tip scrubbing the layers of organic contamination from the top of the lead carbonate layer. The tip would stop scrubbing at the lead carbonate layer and this would mean that the contrast obtained from scans would be the same no matter what tip pressure was applied. This theory is supported by Figure 2 showing the samples after cleaning, removing the organic and loose material from the top surface of the lead carbonate film, the images have the same contrast from the first soft scan to last hard scan.

Electrically, both the inhomogeneous E-field due to the PFM scanning tip and the dielectric properties of the lead oxide / lead carbonate layers would influence the observed piezoresponse during PFM. It has been observed that due to the shape of the PFM tip the E-field would be inhomogeneous and that the greatest flux density would be within the first few nanometres of the film [6],[15]. The thickness of the contamination layer would therefore dictate what strength of E-field the ferroelectric film would experience. If the contamination is thick then the ferroelectric will experience a greatly reduced E-field compared to that of one with a very thin contamination layer. Finite element models are being set-up to examine the effect of the PFM tips inhomogeneous E-field within the film. The different dielectric properties of the contamination layers would also add to this variation in E-field.

5 CONCLUSIONS

This work has shown that the condition of the ferroelectric sample surface has a great impact on the quality of image obtained using Piezo Force Microscopy. It was also found that some samples gave a PFM contrast that was tip to surface pressure dependent and others tip pressure independent. By testing various samples of thin film PZT on different substrates of a variety of ages it was shown that age, handling and fabrication methods has an effect on the PFM image quality.

During investigation it was found that samples that had been extensively handled since fabrication showed a distinct layer of contamination on their surfaces. This contamination had the effect of reducing the effective piezoresponse of the thin film possibly due to a masking of the applied E-field from the PFM tip. The contamination was probably organic and inorganic material from sources handled prior to the sample being touched. This layer was easily scrubbed away with the AFM tip operated at a high tip to surface pressure, the resulting PFM image becoming clearer and showing a contrast that became tip pressure independent.

The hypothesis was put forward that the pressure dependency was related to contamination layers that are native to the surfaces of all the samples, namely lead oxide and lead carbonate, not the contamination due to handling. It is thought that the contrast or effective measured piezoresponse is influenced by the degree of electrical and mechanical coupling that exists between the PFM tip and the sample surface. Future work is aimed at exploring this pressure dependence of PFM measurements on ferroelectric thin films.

REFERENCES

[1] O.Kolosov et al., "Nanoscale Visualisation and control of ferroelectric domains by atomic force microscopy", Physical Review letters, Vol.74, No. 21, May 1995.

[2] A.Gruverman, "Domain structure and polarisation reversal in ferroelectrics studied by atomic force microscopy", Journal of Vacuum Science and Technology, B13, Issue 3, 1995.

[3] S.V.Kalinin et al., "Temperature dependence of polarisation and charge dynamics on the BaTiO3 (100) surface by scanning probe microscopy", Applied Physics Letters, Vol 78. No.8. Feb 2001.

[4] A.Gruverman et al., "Nanoscale imaging of domain dynamics and retention in ferroelectric thin films",

Applied Physical Letters, 71(24), Pgs 3492-3494, Dec 1997.

[5] A.L.Kholkin et al., "Electromechanical properties of lead based ferroelectric thin films", Proceedings of the Materials Research Society Symposium, Vol. 541. Pgs 623-628, 1999.

[6] B.J.Rodriguez et al., "Piezoresponse force microscopy for piezoelectric measurements of III-nitride materials", Journal of crystal growth, Vol 246, pgs 252-258, 2002.

[7] A.L.Kholkin et al, "Seeding effect on micro- and domain structures of sol-gel derived PZT thin film", Materials Letters. Vol. 50, pgs 219-224, 2001

[8] C.Durkan et al., "Investigation into local ferroelectric properties by atomic force microscopy", Ultramicroscopy Vol. 82, pgs 141 – 148, 2000.

[9] S.Hong et al., "High resolution study of domain nucleation and growth during polarisation switching in Pb(Zr,Ti)O3 ferroelectric thin film capacitors", Journal of Applied Physics. Vol. 86, No. 1. July 1999.

[10] S.Dunn et al., "Substrate effect on domain structures of PZT 30/70 sol-gel films via PiezoAFM", Journal of European Ceramic Society Vol. 22, pgs 825-833, 2002.

[11] P.Lehnen et al., "Ferroelectric domain structures of PbTiO3 studied by scanning force microscopy", Journal of Applied Physics, Vol 33, pgs 1932-1936, 2000.

[12] Desheng Fu et al., "Local Piezoelectric response in Bismuth-Base Ferroelectric thin films investigated by scanning force microscopy", Japanese Journal of Applied Physics, Vol. 41, pgs 1103-1105, 2002.

[13] S.S.Roy et al., "Growth and characterisation of Lead Zirconate Titanate (30/70) on indium tin oxide coated glass for oxide ferroelectric-liquid crystal display application", Integrated ferroelectrics, Vol.29, pgs 189-213, 2000.

[14] Q.Zhang et al., "Sol-gel PZT and Mn-doped PZT thin films for Pyroelectric applications", Journal of Physics D: Applied Physics (0022-3727), Vol. 34, No. 15, pgs. 2296-2307, 2001

[15] M.Abplanalp et al., "Imaging of Ferroelectric domains with sub micrometer resolution by scanning force microscopy", ISAF 98. Proceedings of the Eleventh IEEE International Symposium on applications of ferroelectrics, pgs 423 –426 Aug 1998

Improved Reproducibility in Porous Silica Sol-Gel Processing

Using Tertiary Butanol Solvent

A. Rissanen, M. Blomberg and H. Kattelus

MICRONOVA Centre for Micro- and Nanotechnology, VTT Information Technology
P.O. Box 1208, FIN-02044 VTT, Espoo, Finland, Anna.Rissanen@vtt.fi

ABSTRACT

Silica aerogels and xerogels are porous materials, which possess many exceptional properties. The research for the use of aerogel thin films in MEMS applications, such as humidity sensors, biosensors and calorimetric devices, is still at an early stage. Therefore the aerogel thin film process development is prerequisite. Previously, highly volatile liquids like ethanol or isopropanol (IPA) have been used for the solvent in the traditional two-step acid-base sol-gel process, which may deteriorate the reproducibility of the process. In this paper, a novel method to produce porous films with improved reproducibility is presented. It is also demonstrated that aerogel films act as a source for particles during the supercritical drying process. Aerogel patterning with lithography and plasma etching is investigated.

Keywords: sol-gel, aerogel, xerogel, TBA, porous film patterning

1 INTRODUCTION

Aerogels are extremely porous materials, which consist of nanometer-size pores and particles. They are made using sol-gel chemistry to obtain gels consisting of a solid silica part and a liquid solvent part. The gel can be dried by removing the solvent from the pores so that the solid matrix does not collapse, creating a highly porous material.

Usually the sol consists of tetraethyl orthosilicate (TEOS) or tetramethyl orthosilicate (TMOS), alcohol, water and a base/acid catalyst. The sol turns into a gel through hydrolysis and condensation reactions. The solid silica network needs time to become stronger in a process called aging. After this, the gel can be dried to form a porous silica film.

The highest porosities can be obtained by supercritical drying. The solvent liquid in the pores is exchanged with liquid CO_2. When CO_2 is evaporated the pores will not collapse due to the lack of surface tension. If other drying methods are used, like conventional evaporation by heating, the pores will lose some of their size and the obtained material is called xerogel.

The porous silica materials possess many exceptional properties: they have a very low dielectric constant, an adjustable refractive index, low thermal conductivity, low density and a high surface area. The properties of the film can be modified by tailoring the sol-gel process. Because of these unique properties, a lot of research has been done concerning the applications of these porous materials in IC industry and optical devices. However the research for the use of aerogel thin films in MEMS applications, such as humidity sensors, biosensors and calorimetric devices, is still at an early stage. In order to be used as a part of a MEMS process, the film preparation must be well repeatable.

2 EXPERIMENTAL

2.1 Film Preparation

In this work, films were synthesized using a two-step acid-base sol-gel process using tertiary butanol (TBA) as the solvent. The sol composition used in the hydrolyzation step was TEOS:TBA:H_2O:HCl = 1: 3: 2: 7.3 x 10^{-4} in molar ratio. Films were also prepared using several variations of the traditional two-step acid-base sol-gel process [1-3] using IPA as the solvent.

Film preparation process is presented in Figure 1. The solution was hydrolyzed at 60 °C for 1.5 hours. This stock solution was stored below room temperature and before the use it was heated to 30 °C. To catalyze the condensation reaction, 0.6 ml of 0.5 M NH_4OH and 0.4-0.6 ml of DI water was added to 10 ml of the warm stock solution. The gelation time was 11-13 minutes.

The solution was spun on Si substrates in a solvent saturated atmosphere in the optimized viscosity range at $t < 0.4\ t_{gel}$. Spinning was conducted at 2000 – 3000 rpm for 10 s. After coating, 5 ml of IPA was pipetted onto the substrates through the lid of the spinner to prevent evaporation and the wafers were transferred to petri dish for aging. The films were aged using IPA/ ammonia solution (pH 10) from 18 h to 4 d. The solvent was changed to isopropanol and removed with CO_2 supercritical drying.

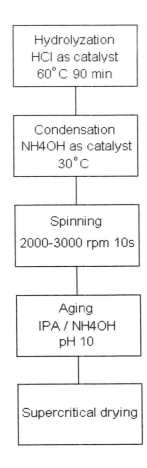

Figure 1: Process flow of silica aerogel film preparation.

2.2 Characterization

Films were characterized with ellipsometry, SEM (see Fig. 2 and Fig. 3) and RBS measurements. The changes in the film properties during lithography and plasma etching in RIE with CF_4 were monitored. IC compatibility of aerogel films was evaluated by detecting particles transferred over a blanket silicon reference sample during supercritical drying.

Film porosity was determined from refractive index measurements and RBS. The density and porosity (P) can be determined from the refractive index n using the empirical equation [4, 5]:

$$P = \frac{1.458 - n}{0.458} \tag{1}$$

The density and porosity of the films were also determined using Rutherford backscattering spectrometry (RBS) with collimated 2 MeV He^+ ions. The areal density value (number of atoms per unit area of film) can be determined from the spectra by calculating the energy gap between the leading and trailing edges of the signals related to each element. The density of the film can be calculated

by dividing the total areal mass by the thickness (d) of the film.

IC compatibility of aerogel films was evaluated by detecting particles created during supercritical drying in Tousimis automegasamdri –915B. This was done by processing two wafers in Tousimis, one blank one (wafer #1) by itself and another blank wafer (wafer #2) with a test wafer coated with aerogel, and comparing the particle counts on these two blank wafers with light point defect counter and PC- based software (VTTWAFER) [6]. Tousimis is not an IC clean instrument, but the difference in the particle counts gives an estimate of how many particles are produced by the aerogel film.

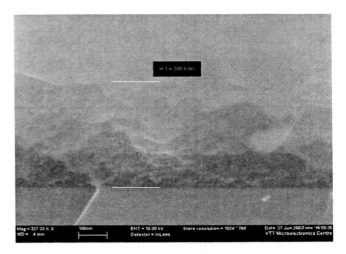

Figure 2: SEM cross section image of a 387 nm-thick aerogel film.

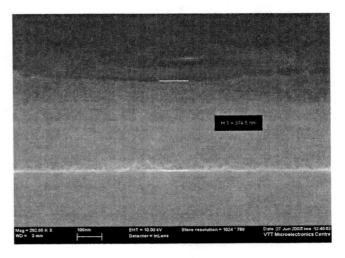

Figure 3: SEM cross section image of 375 nm-thick aerogel film.

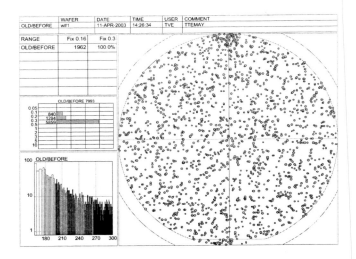

Figure 4: Particles sized 160-300nm on wafer #1.

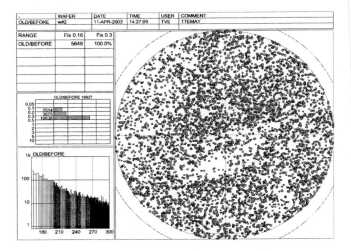

Figure 5: Particles sized 160-300nm on wafer #2.

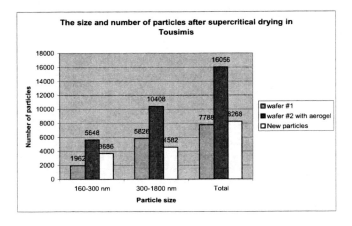

Figure 6: The particle counts on wafers processed with and without an aerogel-coated wafer.

2.3 Patterning

The wafers were primed in HMDS vapor at 115 °C for 30 minutes. Photoresist EZ5214 was manually spun on at 5000 rpm for 35 s. The samples were then prebaked at 90 °C for 1 min on a hot plate.

UV exposure was done with SUSS Mask Aligner MA 6 with a resolution mask. The exposure time was 5 s. The wafers were developed using a mixture of developer AZ351 and DI water (1: 3.5). The developing time ranged from 35 s to 70 s. All the aerogel films remained undamaged during lithography. The films did not suffer from the liquid contact during photoresist development and movement during the development had no effect.

The samples were etched with CF_4 in RIE (reactive ion etching) Electrotech under following conditions:

- Pressure = 20 mTorr
- CF_4 flow rate = 20 sccm
- P (RF) = 100 W
- Time = 2 – 6 min

After etching the photoresist was stripped with O2 plasma in RIE under the following conditions:

- Pressure = 125 mTorr
- O_2 flow rate = 40 sccm
- P (RF) = 150 W
- Time = 3.5 – 10 min

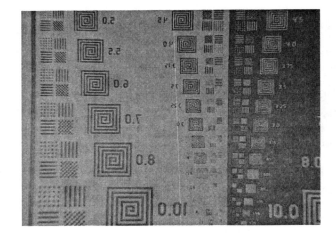

Figure 7: Patterned aerogel test structures.

From Figure 7 it can be seen that it is easier to pattern a hole and trenches into the film than form dots and lines of the film. In the latter case adhesion problems may occur.

3 RESULTS AND DISCUSSION

Films were prepared using several variations of the two-step acid-base sol-gel process to produce films with porosity in the range between 60-80%. It was observed that films made with high volatility solvents like IPA were subject to serious reproducibility problems caused by evaporation during spinning.

The desired porosity and reproducibility were obtained by using tert-butanol (TBA) as a solvent instead of ethanol or IPA. Tert-butanol is a low-volatility solvent and has previously been used in film synthesis through sublimation [7] because of its high freezing point (25.6 °C). Excess amount of TBA during freeze-drying causes film cracking [7], so to obtain the best reproducibility, which is required for MEMS/IC- processing, supercritical drying is preferable. The obtained refractive indices were in the range of 1.10-1.18 for the aerogel films and 1.22-1.26 for the xerogel films, depending on the aging time. The porosities for the aerogel films were in the range of 60% to 78% and for xerogel films in the range of 43% to 52%. The values obtained from the RBS measurements showed a slightly higher value for the porosities than the values obtained indirectly from the refractive indices.

Films made with TBA showed a greatly improved reproducibility since solvent evaporation during processing was significantly decreased. Process reproducibility is important when planning to use aerogel films as a part of commercial MEMS structures for example as semipermeable etching layers or as the active film in humidity sensors. With long device processes, the aerogel coating quite often may take place as one of the last steps and if the coating process is not stable, failing to obtain the desired porosity can be quite expensive.

Aerogel films still have obstacles in becoming a part of commercial products because of the particle problem demonstrated in this work. From Figures 4 and 5 and the summary in Figure 6 it can be seen that the particle counts are much higher on wafer #2, which was processed together with the aerogel wafer. Small particles (160-300 nm) were increased by 3686 particles and big particles (300-1800 nm) were increased by 4582 particles. Introducing a source of thousands of particles to an IC process is not acceptable. In MEMS processing the particles are not necessarily as bad as a problem, especially if the film is deposited at a later stage in the processing chain. However if particles are introduced to the process at an early stage, that can lead to problems in such steps as structure releasing. Other layers such as photoresist might be useful in inhibiting the particles from the aerogel layer to spread in other processing steps.

Aerogel film patterning has previously been demonstrated with surfactant templated films [8], but based on our work patterning sol-gel-based films seems to be more challenging. We found that the porosity suffered somewhat during patterning, especially with films using IPA as the solvent. Films made with TBA showed also in this respect to be superior; they had lower refractive indices after patterning than the films made with IPA. Photoresist trapped in the pores showed also to be a problem with some of the films.

It was also demonstrated that aerogel films do not adhere to some metals, like Mo-N and aluminum films. In processing this can be solved by depositing an adherence layer. However, this feature can also be used as a benefit to avoid lithography steps and pattern aerogel selectively on areas not covered by the metal film.

4 CONCLUSIONS

It was shown that the reproducibility and the porosity of the films improved using TBA instead of IPA. Porosity was about 76% and refractive index 1.11 for 380 nm thick films. Carbon content was about 12-24 at-% in films prepared with IPA. Particle transfer between wafers during supercritical drying was recorded to be a problem, thousands of added particles were counted on a 100 mm wafer, Table 1. Films were patterned using standard lithography processes, but in some cases the photoresist clogged in the aerogel pores caused problems.

REFERENCES

[1] C.J. Brinker, G.W. Scherer, "Sol-Gel Science: The Physics and Chemistry of Sol-Gel Processing", Academic Press, San Diego, CA, 1990.

[2] S.S. Prakash, C.J. Brinker, A.J. Hurd, J. of Non-Crystalline Solids 190, 264-275, 1995.

[3] S.V. Nitta, V. Pisupatti, A. Jain, P.C. Wayner Jr., W.N. Gill, J.L. Plawsky, J. Vac. Sci. Technol. B 17(1), 204-212, 1999.

[4] S. Henning, L. Svensson, Phys. Scripta 23, 697, 1981.

[5] L.W. Hrubesh, J.F. Poco, "Thin aerogel films for optical, thermal, acoustic and electronic applications", J. of Non-Crystalline Solids 188 46-53, 1995.

[6] N. Henelius, O. Anttila, H. Ronkainen, J. Molarius, "A new method for simultaneous characterization of process cleanliness and true particle removal efficiency ", 184th Electrochemical Society Meeting, 514 – 515, 1993.

[7] S.H. Hyun, T.Y. Kim, G.S. Kim, H.H. Park, "Synthesis of low-k porous silica films via freeze drying", J. of Mat. Sci. Let. 19(20), 1863-1866, 2000.

[8] S.-K Fan., C.-J. Kim, "MEMS with thin-film aerogel", 14th Int. IEEE conf. MEMS 2001, 122-125, 2001.

Controlling the Photoactivity of Nanoparticles

P. Casey[*], S. Boskovic, K. Lawrence and T. Turney

Commonwealth Scientific and Industrial Research Organisation
Division of Manufacturing and Infrastructure Technology
Locked Bag 33, Clayton South, Victoria 3169, Australia
* phil.casey@csiro.au

ABSTRACT

The relative photoactivity of a series of nanophase ZnO powders with varying crystallite size (20-140nm) was determined using a chemical method based on the photobleaching behaviour of the stable radical 1,1-diphenyl-2-picrylhydrazyl (DPPH). It was confirmed that photoactivity behaviour increased (one to twofold) with decreasing crystallite size for crystallites <100nm. A further series of nanophase ZnO powders doped with metal ions showed that, depending on the type and/or level of dopant, the photoactivity behaviour of nanoparticles could be reduced by more than an order of magnitude.

Keywords: nanoparticle, photoactivity, zinc oxide, doped

1. INTRODUCTION

Nano-particulate oxides such as ZnO and TiO_2 are increasingly being used as pigments and UV absorbers in personal care products (e.g. sunscreens), coatings and paints, predominantly because their absorbance efficiency increases with decreasing particle size. However, whilst protection against UV may be maintained at smaller particle sizes, the rate at which hydroxyl radicals are generated increases, due to the inherent photo-activity of these materials. Consequently, in the case of personal care products, there may be a negative effect on skin cells due to this increased photocatalytic activity. There is little published data that correlates photo-activity of nanoparticles with biological effects, however in vitro work has shown that supercoiled DNA is indeed damaged in the presence of nano-particulate metal oxides and the rate of unwinding itself can be used as a measure of the photoactivity of the metal oxide [1,2].

In personal care applications, to prevent possible damage to skin cells, the photoactivity of these materials needs to be quenched.

Methods by which this may be achieved are by altering the band gap, capping the sites where surface hydroxyl radicals are generated or encapsulating the material by coating the particles with an insulating layer.

Alteration of the band gap is of increasing interest and may be achieved by lowering the energy gap relative to the original system by the introduction of dopants. The majority of published data on the use of dopants to control photoactivity has been with titania [3,4]. A series of undoped and doped nanosized ZnO materials designed to quench photoactivity has been tested for photoactivity (using chemical methods) when exposed to UV. Parameters such as crystallite size, dopant type and level have been considered.

The test method used to evaluate the photoactivity was a modified method of that reported by Dransfield et al. [5] that follows the photobleaching of the stable radical 1,1-diphenyl-2-picrylhydrazyl (DPPH). The method is relatively simple and requires little expensive equipment compared to conventional methods based of the photocatalytic oxidation of propanol [6] or the more recently reported unwinding of supercoiled DNA [1,2]. It is quite sensitive to parameter changes, highly reproducible and is able to be performed in a short period of time as compared to other simple methods looking at pigment colour degradation over months of exposure to UV.

2.   EXPERIMENTAL
2.1   Characterisation of Materials

A series of undoped and doped zinc oxide was prepared by proprietary methods and characterised chemically and physically. Chemical analysis was performed using Inductively Coupled Plasma Atomic Emission Spectroscopy (ICP-AES) methods. Crystallite phase was determined by using a Bruker ASX-D8 X-Ray Diffractometer using CuKα radiation over a 2θ range of 5° to 85° with a step size of 0.02°. Crystallite size was determined by performing a Rietveld refinement of the diffraction data using Siroquant™ Version 2.5 software.

2.2   Test Protocol

The method established and implemented for determining the photoactivity of metal oxide particles such as ZnO and TiO_2 was that initially proposed by Dransfield et al. [5]. It is a colorimetric test that follows the photobleaching of the stable radical 1,1-diphenyl-2-picrylhydrazyl (DPPH). A measure of the photoactivity is determined by the time it takes to reduce the DPPH radical, which is initially purple in colour, to its reduced form, which is yellow.

Test Procedure

A 1:1 weight mixture of mineral oil (white, heavy) and capric caprylic triglyceride was prepared. This was the dispersing medium in which two further mixtures were prepared:

Mixture 1   0.031 g of the metal oxide (ZnO or doped ZnO) was added to 62.5 ml of the dispersing medium and the mixture was stirred magnetically for over 30 minutes to disperse the metal oxide nanoparticles.

Mixture 2   0.0052 g of the DPPH was added to 62.5 ml of the dispersing medium and the mixture stirred magnetically for over 30 minutes to dissolve the DPPH radical.

These two mixtures were then combined in an open glass vessel and stirred magnetically. The vessel was placed 15 cm directly under an irradiation source (Spectroline BIB-150P UV lamp) which emitted predominantly in the UVA region with a steady state output of 6 mW/cm² at an irradiation distance of 38 cm. The irradiation source could be isolated from the vessel by means of a removable barrier. This allowed the UV lamp to be turned on an hour before experimental irradiation exposure so as to let the lamp reach a steady state output. Prior to removal of the barrier and exposure to the irradiation source, 5 ml of the test sample was withdrawn and an absorption spectrum was recorded in a 10 mm quartz cuvette

using a Cary UV-Vis spectrophotometer. The sample was then returned to the bulk dispersion before the isolating barrier was removed to begin the experiment. At regular intervals the isolating barrier was replaced and further absorption spectra were recorded to monitor the photocatalysed reduction and colour change of the DPPH radical.

To quantify photoactivity the value of the absorbance was monitored at 520 nm (the position of a peak in the absorbance spectra due to the purple DPPH radical). Photoactivity was determined by the time it took to change from an initially high absorbance (due to the purple DPPH radical) to a lower baseline absorbance (the yellow colour of the reduced DPPH radical). The reciprocal of the time taken for radical decomposition was defined as the Photoactivity Index (PI) whose units are reciprocal time (min^{-1}).

3.   RESULTS

Comparative XRD results of some selected doped and undoped ZnO samples are presented in Figure 1. Results of chemical analyses, crystallite size and chemical photoactivity rating for the series of oxides examined are presented in Table 1 and Figures 2.1a and b.

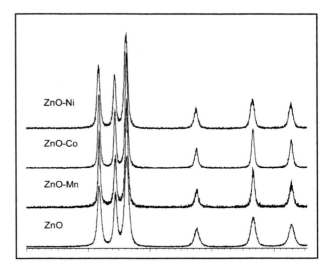

Figure 1. Comparison of XRD traces for doped and undoped ZnO

Compound	Crystallite Size (nm)	Time taken for decay of DPPH radical (min)	P I (min^{-1})
ZnO	24	12	0.083
Dopant			
Fe (0.44wt%)	14	30$^+$	<0.033
Ni (0.64wt%)	20	100	0.01
Co (0.7 wt%)	16	150$^+$	<.0066
Mn (1.1wt%)	21	150	0.0066
Dopant Level (wt%Mn)			
0.2		30	0.0333
0.3	23.3	38	0.0263
0.6	23	95	0.0105
0.8	18.3	220	0.0045

Table 1. Summary of Chemical analyses, Crystallite Size and Photoactvity Index (PI)

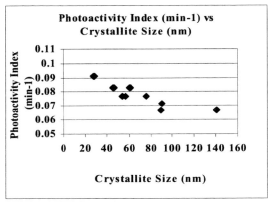

Figure 2.1a Photoactivity Index versus Crystallite Size

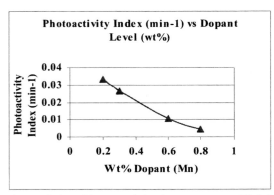

Figure 2.1b Photoactivity Index versus Weight % of dopant (Mn)

4. DISCUSSION

The characteristic XRD pattern for all the materials examined was that of ZnO (see Figure 1). An average crystallite size, determined by Rietveld refinement, varied between ~20 and 140nm. For ZnO this crystallite size may be effectively interpreted as primary particle size if efficient dispersion in a hosting medium can be achieved.

Photoactivity is expected to vary directly with surface area such that as surface area increases so too does photoactivity. A change in particle diameter from 80 to 40nm represents a 2-3 fold increase in surface area /unit weight. A consequent increase in photoactivity of the same order would be expected.

In the case of undoped zinc oxide there appeared to be two distinct regions of photoactivity behaviour. For crystallite sizes < ~ 80nm photoactivity was linearly dependant on crystallite size but above this it became independent of crystallite size. For example, the photoactivity index for crystallite sizes between 40 and 90nm decreased from ~.092 to ~0.07 whilst from 90-140nm it remained essentially unchanged (0.068-0.071 *cf* 0.068). The increase in photoactivity for particles from 80-40nm was in the order of 1-2 fold, far less than the expected 2-3 fold for the associated crystallite size change (based on surface area).

While these experimental results reflect the expected relationship between surface area and photoactivity the changes are less substantial than predicted. Possible reasons for this are that (i) other intrinsic or extrinsic material factors, either physical and/or chemical, besides surface area diminish or influence photoactivity (ii) the conditions under which tests were performed affect either the photoactivity or the measurement of photoactivity.

It is unlikely that intrinsic material factors vary significantly between samples of undoped ZnO and any subsequent effect on photoactivity due to these factors would be insignificant. Chemical analysis, XRD and inspection by SEM showed no changes in chemical composition, crystallite phase or particle shape or morphology between samples. Extrinsic factors may be a more likely cause. Nanoparticles, because of their high surface area, have a great propensity to agglomerate and their successful dispersion as discrete particles and non-reagglomeration in hosting media for targeted applications remains one of the greatest challenges in harnessing the efficient use of their functionality. At greater degrees of dispersion the effective surface area increases and consequently photoactivity would be expected to rise. To determine whether extrinsic factors such as dispersion effects could influence photoactivity measurements, a trial was performed where a sample of undoped ZnO (~30nm) was dispersed using a high shear mixer (Ultra Tarax T50) as opposed to magnetic stirring. The Photoactivity Index rose from 0.083 (magnetically stirred) to 0.125min^{-1} (high shear mixing). This suggests that the degree of dispersion and/or agglomeration influences photoactivity measurements. To compare in house dispersed material to that independently dispersed, two samples of similar size ZnO (~30nm), supplied as sunscreen concentrate dispersions from different commercial suppliers, were also tested. In these cases the photoactivity index (dispersed using magnetic stirring) was the similar to that of ZnO mixed in house using the high shear mixer i.e ~0.125min^{-1}.

Whilst photoactivity measurements will be influenced by the efficiency of dispersion, as suggested here, the degree by which it rises under these test conditions is still below that expected based on surface area. Possible reasons for this are (i) there is a degree of agglomeration remaining that diminishes photoactivity (ii) the relationship between photoactivity and crystallite size cannot be strictly applied but can only be used as a guide for interpreting the experimental measurements. It is unlikely that the ZnO samples used here are of a singular particle size but are more likely to be distributed across a range of sizes. Rietveld refinements calculate an average crystallite size but make no determination of size distribution or population balance.

In considering whether there were variations in the conditions under which tests were performed that affect either the photoactivity or the measurement of photoactivity an experiment on a selected sample was repeated. In this case an identical Photoactivity Index was obtained. When photoactivity was low (doped ZnO), dispersions could be exposed for long periods of time (>> 30 minutes cf 8-12mins for undoped ZnO) to the UV lamp with a consequent possibility that temperature rises may influence the stability of the radical. An experiment was performed where a dispersion of the radical alone was exposed to the UV lamp for an extended period. No degradation occurred even though the temperature increased by ~ 5-10°C.

The photoactivity of ZnO may be quenched by the incorporation of a dopant material. The main factors, besides crystallite size, that influence the subsequent

photoactivity reduction produced in doped materials are the type and level of dopant.

All dopants used in this study were effective in significantly reducing photoactivity (see Figure 3). The most effective appeared to be Mn, Co and Ni with Fe being less effective than these. Depending on the level of dopant, photoactivity could be reduced by 0.5-1.0 order of magnitude from that exhibited by ZnO at a similar crystallite size. Other dopants (not reported here) were tested and displayed less significant changes in photoactivity than Mn, Ni, Co and Fe.

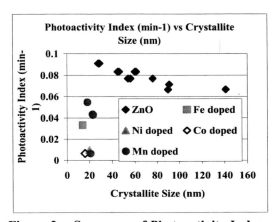

Figure 3   Summary of Photoactivity Index versus Crystallite Size

To determine whether the level of doping is important, a series of Mn doped ZnO was prepared at dopant levels ranging from 0.2wt% to 0.8wt%. From this series, the level of dopant appeared to be determinant in the degree of subsequent photoactivity with a clear relationship indicated between increasing dopant level and decreasing activity. This is illustrated in Figure 2.1b where, as the level of dopant is increased from 0.2 to 0.8wt%, the photoactivity index drops from 0.0333 to 0.00454min^{-1} (almost and order of magnitude) whilst crystallite size remains relatively constant ~ 20 nm (see Table 1).

When the level of dopant is determined on an atomic ratio basis (dopant: zinc) what appears to emerge is a linear relationship between at% doping and photoactivity up to a ratio of 1 but above which further reductions in photoactivity appear independent of at% (see Figure 4). For the series of dopants used here, this raises the question whether it matters which of these is used. Mn, Fe, Co and Ni are all transition elements with d orbital electrons, similar ionic charge and atomic radii and so would be expected to behave in a similar manner. It is not surprising that they appear to follow a general similarity in chemical and photochemical behaviour.

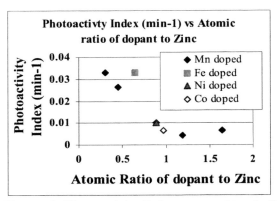

Figure 4. Photoactivity Index versus Atomic Ratio of dopant to Zinc

In general then, the photoactivity of nano phase zinc oxide has been found to increase with decreasing particle size but may be controlled by the incorporation of selected dopants. The current study has not reported on how or where the dopant has been incorporated into ZnO but of the three possibilities suggested in the introduction it is assumed that the band gap has been altered.

There is some data in the literature that correlates photoactivity and biological effects. In particular, photodynamic DNA strandbreaking activity of metal oxides on supercoiled DNA which has been found to correlate with the photocatalytic oxidation of propan-2-ol [4,5]. These methods have focussed mainly on titania and at particle sizes much larger than the ones examined here.

The method presented here is useful as screening guide in assessing the photoactivity behaviour of nanoparticles before they are considered for use in personal care and other applications. Preliminary biological trials (the uncoiling behaviour of DNA under exposure to UV) using materials selected from this study show good correlation between the level of photoactivity reported here and biological behaviour.

It is proposed therefore that the protocol of performing simple chemical photoactivity tests such as that presented here be used as a screening technique for the development of designer nanoparticles in personal care

applications. Correlation with biological activity such as DNA unravelling can then be determined.

REFERENCES

1. Takao Ashikaga, Masayoshi Wada, Hiroshi Kobayashi et al. Effect of photocatalytic activity of TiO_2 on plasmid DNA. Mutation Research 466 (2000) p1-7
2. Hisao Hidaka, Satoshi Horikoshi, Nick Serpone, John Knowland In vitro photochemical damage to DNA, RNA and their bases by an inorganiuc sunscreen agent on exposure to UVA and UVB radiation. Journal of Photochemistry and Photobiology A: Chemistry 111 (1997) p 205-213
3. Sivalingam, G; Nagaveni, K; Hegde, MS, et al. Photocatalytic degradation of various dyes by combustion synthesized nano anataseTiO2. Appl Catal B-Environ, 45 (1): 23-38 Sep 25 2003
4. Hirano, M; Nakahara, C; Ota, K, et al. Photoactivity and phase stability of ZrO2-doped anatase-type TiO2 directly formed as nanometer-sized particles by hydrothermal conditions. J Solid State Chem, 170 (1): 39-47 Jan 2,2003
5. Dransfield, G.; Guest, P.J.; Lyth, P.L.; McGarvey, D.J.; Truscott. T.G. *Journal of Photochemistry and Photobiology B: Biology* 2000, 59, 147.
6. Irick,G,Jr. Determination of the Photocatalytic Activities of Titanium Dioxides and other White Pigments. J. Applied Polymer Science., 16,2387 (1972)

Phase Morphology Control of the Electrospun Nanofibers from the Polymer Blends

M. Wei*, B. Kang**, C. Sung**and J. Mead*

*Department of Plastics Engineering
**Department of Chemical Engineering
University of Massachusetts Lowell
One University Ave, Lowell, MA 01854, USA, Joey_Mead@uml.edu

ABSTRACT

Nanofibers of polybutadiene/polycarbonate blends were electrospun from polymer solutions. The effect of composition on the resulting internal morphology of the nanofibers was studied using Transmission Electron Microscopy (TEM). When the composition ratio of PB/PC blends was larger than 75/25, a co-continuous structure was formed, but when composition ratio of PB/PC blends was changed to 25/75, a core-sheath structure was formed. The effect of type of polymer in the blend on the phase morphology was also investigated. The solubility of the components in the electrospinning solvent was found to play a critical role in phase morphology of nanofibers. Larger solubility differences in components of the blend favor the formation of core-sheath structures, while smaller solubility differences lead to co-continuous structures.

Keywords: electrospinning phase separation blend morphology

1 INTRODUCTION

Electrospinning offers the ability to produce polymer fibers with diameters in the nanometer range.[1,2,3] The high surface area/volume ratio endows nanofibers with enhanced properties for a variety of applications including tissue growth, protective clothing, thermal insulation, filters for fine particles, reinforcing fibers, wound dressings, and artificial blood vessels.[4,5] Most of the past and current research efforts have focused on fiber formation mechanisms, electrospinning of different polymer types and blends, and control of mat properties, such as fiber size, type, process, etc.[6,7] The electrospinning of polymer blends offers the potential to prepare unique materials, but has received more limited investigation. The focus has been primarily on the properties of the overall mat. Blends of polyaniline and polyethylene oxide or sulfonated polystyrene and polyethylene oxide have been used to produce conductive and photo-responsive nanofibers for electronic devices. Nanofibers prepared from Styrene-Butadiene-Styrene (SBS) triblock copolymers showed a two-phase morphology with domains elongated along the fiber axis.[8] Annealing the nanofibers improved the ordering of the domains of the nanofibers. Co-continuous phase morphologies were observed in electrospun polylactide /polyvinylpyrrolidone blends and specific surface topologies or fine pores were generated by selective removal of one of the components.[4] There has been little effort made towards control of the internal morphology of the nanofiber itself. Therefore, the focus of this paper was to study the effect of composition ratio of polymer blend and polymer types in the blends on the resulting morphology within the electrospun fiber. A number of unique morphologies were found, including core-sheath and co-continuous structures.

2 EXPERIMENTAL

2.1 Materials

Polycarbonate (PC) with molecular weight 21.9 kD was supplied by GE Plastics. Polybutadiene (PB) with molecular weight 420 kD, Polymethyl methacrylate (PMMA) with molecular weight 120 kD, Polystyrene (PS) with molecular weight 280 kD, tetrahydrofuran. (THF), osmium tetroxide (OsO_4) and ruthenium tetroxide (RuO_4) were purchased from Aldrich. Table 1 presents the physical properties of the polymers used in this research. Polymer blend solutions of PB/PC with 90/10, 75/25, 65/35, 50/50, 65/35, 25/75 and 10/90 wt% in THF, as well as PMMA/PC, PS/PC, PB/PMMA and PB/PS blends with 25/75 wt% were prepared for electrospinning experiments.

2.2 Electrospinning

The electrospinning apparatus consisted of a high voltage power supply (Gamma High Voltage Research Co.), a digitally adjusted syringe pump (Harvard Apparatus, PHD 2000, South Natick Co.), a steel syringe needle, and an electrically grounded aluminum foil target. Polymer solution was fed through the needle tip by the syringe pump at a controlled flow rate. As the electrical field between needle tip and target increased, nanofibers were formed and collected on the target. The applied voltage was 15 kV, the flow rate of the polymer solution was 0.02 ml/min, and the distance between the needle tip and the target was 20 cm in all cases.

	PB	PC	PMMA	PS	THF
Critical surface tension (γ_c) (mN/m)	42	32	39	33	
Solubility parameter (δ) (MPa$^{1/2}$)	17.2	19.4	19.0	18.6	18.6
Brookfield viscosity (η) at 6% in THF (mPa·s)	320	13	11	24	
Mw (kD)	420	21.9	120	280	

Table 1: Physical properties of polymers.

2.3 Characterization Methods

The morphology of nanofibers was observed using a Transmission Electron Microscope (TEM, Philips EM 400 T). For the preparation of TEM samples, the nanofibers were collected on a carbon coated copper specimen grid, followed by staining in OsO$_4$ or RuO$_4$ vapor by suspending them over a 4.0% aqueous solution of OsO$_4$ or RuO$_4$ for 30 minutes.

3 RESULTS AND DISCUSSION

The phase morphology of polymer blends depends on a number of factors. The composition, interfacial tension, processing conditions, and rheological properties of the components will all influence the resulting morphology of the blends.[9] In the electrospinning process, solvent removal takes the mixture from a homogeneous solution across the phase boundary, until two phases are formed. This process of solvent evaporation occurs at time scales in the millisecond range, during the whipping motion of the electrospinning jet.[2,4] As a result, the rapid fiber structure forming process, coupled with the phase separation process may result in the formation of unique fiber structures.[2] Due to the high surface to volume ratios in nanofibers, interfaces play a key role. In this paper, polycarbonate and polybutadiene, which have large solubility parameter differences, as well as large viscosity differences, were selected for initial studies. By varying the composition ratio of PB/PC blends, we can study the effect of composition ratio on the resulting phase morphology of PB/PC blends within the electrospun nanofibers. Variations in the compositions of PB/PC blends, as well as blends of PB/PMMA, PB/PS, PMMA/PC, and PS/PC were electrospun to study the factors that had the greatest effect on the phase morphology of nanofibers.

Figure 1 shows the TEM images of PB/PC blends with different weight ratios from 90/10, 75/25, 65/35, 50/50, 65/35, 25/75 to 10/90 after staining in OsO$_4$ vapor for 30 minutes. OsO$_4$ is known to preferentially stain the polybutadiene phase, so that the dark regions in the TEM images are identified as the polybutadiene phase, and the light regions are the

polycarbonate phase.[4] It was observed that there existed a change in morphology around 25/75 weight ratio of PB/PC blends. When the weight ratio of PB/PC blends was larger than 25/75, a co-continuous structure with interconnected PB and PC nanolayers or strands was formed and the nanolayers were oriented along the fiber axis. When the weight ratio of PB/PC blends was increased to 25/75, the nanofibers presented large layers of only one material and evidence of core-sheath structures with PB located in the center and PC located outside.

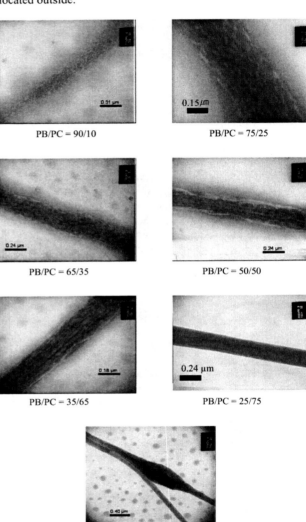

PB/PC = 90/10

PB/PC = 75/25

PB/PC = 65/35

PB/PC = 50/50

PB/PC = 35/65

PB/PC = 25/75

PB/PC = 10/90

Figure 1. TEM images of electrospun PB/PC nanofibers with different composition ratios.

The formation of these structures is dependent on the ability of the two materials to phase separate during the rapid solvent evaporation process, which may not allow coarsening of the phase separated regions and thus preserving a very fine phase morphology.[4] Orientation of the nanolayers is a result of the elongation of the fiber during the whipping process.

The phase separation mechanism involved with rapid solvent evaporation is very similar to that involved with spinodal decomposition.[10] The formation of the unique core-sheath structure may be due to a combination of factors, including viscosity and solubility of the polymers. Although the viscosity of the starting solutions are the same, as the solvent evaporates the viscosity of the system will be lower for the blends with high PC content, because of the increasing fraction of the lower viscosity PC. This decreased viscosity promotes the development of a core sheath structure by two processes. The first is involved with rheological processes and the effect of the viscosity ratio of compositions. Typically, in capillary flow of polymer blends, the lower viscosity material will tend to migrate to regions of highest shear rate. For the capillary flow of a polymer solution in the needle of the electrospinning equipment, the lower viscosity PC component would like to encapsulate the higher viscosity PB component according to the minimum energy dissipation theory of a flow system.[3] With increasing PC weight fraction, the decreased viscosity of the whole blend system will give sufficient mobility for this rheological process to occur. Thus, the PC phases will be able to migrate to the outside of the fiber, while the PB phases are located at the center of the fiber forming a core-sheath structure. The second effect is due to solvent effects, or the solubility of the components in the common solvent. THF is a poorer solvent for PC than PB because both PC molecules and THF are nucleophilic, although PC has a lower solubility parameter difference than PB. Therefore, with the evaporation of solvent, the different solubilities of PB and PC in THF would be likely to cause the PC to coalesce, leading to the formation of a core-sheath structure.[11] The process of phase separation of PC from PB will also be dependent on the viscosity of the whole blending system, i.e. the mobility of the molecules at high solution concentration as the solvent evaporates.

To further study the influence of the viscosity ratio, solubility of compositions in the common solvent, and the viscosity of the whole blending system on the phase morphology, we investigated several other polymer blend combinations. In the following experiments, we changed the composition of blends to investigate the effect of viscosity ratio and solubility of compositions on the phase morphology. In PMMA/PC and PS/PC blends, PMMA and PS, with much lower viscosity than PB, were used to replace PB in order to study the effect of viscosity ratio on the phase morphology. In PB/PMMA and PB/PS blends, PMMA and PS with better solubility in THF than PC were used to replace PC in order to study the effect of the solubility of components on the phase morphology.

Figure 2 shows TEM images of PMMA/PC, PS/PC, PB/PMMA and PB/PS at 25/75 wt%. PMMA/PC and PS/PC blends were stained by RuO_4 vapor for 30 minutes to darken the PC phase, and PB/PMMA and PB/PS blends were stained by OsO_4 for 30 minutes to darken the PB phase. It was observed that PMMA/PC and PS/PC present core-sheath structures, but PB/PMMA and PB/PS blends present co-

continuous structures. Table 2 presents the molecular weight, approximate interfacial tension, solubility parameter difference, and resulting phase morphology of the electrospun fibers. We found that even with the large viscosity ratio in the PB/PMMA and PB/PS blends, they still presented co-continuous structures, indicating that the viscosity ratio was not the dominant factor for the formation of core-sheath structures. The good solubility of PMMA and PS in THF solvent, leads to the development of co-continuous structures. On the other hand, we observed that any blends having PC in the composition presented core-sheath structures, even with small differences in the viscosity ratio. The interfacial tension and solubility parameter differences between the polymers appear to have little effect on the morphology. Clearly, PC played a key role in the formation of core-sheath structures at certain composition ratios.

4 Conclusions

Core-sheath structures were observed in PB/PC, PMMA/PC, and PS/PC blends and co-continuous structures were observed in PB/PMMA and PB/PS blends at 25/75 wt% by TEM. The effect of interfacial tension, viscosity ratios, and solubility on nanofiber internal morphology were studied using different polymer blend systems. It was found that solubility differences of the polymer components in the electrospinning solvent played a key role in the resulting phase structure of the nanofiber. Larger solubility differences led to the formation of core-sheath structures. Smaller solubility differences for the compoents of the blends leads to the formation of co-continuous structures.

5 Acknowledgements

The authors of this paper would like to thank the support of the National Science Foundation under grant number DMI-0200498.

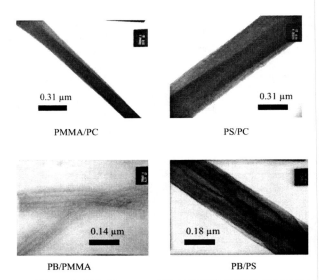

PMMA/PC   PS/PC

0.31 μm   0.31 μm

0.14 μm   0.18 μm

PB/PMMA   PB/PS

Figure 2. TEM images of PMMA/PC, PS/PC, PB/PMMA and PB/PS blends at 25/75 wt%.

	PB/PC	PB/PMMA	PB/PS	PMMA/PC	PS/PC
Mw (kD/kD)	420/21.9	420/120	420/280	120/280	280/21.9
$\Delta\gamma c$ (mN/m)	10	3	9	7	1
$\Delta\delta^*$ (MPa$^{1/2}$)	2.3	1.8	0.6	0.4	1.6
$\Delta\delta^{**}$ (MPa$^{1/2}$)	1.4/0.8	1.4/0.8	1.4/0	0.4/0.8	0/0.8
$\Delta\eta$ (mPa·s)	307	309	296	2	11
Morphology	Core-sheath	Co-continuous	Co-continuous	Core-sheath	Core-sheath

*: The solubility parameter difference between the components of blends
**: The solubility parameter difference between the components of blends and THF solvent.

Table 2 The parameters of for blend compositions and phase morphology.

References
[1] Pawlowski, K. J.; Belvin, H. L.; Raney, D. L.; Su, J., Polymer, 2003, 44, 1309.
[2] Walters D. A.; Ericson L. M.; Casavant M. J., Appl. Phys. Lett. 1999, 74, 3803.
[3] Salem, D. R., Structure Formation in Polymeric Fibers, Hanser Publishers, Munich 2000.
[4] Bognitzki, M.; Frese, T.; Steinhart, M., Polym. Eng. and Sci., 2001, 41, 982.

[5] Norris, I. D.; Shaker, M. M.,; Ko, F. K., MacDiarmid A. G., Synthetic Metals, 2000, 114, 109.
[6] Doshi, J.; Reneker, D. H., J. Elec., 1995, 35, 151.
[7] Baumgaarten, P. K., J. Coll. And Int. Sci., 1971, 36, 71.
[8] Fong, H.; Reneker, D. H., J. Polym. Sci B: Polym. Phys., 1999, 37, 3488.
[9] Paul, D.R., "Fibers from Polymer Blends" in Polymer Blends, Vol. 2., ed. by Paul, D.R., and Newman, S., Academic Press, New York, NY 1978, Ch. 16.
[10] Lee, J. K., Han, C. D., Polymer, 1999, 40, 2521
[11] Walheim, S., Boltau, M., Mlynek, J., Krausch, G., Steiner, U., Macromolecules, 1997, 30, 4995.

Conjugation of DNA to Streptavidin-coated Quantum Dots for the Real-time Imaging of Gene Transfer into Live Cells

J. K. Kim[*], S. H. Lim[*], Y. Lee[**], Y. S. Shin[**], C. Chung[*], J. Y. Yoo[**] and J. K. Chang[*, ***]

[*]Digital Bio Technology, Co., Seoul 151-742, Korea
[**]School of Mechanical and Aerospace Engineering, [***]School of Electrical Engineering and Computer Science, Seoul National University, Seoul 151-742, Korea

ABSTRACT

We have developed the method for the conjugation of biotinylated DNA to streptavidin-coated quantum dots (QDs). QD-DNA conjugates and a highly sensitive fluorescence imaging technique are adopted to visualize gene transport across the membrane of the live cell in real time. Endocytotic cellular uptake of oligonucleotide is monitored by a real-time confocal imaging system. Long-term kinetic study enables us to reveal the unknown mechanisms and rate-limiting steps of extracellular and intracellular transport of DNA. Gel electrophoresis is used to verify the effect of incubation time and the molar ratio of QDs to DNA on the conjugation efficiency. It is possible to fractionate the QD-DNA conjugates according to the DNA concentration and obtain the purified conjugates by a gel extraction technique. QD-DNA conjugates have a potential to be nanoscale building blocks by self-assembly process as well as a versatile tool for fluorescence imaging and monitoring of biological systems.

Keywords: quantum dot, conjugation, gene transport, live cell imaging, gel electrophoresis

1 INTRODUCTION

Fluorescent semiconductor nanocrystals, quantum dots (QDs), have significant advantages over the classical organic dyes with their unique properties and they are expanding application fields in life sciences [1]. DNA labeling by conjugation of QDs to DNA is the first step for studying kinetics of DNA transfer from the extracellular to the intracellular space through live cell imaging techniques. Strategies for the conjugations of DNA to gold nanocrystals [2] or silanized semiconductor nanocrystals [3] have been developed by other groups. Jaiswal et al. [4] have developed procedures for using QDs to label live cells and applied them for long-term multiple color imaging. We have developed the method for the conjugation of biotinylated DNA to streptavidin-coated QDs. Optimized reaction condition is determined from an analogous scale-up conjugation model using micron-sized beads. QD-DNA conjugates and a high-sensitive fluorescence imaging technique are adopted to visualize gene transport across the membrane of the live cell in real time. Endocytotic cellular uptake of oligonucleotide, electrically-mediated plasmid DNA transfer into the live cell, and localization of DNA in the intracellular compartments are monitored by a real-time confocal imaging system. Elucidating the rate-limiting steps and deciphering the mechanism of gene delivery to the cell nucleus are crucial for the understanding of transfection as well as for the development of gene therapy tools.

2 MATERIALS AND METHODS

2.1 QD-DNA Conjugation

To obtain purified conjugates, careful considerations are required for attaching fluorescent nanocrystals to biological macromolecules covalently. We designed experimental protocols to conjugate the biotinylated oligonucleotide or the plasmid DNA to commercially available streptavidin-coated QDs (Qdot™ 605 Streptavidin Conjugate, Quantum Dot Corp., CA, USA). Biotinylated oligonucleotides (ACE/F-biotin) were purchased from Bioneer (Daejeon, Korea). Each strand has a biotin group at the 5′ end. We used angiotensin-converting enzyme (ACE) gene. (5′-CTG GAG ACC ACT CCC ATC CTT TCT-3′) which has 24 base pairs and its estimated total length is about 8 nm. Biotinylation of 4.7 kb pEGFP-N1 plasmid (Clontech, CA, USA) was realized by three different approaches; biotin labeling by nick translation with Klenow enzyme and random hexamer, by end-filling with Klenow enzyme, and by adding one or more deoxynucleotide on to the 3' terminus of a DNA molecule with terminal deoxynucleotidyl transferase (TDT). Schematic of QD-DNA conjugation is shown in Figure 1. Streptavidin-coated QDs diluted by 1/100 were mixed with biotinylated DNA and incubated at room temperature.

2.2 Electrophoretic Fractionation of QD-DNA Conjugates

Inappropriate reaction condition for QD-DNA conjugation resulted in complex formation by aggregation. Gel electrophoresis was used to verify the effect of incubation time and the molar ratio of QDs and DNA on the conjugation efficiency. The concentration ratio of QD to DNA solutions was varied from 1:20 to 1:0.025 resulting in 10 mixtures. The QD-DNA conjugates were diluted in loading buffer and were run in 0.5 × tris-borate-EDTA buffer on a 2% agarose gel at 10 V/cm. The gels were post stained with ethidium bromide to illuminate the DNA. A polaroid camera was used to acquire the fluorescence images of QDs and DNA illuminated with an ultraviolet transilluminator. Fractionated conjugates were obtained by standard gel extraction technique.

2.3 Scale-up Model of Avidin-Biotin Conjugation

To avoid drying effects occurring in the use of transmission electron microscopy (TEM) for characterizing the nanoscale structures, we investigated DNA-mediated assemblies of 2-μm beads coupled with avidins and 0.5-μm beads conjugated with ACE/F-biotin using confocal microscopy. Carboxylate-modified FluoSpheres beads purchased from Molecular Probes (OR, USA) were diluted to final concentration of 0.5% with one ml PBS buffer. They were coated by incubation with 0.2 mg/ml EDC and 0.2 mg/ml NHS at room temperature to facilitate coupling with avidins. The molar ratios of avidins coupled with 2-μm beads to ACE/F-biotin conjugated with 0.5-μm beads were 1.67 μM:1.6 μM, 1.67 μM:16 nM, 16.7 nM:1.6 μM, and 16.7 nM: 16 nM. A control system was made by mixing of the bidisperse beads that were not functionalized.

2.4 Cell Preparation

SKOV and 293 cells were grown in Dulbecco's modified Eagle's medium supplemented with 10% (v/v) heat-inactivated fetal bovine serum (FBS, Sigma), penicillin (100 units/ml), streptomycin (100 μg/ml) and L-glutamine (4 mM) at 37°C in a humidified 5% CO_2 incubator. Cells were dissociated from the 25-cm^2 tissue culture flask by using Trypsin-EDTA. The final cell suspension density was adjusted to 1×10^4 cells/ml. The cells were transferred to 8-well chambered coverglass (Lab-Tek II, Nalge NUNC International, IL, USA) and monitored by an imaging system after QD-DNA conjugates were added to each well.

2.5 Confocal Imaging System

Confocal fluorescence images of QD-DNA conjugates translocated into live cells were taken using CARV spinning-disk confocal system (Atto Bioscience, MD, USA) equipped with an Olympus IX71 fluorescence microscope and a Sensicam cooled CCD camera (Cooke, Mich, USA) as shown in Figure 2. Color images were acquired by a 3-CCD camera (AW-E300, Panasonic, CA, USA). The QDs fluoresce at any excitation wavelength in the visible range and emit at 605 nm. Excitation light from a 100 W mercury lamp was filtered by a 460-500 nm bandpass filter and then reflected by a dichroic mirror and focused through a 100×/1.4 NA objective into the sample. The fluorescence emission from the QDs was filtered with a longpass filter which transmits light above 510 nm. 640 × 480 pixel images were obtained by the cooled CCD camera in 2 × 2 binning mode with appropriate background rejection.

3 RESULTS AND DISCUSSION

Fig. 3 shows the gel electrophoresis migration pattern. We can observe the colocalized patterns of QDs and DNA by comparing the fluorescent gel images of QD and QD-DNA conjugates, which indicates that most DNA is specifically bound to the QDs. There was almost no existence of the complex formation even after overnight incubation. It is possible to fractionate the QD-DNA conjugates according to the DNA concentration and obtain the purified conjugates by a gel extraction technique.

Figure 4 represents confocal fluorescence images of assembly structures in suspensions comprised of 2.0 and 0.5-μm beads coupled with avidins and ACE/F-biotin, respectively. Self-assembled structures were compared with aggregates in the control system. The DNA-driven assemblies in the two-bead system were significantly affected by the concentration of avidins and highly aggregated patterns were observed at the control system.

Z-series confocal images of the cultured SKOV cells are shown in Figure 5. Lower right corner image is overlay of the bright-field and the confocal fluorescence images. Due to the high photo stability and high brightness of QDs we could acquire time-lapse fluorescent DNA images for a long time and analyze them quantitatively by appropriate image processing schemes. Figure 6 shows bright-field (upper row) and confocal fluorescence (lower row) images of the cultured cells uptaking QD-oligonucleotide conjugates by endocytosis. Highly localized patterns of QDs are observed predominantly in perinuclear region. It should be noted that the distribution of QD-DNA conjugates was asymmetric around the cell nucleus. Figure 7 shows z-series confocal fluorescence images of the 293 cell. The complicated cellular structures, such as a nucleus and vesicles, could be discriminated by quantum dot labeling and confocal imaging.

4 CONCLUSION

We have designed efficient protocols for the conjugation of biotinylated oligonucleotides or plasmid to streptavidin-coated QDs and have observed the localization patterns of DNA in live cells using real-time confocal fluorescence microscope system. Long-term kinetic study by monitoring multiple colored images of high spatial and temporal resolution will enable us to reveal the unknown mechanisms and rate-limiting steps in extracellular and intracellular gene delivery process.

REFERENCES

[1] W. J. Parak, D. Gerion, T. Pellegrino, D. Zanchet, C. Micheel, S. C. Williams, R. Boudreau, M. A. Le

Gros, C. A. Larabell and A. P. Alivisatos, Nanotechnology, 14, R15-R27, 2003.

[2] D. Zanchet, C. M. Micheel, W. J. Parak, D. Gerion and A. P. Alivisatos, Nano Letters, 1, 1, 32-35, 2001.

[3] W. J. Parak, D. Gerion, D. Zanchet, A. S. Woerz, T. Pellegrino, C. Micheel, S. C. Williams, M. Seitz, R. E. Bruehl, Z. Bryant, C. Bustamante, C. R. Bertozzi, and A. P. Alivisatos, Chem. Mater., 14, 2113-2119, 2002.

[4] J. K. Jaiswal, H. Mattoussi, J. M. Mauro and S. M. Simon, Nat. Biotech., 21, 47-51, 2003.

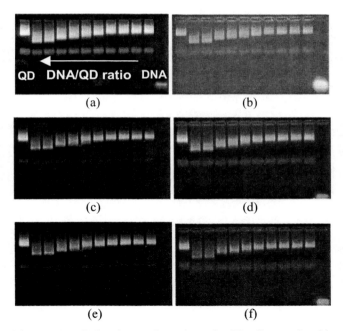

Figure 3: Gel electrophoresis of QD-oligonucleotide conjugates. (a) 1 h incubation, (b) 1 h incubation, EtBr stained, (c) 1 h incubation, (d) 2 h incubation, EtBr stained, (e) 24 h incubation, (f) 24 h incubation, EtBr stained.

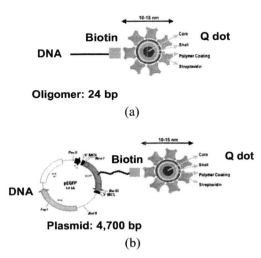

Figure 1: Schematic diagram of the conjugation of streptavidin-coated quantum dots to biotinylated DNA: (a) QD-oligonucleotide, (b) QD-plasmid.

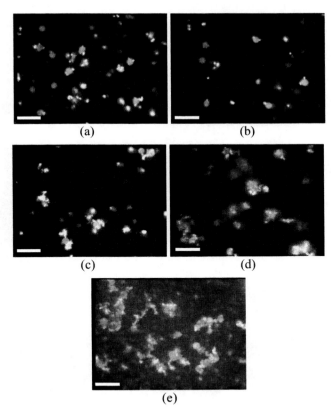

Figure 4: Confocal fluorescence images of assembly structures (a-d) in suspensions comprised of 2.0 and 0.5 μm beads coupled with avidins and ACE/F-biotin, respectively. (e) control system. Bar, 10 μm.

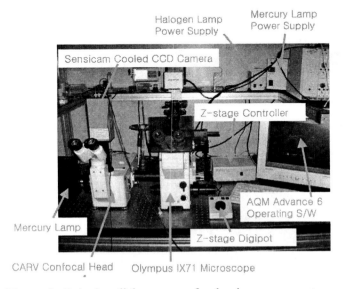

Figure 2: Spinning-disk type confocal microscope system for the real-time imaging of gene transfer into live cells.

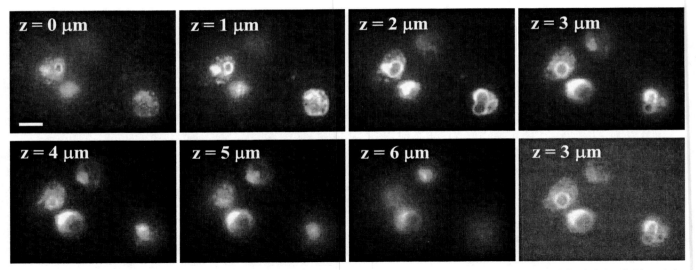

Figure 5: Z-series confocal images of the cultured SKOV cells. Lower right corner image is overlay of the bright-field and the confocal fluorescence images. Bar, 10 μm.

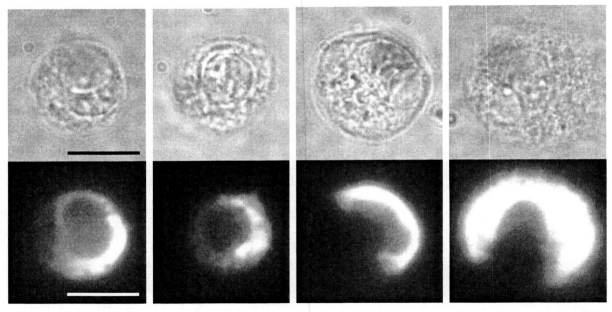

Figure 6: Bright-field (upper row) and confocal fluorescence (lower row) images of the cultured cells uptaking quantum dot-oligonucleotide conjugates by endocytosis. Highly localized patterns of quantum dots are observed in perinuclear region. Bar, 10 μm.

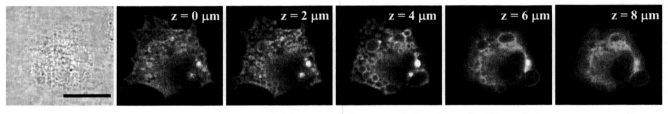

Figure 7: Bright-field (left) and z-series confocal fluorescence images of the 293 cell. The cellular structures, such as a nucleus and vesicles, are discriminated by quantum dot labeling. Bar, 10 μm.

Adsorption of PTCDA on Si(100) and Si(100)-H

T.Soubiron, F.Vaurette, B.Grandidier, X.Wallart, J.P.Nys, D.Stiévenard,

Institut d'Electronique , de Microélectronique et de Nanotechnologie, IEMN (UMR 8520, CNRS), Dpt ISEN, 41 Bd Vauban, 59046 Lille Cédex (France)

ABSTRACT

Submonolayer coverages of perylene-tetracarboxylic-dianhydride (PTCDA) molecules deposited on Si(100) and Si(100)-H surfaces have been investigated with scanning tunneling microscopy (STM) and monolayer coverages by X-ray photoelectron spectroscopy (XPS). At monolayer coverage, X-ray measurements indicate that both on Si(100) and Si(100)-H surfaces, most of the molecules do not react with the surface. At lower coverage, the STM images show that the molecules adsorb mostly in a planar configuration with its axis parallel to the Si dimers. Only a few molecules appear to be chemisorbed. Finally, we compare one ML deposition on both surfaces. A complete coverage is observed on Si(100), associated with a low mobility of the molecules, whereas islands are formed on Si(100)-H, associated with a high mobility of the molecules.

Keywords : Silicon surface, PTCDA, STM, XPS

1 INTRODUCTION

In the general development of the molecular electronic devices, the growth of two dimentionnal molecular films on inorganics substrates is of major interest because of possible application as organic LEDs [1], organic field effect transistors [2], non-linear optical devices [3], …Thin layers of 3,4,9,10-perylene-tetracarboxylic dianhydride (PTCDA) deposited on metals (Cu(110) [4,5], Au(111) [6], Au(100) [7], Ag(111) [8] and semiconductors surfaces (GaAs(100) [9] and InAs(001) [10] have been mainly studied by means of scanning tunnelling microscopy (STM) or low energy electron diffraction (LEED). PTCDA is generally found to grow in a herringbone fashion [7].

In this paper, we investigate, using both STM and X photoemission spectroscopy (XPS) the interaction of the PTCDA molecules with Si(100)-2x1 and hydrogenated Si(100)-2x1 surfaces. The XPS technique allows to analyse the chemical nature of the molecule-surface bounding and the STM gives structural and conformational details of the molecules lon the surfaces.

2 EXPERIMENTAL and DISCUSSION

Experiments were performed in an UHV system containing different chambers with base pressure less than 10^{-10} Torr. The Si(100) surfaces, doped with P in the range of 0.004 to 0.006 Ωcm, were prepared by standard procedures [11]. To prepare hydrogen-terminated Si(100)-2x1 surface, the surface is exposed to atomic hydrogen obtained by cracking molecular hydrogen with a 1500°C W filament placed 5 cm away from the sample surface. The UHV chamber is backfilled with $2x10^{-6}$ T of H_2 during 3 min which corresponds approximativety to 360 L. During the hydrogen exposure, the sample is heated at 350-370°C to prevent dihydride formation and subsequent etching of the Si surface [12]. The PTCDA molecule (see figure 1) were deposited from a *home made* Knudsen cell held at 380 °C, at a rate of about ??? molayer per second. Prior to the STM experiments, the W tips were electrochemically etched, cleaned in UHV and their radius of curvature was checked in field emission. All STM images were taken in constant current mode. For XPS measurements, the samples were transferred to a XPS chamber using a clean box, under overpressure of dry nitrogen.

The chemical structure of PTCDA molecule is shown in Figure 1.

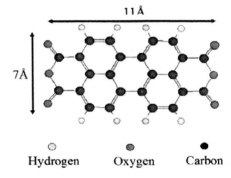

Figure 1 : Chemical representation of the PTCDA molecule

The XPS technique was first used in order to determine the chemical interaction between the molecules and the silicon surfaces (Si or Si-H). Figure 2 gives XPS spectra of the C1s levels for a monolayer coverage.

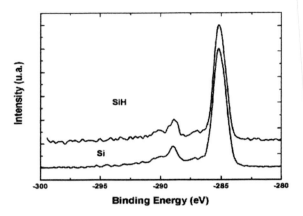

Figure 2 : XPS spectra of the C1s level associated with PTCDA deposited on Si and Si-H surfaces.

The main peak, located at 285.1 eV is associated with the carbon atoms of perylene group. The peak positioned at 289 eV is associated with the carbon atoms of the anhydride group. The main results are : i) the XPS spectra are similar, for both surfaces, ii) no Si-C peak is observed , showing that whatever the silicon surfaces, the molecules are mainly physisorbed, i.e. in low interaction with the surface.

In order to characterize individual molecules with the STM, sub-monolayer of PTCDA were sublimated on Si(100)-2x1 surfaces. Figure 3 is a typical STM image (V=-3.5 volts, I = 40 pA, 200 nm x 200 nm). The rows of silicon dimers are clearly apparent and the bright hillock are associated with the molecules. Taking the silicon surface as a reference for the dimension, the observed adsorbates have the size of the PTCDA molecules (typically 11 Å x 7 Å). Figure 4 and 5 are high resolution STM image (size 6.8 nm x 6.8 nm and 5.2 nm x 5.2 nm respectively) of individual molecule on Si(100)-2x1 surface. In Figure 4, the molecule is perpendicular to the silicon dimer rows (as illustrated on the schematic representation associated with the STM image) and is located over two adjacent dimers. Two bright lobes are observed , associated with the two perylene groups. At the corners of the adsorbate, four spots are observed, with a lower constrast than the main part of the hillock. These spots can be associated with the carboxylic groups. The chemical structure of the molecule seems to be unperturbated and we propose that this molecule is in low

interaction with the surface, i.e. is physisorbed, in agreement with the XPS results.

Figure 3 : STM image (V=-3.5 volts, I = 40 pA, 200 nm x 200 nm) of individual PTCDA molecules on Si(100)-2x1 surface

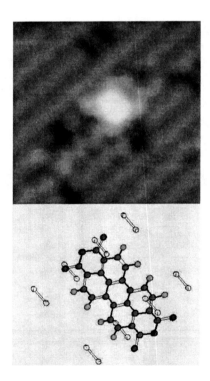

Figure 4 : High resolution STM image (6.8 nm x 6.8 nm, V = -2.5 volts, I = 40 pA) of individual PTCDA molecule

For a few percent of the molecules, the STM image is completely different, as shown in Figure 5.

In such case, four bright lobes are clearly observed. The spots aligned on a silicon dimmer row are separated by a distance of typically 1 nm, whereas the distance between two lobes perpendicular to a silicon dimer row is typically 7 Å. The center of the molecule exhibits a low contrast.

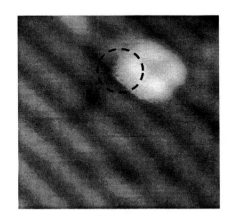

Figure 5 : High resolution STM image (5.2 nm x 5.2 nm, V = -3.5 volts, I = 80 pA) of individual PTCDA molecule

As shown in the schematic representation in Figure 5, below the STM image, we propose that the molecule is parallel to the silicon dimer rows and located between two adjacent rows. The molecule is chemisorbed (cycloaddition reaction [14])on the surface *via* the atomic binding of the oxygen atom of the anhydride groups and silicon surface atom (creation of a Si-O bond). Consequently, there is a modification of the chemical structure of the molecule (the π C=O bonds of the anhydride groups are broken and a new C=C is formed between the two carbon atoms of the anhydride group). The four associated silicon dimmers are also modified.

The π Si=S bond is broken. One silicon atom is bonded to an oxygen atom and a dangling bond is associated with the second silicon atom (dotted circles in Figure 5). As the energy level of the dangling bond (sp^3 level) is near the Fermi level of the surface [13], the dangling bond appears bright in the STM image.

Finally, we first achieve the deposition of one monolayer PTCDA films, on Si(100)-2x1 surfaces. Figure 6 is an STM image (85 nm x 85 nm, V = -3.5 volts, I = 40 pA) of a Si(100)-2x1 surface.

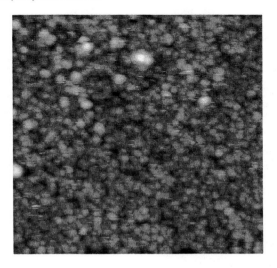

Figure 6 : STM image of Si(100)-2x1 surface after the deposition of one ML of PTCDA (V= -3.5 volts, I = 40 pA)

There is a uniform coverage of the surface, without any islands. Only few small clusters of molecules are observed. We deduce that the physisorbed molecules have a low diffusion behaviour on the silicon surface. As demonstrated by XPS results, there is no strong molecule-surface interaction, but the interaction is high enough to prevent any diffusion of the molecules. After the deposition of a 2 ML of molecules, there is the formation of islands, with a crystal - like structure as shown in Figure 7 where the top surface of an island exhibits a regular structure.

On the contrary, the deposition of one ML of PTCDA molecules on a hydrogenated Si(100) - 2x1 exhibits a complete different behaviour, as shown in Figure 8. In such a case, we observe the spontaneous formation of islands, associated with the two main bright features. They have an average lateral size of 12 nm and a height of 1.2 nm. The small features (a few nm size) are small clusters of molecules. We thus deduce that on such a surface, the mobility of the molecules is high. A detailed study of the kinetics of formation of the islands is under study.

Figure 7 : STM image (15 nm x 15 nm, V = -3.8 volts, I = 35 pA) of the top face of a PTCDA island after a 2 ML deposition on a Si(100)-2x1 surface.

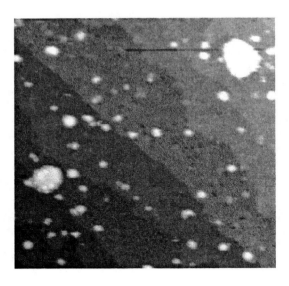

Figure 8 : STM image (85 nm x 85 nm, V = -3.5 volts, I = 40 pA) of a H-Si(100)-2x1 surface after the deposition of one monolayer of PTCDA molecules.

3 CONCLUSION

Using both XPS and STM techniques, we have shown that PTCDA molecules are mainly physisorbed on Si(100)-2x1 or H-Si(100)-2x1 surfaces. However, the mobility of the molecules is higher on H-Si surface compare to Si surface, leading to the creation of islands on H-Si surface even for one monolayer PTCDA deposition. The kinetics of formation of these islands is under study.

REFERENCES

[1] J.Blochwitz, M.Pfeiffer, T.Fritz and K.Leo, App.Phys.Lett. 73, 729, 1998
[2] Z.N.Bao, A.J.Lovinger, A.Dodabalapur, Adv. Mater. 9 (1), 42, 1996
[3] H.S. Nalwa, J.S. Shirk, in : Phthalocyanines, Properties and Applications", A.B.P. Lever (Eds), VCH, New York, 79, 1996
[4] M.Stöhr, M.Gabriel and R.Möller, Surf. Sci. 507, 330, 2002
[5] M.Stöhr, M.Gabriel and R.Möller, Europhys. Lett. 59, 423, 2002
[6] T.Schmitz-Hübsch, T.Fritz, F.Sellam, R.Staub and K.Leo, Phys.Rev. B 55, 7972, 1996
[7] T.Schmitz-Hübsch, T.Fritz, R.Staub, A.Back, N.R. Armstrong and K.Leo, Surf. Sci. 437, 163, 1999
[8] E.Umbach, K.Glöcker and M.Sokolowski, Surf.Sci. 402, 20, 1998
[9] C.Kendrick and A.Khan, Surf. Rev.Lett. 5, 289, 1998
[10] C.Kendrick and A.Khan, Appl.Surf.Sci. 123, 405, 1998
[11] K. Hata, J. Vac. Sci. Technol. A 18, 1933, 2000.
[12] J.J.Boland, Phys. Rev.Lett.65, 3325, 1990.
[13] J.J.Boland, Science 262, 1703, 1993.
[14] L.Fang, J.Liu, S.Coulter, X.Cao, M.P.Schwartz, C.Hacker and R.J.Hamers, Surf. Sci., 514, 362-375, 2002.

Nanoparticle Dispersions for Polishing Applications

Patrick G. Murray

Nanophase Technologies Corporation, Romeoville, IL, pmurray@nanophase.com

ABSTRACT

Increasingly, improvements in surface roughness and planarity are being sought in a variety of technologically advanced polishing applications related to the electronics industry. From the production of photomask blanks to the fabrication of microelectronic memory and process devices, improvements in global surface planarity and roughness, coupled with reduced incidents of defectivity, are critical to achieving the specifications of future technology nodes. Nanophase Technologies Corporation (NTC) uses a patented plasma arc synthesis technique to produce nano-sized cerium oxide (ceria) and aluminum oxide (alumina) particles in commercial quantities and quality. The small particle size, controlled particle size distribution, and unique crystal morphology of these materials are allowing new performance benchmarks to be realized in the planarization of silicon wafers and the polishing of photomask blanks, lithography optics and rigid memory substrates. However, the success or failure of a nanocrystalline material in a particular polishing application depends not only on the characteristics of the particle but also on the ability to form and maintain a stable dispersion of those particles with controlled rheology. Thus, both the attributes of the particles which provide utility in the above mentioned polishing applications as well as observations relevant to the production of stable dispersions are discussed.

Keywords: nanoparticle, ceria, alumina, polishing, CMP

1. INTRODUCTION

Nanophase Technologies Corporation produces nano-crystalline ceria and alumina in multi-ton quantities using a plasma-based vapor-phase synthesis (VPS). In the VPS process, a metal or metal oxide precursor is vaporized in a plasma arc and the evaporated precursor is allowed to combine with a reactive gas, usually oxygen. As the molecules of the newly formed metal oxide begin to associate, nanoparticles condense from the vapor phase and are then rapidly cooled so as to preclude sintering. Fortunately, the nanoparticles associate with each other through weak electrostatic forces in the dry state, which permits collection of the product as a low bulk density powder.

Many primary particle attributes which are derived from the VPS process influence the behavior and performance of these particles when they are used for demanding polishing applications typically associated with advanced electronic applications. The primary particle size, the particle morphology, the particle density and the particle surface chemistry are all set during the VPS process. These particle attributes, along with specific downstream dispersion processing operations, render these nanocrystalline materials particularly suitable for the following polishing applications: Chemical Mechanical Planarization (CMP), including Shallow Trench Isolation (STI), copper interconnect and interlayer dielectric planarization, and final polishing of extreme ultraviolet lithography (EUVL) photomask blanks, optics for lithographic processes, and rigid memory storage substrates with increased aerial density.

2. PARTICLE ATTRIBUTES

The primary particle size is set during the vapor phase synthesis. A TEM image of the VPS nanocrystalline ceria is shown in Figure 1, and a TEM image of the VPS nanocrystalline alumina is shown in Figure 2.

Figure 1. TEM image of VPS produced ceria

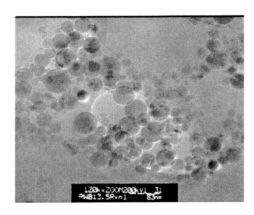

Figure 2. TEM image of VPS produced alumina

As can be observed in the TEM images, the nanocrystalline ceria and alumina are equiaxed and nanoscale in every dimension. The VPS process does not produce any high aspect ratio particles, such as needles or platelets, which may be nanoscale in one dimension, but sub-micron in another. The morphology obtained for a given nanocrystalline material is determined by thermodynamics; ceria forms sharp-edged, multi-faceted, cubic crystals while the alumina particles are spherical.

The chemical purity of the particles produced by the VPS process is extremely high, since high purity raw materials can be sourced, nothing else is added except high purity oxygen, and no by-products or waste streams are generated. This situation may be contrasted with particles produced by fumed or precipitated processes, wherein the products are often contaminated with various rogue ions such as chlorides or sulfates.

The density of the nanocrystalline particles obtained from the VPS process is near theoretical. For polishing applications, this lack of porosity results in non-friable particles which do not change size or shape during a polishing operation, resulting in consistency and reproducibility in removal rates and other performance figures of merit.

The VPS process allows for a great degree of control over the chemistry at the particle surface. For example, as will be discussed subsequently, the particle surface chemistry can be "tuned" with the plasma process so that a high positive zeta potential (+45 mV is typical) results when the particles are dispersed in water. This high positive zeta potential allows concentrated, stable dispersion to be prepared which can be used directly for polishing or subjected to further chemical or physical treatments.

Perhaps most importantly, the particles which form under the VPS synthesis conditions are discrete, individual nanocrystals which, once properly dispersed, do not form secondary aggregate or agglomerate structures as is common with silica and milled ceria. This property translates directly into consistent removal rates and reduced defectivity and sub-

surface substrate damage during polishing. For example, dispersions of the VPS nanocrystalline ceria at 20 weight percent which are over two years old show no evidence of particle size growth.

3. DISPERSION ATTRIBUTES

The success or failure of a nanocrystalline material in a particular polishing application is highly dependent upon the ability to prepare and maintain a stable dispersion. This enables improvements in surface roughness to be achieved along with a corresponding reduction in surface defects and residual particles, both of which are necessary whether the application is microelectronic wafer planarization or lithographic optics polishing.

In moving toward this objective, it is necessary to tailor the zeta potential of the particles to accommodate a wide range of pH environments to ensure that the particles provide sufficient removal rate while being strongly resistant to flocculation during polishing.

Nanocrystalline ceria and alumina made by the VPS process can be dispersed in water such that the particles carry a very high positive zeta potential because of the unique surface chemistry imparted to the particles during VPS synthesis. In these cases, the natural pH of the dispersions is acidic and well below the isoelectric pH of the materials. These dispersions are most widely used in CMP applications and other metal polishing applications. Because of the high positive zeta potential, stable dispersions at concentrations of 20 to 25 weight percent are readily prepared. Even dispersions as concentrated as 50 weight percent have been prepared for specific applications.

Once the nanocrystalline alumina or ceria has been dispersed down to individual, fully solvated primarily particles, a variety of further complimentary physical and chemical processing can be accessed. For example, oftentimes in sensitive polishing applications it is desirable to remove a small fraction of the larger particles which can contribute to scratches or defects on the substrate. This truncation operation, virtually impossible to perform in the dry state, is straightforward and controllable with the dispersion given the proper equipment. From a polishing perspective, this means that the particle size distribution can be controlled to a performance specification. Normally, since only a fraction of the whole lognormal particle size distribution is removed in such an operation, the mean particle size does not change, but instead only the standard deviation about the mean particle size is reduced. In polishing, this means that the removal rate is unaltered, but typically the defectivity is reduced and the surface roughness which can be achieved is improved.

An additional benefit of a stable dispersion is that a variety of performance additives such as buffers, chelants, dispersants, and lubricants can be added. Depending on the polishing application, such additives are often necessary to convert a dispersion of abrasive particles to a ready-to-use polishing slurry.

As an example, a typical particle size distribution for an aqueous dispersion of NTC's VPS produced ceria is shown in Figure 3 at pH 4.0. As shown, the particle size distribution is narrow and centered around 100 nm (volume-weighted average).

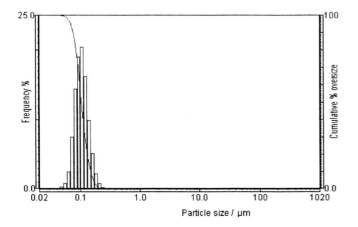

Figure 3. Mean volume weighted particle size histogram for post-processed VPS ceria in water at pH 4.0.

Even though a stable dispersion can be formed at pH values below the isoelectric points without any additional chemical dispersant, formulating with a dispersant under these conditions can extend the life of the slurry in the polishing application by making the dispersion less sensitive to silicates, metal ions, and other species which will increase in concentration during polishing. Moreover, the presence of a dispersant, particularly a polymeric dispersant, can aid in post-polishing cleaning and reduce the amount of residual slurry particles which may remain behind on the substrate.

Most glass polishing, on the other hand, is conducted at an elevated pH, generally with ceria, since glass is softened at higher pH. Examples of this would include the preparation of photomask blanks and the finish polishing of lithography optics and rigid memory storage media discs. However, as the pH of the dispersion is raised, and the isoelectric point of the particle is approached, flocculation of the dispersion is most likely. This is to be avoided at all costs, since poor surface finish and high levels of defects will be observed on the substrate. An example of this is shown in Figure 4, where the larger, agglomerated particles evident in the distribution could be expected to negatively impact all important process metrics.

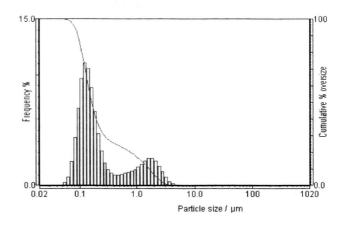

Figure 4. VPS ceria dispersion after crossing the isoelectric pH without the appropriate particle surface treatment

Fortunately, when the appropriate particle surface treatment is employed, a stable dispersion on the other side of the isoelectric point (at pH 8.0) is achieved, as shown in Figure 5. Under these circumstances, the particles will have a highly negative zeta potential (-45 mV is typical) and the dispersion will be stable based on the electrostatic repulsion of the particles.

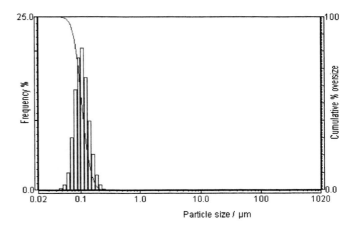

Figure 5. VPS ceria dispersion in water, pH 8.0, surface treated particle

4.   CONCLUSIONS

Nanocrystalline ceria and alumina produced using Nanophase Technologies Corporation's VPS process have unique attributes which render them especially suitable for use in polishing applications. As examples, the morphology, density and primary particle size of these materials have been instrumental in achieving angstrom level surface roughness with no subsurface damage in the polishing of photomask blanks [1] and in setting new industry benchmarks for selectivity and reduced defectivity the fabrication of microelectronic devices [2].

However, the successful application of these nanocrystalline materials in high performance polishing applications requires an understanding of those factors which influence dispersion stability in the environment in which the particles are to be used. The ability to control zeta potential, both at particle inception and in dispersion, has proven to be a key enabling capability in this regard.

5.   REFERENCES

[1] Murray, P. G., *et. al.*, "Nanocrystalline Ceria Dispersions for Ultrafine Polishing and Defect Reduction in EUVL Photomask Blank Manufacture," Optifab 2003, May 19-23, 2003, Rochester, NY.

[2] "Rodel Solidifies Position in Direct STI Market," Press Release, Rodel, Inc., December 3, 2003.

Growth Kinetics of Gold Nanoparticles in Reverse Micelles

A.P. Herrera*, E. Vicu a*, C. Rinaldi*, M. Castro**, L. Sol ***, R. Irizarry*** and J.G. Briano*

*Chemical Engineering Department, University of Puerto Rico
Mayagüez PR 00680, crinaldi@uprm.edu, jbriano@uprm.edu
**Chemistry Department, University of Puerto Rico
Mayagüez PR 00680, mcastro@uprm.edu
*** DuPont Electronics Microcircuits Industries, Ltd.
Manatí PR 00674-3000, Roberto.Irizarry@PRI.dupont.com

ABSTRACT

Gold nanoparticles were formed from the reduction of gold (III) by sulfite in reverse micelles and mo nitored with a stopped flow analyzer. The gold and the reducing agent micelles were formed in isooctane and AOT (as surfactant). The overall reduction of gold (III) occurred in less than 20 [ms]. The growth kinetics was followed from measurements of the time-dependent wavelength absorption spectra, and by analyzing the evolution of the dipole resonance. The dielectric constant of the medium was calculated, and the growth results were modeled with Mie's scattering theory. The modeling assumes the mean size of the particles to be characterized by the position of the plasmon. Excellent agreement between theory and experiment are found at the peak maximum and long wavelength tail, for particle sizes up to 90 [nm]. The dependence of the gold dielectric constant on particle size and the effect of multi-pole interactions were taken into consideration.

Keywords: micelles, growth kinetics, gold nanoparticles

REVERSE MICELLES

A wide range of techniques for the synthesis and characterization of gold nanoparticles have been developed in the past years with the objective of controlling the size and shape of the resulting nanoparticles. Examples include reduction of Au^{+3} in aqueous solution [1, 2], ultrasonic mixing of the reactants [3], ultraviolet photo-activation techniques [4], and preparation in reverse micelles [5-10], among others. Characterization of the resulting nanoparticles can be performed using UV-VIS absorption spectroscopy, Transmission Electron Microscopy (TEM), and also Surface Second Harmonic Ge neration (SSHG).

Reverse micelles consist of aqueous nanometer-sized droplets that are separated from the bulk organic phase by a surfactant layer (figure 1). This is an isotropic and thermodynamically stable single-phase system that consists of three components: water, organic solvent, and surfactant [11].

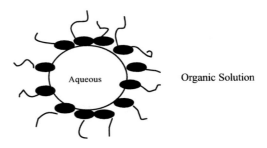

Figure 1: Illustration of a reverse micelle.

A surfactant molecule lowers the interfacial tension between water and oil resulting in the formation of a transparent solution as the two phases intermingles. Using reverse micelles for nanostructure synthesis takes advantage of the small size of the micellar water pools, which essentially act as nanoreactors. These nanoreactors collide continuously due to thermal Brownian motion, hence a small fraction of droplets exist in the form of short-lived dimmers which exchange their water content. After a short time these dimmers separate to form two droplets (as illustrated in Figure 2). As a result of this coalescence and separation mechanism, molecules in the water pools are redistributed over the micellar population [12].

The size of the water pool inside the reverse micelle greatly influences the size of the resulting nanoparticles. Thus, changing the size of the water pool provides a mechanism to control the final nanoparticle size [13]. The final size of the water pool has been described to depend on the water-surfactant molar ratio, $w = [H_2O]/[AOT]$. In order to obtain reverse micelles in the water/AOT system, this ratio must be below 15 [14, 15]. By using Aerosol OT [Sodium bis(2ethylhexyl) sulfosuccinate], Na(AOT), as the surfactant, and isooctane as solvent, it has been demonstrated that the water pool diameter R is related to the water content w of the droplet by [14, 15]:

$$R_w \quad 3V_{Aq} \frac{H_2O}{S} \qquad (1)$$

Where V_{Aq} is the volume of water molecules, s is the area per polar head, and S is the concentration of the surfactant AOT. With the definition $w = [H_2O]/[AOT]$, equation (1) can be expressed in the form:

$$R_w(\text{Å}) \quad 1.5w \qquad (2)$$

As illustrated in Figure 2, synthesis of nanoparticles by using reverse micelles, involves the preparation of two identical water-in-oil microemulsions. Reactants A and B, are dissolved in the aqueous phases of these two microemulsions. Upon mixing, collisions and droplet coalescence bring the reactants A and B into contact to form the AB precipitate. This precipitate is confined to the interior of the reverse micelle droplets and the size and shape of the particle formed reflects the interior of the droplet [16]. Since the compositions of the two microemulsions (I and II) are identical, differing only in the nature of the aqueous phase, the microemulsion is not destabilized upon mixing. One of the disadvantages of using this method is that it is necessary to use large quantities of solvent and surfactant to produce a relatively small amount of nanoparticles. Also, the formation of the reverse micelle depends on many factors including temperature, mixing time, mixing type and order, and the way in which precursors are added to the reaction vessel.

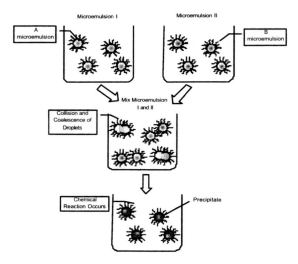

Figure 2: Schematic representation of synthesis of nanoparticles in reverse micelle.

OPTICAL PROPERTIES

Optical properties of gold nanoparticles have been intensively investigated during recent years, particularly their dependence on size [1, 2]. Absorption spectra of colloidal dispersions of metals exhibit broad regions in the UV-Visible range due to the excitation of the plasmon resonance or inter-band transitions, characteristic of the metallic properties of the material [17]. The UV-Visible absorption spectra of a fairly dilute dispersion of colloidal particles can be calculated from Mie's Theory [2, 18, 19]. This theory was developed for a single spherical particle, assuming the dielectric constant to be independent of the particle size [18].

The extinction coefficient C_{ext} is related to the Mie scattering coefficients a_n and b_n through

$$C_{ext} \quad \frac{2}{x^2} \sum_{n=1}^{\infty} (2n \quad 1)[\text{Re}(a_n \quad b_n] \qquad (3)$$

where x = $dn_0/$, n_0 is the refractive index of the host medium, is the wavelength of the incident light *in vacuo*. a_n and b_n are the scattering coefficients, which are functions of the particle diameter, d, and of the wavelength in terms of Ricatti-Bessel functions. For particles with diameters less than 100 [nm], only the first three terms in equation (3) are usually needed. In the case of gold, an excellent agreement is obtained for particle sizes higher than 40nm in aqueous medium The discrepancy for smaller sizes is due to the unknown dependency of the dielectric constant with size. The position of the maximum absorption is directly related [1, 2] to the mean size of the particles as showed on Figure 3. Our experimental data were obtained from commercial colloids which have narrow size distributions, but important enough to explain any deviations from Mie's model.

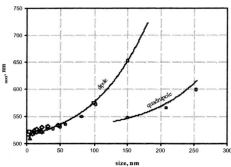

Figure 3. Position of the maximum absorption peak, versus particle size, for gold particles in aqueous medium. The solid lines are simulations using Mie's equation and the symbols are absorption measurements. Experimental data is shown as: squares [Galleto, P., et al., 1999 [20]], triangles up [Logunov, S.T., et al., 1997 [21]], triangles down [Stephan, L., and M. A. El-Sayed, 1999 [22]], diamonds [Sau, T.K., et al., 2001 [23]], and circles [Viera, O., 2002 [24]].

When using micelles, the optical properties of the metallic particle will be influenced by the dielectric function of the medium which depends on the nature of the organic solvent, and of the surfactant, as well as on the size of the micelles. To estimate the effect of the dielectric constant on the position of the plasmon resonance, we prepared reverse micelles using commercial gold colloids at w ranging from 5 to 20. Our preliminary study on this effect is summarized on Figure 4, where the shift in the plasmon position may be seen as function of w.

In this preliminary study, we see that the plasmon peak seems to move to the lower energy range as the micelles get smaller. For $w = 5$, there is a shift of about 9 nm to the right for all particle sizes. As micelles get larger, the optical properties approach that of the aqueous medium case.

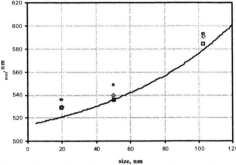

Figure 4. Position of the maximum absorption peak versus particle size of gold particles, in reverse micelles. The solid line is a numerical simulation using Mie's equation for aqueous medium. Small circles correspond to $w = [H_2O]/[AOT] = 5$, diamonds to $w = 10$, and squares to $w = 15$.

GROWTH KINETICS

L.E Murillo et al [1] have studied the formation of gold nanoparticles from aqueous solution by reducing $AuCl_4$ with sulfite, the reduction/nucleation/growth process took less than 1 s, and required a particularly fast technique to follow the optical changes; for this reason, a Stopped Flow Reactor (SFR) was used. In this work, the specific optical properties of the metallic particles, and the gold plasmon resonance were followed in the range of 10 to 100 nm, by using Mie's theory, as developed for a single spherical particle.

A typical time dependent spectrum for reduction and growth of nano-sized gold particles is presented in Figure 5. The initial reduction stage is characterized by the depletion of the 300 nm band (characteristic of gold III). Complete reduction of gold (III) was obtained in 5 ms. Plasmon resonance due to metallic behavior is associated with the peak in the 500 nm region. Figure 7, shows the computation of size evolution assuming the suspension to be statistically narrow [2].

To follow the formation of gold nanoparticles into reverse micelles, we used the same equipment than for the aqueous medium case. Since the system is confined into

these "nanoreactors", the process is slower and the particles growth up to a given smaller size. In fact, we were able to stabilize gold particles of any size by just changing w, or the relative concentrations of the reactants.

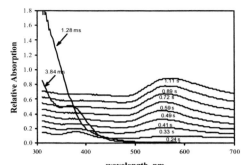

Figure 5: Gold particle growth in aqueous solution [1]

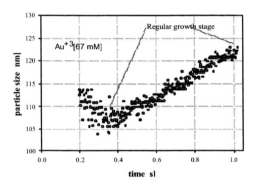

Figure 6: Growth kinetics of gold particles [1].

For similar concentrations in the water phases, the growth process to form particles of about 100 nm, goes for less than 1s, for the aqueous medium process, represented in Figures 5 and 6, to several minutes, for the process with reverse micelles, represented in Figures 7 and 8. Coalescence of micelles after reduction seems to be the limiting step in this process.

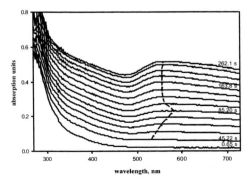

Figure 7: Gold Particle Growth in reverse micelles; Au^{+3} [0.038 mol/l], SO_3^{-2} [0.126 mol/l], $w = [H_2O]/[AOT] = 10$.

In addition, particle size reaches a maximum and then decreases (see Figures 7 and 8). This is apparently due to aggregation of particles that are covered by surfactant, and that easily separate later on.

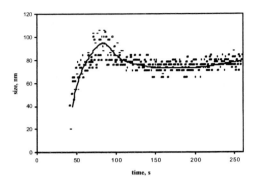

Figure 8: Gold Particle Growth in reverse micelles; Au^{+3} [0.038 mol/l], SO_3^{-2} [0.126 mol/l], $w= [H_2O]/[AOT] = 10$.

In the particular case indicated on the above figure, the particles reach a final size of about 80 nm. This microemulsion of gold particles remained stable for several weeks.

REFERENCES

1. Murillo, L.E; et al. Computational Nanoscience and Nanotech, 435-438, 2002

2. Vicu a, E.; et al. Nanotech, 3, 191, 2003.

3. Okitsu, K.; et al. Langmuir 17: 7717-7720, 2001.

4. Mandal, M.; et al. Bull. Mater. Science 25: 509-511, 2002.

5. Mandal, M.; et al. Langmuir 18: 7792-7797, 2002.

6. Feldheim, D., and C. Foss Jr., *Metal nano-particles: synthesis, characterization and applications.* Marcel Dekker Inc, New York, 2002.

7. Hainfield, J., and R. Powell. Journal of Histochemistry and Cytochemistry 48: 471-480, 2000.

8. Kreibig, U.; et al. Physical Stat. sol (a) 175:351-367, 1999.

9. Chen, F.; et al. Materials Letters 4325: 1-5, 2003.

10. Liu, S; et al. Langmuir 18: 8350-8357, 2002.

11. Feltin, N., and M.P. Pileni. Langmuir 13: 3927-3930, 1997.

12. Pileni, M.P. Journal of Physical Chemistry 97: 6961-6973, 1993.

13. Petit, C.; et al. Journal of Physical Chemistry (B) 103: 1805-1810, 1999.

14. M.P Pileni. Crystal Research Technology 33: 1155-1186, 1998.

15. Pileni, M.P; et al. Chemical Physics Letters 118: 414-420, 1985.

16. LLai, R.; et al. Journals of Magnetism and Magnetic Materials 163: 243-248, 1996.

17. Sau, K.T.; et al. Journal of Physical Chemistry B. 105: 9266-9272, 2001.

18. Mie, G, Ann. Phys., 25: 377, 1908.

19. Kreibig U. and M. Vollmer, *Optical Properties of Metal Clusters.* Springer, Berlin, 1995

20. Galletto P., Brevet, P., Girault, H., Antoine, R., and M. Broyer. J. Phys. Chem. B, 103: 8706-8710, 1999

21. Logunov S., Ahmadi,T., El-Sayed, M., Khoury, J., and R. Whetten,. J. Phys. Chem B, 101:3713, 1997.

22. Stephan, L., and M.A. El-Sayed, J. Phys.Chem. B, 103: 4212, 1999

23. Sau T.K., Pal A., Jana N.R., Wang Z.L., and T. Pal., J. Nanoparticle Res., 3: 257, 2001

24. Viera O., *Growing of Gold Nano-particles from Aqueous Solution.* MS thesis, University of Puerto Rico at Mayagüez, 2002.

Finite-Size Effects and Surface-Enhanced Raman Scattering from Molecules Adsorbed on Noble-Metal Nanoparticles

V. N. Pustovit, K. M. Walker and T. V. Shahbazyan

Department of Physics and Computational Center for Molecular Structure and Interactions,
Jackson State University, Jackson MS 39217

ABSTRACT

We study the role of a strong electron confinement on the surface-enhanced Raman scattering from molecules adsorbed on small noble-metal nanoparticles. We describe a novel enhancement mechanism which originates from the different effect that confining potential has on s-band and d-band electrons. We demonstrate that the interplay between finite-size and screening efects in the nanoparticle surface layer leads to an enhancement of the surface plasmon local field acting on a molecule located in a close proximity to the metal surface. Our calculations show that the additional enhancement of the Raman signal is especially strong for small nanometer-sized nanoparticles.

Keywords: nanoparticles, Raman scattering, surface plasmon

1 INTRODUCTION

An recent interest in single-molecule surface-enhanced Raman scattering (SERS) stems from the discovery of enormously high (up to 10^{15}) enhancement of Raman spectra from certain (e.g., Rhodamine 6G) molecules fixed at the nanoparticle surfaces in gold and silver colloids [1]–[3]. Major SERS mechanisms include electromagnetic (EM) enhancement by surface plasmon (SP) local field near the metal surface [4]–[8] and chemical enhancement due to dynamical charge transfer between a nanoparticle and a molecule [9]–[12]. Although the origin of this phenomenon has not completely been elucidated so far, the EM enhancement was demonstrated to play the dominant role, especially in dimer systems when the molecule is located in the gap between two closely spaced nanoparticles [10]–[13].

An accurate determination of the SERS signal intensity for molecules located in a close proximity to the nanoparticle surface is a non-trivial issue. The classical approach, used in EM enhancement calculations [4]–[8], is adequate when nanoparticle-molecule or interparticle distances are not very small. For small distances, the quantum-mechanical effects in the electron density distribution can no longer be neglected. These effects are especially important in noble-metal particles where the SP local field is strongly affected by highly-polarizable

(core) d-electrons. In the bulk part of the nanoparticle, the (conduction) s-electrons are strongly screened by the localized d-electrons. However, near the nanoparticle boundary, the two electron species have different density profiles. Namely, delocalized s-electrons spill over the classical boundary [14], thus increasing the effective nanoparticle radius, while d-electron density profile mostly retains its classical shape. The incomplete embedding of the conduction electrons in the core electron background [15]–[18] leads to a reduced screening of the s-electron Coulomb potential in the nanoparticle surface layer. The latter has recently been observed as an enhancement of the electron-electron scattering rate in silver nanoparticles [19], [20].

Here we study the role of electron confinement on SERS from molecule adsorbed on the surface of small Ag particles. To this end, we develop an EM theory for SERS in noble-metal particle, based on the two-region model [15]–[17], [21], which describes the role of the surface-layer phenomenologically. We find that the reduction of screening near the surface leads to an additional enhancement of the Raman signal from a molecule located in a close proximity to the nanoparticle. In particular, we address the dependence of SERS on nanoparticle size and show that the interplay of finite-size and screening effects is especially strong for small nanometer-sized particles.

2 TWO-REGION MODEL

We consider SERS from a molecule adsorbed on a spherical nanoparticle with radius R. For sufficiently small R, the retardation effects are negligible and the frequency-dependent potential is determined from the Poisson equation,

$$\Phi(\omega, \mathbf{r}) = \phi_0(\mathbf{r}) + \int d^3 r' \frac{\delta N(\omega, \mathbf{r}')}{|\mathbf{r} - \mathbf{r}'|}, \qquad (1)$$

where $\phi_0(\mathbf{r}) = -e\mathbf{E}_i \cdot \mathbf{r}$ is the potential of the incident light with electric field amplitude $\mathbf{E}_i = E_i \mathbf{z}$ along the z-axis, and $\delta N(\omega, \mathbf{r})$ is the induced charge density (hereafter we suppress frequency dependence). In noble-metal particles, there are four contributions to $\delta N(\mathbf{r})$ originating from the valence s-electrons, $\delta N_s(\mathbf{r})$, the core d-electrons, $\delta N_d(\mathbf{r})$, the dielectric medium, $\delta N_m(\mathbf{r})$, and the molecule, $\delta N_0(\mathbf{r})$. The density profile of delocalized s-electrons is not fully inbedded in the background of localized d-electrons but extends over that of localized d-electrons by $\Delta \sim 1 - 3$ Å[15]–[17]. Here we adopt the classical two-region model with d-electron density extending up to $R_d = R - \Delta$, where Δ is the thickness of the surface layer without d-electron population. The induced density is expressed as $\delta N(\mathbf{r}) = -\nabla \cdot \mathbf{P}(\mathbf{r})$, where the electric polarization vector $\mathbf{P}(\mathbf{r}) = \mathbf{P}_d + \mathbf{P}_s + \mathbf{P}_m + \mathbf{P}_0$ can be related back to the potential via

$$
\begin{aligned}
\mathbf{P}_d(\mathbf{r}) &= -\frac{\epsilon_d - 1}{4\pi}\,\theta(R_d - r)\nabla\Phi(\mathbf{r}), \\
\mathbf{P}_s(\mathbf{r}) &= -\frac{\epsilon_s - 1}{4\pi}\,\theta(R - r)\nabla\Phi(\mathbf{r}), \\
\mathbf{P}_m(\mathbf{r}) &= -\frac{\epsilon_m - 1}{4\pi}\,\theta(r - R)\nabla\Phi(\mathbf{r}), \\
\mathbf{P}_0(\mathbf{r}) &= -\delta(\mathbf{r} - \mathbf{r}_0)\,\alpha_0 \nabla\Phi(\mathbf{r}_0).
\end{aligned}
\tag{2}
$$

Here the step functions $\theta(x)$ enforce the appropriate boundary conditions, ϵ_d and ϵ_m are the core and medium dielectric functions, respectively, and

$$
\epsilon_s(\omega) = 1 - \omega_p^2/\omega(\omega + i\gamma),
\tag{3}
$$

is the Drude dielectric function of s-electrons, ω_p being the bulk plasmon frequency. The width $\gamma = \gamma_0 + \gamma_s$ includes contribution of the bulk-like electron-phonon scattering, γ_0, and of the electron surface scattering, $\gamma_s = g_s v_F / R$, where v_F is the Fermi velocity and g_s is a numerical factor ($g_s \simeq 1$ in noble-metal particles). The molecule is represented by a point dipole with polarizability α_0 located at $\mathbf{r}_0$ (we chose origin at the nanoparticle center).

To obtain a self-consistent equation for the potential $\Phi(\mathbf{r})$, we substitute the above expressions into $\delta N = \delta N_s + \delta N_d + \delta N_m + \delta N_0$ in the rhs of Eq. (1) and, after integrating by parts, obtain,

$$
\begin{aligned}
\epsilon(r)\Phi(\mathbf{r}) = {} & \phi_0(\mathbf{r}) - \nabla_0 \frac{1}{|\mathbf{r} - \mathbf{r}_0|} \cdot \alpha_0 \nabla_0 \Phi(\mathbf{r}_0) \\
& + \int d^3 r' \nabla' \frac{1}{|\mathbf{r} - \mathbf{r}'|} \cdot \nabla'[\chi_d \theta(R_d - r') \\
& + \chi_s \theta(R - r') + \chi_m \theta(r' - R)]\Phi(\mathbf{r}'),
\end{aligned}
\tag{4}
$$

where $\epsilon(r) = \epsilon_d + \epsilon_s - 1$, ϵ_s, and ϵ_m in the intervals $r < R_d$, $R_d < r < R$, and $r > R$, respectively, and $\chi = (\epsilon - 1)/4\pi$ are the corresponding susceptibilities. Since the incident field has the form $\phi(\mathbf{r}_0) = \phi(r_0)\cos\theta = -E_0 r \cos\theta$, we expand Φ and in terms of spherical harmonics and, after retaining only the dipole term ($L = 1$), obtain,

$$
\begin{aligned}
\epsilon(r)\Phi(r) = {} & \phi_0(r) - \frac{\epsilon_d - 1}{4\pi} R_d^2 \,\partial_{R_d} B(r, R_d)\,\Phi(R_d) \\
& + \frac{\epsilon_m - \epsilon_s}{4\pi} R^2 \,\partial_R B(r, R)\,\Phi(R) \\
& - \alpha_0 \nabla_0 B(r, r_0) \cdot \nabla_0 \Phi(r_0),
\end{aligned}
\tag{5}
$$

where

$$
B(r.r') = \frac{4\pi}{3}\left[\frac{r'}{r^2}\theta(r - r') + \frac{r}{r'^2}\theta(r' - r)\right]
\tag{6}
$$

is the dipole term of the radial component of the Coulomb potential, and we assumed that the molecule is located on the z axis.

The second and third terms in rhs of Eq. (5) originate from the scattering due to change of dielectric function at $r = R_d$ and $r = R$, respectively, while last term represents the potential of the molecular dipole. The values of Φ at the boundaries and at the molecule position can be found by setting $r = R_d, R, r_0$ in Eq. (5). In doing so, one determines a closed-form expression for the self-consistent potential in the presence of molecule, nanoparticle, and dielectric medium.

3 RAMAN SCATTERING

In the electromagnetic mechanism of SERS, the enhancement of the Raman signal comes from the far-field of the radiating molecular dipole and secondary field of this dipole scattered by the nanoparticle. In order to extract the Raman contribution, present the self-consistent potential as a sum $\Phi = \phi + \phi^R$, where ϕ^R is the potential of the radiating dipole. Since the molecular polarizability is very small, we restrict ourselves by the lowest order in α_0; in this case ϕ is the potential in the absence of molecule and ϕ^R determines the Raman signal in the first order in α_0. Inclusion of higher orders in α_0 leads to renormalization of molecular and nanoparticle polarizabilities due to image charges; these effects are not considered here.

Keeping only zero-order terms in Eq. (5), we find

$$
\phi = \varphi_0 + \delta\varphi,
\tag{7}
$$

where $\varphi_0 = \phi_0/\epsilon(r) = -E_0 r/\epsilon(r)$, and

$$
\begin{aligned}
\delta\varphi(r) = \frac{1}{\epsilon(r)}\Big[& -\beta(r/R_d)\phi_0(R_d)\frac{\lambda_d(1 - 2\lambda_m)}{1 - 2a^3\lambda_d\lambda_m} \\
& +\beta(r/R)\phi_0(R)\frac{\lambda_m(1 - a^3\lambda_d)}{1 - 2a^3\lambda_d\lambda_m}\Big].
\end{aligned}
\tag{8}
$$

Here $\beta(r/R) = \frac{3}{4\pi}R^2\partial_R B(r, R)$ is given by

$$
\beta(x) = x^{-2}\,\theta(x - 1) - 2x\theta(1 - x),
\tag{9}
$$

and

$$
\lambda_d = \frac{\epsilon_d - 1}{\epsilon_d + 3\epsilon_s - 1}, \qquad \lambda_m = \frac{\epsilon_m - \epsilon_s}{2\epsilon_m + \epsilon_s}.
\tag{10}
$$

The spatial dependence of the nanoparticle-induced potential, $\delta\varphi$, is thus determined by $\beta(x)$. Inside the particle, the potential linearly increases for $r < R_d$, while exhibiting a more complicated behavior in the surface layer, $R_d < r < R$. Outside of the nanoparticle, the induced potential falls off quadratically,

$$\delta\varphi(r) = \frac{E_i \alpha}{\epsilon_m r^2}, \qquad r > R, \qquad (11)$$

with $\alpha(\omega)$ is the particle polarizability

$$\alpha(\omega) = R^3\left[1 - \frac{(1+\lambda_m)(1-a^3\lambda_d)}{1 - 2a^3\lambda_d\lambda_m}\right]. \qquad (12)$$

In the absence of the surface layer, $R_d = R$, we recover the usual expression for the polarizability in the presence of d-electrons and dielectric medium,

$$\alpha^0 = R^3 \frac{\epsilon_s + \epsilon_d - 1 - \epsilon_m}{\epsilon_s + \epsilon_d - 1 + 2\epsilon_m}, \qquad (13)$$

with resonance peak at the SP energy $\omega_M = \frac{\omega_p}{\sqrt{\epsilon_d + 2\epsilon_m}}$. In the presence of the surface layer, the SP energy experiences a blueshift [15], [16] due an effective decrease in the d-electron dielectric function in the nanoparticle (see Fig. 3). At the same time, an increase in the peak amplitude (see Fig. 3), which accompanies the blueshift, indicates a stronger local field at resonance energy acting on a molecule in a close proximity to nanoparticle surface.

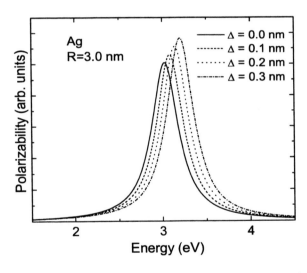

Figure 1: Calculated nanoparticle polarizability for different surface layer thicknesses.

The Raman signal is given by the self-consistent field of the radiating dipole, ϕ^R, determined from Eq. (5) in the first order in α_0,

$$\begin{aligned}\epsilon(r)\phi^R(r) &= \phi_0^R(r) - \frac{\epsilon_d - 1}{3}\beta(r/R_d)\phi^R(R_d) \\ &+ \frac{\epsilon_m - 1}{3}\beta(r/R)\phi^R(R),\end{aligned} \qquad (14)$$

where

$$\phi_0^R(r) = -\beta(r/r_0)\frac{4\pi\alpha_0}{3r_0^2}\frac{\partial\phi(r_0)}{\partial r_0} = S\tilde\phi_0, \qquad (15)$$

is the potential of the molecular dipole in the presence of local field, and we introduced notations

$$S(\omega) = \frac{8\pi\alpha_0}{3E_i r_0^3}\frac{\partial\phi(r_0)}{\partial r_0} \qquad (16)$$

and

$$\tilde\phi_0(r) = -E_i\left[r\theta(r_0 - r) - \frac{r_0^3}{2r^2}\theta(r - r_0)\right]. \qquad (17)$$

Note that while the frequency dependence of the local potential, $\phi(r_0)$, given by Eqs. (7,11), is that of the incident field, ω, the Raman field, $\phi^R(r)$, depends on the Stokes-shifted frequency, $\omega_s = \omega - \omega_0$, where ω_0 is the molecular vibrational frequency determined by α_0.

The last two terms in Eq. (14) describe potential of molecular dipole scattered from the d-electron and s-electron distribution boundaries and can be determined by matching ϕ^R at $r = R_d$ and $r = R_d$. We now notice that, at these values of $r < r_0$, we have $\tilde\phi_0(r) = \phi_0(r)$, so the Raman potential is found as

$$\phi^R = S(\omega)[\tilde\varphi_0 + \delta\varphi(\omega_s)], \qquad (18)$$

where $\tilde\varphi_0(r) = \tilde\phi_0(r)/\epsilon(r)$ and $\delta\varphi(r)$ is given by Eq. (8) but with ω_s instead of ω. For the far field ($r > r_0$), we then obtain

$$\phi^R(r) = \frac{E_0 r_0^3}{2\epsilon_m r^2}S(\omega)\left[1 + g(\omega_s)\right], \qquad (19)$$

where

$$g = \frac{\alpha}{r_0^3} = \left(\frac{R}{r_0}\right)^3\left[1 - \frac{(1+\lambda_m)(1-a^3\lambda_d)}{1 - 2a^3\lambda_d\lambda_m}\right] \qquad (20)$$

with λ given by Eq. (10). The expression for $S(\omega)$ in Eq. (raman-far-field) can obtained in the same manner from Eqs. (7,11) by substituting the nanoparticle electric field at the molecule location,

$$E(r_0) = -\frac{\partial\phi_0}{\partial r_0} = \frac{E_i}{\epsilon_m}\left[1 + 2g(\omega)\right], \qquad (21)$$

into Eq. (16). We finally arrive at

$$\phi^R(r) = -\frac{4\pi\alpha_0 E_0}{3\epsilon_m r^2}\left[1 + 2g(\omega)\right]\left[1 + 2g(\omega_s)\right]. \qquad (22)$$

The above expression generalizes the well-known classical result [4], [5] to the case of noble-metal particle with different distributions of d-electron and s-electron densities. The surface-enhanced Raman field retains the same functional dependence on the nanoparticle polarizability, however the latter contains all the information about the surface layer effect. Note that the form (22) of the Raman field remain unchanged even for a non-classical distribution of the electron densities provided that the electronic wave-functions in the nanoparticle do not oberlap with the molecular orbitals.

Finally, the Raman enhancement factor is given by

$$A(\omega, \omega_s) = \left| 1 + 2g(\omega) + 2g(\omega_s) + 4g(\omega)g(\omega_s) \right|^2. \quad (23)$$

4 DISCUSSION

For large nanoparticle with $R \sim 100$ nm, the classical EM theory provides an enhancement of the Raman signal as large as 10^6-10^7 [6]. In reality, the EM enhancement is inhibited by various factors. In noble-metal particles, the interband transition between d-electron and s-electron bands reduce the SP oscillator strength leading to a weakening of the local fields. For nanoparticle radius below 15 nm, finite-size effects become important. The SP resonance damping comes from the electron scattering at the surface leading to the size-dependent SP resonance width $\gamma_s \simeq v_F/R$. At the resonance frequency, the size-dependence of SERS is quite strong. Indeed, if molecular vibrational frequencies are smaller that the SP width, the enhancement factor decreases as $A \propto R^4$ for small nanoparticles, resulting in several orders of magnitude drop in the Raman signal.

Our main observation is that, in small nanoparticle, the local field enhancement due to reduced screening in the surface layer can provide an additional enhancement of the Raman signal. Although the thickness of the surface layer (0.1-0.3 nm) is small as compared to oveall nanoparticle size [15]–[17], such an enhancement can be condiderable for a molecule located in a close proximity to the surface. In Fig. 4 we show the results of our numerical calculations of the Raman enhancement factor for various surface layer thicknesses, Δ. Although the overall magnitude of the enhancemet increases with Δ, a more important effect is its size-dependence. For finite Δ, the enhancement factor descreases more slowly that in the absence of the surface layer: as nanoparticle size decreases from $R = 5$ nm to $R = 1$ nm, the signals strength ratio between $\Delta = 0.3$ nm and $\Delta = 0$ nm *increases* from $\simeq 1.5$ to $\simeq 4.0$. The reason is that, as the nanoparticle becomes smaller, the fraction of the surface layer increases, and so does the contribution of the unscreened local field into SERS.

This work was supported by NSF under grants DMR-0304036, DMR-0305557 and HRD-0318519, and by ARO under grant DAAD19-01-2-0014.

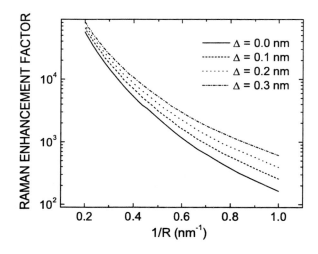

Figure 2: Raman enhancement factor as a function of nanoparticle size for different surface layer thicknesses.

[4] M. Kerker, D.-S. Wang, and H. Chew, Appl. Optics **19**, 4159 (1980).

[5] J. Gersten and A. Nitzan, J. Chem. Phys. **73**, 3023 (1980).

[6] G. S. Schatz and R. P. Van Duyne, in *Handbook of Vibrational Spectroscopy*, edited by J. M. Chalmers and P. R. Griffiths (Wiley, 2002) p. 1.

[7] H. Xu *et al.*, Phys. Rev. Lett. **83**, 4357 (1999).

[8] H. Xu *et al.*, Phys. Rev. B **62**, 4318 (2000).

[9] A. Otto *et al.*, J. Phys. Cond. Matter **4**, 1143 (1992).

[10] M. Michaels, M. Nirmal, and L. E. Brus, J. Am. Chem. Soc. **121**, 9932 (1999).

[11] A. M. Michaels, J. Jiang, and L. E. Brus, J. Phys. Chem. B **104**, 11965 (2000).

[12] Otto, Phys. Phys. Stat. Sol. (a) 4, 1455 (2000).

[13] W. E. Doering and S. Nie, J. Phys. Chem. B **106**, 311 (2002).

[14] W. Ekardt, Phys. Rev. B **31**, 6360 (1985).

[15] A. Liebsch, Phys. Rev. **48**, 11317 (1993).

[16] V. V. Kresin, Phys. Rev. **51**, 1844 (1995).

[17] A. Liebsch and W. L. Schaich, Phys. Rev. **52**, 14219 (1995).

[18] J. Lermé *et al.*, Phys. Rev. Lett. **80**, 5105 (1998).

[19] C. Voisin *et al.*, Phys. Rev. Lett. **85**, 2200 (2000).

[20] C. Lopez-Bastidas, J. A. Maytorena, and A. Liebsch, Phys. Rev. **65**, 035417 (2001).

[21] A. A. Lushnikov, V. V. Maksimenko, and A. J. Simonov, Z. Physik B **27**, 321 (1977).

REFERENCES

[1] S. Nie and S. R. Emory, Science **275**, 1102 (1997).

[2] K. Kneipp *et al.*, Rev. Lett. 78, 1667 (1997).

[3] K. Kneipp *et al.*, Chem. Rev. **99**, 2957 (1999).

Nanoscaled Science and Engineering for Sensing: Quantum Dots Fluorescence Quenching for Organic NO₂ Sensing

S. Nieto*, Alberto Santana*, R. Delgado*, S.P. Hernandez*, R.Thomas Chamberlain**,
Richard Lareau** and Miguel E. Castro*

*Chemical Imaging Center, The University of Puerto Rico at Mayaguez,
Mayaguez, Puerto Rico 00680, mcastro@uprm.edu
** Visiting Scholars, Federal Transportation Security Administration, Federal Aviation Administration
Technical Center, Atlantic City, New Jersey, 08405

ABSTRACT

We report on femtosecond laser off-resonance excitation of quantum dots-TNT assemblies in solution. The fluorescence emission from quantum dots shifts toward lower wavelengths and the relative peak intensity decreases with the amount of TNT added to solutions of the dots. A charge transfer mechanism based on charge injection from the photoexcited electrons in the TNT into the conduction band of the nanoscaled semiconductor is proposed to account for the observed results.

Keywords: femtosecond excitation, quantum dots-TNT assemblies, TNT sensing, nanoscaled sensors, charge transfer

1 INTRODUCTION

Detection of trace amounts of explosive materials is an important task with a broad range of implications, from national security to the cleanup of demilitarized installations. Nanotechnology is ideally suited to serve our Nation needs by providing new materials and methods that can be employed for trace explosive detection.[1-4] Fluorescent nanoparticles, like nanoscaled semiconductors or quantum dots (QD), have the potential to be employed in trace explosive detection in a number of environments. Variations in the particle fluorescence wavelength may serve as a tool to the detection of trace amounts of explosive chemicals. In the ideal case scenario, a QD emitting light with wavelength λ_1 will change its emission line to λ_2 upon bonding to an explosive molecule, as illustrated in schemes 1 and 2 below.

In typical QD, fluorescence electrons and holes are created following excitation with light of energy equal or larger to the band gap or energy difference between the valence and conduction band. Photo-excited electrons in the conduction bands are energetically relaxed to states introduced by traps in the particle. Luminescence emission results from relaxation of electrons from trap states into the valence band or to other empty "trap states".

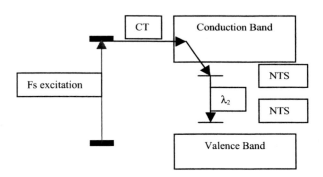

Scheme 2: Proposed charge transfer (CT) or new trap state (NTS) model for the changes in fluorescence emission of QD-TNT assemblies.

We report here on the effect of 2,4,6 trinitrotoluene (TNT) on the fluorescence emission of zinc sulfide (ZnS) protected cadmium selenide (CdSe) quantum particles. The fluorescence is excited off-resonance with respect to the quantum dots, but close to the absorption band edge of the TNT. This is proposed to result in charge transfer (CT) process between TNT and the quantum dots and/or the involvement of new trap states (NTS) in the fluorescence emission of the dots, as illustrated in scheme 2.

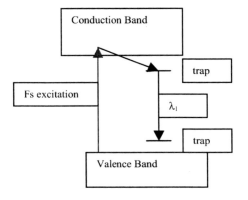

Scheme 1: Typical fluorescence mechanism from QD.

2 EXPERIMENTAL

Solutions of CdSe quantum dots in toluene were purchased from Ocean Optics. The dots are between 2 and 12 nm in diameter. The CdSe core shell is protected by a ZnS layer. The 800 nm line from a Spectra Physics femtosecond laser system operating at a repetition rate of 1 kHz was feeded into a Spectra Physics Wave Generator were it was frequency double by sum frequency methods to provide the 400 nm line employed to excite the fluorescence in the quantum dots. Solutions containing the quantum dots were placed in a standard quartz cell. TNT was added with the use of a calibrated micropipette from a 1 ppm TNT in toluene solution. The fluorescence was collected at 90° degrees with respect to the incident radiation and measured with a JY Triax 550 spectrograph equipped with a UV-visible CCD.

3 RESULTS

Representative fluoresce spectra of quantum dots in toluene and quantum dots exposed to various amounts of the TNT in toluene solution are displayed in figure 1. The fluorescence of the QD in toluene solution used to collect the results displayed in figure 1 is centered at 534 nm. The position of the peak is independent of the amount of toluene added, as verified in independent measurements. The peak shifts toward lower wavelengths and the peak area decreases with the amount of TNT added. So far, we have observe this trend when the QD are exposed to the TNT solution. Independent measurements revealed that the position of the peak does not change upon the addition of highly reactive species, like alkyl thiols or carboxylic acids.

Figures 2 and 3 summarizes results from quantitative analysis of peak intensities and maximum fluorescence wavelength as a function of the amount of added TNT to solutions containing the QD employed for the measurements described on figure 1. The fluorescence relative peak area decreases with the amount of added TNT to the solution. This decrease in peak area is not related to a dilution effect, as verified by independent measurements performed on QD exposed to volumes of toluene free of TNT. Such measurements showed that the fluorescence intensity varied proportional to the concentration of the QD in solution. We conclude then, based on the results presented in figures 1 and 2, that the fluorescence of the QD is quenched by the added TNT.

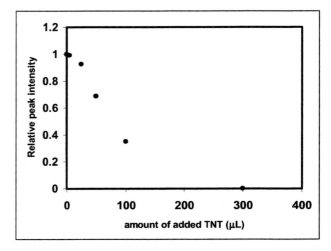

Figure 2: Fluorescence relative peak intensity as a function of the amount of TNT in toluene added to a solution of QD in toluene.

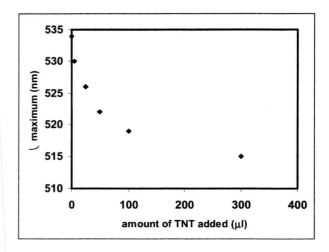

Figure 3: Fluorescence maximum wavelength (nm) as a function of the amount of TNT added (μL).

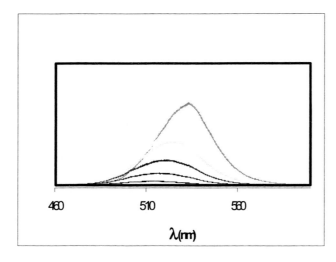

Figure 1: Fluorescence spectra of CdSe/ZnS QD exposed to 0, 5, 25, 50, 100 and 300 mL of a 1 ppm TNT in toluene solution.

The fluorescence maximum wavelength also experiences marked changes upon the addition of TNT to the QD in toluene solution. These are summarized in figure 3. The fluorescence maximum wavelength shifts toward lower energies with increasing the amount of TNT added to the QD in toluene solution. We take this result as evidence for strong interactions among the TNT molecules and the QD.

QD emitting at other wavelengths were also studied. The general trend presented in the previous paragraphs regarding the effect of TNT on the fluorescence maximum wavelength and relative peak intensity was also observed. For instance, the fluorescence maximum of a QD that emits at 567 nm was shifted to 558 nm upon the addition of 100 μL of a 1 ppm TNT solution. The relative peak intensity was observed to decrease by 30 % in the above case. Fluorescence quenching was more marked in a QD that emits at 605.9 nm. In that case, the fluorescence maximum emission wavelength shifted to 598 nm and the fluorescence relative peak intensity decreased by 92 % upon the addition of 50 μL of a 1ppm TNT solution. Thus the results are consistent with a quenching in the fluorescence of the QD and a change in the fluorescence mechanism of the QD upon the addition of TNT.

4 DISCUSSION

The results presented here are consistent with extensive quenching and a change in the fluorescence emission of QD by TNT. The results prove that the QD may find applications in TNT sensing from observations of a second wavelength (λ_2), distinctive from the emission wavelength (λ_1) of the QD in the absence of the TNT.

The changes in the maximum fluorescence wavelength as a function of added TNT are major finding of the work presented here. The changes may be accounted for by a charge transfer mechanism (CT) of TNT electrons into the conduction band of the QD or by the appearance of new "trap states" (NTS) that are not present for the QD, as presented in schemes 1 and 2. Alternatively, a mechanism involving both, CT and NTS, may be needed to explain the observed shifts in the fluorescence of the QD as a function of added TNT.

The arguments presented above assume that the TNT has no effect in the size of the QD. Particle size plays an important role in the eavelength emission of quantum dots, as it is well known that the band gap energy (ΔE) is inversely proportional to the diameter of the particle. If TNT induces a change in the size of the QD, then, ΔE will change and, consequently, the fluorescence emission wavelength. Independent UV-visible absorption measurements (not shown) failed to show a change in the leading edge of the absorption spectra of the QD upon the addition of TNT, indicating that the results presented here

can not be explained by a mechanism that involves a change in the size of the particle. In view of the above arguments, we believe that either a CT and/or NTS mechanism may be responsible for the effect of TNT in the fluorescence emission of the QD. Further work is needed to clearly establish the mechanism of fluorescence of the QD in the presence of TNT.

5 ACKNOWLEDGEMENT

Partial financial support for the operations of the CIC from the Federal Aviation Administration (FAA), grant number 99-G-029 and the National Science Foundation (award number 0304348) are gratefully acknowledged.

REFERENCES

[1] James Baker, Richard Colton, Heidi Schroeder Gibson, Michael Grunze, Stephen Lee, Kenneth Klabunde, Charles Martin, James Murday, Thomas Thundat, Bruce Tatarchuk, and Ketih Ward, Grand Challenge Workshop Series, Nanotechnology Innovation for Biological, Chemical, Radiological and Explosive (CBRE): Detection and Protection; November 2002, Monterrey, California. http://www.wtec.org/nanoreports/cbre/CBRE_Detection_11_1_02_hires.pdf

[2] M.C. Roco, AAAS XXVII: Research and Development FY 2003; National Nanotechnology Investment in the FY 2003 Budget Request; http://www.aaas.org/spp/rd/03pch24.htm

[3] Jehuda Yinon, "Detection of Explosives by Electronic Noses", Analytical Chemistry, March, 2003, pages 99 A – 105 A.

[4] L.A. Pinnaduwage, J.E. Hawk, V. Boiadjiev, D. Yi and T. Thundat, "Use of Microcantilevers for the Monitoring of Molecular Binding to Self Assembled Monolayers", Langmuir, 19, (2003), pages 7841-7844.

[5] Murphy, C. J. Analytical Chemistry. 2002, October 1, 520 A.

Desorption of Surfactant and Sintering of Surface Modified Pd_xNi_{1-x} Nanoparticles

Kuan-Wen Wang [a], Shu-Ru Chung [a], Wei-Hsiu Hung [b], and Tsong P. Perng [a*]

[a] Department of Materials Science and Engineering
National Tsing Hua University, Hsinchu, Taiwan, *tpp@mse.nthu.edu.tw
[b] Synchrotron Radiation Research Center, Hsinchu, Taiwan

ABSTRACT

The desorption of surfactant and sintering of surface modified Pd_xNi_{1-x} nanoparticles are studied. The surface energy of the nanoparticles depends on the composition and is related to the desorption temperature of surfactant from the nanoparticle surface. For the alloys with higher Ni content, the surface energy is higher and the desorption temperature of surfactant is also higher. With the same composition, the desorption temperature of stearic acid (SA) is always higher than that of polyethylene glycol (PEG) because SA has a higher thermal resistance. After surface modification, the surface energy of the nanoparticles is changed. During heating, the more stable nanoparticles have less sintering and grain growth. Surface modification not only changes the surface energy of the nanoparticles but also prevents the nanoparticles from sintering. SA has better thermal resistance than PEG and shows better suppressing of sintering of the nanoparticles.

Keywords: PdNi, desorption, TPD, surface modification, sintering

1. INTRODUCTION

Nanostructure materials are very attractive because of their large surface area and surface energy [1]. By controlling the surface area or surface energy, the properties of nanomaterials can be manipulated. One of the most common surface modification methods is to cover a layer of surfactant on the nanoparticle surface to prevent the particles from agglomeration and grain growth during sintering [2], or to increase the dispersibility because of changing of the surface energy [3]. Therefore, surface modification is often used to change the surface energy and in the mean time to control the properties of nanomaterials [4-10].

In this study, a series of the Pd_xNi_{1-x} alloy nanoparticles were prepared. Because Ni and Pd have different surface energy, the surface energies of particles with different compositions are different. The nanoparticles are modified with two surfactants, stearic acid (SA) with a chemical formula $CH_3(CH_2)_{16}COOH$ and polyethylene glycol (PEG) with a formula $H(OCH_2CH_2)_nOH$. The desorption temperature of the surfactants were measured and related to the properties of the materials, such as surface compositions and sintering behavior [12-15].

2. EXPERIMENTAL PROCEDURE

A series of Pd_xNi_{1-x} nanoparticles were prepared by a chemical precipitation method. Solutions of palladium nitrate and nickel nitrate were mixed at desired stoichiometries 80:20, 70:30, 30:70 and 20:80, and then reduced by sodium borohydride in an alkaline environment. The average particle size was about 10 nm. Solutions of SA and PEG were prepared, and 1 wt% SA or PEG was added to the nanoparticles to form a thin layer on the particle surface. The samples modified by SA or PEG are named as Pd_xNi_{1-x}-s and Pd_xNi_{1-x}-p, respectively.

The phase structure and composition of the nanoparticles were examined by X-ray diffraction (XRD). The weight variation of surfactants during heating was analyzed by thermogravimetric/differential thermal analyzer (TG/DTA). Temperature programmed desorption (TPD) was used to examine the desorption temperature of SA or PEG from the Pd_xNi_{1-x} nanoparticles of different compositions. The heating rate was 0.5K/s. The surface morphologies before and after heating were examined by scanning electron microscopy (SEM), and the size and weight of the pellets before and after heat treatment were also measured.

3. RESULTS AND DISCUSSION

The XRD patterns of four Pd_xNi_{1-x} nanoparticles are shown in Fig. 1. The Pd-Ni nanoparticles all form alloys and the particle sizes are all about 10 nm.

The thermal stabilities of the surfactants used in this experiment, SA and PEG, were analyzed by TG. The curves of weight variation during heating are shown in Fig. 2. The decomposition temperature of PEG is lower than that of SA. It starts to decompose at about 100°C, and the slope becomes more rapid at above 170 °C. PEG has almost decomposed at above 270 °C, with less than 1.1 % of residue. SA has a higher decomposition temperature, being at about 140 °C. Above 300 °C, there is still 10 % of residue. As their molecular weights imply, the molecular weight of PEG (200 g/mole) is less than that of SA (284 g/mole), SA is expected to have a higher thermal stability

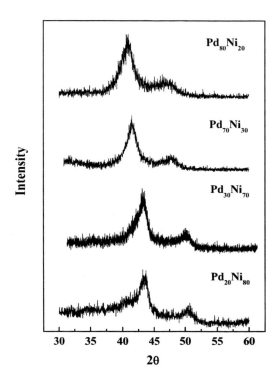

Fig. 1　XRD patterns of Pd_xNi_{1-x} nanoparticles.

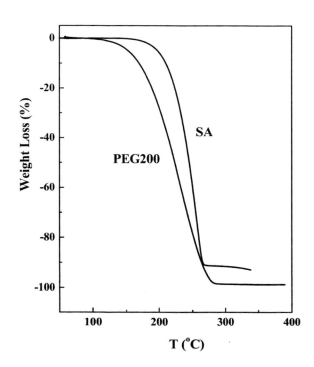

Fig. 2　TG analysis of PEG and SA.

than PEG.

The desorption curves of SA from the nanoparticles by the TPD measurement are described in Fig. 3. According to the mass spectrum in NIST handbook for mass 43, the strongest desorption peak, is used to identify the desorption of SA. The peak position in TPD is defined as the desorption temperature of SA (T_d(SA)). In the TPD results of $Pd_{20}Ni_{80}$, the lower temperature peak at 133 °C is the decomposition or evaporation temperature of free SA. Their temperature is consistent with that from the TGA in Fig. 2. The higher temperature peak indicates the desorption of SA from the nanoparticles. As the Ni content increases, the T_d(SA) increases. Because the surface energy of Ni is higher than that of Pd (2.45 J/m^2 for Ni and 2.05 J/m^2 for Pd) [11], more Ni means a higher surface energy. There is more interaction between the surfactant and the nanoparticle. During desorption, it takes more energy and therefore higher temperature, for SA to leave from the nanoparticle surface. The influence of surface energy is so important that the T_d(SA) is changed from 240 to 394 °C when the Ni content is increased from 20% to 80%. Therefore, the composition or surface energy is a function of desorption temperature.

For the desorption of PEG, the results are similar. The decomposition temperature of free PEG in $Pd_{20}Ni_{80}$ is 105°C.

The T_d(PEG) also increases as the Ni content increases, and is changed from 197 to 377 °C as the Ni content increases form 20% to 80%. The comparison of the T_d(SA) and T_d(PEG) for the samples is shown in Fig. 4. The T_d(PEG) is always lower than T_d(SA) for the same composition. This indicates that PEG on the surface has lower thermal resistance than SA. This property could affect the sintering behavior of surface modified nanoparticles.

In Fig. 5, the sintering parameters φ is displayed for the three $Pd_{70}Ni_{30}$ samples. This $Pd_{70}Ni_{30}$ composition is chosen to study the sintering properties of surface modified nanoparticles. The parameter φ is defined as $(\rho_s-\rho_g)$ / $(\rho_t-\rho_g)$, where ρ_s, ρ_g, and ρ_t are sinter density, green density, and theoretical density, respectively. For all three samples, φ increases as temperature increases. The as-prepared $Pd_{70}Ni_{30}$ sample sinters better than the surface modified samples because it has a higher surface energy and the surface is less stable. SA has a greater thermal resistance than PEG. The surface energy of the nanoparticles is lowered more, and the sintering of nanoparticles is more retarded. This result is coincided with those from TPD and TG. In the TG analysis, at the same temperature, there is more residual SA on the nanoparticle surface to protect the nanoparticles from sintering.

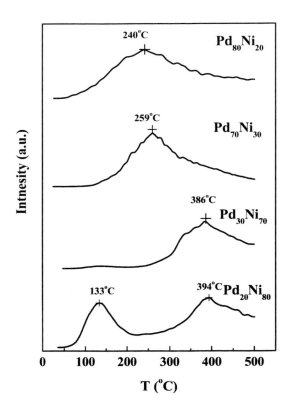

Fig. 3 TPD curves of SA from the Pd_xNi_{1-x} nanoparticles.

The SEM micrographs of the $Pd_{70}Ni_{30}$ nanoparticles sintered in vacuum at $450^{\circ}C$ are shown in Fig. 6. The unmodified sample is well sintered together. Substantial growth of particles is observed. Sintering of the $Pd_{70}Ni_{30}$-p sample is somewhat suppressed because of the presence of PEG. The sample $Pd_{70}Ni_{30}$-s almost remains unsintered. $Pd_{70}Ni_{30}$-s also has smaller particle size, only 40 nm, followed by 100 nm of $Pd_{70}Ni_{30}$-p and several hundred nanometers of $Pd_{70}Ni_{30}$,

4. CONCLUSION

The surface energy of the modified nanoparticles is a function of desorption temperature of the surfactant from the nanoparticle surface. The desorption temperature increases as the Ni content increases. Surface modification by surfactant can change the surface energy and thus changing the sintering behavior of nanoparticles. SA, which has better thermal resistance and higher desorption temperature than PEG, has a higher ability to retard the nanoparticles from sintering and to suppress grain growth.

5. ACKNOWLEDGEMENT

This work was supported by Ministry of Education of ROC under Contract A-91-E-FA04-1-4.

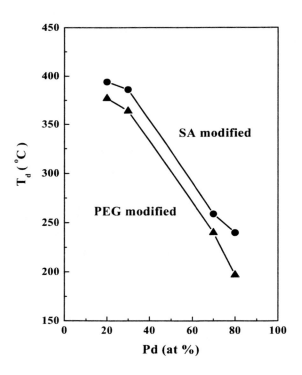

Fig. 4 Desorption temperatures of SA and PEG for the nanoparticles with various contents of Pd.

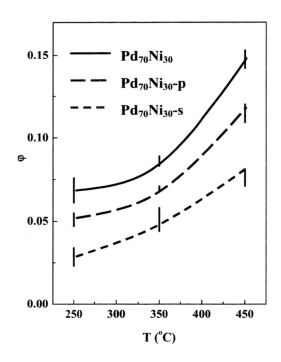

Fig. 5 Sintering curves in vacuum for $Pd_{70}Ni_{30}$ with various surface conditions.

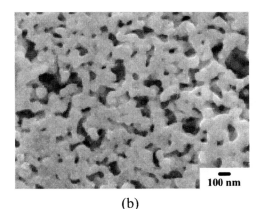

(a)

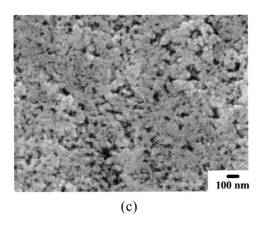

(b)

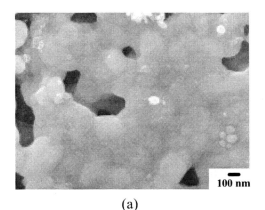

(c)

Fig. 6   SEM morphologies of various $Pd_{70}Ni_{30}$ samples sintered at 450°C in vacuum.   (a) $Pd_{70}Ni_{30}$,   (b) $Pd_{70}Ni_{30}$-p, and (c) $Pd_{70}Ni_{30}$-s.

REFERENCES

[1] M. J. Weins, H. Gleiter, and B. Chalmers, J. Appl. Phys., 42, 2639   (1971).

[2] N. L. Wu, S. Y. Wang, and A. Rusakova, Science, 285, 1375 (1999).

[3] S. F. Lomayeva, E. P. Yelsukov, G. N. Konygin, G. A. Dorofeev, V. I. Povstugar, S. S. Mikhailova, A. V. Zagainov, and A. N. Maratkanova, Colloid Surf., 162, 279 (1999).

[4] P. Milani, M. Ferretti, A. Parisini, C. E. Bottani, M. A. Malvezzi, and A. Cavalleri, Carbon, 36, 5 (1998).

[5] A. Meier, I. Uhlendorf, and D. Meissner, Electrochimica Acta, 40, 10 (1995).

[6] A. B. Bourlinos, A. Bakandritsos, V. Georgakilas, and D. Petridis, Chem. Mater., 14, 3226 (2002).

[7] J. Ji, Y. Chen, R. A. Senter, and J. L. Coffer, Chem. Mater., 13, 4793 (2001).

[8] H. Dai, J. H. Hafner, A. G. Rinzler, D. T. Colbert, and R. E. Smalley, Nature, 384, 147 (1996).

[9] V. Lordi, N. Yao, and J. Wei, Chem. Mater., 13, 733 (2001).

[10] C. Mao, W. Sun, Z. Shen, and N. C. Seeman, Nature, 397, 144 (1999).

[11] L. Vitos, A. V. Ruban, H. L. Skriver, and J. Kollar, Surf. Sci., 411, 186 (1998).

[12] I. Matolinova, S. Fabik, K. Masek, L. Sedlacek, T. Skala, K. Veltruska, and V. Matolin, Vacuum, 71, 41 (2003).

[13] T. V. Buuren, M. K. Weilmeier, I. Athwal, K. M. Colbow, J. A. Mackenzie, T. Tiebje, P. C. Wong, and A. R. Mitchell, Appl. Phys. Lett., 59, 22 (1991).

[14] Y. H. Lai, C. T. Yeh, J. M. Hwang, H. L. Hwang, C. T. Chen, and W. H. Hung, J. Phys. Chem. B 105, 10029 (2001).

[15] Y. H. Lai, C. T. Yeh, H. J. Lin, C. T. Chen, and W. H. Hung, J. Phys. Chem. B, 106, 1722 (2002).

Effect of Various Fuel Conditions on the Synthesis of Nano Crystalline LiNiVO$_4$ Powders for Lithium Battery Applications

S. Vivekanandhan, M. Venkateswarlu[*]**, N. Satyanarayana**[+]

Raman School of Physics, Pondicherry University, Pondicherry- 605 014, India
[+]**Corresponding author**: E-mail: nallanis2000@yahoo.com
[*]**Present address:** Department of Chemical Engineering, NTUST, Taipei, Taiwan

Abstract

Nano crystalline inverse spinel LiNiVO$_4$ powders were prepared by the gel combustion route with citric acid and glycerol under three different fuel conditions as (i) total metal ion to citric acid to glycerol ratio (M/CA/G) = 1:1:1 (ii) M/CA/G = 1:1:2 and (iii) M/CA/G = 1:2:1. Thermal behavior, structure and phase formation of the polymeric intermediates as well as LiNiVO4 powders, were investigated by DSC, FTIR, SEM and XRD measurements. The pure crystalline phase of LiNiVO$_4$ powders with smallest crystallite size of 39nm were obtained from the polymeric intermediate, calcined at 450°C for 12 h, synthesized by Gel combustion route performed under the fuel condition of M/CA/G = 1:1:1.

Key words: Nano - crystalline LiNiVO$_4$, glycerol, polymeric gel, DSC, FTIR, SEM, XRD.

1. Introduction

Lithium nickel vanadate (LiNiVO$_4$), having inverse spinel structure, has been used as the cathode material for high voltage (4.8 V) lithium battery applications [1-4]. LiNiVO$_4$ type of high voltage cathode material offers a great promise for increasing the energy density of the lithium batteries. Better performance of the cathode material is due to its physical and chemical properties like crystallite size, homogeneity, stichiometry etc. which are mainly depend on the synthesis process [5]. Conventional solid state reaction method for preparing the LiNiVO$_4$ powders haven been reported, which showed the disadvantages as bigger grain size, high processing temperature, poor homogeneity etc [6,7]. In order to overcome these disadvantages, a variety of solution chemistry (wet chemistry) methods like sol- gel, hydrothermal, co-precipitation, gel combustion etc. have been developed and reported [8-12].

Among them, gel combustion method is a versatile, simple and cost effective process for the synthesis of multicomponent oxide powders in nano scale [13-17].Some of the reports are available on the synthesis of LiNiVO$_4$ powders but smallest the reported grain size of LiNiVO$_4$ powders were found to be micro meter range only [2,7,13, 18]. Even the latest paper by Jina –Rui Liu et al

reported the particle size of as small as 0.5µm for the LiNiVO$_4$ powders prepared using citric acid at 500°C [18].

Recently, we have successfully developed a new gel combustion method using the combination of citric acid and glycerol (poly hydroxyl alcohol with three -OH groups in a molecule) for the synthesis of nano crystalline materials [20]. It is found that the fuel conditions can affect the formation of the polymeric network, which modifies the microstructure of the oxide powders. Hence, in the present work, gel combustion process was used, with three different fuel conditions as (i) M/CA/G = 1:1:1, (ii) M/CA/G = 1:1:2 and (iii) M/CA/G = 1:2:1, to synthesis nano crystalline LiNiVO$_4$ powders.

2. Experimental

Required stoichiometric amounts (metal ions to citric acid to glycerol ratio (M/CA/G) of 1:1:1, 1:1:2 and 1:2:1) of the lithium nitrate (SD-Fine, AR grade), nickel nitrate (MERCK India, Extra Pure) and ammonium vanadate (TRAC, AR grade) solutions (mole ratio of 1:1:1) were mixed together under the continuous stirring condition. Resulting transparent blue colour solutions were evaporated at 75°C under constant stirring condition and the continuous evaporation lead the formation of block colored gel (Stage-I. The block colored gel was then heated at 175°C for 12 hours to obtained polymeric intermediate (Stage-II). Thermal decomposition of the polymeric intermediates led to the formation of nano crystalline LiNiVO$_4$ powders, which were investigated using DSC, FTIR, SEM and XRD measurements.

3. Results and discussion

The DSC curves, measured between 30°C to 500°C, of the polymeric intermediates (Stage –II products) synthesized under three different fuel conditions are shown in fig. 1. From fig. 1, a single exothermic peak appears between 370 and 400°C for each sample indicates the combustion reaction of the polymeric intermediates. The combustion reaction observed at different temperatures is attributed to the nature of polymeric network exists in the intermediates, which was formed under different fuel conditions. The DSC thermogram of exothermic peak due to the combustion reaction indicate that the organic derivatives are decomposed at 400°C, which lead the formation of crystalline LiNiVO$_4$ phase and it is further confirmed by FTIR and XRD analysis. DSC curve for the polymeric intermediate synthesized under the fuel condition of M/CA/G = 1:1:2 exhibits lower ignition temperature and high heat evaluation. The higher heat evaluation may cause the formation higher crystallinity as well as the formation of improper phase in LiNiVO$_4$ powders, which is confirmed by XRD spectra.

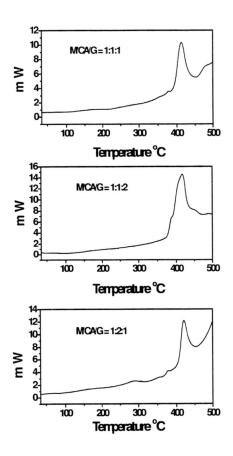

Fig. 1. DSC curves of the polymeric intermediates prepared with three different fuel conditions.

Fig 2 show the scanning electron micrographs of the polymeric intermediates. The SEM picture of fig. 2a and 2c showed the polymeric nature with large voids for the stage – II products prepared under the fuel conditions of M/CA/G = 1:1:1 and 1:2:1. Increase of the glycerol over the citric acid ratio (i.e. M/CA/G =1:1:2) lead the formation of the cluster structure as shown in fig. 2b, which indicates the poor polymerization due to the insufficient of the –COO group (from the citric acid) to form the network through the esterification.

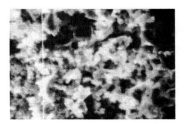

2a. M/CA/G = 1:1:1

2b. M/CA/G = 1:1:2

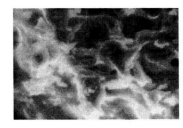

2c. M/CA/G = 1:2:1

Fig. 2. SEM images (2000 X) of the polymeric intermediates prepared with different fuel conditions.

FTIR spectra of the polymeric intermediates are shown in fig .3. From fig. 3, the IR peak at 3392 cm^{-1} for the polymeric intermediate is due to the O-H stretching associated with the hydroxyl group of the citric acid and the glycerol present in the polymeric intermediates. The peak at 1570 cm^{-1} is assigned to the asymmetric stretching of carboxylates (-COO$^-$), which confirms the chelation of metal ions through the formation of metal carboxylates [22]. The IR peak observed at 1255cm^{-1} is due to the –C-O-C- vibration shows the evidence for the esterification between the carboxylic and hydroxyl groups in citric acid and glycerol [23]. The intensity of the 1225cm^{-1} peak decreases with increases of the glycerol amount (M/CA/G = 1:1:2) and the same peak was clearly observed for the stage- II product with M/CA/G = 1:1:1 and 1:2:1. Thus, confirms the decrement of the ester formation between hydroxyl and carboxyl groups with increase of glycerol amount, which caused the poor polymerization and it is also clearly confirmed by SEM analysis. From fig. 4, FTIR spectra showed the formation of two new peaks at 811 cm^{-1}

and 690 cm^{-1} for the Stage – II powders calcined at 450°C, which are attributed to the stretching vibrations of V-O bonds of VO$_4$ tetrahedra in LiNiVO$_4$ [24, 25].

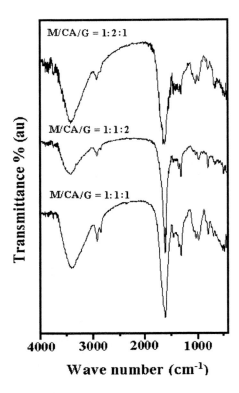

Fig. 3. FTIR spectra of the polymeric intermediates prepared with three different fuel conditions

Fig. 5 shows the typical XRD spectra for as prepared as well as calcined polymeric intermediates. From fig.5, the peak free XRD spectrum for the polymeric intermediates confirmed the amorphous/ polymeric nature. From fig.5 the crystalline LiNiVO$_4$ phase begun to form at 300°C and the complete phase was obtained 450°C and above. Fig 6 shows the XRD spectra of polymeric intermediates calcined at 450°C alongwith the respective JCPDS data. Crystallite size for the prepared crystalline LiNiVO$_4$ powders was calculated using the Scherrer's formula and the FWHM data obtained by Lorentz fit for the strongest XRD peak observed at 35°54'. The crystallite size for the LiNiVO$_4$ powders obtained from the dried gels, calcined at 450°C, prepared with three different fuel conditions as M/CA/G = 1:1:1, 1:1:2 and 1:2:1 is 39, 55 and 40 nm respectively.

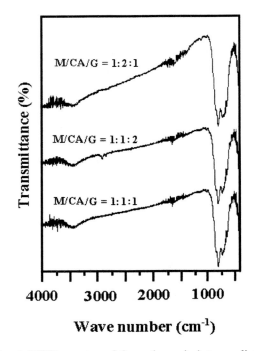

Fig. 4. FTIR spectra of the polymeric intermediates calcined at 450°C

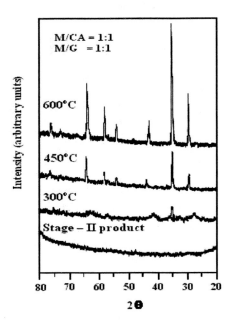

Fig. 5. Typical XRD spectra of the polymeric intermediates (M/CA/G/ = 1:1:1) calcined at different temperature:

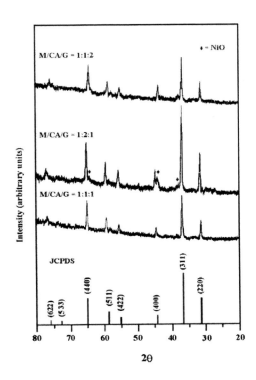

Fig. 6. XRD patterns of LiNiVO4 prepared by calcining the polymeric intermediates at 450°C for 12 hours

4. Conclusions

Gel combustion process was systamatically investigated for the preparation of nano crystalline LiNiVO$_4$ powders under three different fuel conditions. DSC, SEM, FTIR and XRD results confirm that the smallest crystallite size of 39 nm for LiNiVO$_4$ powders with pure crystalline phase obtained from the dried gels, calcined at 450°C for 12 hrs, synthesized using the gel combustion route with the fuel condition of M/CA/G= 1:1:1.

Acknowledgements

Authors are gratefully acknowledged DRDO, DST, CSIR and AICTE for utilizing the research facilities available from the major research projects.

References

[1] G. T. K. Fey, W. Li, J. R, Dahn, J. Electrochem. Soc. 141, 2279, 1994.
[2] G. T. K. Fey, J. R. Dahn, W. Li, M. J. Zhang, J. Power Sources 68 - 2, 549, 1997.
[3] J. M. Tarascon, W. R. Mckinnon, F. Coowar, et al., J. Electrochem. Soc. 141, 1421, 1994.
[4] G. T. K. Fey, K. S. Wang, S. M. Yang, J. Power Sources 68 - 1, 159, 1997.
[5] Zhiping Jiang, K.M Abraham, J. Electrochem. Soc., 143, 1591, 1996.
[6] G. T. K. Fey, W. B. Perng, Mater. Chem. Phys. 47, 279, 1997.
[7]. George Thing- Kuo Fey, Way –bing Perng, Ma, Mater. Chem. Phys. 47, 279, 1997.
[8] Yun-sung Lee, Yang-Kook Sun, Kee-Suk Nahm, Solid State Ionics, 109, 285, 1998.
[9] Chung – Hsin Lu, Shaw – Jang Liou, Mater. Sci. Eng. – B, B75, 38, 2000.
[10] Y. M. Hon, S. P. Lin, K. Z. Fung, M. H. Hon, J. Eur. Ceram. Soc., 22, 653, 2002.
[11] S. R. Sahaya prabaharan, M. Siluvai Michael, T. Prem Kumar, A. Mani, K. Athinarayanaswamy, R. Ganagadharan, J. Mater. Chem., 5(7), 1035, 1995.
[12].B. J. Hwang, R. Santhanam, D.G. Liu, J. Power Sources, 97-98, 2001, 443.
[13] P. Kalyani, N. Kalaiselvi, N. Muniyandi, Mater. Chem. Phys. 77, 662, 2002.
[14] Sang Woon Kwon, Seung Bin Park, Gon Seo, Sung Tai Hwang, J. Nuel. Mater. 257, 172, 1998.
[15] S. Sundar Monoharan, N. R. S. Kumar and K. C. Patil, Mat. Res. Bull., 25, 731, 1990.
[16] Sukumar roy, Liwu Wang, Wolfgang Sigmund, Fritz Aldinger, Mater. Lett. 39, 138, 1999.
[17] W. Liu and g. C. Farrington, F. chaput, B. Dunn, J. Electrochem. Soc., 143, 879, 1996.
[18] Jina – Rui Liu, Meng Wan, Xin Lin, Da – chuan Yin, Wei – Dong Huang, J. Power Sources, 108, 113, 2002.
[19] Ch. Laberty- Robert, et al, Mat. Res. Bull., 36, 2083, 2001.
[20] S. Vivekanandhan, M. Venkateswarlu, N. Satyanarayana, Materials letters (2003) In press.
[21] Klug, H. P., Alexander, L. E., X-ray Diffraction Procedures for Polycrystalline and Amorphous Materials, Wiley, New York, 1954.
[22] Xionghiu Zeng, et al. Mater. Chem. Phys. 77, 209, 2002.
[23]. George Socrates, Infrared and Raman Characteristic Group Frequencies, John Wiley and Sons, New York, 2001.
[24]. C. Julien, M. Massot, C. Perez – Vicente, Mater. Sci. Eng – B, B75, 6, 2000.
[25]. S. Chitra, et al., Mater. Chem. Phys. 65, 32, 2000.

Atomic-scale structure of nanocrystals by the atomic pair distribution function technique

V. Petkov

Department of Physic, Central Michigan University
Mt. Pleasant, MI 48858, USA, petkov@phy.cmich.edu

ABSTRACT

The approach of the atomic pair distribution function technique to determine the atomic-scale structure of nanocrystalline materials is introduced and illustrated with results of studies on V_2O_5 nanotubes and clusters of Cs atoms intercalated inside the pores of the zeolite ITQ-4. We find that V_2O_5 nanotubes are built of double layers of V-O5 and V-O4 units. Inside the channels in ITQ-4 Cs atoms are found to assemble in short-range ordered zigzag chains.

Keywords: structure determination, x-ray diffraction

1. INTRODUCTION

Knowledge of the atomic-scale structure is an important prerequisite to understand and control the properties of materials. In the case of crystals it is obtained solely from the Bragg peaks in their diffraction pattern and is given in terms of a small number of atoms placed in a unit cell subjected to symmetry constraints. However, many materials of technological importance, including nanocrystalline materials, do not possess the long-range order of conventional crystals and often it is this deviation from perfect order that makes them technologically and/or scientifically important. The diffraction patterns of such materials show only a few Bragg peaks and a pronounced diffuse component. This poses a real challenge to the usual techniques for structure determination. The challenge can be met by employing the so-called atomic pair distribution function (PDF) technique. The frequently used atomic PDF, G(r), is defined as $G(r)=4\pi r[\rho(r)-\rho_o]$, where $\rho(r)$ and ρ_o are the local and average atomic number densities, respectively. It peaks at real space distances where the local atomic number density deviates from the average one, i.e. where most frequent interatomic distances occur, and thus reflects the structure of materials. The PDF G(r) is the sine Fourier transform of the so-called total scattering structure function, S(Q),

$$G(r) = \frac{2}{\pi} \int_{Q=o}^{Q_{max}} Q[S(Q)-1]\sin(Qr)dQ, \qquad (1)$$

where Q is the magnitude of the wave vector. The structure function $S(Q)$ is related to the coherent part, $I^{coh}(Q)$, of the powder diffraction intensities as follows:

$$S(Q) = 1 + \left[I^{el.}(Q) - \sum c_i \left| f_i(Q) \right|^2 \right] \Big/ \left| \sum c_i f_i(Q) \right|^2, \qquad (2)$$

where c_i and f_i are the atomic concentration and atomic scattering factor respectively, for the atomic species of type i [1]. As can be seen the PDF is simply another representation of the diffraction data. However, exploring the experimental data in real space is advantageous, especially in the case of materials with reduced long-range order such as nanocrystals. First, as *eqs.1* and *2* imply the *total*, not only the Bragg diffracted, intensities contribute to the PDF. In this way both the *average*, long-range atomic structure, manifested in the sharp Bragg peaks, and the *local*, non-periodic structural features, manifested in the diffuse components of the diffraction pattern, are reflected in the PDF. Note the conventional crystallographic studies take only the Bragg peaks into account. Second, the PDF is obtained with no assumption of periodicity and is barely affected by diffraction optics and experimental factors since these are accounted for in the step of normalizing the raw diffraction data and converting it to PDF data. This renders the PDF a structure-dependent quantity giving directly relative positions of atoms in materials and enables convenient testing and refinement of structural models. Here we demonstrate how the PDF technique works by employing it to determine the structure of two nanocrystalline materials: V_2O_5 nanotubes and nanoscale clusters of Cs atoms.

Crystalline vanadium pentoxide, V_2O_5, is an important technological material widely used in applications such as optical switches, chemical sensors, catalysts and solid-state batteries (2). The material possesses an outstanding structural versatility and can be manufactured into nanotubes that have many of the useful physicochemical properties of the parent V_2O_5 crystal significantly enhanced. For example, the high specific surface area of the nanotubes renders them even more attractive as positive electrodes in secondary Li batteries (3). Also, the nanotubes show significantly increased capability for redox reactions (4). The structure of the nanotubes is an important issue since good knowledge of it is a key to understanding material's properties.

Cs-intercalated $Si_{32}O_{64}$ belongs to the class of novel low-dimensional electron systems known as "electrides" [5]. In these materials a low-density gas of correlated electrons is confined in the cavities of an inert host and coexists with a nanoscale array of alkali ions that provide charge balancing. The confined electron gas exhibits Mott insulation behavior and Heisenberg antiferromagnetism. Recently, the first inorganic electride that is stable at room

NSTI-Nanotech 2004, www.nsti.org, ISBN 0-9728422-9-2 Vol. 3, 2004

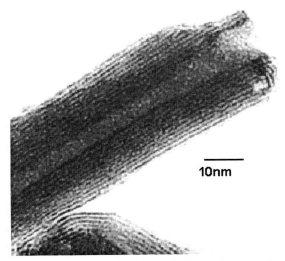

Figure 1. TEM image of vanadium oxide nanotubes used in the present study.

temperature has been synthesized by intercalating Cs into the zeolite ITQ-4 ($Si_{32}O_{64}$). Here we use the PDF technique to find the arrangement of Cs^+ ions inside the nanopores of ITQ-4.

2. EXPERIMENTAL

Vanadium oxide nanotubes were synthesized by mixing crystalline vanadium oxide and dodecylamine in absolute ethanol, stirring in air for two hours and then hydrolyzing with osmosis reversed water. After aging for 3 days and hydrothermal treatment in an autoclave at 180 °C for 7 days, a black precipitate was obtained. The product was washed with water, ethanol, hexane and diethyl ether to remove the unreacted amines and decomposed products, and dried overnight under vacuum at 80 °C to prevent oxidation of vanadium +4 to +5. Thus prepared nanotubes and crystalline V_2O_5 were sealed between Kapton foils and subjected to x-ray diffraction experiments.

Electrides $Cs_xSi_{32}O_{64}$ (x=3.6, 4.6) were prepared by mixing dehydrated $Si_{32}O_{64}$ and a weighted amount of cesium metal in sealed borosilicate flasks and heating to 60° C. The amount of absorbed Cs was determined by collection of hydrogen evolved upon reaction with water, and by titration of the CsOH formed. As prepared electrides and pristine ITQ-4 were sealed in capillaries and subjected to diffraction experiments.

The diffraction experiments on the vanadium oxide nanotubes were carried out at the beamline 1-ID at the Advanced Photon Source, Argonne National Laboratory and those on the electrides at the beam line X7A of the NSLS, Brookhaven National Laboratory. Several runs were conducted with each of the samples measured and the resulting XRD patterns were averaged to improve the statistical accuracy and to reduce any systematic effect due to instabilities in the experimental set-up. The raw diffraction data were corrected for the decay of the incoming synchrotron radiation beam, for background and

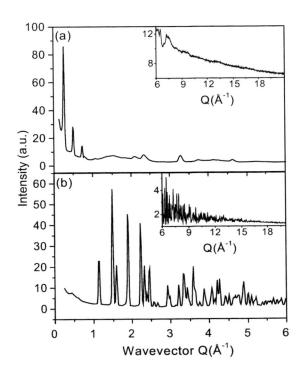

Figure 2. Experimental powder diffraction patterns for V_2O_5 nanotubes (a) and crystalline V_2O_5 (b). The high-Q portion of the patterns is given in the insets on an enlarged scale.

Compton scattering, for sample absorption, normalized, i.e. converted into electron units, and then reduced to structure functions $S(Q)$ and the corresponding PDFs G(r). All data processing was done using the program RAD [6].

3. RESULTS AND DISCUSSION

Vanadium oxide nanotubes are typically up to 15 μm long and have inner diameters between 5 and 15 nm while the outer diameters range from 15 to 100 nm. Their walls consist of several vanadium oxide layers as can be seen in Fig. 1. The lack of long-range order due to the curvature of the tube walls has a profound effect on the diffraction pattern of the materials. While the diffraction pattern of crystalline vanadium oxide shows sharp Bragg peaks (see Fig. 2b) that of the nanotube counterpart shows a pronounced diffuse component and only a few Bragg peaks (see Fig. 1a). This renders the traditional techniques for structure determination inapplicable. However, the atomic PDF of nanocrystallineV_2O_5 is, just like that of the corresponding crystal, rich in distinct, structure related features (see Fig. 3) and lends itself to structure determination. The only difference is that the PDF of the crystalline material persists to very high real space distances as it should be with a material exhibiting a long-range order. The PDF of the nanotubes decays to zero already at 50 Å (see Fig. 3b) due the limited structural coherence in this nanocrystalline material.

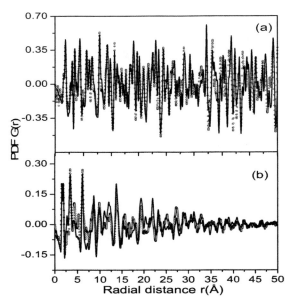

Figure 3. Comparison between experimental (circles) and model (solid line) PDFs for: a) crystalline V_2O_5 and its well known orthorhombic structure [7], b) V_2O_5 nanotubes and the triclinic $BaV_7O_{16}.nH_2O$-type structure [9] refined against the experimental PDF data.

The well-known 16-atom unit cell of crystalline V_2O_5 [7] was fit to the experimental PDF observing the symmetry constraints of the Pmmn space group (see Fig. 3a). The fit was done with the program PDFFIT [8]. It yielded the following cell constants for V_2O_5: a = 11.498(3) Å, b = 3.545(3) Å, c= 4.345(3) Å. They are in good agreement with those obtained by traditional diffraction experiments: a = 11.519 Å, b = 3.564 Å, c= 4.373 Å [7]. The refined atomic positions are also in good agreement with previous results. The agreement well documents the fact that the atomic PDF provides a reliable basis for structure determination. Several structural models were attempted for the vanadium oxide nanotubes [9]. Best fit to the PDF data was achieved with a model based on a triclinic unit cell (S.G. P$\bar{1}$) with parameters a=6.020(3) Å, b=6.1305(25) Å, c=18.973(10)Å, α=93.235(9)°, β=91.067(9)°, γ=90.067(9)° (see Fig. 3b). The model features the walls of the nanotubes as an ordered assembly of double layers of octahedral $V-O_6$ units with a small number of $V-O_4$ tetrahedra embedded into the layers [9]. The outcomes of the structure determination unambiguously show that even a nanocrystal with the complex morphology of vanadium pentoxide nanotubes possesses an atomic-scale structure very well defined on the nanometer length scale and well described in terms of a unit cell and symmetry. The new structural information allows to construct a real-size model of the nanotubes and understand their peculiar morphology [9].

Experimental powder diffraction patterns for the electride samples are shown in Fig. 4. Sharp Bragg peaks are present in the pattern of pristine $Si_{32}O_{64}$. The corresponding G(r)

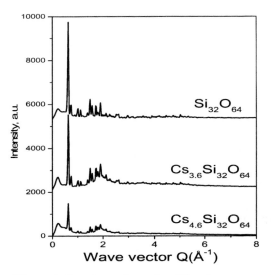

Figure 4. Experimental powder diffraction patterns for $Cs_xSi_{32}O_{64}$ (x=0,3.6, 4.6).

(see Fig. 5) also features sharp peaks reflecting the presence of a long-range ordered network of SiO_4 units in this crystalline material. As expected a model PDF calculated on the basis of the well-known 96-atom unit cell (S. G. I2/m) of $Si_{32}O_{64}$ [10] reproduces the experimental PDF data quite well. Bragg peaks in the diffraction patterns of the samples loaded with Cs are significantly attenuated and already at 2 Å^{-1} merge into a slowly oscillating diffuse component. This reflects the tendency of the zeolite host to become structurally disordered upon absorbing alkali metals. The significant diffuse scattering present in the diffraction patterns of $Cs_xSi_{32}O_{64}$ sample contains information about the atomic ordering of the intercalated Cs atoms. This information is lost in a usual crystallographic studies relying on the Bragg scattering only but is taken into account in the corresponding atomic PDFs shown in Fig. 5. Careful inspection of the figure reveals a new feature at approximately 4 Å in the experimental PDFs of the Cs-loaded samples. The feature grows in line with the number of Cs atoms absorbed in the nanopores and thus can be unambiguously ascribed to atomic correlations involving Cs atoms. To identify the atomic ordering in the assembly of intercalated Cs atoms/ions we considered several structural models and tested them against the experimental PDFs. The models were constructed by placing cesium atoms and/or ions inside the nanopores in $Si_{32}O_{64}$ [11]. Best agreement with the experimental PDF data (see Fig. 5) was achieved with Cs^+ ions occupying positions 4h (1/2, 0.870(5), 0) and 4g (0,0.370(5),0) of the Space group I2/m. According to the model, Cs^+ ions sit close to the walls of the sinusoidal channels in ITQ-4. Since the free space is limited, the intercalated metals are forced to come as close together as possible with increasing doping. As expected, this trend is opposed by an increase in the Coulomb repulsion between the positively charged Cs+ ions. The balance is achieved when most of Cs ions occupy position 4h and 4g

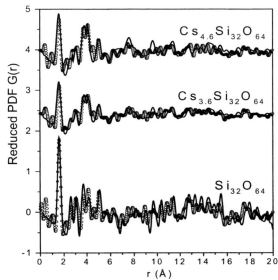

Figure 5. Experimental (circles) and model (line) PDFs for $Cs_xSi_{32}O_{64}$ (x=0,3.6, 4.6). The experimental PDFs are obtained from the diffraction data of Fig. 4.

alternatively forming an assembly of short-range ordered zigzag chains as shown in Fig. 6. Such an extended cationic sublattice will necessitate that its counterpart, the confined electron gas, spreads out along the encapsulating channels rendering the material an electride.

4. CONCLUSION

In summary, the atomic-scale structure of nanocrystals and nanoscale clusters can be determined by employing a non traditional approach such as the atomic PDF technique. The technique relies on diffraction data obtained from the nanomaterial, it is sensitive to fine structural features including the immediate atomic ordering and could easily differentiate between competing structural models. It yields the atomic structure in terms of a small number of sensible parameters such as a unit cell and atomic coordinates that may be used to explain, predict and possibly improve the structure sensitive properties of the material. This a major advantage over the traditional crystallographic techniques that may identify but may not refine the structural characteristics of a nanocrystal. Also, this is a major advantage over employing a statistical description of the structure as it is done with completely disordered materials such as glasses. Thus the PDF technique has all the potential to become the tool for structure determination that is highly needed in the newly emerging field of nanoscience and technology.

ACKNOWLEDGMENTS

The work was supported in part by NSF through grant DMR 0304391(NIRT). APS is supported by DOE under Contract W-31-109-Eng-38. NSLS is supported by DOE under Contract No. DE-AC02-98CH10886. The author thanks S.J.L. Billinge, M.G. Kanatzidis, J. Dye, P. Zavalij, T. Vogt and S. Shastri for the fruitful cooperation.

Figure 6. Fragment of the tetrahedral network in $Cs_xSi_{32}O_{64}$ with Cs ions (circles) in the nanopores assembled in zigzag chains.

REFERENCES

[1] H.P. Klug and L.E. Alexander, *X-ray diffraction procedures for polycrystalline and amorphous materials* (Wiley, New York, 1974).

[2] J. Livage, *Chem. Materials* **3** (1991) 578.

[3] S. Lutta, A. Dobley, K. Ngala, S. Yang, P. Zavalij & M. S. Whittingham, *Mater. Res. Soc. Proc.* **703** (2002) 323.

[4] F. Zhang & M. S. Whittingham, *Electrochem Commun.*, **2** (2000) 69.

[5] J. L. Dye, *Science* **247** (1990) 663.

[6] V. Petkov, *J. Appl. Cryst.* **22** (1989) 387.

[7] R. Wyckoff in *Crystal Structures* (Wiley, New York) 1964.

[8] Th. Proffen & S.J.L. Billinge, *J. Appl. Cryst.* **32** (1999) 572.

[9] V. Petkov, P. Y. Zavalij, S. Lutta, M. S. Whittingham, V. Parvanov & S. Shastri, *Phys. Rev. B.* (2003), submitted.

[10] P.A. Barrett, M.A. Camblor, A. Corma, R.H. Jones & L.A. Villaescusa, *Chem. Mater.* **9** (1997) 1713.

[11] V. Petkov, S.J.L. Billinge, T. Vogt, A.S. Ichimura and J.L. Dye, *Phys. Rev. Lett.* **89** (2002) 075502.

Burned Metal Phenomenon: A Study of Critical Factors and Their Effects on IC Devices during Parallel Lapping

P. M. Batteate and J. Y. Liao

NVIDIA Corporation
2701 San Tomas Expressway, Santa Clara, CA 95050
jliao@nvidia.com
pbatteate@nvidia.com

ABSTRACT

Parallel lapping is widely employed in destructive physical analysis on semiconductor integrated circuits, to reveal defects isolated by electrical failure analysis tools and techniques. A clean sample surface is critical so that the analysts can accurately identify the real defects and not be misled by anomalies possibly introduced during deprocessing. Burned metal has been identified as one type of failure mechanism, caused by electrically overstressing the device. Its presence as failure mechanism is usually validated by electrical and circuit analysis. However, in some cases additional burned metal was also observed on areas unrelated to suspect locations. To understand this phenomenon, we carefully examined the procedure of parallel lapping and conducted a series of experiments to determine the cause and effects of critical factors on IC devices during parallel lapping.

Keywords: IC failure analysis, parallel lapping

1 INTRODUCTION

With the continued acceleration of feature size reduction in semiconductor integrated circuits (ICs), fault isolation complexity is expected to grow exponentially. Development of novel failure analysis technologies, and perfection (and creative approaches) of the existing failure analysis techniques and tools are equally important in identifying the root cause of the failure in an accurate and timely manner [1].

IC failure analysis consists of two categories: electrical failure analysis (EFA) and physical failure analysis (PFA). The failed device is first subjected to a series of vigorous electrical tests and failure analysis techniques to isolate the defect to a suspect area. The unit is then sent to IC device analytical laboratory for deprocessing to reveal the physical anomalies at the suspect fault locations. Semiconductor deprocessing involves a number of critical chemical and mechanical procedures, and one of the most widely employed is parallel lapping. Parallel lapping is performed by mechanically polishing off one process layer at a time in order to observe physical anomalies that may be obscured by the upper process layers of the device. In any destructive physical analysis, a clean sample surface is critical so that the analysts can accurately identify the real defects, and not be misled by anomalies possibly introduced during deprocessing.

Burned metal has been identified as one type of failure mechanism, caused by electrically overstressing the device. The burned metal is usually found in the vicinity of the photon or thermal emission site, or on the signal lines connecting to the light emitting location. Electrical and circuit analyses validate and confirm the physical presence of burned metal as the cause of failure. However, in some cases additional burned metal was also observed through out the device. This raised several questions: Was the defect real? If so, why didn't the device reveal more emission sites during light emission or thermal emission microscopy analyses? Did we introduce this anomaly during parallel lapping?

To understand this burned metal phenomenon, we conducted a series of experiments focusing on parallel lapping procedures. The samples were all taken from the same wafer of graphics ICs.

2 EXPERIMENTS AND RESULTS

In a careful examination of the procedure of parallel lapping, we concluded that the varying factors were elevated temperature and the quality of water; e.g., city water versus de-ionized (DI) water. Temperature (and time duration of the samples exposed to heat) was chosen due to the fact that almost all jobs submitted for parallel lapping are packaged units, and heat must be introduced in order to extract the device from its package. Water was to be considered because of the possible high chlorine content in city water, and chlorine is known to be highly corrosive when in contact with aluminum [2] [3].

Experiment results are presented in Table 1. Temperature and time duration of the samples exposed to heat were chosen according to procedural and experimental data to successfully remove the die from its ceramic package. The temperature is set at 526°C and the duration is 20 minutes.

Optical micrographs and Scanning Electron Microscopy (SEM) photographs in Figures 1 and 2 demonstrate the burned metal phenomena when city water was utilized during parallel lapping. Experiment II was a repeat of Experiment I to confirm the findings. Devices parallel-lapped with DI water did not reveal any burned metal. This observation was also shown on known failed devices subjected to parallel lapping with DI or city water (Experiment III).

After analyzing the results of these three experiments, we conclude that heat has no effect on the metal. However, the samples that were subjected to parallel lapping with city water revealed corroded metal. To study the effect of water during parallel lapping, we engaged Balazs Analytical Laboratory to perform the analysis of water quality. Upon receipt of the results, a fourth experiment was conducted to better understand the corrosiveness of chlorine to aluminum metal lines on ICs.

City water was not utilized in Experiment IV; however, Clorox was added to a beaker of DI water used in the rinse step after polish. Energy dispersive X-ray (EDX) analysis was performed on a sampling of Clorox to determine its elemental composition. Burned metal was observed on both samples. Chlorine and chloride percentages in the high and low concentrations of Clorox used in this experiment were 21,000 µg/g and 1,500 µg/g, respectively.

Experiment I (4 samples)

S/N	Heat	Water	Results
1	No	DI	Clean metal
2	No	City	Burned metal
3	Yes	DI	Clean metal
4	Yes	City	Burned metal

Experiment II (4 samples, repeat of Experiment I)

S/N	Heat	Water	Results
1	No	DI	Clean metal
2	No	City	Burned meta
3	Yes	DI	Clean metal
4	Yes	City	Burned metal

Experiment III (2 samples, known failed devices)

S/N	Heat	Water	Results
1	No	DI	Clean metal
2	No	City	Burned metal

Table 1: Experiments to identify critical factors during parallel lapping

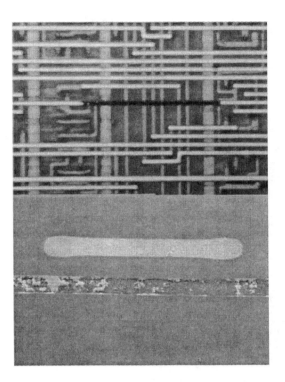

Figure 1: Optical (upper) and SEM (lower) micrograph showing burned metal. The sample was prepared at room temperature and city water was utilized in parallel lapping.

Figure 2: Optical (upper) and SEM (lower) micrograph showing burned metal. The sample was prepared at elevated temperature and city water was utilized in parallel lapping.

City Water: Concentration in *ppm* (*µg/g*)

*Analysis revealed that the ion was not found above the detection limit.

Anions	Detection limit	City Water
Fluoride (F$^-$)	1.0	*
Chloride (Cl$^-$)	0.5	70
Nitride (NO$_2^-$)	0.5	3.6
Phosphate (HPO$_4^=$)	1.0	*
Bromide (Br$^-$)	1.0	*
Nitrate (NO$_3^-$)	1.0	2.7
Sulfate (SO$_4^=$)	1.0	45

Di Water: Concentration in *ppb* (*ng/g*)

Anions	Detection limit	City Water
Fluoride (F$^-$)	2.0	*
Chloride (Cl$^-$)	0.02	0.10
Nitride (NO$_2^-$)	0.02	*
Phosphate (HPO$_4^=$)	0.02	*
Bromide (Br$^-$)	0.02	*
Nitrate (NO$_3^-$)	0.02	*
Sulfate (SO$_4^=$)	0.05	0.08

Table 2: Water laboratory results: anions by ion chromatography performed by Balazs Analytical Laboratory.

Experiment IV (2 samples)

S/N	Heat	Water	Results
1	No	DI (low Clorox)	Burned metal
2	No	DI (high Clorox)	Burned metal

Table 3: Experiments to confirm chlorine's effect on parallel lapping

3 DISCUSSION AND SUMMARY

These experiments demonstrated that additional burned metal observed on areas of the die not pertaining to emission locations was caused by the high concentration of chlorine in city water. Consultation with representatives of Balazs Analytical Laboratory indicated that a single digit reading of chlorine is enough to cause corroded metal. City water contains 70ppm, well over the limit, while DI water contains only 0.10ppb.

It must be noted that the results of the experiment do not imply that burned metal would not occur if DI water were used during parallel lapping. The burned metal witnessed in the cases correlated with electrical failure analysis techniques is known to be real and true, caused by electrical overstress of the devices. Figure 3 shows an example of burned metal resulting from electrical overstress.

The samples subjected to chlorinated DI water revealed corroded metal identical to that subjected to city water. We also observed that devices that failed for high level leakage current were more susceptible to the burned metal response to city water than normal leakage units.

Based on the experiment results and analyses, we have converted all polishing wheels to use DI water only. Hence, burned metal phenomenon on areas not confirmed and validated with electrical fault isolation procedure was no longer experienced.

Figure 3: Optical (upper) and SEM (lower) micrograph showing burned metal resulting from electrical overstress.

REFERENCES

[1] The international technology roadmap for semiconductors, 2001.

[2] Technical Data on Corrosion, Lithonia Lighting, http://www.lithonia.com/industrialcd/TechnicalData/corrosion.htm

[3] Technical Report on Corrosion of Electronics, CorrosionCost, www.corrosioncost.com/pdf/electronics.pdf

[4] Report on Water Quality from Balazs Analytical Laboratory

Elasto-plastic properties of Electroplated and Polycrystalline Nickel using numerical modelling of the nanoindentation test coupled with an inverse method.

M. Qasmi, P. Delobelle and F. Richard,

Applied Mechanics Laboratory, LMARC, UMR CNRS 6604
University of Franche-Comté
24, chemin de l'Epitaphe, 25000 Besançon, France
E-mail : patrick.delobelle@univ-fcomte.fr

ABSTRACT

The aim of this paper is a wide investigation at the nano scale of the elastoplastic properties of electroplated Nickel, which is extensively used in the LIGA techniques and polycrystalline hardened and annealed Nickel with large grains diameter. Using instrumented nanoindentation technique, we have obtained the Berkovich's Hardness H_b and the Young's modulus E. These mechanical properties are affected by the current densities used during the electroplating as well as the nature of the substrates and ion chloride density. Experiments were made on Ni samples having a colony diameter ϕ between 0.2 up to 3µm and a polycrystalline of 150µm of diameter. Results obtained through nanoindentation let us to show that it exists a Hall & Petch relationship between the hardness H_b and the colony diameter. Numerical modelling of the nanoindentation test using a Finite Element Method (FEM) was done in order to identify the elasto-plastic parameters of an isotropic behaviour law of the material using an inverse method.

Keywords : Nanoindentaion, Identification, Young's modulus, Hardness, Grain diameter, Nickel.

1 EXPERIMENTAL TECHNIQUES

The tested materials were on the one hand a pure hardened and annealed Nickel with large grains diameter ($\approx$ 150µm) and in the other hand a very small grains diameter widely used in the development of the MEMS trough LIGA technology [1]. The plating was done on four different layers, Si, Cu, Ti-Cu and Stainless steel and was obtained from a bath of Nickel sulphamate (900gr/l of Ni), Nickel chloride (20gr/l or 5gr/l), boric acid (40gr/l) and wetting agent. The pH of the solution was 4 $\pm$ 0.1 and the temperature was fixed to 55±0.5°C. All the experiments were carried out with a current densities J between 0.16 and 4.1 Adcm^{-2}. The colonies size diameter obtained varied from 0.2 to 3µm according to the current densities and to the Nickel chloride densities. We notice that inside these colonies, there exist a small fine grains of diameter d

($d = 0.21\phi$). Measurement of the grain orientation was investigated on a pure Nickel with large grain and for two samples having a small colonies diameters. The EBSD technique was applied for that. Results show that it exist a light texture <110> for the colonies, the very fine grains are lightly disoriented inside the same colony.

The Berkovich's indentations were performed with NanoInstrument (II-S). The hardness H_b and the Young's modulus E are deduced using the classical or the continuous stiffness methods described as follow. One of the more commonly used methods for analysing nanoindentation data is the Oliver and Pharr [2], which expands on earlier ideas developed by Loubet *et al* [3] and Doerner and Nix [4]. In the Oliver and Pharr method, the hardness H_b and the reduced modulus E_r are derived from:

$$H_b = \frac{P_{\max}}{A} \qquad (1)$$

and
$$\left(\frac{dP}{dh}\right)_{unload} = S = \frac{2\beta}{\sqrt{\pi}} E_r \sqrt{A} \qquad (2)$$

where $P_{\max}$ is the maximum indentation load, A is the projected contact area, and S the unloading stiffness measured at the maximum depth of penetration h and $\beta = 1.034$ for the Berkovich's tip. The reduced modulus is used in the analysis to take into account that elastic deformation occurs in both the indenter and the specimen and it is given by:

$$\frac{1}{E_r} = \frac{1-v^2}{E} + \frac{1-v_i^2}{E_i} \qquad (3)$$

where E is the elastic modulus and v is the Poisson's ratio of the specimen, E_i and v_i are the same parameters for the indenter. The contact area A is given as a function of the contact depth h_c. The contact depth at maximum load is then:

$$h_c = h_{\max} - \varepsilon(h_{\max} - h_f) \qquad (4)$$

where $\varepsilon = 0.72$ for a conical indenter as given by Sneddon [5] and h_f is the final contact depth. The expression for the contact area for a Berkovich indenter is usually approximated by the following formula:

$$A(h_c) = 24.5h_c^2 + \sum_{n=1}^{4}(a_n h_c^{1/n}) \qquad (5)$$

where the first term gives the contact area of an ideal indenter whereas the term present in the sum takes into account geometrical deviations due to the tip rounding. In our study, the coefficients a_n (for n=1-2) where established by using a calibration procedure that takes into account the load frame compliance and allowed calculation of the indenter geometry by reference to a material of well-known Young's modulus. Nanoindentation experiments were made with a fixed maximum depth, which are 125 and 900 nm.

2 FEM MODELLING

In this part numerical modelling of the nanoindentation process is done using the axisymmetric two dimensional finite element method (2D) with uniaxial stress-strain data as input. Lichinchi et al. [6] demonstrate that there is no difference between the (2D) axisymmetric simulation with a conical indenter and the (3D) simulation using a pyramidal indenter. Thus, the indenter and specimen were treated as revolution bodies as shown in (Fig.1). The indenter was modelled as an elastic cone with a semivertical angle $\varphi = 70.3°$ that gives the same contact area to depth ratio as a perfect Berkovich pyramid.

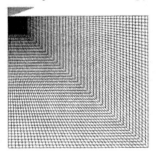

Fig.1: Finite element meshing of half-space

The tip radius of this cone was investigated for the numerical simulation according to the contact depth h_c. In order to identify the radius R, Eq. (5) with $a_1 = 1093$ and $a_2 = 660$ was solved as follow:

$$A(h_c) = 24.5h_c^2 + a_1 h_c + a_2 h_c^{1/2} = S \qquad (6)$$

where S is the cone geometrical surface described by:

$$S = S_1 + S_2 \,, \quad S_1 = \pi a(r+r') \text{ and } S_2 = 2\pi R \delta \qquad (7)$$

where $\delta = R(1 - \cos\alpha)$ is the distance between the spherical apex and the plane passing by the points where the sphere is tangent to the cone (Fig.2). Results obtain for the variation of the radius R is presented in (Fig.3). As we explain before, our investigation was done with maximum depth of 125 and 900 nm. Thus, according to result founded, the radius R is respectively equal to R= 300nm for 125nm maximum depth and R=0 for the other one.

This numerical modeling was established using three finite element codes (CAST3M, ZEBULON and LSDYNA) and then compared to the result in the literature [7]. The friction coefficient at the indenter-material contact was considered constant and equal to 0.2 [8]. To insure the numerical convergence problem, we adopted a progressive meshing increasing from the indented area to the boundary conditions. The dimensions of the substrate are thirty times higher than the maximum indentation depth.

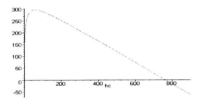

Fig.2 : Indenter modelled with a spherical extremity of radius R

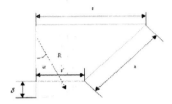

Fig.3: Variation of R depending of hc

2.1 Constitutive law

For the numerical simulation we were considering an isotropic elasto-plastic law in accordance with equations (8).

$$\varepsilon^t = \varepsilon^e + \varepsilon^p \,, \quad d\varepsilon_{ij}^t = d\varepsilon_{ij}^e + d\varepsilon_{ij}^p \,, \quad d\varepsilon_{ij}^e = \frac{1+\nu}{E}d\sigma_{ij} - \delta_{ij}\frac{\nu}{E}d\sigma_{kk}$$

$$d\varepsilon_{ij}^p = d\lambda\frac{\partial f}{\partial \sigma_{ij}} = \frac{3}{2}d\lambda\frac{S_{ij}}{J_2(\sigma)} \,, \quad f = J_2(\sigma) - R - R_0 \qquad (8)$$

$$dR = b(Q-R)d\varepsilon^p \,, \quad R(0) = 0 \,,$$

$$d\lambda = \overline{d\varepsilon^p} = (\frac{2}{3}d\varepsilon_{ij}^p d\varepsilon_{ij}^p)^{1/2} = H(f)\frac{dJ_2(R)}{b(Q-R)}$$

In these equations, E and ν are respectively the Young's modulus and Poisson's ratio, δ_{ij} is the Kronecker symbol, S_{ij} represents the stress deviatoric stress tensor ($S_{ij} = \sigma_{ij} - \frac{\delta_{ij}}{3}\sigma_{kk}$), $J_2(\sigma)$ is the Von Mises equivalent stress [$J_2(\sigma) = (\frac{3}{2}S_{ij}S_{ij})^{1/2}$], R_0 is the initial yield stress, R the scalar hardening variable. After integrating this law (1D), we obtain the classical relation as follow:

$$\sigma = R_0 + Q(1 - \exp(-b\varepsilon^p)) \qquad (9)$$

2

3 INVERSE METHOD

3.1 Introduction

In nanoindentation, the direct identification of the flow stress curve using an analytical model is quite impossible. So to improve this identification an inverse method is used to identify the parameters of the constitutive behavior. The inverse identification is based on optimization techniques to adjust material parameters so that the calculated response obtained by FEM matches the measured one.

3.2 Inverse method

The problem is formulated as follow:

$$\text{Find } P^{opt} \in \Gamma \text{ such as } \psi(P^{opt}) = \underset{\Gamma}{Min}\psi(P)$$

$$\text{With } \psi(P) = \sum_{n=1}^{n=Nm} \frac{1}{2}(F_i^{exp} - F_i^{comp})^2 \quad (10)$$

Where ψ is the objective function, F_i^{comp} the ith computed data, F_i^{exp} the ith experimental data, Nm the number of measurements, P^{opt} the optimal parameters and Γ a constraint space. A Levenberg-Marquardt [9-10] method is used to minimize the objective function.

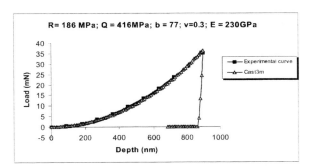

R= 186 MPa; Q = 416MPa; b = 77; v=0.3; E = 230GPa

Fig.4: Loading-unloading curve, comparison between experimental data and simulation results from identification.

Do while $\quad |\nabla\psi| > \varepsilon_{stop}$

(a) $\quad H(P^k) = \nabla F^{comp^T} \nabla F^{comp} + \lambda_{LM} I$

(b) $\quad H(P^k)\Delta P^k = -\nabla\psi(P^k)$

$\quad\quad P^{k+1} = P^k + \Delta P^k$

If $P^{k+1} \in \Gamma$ and $\psi(P^{k+1}) < \psi(P^k) \Rightarrow \lambda_{LM} = \lambda_{LM}/10$

End of iteration k

Else $\lambda_{LM} = \lambda_{LM} *10$ go back again to Eq.(b).

The identification of the parameters (R_0, Q and b) was made for all experimental data and results (Fig.4) are in good agreement between experimental and numerical curves.

4 RESULTS AND DISCUSSIONS

4.1 Influence of the grain size diameter

From measurement of all experimental data of the hardness on different specimens we were able to draw the evolution of H_b according to the colony size diameter ϕ. We report in Fig.5 the variation of the hardness H_b versus $1/\sqrt{\phi}$, which is Hall & Petch relationship given by equation (11).

$$H_b = H_{b_0} + \frac{B}{\sqrt{\phi}} \text{ with } H_{b_0} = 2.0GPa \text{ and } B = 2.5GPa/\mu m^{1/2} \quad (11)$$

This physical parameter ϕ (or d) depends mainly on the manufacturing conditions: nature of the specimens, chloride density, current density J (Fig.6). We also can notice the influence of 30% of strain hardening on the hardness on large polycrystalline grains.

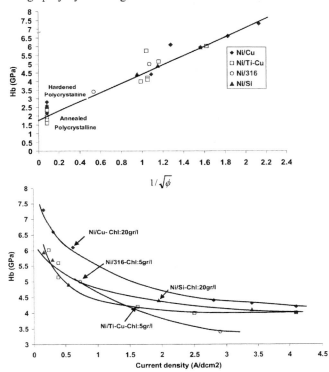

Fig.5 : Linear relation of Hb versus $1/\sqrt{\phi}$.

Fig.6: Evolution of Hb versus the current density for different layer and different ion chloride density.

4.2 Optical profilometer measurement

Measurements were made on hardened and annealed Nickel samples having a grain diameter of 150µm. The mark left by the indenter after the unloading was measured

3

by an optical profilometer (Fig.7). This procedure allowed to determine the average value of the pile-up found around the indents. This allowed to calculate the total projected contact area, which is used to determine the true hardness and the Young's modulus of the specimen. For the different grains orientations the values of experimental modulus $M_{hkl} = \beta_{hkl}(\frac{E}{1-\nu^2})_{isotr}$ are equal to the theoretic values [11-12]: $M_{<111>} = 238\text{GPa}$; $M_{<110>} = 232\text{GPa}$; $M_{<001>} = 214\text{GPa}$ and $M_{isotr} = 230\text{GPa}$.

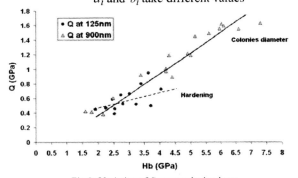

Fig.7: Measurement of the pile-up along the indicate line

4.3 Evolution of the parameters according to the microstructure

Figures 8, 9, 10 show the evolution of the parameters $R_0 + Q$, Q and R_0 versus the hardness H_b. The identification method was done on the indents of 900nm and 125nm of depth. We can notice that all results follow relationships given by:

$$R_0 + Q, Q, R_0 = a_i H_b + b_i \quad \text{and} \quad R_0 + Q = 0.32 \times H_b \quad (12)$$

a_i and b_i take different values

Fig.8: Variation of Q versus the hardness.

The hardness that is depended on plating conditions can give an approximation of the hardening parameters R_0 and Q of the material, E and ν being known. All these parameters allowed plotting the stress-strain relation given by equation (9). Results are in good agreement with those obtained by Dalla Torre et al [13] on solid electroplated samples using tensile test. From these results we are also able to distinguish two types of hardening from grains size effect and from hardened grains (coefficients a_i and b_i are different in Fig.8-9 but it respect Eq(12)). We can also report the influence of the microstructure dislocations, existence of dislocations cells on the hardened material, on

the local behavior law on a small indentation depth and to connect them to the local density of the dislocations [12].

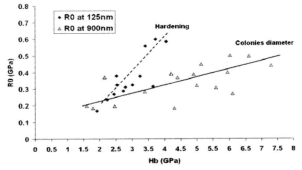

Fig.9: Variation of R₀ versus the hardness.

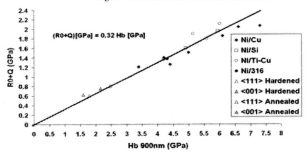

Fig.10: Relationship between R₀+Q and Hb.

5 CONCLUSION

We have presented in this paper a wide investigation trough the nanoindentation test of the mechanical properties of the electroplated Nickel used in LIGA techniques, we studied as well the influence of the grains orientation and of the hardened grains. We have shown that the characterization methods used for the elasto-plastic properties must be carefully analyzed. An identification technique from nanoindentation experimental data is proposed. This identification scheme is based on the combination of an optimisation algorithm and Finite Element Method. Results from experiments give lots of information about the samples microstructures, which depends on the manufacturing conditions.

6 REFERENCES

[1] L. Robert, S. Basrour, A. Sergent, P. Delobelle, M. Schmidt et H. Lowe, Harmst '97, Worldwide LIGA Forum Madison USA (1997).
[2] W. C. Oliver, G.M. Pharr, J. Mater. Res vol 7 (1992) 1564.
[3] J.L. Loubet, J.M Georges, O. Marchesini, J. Tribology 106, 43 (1984).
[4] M.F. Doerner and W.D. Nix, J. Mater.Res. 1, 601 (1986).
[5] I.N. Sneddon, Int. J. Eng. Sci. 3 (1965) 47.
[6] M. Lichinchi, C. Lenardi, J. Haupt, R. Vitali, Thin film solid 312 (1998) 240-248.
[7] H. Pelletier, J. Krier, A. Cornet, P. Mille, Thin film solid 379 (2000) 147-155.
[8] B. Taljat, T.Zacharia, F.Kosel, Int. J. Solid Struct. 35 (33), 4411-4426 (1998).
[9] K. Levenberg, Quart. Appl. Math.. 2, 164-168, 1944.
[10] D.W. Marquardt, SIAM J. Appl. Math., 11, 431-441, 1963.
[11] J. Vlassak, W.D. Nix, Mech. Phys. Solids, vol 42 n°8, 1223-1245 (1994).
[12] P. Delobelle, H. Haddou et X. Feaugas Matériaux 2002 Tours (France).
[13] F. Dalla Torre, H. Van Swygenhoven, M. Victoria, Acta Materiala, 50 (2002) 3957-3970.

4

MEMS Resonator Tuning using Focused Ion Beam Platinum Deposition

S. Enderling*, Liudi Jiang*, A.W.S. Ross*, S. Bond*, J. Hedley**, A. J. Harris**, J. S. Burdess**,
R. Cheung*, C. A. Zorman***, M. Mehregany*** and A. J. Walton*

*Scottish Microelectronics Centre, University of Edinburgh,
Edinburgh EH9 3JF, UK, Stefan.Enderling@ee.ed.ac.uk
**Newcastle University, Newcastle-upon-Tyne NE1 7RU, UK, John.Hedley@ncl.ac.uk
***Case Western Reserve University, Cleveland, Ohio, USA, caz@cwru.edu

ABSTRACT

This paper presents a novel post-fabrication tuning method which changes the resonant frequency of micromechanical beam resonators using Focused Ion Beam (FIB) deposition and removal of platinum. Tuning was achieved by depositing platinum on a $13 \times 5\mu m$ surface area at the tips of cantilever and the middle sections of bridge resonators in thicknesses ranging from 0.3 to $3.1\mu m$. Measurements on both types of resonator structures showed a maximum frequency change of -12% for 2.4 and $3.1\mu m$ thick deposition. A decrease was observed in the quality (Q) factor due to the damping effect of the platinum material and the increased surface roughness of the resonator. After deposition, the change in resonant frequency was re-adjusted by precise milling of the deposited platinum.

Keywords: resonators, frequency tuning, focused ion beam

1 INTRODUCTION

Characteristics such as high-Q factors, good temperature stability and favourable aging properties make micromechanical resonators attractive components for RF filter [1] applications. In order for these frequency-selective devices to deliver optimum performance, a precise resonant frequency is often required. This is not always possible, since fabrication tolerances, stresses and defects result in a resonant frequency which is different from the designed value [2]. In this situation a frequency tuning method can be used to adjust the resonant frequency. Other researchers have achieved this by changing the stiffness; using localised filament annealing [3]; or by altering the mass of the resonator employing either a localised Chemical Vapour Deposition process [2] or a laser [4]. However, when high temperatures are employed for frequency tuning, stress is generated in the material and results in deformation of the resonator structure. Additionally, laser deposition processes are often limited to the minimum laser spot size ($10\mu m$) which is critical when resonator sizes become smaller. Most of the tuning methods are not able to re-adjust the frequency after tuning. This can be problematic as the resonant frequency changes over time due to stress release in the device material after packaging [5]. In this paper, a new frequency tuning method is demonstrated which uses FIB induced platinum deposition onto silicon carbide beam resonators. A FIB system has been employed before to change the resonant frequency by altering the resonator stiffness [6]. However, this paper uses the system to alter the resonator mass. The tuning process presented has the advantage of precise material deposition at room temperature and allows the frequency shift to be re-adjusted through removal of the previously deposited platinum.

2 FIB TUNING PROCESS

A FIB system uses a beam of gallium ions to physically sputter material from the sample surface or to deposit different materials such as platinum. In the deposition mode, an organometallic precursor gas (trimethyl platinum) is decomposed under a high energy (30kV) gallium ion beam leaving platinum on the surface of the resonators (Fig. 1).

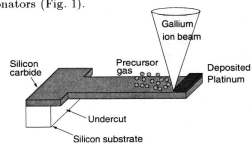

Figure 1: Focused ion beam platinum deposition.

2.1 SIMULATION

In order to evaluate FIB platinum deposition for resonator tuning, structures were first simulated using a Finite Element (FE) model, which predicted the change in resonant frequency due to the deposited material. The structures modelled were $200\mu m \times 15\mu m \times 2\mu m$ 3C silicon carbide cantilever and bridge resonators which included an undercut shown in the attachment region in Fig. 1 [7]. The software employed for the FE-analysis was Coventorware 2001.3. The physical material properties used for the simulations are summarised in Table 1.
An ideal rectangular cross section and a uniform resonator thickness were assumed during the simulations.

Property	3C SiC	Platinum
Young's modulus, E [GPa]	400	168
Density, ρ [$kg\times m^{-3}$]	3230	21500
Poison ratio, η	0.377	0.22
Tensile stress, T [MPa]	400	0

Table 1: Material properties of 3C SiC [7], [9], [10] and platinum [11].

In order to evaluate the mass tuning effect, a platinum layer with a fixed $13\times5\mu$m surface area and a range of thicknesses from 0.3 to 3.1μm was placed on the tips of cantilevers and the middle sections of bridges. Fig. 2 presents the simulated frequency change due to the platinum deposited on the silicon carbide resonators. There is a linear decrease in frequency with increasing material thickness. The simulated maximum frequency change was -23% for the cantilevers and -19% for the bridges with 2.4μm and 3.1μm deposited platinum, respectively.

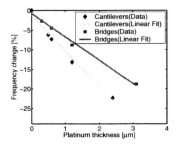

Figure 2: Simulated frequency change of silicon carbide cantilever and bridge resonators versus deposited platinum thickness.

3 EXPERIMENTAL RESULTS

A FEI FIB 200 workstation was employed to deposit platinum onto silicon carbide cantilever and bridge resonators. For optimum control over the tuning process, a platinum deposition rate of 120nm$\times$min^{-1} was experimentally determined for a beam current of 150pA. Platinum was deposited on a $13\times5\mu$m surface area of the resonators in thicknesses ranging between 0.3 and 3.1μm. Fig. 3 shows deposited platinum on a silicon carbide cantilever and bridge resonator.

The tuning process was characterised using the optical 'workstation' shown in Fig. 4 [8]. The resonator samples were attached to a piezoelectric disc and inertially actuated in a vacuum.

Their resonant peaks were detected using a laser vibrometer and the deposited platinum thicknesses measured by a surface profiler. The change in resonant frequency was obtained by measuring the fundamental resonant frequency of each resonator before and after the depos-

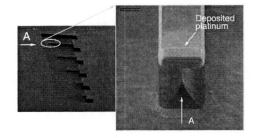

(a) Platinum on the tip of a cantilever.

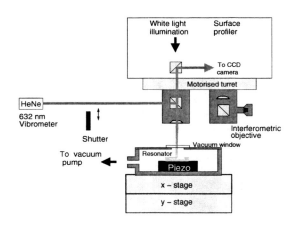

(b) Platinum on the middle of a bridge.

Figure 3: Selective deposition of 1.2μm platinum on a $13\times5\mu$m section of the silicon carbide resonators. The arrow 'A' indicates the the viewing direction for the enlarged picture of the cantilever.

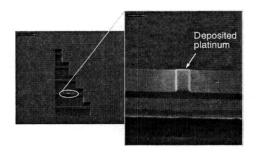

Figure 4: Optical 'workstation' for tuning characterisation.

ition process. The measured change in resonant frequency as a function of platinum thickness is presented in Fig. 5.

It can be observed that the resonant frequency of the cantilevers decreases with thickness in a linear fashion as predicted by the simulation and in the literature for laser deposition [4]. Bridges show a deviation from a lin-

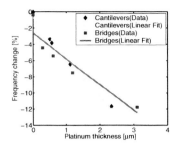

Figure 5: Measured resonant frequency change of silicon carbide cantilever and bridge resonators as a function of deposited platinum thickness.

ear decrease in frequency change. It is believed that this is due to slightly off-centered platinum position and temperature variations between measurements before and after deposition. The maximum measured change in resonant frequency was found to be -12% for both cantilevers and bridges with a platinum thickness of 2.4μm and 3.1 μm, respectively.

The simulation results differ markedly from the measurement data by up to 92%, because properties of pure platinum and a specified value for the silicon carbide thickness were assumed for the FE-analysis. In order to calibrate the simulations with the measurement data, the thickness of the silicon carbide was determined using an ellipsometer (SOPRA SE-5). Measurements gave a material thickness of 2.5μm instead of the 2μm assumed for the previous FE-simulation. After the resonator thickness was known, the value of bulk platinum density was changed during the simulation from $\rho_1 = 21500 kg \times m^{-3}$ to $\rho_2 = 13400 kg \times m^{-3}$. Fig. 6 presents the re-simulated frequency change using the modified silicon carbide thickness and platinum density.

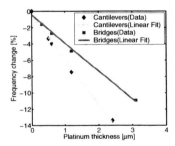

Figure 6: Re-simulated frequency change of silicon carbide cantilever and bridge resonators versus deposited platinum thickness.

Using these parameters, the maximum frequency change was -13% for the cantilevers and -11% for the bridges which brings the simulation and measurement results

into close agreement. The small discrepancy is probably due to the assumed ideal rectangular cross section of the resonators and temperature variations between measurements. Fig. 7 shows an example of the normalised amplitude of a cantilever resonator as a function of frequency before and after platinum deposition. A clear decrease can be observed in the resonator's quality factor (Q) after the deposition process. In this case, 0.5μm of platinum has been deposited onto the resonator.

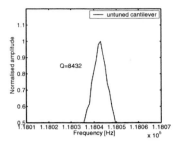

(a) Before platinum deposition.

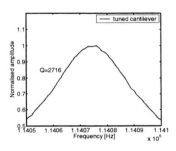

(b) After platinum deposition.

Figure 7: Normalised amplitude as a function of frequency for a silicon carbide cantilever before and after platinum deposition.

The decrease in Q can be caused by: 1) the increased resonator surface roughness [12] due to the ion bombardment during FIB operation and 2) the damping effect of the deposited platinum film. It is believed that a lower beam current for sample imaging will reduce the surface roughness and retain a higher Q. However further investigations are required. In order to demonstrate the ability to re-adjust the change in resonant frequency, 375nm of platinum was removed from a cantilever and 600nm from a bridge resonator. For the platinum removal, a beam current of 350pA was used. The resonant frequency decreases after FIB platinum deposition (FIB step 1) and increases after platinum removal (FIB step 2) as shown in Fig. 8.

The resonant frequency of the bridge after milling is

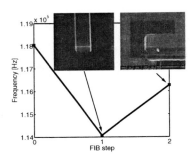

(a) Re-adjustment of silicon carbide cantilever tuning.

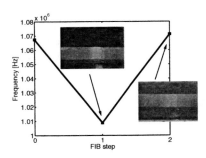

(b) Re-adjustment of silicon carbide bridge tuning.

Figure 8: Stages of silicon carbide resonator tuning. FIB step 1: platinum deposition and FIB step 2: platinum milling.

slightly higher than before deposition. This is because a small amount of silicon carbide was removed during milling. The cantilever has a lower resonant frequency than before deposition, because not all platinum was removed from the resonator.

4 CONCLUSIONS

A novel post-fabrication frequency tuning method for micromechanical resonators using FIB platinum deposition and removal has been demonstrated in this paper. The resonant frequency was decreased by platinum deposition on a $13\times5\mu m$ surface areas of cantilever and bridge resonators with thicknesses ranging from 0.3 to $3.1\mu m$. There is a slight difference in the maximum frequency change obtained from the simulations and the measurements. Modelling can only serve as a guide to predict the frequency tuning of resonators. For accurate tuning control, it is therefore necessary to monitor the change in resonant frequency during the tuning process. The Q-factor of the resonators decreases after deposition probably due to the increased surface roughness caused by the gallium ion beam and damping of the platinum material. In addition, the resonant frequency was re-adjusted by removal of some or all of the platinum

from the resonators. FIB can deposit smaller amounts of platinum than other published methods [4]. Therefore, this method has major attractions for changing the resonant frequency of small (high frequency) resonators. Furthermore, the method offers the flexibility of being able to remove platinum after deposition to compensate for changes in the tuned resonant frequency over time. Hence, FIB induced platinum deposition provides a precise and re-adjustable tuning process.

ACKNOWLEDGEMENTS

The authors would like to acknowledge the financial support of EPSRC (GR/R38019), and thank MIAC for access to the FIB system, P. Heinrich from SOPRA for help on the ellipsometer work and Dr. Jon G. Terry for constructive comments.

REFERENCES

[1] Frank D. Bannon, III, John R. Clark and Clark T.-C. Nguyen, *IEEE Journal of Solid-State Circuits*, vol. 35, no. 4, pp. 512 - 526, 2000.

[2] Daphne Joachim and Liwei Lin, *IEEE Journal of Microelectromechanical Systems*, vol. 12, no. 2, pp. 193 - 200, 2003.

[3] Kun Wang, Ark-Chew Wong and Clark T.-C. Nguyen, *Transducers'97*, vol. 1, pp.109-112 , 1997.

[4] Mu Chiao and Liwei Lin, *Transducers'03*, pp. 1820 - 1823, 2003.

[5] R. Ramesham and R. Ghaffarian, *Electronic Components and Technology Conf.*, pp. 666 - 675, 2000.

[6] R.R.A. Syms and D.F. Moore, *IEE Electronics Letters*, vol. 35, no. 15, pp. 1277 - 1278, 1999.

[7] Liudi Jiang, R. Cheung, M. Hassan, A.J. Harris, J.S. Burdess, C.A. Zorman and M. Mehregany, *accepted for publication in Journal of Vacuum Science and Technology B*, 2003.

[8] John Hedley, Alun Harris, Jim Burdess and Mark McNie, *Proceedings of the SPIE*, vol. 4408, pp. 402 - 408, 2001.

[9] Lijun Tong, Mehran Mehregany and Lawrence G. Matus, *Applied Physics Letters*, vol. 60, no. 24, pp. 2992 - 2994, 1992.

[10] S. Roy, A.K. McIlwain, R.G. DeAnna, A.J. Fleischman, R.K. Burla, C.A. Zorman and M. Mehregany *Solid-State Sensor and Actuator Workshop, Hilton Head Island*, pp. 22 - 25, 2000.

[11] G.W.C. Kaye and T.H. Laby, "Tables of Physical and Chemical Constants," *Longman Inc., London and New York*, 15 th edition, 1986.

[12] V.B.Braginsky, V.P.Mitrofanov and V.I.Panov, "Systems with Small Dissipation," *The University of Chicago Press*, 1985.

Uniform Polycarbonate Nanofibers Produced By Electrospinning

N. Kattamuri and C. Sung

Center for Advanced Materials, Department of Chemical and Nuclear Engineering,
University of Massachusetts, Lowell, MA 01854. Changmo_sung@uml.edu

ABSTRACT

Electrospinning is a process of generating very fine fibers in the sub micron to nanorange by applying a voltage to a polymer in solution. The high electrostatic potential overcomes the surface tension of the polymer and makes it stretch into fibers. In the present work electrospinning of polycarbonate has been attempted to generate uniform nanofibers with minimal bead densities. Electrospinning was performed at various concentrations ranging from 14 to 18%, voltages of 25-30 kV, flow rates of 0.01 to 0.5 ml/min and distances 4 to 10 inches. The ambient parameters like temperature and humidity were monitored throughout the experiment. After imaging the spun fibers with SEM, image processing was done to find the fiber thickness and bead densities. The Design of experiment was performed. It was concluded that a 16% PC with 0.01ml/min, distance of 10 inches, voltage of 30 kV, temperature of 28°C and humidity around 32% could generate uniform PC nanofibers with minimal bead densities.

Keywords: Electrospinning, Polycarbonate, Nanofibers, Design of Experiment

1. INTRODUCTION

Electrospinning is an emerging technology being used for the production of fibers in the nano diameter size. Fibers of such small size have large surface area to volume ratios and very small pore sizes. Due to these advantages, the nanofibers are being used in many commercial applications like manufacture of filtration membranes, fabrication of protective clothing, nanotubes etc. However the process itself is not as complicated. A polymer is chosen and is dissolved by means of a suitable solvent. The complete dissolution is done through an ultrasonicator. The dissolved solution is taken in a syringe attached to a syringe pump. The pumping of the flow meter ejects a drop of liquid from the tip of the syringe. A voltage is supplied to the polymer from a voltmeter via an electrode. The surface tension of the polymer is overcome because of this electric potential. The polymer undergoes a whipping process and gets stretched into fibers which collected as a mat on an aluminum target placed opposite to the syringe.

2. EXPERIMENTAL CONDITIONS

The polymer polycarbonate of bisphenol A of molecular weight 28,000 Daltons was purchased from Aldrich Chemical Company Inc. The solvents used were Tetrahydrofuran (THF) and Dimethylformamide (DMF) which were also purchased from Aldrich. A solvent mixture ratio of THF: DMF = 60:40 was used throughout the experiment. The experimental set-up of electrospinning is shown in Figure 1. At the beginning of the work a design of experiment was planned with suitable parameters with Minitab software. Experiment was conducted for these combinations of parameters to get nanofibers. The spun fibers were then imaged with an Amray 1400 Scanning Electron Microscope and Philips EM 400 Transmission Electron Microscope.

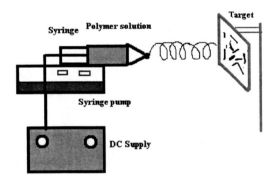

Figure 1: Experimental set-up of electrospinning

3. RESULTS AND DISCUSSION

3.1 Effect of concentration

Electrospinning was performed for polycarbonate concentrations varying from 14% to 18%. It was found that as the concentration increased the fiber thickness also increased, thereby making the fibers at 18% the thickest. It can be observed from the SEM micrographs of Figure 2 that there was an increase of fiber thickness from 75.4 nm at 14% concentration to 179.6 nm at 18% concentration. The reason for increased fiber thickness was that the increased solution viscosity made it difficult for the polymer to be whipped into fibers.

14%   18%

Figure 2: Increase of fiber thickness with concentration

During electrospinning, fibers may not be formed at all times. Due to poor processing parameters, sometimes the polymer might not spin properly and form beads instead of fibers. Our aim was to try to optimize parameters and reduce the beads to a minimal value. It was also found that as the concentration increased there was a decrease in bead density [2]. This is attributed to the increase in viscosity of the solution.

Figure 3 shows the decrease of bead density from 29% at 14% concentration to 9% at 18% concentration. But, since the aim of the present work was to obtain nanofibers with minimal bead density a optimized concentration had to be chosen which had thin fibers and yet very few beads.

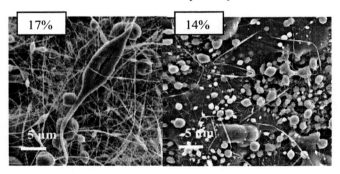

Figure 3: Effect of concentration on bead density

3.2 Effect of flow rate

The rate at which the polymer was being ejected from the tip of the syringe was controlled with the help of a syringe pump. Flow rates from 0.01 to 0.5 ml/min were applied to study the effect of changing flow rates on fiber morphology. It was found that too high flow rates did not yield fibers as the polymer shot out of the syringe so fast that it did not have enough time for whipping. However a very low flow rate was also not preferable due to the time taken to eject the polymer. So, an optimized value between the two extremes had to be found out.
It was found that as the flow rate increased, the fiber thickness and bead density also increased. It can be seen

from the SEM micrographs of Figure 4 that smaller beads were obtained at 0.05 ml/min but very large beads were obtained at 0.09 ml/min

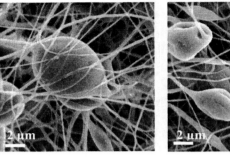

0.09 ml/min   0.05 ml/min

Figure 4: Effect of flow rate on bead density

4. IMAGE ANALYSIS

4.1 Measurement of bead density

Bead density measurement was performed with the help of Scion imaging software. A high contrast SEM image was taken. The software recognized the difference in contrast in the image after following a series of steps of inverting and thresholding. The beads were counted and the area of each bead measured with the help of the software. The bed density was then calculated by the formula,

Bead density = [Area of beads in image/Area of image]*100.

The image in Figure 5 shows the image processed image for bead density measurement

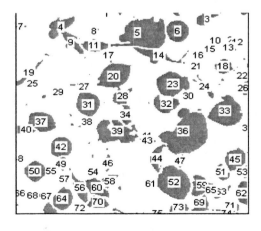

Figure 5: Bead density measurement

4.2 Fiber thickness Measurement

Fiber thickness was measured with the help of GAIA Blue software. Lines were drawn on each fiber in the image and the software gave the thickness of the fiber according to the pre-calibrated scale. Approximately 60 fibers were marked on each fiber and three images were image processed to generate an average fiber thickness value for good approximation.

5 DESIGN OF EXPERIMENT

The design of experiment (DOE) was performed by using Minitab software to find the most significant parameter out of all the parameters involved in the experiment. After the image processing the fiber thickness and bead density values were applied into the worksheet given by Minitab. The software then analyzed the data and produced Pareto charts which showed the most significant parameters for this system. From the Pareto chart in Figure 6, concentration was found to be the most significant parameter to reach the goals of fiber thickness and bead density.

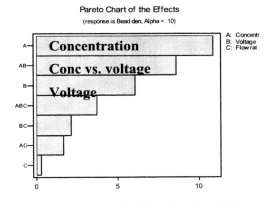

Figure 6: Pareto charts of effects for fiber thickness

The software also gave the main effects plot shown in Figure 7 which interprets that as the concentration increases the fiber thickness increases. From the cube plot shown in Figure 8 we were able to optimize the concentration at a value of 16%. The plot served as a map to show the direction to follow for the rest of the experiment. So, in order to reach the value of 54.2 nm it showed that the voltage and flow rate have to be increased and a concentration of 16% has to be used.

The DOE also helped to predict a theoretical model to the experiment. This is useful because we are able to develop some form of model to a completely unknown system.

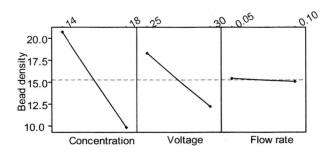

Figure 7: Main effects plot

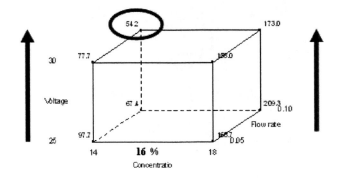

Figure 8: Cube plot

6 EFFECT OF DISTANCE

Since concentration had been optimized at 16%, in the next part of the experiment the flow rate and the distance were varied to see the effect on the fiber morphology. Distance was varied from 6 to 10 inches and flow rate was varied from 0.01 to 0.1 ml/min. It was found that uniform nanofibers were obtained for 10 inches and not for 6 and 8 inches. The reason was that the 6 and 8 inches were studied during the winter time when the temperature was less, about 20.1°C and humidity was about 26%. But when the spinning was done for 10 inches, the temperature had risen considerably to 28.2°C and humidity was 34%. Since all other parameters were constant, we suspect that the humidity and temperature have a major effect on the solvent evaporation rate. It can be seen from the SEM micrographs of Figure 9 that the fibers obtained at 6 inches have a lot of beads and are non-uniform. But the fibers obtained at 10 inches are more or less uniform and have very few beads.

6 inches

2 μm

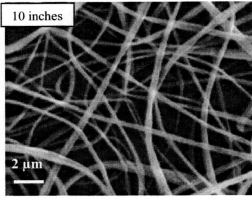

10 inches

2 μm

Figure 9: Fibers formed at 6 and 10 inches target distance

On comparing the fibers that were obtained from previous studies we found that smoother and more uniform fibers were obtained now. Previously due to the high solvent mixture ratio of THF: DMF, the DMF evaporated slowly from the fibers after they were formed and caused pores on them. But in the present study, since a right solvent mixture ratio has been used we obtained uniform and smooth nanofibers as seen in the TEM micrograph of Figure 10

0.39 μm

Figure 10: TEM micrograph of smooth polycarbonate nanofibers

7 CONCLUSIONS

From the present work it has been shown that nanofibers of polycarbonate can be produced with minimal bead densities at a concentration of 16%, voltage of 30 kV, flow rate of 0.01 ml/min, distance of 10 inches, solvent mixture ratio of THF:DMF = 60:40, temperature of 30°C and humidity of around 35%. From the statistical analysis of the data it has been concluded that concentration is the most important parameter to reach our goals.

REFERENCES

[1] Shin, Y.M. "Experimental characterization of electrospinning: the electrically forced jet and instabilities". Polymer, Volume 42, Issue 25, Pages 09955-09967, December 2001.
[2] Deitzel, J. M. "The effect of processing variables on the morphology of electrospun nanofibers and textiles", Polymer, Volume 42, Issue 1, Pages 261-272, January 2001
[3] Viswanathamurthi.P, "Vanadium pentoxide nanofibers by electrospinning", Scripta Materialia, Volume 49, Issue 6, Pages 577-581, September 2003.
[4] MacDiarmid, A.G. "Electrostatically-generated nanofibers of electronic polymers". Synthetic metals, pg 27-30, 119 (2001).
[5] Fong, H. "Beaded nanofibers formed during electrospinning". Polymer, Volume 40, Issue 16, Pages 4585-4592, July 1999.

DOE Optimization and Phase Morphology of Electrospun Nanofibers of PANI/PMMA Blends

K. Desai, C.Sung

Center for Advanced Materials,
Department of Chemical & Nuclear Engineering,
University of Massachusetts, Lowell, MA, Changmo_Sung@uml.edu

ABSTRACT

Electrospinning derived from electro spraying is a process by which sub-micron polymer fibers can be produced using an electrostatically driven jet of polymer solution. The fibers are collected as non-woven mat and offer a high surface to volume ratio. Polyaniline (PANI) is an organic conducting polymer and can be used to fabricate conducting nano fibers by blending with suitable polymers like poly methyl methacrylate (PMMA). In this present work we have explored the effects of electrospinning parameters on the formation of PANI/PMMA fibers using design of experiments (DOE) and the phase morphology of the electrospun fibers using transmission electron microscopy (TEM) and electrostatic force microscopy techniques (EFM).

Keywords: Electrospinning, Polyaniline, TEM Staining, Electrostatic Force Microscopy, Design of Experiments.

1 INTRODUCTION

Electrically conducting organic polymers are a novel class of 'synthetic metals' that combine the chemical and mechanical properties of polymers with the electronic properties of metals and semiconductors. Electronically conducting polymers have been studied extensively owing to their applications in energy conversion devices, sensors, electro chromic devices, electromagnetic interference shielding (EMI), electronic circuits etc. Polyaniline is an organic polymer, but is totally unmoldable and insoluble. It is a conducting polymer and exists in three oxidation states [1]. It has varied applications; its first applications were to make conducting coating of plastic materials i.e. printed circuit board production and also corrosion protection.

Electrospinning is a process by which sub-micron polymer fibers can be produced using an electrostatically driven jet of polymer solution [2]. The fibers are collected as a non-woven mat or membrane with high surface area to volume ratio. Since polyaniline is insoluble in most organic solvents it cannot be spun by itself, as for electrospinning a polymer solution is required. Therefore it has to be blended with other polymers to form a polymer solution for spinning. In this study we have blended polyaniline (PANI) with poly methyl methacrylate (PMMA) to form electrospun fibers.

2 EXPERIMENTAL

2.1 Materials

Polyaniline (PANI) with molecular weight 65,000, Campor Sulphonic Acid (CSA), and poly methyl methacrylate (PMMA) with molecular weight 120,000 were purchased from Aldrich and used. The solvent chloroform was also purchased from Aldrich. Solution of polyaniline / poly methyl methacrylate at varying compositions were prepared and electrospun at various conditions to obtain nanofibers.

2.2 Electrospinning Set-Up

The electrospinning apparatus consisted of a DC power source (Gamma High Voltage Research, Inc. Model HV ES 30P/100), where the charged electrode wires were connected to the polymer solution containing syringe. The polymer solution was fed into a syringe pump so that the flow rate of the solution could be controlled. Electrospun fibers were collected on an electrically grounded target. Spinning potentials ranged from 15 to 25 kV. **Figure 1** shows the basic layout of the electrospinning apparatus.

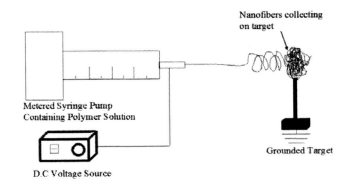

Figure 1: Electrospinning Set-Up

2.3 Design of Experiments (DOE)

Electrospinning is a process by which sub-micron sized polymeric fibers can be easily obtained as a non-woven mat when the polymer solution is sufficiently charged such that the polymeric molecules overcome their surface tension and viscous forces become dominant such that fibers are formed.

The formation of fibers depends on the concentration of the polymers, voltage applied to polymeric solution, the distance between tip and target and the flowrate of the solution.

With so many processing parameters involved in electrospinning to study the effects of individual parameters and the interplay between the key parameters a design of experiments on our process was performed using the Minitab DOE software. A full factorial two level design was used wherein the factors were the polymer concentrations, spinning voltage, tip-target distance and solution flow rates. The responses were characterized in terms of fiber thickness and bead density.

2.4 Phase Characterization

The electrospun fibers were characterized using Scanning electron microscope (SEM, Amray 1400 LaB6) to study fiber morphology, Transmission electron microscope (TEM, Philips EM 400 T) to study phase morphology by staining and Electrostatic force microscopy (EFM, PSIA model XE-100) to study phase morphology of the electrospun fibers. The photomicrographs obtained were then image processed by image processing software (GAIA Blue) to calculate fiber thickness, bead density.

For phase characterization by TEM, the fibers were spun onto a 3mm copper grid which was stained using OsO_4 (Osmium tetroxide), the stain attacks the amide bond in polyaniline giving it a darker contrast compared to the PMMA region.

Electrostatic force microscopy was also used for the phase characterization of the fibers. Since, polyaniline is a conducting polymer when a charged EFM tip is brought in contact with the PANI rich phase it will exhibit a repulsive or attractive force depending on the charge carried by the tip and the material. Polyaniline is positively charged so when a positively charged EFM tip comes in contact with the PANI rich phase would appear darker and if the EFM tip is negatively charged then the PANI rich phase would have a brighter contrast.

3 RESULTS AND DISCUSSION

When a polymeric solution is sufficiently charged it starts spinning into fibers, as the goal was to make nanofibers we have studied the influence of processing parameters in terms of the fiber thickness and bead density. Beads are formed either when the solution is not sufficiently charged or the viscosity of the solution is not sufficient, they should kept minimal.

3.1 Design of Experiments

The experimental runs were generated using Minitab design of experiments software. A two level full factorial design was created involving the four basic electrospinning parameters, the polymer concentrations, the spinning

voltage, tip-target distance and solution flow rate. The experiments were performed according to the runs generated and the response values namely the fiber thickness and the bead density were input into the software so as to perform the statistical analysis.

The software generated a set of graphs showing the importance of individual processing parameters on the response values known as the pareto chart. The pareto chart of effects for fiber thickness is shown in Figure 2.

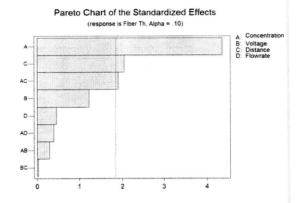

Figure 2: Pareto chart of effects for fiber thickness.

As can be seen from figure 2 concentration and voltage have the most effect on the fiber thickness as we increase the concentration of the polymers in the solution thicker fibers but at the same time a minimal concentration of polymers is needed to generate electrospun fibers otherwise only polymer spraying occurs. [3]. As the spinning voltage is increased thinner fibers are formed as more charge is applied to polymer molecules and they electrospun jet is stretched. Figure 3 shows the plot of effect of processing parameters on the bead density.

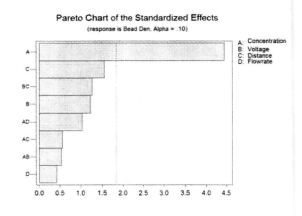

Figure 3: Pareto chart of effects for bead density.

It can be seen from figure 3 that controlling the bead density also polymer concentration and distance between the tip and target are the most important parameters. With increasing polymer concentrations the viscosity of the solution increases and overcomes the surface tension effects and beads are stretched into fibers [3]. It can also be observed that interaction of two parameters also has a significant effect on the response like for fiber thickness interaction of the polymer concentration and the tip-target distance has a significant effect on the fiber thickness thus emphasizing the complexity in understanding the electrospinning process.

Thus by performing a design of experiments we can understand the importance of the effects of the individual processing parameters and further optimize our process to get the desired output. From our experimental and statistical analysis we found for electrospinning of polyaniline/poly methyl methacrylate blends a polymer concentration of 4 wt% PANI-CSA and 16 wt% PMMA and 8 wt% PANI-CSA and 12 wt % PMMA, spinning voltage of 25 kV, tip-target distance of 5 inches and flowrate of 0.3 ml/min is suitable for fabricating of fibers with thickness ranging in the sub-micron range (120 – 800 nm) and bead density of 8% - 28 %.

3.2 Phase Characterization

To understand the phase morphology of the electrospun fibers of PANI/PMMA blends we used staining of polymers technique and electrostatic force microscopy (EFM).

The PANI/PMMA fibers were stained using OsO_4 wherein the stain attacked the amide bond in polyaniline to give it a darker contrast in the TEM images. Figure 4 shows the TEM image of unstained fibers of 8 wt% PANI-CSA and 12 wt% PMMA. Figure 5 shows the TEM image of stained 8 wt% PANI-CSA -12 wt% PMMA fibers showing the presence of polyaniline particles in the PMMA matrix. From TEM images we can see that polyaniline does not form a homogenous mixture with PMMA but aggregates along length of PMMA fibers.

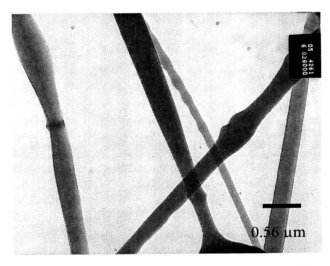

Figure 4: Unstained TEM image of 8 wt% PANI-CSA and 12 wt % PMMA.

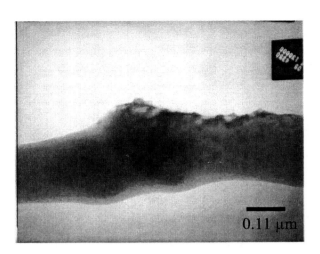

Figure 4: Stained TEM image of 8 wt% PANI-CSA and 12 wt % PMMA.

Polyaniline is a conducting polymer therefore to validate the phase information obtained by TEM we also tried electrostatic force microscopy on the formed fibers and PANI/PMMA thin films. In EFM the AFM cantilever is given a positive or negative bias such that when it scans over a conductive area of the sample the tip will vibrate differently depending on the electrostatic force between the charged tip and conducting surface. If the tip and the sample are of the same bias then the conducting area would appear dark in the EFM image and if both are oppositely charged then it would appear bright. Polyaniline is positively doped with CSA thus giving it a positive charge, therefore when a negatively biased AFM cantilever is brought in contact with the PANI region a brighter contrast is observed in the EFM image.

The difficulty in obtaining a good EFM image of the formed fibers initially propelled us to study the phase morphology of the PANI/ PMMA thin films by EFM. Figure 5 shows the EFM image of the PANI/PMMA thin films it can be seen from them that PANI forms aggregates in the PMMA matrix with some interconnectivity observed within the distinct PANI domains. This is consistent with phase morphology of PANI/PMMA thin films reported earlier [4]. The EFM images of the PANI/PMMA fibers as shown in figure 6 also shows similar information wherein PANI is present as aggregates in the PMMA region and some interconnectivity observed between the distinct PANI domains.

Thus from the phase characterization of the PANI/PMMA fibers by both EFM and TEM staining we can conclude that PANI forms aggregates along the length of the PMMA fibers. However, to obtain conducting nanofibers out of these blends we would have to obtain inter connected network of PANI particles in the PMMA matrix and this could be achieved using a higher molecular weight of PANI.

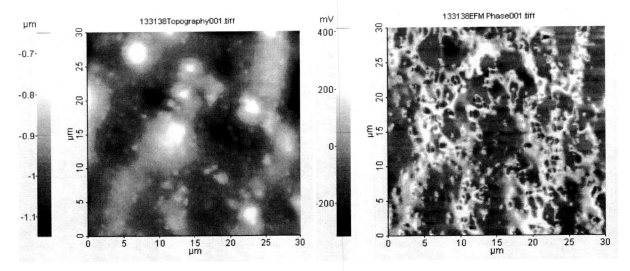

Figure 5: EFM image of 8 wt% PANI-CSA and 12 wt% PMMA thin films (bright area is PANI phase).

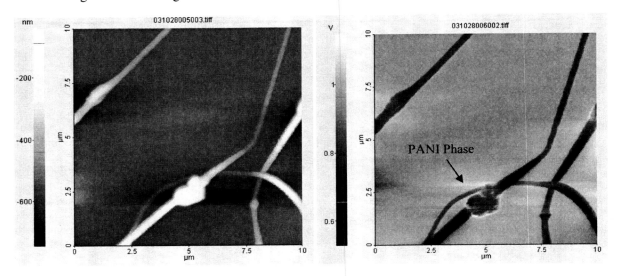

Figure 6: EFM image of 8 wt% PANI-CSA and 12 wt% PMMA fibers.

4 CONCLUSIONS

Electrospinning is a relatively easy method of fabricating nanofibers of polyaniline blends. There is a strong dependence of the processing parameters on the obtained fiber morphology and the desired fiber size can be obtained by controlling the processing parameters as has been observed by design of experiments. The phase characterization of the fibers reveals the presence of PANI aggregates along the length of the fibers, thereby to form conducting nanofibers the process needs to be further optimized so as to get more dispersion of PANI in PMMA perhaps by using a higher molecular weight of PANI in the blend.

REFERENCES

[1] B.Wessling, Synthetic Metals, 85,1313-1318, (1997).
[2] J. M. Deitzel, J. Kleinmeyer, D. Harris and N.C. Beck Tan, Polymer **42**, 261-272 (2001).
[3] K.Desai and C.Sung, Mater.Res.Soc.Proc., 736, 121-126,2002.
[4] C. Y. Yang, Y. Cao, Paul Smith and A. J. Heeger, Synthetic Metals 53, 293-301 (1993).

Phase Identification and Elastic Property of Blend Copolymer Characterized by Force Modulation Microscopy and Force-Distance Curve

Bongwoo Kang, Jun S. Lee, Lan Pham, and Changmo Sung

Center for Advanced Materials, Department of Chemical Engineering and Nuclear Engineering, University of Massachusetts, One University Avenue, Lowell, MA 01854
Bongwoo_kang@uml.edu

ABSTRACT

The technique of polymer blending has been used to create new types of polymers with desirable properties in the past decade. Force Modulation Microscopy (FMM) provides a local contrast due to the local elasticity difference of a sample surface in addition to topography information [1]. The mechanical properties are examined continuously over the extended area and force modulation mode is utilized to identify two phases to measure local elastic properties [2]. Poly(styrene-isobutylene-styrene) (SIBS) blended with poly(styrene-maleic anhydride) (SMA) was characterized by Atomic Force Microscopy (AFM) to understand phase identification. Young's moduli of bulk materials were obtained by tensile testing machine from the slope of stress-strain curve [3]. The localized elastic modulus of each phase is obtained by F-D curve through adopting the mathematical theory [4] and image analysis was employed to measure the volume fraction from FMM images at different blending ratios.

Keywords: Force Modulation Microscopy (FMM), Elastic Modulus, Phase Identification, Polymer blends

1. INTRODUCTION

Polymer blending has been a technique used to create new types of polymers with desirable properties. But most polymers are not miscible and as a result will phase separate into their own structural domain. Unfortunately most polymer pairs exhibit poor adhesion between these domains and will ultimately lead to inferior mechanical behavior of the mixture. Therefore understanding the factors that affect morphology and mechanical properties will assist in developing superior plastic materials.

Force Modulation Microscopy (FMM) is an extension of AFM imaging that includes powerful technique for scientific research of the sample's mechanical properties. The FMM image is from the amplitude of the vertical vibration of the tip. It has great resolution for sample features that are difficult in the contact mode of AFM and Electron Microscopy. It is a continuous 2-D mapping of local mechanical properties on nanometer scale. The probe is modulated into contact with a sample and the sample's surface resists the oscillation and cantilever bends.

With Force vs. Distance curves, AFM can distinguish surface regions of different stiffness and adhesion characteristics. In this technique, force applied on the surface is measured by the deflection of the cantilever while approaching and retracting from the surface [7].

2. EXPERIMENTAL

2.1 Sample preparation

SMA   **SIBS**

The SIBS1027 contained 26 wt% styrene and SMA, random copolymer, contained 14 wt% maleic anhydride and 86wt% styrene. Polymer blends were made with various compositions as described in Table 1. Each composition was blended from a solution consisting of 10% solids with 65% tetrahydrofuran (THF) and 25% toluene. Toluene was added to prevent moisture or water buildup that may cause the SIBS to precipitate during the mixing stage. The solution was stirring for at least 24 hours. Immediately after the solution was stopped, methanol was used to precipitate the blended polymers. The precipitated material dried for at least 2 days at room temperature and final solvent removal occurred in a vacuum oven at 70°C for at least another 2 days.

In order to perform further testing, the blended material was then compression molded into thin flat sheets. Teflon coated aluminum sheets were used to sandwich the material to produce a flat, smooth surface and prevent the material from sticking to the heated plates. An aluminum mold was used to make a uniform film thickness of approximately 1.3mm. The material was heated at a temperature of 160°C for 25 minutes. Afterwards, approximately 6-7 metric tons of force (1300-1600psi) was applied onto the sample for another 25 minutes. In order to complete the process the material was set aside to cool at room temperature.

Table 1: Compositions of Blends (weight %)

Sample #	SIBS1027	SMA
1	100	0
2	80	20
3	60	40
4	40	60
5	20	80
6	0	100

2.2 Force Modulation Microscopy

This study examined the morphology of SIBS and SMA blends at different compositions by FMM as the two polymers have different elastic properties and these images were compared with Scanning Electron Microscopy (SEM) images. To distinguish the two phases by FMM, samples were embedded in epoxy and polished to produce a smooth flat surface. FMM images were obtained by using on XE-100 microscope manufactured by PSIA Corp. A non-contact tip, stiffer than contact tip, with on silicon cantilever was used. The cantilever's length and width were 125μm and 35μm, respectively. The cantilever had a resonant frequency of 325 kHz and a force constant of 40N/m. Choice of the cantilever is a very critical factor in FMM. Phase and Force Modulation images were obtained at optimized scanning conditions. Image analysis has been carried to measure volume fraction from FMM images at different blending ratios.

2.3 Image Analysis

Image analysis has been carried to measure volume fraction from FMM images at different blending ratios. The volume fraction of SMA phase was obtained by dividing the area of that phase by the total area of image.

Volume fraction of SMA phase

$$= \frac{\text{Total area of one phase}}{\text{Total area of image}} \times 100 \qquad (1)$$

2.4 Elastic Modulus from F-D Curve

Based on the FMM and modulus calculation by the F-D curves, the phase identification can be efficiently verified between the experimental results and the presented values by the F-D data. The slope of curve is a representative of the stiffness of the material. The hard segment, high elastic modulus, has steeper slope than the soft segment.

The cantilever chosen for F-D analysis had a resonant frequency of 105kHz and a force constant of 0.9N/m. The data were acquired at 1.0μm/s down speed and up speed. The regions decided by each phase were measured at the one point and plotted by z-distance and vertical force in the automated computer interface. After analyzing force and distance data, we can obtain the slope of the approaching line which indicates a compression modulus.

3. RESULTS AND DISCUSSION

To characterize the surface morphology of the blend copolymer at the different mixing ratios, SEM and FMM were used. Figure 1 shows that particles ranged from 1-3μm were introduced at the 80wt% SMA in Figure 1(C). But at a concentration of 40wt% SMA in Figure 1(B), the agglomerated particles were no longer present and a co-continuous phase of materials had formed with no defining features. From Figure 1, two phases were difficult to detect with SEM.

However, the FMM results provided further evidence of phase separation between the SIBS/SMA force modulation microscopic images. FMM images revealed differences between the two phases clearly, even though SEM did not show any phase separation images due to poor contrast of two phases.

From the image analysis by GAIA Blue software with FMM image, the volume fraction of SMA was calculated as 18.8% for Figure 1(a), 42.5% for Figure 1(b), and 77.0% Figure 1(c). These values are in good agreement with blending ratio used to formulate the blends.

The microstructure depended on the composition of the blend, compatibility of the two components, fabrication of the blended materials, and the physical properties of the two polymers. Figure 2 shows the Young's Modulus of bulk materials at the different mixing ratio. Small amounts of SMA do not greatly affect the modulus. From 26.7wt% SMA, the modulus increased dramatically.

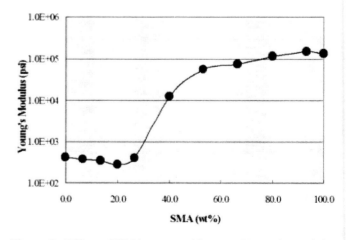

Figure 2: Effect of SMA compositions on Young's modulus from composite materials.

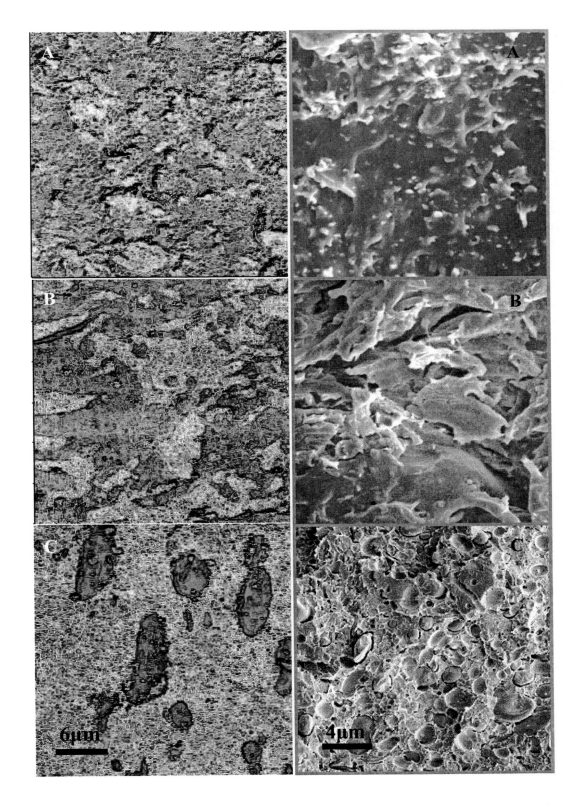

Figure 1: FMM images for phase identification of (a) 20wt% SMA/80wt% SIBS1027, (b) 40wt% SMA/60wt% SIBS1027, and (c) 80wt% SMA/20wt% SIBS1027.

The evaluation of mechanical properties with the AFM has been purely qualitative. The operation of the AFM in force modulation mode allows for a more quantitative characterization of polymer behavior under mechanical loads. This quantitative information gives more confident for phase identification.

The verification of phase identification in the FMM images is important for polymer blends. Even though FMM images show the information of phase separation, we cannot identify each region clearly. In terms of F-D analysis, the comparison of stiffness is able to identify those phases of polymer blends.

In Figure 3, the F-D curve represents the vertical trail of AFM tip's movement. According to this analysis, the data for approaching the tip were collected to analyze the comparison with 1024 data points. The local elastic modulus of the sample is determined from the slope of the initial portion of the force-distance curve as the AFM tip comes into contact with the sample surface.

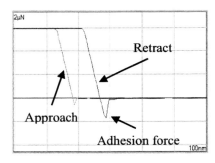

Figure 3: Typical force versus distance curve of SMA / SIBS blends

By means of the linear fitting from the portion of approaching line, the moduli were obtained for each phase. Figure 4(a) shows that the slope of F-D curve, which means the modulus, is 1.785 N/m for the 80 wt% of SMA. The modulus of 20 wt% SIBS is 1.040 N/m. The difference of each modulus is 0.75 N/m. In other words, the SMA phase is 1.7 times bigger than SIBS phase. Consequently in the FMM images, the brighter contrast shows SMA phase, hard segment, and the darker contrast reveals SIBS phase, soft segment. Therefore, we can clearly identify each region of SMA and SIBS qualitatively. FMM images could be easily analyzed due to the stiffness of SMA and SIBS.

4. CONCLUSIONS

The microstructure depended on the composition of the blend, compatibility of the two components, fabrication of the blended materials, and the tensile properties of the two polymers. FMM images clearly revealed differences between the two phases because of various elastic moduli, while secondary electron images of SEM was not clear to show phase differences due to poor contrast. FMM is

proved to be a unique and powerful method to characterize phase separation of polymer blends. From F-D curve of each phase, the local mechanical properties such as modulus were compared between the two phases. Therefore, the phase identification in the FMM images can be proved by combining the force analysis of AFM.

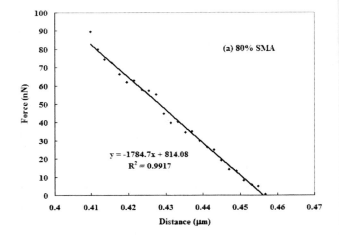

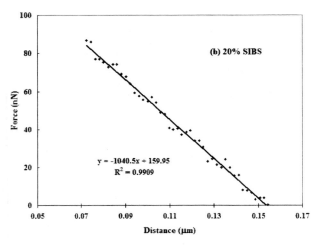

Figure 4: Linear fitting of tip approaching region, blend ratio of SMA/SIBS = 80/20

REFERENCES

[1] D. R. Paul and S. Newman (eds.), "Polymer Blends, Vol. 1 and 2," Academic Press, New York, 1978.
[2] D. R. Paul, Ch. 12 in "Polymer Blends, Vol. 2," D. R. Paul and S. Newman (eds.), Academic Press, New York, 1978.
[3] N. Dharmarajan and S. Datta, Polymer, 33, 3848 (1992).
[4] S. Datta, N. Dharmarajan, G. Verstrate, and L. Ban, Polym. Eng. Sci., 33, 721 (1993).
[5] J.H. Kim, H. Keskkula, and D. R. Paul, J. Appl. Polym. Sci., 40, 183 (1990).
[6] B. Krache and B. Damaschke, Appl. Phys. Lett. 77, 361 (2000)
[7] N.K Sahoo, S.Thakur, M.Senthilkumar, N.C.Das, Appl. Surface Sci., 206, (2003) 271

Mechanical Parameter Extraction of Thin-Film Coated Materials

Yiquan Li and Kevin K. Tseng[*]

Vanderbilt University, Dept. of Civil & Environmental Engineering
VU Station B 351831, Nashville, TN, USA
[*]corresponding author, e-mail: Kevin.Tseng@Vanderbilt.Edu

ABSTRACT

Materials coated with thin-films have become popular in many advanced applications. The thin layer of high-strength coating can improve the performance of the material. However, challenges arise when characterizing or extracting mechanical parameters for this class of material due to the presence of one or multiple layers of coating. Experimental methods such as nanoindentation and scratch test can be used to determine the mechanical properties for a given specimen. This paper presents a modeling framework for predicting the mechanical property of thin-film coated materials based on the image method. Analytical closed-form expressions for the stress distribution on a coated material subjected to a distributed loading on the surface of the coating material are derived in recursive forms. These recursive equations are implemented numerically to calculate the solutions of the problem. Theses solutions can be used to validate the experimental results of nanoindentation and scratch tests.

Keywords: nanoindentation, thin-film, image method, interface, distributed force

1 INTRODUCTION

Materials coated with thin-films have become popular in many advanced applications including semiconductors and optical storage devices for modern computers. The thin layer of high-strength coating improves the overall mechanical and electro-mechanical performance of the material. However, challenges arise when characterizing or extracting mechanical parameters for this class of material due to the presence of one or multiple layers of coating. Experimental methods involving contact probes such as nanoindentation and scratch test can be used to determine the mechanical properties for a given specimen. This can be used to calibrate a characterization and parameter extraction framework comprising a combination of analytical formulations and numerical algorithms. Such a modeling framework will provide a useful tool for predicting the mechanical property of thin-film coated materials. This paper will present a modeling framework for predicting the mechanical property of thin-film coated materials based on the image point method. Analytical close-form expressions for the stress distribution on a thin-film coated material subjected to a distributed loading on the surface of the coating material are derived in recursive forms. These recursive equations are implemented numerically to calculate the solutions of the problem. These solutions can be used to validate the experimental results of nanoindentation and scratch tests.

2 FOUNDAMENTAL SOLUTION

Consider a semi-infinite plane substrate deposited by a thin film, shown as Fig. 1, subjected to a concentrated force. A concentrated point force applies on the free surface. The upper thin layer is denoted by material 'I' and is with the thickness 'h', shear modulus μ_1, and Poisson's ratio ν_1. The substrate, denoted by material 'II', is a half-infinite plane, of which the shear modulus is μ_2 and Poisson's ratio ν_2. The x-axis is set to lie along the interface. Image points upon the free surface (including surface) are denoted by O_i, and the corresponding local coordinates are expressed by the complex form as $z_k = x + iy_{k1}$. The image points underneath the interface are denoted by C_i with the corresponding local coordinates: $\varsigma_k = x + iy_{k2}$. The relationship between the local coordinates and the global coordinates are:

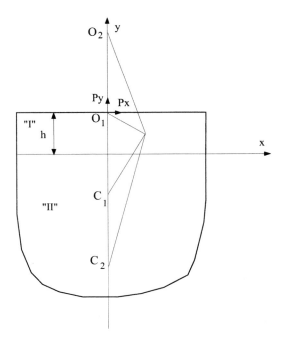

Fig. 1 Analysis model

$$z = x + iy, \quad \overline{z} = x - iy, \quad i = \sqrt{-1}$$
$$z_k = z - 2(k-1)hi \tag{1}$$
$$\varsigma_k = z + (2k-1)hi$$

The elastic fields should satisfy the Goursat formula:

$$\sigma_y + i\tau_{xy} = \varphi' + \overline{\varphi}' + \overline{z}\varphi'' + \psi', \quad \sigma_x + \sigma_y = 4\operatorname{Re}(\varphi') \tag{2}$$
$$2\mu(u+iv) = \kappa\varphi - z\overline{\varphi}' - \overline{\psi}$$

where

$$\begin{cases} \kappa = 3 - 4\nu & \text{for plane strain} \\ \kappa = (3-\nu)/(1+\nu) & \text{for plane stress} \end{cases} \tag{3}$$

Assuming stress functions take the following form:

$$\begin{cases} \varphi_I = \sum_{k=1}^{\infty}[A_k(z_k) + \Phi_k(\varsigma_k)], \quad \varphi_{II} = \sum_{k=1}^{\infty}a_k(z_k) \\ \psi_I = \sum_{k=1}^{\infty}[B_k(z_k) + \Psi_k(\varsigma_k)], \quad \psi_{II} = \sum_{k=1}^{\infty}b_k(z_k) \end{cases} \tag{4}$$

Substituting Eq. (4) into Eq. (2) yields the following set of equations for the stresses in the problem:

$$\sigma_{yI} + i\tau_{xyI} = \varphi'_I + \overline{\varphi}'_I + \overline{z}\varphi''_I + \psi'_I$$
$$= \sum_{k=1}^{\infty}[A'_k(z_k) + \Phi'_k(\varsigma_k)] + \sum_{k=1}^{\infty}[\overline{A'_k}(z_k) + \overline{\Phi'_k}(\varsigma_k)] \tag{5a}$$
$$+ \overline{z}\sum_{k=1}^{\infty}[A''_k(z_k) + \Phi''_k(\varsigma_k)] + \sum_{k=1}^{\infty}[B'_k(z_k) + \Psi'_k(\varsigma_k)]$$

$$\sigma_{xI} + \sigma_{yI} = 4\operatorname{Re}(\varphi'_I) = 4\operatorname{Re}\left(\sum_{k=1}^{\infty}[A'_k(z_k) + \Phi'_k(\varsigma_k)]\right) \tag{5b}$$

$$2\mu_I(u_I + iv_I) = \kappa_I\varphi_I - z\overline{\varphi}'_I - \overline{\psi}_I$$
$$= \kappa_I\sum_{k=1}^{\infty}[A_k(z_k) + \Phi_k(\varsigma_k)] - z\sum_{k=1}^{\infty}[\overline{A'_k}(z_k) + \overline{\Phi'_k}(\varsigma_k)] \tag{5c}$$
$$- \sum_{k=1}^{\infty}[\overline{B_k}(z_k) + \overline{\Psi_k}(\varsigma_k)]$$

$$\sigma_{yII} + i\tau_{xyII} = \varphi'_{II} + \overline{\varphi}'_{II} + \overline{z}\varphi''_{II} + \psi'_{II}$$
$$= \sum_{k=1}^{\infty}a'_k(z_k) + \sum_{k=1}^{\infty}\overline{a'_k}(z_k) + \overline{z}\left[\sum_{k=1}^{\infty}a''_k(z_k)\right] + \sum_{k=1}^{\infty}b'_k(z_k) \tag{6a}$$

$$\sigma_{xII} + \sigma_{yII} = 4\operatorname{Re}(\varphi'_{II}) = 4\operatorname{Re}\left(\sum_{k=1}^{\infty}\overline{a'_k}(z_k)\right) \tag{6b}$$

$$2\mu_{II}(u_{II} + iv_{II}) = \kappa_{II}\varphi_{II} - z\overline{\varphi}'_{II} - \overline{\psi}_{II}$$
$$= \kappa_{II}\left[\sum_{k=1}^{\infty}a_k(z_k)\right] - z\left[\sum_{k=1}^{\infty}\overline{a'_k}(z_k)\right] - \sum_{k=1}^{\infty}\overline{b_k}(z_k) \tag{6c}$$

The continuity conditions at the interface are:

$$\sigma_{yI} + i\tau_{xyI} = \sigma_{yII} + i\tau_{xyII} \quad \text{at} \quad y = 0 \tag{7}$$
$$u_I + iv_I = u_{II} + iv_{II}$$

Tractions free condition at the free surface is:

$$\sigma_{yI} + i\tau_{xyI} = 0 \quad \text{at} \quad y = h \tag{8}$$

Substituting Eqs. (5a)–(6c) into Eqs. (7) and (8), one can obtain the recurrence relationships of the stress functions:

$$\begin{cases} \Phi_k = m_1\left[(\varsigma_k - (2k-1)hi)\overline{A'_k} + \overline{B_k}\right] \\ \Psi_k = -m_2\overline{A_k} - m_1\left\{\begin{array}{l}(\varsigma_k - (2k-1)hi)^2\,\overline{A''_k} \\ + [\varsigma_k - (2k-1)hi](\overline{A'_k} + \overline{B'_k})\end{array}\right\} \end{cases} \tag{9}$$

$$\begin{cases} a_k = (1 - m_2)A_k \\ b_k = (m_1 + m_2)(z_k + (2k-1)hi)A'_k + (m_1 + 1)B_k \end{cases} \tag{10}$$

$$\begin{cases} A_{k+1} = -[z_{k+1} + (2k+1)ih]\overline{\Phi'_k} - \overline{\Psi_k} \\ B_{k+1} = (z_{k+1} + (2k-1)hi)(\overline{\Phi'_k} + \overline{\Psi'_k}) - \Phi_k \\ + (z_{k+1} + (2k+1)hi)[z_{k+1} + (2k-1)ih]\overline{\Phi''_k} \end{cases} \tag{11}$$

where:

$$m_1 = \frac{\beta - \alpha}{1 - \beta}, \quad m_2 = \frac{\alpha + \beta}{1 + \beta} \tag{12}$$

Here α and β are Dundur's parameters[2]:

$$\alpha = \frac{\mu_1(\kappa_2 + 1) - \mu_2(\kappa_1 + 1)}{\mu_1(\kappa_2 + 1) + \mu_2(\kappa_1 + 1)}, \beta = \frac{\mu_1(\kappa_2 - 1) - \mu_2(\kappa_1 - 1)}{\mu_1(\kappa_2 + 1) + \mu_2(\kappa_1 + 1)} \tag{13}$$

For a homogenous semi-infinite plane subjected to a concentrated force on the surface, the stress functions are:

$$A_1 = C\log z_1, \quad B_1 = -\overline{C}\log z_1 + \frac{ihC}{z_1} \tag{14}$$

where

$$C = -\frac{P_x + iP_y}{2\pi} \tag{15}$$

Starting from the stress functions in Eq. (14), the other functions can be derived using the recurrence relationships in Eqs. (9), (10), and (11). The symbolic computation software such as Mathematica or Maple can be used to calculate the stresses.

3 INTEGRAL SOLUTION FOR DISTRIBUTED FORCES

When arbitrarily distributed forces acting on the surface, shown as Fig. 2, the stress and displacement fields can be obtained by superposition, through the integration of the fundamental solutions.

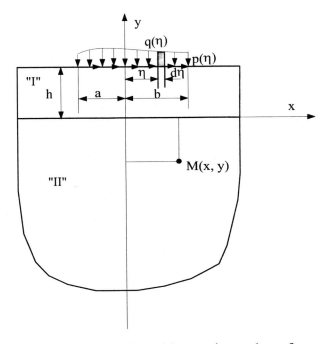

Fig. 2 Arbitrarily distributed force acting on the surface

In Fig. 2, the distribution of the shear and normal forces are $p(\eta)$ and $q(\eta)$, respectively. At the distance η away from the load point along the surface, the equivalent differential loadings are $dP_x = p(\eta)d\eta$ and $dP_y = q(\eta)d\eta$. The stresses and displacements at any point $M(x, y)$ under dP can be calculated according to Eqs. (5) and (6), using $z-\eta$, $z_k - \eta$, and $\varsigma_k - \eta$ instead of z, z_k, and ς_k in the formulas. The results are as following:

$$d(\sigma_{yI} + i\tau_{xyI}) = \{ \sum_{k=1}^{\infty} [A'_k(z_k - \eta) + \Phi'_k(\varsigma_k - \eta)]$$

$$+ \sum_{k=1}^{\infty} [\overline{A'_k}(z_k - \eta) + \overline{\Phi'_k}(\varsigma_k - \eta)]$$

$$+ (\overline{z} - \eta)\sum_{k=1}^{\infty} [A''_k(z_k - \eta) + \Phi''_k(\varsigma_k - \eta)]$$

$$+ \sum_{k=1}^{\infty} [B'_k(z_k - \eta) + \Psi'_k(\varsigma_k - \eta)]\}d\eta \qquad (16a)$$

$$d(\sigma_{xI} + \sigma_{yI})$$
$$= \left\{ 4\,\mathrm{Re}\left(\sum_{k=1}^{\infty} [A'_k(z_k - \eta) + \Phi'_k(\varsigma_k - \eta)] \right) \right\}d\eta \qquad (16b)$$

$$2\mu_I d(u_I + iv_I) = \{ \kappa_I \sum_{k=1}^{\infty} [A_k(z_k - \eta) + \Phi_k(\varsigma_k - \eta)]$$

$$- (z - \eta)\sum_{k=1}^{\infty} [\overline{A'_k}(z_k - \eta) + \overline{\Phi'_k}(\varsigma_k - \eta)]$$

$$- \sum_{k=1}^{\infty} [\overline{B_k}(z_k - \eta) + \overline{\Psi_k}(\varsigma_k - \eta)]\}d\eta \qquad (16c)$$

$$d(\sigma_{yII} + i\tau_{xyII}) = \{ \sum_{k=1}^{\infty} a'_k(z_k - \eta) + \sum_{k=1}^{\infty} \overline{a_k}(z_k - \eta)$$

$$+ (\overline{z} - \eta)\left[\sum_{k=1}^{\infty} a''_k(z_k - \eta) \right] + \sum_{k=1}^{\infty} b'_k(z_k - \eta)\}d\eta \qquad (17a)$$

$$d(\sigma_{xII} + \sigma_{yII}) = \left\{ 4\,\mathrm{Re}\left(\sum_{k=1}^{\infty} \overline{a'_k}(z_k - \eta) \right) \right\}d\eta \qquad (17b)$$

$$2\mu_{II} d(u_{II} + iv_{II}) = \{ \kappa_{II}\left[\sum_{k=1}^{\infty} a_k(z_k - \eta) \right]$$

$$- (z - \eta)\left[\sum_{k=1}^{\infty} \overline{a'_k}(z_k - \eta) \right] - \sum_{k=1}^{\infty} \overline{b_k}(z_k - \eta)\}d\eta \qquad (17c)$$

The coefficients of the stress functions in Eqs. (16) and (17) are determined by the recursive relations expressed in Eqs. (9), (10) and (11), and each term includes C or it's complex conjugate $\overline{C}$. Since $C = -\dfrac{P_x + iP_y}{2\pi}$ in the fundamental solutions, for the differential force dP C will be $C = -\dfrac{p(\eta) + iq(\eta)}{2\pi}d\eta$. Therefore, the stresses and displacements can be integrated with respect to the parameter η using Eqs. (16) and (17) for the case of distributed forces. For instance, integrating equation (16a) we can obtain the normal and shear stresses in material I, the thin layer. The integration can be carried out in complex form and then separate the solutions into real and imaginary parts, corresponding to the normal stress and the shear stress, respectively.

4 COMPARISON WITH NUMERICAL RESULTS

In order to validate of the analytical results presented in the previous sections, some numerical analysis based on finite element method are carried out for the case of uniform distributed normal force along the surface. In such case, $p(\eta) = 0$ and $q(\eta) = $ constant, the region of the distributed loading is taken from -1 to 1, and the thickness of the film is 1mm. The material constants for the coating and substrate are summarized in Table 1. The results are shown in Figs. 3(a) and 3(b).

Material	I	II
Young's Modulus E (GPa)	546	206
Poisson's Ratio ν	0.3	0.3

Table 1 Material constants

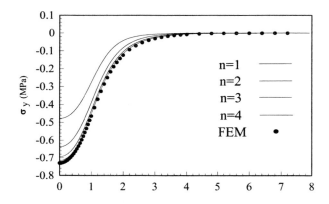

(a) Normal stress σ_y for $p(\eta) = 0$, $q(\eta) = 1 N / mm^2$

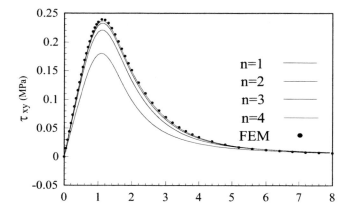

(b) Shear stress τ_{xy} for $p(\eta) = 0$, $q(\eta) = 1 N / mm^2$

Fig. 3 Stress distributions along the interface for the case of normal loading

In the legends of Figs. 3(a) and 3(b), n=k denotes the order of image point. With comparisons to the numerical results from finite element analysis, it can be found that the stresses increments decrease rapidly with the order of the image points increasing. And the stresses superposed with 5 image points have enough accuracy for the material combination listed in the Table 1. These results demonstrate that the convergence rate of the theoretical solutions is very fast.

5 SUMMARY

In this paper, we have formulated a framework for solving the problem of film-substrate material system subjected to distributed loads. In fact, in indentation or nanoindentation tests, the contact distributed forces are not uniformly distributed normal forces. However, many kinds of continuous distributed loads can be simulated by polynomial expression. Noting that all the stress functions are polynomial, except the stress functions corresponding to the first order image point. Therefore, Eqs. (16a) through (17c) can be integrated in closed-form.

It should be pointed out that the displacement fields are needed for validating the results in nanoindentation tests. The authors are conducting on-going research program to solve the displacement fields of this problem and to further extend the loading case to more general forms.

ACKNOWLEDGEMENT

This research is supported in part by a nano-initiative fellowship from Vanderbilt Institute for Nanoscale Science and Engineering.

REFERENCES

[1] J.Q. Xu, Y.Mutoh and L.D.Fu, JSME, A68-672(2002), 1259-1265.
[2] Dundurs, J., J. Composite Mater., 1(1967), 310-322

Nano-Structured C_{60}-SAM Formed on Ultrathin Au Films/MgO Single Crystal

H. Yamamoto, N. Iwata and Y. Shimizu

College of Science & Technology, Nihon University
Funabashi-shi, Chiba 274-8501, Japan, hyama@ecs.cst.nihon-u.ac.jp

ABSTRACT

Exotic electric conducting properties have been expected in organic molecules/metals interfaces with low-dimensional nano-structures. A self-assembly monolayer of a C_{60} derivative, C_{60}-O-C_8SH, (C_{60}-SAM) was synthesized on ultrathin Au films which were deposited on MgO (100) single crystal substrates. Since the surfaces of annealed MgO were atomically flat and revealed step-terrace structures, the ultrathin Au films were deposited along the edges of the steps with the length of a few micrometer. As the result of the characteristic nano-structure of the Au films, the C_{60}-SAM adsorbed on them showed also one-dimensional alignments with the width of few tens nm. The surface morphology of the C_{60}-SAM/Au was observed and analyzed by AFM. In particular samples the resistivity deviated from a normal linear dependence on temperature and revealed an anomalous decrease in the temperature range of 150 - 250K.

Keywords: C_{60}, self-assembly-monolayer, nano-structure, resistivity anomaly

1 INTRODUCTION

Exotic electric conducting properties have been expected in organic molecules/metals interfaces with low-dimensional nano-structures. It is well known that Little[1] or Ginzburg[2] proposed the room temperature superconductivity by an excitonic mechanism. They discussed the possibility of the room temperature superconductivity by the exciton in the 1-dimensional system and the 2-dimensional system, respectively. Then the expected T_c is above room temperature.

As shown in Fig. 1 the 2-dimensional model is consisted with the exciton parts (the side plane) which are adjacent to electron conducting layer (the main conducting plane). The side plane must have enough high energy of elementary excitation and interactions with free electrons in the conducting plane.

The candidate of the conducting layer for the 2-dimensional system is a metal thin film. Since the shielding length of the exciton in metals is in the order of nm, the metal layer must be continuous and/or ultrathin.

As the side plane π-electronic molecules layers may be available because of the high excitonic energy in the order of eV. In this work we have noticed a C_{60} derivative

monolayer as the side plane because of its physical and chemical stability or easiness for handling. Recently Shi *et al*[3]. reported the preparation of highly closed packing alignments of the C_{60} derivatives self-assembled monolayer (C_{60}-SAM) on Au (111) / mica substrates. They used the C_{60} derivative (C_{60}-O-C_6SH) with alkanethiol at the end of long alkyl chain. Such the monomolecular layer of the C_{60} derivative may be adopted for the side chain.

We prepared a similar C_{60} derivative. Au ultrathin films were prepared on MgO substrates which had atomically flat surfaces. When the C_{60}-SAM's are synthesized on Au ultrathin films, we expect that the substance system becomes the model of the excitonic superconductor.

The detailed experimental conditions are explained. The surface structure of the specimen films are evaluated and the observed temperature dependences of resistance are discussed.

SIDE PLANE

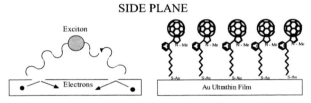

MAIN CONDUCYING PLANE

Figure 1: The schematic model of a 2-dimensional excitonic superconductor and the materials system proposed in this work.

2 EXPERIMENTAL

As the substrates MgO (100) single crystals were annealed in a gold furnace. The two MgO (100) substrates were faced to each other surface and piled on a Nb-STO crystal which effectively absorbed infra-red rays.[4] The annealing temperature was increased with the speed of 450 K/h and was kept at 1173-1273 K for 12 hour. The temperature was cooled down with the speed of 450 K/h and the specimen was taken out at room temperature.

Au films were prepared simultaneously on the MgO substrate by RF magnetron sputtering through a metal mask in order to obtain the both the electrodes for resistance measurements and the area of ultrathin metallic films.

Sputtered Au flux passed around behind the holes of the mask. The diameter of the hole was determined by chemical etching time and the distance between the holes was about 190μm. By adjusting the distance between the mask and the substrates the diameter of the Au electrodes was changed to obtain one of typically about 170μm.

The used target was a Au plate with a diameter of 80mm. The conditions of sputtering were: distance between target and substrates of 10 cm, substrate heater temperature of 673 K, Ar gas pressure of 1.2 Pa, sputtering power of 100 W, and sputtering time of 10 min.

The C_{60} derivative (C_{60}-O-C_8SH) with a long alkyl chain was synthesized by following the similar method previously reported[3]. The molecule was schematically shown in Fig. 2. The synthesized powder was purified by the HPLC with Buckyprep column (Nacalai Tesque Co.) and toluene, and characterized by FAB-MS and H-NMR spectroscopy. The length of the C_{60} derivative is calculated to be 2.2 nm by a space-filling model.

In the SAM process the C_{60} derivatives strongly adsorb on Au with alkanethiol and form monomolecular layer in highly dense structures. The SAM process is illustrated in Fig. 2. The Au/MgO substrates were soaked in benzene solution of 0.01 mM C_{60} derivatives. The soaking cell was filled with N_2 gas. After the soak non-adsorbed C_{60} derivatives were washed out by benzene and the specimen was dried by dry N_2 gas.

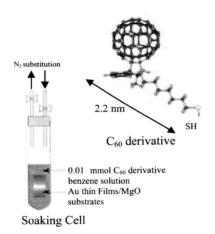

Figure 2: Schematic equipment of the SAM process and the used C_{60} derivative (C_{60}-O-C_8SH). The soaking cell was filled with N_2 gas. The soak was done at room temperature for 20 hr.

The depth profile of XPS C_{1s} peak intensity indicated the existence of C_{60} derivatives monomolecular layer on Au films. The covering ratio of the C_{60} derivatives was evaluated by comparing the intensity of the XPS C_{1s} peak of SAM with that of evaporated C_{60} films. The value of the covering ratio increased as increasing the soaking time until

about 10 hr and almost saturated for 15 hr at room temperature.[5] So the specimen were soaked typically for 20 hr in this work.

The surface morphology of the specimen films was observed by an Atomic Force Microscopy (AFM) (Seiko Instruments Inc.). The resistance of the specimen was measured by a four probe technique in the temperature range from 300 K to 77 K.

3 RESULTS AND DISCUSSION

The typical cross section of the prepared specimen is shown schematically in Fig. 3. The diameter of the Au electrodes was about 170μm. The width of the valley in which ultrathin Au films were deposited was less than 20μm length. The C_{60}-SAM was formed all over the specimen surface.

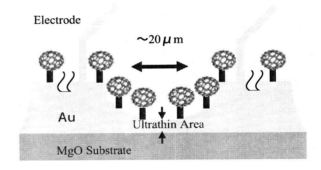

Figure 3: Schematic cross section of the SAM specimen and the Au configuration of electrodes.

The AFM image of the C_{60}-SAM formed on ultrathin Au films is shown in Fig. 4. The enlarged image is shown in the inset. The line profile is also shown in the attached figure. It was found that C_{60}-SAM's aligned one-dimensionally along the edge of the steps of the MgO. The bright lines in the figure revealed that the height of the accumulated substances was 1.0-1.5nm and the width of the SAM region was narrow, a few tens nm. Since the length of the C_{60} derivative is 2.2 nm, C_{60}-SAM's may be accumulated in canting.

The temperature dependence of the resistance was measured. In normal Au films resistance decreased linearly as decreasing temperature. Several specimens showed anomalous decreases of the resistance. Figure 5 shows such the typical temperature dependence of resistance. It revealed a ⌐⌐-shape and slightly deviated from the linear dependence in the temperature range below 150 K. The value of the differentiation coefficient of resistance vs. temperature increases rapidly from 5.0×10^{-3} to 8.0×10^{-3} ohm/K at about 150K. It means that resistance decreased rapidly in the temperature range by some other mechanism rather than usual lattice vibrations.

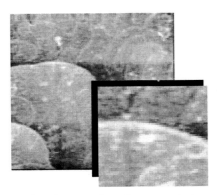

(a) The AFM image of surface morphology of the SAM specimen of 5μm square. The inset is the enlarged image of 1 μm square.

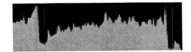

(b) The line profile of AFM in the length of 5μm. The maximum height of the rectangular is 3.47 nm.

Figure 4: AFM image and line profile of the SAM formed on ultrathin Au films/MgO (100) single crystal.

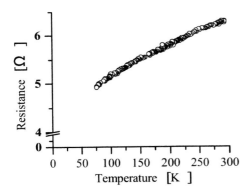

Figure 5: The typical temperature dependence of resistance of the SAM specimen.

This type of resistivity anomaly was observed only in the specimen with ultrathin Au films. As the evidence of the effect of Au film thickness the temperature dependences of the resistance were measured at three different points in one specimen for comparing and discussing the effect of the thickness of the Au films. The results is shown in Fig. 6.

In the two points where Au film was comparatively thick, resistance revealed the normal metallic temperature

dependence. However, the part of ultrathin Au film showed remarkable anomalous decreases of resistance in the temperature range of 150-250K.

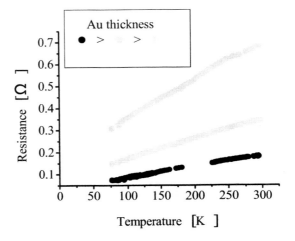

Figure 6: Temperature dependences at three different parts of the same SAM specimen. The upper line corresponds the thinner Au film.

The anomalous temperature dependence of resistance was fitted in the two temperature regions with the linear lines. If we assume that only the limited region of the specimen changes into a zero-resistance state at the peculiar temperature, the linear line bends accompanying with a drop of resistance. Then the other regions of the specimen were normal metallic state and give finite resistance.

We think that the observed phenomenon was caused by the influence of excitonic interactions between free electrons in the metal and electric dipoles excited in the molecules. The anomalous effect may be suppressed in the region of thick Au films, since the shielding length in metals was so small, nm order.

The transition width of resistance was very broad. This may be interpreted as the result of large fluctuations which usually take place in a low dimensional system.

4 SUMMARY

The characteristic interface of ultrathin Au films/ C_{60} derivative monomolecular layer was proposed as the novel substance model for an excitonic superconductor.

The Au ultrathin films were deposited by RF magnetron sputtering. The C_{60}-SAM adsorbed on the Au films was accumulated in 1-dimensional alignment along the edges of the steps of MgO(100) substrates. The surface morphology and structure of the specimen were observed. The some specimen showed the anomalous decreases of resistance which were observed in the temperature range of 150K-250K. These exotic behaviors may be caused by the change of conductivity in the interface of ultrathin Au/C60 derivatives through an excitonic mechanism.

The detailed measurement and analyses of resistance in the micro-scaled area of the specimens are now proceeded and the future subject.

ACKNOWLEDGEMENTS

The authors would like to thank Prof. K. Kikuchi (Tokyo Metropolitan University) and Dr. M. Chikamatsu for preparation of the C_{60} derivatives. This work was supported by the Ministry of Education, Culture, Sports, Science and Technology through a Scientific Grant-In-Aid (No, 12650322 & 15360173)

REFERENCES

[1] W. A. Little, *Phys. Rev.*, **134**, A 1416, 1964.

[2] V. L. Ginzburg, *Contemp. Phys.*, **9**, 355, 1968.

[3] X. Shi, W. Brett Caldwell, Kaimin Chen, and Chad A. Mirkin ; *J. Am. Chem. Soc.*, **116**, 11598, 1994.

[4] Hiroyuki Imai, Nobuyuki Iwata and Hiroshi Yamamoto, *Nanotechnology.*, **13**, 768, 2002.

[5] H. Furusawa, K. Sakaguchi, T. Shimizu, T. Watanabe, H. Yamamoto, M. Chikamatsu and K. Kikuchi; *Trans.Mate.Res.Soc.Jpn.*, **25**, 413, 2000.

Single-Dot Spectroscopy of Low Density GaAs Quantum Dots Grown by Modified Droplet Epitaxy

M. Yamagiwa*, F. Minami* and N. Koguchi**

* Department of Physics, Tokyo Institute of Technology
Meguro-ku, Tokyo, Japan, yamagiwa@lindberg.ap.titech.ac.jp, minami@ap.titech.ac.jp
**Nanomaterials Laboratory, National Institute for Materials Science
Tsukuba, Ibaraki, Japan, KOGUCHI.Nobuyuki@momokusa.nims.go.jp

ABSTRACT

Low density GaAs/AlGaAs quantum dots (QDs) have been fabricated using Modified Droplet Epitaxy (MDE) [1] for the spectroscopic study of single QDs using micro-photoluminescence (μPL). In μPL measurements, the excitation/observation area is focused by a 50× objective to a spot size of approximately $0.6\,\mu$m. Thus, a low density sample was necessary to limit the number of QDs in such a spot size to no more than one. This work made possible the first broad-spectrum single QD spectroscopy of GaAs/AlGaAs QDs fabricated by MDE. The excitation intensity dependence of the μPL spectrum of this sample shows multiple spectral lines which appear at higher excitation intensities. These lines appear from the recombination of electrons and holes from higher energy levels.

Keywords: modified droplet epitaxy, quantum dot, micro-photoluminescence, GaAs

1 INTRODUCTION

With the advent of high-quality crystal growth technology, the controlled fabrication of various semiconductor heterostructures has become possible. Stranski-Krastanov (SK) epitaxy made possible the appearance of self-aggregated quantum dots grown on a two dimensional wetting layer. However, self-assembling of QDs with the SK method has its drawback in the necessary presence of strain for triggering the island formation in the epitaxial growth. Recently, self-assembling of strain- and defect-free GaAs/AlGaAs QDs with no wetting layer has been achieved by MDE. This material is a promising candidate for investigating the interactions among confined carriers without any interactions with carriers confined in a wetting layer, as well as the undesirable effects of strain. In this work, efforts have been made to fabricate QDs with a density two to three orders of magnitude smaller than those used in a previous work [2] to make single-dot spectroscopy possible.

2 SAMPLE GROWTH

2.1 Growth Procedure

The sample is grown by MDE using a Riber-32P molecular beam epitaxy (MBE) system with elemental sources and an EPI (Veeco)-valved As cracking source, which enables the rapid irradiation of a high As$_4$ flux. After native oxide desorption, a 300 nm-thick GaAs buffer layer and a 500 nm-thick AlGaAs barrier are grown on a GaAs (001) wafer at 580°C. The substrate temperature is then reduced to 330°C. The As supply is stopped, and the minimum Ga supply (one monolayer) necessary for droplet formation on a c(4×4) surface is irradiated with a Ga flux equivalent to a GaAs growth rate of 0.05 monolayers/s. The substrate temperature is lowered to 150°C, and these droplets are crystallized by the irradiation of a high As$_4$ flux. The substrate temperature is raised to 200°C, and a 10 nm-thick AlGaAs barrier layer is grown over the QDs using migration enhanced epitaxy (MEE) [3] to avoid the two-dimensional regrowth of the QDs. The temperature is raised to 580°C, and a 90 nm-thick AlGaAs barrier and a 10 nm-thick GaAs capping layer are grown by standard MBE. The sample is consecutively annealed at 680°C for one hour.

2.2 Sample Surface Analysis

The reflection high-energy electron diffraction (RHEED) patterns, which give important information concerning the surface reconstruction, are observed during the growth process. A change in the RHEED pattern from that of an As-rich surface (c(4×4)) to that of a mostly Ga-terminated surface ((4×1)) with very weak 4× lines is seen during Ga droplet formation. The supplied Ga corresponds to the As coverage of the As-stabilized GaAs (001) c(4×4) surface [4]. These facts suggest that most of the Ga adatoms contribute to the formation of a Ga-terminated surface. The relatively low number of remaining Ga adatoms which coalesce into droplets, combined with the low supersaturation pressure (i.e. low nucleation rate) due to the relatively low Ga flux, contributes to the lowering of the QD density. The RHEED patterns after crystallization show weak {111} facet patterns, suggesting the formation of pyramidal QDs.

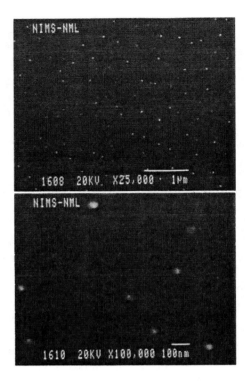

Figure 1: HRSEM image of the GaAs QD formation surface

High-resolution scanning electron microscope (HRSEM) measurements of samples fabricated under the same conditions but without a barrier and capping layer over the QDs show that the sample has an average QD base size of 40 nm and a density of $7 \times 10^8 \, \text{cm}^{-2}$, as shown in Figure 1. Atomic force microscopy (AFM) images of a QD sample without a barrier and capping layer show that the QDs are of a pyramidal shape. In previous works, it has been observed from cross-sectional HRSEM images of GaAs QDs made by MDE that the wetting layer possibly washes out due to group III interdiffusion during the final annealing process, and that the density of the QDs is retained after annealing. A small size reduction is also seen [5].

3   EXPERIMENTAL SETUP

In the PL measurements, the sample is cooled to 4 K by a He gas dilution refrigerator, and an Ar^+ laser (514 nm) is used as an excitation light source. The PL spectra are observed using a spectrometer and a GaAs photodetector.

In the single QD μPL measurements, the sample is placed on a continuous flow Janis ST-500 Microscopy Cryostat and cooled down to 10 K. A frequency-doubled YAG cw laser (532 nm, 20 mW) is used as an excitation light source and the light is coupled into a single mode fiber. The light from the fiber is collimated by a fiber coupler and focused onto the sample by a 50× objec-

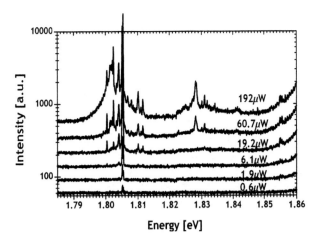

Figure 2: Excitation intensity dependence of the μPL spectrum

tive lens to a spot size of $0.6 \, \mu$m. The position of the excitation beam spot on the sample is controlled by adjusting the position of a concave lens placed between the fiber coupler and the objective lens. The μPL is picked up by the same objective and coupled into a different single mode fiber. The μPL spectra from the fiber are observed using a 50 cm spectrometer with a 1800g/mm grating and a liquid nitrogen-cooled CCD detector. The excitation power is changed by a neutral-density filter.

4   RESULTS AND DISCUSSION

The PL of this sample at 4 K shows a QD PL peak centered at 1.766 eV with a full-width half-maximum of 109 meV. The intensity of the PL peak of the QDs is low compared to that of the AlGaAs PL peak, and this reflects its low density.

The excitation intensity dependence of the μPL spectrum of this sample shows multiple spectral lines which appear at higher excitation intensities, as shown in Figure 2. The main QD peak, centered at 1.805 eV, first appears at the lowest excitation power. Various peaks surrounding the main peak begin to appear with increased excitation power. By observing the PL spectrum at high excitation power while changing the position of the excitation/observation spot in various directions on the sample with the concave lens, it is possible to differentiate the spectra of the QD in question from the spectra of separate QDs (if any exist) from the intensity change of the peaks. All the peaks' intensity in Figure 2 change in unison, thus it can be said that the observed spectral lines are all from the same QD. At 1.828 eV and its vicinity, separate peaks appear. The energy separation from the main peak is approximately 20 meV. From high pressure PL measurements of higher-density GaAs/AlGaAs QDs made by MDE, it has been determined that the heavy hole ground state and heavy hole first excited state have an energy difference of 24 meV. Thus, the peaks at 1.828 eV are possibly from the recombination of the QD conduction band ground state electron with the valence band first excited state hole. Such a recombination is possible by the coupling between optically allowed and forbidden transitions induced by Coulomb interaction among the carriers [6]. The multiple peaks which are observed several meV lower-energy side of the main peak are possibly caused by such an interaction (i.e. the electron-electron and hole-hole exchange interaction).

5 CONCLUSION

A GaAs/AlGaAs QD sample with an average QD base size of 40 nm and a density of 7×10^8 cm^{-2} has been fabricated by MDE. From the RHEED patterns during growth, it can be seen that most of the Ga supply contributes to the formation of the Ga-terminated surface, leaving a low number of Ga adatoms to coalesce into droplets. This is the main factor in the reduction of the QD density. From μPL measurements of this sample, single QD spectroscopy has been accomplished. Most of the QD bandgap were bulk-like due to their large sizes, and their μPL spectra were indistinguishable from that of bulk GaAs. The smaller QDs with strong confinement, however, had μPL peaks at a much higher energy than that of the bulk, and the small number of small QDs in the ensemble contributed greatly to the reduction of the effective "confining" QD density, making single QD μPL observation possible. From the excitation intensity dependence of the single QD-μPL spectra, multiple spectral lines can be seen. These lines appear from the recombination of the QD conduction band ground state electron with the valence band first excited state hole.

This work was supported partially by the Grant-in-Aid for Scientific Research from the Ministry of Science, Education and Culture of Japan, and by a 21 st Century COE Program at TokyoTech "Nanometer-Scale Quantum Physics" by the Ministry of Education, Culture, Sports, Science and Technology.

REFERENCES

[1] K. Watanabe, N. Koguchi, and Y. Gotoh, Jpn. J. Appl. Phys. 39, L79, 2000.

[2] S. Sanguinetti, K. Watanabe, T. Kuroda, F. Minami, Y. Gotoh, and N. Koguchi, J. Crystal Growth 242, 321, 2002.

[3] Y. Horikoshi, M. Kawashima, H. Yamaguchi, Jpn. J. Appl. Phys. 27, 169, 1988.

[4] A. Ohtake, J. Nakamura, S. Tsukamoto, N. Koguchi, and A. Natori, Phys. Rev. Lett. 89, 206102, 2002.

[5] K. Watanabe, S. Tsukamoto, Y. Gotoh, and N. Koguchi, J. Crystal Growth 227-228, 1073, 2001.

[6] U. Hohenester, R. Di Felice, E. Molinari, and F. Rossi, Appl. Phys. Lett. 75, 3449, 1999.

Self-assembled Alternating Nano-scaled Layers

Wan-Yu Wu and Jyh-Ming Ting

Mina Materials Laboratory, Department of Materials Science and Engineering,
National Cheng Kung University,
1 University Road, Tainan, Taiwan, jting@mail.ncku.edu.tw

ABSTRACT

This paper reports the use of a conventional sputter deposition technique to produce self-assembled alternating nano-scaled layers consisting of diamond-like carbon (DLC) and metal. A conventional sputter deposition technique produces alternating layers only when multiple sputtering targets are employed. However, we have successfully demonstrated that the use of only one single sputtering gun in a conventional sputter deposition process could lead to the formation of alternating nano-scaled layers through self-assembling. The microstructure of the alternating layers was examined as a function of the deposition parameters.

Keywords: self-assembling, layered structures; diamond-like carbon; thin films, sputter deposition

1. INTRODUCTION

It is known that diamond-like carbon (DLC) exhibit excellent tribological properties especially against steel [1,2]. Many techniques have been used for the preparation of Me-DLC thin films [3-10]. However, high internal stresses and insufficient adhesion of DLC films to the substrate have also been found to limit their applications. Therefore, metallic elements were introduced into DLC films reduce the stresses and enhance the adhesion [5,6]. It has been shown that the metallic component in an Me-DLC thin film forms nanosized clusters that distribute homogeneously in the DLC matrix [11,12]. Recently, alternating layers of carbon and metal or metal carbide were observed in W-DLC thin film [13,14]. Such an alternating layered structure was found to have better wear and fatigue resistance. A reactive dc unbalanced multiple-target magnetron sputter deposition technique with moving substrate was used for the growth of the W-DLC thin films. A total of four magnetron guns with two different types of target materials were used in the studies reported in Refs. 13 and 14. In this study, we report the use of a dc reactive sputter deposition system equipped with only one single magnetron gun to deposit self-assembled alternating layers of Me-DLC thin films.

2. EXPERIMENTAL

The target materials used in the sputter

deposition system include nickel (99.99%), platinum (99.99%), and copper (99.95%). Single crystal wafers of (100) Si were used as the substrates. For the film deposition, the sputter deposition chamber was first evacuated to a pressure lower than 5×10^{-5} torr and then back-filled with an argon/methane gas mixture to a desired deposition pressure of 1×10^{-2} torr. The electrode distance was 40mm. The deposition time varied from 1min to 30min. The substrates were not heated during the deposition under all the conditions. Compositions of the resulting Me-DLC thin films were examined using energy dispersive spectrometry (EDS). Surface morphologies and cross-section views of Me-DLC thin films were examined using scanning electron microscopy (SEM). Deposition rates of Me-DLC films were determined from the thickness measured on the SEM cross-sectional images. Crystallinity and microstructure of the Me-DLC thin films were investigated using grazing incident x-ray diffraction (GID), transmission electron microscopy (TEM), and micro-Raman microscopy.

3. RESULTS AND DISCUSSION

Me-DLC (Me = Ni, Cu, or Pt) thin films were obtained. The crystallinity was first examined using GID. It was found that in general the carbons in the films are amorphous while the metals exhibit polycrystalline structure, as shown in Fig. 1. However, it appears that Ni has a worse degree of crystallinity, and both copper and platinum have a better degree of crystallinity. Occasionally, a minor amount of polycrystalline nickel carbide was also found in Ni-DLC. By increasing the deposition time, i.e., increasing the thickness, the crystalline of the metals can be improved.

The thickness of these thin films is in the order of 10^2 nm. Most of the films obtained exhibit alternating layered structures, which can be seen in TEM micrographs. Regardless of the film thickness, the alternating layered structure starts at the film/substrate interface and continues to the film surface. This is shown in Fig. 2. In a period, there are a metal-rich layer and a carbon-rich film. The occurrence of alternating layered structure and the periodicity, however, depend on the type of metals and the deposition parameters used. TEM analysis was performed to examine the cross sections of Ni-DLC, Cu-DLC, and Pt-DLC films prepared using a power of 100 W, an Ar/CH$_4$ ratio of 1, and a pressure of 1×10^{-2} torr. It was found that there is no alternating layered structure in the Ni-DLC film. The film exhibits a microstructure in which the nano-sized Ni particles are embedded homogeneously in the DLC matrix. On the other hand, alternating layered structure is clearly seen in both Cu-DLC film and Pt-DLC film prepared using the same conditions. The periodicities are 13 nm and 10 nm, respectively, for Cu-DLC film and Pt-DLC film. However, as the Ar/CH$_4$ ratio increases, alternating layered structure appears in Ni-DLC film. It was then observed that all the Me-DLC thin films prepared using a power of 100 W, an

Ar/CH$_4$ ratio of 3, and a pressure of 1 x 10^{-2} torr exhibit alternating layered structure. The periodicities are 15 nm, 15 nm, and 22 nm, respectively, for Ni-DLC film, Cu-DLC film, and Pt-DLC film.

4. CONCLUSION

Thin films with alternating layered structures are normally produced in a conventional sputter deposition technique by using multiple sputtering targets. However, we have successfully demonstrated that the use of only one single sputtering gun in a conventional sputter deposition process could lead to the formation of alternating nano-scaled layers through self-assembling. The thickness of these thin films is in the order of 10^2 nm. Regardless of the film thickness, the alternating layered structure starts at the film/substrate interface and continues to the film surface. The occurrence of alternating layered structure and the periodicity, however, depend on the type of metals and the deposition parameters used.

Finally the Raman spectra were investigated for Cu-DLC specimens. Both the D-band and G-band were detected a 41.6 at% Cu-DLC thin film. It was further found that the I_D/I_G ratio increased with Cu concentration, indicating a smaller carbon cluster size at a higher Cu concentration. This would have given a larger full width at the half maximum (FWHM). However, the FWHM decreases with the Cu concentration, suggesting that through the addition of copper the film stresses are greatly released. The stress relieve is strong enough to overcome the effect of reducing carbon cluster size on the FWHM.

ACKNOWLEDGEMENT

This work was supported by the National Science Council in Taiwan under Grand Nos. NSC 91-2120-E-006-002 and NSC 91-2216-E-006-033.

REFERENCES

[1] K. Enke, Thin Solid Films 80 (1981) 227.

[2] T. Mori, Y. Namba, J. Appl. Phys. 55 (1984) 3276.

[3] K. Bewilogua, C.V. Cooper, C. Specht, J. Schroder, R. Wittorf, M. Grischke, Surf. Coating Technol. 127 (2000) 224.

[4] D.P. Monaghan, D.G.. Teer, P.A.. Logan, I. Efeoglu, R.D. Arnell, Surf. Coating Technol. 60 (1993) 525.

[5] K. Oguri, T. Arai, Surf. Coating Technol. 47 (1991) 710.

[6] Y.H. Wu, C.M. Hsu, C.T. Chia, I.N. Lin, C.L. Cheng, Diamond and Related Materials 11 (2002) 804.

[7] T. Hioki, Y. Itoh, A. Itoh, S. Hibi, J. Kawamoto, Surf. Coating Technol. 46 (1991) 233.

[8] Q. Wei, J. Sankar, J. Narayan, Surf. Coating Technol. 146-147 (2001) 250.

[9] G.Y. Chen, J.S. Chen, Z. Sun, Y.J. Li, S.P. Lau, Applied Surface Science 180(2001) 185.

[10] V. Rigato, G. Maggioni, D. Boscarino, G. Mariotto, Surf. Coating Technol. 116 (1999) 580.

[11] J.B. Pethica, P. Koidl, J. Vac. Sci. Technol. A3 (1985) 239.

[12] H. Dimigen, H. Hubsch, R. Memming, Appl. Phys. Lett. 50 (1987) 1056.

[13] K. Bewilogua, C.V. Cooper, C. Specht, J. Schroder, R. Wittorf, M. Grischke, Surf. Coating Technol. 127 (2000) 224.

[14] C. Strondl, N.M. Carvalho, J.Th.M. De Hosson, G.J.

van der Kolk, Surf. Coating Tech. 162 (2003) 288.

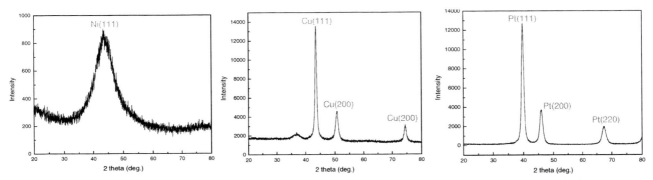

Fig. 1. Low angle XRD patterns of 1-min samples: Ni-DLC (top), Cu-DLC (middle), and Pt-DLC (bottom).

Fig. 2. TEM cross sectional image showing alternating layered structure in a Ni-DLC film.

Fabrication of Metallic Nano-stamper and Replication of Nano-patterned Substrate for Patterned Media

Youngkyu Kim[*], Namseok Lee[**], Young-Joo Kim[**] and Shinill Kang[*,**]

[*]School of Mechanical Engineering, Yonsei University,
[**]Center for Information Storage Device, Yonsei University,
134 Shinchon-dong, Seodeamun-ku, Seoul, Korea, snlkang@yonsei.ac.kr

ABSTRACT

With increasing demand for ultra high-density information storages, patterned media has received much attention as a solution to overcome the limits of conventional continuous magnetic media. Current methods to fabricate the patterned media are to use direct patterning and etching. However, those procedures are very costly and are not suitable for mass production. In this study, we investigate the possibility of mass production of patterned media by molding technology with the metallic nano-stapmer. The metallic nano-stamper was fabricated using an electroforming process, and then the nano-patterned substrate was replicated using a nano-molding process. To evaluate the replication quality of the nano-patterned substrate, the surface profile and the roughness of the patterns on the molded substrate were measured and analyzed. The magnetic layer was deposited on the nano-patterned substrate, and a single magnetic domain state was observed on the nano-patterned substrate.

Keywords: nano-molding, metallc-nano stamper, nano-patterned substrate, patterned media

1 INTRODUCTION

As the need for information storage with high storage density increases, patterned media has received much attention as a solution to overcome limits of conventional continuous magnetic media, such as superparamagnetic effect and medium noise. Patterned media is regarded as a forefront candidate to achieve high storage density up to a few Tbits/in^2. Data can be stored in a bit on the discrete nanoscale patterns of single magnetic domains on the media.

Patterned media with nanoscale patterns can be produced using various micro-electro-mechanical system (MEMS) and nano-electro-mechanical system (NEMS) technologies. Aoyama and Hao et al. fabricated the patterned media using electron beam (e-beam) and holographic lithography, and etching [1-2]. Also, a focused ion beam (FIB) was proposed to directly fabricate discrete magnetic patterns [3]. However, those procedures are high-costed and low-throughput techniques. Chou et al. has proposed a nano-imprint lithography to produce nanoscale patterns at low-cost [4]. In a nano-imprint lithography process, a mold containing initial patterns is required,

silicon or quartz is used as a mold due to its ease of fabrication [5-6]. However, the silicon or quartz mold has a short lifetime, because they are too brittle to be used for nano-imprint lithography, which requires the process under repeated pressure and temperature. McClelland et al. applied the nano-imprint lithography to fabricate the patterned media, and a photopolymer was used as a mold material [7]. However, the polymer mold is not suitable to use a mold repeatedly for mass production in imprint process because of its flexibility. A metallic nano-mold can solve this problem, and Heydermann et al. made metallic nano-structures using numerous steps; deposition of the seed-layer on the substrate using a evaporation or sputtering, fabrication of patterns using lithography processes, fabrication of metallic structures using a electroplating, and reactive ion etching (RIE) [8]. However, when a metallic mold was fabricated, RIE is applied to each cycle to these processes.

This paper describes the methodology of fabricating a nano-patterned substrate for patterned media using a molding process, and application of this methodology. The metallic nano-stamper was fabricated using a electroforming, and then the nano-patterned substrate was replicated using a nano-molding with the metallic nano-stamper as illustrated in Figure 1.

As a first step, the master nano-patterns were fabricated by e-beam and holographic lithography. After the seed-layer was deposited on the master by e-beam evaporation and sputtering, the electroforming process followed to

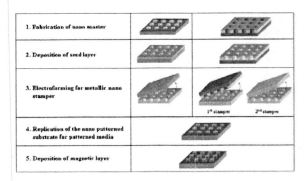

Figure 1: Fabrication procedures for patterned media using the present method

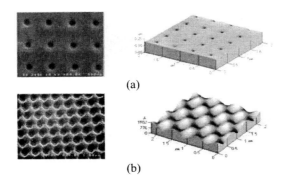

(a)

(b)

Figure 2: SEM and AFM images of the master nano-
patterns by (a) e-beam lithography and (b)
holographic lithography, respectively.

make the nickel nano-stamper. And then, the polymeric
nano-patterned substrate was replicated by nano-molding.
The surface profile and the roughness of the patterns on the
polymeric substrate were measured and analyzed. Finally,
the Cr underlayer and the Co magnetic layer were deposited
on the patterned polymeric substrate, and a single magnetic
domain state was observed on the nano-patterned substrate.

2 EXPERIMENTS AND RESULTS

2.1 Fabrication of Master Nano-patterns

To replicate patterns of nanoscale sizes and pitches by a
nano-molding process, we fabricated the master nano-
patterns using e-beam and holographic lithography.

E-beam lithography can fabricate the arbitrary patterns
with various hole and pillar patterns in nanoscale diameters,
pitches and depths. Also, holographic lithography can
generate the periodically repeated nano-structures in large
area. Figure 2 shows scanning electron microscope (SEM)
images and atomic force microscope (AFM) images of the
master nano-patterns by e-beam and holographic
lithography, respectively. Hole patterns with a diameter of
200 nm, a pitch of 500 nm and a depth of 100 nm, were
fabricated by e-beam lithography. To produce two-
dimensional crossed gratings with a period of 500 nm and a
depth of 60 nm on 4-inch wafer, double exposures with half

Parameters		Range of values
Temperature (°C)		43-45
pH		4-4.2
Current density (mA/cm^2)		10-20
Concentration of Electrolytes (g/l)	Ni·(NH$_2$SO$_3$)$_2$ 4H$_2$O	300-450
	Ni·Cl$_2$6H$_2$O	15
	H$_3$BO$_3$	45

Table 1: Processing parameters for nickel electroforming

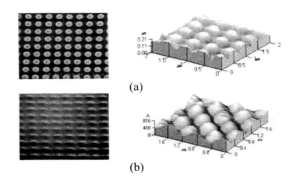

(a)

(b)

Figure 3: SEM and AFM images of the electroformed
metallic nano-stampers using (a) the e-beam
patterned master and (b) the holographic
patterned master, respectively.

of exposure energy along different directions by 90°
rotating the sample, were performed.

2.2 Fabrication of Metallic Nano-stamper

The seed-layer was deposited on the master by e-beam
evaporation and sputtering to enhance the coverage and to
uniformize the deposition states of metallic particles.
Because the seed-layer functions as not only nano-patterns
on the metallic nano-stamper but also as a conduction layer
in the electroforming process, control of the deposition
process and selection of the material for the seed-layer are
important. The nickel was chosen as the seed-layer because
it can satisfy desirable requirements, such as high hardness
and thermal stability [9]. Also, processing conditions should
be properly controlled during the process not to separate the
metallic layer from the patterned resist-layer on the master.

After the seed-layer was deposited, the electroforming
process using a nickel sulfamate solution was peformed to
make the nickel nano-stamper with a thickness of 0.5 mm
[10]. As shown in Table 1, we controlled an electroforming
process to suppress the separation of seed-layer and the
development of residual stress in the electroformed layer. In
particular, an electric current was maintained at low value
to slow the electroforming rate until the metallic layer was
stable. After the electroforming, the silicon wafer and
residue resist were removed. The nanoscale pillar and hole
shapes of the metallic nano-patterns, which were
transferred from the shapes of master nano-patterns, were
formed. Figure 3 shows SEM and AFM images of the
electroformed metallic nano-stampers.

2.3 Replication of Polymeric Nano-patterns

Metallic nano-stamper was used for the nano-molding
process, such as compression molding, ultra-violet (UV)
molding and injection molding. Nano-molding process can
be regarded as a suitable mass production process for
replicating nano-patterns. The present molding system is
modified from the micro-compression molding system [11].

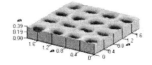

Figure 4: SEM and AFM images of the polymeric nano-patterns with a diameter of 200 nm, a pitch of 500 nm and a depth of 100 nm

A polymethyl methacrylate (PMMA) powder was used as the molding material. In nano-molding process, the molding temperature and compression pressure are the most dominant governing processing conditions that affect the replication quality of the molded parts. If the molding temperature and compression pressure are too low, molded patterns can not be made because PMMA powder particles can not bond, and the polymeric nano-patterns can not fill the metallic nano-stamper. Also, if the molding temperature and compression pressure excessively increased, they can deteriorate the replication quality because of the sticking effect. Therefore, it is important to control the molding temperature and compression pressure conditions to improve the quality of replication in the molded parts.

After the proper processing conditions were determined from experiments, the molding process was carried out at a molding temperature from 210-220 °C and compression pressure 2-4 MPa. To evaluate the replication quality of the polymeric nano-patterns, the surface profile and the surface roughness of the molded parts were measured and analyzed by SEM and AFM. Figure 4 shows SEM and AFM images of hole patterns with a diameter of 200 nm, a pitch of 500 nm and a depth of 100 nm in the molded parts. And Table 2 shows the surface roughness of the patterns in the master, the stamper and the molded part. As shown in Figure 4 and Table 2, the pillar shapes in the metallic stamper were properly transferred to the polymeric patterns. Also, the surface roughness of molded patterns was less than 10 Å.

These experimental results show that our replication method with the metallic nano-stamper can be applied to fabricate the nano-patterned substrate for patterned media.

2.4 Application to Patterned Media

To apply our replication method to patterned media, the nano-patterned substrate with pillar patterns should be replicated. After the metallic nano-stamper with hole

	Ra (Å)
Master nano-patterns	3.29
Metallic nano-stamper	6.42
Polymeric nano-patterns	7.55

Table 2: Surface roughness of the master nano-patterns, the metallic nano-stamper and the polymeric nano-patterns

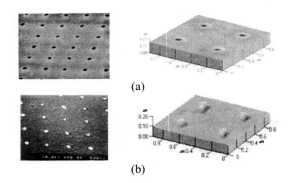

(a)

(b)

Figure 5: SEM and AFM images of (a) the metallic nano-stampers and (b) the corresponding nano-patterned substrate with a diameter of 200 nm and a pitch of 500 nm

patterns was fabricated using e-beam evaporation and electroforming, the corresponding nano-patterned substrate with pillar patterns was replicated. The nano-patterned substrate had pillar patterns with a diameter of 200 nm and a pitch of 500 nm, and a diameter of 100 nm and a pitch of 250 nm, respectively. Figure 5 and Figure 6 show SEM and AFM images of the metallic nano-stamper and nano-patterned substrate for patterned media. Comparing with the molded substrate with hole patterns, it was observed that molded patterns were not stable for the case of pillar shape. In particular, when the pillar diameter was 100 nm, some defects were founded in the substrate, and patterns in shape and size were not uniformly distributed as desired. From these results, we believe that the replication of pillar patterns is more difficult than that of hole patterns, because non-filling and sticking problems can occur. To solve these problems, it is necessary that more precise processing condition is optimized and the releasing method is developed in the nano-molding process.

Finally, the magnetic layer was deposited on the molded patterns for patterned media. The magnetic material and

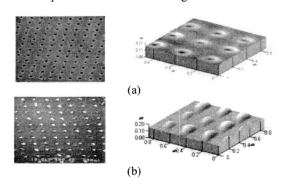

(a)

(b)

Figure 6: SEM and AFM images of (a) the metallic nano-stampers and (b) the corresponding nano-patterned substrate with a diameter of 100 nm and a pitch of 250 nm

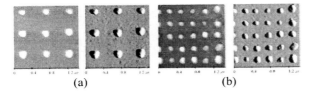

(a)   (b)

Figure 7: AFM and MFM images of the magnetic layer with (a) a diameter of 200 nm and a pitch of 500 nm, and (b) a diameter of 100 nm and a pitch of 250 nm on the nano-patterned substrate

thickness were designed for the polymeric nano-patterns with diameters of 200 nm and 100 nm. The Cr underlayer was first deposited on the molded patterns with a thickness of 100 Å, and then the Co magnetic layer was deposited on it with a thickness of 200 Å in an ultra-high vacuum (UHV) system. To evaluate the magnetic characteristics, the Co magnetic layer was saturated along the longitudinal direction of the patterns. The magnetic domain structure was analyzed by magnetic force microscope (MFM). Figure 6 shows AFM and MFM images of the magnetic layer with a diameter of 200 nm and a diameter of 100 nm. Single magnetic domain states were observed on the nano-patterned substrate with a diameter of 200 nm. On the other hand, the pillars with a diameter of 100 nm show multi-domain states, probably due to non-uniform distributions of shapes and sizes of the nano-patterned magnetic pillars.

3 CONCLUSION

In this study, it was showed that our molding process with the metallic nano-stamper could be applied to mass production of patterned media. The master nano-patterns, which contain nanoscale hole and pillar patterns, were fabricated by e-beam and holographic lithography. Electroforming process was followed to make the nickel nano-stapmer, and then the nano-molding was performed to replicate the polymeric nano-patterns. The replication quality of the molded parts was analyzed by SEM and AFM, and these results showed that our replication method could be applied to fabricate the nano-patterned substrate for patterned media. Finally, the magnetic layer was deposited on the nano-patterned substrate; results showed that a single magnetic domain was established on the nano-patterned substrate. However, the molded and magnetic domain states have non-uniform distributions when the pillar diameter is 100 nm. Research on the improvement of replication quality and uniformity of the molded patterns, and on the development of injection molding and UV molding with the metallic nano-stamper are ongoing.

ACKNOWLEDGEMENTS

This research was supported by a grant (M102KN 010001-03K1401-00610) from Center for Nanoscale Mechatronics & Manufacturing, one of the 21st Century Frontier Research Programs, which are supported by Ministry of Science and Technology, KOREA

REFERENCES

[1] T. Aoyama, K. Uchiyama, T. Kagotani, K. Hattori, Y. Wada, S. Okawa, H. Hatate, H. Nishio and I. Sato, "Fabrication and properties of CoPt patterned media with perpendicular magnetic anisotropy," IEEE Transactions on Magnetics, Vol. 37(4), 1646-1648, 2001.

[2] Y. Hao, M. Walsh, M. Farhoud, C. A. Ross, H. I. Smith, J. Q. Wang and L. Malkinski, "In-plane anisotropy in arrays of magnetic ellipses," IEEE Transactions on Magnetics, Vol. 36(5), 2996-2998, 2000.

[3] J. Lohau, A. Moser, C. T. Rettner, M. E. Best and B. D. Terris, "Writing and reading perpendicular magnetic recording media patterned by a focused ion beam," Appl. Phys. Lett., Vol. 78(7), 990-992, 2001.

[4] S. Y. Chou, P. R. Krauss and P. J. Renstrom, "Imprint lithography with 25-nanometer resolution," Science, Vol. 272, 85-87, 1996.

[5] S. Lebib, Y. Chen, J. Bourneix, F. Carcenac, E. Cambril, L. Couraud and H. Launois, "Nanoimprint lithography for a large area pattern replication," Microelectronic Engineering, Vol. 46, 319-322, 1999.

[6] Y. Hirai, N. Takagi, H. Toyota, S. Harada, T. Yotsuta and Y. Tanaka, "Nano chamber fabrication on an acrylic plate by direct nano imprint lithography using quratz mold," Microprocesses and Nanotechnology Conference 2001 International, 104-105, 2001.

[7] G. M. McClelland, M. W. Hart, C. T. Rettner, M. E. Best, K. R. Cartner and B. D. Terris, "Nanoscale patterning of manetic islands of imprint lithography using a flexible mold," Appl. Phys. Lett., Vol. 81(8), 1483-1485, 2002.

[8] L. J. Heydermann, H. Schift, C. David, B. Ketter, M. Auf der Maur and J. Gobrecht, "Nanofabrication using hot embossing lithography and electroforming," Microelectornic Engineering Vol. 57-58, 375-380, 2001.

[9] S. Moon, N. Lee and S. Kang, "Fabrication of a microlens array using mirco-compression molding with an electroformed mold insert," Journal of Micromechanics and Microengineering, Vol. 13(1), 98-103, 2003.

[10] N. Lee, S. Moon and S. Kang, "The effect of wettability of nickel mold insert on the surface quality of molded microlens," Optical Review, Vol. 10(4), 290-294, 2003.

[11] S. Moon, S. Kang and J. Bu, "Fabrication of polymeric microlens of hemispherical shape using micromolding," Opt. Eng., Vol. 41(9), 2267-2270, 2002.

Nano-Lithography in Ultra-High Vacuum (UHV) for Real World Applications

J. Gilsinn, H. Zhou, B. Damazo, J. Fu, R. Silver

National Institute of Standards & Technology (NIST)
Manufacturing Engineering Laboratory (MEL)

ABSTRACT

As nano-lithography technology improves, more companies and research groups have the capability to create nano-scale structures. Scanning tunneling microscopes (STMs) are commonly used to create these structures and evaluate them afterward. One difficulty is that these nano-structures are difficult to find on a centimeter-sized sample without very specialized hardware and post-processing. The National Institute of Standards and Technology (NIST) is conducting research into developing an integrated system consisting of a high-precision STM, course motion system, interferometer, and vision system that would allow these structures to be accurately created, recognized, and evaluated in ultra-high vacuum (UHV). The fast scan direction in this high-precision STM uses a piezo-driven flexure stage with an interferometer system that allows the motion to be tracked very accurately. A course motion system allows the sample to be moved in larger steps than the STM could move on its own. A vision system capable of locating micro-scaled structures on the sample is used to map larger features on the sample and aid the STM in writing and locating nano-scale features. By combining all these components into one integrated system, NIST hopes to develop the capability to combine nano-scale features with large-scale structures.

Keywords: nano-lithography, STM, interferometer, microscopy, etching

1 INTRODUCTION

With electronics manufacturing driving commercial devices smaller and smaller, the demand for building nano-scale devices (50 nm to 1000 nm in size) is seen as a future industry with great potential. Nano-lithography is one promising technology for building these structures. Although there are multiple techniques for nano-lithography and the theory behind these techniques has been around for years, nano-lithography has not moved far from research and development facilities. Some barriers preventing broader application of these technologies are the high startup costs and the ability to perform relevant analysis on these nano-structures once created.

One technique for creating and inspecting nano-structures uses a scanning tunneling microscope (STM), but a problem is that these nano-structures are difficult to locate on a centimeter-sized sample without very specialized hardware and post-processing. The National Institute of Standards and Technology (NIST) is conducting research into developing a system that would allow these nano-structures to be more easily manufactured, located, and identified.

2 SYSTEM OVERVIEW

The heart of the new system is a modified commercial Scanning Tunneling Microscope (STM). Commercial STMs normally uses a lead zirconate titanate (PZT) piezo-electric tube to drive the STM tip in two directions. The tip scans fast in the X direction (left to right) and scans slow in the Y direction (top and bottom). While STMs produce very precise results after post-processing, their scanning motion is not easily controllable. The NIST system replaces the fast scan direction of the STM with a PZT-driven flexure stage with an interferometer system that allows the motion of the STM to be monitored very accurately.

The new system will also incorporate a millimeter-scale motion system for course positioning, allowing larger scale samples to be used and larger features to be written and imaged. Two PZT-driven linear stages mounted perpendicular to one another are used to move the sample in both the X and Y direction relative to the STM tip allowing the entire sample to be utilized [1].

While nano-scale devices have been written for many years, there have been few or no commercial applications of them yet. One of the difficulties has been to place them onto a specific portion of a sample and locate them again once they have been created. Typical STMs cannot image more than a few hundred nanometers in high-precision mode, which is not large enough to locate nano-scale devices on a centimeter-sized sample. A vision system is being developed for the new nano-manufacturing system that will allow the location of the STM tip to be known relative to other sample features. While the features on the current samples are calibrated lines and grids, the technology should allow for circuit lines and other structures to be viewed just as easily.

The technologies discussed above are a key part of writing nano-scale structures on a sample, but proper sample preparation and pattern generation techniques are also necessary. The samples used in the NIST system have been prepared using a silicon etching technique and written using selective modification of a hydrogen-terminated surface.

3 STM SYSTEM

The heart of the NIST nano-lithography system is the STM system. It combines a commercial STM tube, flexure stage, and an interferometer into one complete system allowing the movement of the STM tip to be very accurately known.

3.1 Flexure Stage

The fast scan direction of the commercial STM used in the NIST system can be replaced by the motion of a flexure stage driven by a PZT actuator. The motion of the stage along the X-axis is approximately 1 µm and constrained along the axis by the design of the flexure joints. A sample holder station is mounted to the face of the stage carrying a sample holder and plane mirror for the interferometer. The sample and plane mirror are moved along the fast scan direction with respect to the STM scanning tip. A schematic drawing of the flexure stage and STM system design can be seen in Figure 1.

3.2 Interferometer

A new implementation of a Michelson interferometer capable of approximately 20 pm resolution has been developed [2]. This new method uses a tunable diode laser as a light source with the diode laser wavelength continuously tuned to fix the number of fringes in the measured optical path. High-speed, accurate measurements of the beat frequency created by beating the diode laser against a reference laser enable the diode laser wavelength to be measured.

The basic principle of this measurement system is that the output frequency of a tunable diode laser is adjusted so that as the measurement arm of a Michelson interferometer is scanned, the signal remains locked on a fringe [3]. Figure 2 shows a schematic of the optical apparatus with the measurement and control strategy and a CAD drawing of the in-vacuum equipment. The fringe signal arises in this apparatus from a differential measurement of the reference arm signal against the measurement arm signal, where the measurement arm is the distance from the birefringent crystal to the plane mirror on the flexure stage. To track the laser diode frequency, a portion of the beam is split off and beat against a reference, a stabilized Helium-Neon (HeNe) laser [4]. This allows the diode laser wavelength to have direct, unbroken traceability.

This is a homodyne interferometer system meaning only a single frequency is used in the interferometer portion of the system. The two-frequency portion of the measurement system is independent of the interferometer and measurement mirrors. In addition, no fringe interpolation is required since the measurement is based on holding constant the number of fringes in the optical-path difference between the reference arm and the measurement.

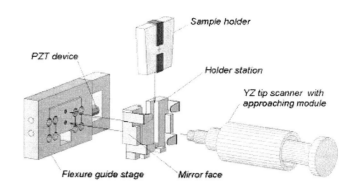

Figure 1 - Schematic Diagram of the Flexure Stage and STM System Design.

The diode laser frequency is continuously controlled by a PZT actuator to lock to a null point in the fringe signal measured at the bicell detector. As the plane mirror is scanned, the diode laser frequency is changed to maintain a constant zero in the difference signal between the two polarization states exiting the 45° rotated birefringent crystal. The diode laser is operated with a control bandwidth of a few hundred hertz since the PZT actuator has a maximum operation speed of 2 kHz.

A stabilized HeNe laser is used as the reference signal to measure the diode laser frequency. The tunable diode laser beam is split with a polarizing beam splitter. Half the signal goes to the reference measurement system and the other half goes to the interferometer. The HeNe reference beam and the diode laser beam are mixed in the reference measurement path so that they cause a beat frequency signal. This nominally 2 GHz signal gives a direct measure of the diode laser frequency relative to the stabilized HeNe reference signal, which has been appropriately calibrated. A computer logs the frequency measurement synchronized with the STM Z height data, providing traceability for the STM data.

4 MILLIMETER-SCALE MOTION

While the STM used in the NIST system is capable of atomic resolution scans and nano-scale writing, it is not capable of doing either of these over an entire sample. The samples used for the NIST experiments are approximately 1 cm square, while the largest motion of a high-precision STM is typically on the order of a few hundred nanometers. In order to increase the motion capabilities of the system, a millimeter-scale motion system is required.

The millimeter-scale motion system consists of two PZT-driven linear stages mounted perpendicular to one another and oriented along the X and Y axes of the STM system. The stages are PZT-driven linear stepper stages that move with 4 nm steps over a distance of 5 mm. While the current stages cannot image an entire 1 cm square sample, they are capable of proving the concept of large-scale nano-structure design and fabrication.

Since the interferometer is not capable of measuring the distances required to scan or write on an entire 1cm square sample, a method to increase the measurement capability of the interferometer is needed. A possible solution that utilizes the existing interferometer system is called fringe hopping. During operation, the interferometer is locked into the null-point on one fringe of a beat signal. Fringe hopping causes the interferometer to lock into the null point from the next fringe, which increases the interferometer's measurable distance.

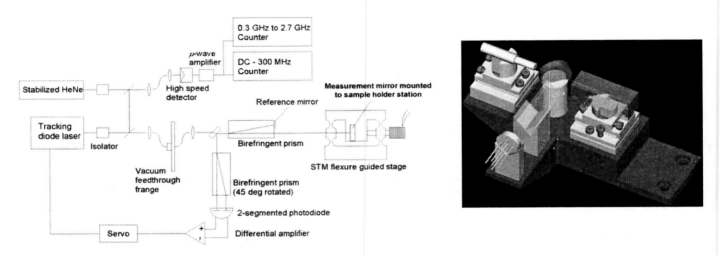

Figure 2 - Schematic & CAD Drawing of the Interferometer Beam Path & System Design

5   VISION SYSTEM

The vision system being designed should be capable of viewing the silicon surface with a resolution of 1 μm using optical technology. While this will not be capable of viewing the nano-scale objects themselves, it will be able to identify a region of the sample surface to be written on and to search for the object in future scans. This is important when trying to connect things like circuit lines or pads with the nano-scale objects.

The main design goals for this vision system are UHV compatibility and long working distance. None of the components of the vision system are vacuum compatible, so a retractable tube with a quartz window is being designed to extend into the UHV chamber. An optical microscope designed with a working distance of approximately 3 cm will be inserted into this tube. The vision system will incorporate both through the lens illumination as well as offset lighting in order to get the best contrast for viewing the micro-scale objects on the sample. The vision system also incorporates interchangeable 10x and 20x lenses that give the appropriate magnification needed for a particular experiment. During bake-out of the UHV chamber, the vision system will be removed from the tube to avoid damaging any of the sensitive components.

6   NANO-LITHOGRAPHY TECHNIQUE

6.1   Sample Preparation

To prepare nanometer scale features in a vacuum environment, nearly atomically flat and ordered surfaces are required. Silicon is a highly desirable substrate for this work because of its widespread usage in semiconductor manufacturing. One difficult challenge several research groups have encountered in attempting to use Si in nanotechnology applications has been in finding a robust method for the repeatable preparation of large, ordered, flat surfaces.

Wet chemical etch processing can produce high quality, atomically flat Si(111) surfaces without requiring high temperature [5]. The etching process results in hydrogen-terminated Si (1x1) surfaces that are quite stable in an ambient environment [6]. The hydrogen termination can be locally removed through electron stimulated desorption [7][8], making it possible to fabricate nano-scale structures using a STM.

Currently, most research groups working on this subject are using a common RCA [9] type pre-cleaning followed by 40% NH4F etching. However, details in the methods used vary widely. In order to better understand the mechanisms of the wet etching process and to develop more repeatable procedures, NIST has systematically studied various etching conditions on several different control wafers. We have found that the etching time and wafer miscut play a key role in the determination of the final morphology of the Si(111) surfaces.

Figures 3 and 4 show example results for wafers with a $0.12°$ miscut in the $\left[1,1,\overline{2}\right]$ direction. This miscut angle results in an average terrace width of 150 nm. The images show a 2μm square area of the wafer surface with the

$\left[1,1,\overline{2}\right]$ direction pointing to the right and $\left[\overline{1},1,0\right]$ direction pointing up. After 2 minutes etching, shown in Figure 3, the terraces become the dominant features on the surface. There are still many triangular pits, but they only have an average size of 50 nm on the terraces. After 6 minutes etching, shown in Figure 4, no triangular pits are observed and the step edges are much straighter. Samples prepared in this way are the substrates for the writing technique discussed below.

Figure 4 - AFM Image, 2 Minutes Etch

Figure 3 - AFM Image, 6 Minutes Etch

6.2 Pattern Generation

One of the major barriers preventing the widespread application of UHV patterning is the lack of a robust pattern transfer technique or the ability to locate and subsequently image the patterns in an external tool. NIST has developed a robust method to transfer patterns into the substrates, with known feature positions, so that they can be accessed by a variety of external metrology tools.

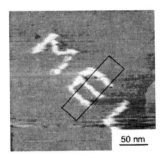 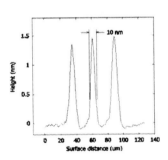

Figure 5 - STM Image of Written Sample

Figure 6 – Cross Section of Sample Image

Figures 5 and 6 show some UHV STM pattern generation results. The pattern generation technique uses selective modification of the hydrogen termination on a surface using a STM. These results were obtained with the writing parameters of 800 pA constant current and +6 V bias voltage. A cross section of the image, showing averaged data from the area delineated, is shown in Figure 6. These patterns have an average line width of nominally 10 nm and, as measured by the STM, have a height of approximately 1 nm.

7 SUMMARY

NIST is integrating commercial and newly developed components and techniques to develop a nano-lithography system capable of creating large-scale nano structures. While none of the components or techniques described in this paper are ground breaking, their combined application in a complete system should demonstrate a capability that has been difficult for many to achieve. Once the motion of the STM tip can be measured and moved accurately, it can be controlled. Also, a vision system capable of viewing micro-scale features allows for writing nano-scale structures in a known location on a sample. These results would allow nano-structures to be integrated with larger features created by standard semiconductor techniques.

8 REFERENCES

[1] J. Gilsinn, B. Damazo, R Silver, H. Zhou, "A Macro-Micro Motion System for a Scanning Tunneling Microscope", *Robotics, Automation, Control, and Manufacturing: Trends, Principles, and Applications*, World Automation Congress, v. 14, pp. 281-289, 2002.

[2] R. M. Silver, H. Zhou, S. Gonda, et. al., "Atomic-resolution Measurements with a New Tunable Diode Laser-based Interferometer", Optical Engineering, to be published January 2004.

[3] L.P. Howard, J. Stone, and J. Fu, "Real–time Displacement Measurements with a Fabry-Perot Cavity and a Diode Laser", Precision Eng., v. 25, p. 321, 2001.

[4] S. Wetzels, P. Schellekens, "Development of a traceable laser-based displacement calibration system with nanometer accuracy", CIRP Annals, 46/1, 1997.

[5] G.S. Higashi, R.S. Becker, Y.J. Chabal, and A.J. Becker, Applied Physics Letter, vol. 58, p. 1656, 1991; R.S. Becker, G.S. Higashi, Y.J. Chabal, and A.J. Becker, Physics Review Letter, vol. 65, p. 1917, 1990.

[6] Y. Kim and C. Lieber, Journal of the American Chemical Society, vol. 113, p. 2333, 1991.

[7] M. Schwartzkopff, P. Radojkovic, M. Enachescu, E. Hartmann, and F. Koch, Journal of Vacuum Science Technology, B 14, p. 1336, 1996.

[8] J.A. Dagata, J. Schneir, H.H. Harary, C.J. Evans, M.T. Postek, and J. Bennet, Applied Physics Letter, vol. 56, p. 2001, 1990.

[9] W. Kern, Semiconductor International, vol. 94, April, 1984.

Investigation of Tooling Surfaces on Injection Molded Nanoscale Features

Sung-hwan Yoon[*], Chinnawat Srirojpinyo[*], Junseok Lee[**], Changmo Sung[**], Joey L. Mead[*], Carol M. F. Barry[*]

[*] Department of Plastics Engineering, University of Massachusetts Lowell, Carol_Barry@uml.edu
[**] Center for Advanced Materials, University of Massachusetts Lowell
1 University Avenue, Lowell, MA 01854, USA

ABSTRACT

The composition and surface properties of tooling materials become more critical as the size of the molded features decreases [1, 2]. This work investigates the effect of tooling surfaces with micro and nanoscale features. These tooling surfaces were employed as inserts for micro injection molding. Insert materials included etched and coated silicon wafers with pattern depths of 600 nm and minimum features of 200 nm. Electroformed nickel-based digital versatile disk (DVD) masters were employed as a control because this tooling currently can reproduce features that are 140 nm in depth. The micro and nano-featured parts were molded with high flow polycarbonate over a range of processing conditions. Atomic force microscopy (AFM) and root mean square (RMS) roughness were used to characterize the surface topography of molded samples. The goal of this study was to explore the effect of different tooling materials on molded plastic parts with nanoscale features in terms of replication quality and durability of mold surface.

Keywords: nanoinjection molding, nanoscale features

1 INTRODUCTION

In recent years, various kinds of nanotechnology have been developed for replication of nanoscale features, including nano imprint lithography [3, 4]. Few of these approaches, however, are economically viable for manufacturing. Among the many molding techniques, injection molding technology offers the potential to rapidly and economically produce components with nanoscale features. Injection molding can produce complex geometries in a single production step using automated processes [5].

Due to the high heat and pressure developed during injection molding, durable materials, such as steel, aluminum, and beryllium copper alloys, are typically used as mold tooling materials [6]. Conventional machining technologies for these materials are not suitable for nanoscale features. As a result, toolmakers have turned to micromachining techniques, such as UV or X-ray lithography, LIGA, laser ablation, or high precision electro discharge machining (EDM) [7]. Some of these technologies are widely used in the semiconductor industry for fabrication of integrated circuits from silicon wafers. Thus, the use of silicon wafers as a tooling material is an attractive alternative. Shah et al. [8] introduced silicon wafers as a tooling material for injection molding in 1999. Becker and Heim [9] utilized silicon wafers for hot embossing, describing the potential advantages of silicon tooling in terms of hardness, tensile strength, linear thermal expansion coefficient, thermal conductivity, and good mold release characteristics due to the flat surfaces.

When molding at the nanoscale, contamination of the surface becomes a critical issue. Heat transfer may also play a much larger role in terms of replication of the surface features. Heat transfer through the insert and the insert features was one of the main concerns for micro injection molding and thin wall injection molding processes [10]. Brittleness of the silicon wafer is also a potential concern.

The goal of this research was to explore the effect of different tooling materials on molded plastic parts with nanoscale features in terms of replication quality and durability of the mold tooling surface. Molding was carried out in the absence of a clean room to assess the robustness of the process. RMS roughness measurement was used as a way to measure the replication quality.

2 EXPERIMENTAL

2.1 Tooling

Two kinds of mold inserts were used for molding. One was an electroformed nickel-based DVD master, and the other was a micro/nano scale patterned silicon wafer produced by e-beam lithography. The silicon wafer and DVD master were cut into small pieces to fit as an insert (9 mm x 4.6 mm) in a larger mold. The small piece was carefully placed into a vial with various cleaning solutions and cleaned in a Branson ultrasonic sonicator following the procedure shown in Table 1.

Step	Solution	Time	Repeat
1	Detergent (Lysol) (1:3) solution	15 min	1 time
2	Deionized water	15 min	2 times
3	Acetone	15 min	2 times
4	Methanol	15 min	2 times

Table 1. Silicon Wafer Cleaning

The detergent solution (Lysol) was used for removing general dirt and grease from the surface, acetone for non-polar impurities, and methanol for polar impurities, respectively. Most of the cleaning process was done at room temperature. A Teflon sheet was inserted between the insert and mold to protect the insert from the high forces generated during the molding process.

2.2 Compression Molding

Compression molding was conducted to study the feasibility of using a silicon wafer (2.54 cm x 2.54 cm) as a tooling material. A nickel-based electroformed DVD master (2.54 cm x 2.54 cm) was also used for compression molding as a reference material. Polymer pellets were fed into a single screw extruder (C.W. Brabender Instrument Inc., Type 2503) at a melt temperature of 288°C. Optical grade high flow polycarbonate (GE Plastics) was used for the compression molding process to allow better filling of small features due to the low viscosity of PC. After stabilizing the extrusion process, small amounts of the extrudate were collected on the DVD master, placed between two Teflon sheets and then placed in the pre-heated (315°C) compression molding machine (Carver Laboratory, Model: C). A clamping pressure of 6.89 MPa was applied for 5 minutes, the pressure was released, then the DVD master was allowed to cool to ambient temperature. The same procedure was followed for polypropylene (Accpro 9433) on the silicon wafer, but at lower processing temperatures (230°C).

2.3 Injection Molding

A micro injection molding machine (Nissei AU3E) was employed for the molding process. Optical grade high flow polycarbonate (GE Plastics) was used for the molding process after drying in the oven at 120°C. The mold temperature was set at 80°C, and the nozzle, joint, and plunger temperature were set to 315°C. The injection velocity was set from 50 to 100 mm/s, and the shot size was gradually increased to 10.50 mm. Packing pressure was not applied in order to avoid fracture of the silicon wafer. Due to the brittle nature of the silicon wafer, failure of wafer was observed whenever packing pressure was applied. A packing pressure of 50 MPa was applied for the nickel-based DVD master.

2.4 Characterization

A high-precision XE-100 atomic force microscope (AFM) manufactured by PSIA Corp. was used to characterize the tooling and molded surface features. The scan rate was 0.5 Hz and tapping mode was used for the scanning. The RMS roughness was measured by XEI 1.1 of PSIA Corp. The location of measurement was the top surface excluded each pit (pink-colored area). The

roughness were measured and analyzed from the same AFM scanned area.

The histogram of each profile shows the distribution of depth of DVD surface.

3 RESULTS AND DISCUSSION

Compression molding trials were performed to assess the overall viability of different tooling materials for injection molding. Figure 1 shows the scanning electron microscope (SEM) image of compression molded polypropylene using the silicon wafer as a tooling surface.

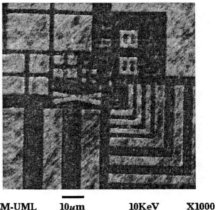

SEM-UML   10μm   10KeV   X1000

Figure 1. SEM image of molded polypropylene from silicon wafer by compression molding.

Demolding from the silicon wafer was extremely difficult, most likely caused by the high shrinkage of the semi-crystalline polypropylene and adhesion of polymer to impurities on surface, which was not cleaned prior to use. Figure 2 shows the AFM image of molded polycarbonate (PC) from a DVD master by compression molding.

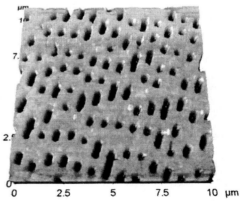

Figure 2. AFM image of molded polycarbonate from nickel-based DVD master by compression molding.

RMS analysis of DVD master measured by XEI 1.1 of PSIA Corp. The flat surface area of the DVD master was measured and used for replication quality. Table 2 shows the rms values of DVD master tooling surface and that of compression molded polycarbonate and injection molded polycarbonate surface. The rms roughness value of the compression molded PC was 12.8 nm, and it was marginally rougher than injection molded PC of 11.1 nm. Injection molding showed better replication quality because of the better thermal control and pressure distribution as compared to compression molding.

Surface	RMS
DVD master	10.7 nm
compression molded PC	12.8 nm
injection molded PC	11.1 nm

Table 2. Root mean square roughness comparison between compression molded PC and injection molded PC

Gallium arsenide (GaAs) wafers were also used as mold inserts for compression molding. These inserts were too fragile and fractured while being placed in the mold. The hardness of silicon wafers (110) measured by nanoindentation techniques is 12 to 14 GPa whereas that of GaAs (001) is only 6.7 to 7.0 GPa [11]. Due to the low fracture toughness, gallium arsenide was not practical for tooling. Based on the compression molding trials both silicon wafers and nickel-based DVD masters were further investigated in the injection molding trials.

Figure 3 shows the AFM three-dimensional topography of the silicon wafer tooling surface used in the injection molding trials.

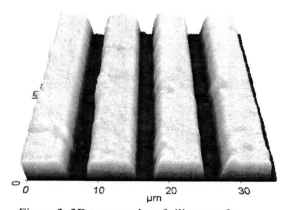

Figure 3. 3D topography of silicon wafer.

Figure 4 presents the polycarbonate surface prepared using the silicon wafer tooling. The well-defined trenches of the molded polycarbonate show that replication of the pattern from the wafer is excellent. The bottom values for the silicon wafer should be compared to the top values for the polycarbonate, as the tooling produces the reverse of the surface features. The data in Table 3 indicate the excellent replication of the tooling surface by the polycarbonate.

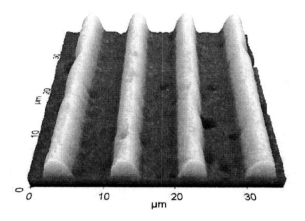

Figure 4. 3D topography of molded polycarbonate from silicon wafer.

	TOP		BOTTOM	
	AVG.	STD	AVG.	STD
Si wafer	25.7	7.2	66.7	9.7
Molded PC	52.4	2	22.4	4.4

Table 3 lists the root mean square roughness (nm) comparison between silicon wafer and molded PC.

Figure 5 shows the three-dimensional topography of the electroformed nickel-based DVD master surface and molded polycarbonate from the DVD master.

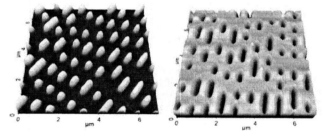

Figure 5. (a) 3D topography of nickel-based DVD master and (b) the molded polycarbonate from the DVD master

As the DVD master tool was a metallic part, 50 MPa of packing pressure could be applied for molding. This packing pressure resulted in excellent replication of the tooling using the DVD master.

The heat transfer problem was not as serious as expected. Since the molded part was neither a thin wall molded part nor micro-injection part in terms of shot size, filling the part was not difficult [10]. High pressure and high injection speed were not required for molding. The thermal conductivity of silicon is 148 W/m-K at 300 K and 90.7 W/m-K for nickel [12]. In both cases, these surfaces produced excellent replication.

Cleaning of the tooling was also not as difficult as anticipated. Once the first cleaning was complete, the collection of dirt on the surface was very low, most likely because of the short time of mold opening during the molding process and the small dimensions of the cavity. The tooling surface seemed to be kept clean by some type of self-cleaning effect. In fact, the surface quality appeared to improve as more parts were made. This may be caused by small particles of dirt being removed by adhering to the polymer melt, rather than remaining on the surface. A more detailed stuffy of this effect will be required.

The silicon wafer showed potential as a possibility for a tooling material under realistic injection molding conditions, however, the brittleness of the silicon wafers may be a limiting factor in the use of silicon wafers as a tooling material. In addition to maintaining low pressure in the cavity, the reduction of micro flaws on wafer surface is critical. Although the wafers used in this research were prepared by manual cutting, machine cutting would reduce the number of flaws significantly.

Wafer cleaning and handling under non-clean room conditions are also difficult problems to solve, but the self-cleaning tendencies could be utilized as a solution. Molding quality could also be enhanced by either better heat transfer from the mold insert to the mold or by a metallic coating on the silicon wafer. The metallic coating would not only enhance heat transfer, but it would also reduce the number of micro flaws on wafer surface. Higher mold temperatures would be also helpful for maintaining the fluidity of the polymer melt to increase the ability to flow around and into nanoscale features during injection molding.

4 CONCLUSIONS

The effect of different tooling materials on molded plastic parts with nanoscale features in terms of replication quality and durability of mold surface were investigated. Compression molding and injection molding were performed with electroformed nickel-based DVD master and silicon wafers as the tooling surfaces. Molding quality of the surface was assessed by AFM image analysis and RMS roughness. Both silicon wafers and DVD masters produced excellent surface replication. The silicon wafers showed potential as an injection mold tooling material. The brittleness of the silicon wafers could be handled by molding without packing pressure.

5 ACKNOWLEDGEMENTS

This research is supported by the National Science Foundation under grant number DMI-200498. The authors would also like to thank Kyu-Pil Lee at the University of Florida and Charlie Currie for providing the silicon and gallium arsenide wafers.

REFERENCES

[1] M. T. Martyn, B. Whiteside, P. D. Coates, P. S. Allan, G. Greenway, and P. Hornsby, SPE Technical Papers, 61, 2582, 2003.

[2] M. T. Martyn, B. Whiteside, P.D. Coates, P.S. Allan, G. Greenway, and P. Hornsby, SPE Technical Papers, 60, 480, 2002.

[3] H. C. Scheer, H. Schultz, T. Hoffmann, and C. M. Sotomayor Torres, Nanoimprint Techniques, in: H.S. Nalwa (ed.), The Handbook of Thin Films, Academic Press, New York, NY, 2001.

[4] Y. Chou, P.R. Krauss, P. J. Renstrom, Appl. Phys. Lett. 67, 3114, 1995.

[5] G. Potsch and W. Michaeli, "Injection Molding - An Introduction", Hanser/Gardner Publications, Cincinnati. OH, 1, 1995.

[6] R. A. Malloy, Plastics Part Design for Injection Molding, Hanser/Gardner Publications, Cincinnati. OH, 312, 1994.

[7] W. Lutz and E. Wolfgang, SPE Technical Papers, 56, 930, 1998.

[8] J. Shah; Y. Su, and L. Lin, Micro-Electro-Mechanical Systems, 1, 235, 1999.

[9] H. Becker and U. Heim, IEEE international Conference on Micro Electro Mechanical Systems, p229, 1999.

[10] Y. K. Shen and W. Y. Wu, International Communications in Heat and Mass Transfer, 29 (3), 423-31, 2002.

[11] S. E. Grillo, M. Ducarroir, M. Nadal, E. Tournie, and J-P. Faurie, J. Phys. D: Appl. Phys., 3, L6, 2003.

[12] G. N. Ellison, "Thermal Computations for Electronic Equipment," Van Nostrand Reinhold, New York, NY, 3, 1984.

Processing Parameters Affecting Nanoinjection Molding

Chinnawat Srirojpinyo[*], Sung-hwan Yoon[*], Jun Lee[**], Changmo Sung[**], Joey L. Mead[*], and Carol M. F. Barry[*]

[*]Department of Plastics Engineering, University of Massachusetts Lowell, Carol_Barry@uml.edu
[**]Center for Advanced Materials, University of Massachusetts Lowell
1 University Ave., Lowell, MA, 01854

ABSTRACT

Although micro parts and features are routinely molded, the performance of polymer melts is not well understood when the part wall thickness is less than 1 mm. In this study, the effects of molding conditions and material properties were determined for the replication of nanoscale features via injection molding. The nanoscale features were part of a thin insert incorporated into the mold. Polypropylene, polystyrene, polycarbonate, and polymethylmethacrylate (PMMA) were examined using a two-level design of experiments designed to investigate the effects of melt temperature, mold temperature, injection velocity, and packing pressure on depth ratio and surface quality. Atomic force microscopy (AFM) was employed to measure the molded parts. Polypropylene provides the best replication with broad process window while polycarbonate is the best replicated of amorphous materials with narrower process window. As expected, replication is material dependent and better achieved with higher melt temperature and mold temperature.

Keywords: nanoinjection molding, nanoscale features

1 INTRODUCTION

Injection molding on the nanoscale offers the potential for high rate, high volume manufacturing of integrated circuits and micro- and nano-electro-mechanical systems (MEMS and NEMS) [1] due to significant reduction in the infrastructure investment, manufacturing costs, and environmental impact. Polymers are more biocompatible than the silicon-based systems current employed for biological MEMS [2] and can be tailored to have desired properties, including chemical resistance and protein adsorption characteristics. Molding would also match with the very broad product range and relatively short lifetimes of biomedical products. Such manufacturing methods would greatly enhance assays and facilitate tissue engineering applications [3].

Molding of micron-scale parts has shown that heat transfer, specifically, mold temperature, plays a far more important role in mold filling as compared to conventional injection molding [4,5], thereby requiring near isothermal mold filling [6,7]. Surface roughness also influences part

quality and the ejection of micromolded parts [8]. Attempts to mold with silicon wafers rather than metal tooling have demonstrated that the wafers are feasible tooling, but are more fragile than metals [9,10]. Finally, interfacial effects, such as surface tension, become more important at the nanoscale. While large nanoscale features have been molded for digital versatile disks (DVD), the molding system was customized for optical grade polycarbonate and nickel tooling. Therefore, replication of nanoscale features via injection was examined with several materials and a range of processing conditions to determine whether heat transfer was the only factor affecting the nanoscale molding (as it is with micromolding) or whether interfacial effects were influencing the quality of molded parts.

2 EXPERIMENTAL

The high flow materials selected for this study included an optical-grade polycarbonate (PC), which is standard material for DVDs, polystyrene (PS), polymethylmethacrylate (PMMA), and polypropylene (PP). Material properties are listed in Table 1. Melt viscosity was characterized using the viscosity at the highest melt temperature and a shear rate of 10,000 s^{-1}, η, the power law index , n, and the activation energies, E_a, calculated for a shear rate of 10,000 s^{-1}. Thermal properties were condensed to the thermal diffusivity, α.

Material	PC	PS	PMMA	PP
n	~ 1.00	0.40	0.45	0.40
η, Pa-s (T, °C)	30 (316)	18 (238)	18 (288)	12 (260)
E_a, kJ/mol	73	39	52	14
α, mm^2/s	0.11	0.09	0.07	0.07

Table 1: Material properties.

The DVD stamper was cut into 9 x 4 mm^2 pieces, cleaned in a sonicator, and inserted into an injection mold, which was then mounted in a 3-ton two-stage micro-injection molding machine (Nissei, model AU3E). A four-factor two-level design of experiments, DOE, as shown in Table 2, was conducted to investigate the effect of melt temperature, T_m, mold temperature, T_w, injection velocity, v_{inj}, and packing pressure, P_{pack}, on feature replication. Parts were characterized using AFM with a PSIA Corp. (model High-precision XE-100) instrument and a non-contact method at 0.5 Hz.

		PC	PS	PMMA	PP
T_m, °C	Low	293	221	266	204
	High	316	238	288	260
T_w, °C	Low	54	38	49	38
	High	80	66	60	66
v_{inj}, mm/s	Low	100	110	100	21
	High	150	163	150	163
P_{pack}, MPa	Low	30	2.7	20	4.7
	High	110	92	130	44

Table 2: Processing conditions.

3 RESULTS

Figure 1 presents a typical topographic image from the AFM. For each image, three scans were analyzed to produce the "surface roughness" traces illustrated in Figure 1. In this image, the projections of the tooling have produced depressions or features in the molded part. These traces give the depth of the molded part as a function of scan distance. Feature dimensions were measured from these traces. Similar AFM images showed that the projections on the DVD tooling were 140 nm deep. The replication of the tooling features was quantified using the depth ratio and feature definition. The depth ratio, DR, was defined as:

$$DR = \frac{d_p}{d_t} \qquad (1)$$

where d_p is the depth of the depression in the molded part and d_t is the depth of the projection in the tooling (i.e., 140 mm).

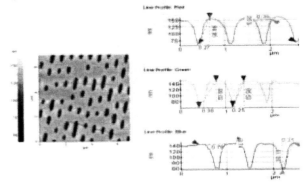

Figure 1: The topography for a molded part. and the typical surface roughness from AFM. The magnification is 8,000 X

Since the geometry of the tooling projections was not uniform, feature definition, FD, was rated on a scale of 1 to 10, where a rating of 10 was the best quality of the molded part replicated from the tooling. Table 3 illustrates the scale used for feature definition. The depth ratio and feature definition analyzed for trends using Minitab.

Scale	Feature Replication	Scale	Feature Replication
2		6	
3		7	
4		8	
5		9	

Table 3: Feature replication scale.

Hypothetically, increasing melt temperature, mold temperature, injection velocity, and packing pressure should improve feature definition and depth ratio. At higher melt temperatures, material can flow longer to complete the part filling, while higher mold temperatures will slow the cooling and allow more material to flow to the end of the fill. Higher injection velocities lower the melt viscosity due to higher rates of shear and elongation, and to shear heating. Lower viscosity melt would be expected to improve the depth ratio and feature definition because the material can more easily flow into the features. Rapid filling also reduces cooling of the molten polymer during filling, and thus, should improve feature replication. For this mold design, packing pressure acts like a compression force and higher pressures should force the melt closer to the cavity walls.

Tables 4 and 5 present the probability values from the ANOVA analysis, with the bold-faced values being significant effects. Melt temperature and mold temperature had the greatest effect on feature replication. The depth ratio was primarily a function of melt temperature and mold temperature while feature definition depended only on mold temperature.

Material	PC	PS	PMMA	PP
T_m	**0.016**	**0.004**	0.219	**< 0.001**
T_w	**< 0.001**	**0.001**	**0.017**	0.061
v_{inj}	**0.002**	**0.001**	0.200	0.075
P_{pack}	**0.029**	0.634	0.567	0.672

Table 4: Probability values for depth ratio.

Material	PC	PS	PMMA	PP
T_m	0.869	0.071	0.669	**0.009**
T_w	**0.021**	0.757	**0.012**	0.500
v_{inj}	**0.015**	0.093	0.526	0.150
P_{pack}	0.424	0.163	0.526	0.081

Table 5: Probability values for feature definition.

As illustrated in Figure 2, depth ratio increased with melt temperature. The depth ratios of polypropylene and polystyrene were most improved by increases in melt temperature and exhibited an increase of 0.45% per °C. Polycarbonate showed an increased of 0.40% per °C whereas PMMA was less affected by temperature and gave an increase of 0.22% per °C. Feature definition, however, was less affected by melt temperature (Figure 3).

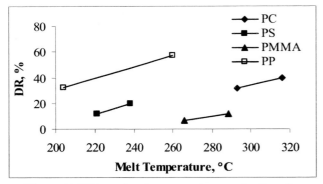

Figure 2: Effect of melt temperature on depth ratio.

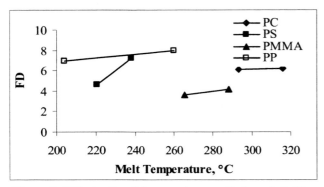

Figure 3: Effect of melt temperature on feature definition.

Only polystyrene showed significant increases in feature definition with higher melt temperatures. In contrast, mold temperature had a greater effect on the feature definition (and depth ratios) of polycarbonate and PMMA. As shown in Figure 4, the definition of polypropylene and polystyrene features did not change significantly with increased mold temperature.

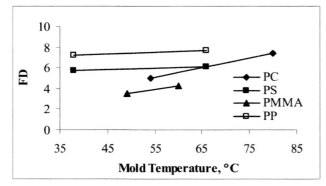

Figure 4: Effect of mold temperature on feature definition.

With respect to material performance, the sensitivity of part reproduction to melt temperatures did not follow trends typical of conventional injection molding. Generally, increases in melt temperature have the greatest effect on materials whose viscosity changes significantly with temperature (i.e., have a high activity energy). At higher melt temperatures, however, feature replication did not improve for polycarbonate (E_a = 73 kJ/mol), showed slight improved for PMMA (E_a = 52 kJ/mol) and polypropylene (E_a = 14 kJ/mol), and increased significantly for polystyrene (E_a = 39 kJ/mol). Although polystyrene, PMMA, and polypropylene exhibited similar levels of shear thinning and similar viscosities, the depth ratio of PMMA was not affected by changes in melt temperature whereas polystyrene and polypropylene showed similar improvements in depth ratio. Melt temperature had the effect on the depth ratio of polycarbonate, which had no shear thinning and a higher nominal viscosity than the other three materials, as it had on polystyrene and polypropylene.

Mold temperature was a significant factor affecting the feature definition in polycarbonate and PMMA parts. Increasing mold temperature, however, had no effect with polypropylene and polystyrene. Polycarbonate and PMMA had the highest temperature sensitivity of melt viscosity and the highest thermal diffusivities. Polystyrene and polypropylene had similar, lower thermal diffusivities, but different activation energies. Although mold temperature is a much more significant factor in the filling of micromolds [11,12], these data suggest that mold temperature completely dominates the replication of nanofeatures.

As expected, the processing window for good replication of nanofeatures was very small. For most materials, the low end of the melt temperature range did not allow adequate filling of the mold or produced parts with incomplete features. Higher temperatures produced black specks in polystyrene. Therefore the melt temperature range was about 20°C for the amorphous materials and 55°C for polypropylene. The interaction of low melt temperatures and all mold temperatures also caused problems. Low melt temperature PMMA exhibited a decrease in feature definition at higher mold temperatures, probably due to this interaction.

Injection velocity had far less effect on feature replication than was expected. Higher speeds provided slight improvements in depth ratio and feature definition, but these effects were material sensitive. Packing pressure initially had little effect on feature replication because the applied effectively compressed the molten polymer. When polycarbonate was molded again using a 316°C melt temperature, 77°C mold temperature, and 20 mm/s injection velocity, packing pressures of 20, 35, and 50 MPa produced depth ratios of 75% and similar feature definition ratings. Since the switchover pressure was 35 MPa, this behavior confirms that the features have cooled during filling.

Table 6 presents the "optimum" processing conditions and the feature replication associated with those conditions. High melt and mold temperatures, rapid injection, and the higher pack pressure produced the best replication of the nanoscale features. All four materials exhibited similar levels of feature definition, but very different depth ratios. Polypropylene achieved a depth ratio of 85% of the tooling depth while the depth ratio of the control (polycarbonate) was 54%. Polystyrene and PMMA produced very low depth ratios of 29 and 36%, respectively. This behavior suggests that the feature definition may be tied to melt viscosity while depth ratio may be influenced by interactions between the tooling surface and the polymer melt. Polypropylene had the lowest viscosity, a low thermal diffusivity, and is non-polar. Polystyrene and PMMA had similar viscosities, but PMMA had a higher thermal diffusivity and is more polar than polystyrene. Polycarbonate had much higher viscosity than the other three polymers, the highest thermal diffusivity, and a nominal polarity that is similar to that of PMMA. Optical grade polycarbonates, however, are formulated to achieve the correct level of flow, and therefore, may be less polar than expected. As a result, the ability to reproduce 100% of the tooling depth may be related to surface tension-related interactions between the tooling surface and the polymer melt. Greater differences in polarity of the tooling surface and melt allow for better reproduction. In addition, the transition temperature of polycarbonate changes as the polymer approaches a surface. Therefore, depth ratio differences in the polymer melts may be related to possible changes in the transition temperatures. Since the surfaces significantly affect reproduction of the nanofeatures, changes in point of solidification would influence the depth ratio.

Material	T_m, °C	T_w, °C	v_{inj}, mm/s	P_{pack}, MPa	DR, %	FD
PC	293	80	150	110	53	9
PS	238	38	163	92	36	9.5
PMMA	288	49	150	130	29	8
PP	260	66	163	44	86	9

Table 6: Optimal processing conditions and resultant feature replication levels.

4 CONCLUSION

Molding of parts with nanoscale features resulted in a very small processing window. Melt temperature and mold temperature affected the ability of the melt to flow around projections in the tooling, while mold temperature was the only processing parameter affecting feature definition. A comparison of material properties and feature replication results suggests that the mold temperature completely dominates polymer flow in nanoscale features and that melt temperature affects only exist as interactions with mold temperature. Injection velocity had little effect on replication. Melt viscosity had a major influence on feature definition, but depth ratio seems to be related to the interactions between the tooling surface and the polymer melt. Since depth ratio increased as the difference in polarity between the tooling and polymer melt increased, this suggests that surface tension may influence replication. Alternatively, changes in transition temperatures may be affecting melt solidification. Both causes are still under investigation.

5 ACKNOWLEDGEMENTS

This research was generously supported by the National Science Foundation under grant number DMI-200498. The authors also would like to thank Nissei America, Inc. for the donation of the microinjection molding machine and Greg Rathbone for the donation of DVD stampers.

REFERENCES

[1] Greiner, R, Kunstoffe Plast Europe, 88, 29-32, 1998.
[2] Lai, S.; Lee, J. L.; Yu, L.; Koelling, K. W and Madou, and. M. J., Materials Research Society Symposium Proceedings, 729, 17-27, 2002.
[3] Madou, M. J.; Lee, J. L.; Koelling, K. W.; Daunert, S. Lai, S.; Koh, C. G.; Juang, Y.; Yu, L.; and Lu, Y., SPE Technical Papers, 59, 245-254, 2001.
[4] Shah, R. K.; London, A. L., Advances in Heat Transfer, 1978.
[5] Yu, L.; Lee, L. J.; Koelling, K. W., SPE Technical Papers, 61, 602-606, 2003.
[6] Martyn, M. T.; Whiteside, B.; Coates, P. D.; Allan, P. S.; and Hornsby. P., SPE Technical Papers, 60, 476-480, 2002.
[7] Yao, D.; and Kim, B., SPE Technical Papers, 61, paper 521-525, 2003.
[8] Martyn, M. T.; Whiteside, B.; Coates, P. D.; Allan, P. S.; Greenway, G.; and Hornsby. P., SPE Technical Papers, 60, 2582-2586, 2003.
[9] Yu, L.; Koh, C. G.; Koelling, K. W.; Lee, J. L.; and Madou, M. J., SPE Technical Papers, 59, 785-789, 2001.
[10] Yu, L.; Koh, C. G.; Lee, J. L.; and Koelling, K. W., Polymer Engineering and Science, 42, 871-888, 2002.
[11] Wimberger, R. F., SPE Technical Papers, 57, 476-480, 1999.
[12] Bushko, W. C.; Dris, I.; Avagliano, A., SPE Technical Papers, 59, 780-784, 2001.

Interfacial instabilities in Multilayer Extrusion

Kicherl Ho, Jun S. Lee, Nantiya Viriyabanthorn,
Changmo Sung, Carol M. F. Barry and Joey L. Mead

University of Massachusetts Lowell, 1 University Ave., Lowell, MA, 01854

ABSTRACT

Polymeric alternating multilayer laminates have been prepared by coextrusion, utilizing a specially designed feedblock to combine the two incoming materials from different extruder. Layer multiplying elements (LME) were used to split the melt stream, divide it, and then bring the two polymeric melt streams together, increasing the number of layers, while decreasing the thickness of each layer.

Polycarbonate (PC) and polypropylene (PP) were extruded to form alternating structures by repeated layer multiplications. The materials were characterized by optical microscopy (OM) and atomic force microscopy (AFM) to study layer instabilities.

Zig-zag type instabilities were found at the interfaces, predominantly in the center of the samples, which originated from the interfaces near the walls. Instabilities could be reduced by increasing the temperature of the LMEs and decreasing the total flow rate of both materials.

Keywords: multilayer, interfacial instabilities, coextrusion, feedblock, AFM, layer multiplying element.

INTRODUCTION

Numerous applications have appeared for nanolayered materials, which are often produced using layer by layer deposition. One attractive method to create nanolayered materials employs coextrusion technologies. [1, 2, 3] This method offers the potential to develop extruded multicomponent thin films for conformable, high-density data or energy storage.

The coextrusion technique has been used in many applications, including those utilizing transparent materials. In the manufacture of films or sheets where appearance is important, product uniformity is critical. Coextrusion techniques, using two incompatible materials, typically require adhesive materials in between the layers or interfaces [4]. These adhesive materials act as a bonding layer and stress reliever at the interface, so that the layers show good uniformity. Without the use of adhesive layers in incompatible compositions, interfacial stresses will be increased as the number of layers increases, generating interfacial irregularities, especially near the walls in the flow channel. Interfacial instabilities result in decreased performance and aesthetics.

Thus, multilayer coextrusion requires the manufacture of uniform layered structures, which can be problematic as a result of these interfacial instabilities caused by differences in chemical or rheological properties between materials. In addition, as the number of layers increases and the thickness of each layer decreases, the individual layers often became discontinuous due to these flow instabilities. This change in behavior as the thickness decreases is an example of the issues unique to maintaining layer identity and uniformity at the nanoscale.

Much of the multilayer instability studies have been based on systems with three to five layer structures, using two or three different polymeric materials [13, 14, 15]. Two major interfacial instabilities have been found; zig-zag and wave instabilities [5]. The zig-zag type (high frequency, low amplitude) instability is driven by high interfacial shear stresses, while the wave type (low frequency, high amplitude) instability is generated by an extreme extensional deformation of the minor layer at the merge point [6 – 12].

This work investigates the role of polymer properties in the development of flow instabilities in mulitlayered films. In this research, multilayer coextrusion experiments were carried out with a semi crystalline and an amorphous material to produce multilayered structures containing tens or thousands of alternating layers.

EXPERIMENTAL

1.1 Materials

Polycarbonate (PC) (ECM Plastics, molding grade) and polypropylene (PP) (Huntsman WL-313) were used in this study. Prior to processing the PC was dried at 120°C. Shear viscosities of the materials were measured using a capillary rheometer.

1.2 Processing

Multilayer layer coextrusion equipment with two single screw extruders having a conventional 25.4 mm single screw (L/D = 24) were used to produce sheets, 25.4 mm wide and 3mm thick. These sheets consisted of three to several thousands of alternating layers, using specially designed layer multiplying elements as shown in Figure 1. Each layer multiplying element (LME) was composed of three plates, where the melt stream was split into two parts, each part was compressed, restretched, and then restacked together. These LMEs were added, one by one, up to 10 sets, until nanoscale layers were obtained. The extrusion temperature was 280°C for PC/PP compositions in most experiments because these two materials showed similar viscosities at this temperature. PC was used as the skin layer as seen in Figure 2, which resulted in PC being 80 % by volume of the total composition in the samples. Extruded samples were quenched in a water bath immediately after processing.

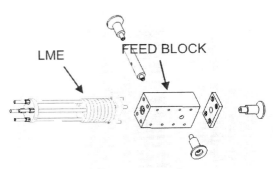

Figure 1. Multilayer coextrusion system.

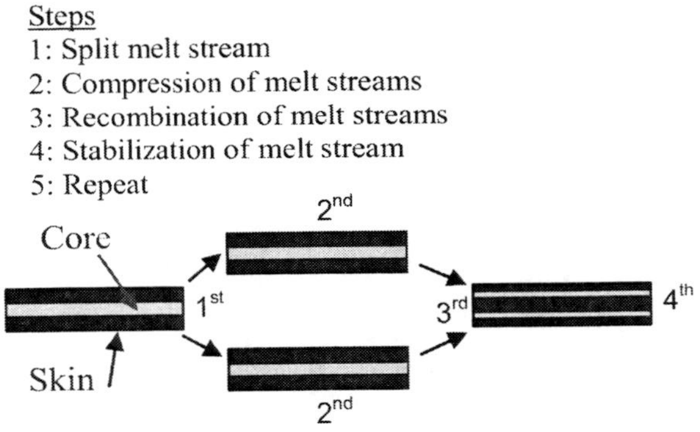

Steps
1: Split melt stream
2: Compression of melt streams
3: Recombination of melt streams
4: Stabilization of melt stream
5: Repeat

Figure 2. Schematic diagram of skin-core structure.

To prevent the formation of wave type instabilities, the volumetric flow rates were adjusted to give the same velocities for the two materials at the first merge point in the feedblock. The output ratio of skin to core layers from the extruders was set at 4 to 1 based on the ratio of the cross sectional areas of the two flow channels in the feedblock. Since the skin layer material was split and placed on either side of the core layer, the actual velocity ratio of skin and core layers in the feedblock was 2 to 1. Prior to taking samples at least 20 minutes of purging was performed.

1.3   Characterization

Multilayer laminate thicknesses were measured using optical microscopy (OM) and atomic force microscopy (AFM). The samples were prepared by embedding the extrudate in epoxy resin to hold each layer tightly and avoid delamination. After curing the epoxy, the surface of the sample was polished using a mechanical grinder with fine-mesh sandpaper (1200 grit) to reduce the surface roughness for AFM measurements. All samples were examined by optical microscopy, and for selected samples, followed by AFM (XE-100 AFM, PSIA Corp) to study the interface using a scan rate of 1Hz [16]. A non-contact ULTRASHARP silicon cantilever (NSC15 series) with Al coating on the laser-reflection side was used in this study. The dimensions of the cantilever were 125 µm × 35 µm × 4 µm. Non-contacting mode was used for the AFM scan, followed by image processing to measure the thickness of each layer.

RESULTS AND DISSCUSSION

In order to avoid the viscous encapsulation behavior [17], PC was extruded as the skin layer and PP as the core layer, due to the lower viscosity of PC over PP at the processing temperatures. Since the PP and PC were incompatible with each other, they delaminated easily, which aided in the investigation of layer uniformity.

As the number of layer multiplications increased, the thickness of layers decreased as seen in Figure 3. When five and seven LMEs were used, the thicknesses of PP layers were about 19 and 4 µm, respectively.

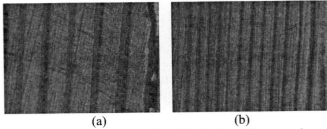

(a)   (b)

Figure 3. Optical micrographs (600x) of multilayers using five (a) and seven (b) LMEs with PC and PP.

As the number of layers increased, interfacial stresses near the wall would be higher than those near the center of samples. Thus, interfacial deformation caused by instabilities was greater in layers nearest the wall, as compared to those in the center as seen in Figure 4.

Wall ◄——————— Center ——————► Wall

Figure 4. Digital photo image of delaminated layers for three LMEs; flow rates PC (50.3 g/min), PP (30 g/min).

This increase in interfacial stresses would be expected to cause zig-zag type instabilities [11]. To avoid these instabilities, interfacial stresses should be decreased by either increasing the temperature of LMEs or decreasing the total flow rate of materials, while keeping the flow rate ratio constant.

After five layer multiplying elements were used, instability appeared, showing merged layers in the center of the cross section of the extrudates. The merged layers were found to originate from the layers closest to the walls, which were brought to the center of the sample during the splitting and recombination of the melt streams. These instabilities were reduced by increasing the temperature of the layer multiplying elements, which would act to decrease the critical interfacial shear stress [18]. This is seen in Figure 5, where delamination occurs more easily in samples with more uniform layers (Figure 5b).

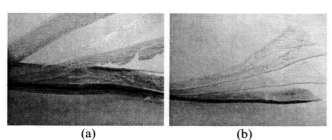

(a)   (b)

Figure 5. Digital photo images of alternating layered structures produced by five LMEs for LME temperatures of (a) 280°C and (b) 302°C.

As the number of LMEs increased over five, most of interfaces were uniform except for layers near the center of samples where two layers merged, which had originated from interfaces close to the walls. Figure 6 shows an image of a merged layer from the center of a sample, where PP layers appear darker and thinner (due to lower volumetric flow rate of PP compared to that of PC). This merged layer results in mechanical interlocking between layers, reducing the ability to delaminate the extrudates. This behavior was previously observed by Ranjbaran and Khomami [12].

Figure 6. Optical micrograph (600x) of instability using six LMEs at 280°C.

When the two materials were recombined with each other in the feedblock, the velocity difference at this first merge point caused stress at the interface, generating non-uniformity at this interface. Experiments indicated that the flow ratio between PC and PP was a critical factor in the

generation of wave type layer instabilities [8]. Balancing the flow rate ratio of the two materials in the feedblock helped to delay the occurrence of the layer instabilities as seen in Figure 7 and Table 1. When the average velocity ratios of PC/PP were less than 0.4, wave type instabilities occurred. Samples including wave type instabilities were rather easy to delaminate, indicating the amplitude of the wave type instabilities was low.

(a)   (b)   (c)

Figure 7. Digital photo images of instabilities appearing on the surfaces of samples prepared with three LMEs with a temperature of LMEs of 280°C and average velocity ratio of PC/PP, (a) 0.4, (b) 0.2 and (c) 0.1, respectively.

As more than six sets of LME were used, the number of merged layers increased as seen in Figure 8 (OM and AFM images of samples from seven LMEs). In addition, the size of the merged layers was increased. Instabilities located in the center of the sample, formed in the most recent combination of the melt streams, showed the greatest size. Layers in between these instabilities were rather uniform.

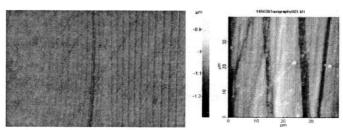

Figure 8. Optical microscope (600x) and AFM image for a sample prepared using seven LMEs at 302°C.

Figure 9 shows a cross sectional view of a sample using 10 LMEs. Most of the interfaces near the center area were disrupted, yielding small PP droplets in the PC domain. Interfaces close to the surface of the samples were stable, and fairly uniform with layers about 500 nm thick.

CONCLUSIONS

We have investigated the effect of layer multiplication on interfacial instabilities in multilayer coextrusion. Interfacial shear stress was found to be a critical parameter controlling the development of zig-zag type instabilities between layers. As the number of layers increased, the amount of instability also increased. Interfacial instabilities could be reduced by increasing the temperature of LMEs, resulting in lowered interfacial stresses. The velocity ratio of materials was also found to be an important factor in wave-type interfacial instabilities. There was a critical value of velocity ratio required to produce uniform layers. The size of instabilities grew as the number of LMEs increased, causing layer breakdown in the center area of samples. After ten LMEs were used, uniform layer structures were only detected near the surfaces.

ACKNOWLEDGEMENTS

This research is generously supported by the National Science Foundation under grant number DMI-200498. The authors also wish to thank the US Army Natick Soldier Center for the donation of the coextrusion line.

REFERENCES

[1] S. Nazarenko, A. Hiltner, E. Baer, J. Mat. Sci., 34, 1461, 1999

[2] S. Nazarenko, M. Dennison, T. Schuman, E. V. Stepanov, A. Hiltner, E. Baer, J. Appl. Polym. Sci., 73(14), 2877, 1999

[3] C. Rauwendaal, Polymer Extrusion, Hanser/Gardner Publications, Cincinnati, OH, 1994

[4] E. Mitsoulis, Advances in Poly. Tech., Vol. 8, No. 3, 225-242, 1988

[5] C. D. Han and R. Shetty, Polym. Eng. Sci., Vol. 18, No. 3, 180, 1978

[6] B. Khomami, J. Non-Newtonian Fluid Mechanics, 37, 19-36, 1990

[7] R. Ramanathan and W. J. Schrenk, Plastics Engineering, 6, 73, 1993

[8] C. Tzoganakis and J. Perdikoulias, Polym. Eng. Sci., Vol. 40, No. 5, 1056, 2000

[9] M. Zatloukal, M. T. Martyn, P. D. Coates, P. Saha, SPE ANTEC Tech. Papers, 295, 2003

[10] W. Kopytko, M. Zatloukal, J. Vlcek, SPE ANTEC Tech. Papers, 7, 2003

[11] R. Ramanathan, R. Shanker, T. Rehg, S. Jons, D. L. Headley, and W. J. Schrenk, SPE ANTEC Tech. Papers, 224, 1996

[12] M. M. Ranjbaran and B. Khomami, Polym. Eng. Sci., Vol. 36, No. 14, 1875, 1996

[13] C. Tzoganakis, J. Perdikoulias, Polym. Eng. Sci., 40, 1056, 2000

[14] M. M. Ranjbaran, B. Khomami, Polym. Eng. Sci., 36, 1875, 1996

[15] R. Ramanathan et al., Proceedings of the Society of Plastics Engineers' Annual Technical Conference, 42, 224, 1996

[16] Joonhyung Kwon, J. Hong, Y. Kim, D. Lee, K. Lee, S. Lee, and S. Park, Review of Scientific Instruments, October 1, 2002

[17] J. Dooley, K. S. Hyun, and K. Hughes, Polym. Eng. Sci., Vol. 38, No. 7, 1061, 1998

[18] W. J. Schrenk and T. Alfrey, Jr., Polymer Blends, Academic Press, Inc., New York, 129, 1978

Screw Speed Ratio (PC/PP)	Flow Rate of PC	Flow Rate of PP	Average Velocity Ratio (PC/PP)	Pressure Drop	Wave Instabilities
[-]	[g/min]	[g/min]	[-]	[MPa]	
45/20	74	18	1.0	1.86	None
30/30	50.3	30	0.4	1.72	None
20/45	34.5	45	0.2	1.79	Yes
20/60	28	60	0.1	2.2	Yes

Table 1. The effect of flow rate ratios on wave type instabilities.

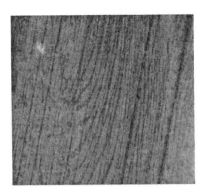
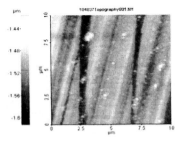
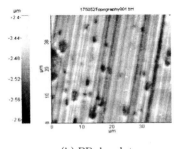

(a) PP layers

(b) PP droplets

Figure 9. Optical microscope (600x) and AFM images for a sample prepared using 10 LMEs at 302°C.

Microcap Selective Packaging through Flip Chip Alignment

C. H. Chao and C. T. Pan

Detp. of Mechanical and Electro-Mechanical Engineering ,
Center for nanoscience and nanotechnology
National Sun Yat-Sen University, Taiwan, R.O.C.

ABSTRACT

In this study, the technique of microcap selective bonding for 3-D microstructures using MEMS processes was presented. The flip-chip assembly was successfully demonstrated that the microcap transferred on the selective area of the host substrate through wafer level alignment. A new packaging technique of microcap with passivation treatment was developed for selective packaging. The metal Ni microcap is superior to those using thin film poly-silicon by surface micromachining technique due to the high stiffness structure. Photo definable material was served as the bonding adhesive layer between the silicon wafer and metal microcap. For the bonding process, several types of photo definable material were explored to characterize for bonding strength. The result shows that excellent bonding strength under bonding temperature can be achieved.

Keywords: flip-chip, passivation, microcap, wafer level, MEMS process

1 INTRODUCTION

One major cost driver in today's MEMS is the packaging process. MEMS includes moving structures, requirements for packaging of such components must take that into account. Some functions also include interaction with the surrounding environment, such as pressure sensors. Packaging should provide mechanical support to the sensitive chip.

MEMS Packaging includes wafer bonding processes and microfabrication. There are several wafer bonding techniques offering permanent wafer protection, such as anodic and fusion bonding. Conventional anodic and fusion wafer bonding techniques can not be employed to all MEMS and IC bonding process due to high temperature or high electric fields during the bonding process, which might damage IC or MEMS devices. These problems have prompted researchers to study low-temperature bonding techniques, and use eutectic bonding .

A process of selective encapsulation with low temperature bonding technique was presented, including localized silicon-gold eutectic bonding, localized silicon-glass fusion bonding, localized solder bonding and localized CVD bonding processes. There are also methods such as the donor/target procedure, using two wafers, one for the micromechanical component and one for the MEMS component. Hermetic packaging of microdevices devices is required in many applications. Resonant, and tunneling devices, infrared and pressure sensors, voltage controlled oscillators and vacuum displays all utilize vacuum sealing to improve performance. Vacuum sealing can be accomplished at the package level using brazing, soldering or welding. Wafer-to-wafer bonding has also been employed to hermetically seal microdevices. However, these approaches require bonding temperature over 220 °C to finish the process. Although the temperature is much lower than that of anodic and fusion bonding, still, it will damage some of the IC and MEMS structures during the bonding process. Another encapsulation method fabricated by MEMS process was developed. Glass wafer was etched to accommodate the microcap. Besides, the transparent characteristics of glass wafer will facilitate the alignment process during bonding. But, glass etching is time consuming and loose of precision. It would be very advantageous if a first-level or device-level packaging were performed in such a way that the remaining packaging procedures for MEMS chips would be the same as those for IC chips using common procedures with existing IC packaging equipment.

In order to accomplish these goals, we use flip chip and wafer level alignment to batch fabrication to lower the manufacturing cost, to fabricating encapsulation with passivation technique to protect MEMS devices, and to offering lower temperature and electric field free bonding process to prevent the IC and MEMS devices from thermal damage and exploring various photo definable material to find the strongest bonding strength. Furthermore, the passivation techniques were utilized in transferring microcap on the selective area of the host substrate. The process flow of wafer level bonding using flip chip packaging is shown in figure 1. Photo definable material with patternable characteristics was served as the bonding adhesive layer between the microcap and host wafer. Bonding experiment with several types of photo definable material were explored for bonding process. It shows that excellent adhesion strength between microcap and host wafer can be obtained.

2 BONDING PROCESS

The wafer bonding process can be divided into following steps: First, the metal microcap is fabricated on the carrier substrate with passivation treatment. Then wafer level and flip chip method was used to transfer the metal microcap selectively on the host substrate. Finally, two wafers were separated. The metal microcap will be transferred on the selected area of the host substrate. The bonder used in the study is a commercially available EV501 (Electronic Visions) with PC controller. AZ-4620 (a positive photoresist from Shipley), JSR-137N (a negative photoresist from Japan Synthesis Rubber Co.), SU-8 (a negative photoresist from Microchem Co.), and SP-341 (a positive photoresist from Toray Co.) were selected as bonding adhesive layer material for the bonding experiments.

2.1   Fabrication of the metal microcap

For the fabrication process for the metal microcap. Four-inch (100) single side polished silicon wafer with 525 μm in thickness was used. Oxide is grown thermally with a thickness of 1.5 μm at 1050 °C. After etching mask was defined on the carrier silicon wafer. The silicon is anisotropically etched using 30 wt% KOH at 70 °C to form a cavity. This etchant does not attack the silicon oxidation. The cavity is formed by the (111) silicon surfaces which etch at a much lower rate than the (100) surfaces in the anisotropic etchant. The exact shape is defined by the edges of the pattern due to orientation-dependent etching, and the (111) surface forms an angle of 54.74 °. A Ni layer of thickness 150nm is then sputtered functioned as seed layer for electroplating, followed by a passivation treatment on its surface. Later the Ni layer with passivation treatment will help the metal microcap to separate from the carrier substrate. Then a thick photoresist was spin coated to form the electroplating template for the metal microcap structures. Ni metal microcap structure with 15 μm in thickness can be formed by electroplating process. Therefore, the metal Ni microcap is superior to those using thin film poly-si by surface micromachining technique due to the high stiffness structure. Then, another photo definable material was coated onto the surface of the silicon wafer. To evaporate the solvent contained in the photo definable materials, the material was baked for a period of time. The sample was then exposed and developed to define the bonding pad on the metal microcap. Finally, the electroplating template was removed.

2.2   Flip chip and wafer level bonding

The flip chip method was used to transfer microcap on the selective area of the host substrate through wafer level alignment. These two wafers were then placed into the bonder. During the bonding processing, the temperature was adjusted based on the adhesive layer material. The bonder allows wafers to contact and be annealed between plane heaters. The final procedure is to remove the carrier silicon from the host silicon wafer. After the carrier silicon wafer is pulled upwards, the microcap will be separated form carrier silicon wafer, and leaving the microcap on the host substrate. It is worth noting that the metal microcap can be easily separated from carrier substrate because the

Ni layer on the carrier wafer has a passivation treatment in advance. The process of passivation treatment is to put the Ni layer into oven flushing with air under a period of time. The chart of time vs. temperature for the treatment is shown in figure 2. And due to the treatment, the carrier substrate can be reused to fabricate metal microcap. In this approach, the bonding process allows transferring microcap on the selective area of the host substrate. MEMS device can be protected by the metal microcap. This method is particularly suitable for the integration of micro-structure with microelectronics involved in MEMS packaging. Once the microcap is transferred on the host wafer, the microdevices are then protected, and can be treated as same as an IC wafer during subsequent dicing. Devices can be batch-fabricated and batch-packaged. Individual packaged microcap-base chip can then be obtained by dicing along dicing line. The individual MEMS device with microcap which can be diced as standard IC dicing process. The microcap from of 50μm   50μm to 1.5mm x 1.5mm in area can be made to protect microdevices.

2.3   Bonding inspection

The common measurement techniques are bond imaging, cross-sectional analysis, and bond-strength measurement. The bond imaging methods are nondestructive and can be used as in-process monitors, while the cross-sectional analysis and bond strength measurements are destructive for characterization.

Infrared (IR) transmission was setup to inspect the bonding structure. It consists of an IR source and an IR sensitive camera with excellent sensitivity in the near-IR range. The bonded wafer pair is located between the source and camera. Any defect after the bonding will show up. Examples of the images obtained by the method for two bonded 4-inch silicon wafer pairs are shown in figure 3. It shows the method can successfully examine bonding result. The bonding pad can be clearly identified as shown in figure 3 (a). On the other hand as shown in figure 3 (b), if Newton's Ring appears, it means void or defect existing in the bonding area. This imaging method generally cannot image voids with a dimension less than one quarter of the wavelength of the IR source.

2.4   Low temperature and electric field free bonding

Photo definable material was applied as the intermediate adhesive layer in this bonding process. It has several features; low bonding temperature, electric field free, strong bonding strength and excellent surface planarization properties. In the experiments, the single polished silicon wafers were bonded with other Ni coated wafer using different photo definable materials. The influences of the bonding material, the bonding force, and the bonding temperature were investigated.

The bonding pad can be patterned on the silicon wafer through photolithography. Therefore, its resolution can reach as small as 5 μm. The result shows that the adhesive layer can adhere on the bonding surfaces firmly to increase the bonding strength.

The bonding temperature around 50 ~ 150 °C was tested. When the adhesive layer material reaches the pre-set bonding temperature, it causes the interface of the two wafers bonded tightly, and forming excellent bonding strength. For the tensile strength test, bonded pairs are cut using dicing saw. The bonding chip is attached using a clamping apparatus in the MTS (Material Test Station) for the bond strength test. The clamping apparatus is pulled apart using a selected pull force and speed.

3 RESULTS AND DISCUSSIONS

The curves for the bonding strengths are shown in figure 4. Each data represents the average of three measurements. Figure 4 shows the tensile test curves as a function of bonding temperature under a constant bonding force (50N). From the result, it can be seen that when the bonding temperature for SU-8, JSR, AZ-4620, and SP-341 was between 80 °C and 100 °C, the bond strength reached their maximum. When the bonding temperatures were higher than 140 °C, AZ-4620 would be scorched. Besides, if the bonding temperatures were higher than 200 °C, SU-8, JSR, and SP-341 exhibited the same bonding strength as at 100 °C. It also shows that SU-8 used as adhesive layer under bonding temperature 90 °C has maximum bonding strength about 213 Kg/cm^2 (20.6 MPa). SU-8 has many attractive properties as an intermediate adhesive layer described as follows. SU-8 has an epoxy feature with very high bonding strength. SU-8 is a negative photoresist and is crosslinked after exposed to UV light. It exhibits increased chemical resistance after UV exposure. In addition, it just requires very low bonding temperature 90 °C to form excellent bonding strength. On the other hand, as for AZ-4620, when the bonding temperature was about 90 °C, the bonding strength reached about 86 Kg/cm^2. When a thinner bonding pad is required, the AZ-4620 photoresist is a good choice for the bonding process. Regarding SP-341, when the bonding temperature about 90°C, the bonding strength of SP-341 would be about 100 Kg/cm^2. When the bonding temperature for JSR was about 90 °C, the bonding strength reached about 88 Kg/cm^2.

The bonding temperatures of 90 °C for SU-8, JSR, AZ-4620, and SP-341 were tested. The result shows that when the bonding force was reduced, the bonding strength became weak between the two wafers because of voids created by the uneven wafer surface and by the voids trapped in photoresist. It is worth noticing that with a lower bonding force, the bonding between the substrate and intermediate adhesive layer can not be formed tightly because the bonding force is not strong enough to eliminate the void and gap in the interface. Besides, when larger bonding force is applied, the adjacent microstructure and IC may be destroyed. The selection of bonding force is an important parameter in the bonding processing. The bonding force is very important factors in wafer bonding technology.

It has been shown that when the bonding strength reached their maximum about 80~210 Kg/cm^2, respectively. The maximum strain of SP-341 is 155E-3, SU-8 is 80E-3, AZ-4620 is 70E-3, and JSR is 60E-3. It is worth noticing that SU-8 will become ductile behavior in the interface after bonding, but the intermediate layer using SP-341 will show elastic characteristics after bonding. The choice of intermediate layer can be based on the requirement of bonding process.

Thickness of adhesive layer less than 20 μm can be achieved (see figure 5). The thickness of the intermediate adhesive layer can be controlled precisely by the spin-coating process. Due to photolithography process, the lateral width of bonding pad can be controlled down to 10μm. Therefore, the bonding has great potential for substrates with higher density of IC and MEMS devices.

4 CONCLUSION

In this study, a silicon wafer bonding technique for 3-D microstructures using photoresist as adhesive layer was presented. The flip chip and wafer level bonding were successfully utilized in demonstrating the transferred microcap on the selective area of the host substrate. Passivation technique can help the metal microcap separating from carrier wafer. Ni metal microcap structure with 15 μm in thickness can be formed by electroplating process. Therefore, the metal Ni microcap is superior to those using poly-si by surface micromachining technique due to the high stiffness structure. The micro encapsulation with passivaton treatment can be used as a new packaging technique. Photo definable material with patternable characteristics can be served as the bonding adhesive layer between the silicon wafer and microcap. Several types of photo definable material were tested for bonding strength. Resolution of bonding pad and thickness can be down to 10 μm and 20 μm, respectively, which is very suitable for high dense IC and MEMS packaging. The results indicated that SU-8 is the best material with bonding strength up to 213 Kg/cm^2 (20.6 MPa) under 90 °C bonding temperature. When bonding is required of a high resolution bonding pad, strong bonding strength, electric field free and low temperature bonding technology, the best solution is to use photo definable material as the bonding adhesive layer. It can offer an excellent bonding result.

5 ACKNOWLEDGEMENT

The author would like to thank the National Science Council of the Republic of China for its support of this work with Grant No. NSC-92-2212-E-110-029.

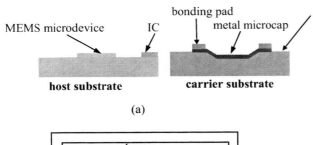

MEMS microdevice   IC   bonding pad   metal microcap

host substrate   **carrier substrate**

(a)

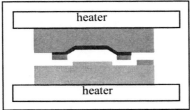

heater

heater

flip chip and wafer level alignment in vacuum environment

(b)

bonding process

(c)

microcap transferred on selective area

(d)

Figure 1 Process flow chart of flip chip and wafer level bonding with microcap protection

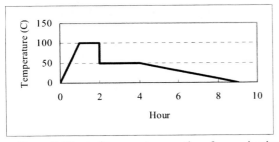

Figure 2 Chart of temperature vs. time for passivation treatment

(a) successful wafer bonding   (b) failed wafer bonding with void

Figure 3 IR transmission image of two bonded wafer pairs

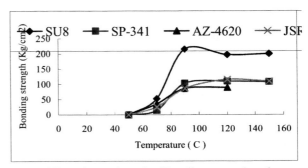

Figure 4 Experimental result of bonding strength and bonding temperature for various intermediate adhesive layers

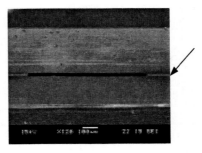

Thickness of adhesive layer less than 20 μm

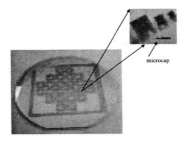

Microcap on selective area

Figure 5 Experimental result of microcap bonding

Miniaturized High Speed Visualization Setup for the Diagnostics of Dynamical Processes in Microsystems

H. Brugger, C. Maier and E.P. Hofer

Dept. of Measurement, Control and Microtechnology
University of Ulm, Albert-Einstein-Allee 41, 89081 Ulm, Germany

ABSTRACT

For the real time visualization of the dynamics of MEMS it is often inevitable to use high speed camera systems because of the optical enlargement. The micro diagnostic system MinVis, which consists of a miniaturized optical setup with a commercial microscope objective and a high performance AlInGaP LED as pulsed light source allows the investigation of these highly dynamic processes in microsystems in real time, and represents an economical alternative to complex and very expensive high speed camera systems. The application of MinVis covers a wide range from diagnostic measurements during the prototype development to inspections for the quality assurance of microsystems. As an easily transportable setup MinVis is also particularly suitable for presentation purposes.

Keywords: high speed visualization, dynamic processes in microsystems

1 INTRODUCTION

Dynamic processes in MEMS (micro electro mechanical systems) are in general very fast due to their small masses. Additional, because of the small sizes, an optical system magnification is required. The consequence of this optical enlargement is a virtual speed-up of the mechanical system. Therefore, for the real time visualization of the dynamics of MEMS it is often inevitable to use multi frame high speed camera systems.

The micro diagnostic system MinVis allows also the investigation of highly dynamic processes in microsystems in real time, and represents an economical alternative to complex and very expensive high speed camera systems. Fast processes on the micrometer scale are visualized in real time with multiple exposure LED (light emitting diode) illumination while the camera shutter is continuously open. Even non-reproducible processes can be captured since the multiple flashing produces a superimposed image containing real time information.

Such a micro diagnostic system can be used for the determination of unknown material parameters on the micrometer scale, for the evaluation of the system performance and for its optimization. Because of its small

size and the modular mechanical-optical structure a single MinVis system or the combination of several MinVis systems can be integrated for e.g. monitoring manufacturing processes [3].

2 COMPONENTS OF MINVIS

MinVis consists of a miniaturized optical setup with a commercial microscope objective and a high performance AlInGaP LED as pulsed light source. LED light sources have already been applied to high speed photomicrography since they have several advantages compared to classical approaches using Xenon flash lamps or spark sources. They have a much higher repetition rate which is necessary for the real time capturing by multi exposures. In order to work with the low light level of the illumination a very sensitive black and white television CCD (charged coupled device) camera is used. Its minimal illumination is specified with 0.00015 *lux*. Moreover, this surveillance camera has a very compact size. To become a truly movable and versatile high speed real time photomicrography system we have also miniaturized the necessary delay and pulse electronic. Figure 1 shows the optical part of MinVis with the microscope objective, an optical system and the small camera. On the left side the LED light source is visible which is composed of a LED and a lens. The whole mechanic optical part of MinVis is about 240 *mm* long.

Figure 1: Mechanic optical part of MinVis.

3 FUNCTIONAL DESCRIPTION

For image capturing a standard video camera is used. Therefore the entire measuring procedure is synchronized concerning the black and white video signal from the camera. The CCD inside the camera have to be illuminated within a defined period of the vertical synchronization signal. With an adjustable time value T0 the starting time of this active time period is defined. This time value can be entered with a resolution of 10 μs. According to this trigger signal two high precise time delays are generated. The time basis for these time delays is a 100 MHz crystal oscillator. Thus, the resolutions for these two time delays are 10 ns. Usually, with time delay A the DUT (device under test) is triggered, whereas time delay B triggers the illumination circuit. Figure 2 illustrates schematically the time delays.

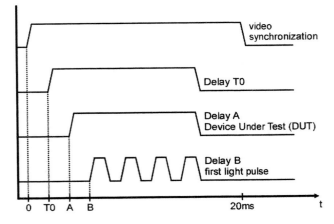

Figure 2: Schematical time delays regarding to the video synchronization.

This illumination circuit produces light pulses with a high performance AlInGaP LED. The number of light pulses, the illumination duration and the time distance between the light pulses are user selectable. The main spectral density of the used LED is at the wavelength of 626 nm. At this wavelength the camera is most sensitive. Within the optical path a matched interference filter is mounted. This interference filter allows the use of MinVis in common laboratories without the need of a shaded environment and prevents an exposure of the CCD during the whole video clock by the background illumination. The resolution for the duration of illumination is also dependent on the used crystal oscillator and is also 10 ns. If there are two MinVis systems necessary, e.g. for the examination of an experiment in two different directions, an optical decoupling with two different wave lengths by the use of different LEDs and interference filters is also possible [2].

For highly reproducible procedures, MinVis is suitable for pseudo cinematographic measurements. Therefore, only one light pulse is used for each captured picture. The dynamic information can be obtained with an increasing time delay B for the light pulse in subsequent pictures. For such measurements, the delay and pulse electronic has implemented an automatic counter

for producing such an increasing series of trigger signals. With a video capturing card the dynamic behavior can be saved like a slow motion recording. The speed of the dynamic variation is defined by the increment value to the time delay B for the light pulse.

MinVis is developed as an easily transportable setup. Thus, the whole pulse and trigger electronic is developed as a small device with a standard RS232 interface to a personal computer. The second version of MinVis includes also a frame grabber for the standard video signal and communicates with the PC with an USB interface. The use of standard interfaces for the communication offers the usage of a Laptop for the control of MinVis.

4 HIGH SPEED VISUALIZATION OF A MICRODOSAGE PROCESS

To demonstrate the performance of MinVis, a commercial micro dosage unit [1] is mounted on a xyz-positioning unit and the ejected droplet is in the focus of the MinVis system. Figure 3 shows this measurement setup with the Microdrop dosage actuator and MinVis.

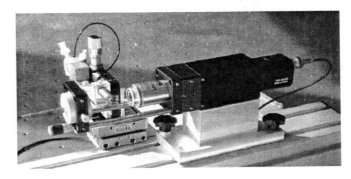

Figure 3: Commercial micro dosage unit as DUT for MinVis.

The Microdrop dispenser heads are designed to process a wide range of different liquids. The inner nozzle diameter is 50 μm and the usual dosing volume is about 65 pl. These values depends strongly on the used liquid. The droplet velocity is in the range of 2.5 m/s with a range of flight of about 20 mm. The Microdrop dispenser uses a glass capillary which is surrounded by a piezo actuator. For different liquids it is necessary to adjust the applied signal to the piezo actuator. There is a rectangular signal with the voltage and width of the signal as parameter. Figure 4 shows on ejection of a droplet, whereas the parameter for the ejection are consciously not optimal. Thus, after the main droplet there exist a tail behind it which alters to an oscillating satellite droplet.

The light pulse for the illumination is 150 μs after the trigger signal for the actuator. The light pulse is 300 ns long. This value is a compromise between a good signal

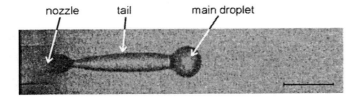

Figure 4: One light pulse for illumination. A main droplet with a tail is ejected by a micro dosage actuator. The bar in each figure denotes a distance of 100 μm.

to noise ratio and blurring effects due to the movement of the object. In Fig. 5 the same ejection is depicted with two light pulses. The second light pulse is shot 50 μs after the first light pulse. The main droplet resides normal and the tail becomes an oscillating satellite droplet. This satellite droplet has a greater velocity and will catch up the main droplet.

Figure 5: Two light pulses. The tail becomes a satellite droplet behind the main droplet.

The contrast in Fig. 5 is inferior to the contrast in Fig. 4 so that Fig. 5 is brighter than the other one. By the use of more light pulses the sensor accumulates more light and defines the maximum number of light pulses. With optimal parameters for the dosage process, a droplet ejection with 5 light pulses is presented in Fig. 6. The first light pulse is 170 μs after the trigger for the actuator and the subsequent light pulses comes with a time delay of 50 μs.

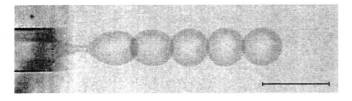

Figure 6: Five light pulses with optimal parameters for the dosage actuator. There are no satellites.

The more highly the number of exposures the smaller becomes the contrast of the captured image. Up to a number of 10 light pulses there is no problem with the contrast. In general, 10 captured moments are usually sufficient, particularly with regard to the limit of 8 pictures of common high speed cameras.

There are also limits for the usage of MinVis. If the dynamic of an object has to be captured in real time, there has to be a characteristic movement in the plane. If there is for example only a rotating device, it's not possible to separate the individual exposures in the captured image. Such an experiment can only be captured by MinVis in a pseudo cinematographic manner. But this means, that the experiment have to be reproducible and in particular that the experiment can be triggered by an electric signal.

In comparison to high speed camera systems, MinVis has several advantages. A high speed camera, like the Imacon468 from DRS Hadland Ltd. consist of eight intensified CCD units which are arranged in a circle around a beamsplitter. There are always discrepancies in the captured images because its not possible to get a perfect optical path for all CCD units. For the illumination a powerful flashlight source is needed and cannot be triggered with high frequencies. Furthermore, for the readout of the captured images an extra I/O-card for a PC is needed. MinVis allows high repetition rates also for non reproducible events and can be watched on a monitor. Furthermore, there is no need for additional hardware.

As mentioned above, MinVis allows also the visualization in a pseudo cinematographic manner. Figure 7 shows 10 pictures of a flying droplet with one satellite. The given time is the delay between the trigger for the dosage actuator and the trigger for one light pulse. The pictures are single frames of an captured video with a PCMCIA frame grabber card. The whole movie shows the flying droplet and the collision of the satellite droplet with the main droplet in a slow motion video. In particular the oscillations of the satellite droplet can be realized very good. In contrast to Fig. 6 its not one and the same droplet. For Fig. 7 ten different droplets are captured with different delay times B for the light pulse. Because of its high reproducibility, it seems to be one moving droplet.

5 GRAPHICAL USER INTERFACE

The graphical user interface is developed with the graphical development environment LabView from National Instruments. This allows the user to enter the necessary parameters for MinVis in a comfort manner and the generation of special series of trigger impulses can be adapted easily to the device under test. The input parameters are sent via the interface (RS232 or USB) to the electronic unit. This circuit generates the trigger impulses for the device and the LED independet from running processes on the computer. Also with the available graphic functions of LabView, some graphic enhancements on the captured image can be carried out. Figure 8 shows the graphical user interface of MinVis with a captured image of the dosage actuator with two light pulses. After a calibrating measurement with a

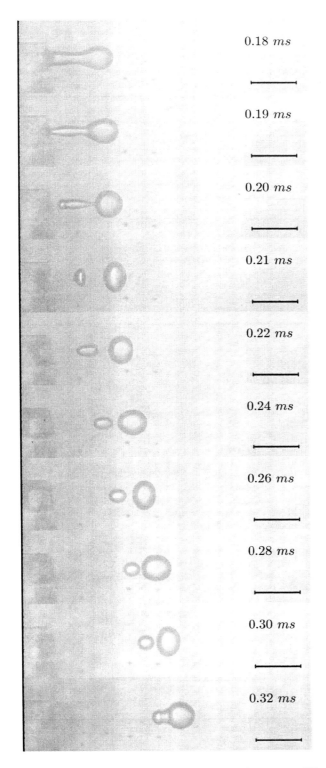

0.18 *ms*

0.19 *ms*

0.20 *ms*

0.21 *ms*

0.22 *ms*

0.24 *ms*

0.26 *ms*

0.28 *ms*

0.30 *ms*

0.32 *ms*

Figure 7: Pseudo cinematographic visualization. The bar in the figure denotes again a distance of 100 μm.

reticle of lines, direct measurements in the image can be carried out. In Fig. 8 the diameter of the ejected droplet is determined with approximately 48 μm.

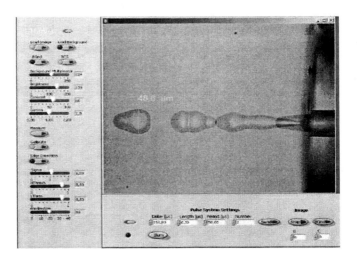

Figure 8: LabView GUI.

6 CONCLUSION

The portable realtime micro diagnostic system Min-Vis is aimed for the inspection of dynamical processes in micro electro mechanical systems (MEMS). Therefore, MinVis visualizes the motion of small devices in real time by flashing the device with one or more short light pulses. The real time information is recorded in one single superimposed picture. The MinVis system with its components and the performance of this measurement setup is demonstrated on a microdosage process. Thus, MinVis facilitates the investigation of dynamic operations and the identification of system parameters like velocities but also dynamic parameters like eigen frequencies.

With the presented characteristics, MinVis is well suited as a cheap alternative for complex and expensive high speed multi frame cameras and can be used as a easily manageable system for a lot of different applications like supervising, identification, optimization, manufacturing, presentation, and so on.

REFERENCES

[1] http://www.microdrop.com

[2] Maier, C. and Hofer, E.P.: Three-dimensional device characterization by high-speed cinematography. In proceedings of SPIE conference: Microsystems Engineering: Metrology and Inspection. Vol. 4400, pp. 16–22, Munich, 20. June 2001, Germany.

[3] Hofer, E.P. and Maier, C. and Brugger, H. and Wecker, T. and Queffelec, X.: MinVis & Stereo: High Speed Visualization for the Diagnostics of Dynamical Processes in Microsystems. Developed at the department of Measurement, Control and Microtechnology, University of Ulm, Hannover Industrial Fair, Germany, 2002.

3D Atomic Holographic Optical Storage Nanotechnology

By

Michael E. Thomas

Colossal Storage Corporation
fedrive@pacbell.net

Abstract

Colossal Storage Inc. has licensed 5 year old patents on new ways of non - contact reading and writing with non destructive reading of information to a ferroelectric molecule.

1. Introduction

The Colossal FE Optical Drive density of 40 gigabits/sq.in. up to 40,000 Terabits/cu.cm.[1] A comparison with harddrives of today is around 4 gigabits/sq.in. maxing at ~200 gigabits.[2] With optically assisted / Blu - Ray drives maxing at ~45 gigabits/sq.in. and contact recording AFM, STM, SPM or SFM, i.e. atomic force microscope and their derivatives, maxing practically out at about ~300 gigabits/sq.in..

2. First Time in the History of Mankind

Colossal Storage uses the Einstein/Planck Theory of Energy Quantum Electrons to control molecular properties by an atoms electron movement/displacement. [3] The Colossal Storage FeDrive - FeHead Semiconductor Integrated Optical Read / Write Head will use Ultraviolet/Blue laser diodes with Voltage transducer to write, photon induced electrical field poling, and UV/Blue laser diode and Nanooptical transistor or Nanofloating gate Mos Fet to read.

2.1 Atomic Switch Controls Optical Data

Molecular dissociation following Thomas' patents cover methods for a non-contact ultraviolet / blue laser photon induced electric field poling using UV at the same wavelength as a molecular transition will create controllable clouds of electrons in harmonic waves (plasmon).

Some organic/inorganic molecules have resonant valence orbit electrons that under the proper Quantum UV/Blue photoexcitation allow conduction band electrons to move freely for a short time. Plasmon known as electric current along with the electric field present providing a mechanism for ferroelectric perovskite molecules to switch binary positions. The unique concept of resonant absorption excitation by UV/Blue light causing molecular dissociation and simultaneous electric field application (Pockels effect) can be used for writing 3D volume data so when it is read back having coherent interference waves in a beam of UV/Blue photon radiation.

The single frequency creates many bright or dark bands from the UV light that are in phase or out of phase with one another. The diffraction by the bistable state nucleus in the center of ferroelectric dipole molecule can therefore be represented as a binary 0 or 1.

Ferroelectric non-linear photonic bandgap crystals offer the possibility of controlling and manipulating light within a UV/Deep Blue frequency. The small size of ferroelectric transparent structures makes it possible to fabricate nano-optical devices like volume holographic storage having both positive and negative index of refraction. The ability to control the diffraction of Ultraviolet photons makes the ferroelectric perovskite NLO photonic materials very attractive for the research and development of 3D volume holographic optical storage. Furthermore, ferroelectric non-linear photonic crystal structures provide the ability for infinite rewritability of a non-volatile holographic storage drive.

The outstanding potentials of ferroelectric molecular materials will revolutionize 3D volume holographic optical storage technologies along with several challenges in design, optimization, fabrication, and characterization an provide for further extensive research and development activities in the field of ferroelectric holographic materials and data storage.

All other known attempts at rewritable holographic storage use electrons clouds to store data and as a result have only been able to achieve write once read many devices. Thomas feels this method can never overcome the Niels Bohr Atomic Theory of electron recapture and therefore this type of Bragg/Compton Scattering recording technique is usually destructive readout and a short data storage shelf life like spatial spectral hole technology.

2 Semiconductor Integrated Optical Read / Write Head and Function

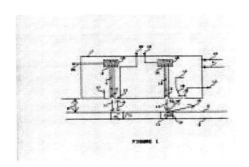

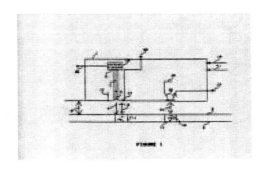

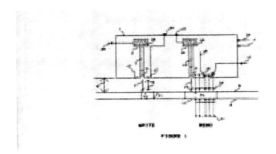

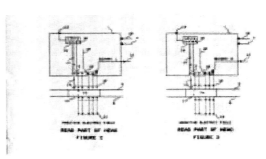

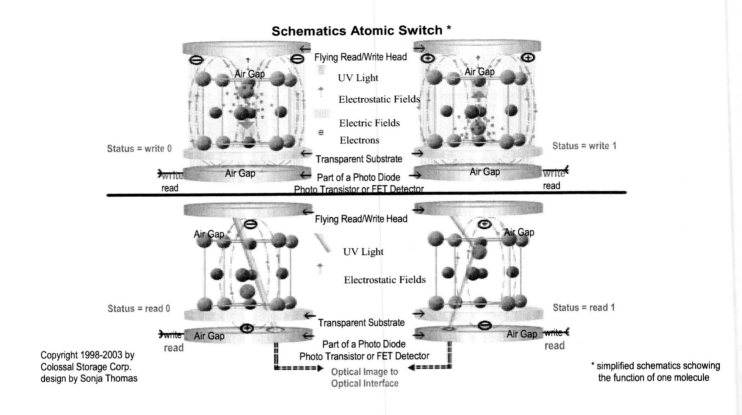

Schematics Atomic Switch *

Flying Read/Write Head
UV Light
Electrostatic Fields
Electric Fields
Electrons
Transparent Substrate
Part of a Photo Diode
Photo Transistor or FET Detector

Status = write 0
Status = write 1
Status = read 0
Status = read 1

Air Gap

write
read

Optical Image to
Optical Interface

Copyright 1998-2003 by
Colossal Storage Corp.
design by Sonja Thomas

* simplified schematics schowing
the function of one molecule

3. Conclusions

The Colossal Storage FE Optical Drive will offer symmetrical non-destructive read and writes for the retention of data storage for 100-years or more. Thomas patents on a semiconductor read/write head for ferroelectric optical storage media memories promises to raise data storage densities by a factor of 1000 or more and will add at least 10,000 times the data storage capacity per peripheral storage footprint.

Colossal Storage Corp.
U.S. Patents, # 6,028,835 2/22/00 and # 6,046,973 4/4/00

Bibliography

[1] Writing on the Fringe - Interfering electrons could lead to atomic data storage
Scientific American *October 1995, p. 40*
Michael W. Noel and Carlos R. Stroud of the University of Rochester

[2] IBM's magnetoresistive and giant magnetoresistive head technologies enable data storage products with the industry's highest areal densities. By Jim Belleson, IBM
Storage Systems Division, & Ed Grochowski, IBM Almaden Research Center.

[3] *Laboratoire de Céramique* (LC) - Matériaux , Prof. Nava Setter,
Research Activities 1996/7 in ferroelectrics.

Cooling and Power Conversion using Nanometer Gaps

A. Tavkhelidze[*] and I. Cox[*]

[*]Cool Chips plc, Gibraltar, exec@coolchips.gi

ABSTRACT

The enormous potential of solid state cooling and the direct conversion of heat into electricity has long been recognized, but despite the many benefits in terms of size, weight, design simplicity and reduced environmental impact, widespread adoption has been limited in large part because of the low efficiency and high cost of available technologies.

Our research has resulted in a new diode design that relies on electron tunneling through a nanometer-scale vacuum gap as its primary operating mechanism. Experiments to date have resulted in consistent production of devices with conformal electrode surfaces, with measured electron tunneling currents in excess of 10A through the vacuum gap. Now, after having successfully resolved the primary technical challenge of creating and maintaining the vacuum gap between the electrodes, research efforts can be focused on the remaining engineering tasks on the path to commercialization. These tasks include integrating low work function materials during electrode fabrication and finalizing package design and assembly.

Potential applications span a broad spectrum ranging from aerospace to automotive, refrigeration and HVAC systems. Once development is complete, these devices are expected to offer high-performance, low-cost replacements for just about any existing cooling or power generation solution at efficiencies far greater than thermoelectric alternatives.

Keywords: electron, tunneling, vacuum, Cool, Power

1 INTRODUCTION

A number of recent research projects have examined the potential of electron tunneling for cooling and power generation applications. Superconductor-Insulator-Normal metal (SIN) junctions have been used for cooling at low temperatures[1]. Other devices have been constructed using a Superconductor – Insulator – Normal metal – Insulator – Superconductor (SINIS) structure[2]. Additionally, calculations of cooling were made for Normal metal-Insulator-Normal-metal (NIN) tunnel junctions[3]. In all of these cases, tunneling takes place through an insulator layer between two metals. Because of the high thermal conductivity of thin insulator layers, the effectiveness of these devices is limited. One solution to the heat backflow problem using multiple tunnel junctions of NIN type in series was offered by Korotkov et. al.[4] but the complex-ity of multiple junction fabrication prevented further development.

Our research has resulted in a new structure, described as a Normal metal-Vacuum-Normal metal (NVN)[5] junction. A key advantage of this structure is the use of a vacuum as the insulator. The vacuum layer offers formally zero heat conductivity between the electrodes, allowing the fabrication of tunnel junctions with extremely low thermal backflow. These tunnel junctions represent a means of creating high-efficiency cooling and power generation, and have begun to draw greater interest. Independent of our own work, theoretical research at Stanford University examined these structures[6] in detail. Other methods of using a vacuum gap, in one case utilizing emission from semiconductor resonant states, have also been investigated[7].

Most cooling and power generation applications require tunnel junctions with a sufficiently large area (on the order of several square centimeters) to provide useful levels of performance. Fabricating NVN tunnel junctions with large areas poses several practical problems. The electrodes for such junctions should be flat within tens of Angstroms across a large area, without any areas of excessive local roughness . State of the art polishing methods allow for the fabrication of surfaces with a flatness of 0.5 micron per centimeter, which is still two orders of magnitude greater than what is required. The local roughness of polished surfaces available today is low enough (as low as 5 Angstroms) to allow tunneling through the vacuum gap, but because of slight deviations in the surfaces it is not possible to bring large areas of two electrodes (polished independently) close enough to each other to allow tunneling to take place. To solve this problem, we have developed a method of fabricating pairs of electrodes in which the topographical features are precisely matched. With a matched electrode pair, proximity can be maintained without perfect flatness.

2 ELECTRODE FABRICATION METHOD

To fabricate the pair of electrodes, we begin with a doped Si wafer as the substrate. A 0.1 micron thick Ti film is first deposited over the Si substrate (fig. 1a). Next, a 1 micron thick Ag film is deposited over the Ti layer. Deposition regimes for Ag are chosen to optimize adhesion of Ag to the Ti film (For our purposes, the optimum adhesion is much lower than typical microelectronics

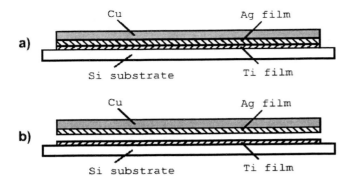

Figure 1. a) Si/Ti/Ag/Cu sandwich, b) Opened sandwich with conformal electrode surfaces.

processes). Finally, a layer of Cu 500 microns thick is grown electrochemically on the Ag film.

The "sandwich" can then be opened on the border of Ti and Ag films (fig.1b). Low adhesion between the Ti and Ag films allows the sandwich to open without significant deformation of the electrodes. After separation, we are left with two conformal electrodes that allow for tunneling over broad areas of the electrodes.

The sandwich is opened in a sealed chamber with the atmosphere evacuated to avoid oxidation of the sample. The sandwich is opened either by cooling or by heating. Because Cu and Si have different Thermal Expansion Coefficients (TEC), the two electrodes separate from the slight mechanical stress induced by the temperature change. As observed earlier, the adhesion between the Ti and Ag films must be low enough to allow the sandwich to open without creating deformation in the electrodes. It is important, however, that the adhesion be high enough to prevent electrochemical liquid from penetrating between the films during the electrochemical growth of the Cu layer. Precise adhesion control between the Ti and Ag films is therefore critical.

Tunneling current between two electrodes increases significantly at distances less than 100 Å[5]. To control the width of the gap, the experimental device uses two-stage regulation. The first stage is mechanical, using a differential screw to regulate the distance within a few microns. The second stage is a piezoelectric cylinder with resolution on the order of 1 Å. Four two-stage regulators are used, with one regulator placed in the center of the round Cu electrode and three regulators placed equilaterally on the perimeter of the Cu electrode. These regulators enable the changing of the distance and angle between the electrodes during the measurements.

Another method for distance regulation is the use of dielectric spacers between the electrodes. Al_2O_3 spacers can be deposited using reactive DC magnetron sputtering of Al. Al_2O_3 is deposited on the Ti film before the deposition of the Ag film (not shown on fig. 1). Porous Al2O3 is

used to minimize the thermal conductivity of spacers. After opening the sandwich, the spacers remain on the Ti film because of their low adhesion to Ag. The spacers prevent the electrodes from short-circuiting.

Capacitance and conductance between the electrodes are monitored during these experiments. The capacitance readout is used to determine the mean distance between the electrodes, and the conductance readout is used to determine the total area of the shorts between the electrodes. I-V characteristics of the junctions are recorded to detect tunneling currents.

3 EXPERIMENTAL RESULTS

The primary obstacle in the creation of conformal sur-

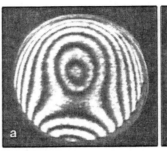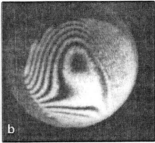

Figure 2. Interferogram showing the curvature of an early electrode pair, a) Si/Ti and b) Cu/Ag . Distance between rings: 317 nm. Electrode diameter: 28 mm.

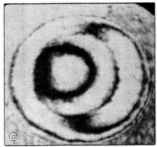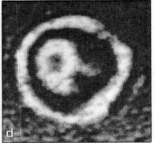

Figure 3. Interferogram showing the curvature of a more recent electrode pair, c) Si/Ti and d) Cu/Ag . Increased flatness is visible in the reduction in contour lines.

faces is the deformation of the electrodes. Interferograms like those shown in figures 2 and 3 (below) were taken at each stage to trace the sources of mechanical tension in the electrodes. The electrode pair in figure 2 shows some conformity, but there is still considerable variation in the interferograms of the Si/Ti and Cu/Ag electrodes. By selecting silicon wafers with reduced surface curvature and reducing the tension in the Cu/Ag electrodes, we were able to produce electrode pairs with significantly lower surface curvature and increased conformity between electrodes. Figure 3 shows a recent electrode pair, demon-

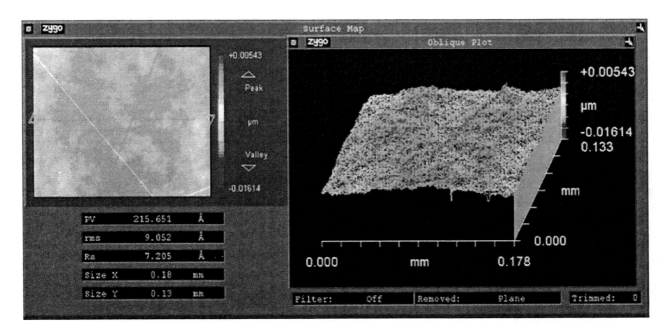

Figure 4. 3-D plot of a 0.18mm X 0.13mm area of the Cu/Ag electrode.

strating both the reduction in curvature and the close conformity of the curvature between the electrodes.

To analyze local roughness of the surfaces of the electrodes we recorded profilograms of the surfaces of the electrodes. Detailed examinations of surface roughness were conducted by a potential commercial partner. Local roughness measurement results of a Cu/Ag electrode are presented in figure 4 (below).

Figure 4 shows that the electrode surface is extremely smooth, with an average roughness of 7.2Å. In fact, because of the extremely low observed roughness of the electrode surface it was suggested that the measuring equipment should be recalibrated to verify the measurements taken. After performing tests with a Silicon Carbide reference flat, measurements were retaken and the results confirmed.

In Table 1, capacitance, conductance, and mean distance between electrodes of six samples are shown. The first two are samples which demonstrated superb C/G ratios. Samples 3-6 have typical C/G ratios. Average distances between electrodes were a few hundred Angstroms.

The I-V characteristics of closed sandwiches were linear, with a resistance of less than 0.5 mΩ. When opened, I-V characteristics became nonlinear as a result of tunneling , with changes to both shape and slope as distance was regulated. All the I-V characteristics we recorded show that the main component of current comes from electron tunneling, with remaining current coming from conduction through any shorts between the electrodes (conducted current being proportional to the applied voltage).

At high applied voltages (1-5 V, depending on the sample) the electrostatic attraction between the electrodes in some cases "closed" the sandwich with a measured attraction force in the range of 4-6 kg. After releasing the sandwich by switching off the applied voltage it remained closed, and only opened after additional distance regulation was applied. In order to prevent unwanted closure and avoid a short circuit between the electrodes, Al2O3 spacers 100 nm and 50 nm in height were tested. With 100 nm spacers a threshold voltage approaching 5 V was obtained, but ridges on the surface of the Al2O3 were observed under an optical microscope. Reducing the thickness of the spacers to 50 nm eliminated the ridges, and yielded a maximum threshold voltage of 3 V.

Device Sample	Capacitance (μF)	Conductance (mS)	Average Distance (Å)
1	0.56	42	107
2	0.49	10	122
3	0.38	445	157
4	0.30	300	200
5	0.23	327	260
6	0.11	214	545

Table 1. Capacitance, Conductance and Average Distance between the electrodes of selected samples.

4 COMPLETING DEVELOPMENT

Work currently underway in several areas must be completed before commercial prototypes can be produced. In addition to further process refinements to reduce the likelihood of electrode deformation, packaging work and integration of cesiated materials into electrode surfaces remains.

In order to get useful cooling and power generation effects at room temperature, the work function of the electrodes must be reduced to 1-1.5 eV by incorporating cesium into the electrodes. The influence of Cs on both Ti and Ag materials is well understood, with work functions of 1-1.5 eV having been achieved[8].

Package designs under consideration use piezoelectric materials both as a means to regulate the gap width and as a seal against atmospheric contamination. For applications with high power generation or cooling requirements, this package design allows for scalability using arrays of devices over a larger area.

The most challenging work, creating and maintaining the gap between electrodes, has been completed. Although refinements may occur to the remaining process steps, we are consistently able to produce electrodes with conformal surfaces and uniformly low local roughness. Once we have completed the integration of cesiated materials, we expect to begin detailed performance analysis. Evaluation devices should then become available to commercial partners once packaging designs are finished.

5 CONCLUSION

After completing development, we expect the impact of this technology to be far-reaching across many areas of thermal management and power generation. Many aerospace applications, for example, have size and weight constraints that would make these compact, high efficiency devices an ideal solution. From esoteric applications like improved radioisotope power generation or cooling sensitive infrared detectors, to more mundane applications like refrigerators and automotive power and cooling, solutions based on electron tunneling across a vacuum gap hold great potential.

Furthermore, because the materials used are both inexpensive and readily available, and because there are relatively few production steps involved, we anticipate that mass-production should be feasible and that mass-produced devices should be inexpensive compared to alternative solutions. Even without the benefits stemming from improved performance and a small, low maintenance form factor, the projected cost advantage alone makes this technology worth pursuing for the next generation of cooling and power generating solutions.

REFERENCES

1. M. Nahum, T. M. Eiles, and M. Martinis, "Electronic refrigeration based on a normal-insulator-superconductor tunnel junction," Appl. Phys. Lett. 65 (24), p. 3123-3125, (1994).

2. A. Luukanen, A.M. Savin, T.I. Suppula, et. al., Integrated SINIS refrigerators for efficient cooling of cryogenic detectors," American Institute of Physics Conference Proceedings v 605 (Low Temperature Detectors), p. 375-378, (2002)

3. A.N. Korotkov, M.R. Samuelsen, S.A. Vasenko, "Effects of overheating in a single-electron transistor," J. Appl. Phys., 76 (6), p. 3623-3631, (1994).

4. F. N. Huffman, "Thermotunnel converter," US patent 3,169,200 (1965).

5. Avto Tavkhelidze, Larisa Koptonashvili, Zauri Berishvili, Givi Skhiladze, "Method for making diode device," US patent 6,417,060 B2, (2001); other patents filed beginning from 1997 are pending.

6. Y. Hishinuma, T.H. Geballe, B.Y. Moyzhes, T.W. Kenny, "Refrigeration by combined tunneling and thermionic emission in vacuum: Use of nanometer scale design," Appl. Phys. Lett. 78, p. 2752-2754, (2001).

7. A.N. Korotkov and K.K. Likharev, "Possible cooling by resonant Fowler- Nordheim emission," Appl. Phys. Lett. 75, P. 2491-2493 (1999)

8. V.S. Fomenko, "Handbook of Thermionic Properties," (Plenum, New York, 1966).

Novel Approach to Circuit Board Testing

R. Glenn Wright*, Larry V. Kirkland**, Marek Zgol*, David Adebimpe*,
Ernest Keenan*, Robert Mulligan*

* GMA Industries, Inc., 20 Ridgely Avenue, Suite 301,
Annapolis, MD, USA, {glenn,marek,david,ern,robert}@gmai.com
** US Air Force, OO-ALC, 7278 4th Street, Bldg. 100,
Hill AFB, Utah, USA, larry.kirkland@hill.af.mil

ABSTRACT

This paper describes research and development efforts
in the application of nanoscale sensors to implement an
original method for bed-of-needles testing for printed
circuit boards testing. This approach performs functional
testing that eliminates the need to have a pre-existing model
of a circuit board while automating much of the
development process. The general concept involves
creating arrays of nanoscale sensor probes , using molecular
electronics for incorporating test instruments and
computing logic for test interpretation directly into a pod
consisting of multiple nails, and distributing these pods
across the face and back of a circuit card using a contact
fixture. The resulting test approach exhibits massive
parallelism combined with extremely compact size that
facilitates novel testing approaches not possible with
current generation or planned test equipment. In
consequence, many existing limitations are expected to be
overcome with test speed and accuracy significantly
improved.

Keywords: carbon nanotube electronics, nanoscale sensors,
circuit board testing

1 INTRODUCTION

A problem exists in circuit board testing whereby
advances in IC and printed circuit board technology have
surpassed existing capabilities with respect to test hardware
and software. The physical size of electronic components
and ICs has decreased over time, making it more and more
difficult to access component parts and suitable test points.
Additional issues that are problematic for today's test
approaches include the increased performance and
complexity of today's ICs, degradation in circuit
performance due to I/O and signal path loss, noise insertion
from long signal paths between unit under test (UUT) and
automatic test equipment (ATE), and increased difficulty in
obtaining comprehensive models of circuit components.
Also, the increased complexity of circuitry on circuit cards
results in longer development times and greater dependency
upon human intervention in test program development. So-
called Automatic Test Program Generators (ATPGs) are
becoming less and less automatic. Furthermore, in

concordance with Moore's Law, the doctrine estimating
that computing power doubles every 18 to 24 months, it is
expected that by year 2015 the dimensions of ICs within
boards will require new testing methodologies as the scale
of their discrete devices will reach sub-micron dimensions
resulting in an increased amount and density of I/O pins [1].
After evaluating alternatives (e.g. non-comprehensive
boundary scan techniques), we took a second look at the
application of new technologies and tools to existing test
methods to see whether the potential exists to improve in
their implementation. Second, we examined this approach
to determine whether these improved techniques could
supplement boundary scan and other existing and future test
approaches.

The result of this investigation is the conceptualization
of a test methodology where we bring nanoscale probes and
measurement equipment (e.g., logic state detection, DC,
AC, frequency, pulse, etc.) directly to the IC and circuit
board signal path. This approach eliminates the potential for
all path loss compensation and interference injection
problems, and dramatically reduces timing problems,
especially those typically encountered during digital testing.
In addition, this approach also reduces the problem of a
lack of test points since it establishes virtually thousands of
test points across the top and bottom of the circuit board.

2 TEST SYSTEM CONCEPT

Figure 1: Individual Sensor Architecture

Figure 1 provides an illustration of a sensor pod architecture including the probe elements chemically bound to sensor circuits in the sensor pod base, as well as in the substrate of the next higher assembly. This combination of sensor filaments, interconnections, and sensor circuitry (where applicable) physically interconnect with the UUT surfaces (e.g., IC pin, circuit path, etc.) from which measurements or other actions are to be made.

Individual sensors are clustered into pods, each having an entire suite of measurement equipment. Sensor pods, in turn, form instrument clusters that combine to create the entire nanosensor pin bed.

During testing, the circuit card is placed between two nanosensor pin beds, whereby the top and bottom surfaces of the card are bathed in sensors that make contact with IC pins and circuit paths.

This approach culminates in the development of a nanosensor pin test system, illustrated in Figure 2. This system is comprised of a test fixture consisting of two nanosensor pin bed assemblies between which the UUT resides, a Test Computer for test program execution, a PXI-bus test fixture facilitating I/O between the NanoSensor Pin Bed assemblies and the test computer, appropriate interface cabling and connectors, and other ATE assets required to exercise the UUT at the card edge connectors and boundary scan port, if available.

Figure 2: Nanoscale Sensors Applied to Both Sides of a UUT During a NanoSensor Pin Test System

Due to their extremely small size, one IC pin or circuit board signal path will extend across multiple sensor pod clusters, ensuring that multiple and redundant measurements may be taken to improve measurement reliability. The test architecture depicted here presents the nanosensor pin beds serving mainly as just an interface device. However, our vision anticipates gradual evolution from the concept of test fixture with molecular circuitry performing just supplementary functions of data collection and enhancing the capabilities of the PXI (or PC) system to the fully independent (stand-alone) arrangement comprised of molecular-test-system-embedded nanosensor pin beds.

The paragraphs that follow describe the challenges of implementing the above-presented concept. Creating the nanosensor-pin test system calls for exploring the truly interdisciplinary field of molecular electronics, which brings together elements of chemistry, physics, and electrical engineering. Therefore, to allow for a broader picture of our idea, before describing our current research progress, we provide background regarding the functionality anticipated to result from our approach.

3 NANOSENSOR-PIN TEST SYSTEM DESIGN AND DEVELOPMENT

The best illustration of our approach is the comparison of our design with well-known concept of bed-of-nails fixtures. We first describe our idea to update this traditional approach where we increase probe density by several orders of magnitude, in addition to placing the measurement instruments directly at the probe point. We follow this discussion with details regarding the methods we use to design and fabricate the nanoscale probe elements and chemically synthesize the electronic circuitry comprising the measurement equipment.

3.1 Concept Background

The first machines using universal fixturing were introduced to the PCB industry in the early '80s. Sometimes called "bed-of-nails" fixtures, the universal grid was a revolution. In 1980, less than 10% of the bare boards produced in the U.S. were electrically tested. In 1988 more than 75% were tested. Today the percentage of boards tops 95% [2]. The use of fixtures for early-dedicated machines was expensive and laborious. Hence, their application was only justified for large volume. A universal grid is what the name implies. It is populated with test points, originally placed on 0.100, or 100 points/in^2, grids. The number of test points, then, is a matter of the desired physical size and the cost. An 18 x 24 grid contains 43,200 test points. When used with a test fixture interface, the number of test points actually in use is only the number required for the board to be tested. To contrast this technology with our approach, we expect to achieve grid density in excess of 10,000+ points/in^2, with densities exceeding 35,000 points/in^2 possible.

To carry out the test it is necessary to gain access to each node on the board. The most common way of achieving this is to generate a "bed of nails" fixture. The board is held in place accurately by the fixture and pulled onto test-fixture pins that make contact with connectors on the board. The board may either be pulled down under the action of a vacuum or it may be achieved mechanically [3].

Because current generation bed-of-nails testers are relatively complex and expensive, this type of testing is most suited to situations requiring testing a large volume of circuits and circuit boards quickly. Furthermore, it will serve to distribute high cost over many tested circuits, and minimizes access requirements to prime ATE assets [2].

3.2 Design Considerations

The following represents our design considerations for the fabrication of all the elements that will constitute the nanosensor-pin test system.

Sensor Pod Cluster

Pod cluster development is divided into two general parts. The first part of the research pertains to the fabrication of the pod cluster panel itself including conductive pod bases and connections providing individual addressing of each pod, as well as carbon nanotube arrays formed on top of the pod bases. The second general part describes the formation of the nano-scale sensor circuits to be created at the back of each pod cluster panel.

The sensor pod clusters consist of a number of individually addressed pods connected to a sensor circuit situated at the back side of the sensor pod cluster panel. The bases of the pods as well as the individual connections are expected to be built using either ink-jet printing-produced conductive-polymer circuitry or metallic circuitry fabricated with use of photolithographic techniques. The production of a mash of the metallic pod bases on silicon substrate will be relatively straightforward, and we anticipate utilizing previously verified sources of photolithography manufacturing in order to accomplish this task [3].

For the production of the sensor pods, we have initially selected carbon nanotubes (CNT) over other types of materials due to their unusual electrical and mechanical properties. Carbon nanotubes exhibit both extraordinary durability and flexibility allowing for multiple usages of created pin beds for various types of circuit boards. Moreover, CNTs are the most studied of all the nanostructures and the most promising for obtaining near term practical results in electronics and electronic devices.

Sensor Circuits

Sensor circuits are to be produced on the backside of the earlier described pod clusters. Due to the relatively small area on which we will have to create fairly complex electronic devices, we predict that in addition to traditional electronic devices, other technologies will have to be applied including nano-scaled molecular electronics.

We are currently exploring methods of creation of simple molecular electronic logic devices, which can then be combined into more sophisticated systems. The molecular systems will include polyphenylene-based devices synthesized using organic chemistry methodologies developed by James Tour [4]. The basic materials for construction of those molecular electronics are Tour wires, which are polyphenylene-based molecules with different substituents incorporated into their structure. Using these structures as simple building blocks, we can fabricate discrete electronic devices including rectifier diodes, tunneling diodes, or transistors. These devices, in turn, can be further combined to create logic gates.

System Software

The primary requirement is the creation of software that will control the nanosensor pin test system hardware, as well as process sensor measurements to create and execute a resulting test program in a LabView environment. The system software will have two primary components that are run sequentially. The first component is the pattern analysis software, whose job is to determine which nanosensor pins of the test bed are relevant to the test of a particular UUT and removes the rest from the test configuration. It also identifies areas of continuity that are representative of IC pins and circuit paths. The second component is the test analysis software itself, comprised of the software used to create a test program through semi-automated means, and the resulting executable test program software hosted on the ATE.

The pattern analysis software examines continuity measurements reported by sensor pods on both NanoSensor Pin Bed assemblies to detect patterns of connectivity illustrative of IC pins and circuit paths. Due to their extremely small size, numerous sensor pod clusters will span the distance of individual IC pins and circuit paths. Areas of continuity on a circuit board represent metallic connections through which electrical signals flow with minimal resistance, i.e., IC pins and circuit paths. Areas where electrical signals do not flow include IC packages and other electronic components, as well as blank areas on a circuit board. Establishment of such areas of continuity on the UUT will effectively result in the creation of a netlist.

This software module is anticipated to use a continuity measurement probe identified in the system requirements specification, consisting of probe elements and circuitry that apply extremely low power stimulus and/or measure very low power signals to determine if a shorted condition may exist between two or more sensor pods through a UUT surface (IC pin, circuit path, etc.). A positive reading would mean two or more sensor pod interfaces are located on the same IC pin or circuit board conductor.

Patterns of continuity will be identified through examination of the measurements obtained from these probes. Positive measurements will be tagged and configured to an active state. Sensor pod clusters that do not depict areas of continuity will be configured to an inactive state. Thus, only the remaining active sensor pod clusters are of interest for the performance of UUT testing and will be queried for this purpose. This step significantly reduces the search space and computational requirements for the test analysis software.

Further processing will then take place to identify specific paths and IC pin connections. This may be accomplished through an automated process in the event a netlist is available from an independent source. In this case, we would be able to compile a list of total connections on a UUT as well as a list of exposed connections. In the event this is not possible, the user may interactively designate connectivity elements presented through a graphical user interface.

Upon completion of sensor pod cluster configuration and netlist generation, the pattern analysis software will compile a record of the test environment for the UUT within which the test analysis software will create a test program set.

For test program set generation, an initial stimulus input set consisting of a uniform distribution of test patterns will be applied. The measurements obtained from the test pods for each input stimulus pattern will be collected. Test patterns will be iteratively generated based on a perturbation function determined from the correlation between the set of input patterns and the resulting measurements. Inconsistent results will appear as singularities in the output measurements, such as a curve of current vs. voltage. The process is similar to a feedback based Monte-Carlo simulation approach, in which an estimate of the joint distribution of a large number of variables can be determined by obtaining samples at intervals on the distribution curve. For traditional Monte-Carlo simulation, values are randomly generated for the variables of interest in order to simulate a model. However for our approach, no simulation is needed, since the actual measurements will be taken for each input stimulus.

4   CONCLUSIONS

Much of the technology identified within this paper is graduating from the theoretical to the practical domain. Improvements in tools such as the atomic force microscope and scanning electron microscope, as well as recent advances in the development of nanoscale manipulation tools, have afforded us unprecedented visibility and capability in directly effecting research at the atomic level – even to the extent of the physical manipulation of individual atoms. As IC technology continues to shrink into the nanometer realm, our nanosensor-pin testing paradigm presents a modern cutting-edge test method, and may one day be the platform through which future generations of test methodologies that address the sub-micrometer dimensions are based.

REFERENCES

[1] G.E. Moore, "Cramming more components onto integrated circuits," Electronics, volume 38, number 8, 19 April 1965.
[2] Advanced Electronic Packaging: With Emphasis On Multi-Chip Modules, Editor: W. D. Brown, Chapter 12, Testing and Qualification, S. Kolluru and D. Berleant, Wiley-IEEE Press, 2001
[3] Lee, et al., Chem. Phys. Lett., 337, 398, 2001
[4] S. Huang, J.M. Tour, J. Am. Chem. Soc., 121, 4908, 1999

Expecting the Public Backlash: Public Relations Lessons for Nanotechnology from the Biotechnology Experience

J. Matsuura*

*University of Dayton School of Law, 300 College Park, Dayton, Ohio, USA,
Jeffrey.Matsuura@notes.udayton.edu

ABSTRACT

This paper discusses the relationship between public reaction to new technology and the legal, regulatory, and public policy environment that develops to provide oversight for the technology. The paper contends that there are useful lessons to be derived for nanotechnology from the relationship between public response and legal oversight that developed with the introduction of biotechnology and other innovative technologies. It identifies the most useful lessons for the nanotechnology industry available in the prior experience. Based on those applicable lessons, the paper describes a regulatory strategy for nanotechnology that is conducive to continued research and effective commercialization in the field.

Keywords: regulation, publicity, promotion

1 PUBLIC RELATIONS AND REGULATION

Public reaction to a field of scientific research or to an advance in technology can have a profound impact on the legal, regulatory, and public policy environment applied to the research or technology. The reaction of the public to a field of research or to a new technology can ultimately shape the laws, regulations, and public policies that govern the research or technology. This interaction between public perceptions and legal oversight seems to exist for any type of new technology. Accordingly, those who would shape the legal context in which new technologies develop and evolve should be acutely aware of the public perception of those technologies, including both public hopes and fears as to the technologies, and they should act to influence the public perception in a constructive manner.

The public influence on the legal climate associated with new technology is exerted in part through political pressure. Political pressure translates into legislation and regulation. Public opinion plays a significant role in the establishment of the policy agenda of government. That agenda is largely implemented through legislation and regulation. Through this process, public perceptions regarding the potential and the risk associated with new technology shape the legal and regulatory framework that oversee that technology.

Public influence is also exerted through private legal actions. Individuals and groups of individuals can initiate private lawsuits enforcing civil law claims. The results of those private legal actions influence future conduct by creating common law principles that interpret legal obligations. In this way, private litigation helps to shape the legal context in which technology develops and is applied. Private enforcement of legal rights provides another key aspect of public influence on technology regulation.

The legal and regulatory climate that evolves for a new technology has both a direct and an indirect effect on development of applications for the technology. The direct effect consists of legal requirements enforced in conjunction with creation, distribution, and use of the technology. Those requirements establish the limits of permissible conduct and they force the parties involved with the technology to bear the costs of compliance. The indirect effect involves potential financial impact on the technology as a result of the level of risk and uncertainty associated with the technology. When there is a clear legal framework associated with a technology, developers, users, and investors involved with that technology are better able to identify and quantify risk. In that environment of greater certainty as to risk assessments, investors and lenders are more willing to offer financial support for the technology, and insurers are more willing to offer coverage for potential liability.

2 THE BIOTECH EXPERIENCE

The public response to the development of advances in biotechnology and genetic research, and the associated legal and regulatory climate that developed, provide a useful model for the field of nanotechnology. By examining the public reaction to the development of biotechnology and the legal framework that evolved for that technology, we can derive a few principles that are likely to be applicable for nanotechnology, and for other new technologies that carry both significant potential promise for public benefit and some risk of potential public harm.

One of the assumptions regarding public acceptance made in the biotechnology context, was that the public would quickly recognize and appreciate the potential value to quality of life presented by biotechnology. Although it is fair to say that the public did recognize many of the potential benefits associated with biotechnology, potential risks associated with the technology captured the attention of segments of the media and important portions of the general public. This attention served to underscore possible adverse consequences, and in the context of limited public understanding of the details of biotechnology, the concern grew into active opposition to various applications of the technology.

Another important assumption was that rational arguments could effectively influence the response to biotechnology. Experience demonstrated, however, that even though much of the public acknowledged the significant benefits of biotechnology, fear as to potential abuses and accidents associated with the technology seemed captivated key segments of the population. The fear and concern surfaced into political and legal action in specific instances, genetically modified foods, for example. Proponents of biotechnology continued to raise rational arguments in defense of the technology, focusing on its vast benefits and explaining that the risks were not as substantial as opponents of the technology suggested. Although the rational arguments expressed in response to the opposition were helpful, they were not able to diffuse entirely the more vocal opposition.

3   LESSONS FOR NANOTECH

Public perception of nanotechnology appears to be developing along a track similar to that followed for biotechnology. Although there is general recognition of the potential benefits from many aspects of nanotechnology, there is also significant public uncertainty regarding the technology and some fear that arises largely from the uncertainty. That combination of support mixed with a lurking fear of possible adverse consequences is very similar to the climate that developed around biotechnology. Proponents of nanotechnology can take several useful lessons from the experience provided by public reaction, and associated legal and regulatory oversight, regarding biotechnology and other technological advances.

One important lesson is to avoid underestimating the ability of public reaction to influence the legal and policy environment. Public reaction to nanotechnology, regardless of the basis for that reaction, can translate into political influence, which in turn affects the form and substance of legislation and regulatory oversight. Both informed and uninformed opposition can affect public perception and have a corresponding impact on the legal environment applied to nanotechnology and its applications.

Another important lesson is to recognize the importance of engaging all concerns regardless of whether they are considered to be based in fact or merely unrealistic speculation. In the context of biotechnology and other forms of new technologies, there has sometimes been a view that concerns or opposition not based in fact can be ignored. This is a mistake. All concerns and opposition should be treated seriously, as all of those concerns can affect the regulatory climate applied to technology.

Another critical lesson is to respond to expressed concerns quickly and with a consistent set of messages. Every effort should be made to engage criticism or concern quickly. Responses should be direct and they should be dynamically presented, in a clear and lively manner. In addition, proponents of nanotechnology should not always wait for the opposition to take the initiative. When possible, supporters of nanotechnology should present the affirmative case for nanotechnology to the public using the most effective available marketing and promotional practices and strategies.

When presenting the affirmative case for nanotechnology, however, advocates of nanotechnology should try hard to avoid overstating the value of the technology. There is often incentive to exaggerate the potential benefits of a new technology, to attract funding, for example. Overstated benefits frequently return to haunt the supporters who oversold the claims, as public disappointment arising from inability to deliver the promised results can easily breed mistrust of the technology and its promoters. When the benefits of the technology are exaggerated, the door is opened for critics to question whether any of the positive assumptions associated with the technology are accurate. It is important for nanotechnology proponents to tell a consistent and realistic story regarding the technology to government, investors, and the general public.

Finally, nanotechnology supporters should recognize that a significant portion of the opposition to nanotechnology is not really inspired by the technology itself, but is instead driven by a combination of fear, frustration, and anger directed toward major institutions in our society. Part of the opposition expressed against nanotechnology, or any other new technology for that matter, is inspired by discontent with respect to government and businesses. To the extent that citizens are frustrated by their government or feel that large businesses have moved beyond the scope of effective control that discontent may surface as opposition to new technologies supported by those institutions, such as nanotechnology. No amount of effort by the nanotechnology industry will address fundamental societal discontent that expresses itself as opposition to nanotechnology.

4   CONCLUSION

This is a critical time for efforts to influence public perception of nanotechnology. Different applications of nanotechnology are emerging into the commercial marketplace, and are becoming more visible to the public. Mass market entertainment, including best selling novels, now occasionally highlight nanotechnology, and often in a highly unflattering light. Controversies over application of biotechnology and other novel technologies are fresh in the public consciousness. Public views on nanotechnology are now being developed.

Past experience with public reaction to other new technologies suggests that a major threat to new technologies is a stampede to regulation that can develop long before the technologies are mature enough to permit effective risk assessment. An effective strategy to reduce the risk of a rush to regulate is the coordinated use of existing legal and regulatory mechanisms. This approach is based on the argument that there is no need for development of an entirely new legal regime for nanotechnology, as its many different applications are already governed by several different regulatory systems. For example, nano-particulates are already governed by existing health and environmental regulations. Nanotechnology applied in the context of medical products and procedures are already regulated by the public health and medical regulatory processes. An effort to prevent premature comprehensive regulation of nanotechnology should begin with the argument that the various applications of different forms of nanotechnology are already effectively controlled through a diverse set of existing regulatory regimes.

The legal and regulatory framework applicable to nanotechnology will be significantly affected by public opinion. The need for proponents of nanotechnology to be mindful of the connection between public perception and legal oversight is underscored by experience with other novel technologies, particularly biotechnology. Supporters of nanotechnology should apply sophisticated promotional and marketing techniques to present to the public the affirmative case in support of nanotechnology. They should also respond quickly and dynamically to opposition to nanotechnology, no matter how rational or irrational the basis for the opposition.

Finally, advocates of nanotechnology should work to persuade the public that the range of nanotechnology applications can be effectively overseen by the current regulatory framework. Each application of nanotechnology will be subject to the regulatory oversight associated with that application. Given that established regulatory regime, the public interest would not be served by implementation of laws or policies directed specifically at nanotechnology. If we learn from past experience regarding public acceptance of new technology, perhaps we can preserved a legal and public policy environment in which innovative and useful applications of nanotechnology can be created and can flourish, for the benefit of society.

REFERENCES

[1] Arnall, A., *Today's Choice: A report for the Greenpeace Environmental Trust*, Greenpeace, London, 2003.

[2] Levidow, L., Carr, S., and Wield, D., "European Biotechnology Regulation: Contested boundaries of environmental risk," *BioSafety*, Vol. 3, Paper 1, 1997, at http://bioline.utsc.utoronto.ca/archive/00000006/01/by977001.pdf.

[3] Malinowski, M., *Biotechnology: Law, Business, and Regulation*, Aspen, New York, 1999.

[4] Philipkoski, K., "Food Biotech Is Risky Business," *Wired News*, Dec. 15, 2003, at http://www.wired.com/news/medtech/0,1286,61096,00.html.

[5] Toi Te Taiao: Bioethics Council, *Nanotechnology Report of the Toi Te Taiao: Bioethics Council to the Minister of the Environment*, New Zealand, 2003, at http://www.bioethics.org.nz/publications/nanotechnology-report-minister.html.

[6] Woods, S., Jones, R., and Geldart, A., *The Social and Economic Challenges of Nanotechnology*, Economic and Social Research Council, 2003.

Nanofabrication via Dip Pen Nanolithography™

R. Eby and J. Leckenby

NanoInk, Inc., 1335 West Randolph Street, Chicago, IL 60607
reby@nanoink.net

Keywords: nanofabrication, nanolithography, dip pen nanolithography, nanostructures, carbon nanotubes, arrays

ABSTRACT:

Dip Pen Nanolithography, DPN™, was discovered in the laboratory of Dr. Chad Mirkin at Northwestern University leading to the founding of NanoInk, Inc. The company mission is to be the worldwide leader in nanoscale manufacturing and applications development through the development of the process of DPN. This process enables researchers demonstrate the viability of building structures at the nanoscale – from the bottom-up. The DPN process is be described along with the parameters essential for reproducibility. The DPN process uses a scanning probe tip as a "pen," which is coated with molecules (the "ink") that are deposited onto a substrate (the "paper") through a controlled diffusion process. Process control is established with the first dedicated DPNWriter™ system – NSCRIPTOR™. Applications using DPN cover a multitude of different materials including organics, inorganics, biomolecules to conducting polymers and carbon nanotubes, thus finding use in the research communities from life sciences to nanoelectronics and fine chemicals. Structures have been "built" in the size range of 15nm to 100nm with as small as 5nm spacing between. Nanostructures are built by a simple three stage process: design, deposition and inspection. To make this a truly viable manufacturing process, the single pen must be replaced by arrays of pens, providing for scalability towards speed and cost reductions.

1.1 Introduction

The process of Dip Pen Nanolithography was first reported in January 1999 in *Science* by Dr Chad Mirkin of Northwestern University. This discovery has lead to the founding of NanoInk in November 2001 and the development of the technology to that provides a means of manufacturing and fabrication on the nanometer scale. In the past two years, the company has developed a range of products and capabilities to assist researchers in academia and industry develop new and exciting methods for working with materials on the nanoscale. A four-step protocol has been implemented whereby the researcher may design a structure, deposit (fabricate) it, inspect (Image with AFM) and investigate its properties within a common platform – i.e., the NSCRIPTOR system. In support of this product development effort, a large investment has been made to build a broad intellectual property portfolio to protect the fundamental technology and potential applications for the use of DPN.

DPN may be easily compared to that of using a quill pen and ink to write onto paper. Imagine the quill pen. It has a tip. To write, it is dipped in ink and may then write directly onto paper. In the case of DPN, the pen is the silicon nitride probe stylus of an atomic force microscope, a pyramidal structure with a fine tip of just 20nm. The DPN pen may be dipped into a variety of different inks where the ink may be one of many different materials: metals, small molecules, polymers, DNA and proteins to name but a few. Once inked, the tip may be brought into contact with a substrate causing the ink to bind to the surface and form a nanostructure. The substrate may also be selected from a wide range of materials. These include silicon, glass and metals, the key criteria being that they are very flat.

The writing process is governed by the rate of diffusion of the ink from the pen to the substrate. Thus, the time in contact between pen and surface, together with such environmental parameters such as temperature and humidity, will affect the size of structure being fabricated. Using this nanoscale pen, it is possible control structures in the 15nm size range, which is approximately 6000 times smaller than the width of a human hair (80 micron diameter).

1.2 Attributes of DPN

DPN offers many key advantages over existing technologies such as photolithography, e-beam lithography and microcontact printing.

Flexibility: DPN is a direct write technique requiring no photomask. It enables the user to write with both hard and soft materials in the same space, none of which are possible with the techniques noted here.

High resolution and accuracy: By using the established pick and shovel technique of the nanotechnology world, scanning probe microscopy and its main technique, atomic force microscopy (AFM), DPN has an incredibly accurate method of controlling the placement of the pen to nanoscale accuracy and to then write structure of nanometer dimensions.

Scalability: To make such a technique viable as a production tool, the methods must be fully scalable. While the initial DPNWriter was a single-pen instrument, the NSCRIPTOR products of 2004 offer multiple pen capabilities to deliver multiple inks using a variety of MEMS fabricated devices, including passive and active arrays of pen systems and associated inkwell designs to meet the increasing demands of the nano-research community.

Low cost: Unlike many "top-down" techniques, DPN uses established low-cost, technologies (SPM, MEMS, etc.) to provide customizable solutions on the nanoscale. A fundamental "bottom-up" tool, NSCRIPTOR offers a cost-effective solution to the challenge of nano-fabrication.

1.3 Applications of DPN

Through the successes of research labs publishing their uses of DPN in the last four years, DPN has seen a tremendous leap in popularity illustrated by the simple search tool on the Internet: the search engine. Employing Google just twelve months ago, you would find perhaps 200 hits under the term of dip pen Nanolithography. A search today would reveal over 2,200 hits.

As the technology has become validated through peer-reviewed publications, more interest for applying this to industrial challenges is being observed. This is because DPN allows companies to produce products that would have been impossible or cost-prohibitive to produce in the past. DPN has been applied to applications in materials discovery, semiconductors and the life sciences. DPN is becoming the tool that allows convergence between these extreme disciplines as companies search for new materials either to build nano circuits or to aid drug discovery through the investigation of combinatorial libraries of materials:

DPN may create biosensors for point-of-care diagnostics that are so sensitive and selective, they are able to lead to earlier and more accurate diagnoses of disease.

DPN has been applied in the direct-write repair of defects in flat panel displays and for photomasks in an entirely new and economical way through the development of specialized ink formulations.

DPN allows the manipulation and placement of nanotubes to create nanoscale circuitry that will offer superior performance over today's technology.

1.4 NanoInk Performance

Since formation in 2001, NanoInk has demonstrated an excellent track record. Funded through VC investment (Galway Partners and the Lurie Group), the company quickly brought products to the market place and has been generating revenues in three areas:

Software: this basic-capability software tool, to enable researchers to learn of DPN, was released in June 2002 and is distributed worldwide by Veeco for their CP-Research platform of AFM tools

Hardware: the first of the NSCRIPTOR family of DPNWriters was released in June 2003 and now has a growing user base now using multi-pen arrays and inkwell systems to demonstrate the scalability of DPN. These products are distributed through both direct and indirect sales channels worldwide.

Consumables: to support the above, NanoInk's own MEMS facility in Campbell, CA, is manufacturing inkwell systems and arrays of pens to meet the growing applications needs of the user base. This is complemented by an in-house team of chemists developing new ink formulations to match these needs.

Furthermore, the product line has been extended to provide a dedicated nanoscale testing platform called NETS™, the nanoscale experimenters test system. This enables utilization of current macroscale test and measurement systems for the measurement of properties on the micro- and nano- scales.

In step with development, a growing intellectual property platform is being established, with the first patents being granted in the US and Taiwan during 2003, and with another 70+ filings in eight jurisdictions worldwide. The company continues to aggressively develop and secure IP worldwide through both in-house and in-licensing agreements.

1.5 Concluding Remarks

NanoInk is founded upon an experienced management team with proven ability to create and commercialize breakthrough products. The technique of DPN is nothing if not such a product. As a truly disruptive technology, NanoInk's technology is emerging as a fundamental platform process for achieving true nanoscale fabrication.

Nanocrystalline Mixed Metal Oxides – Novel Oxygen Storage Materials

Harry Sarkas, Patrick G. Murray, Aaron Fay, R. W. Brotzman, Jr.

Nanophase Technologies Corporation, Romeoville, IL, USA, pmurray@nanophase.com

ABSTRACT

Nanophase Technologies Corporation (NTC) employs a new physical vapor synthesis technique to manufacture ceria-based mixed rare earth oxide nanomaterials. The mixed rare earth oxide nanoparticles are dense, discrete crystals. The compositions are solid solutions that remain thermally stable to above 1050°C. Ceria is an active oxygen storage material. Doping ceria with other rare earth metals enhances the thermal stability of nanocrystalline particle size and increases the oxygen storage capacity (OSC). The mean particle size, and static and dynamic OSC of ceria-based mixed rare earth oxides will be presented as a function of particle composition, morphology, and thermal history. These materials are being engineered for several applications including oxygen storage catalysts and polishing materials.

Keywords: ceria, doped-ceria, nanoparticle, oxygen storage

1 INTRODUCTION

A large relative portion of atoms/molecules reside at a nanoparticle surface. The surface atoms/molecules are unsaturated in bonding and highly reactive. Thus the nanoparticle surface has large interactions with other particles and the application environment.

Nanoparticles produced by physical vapor synthesis in particular have unique properties including high chemical reactivity per mass, wide composition palette, high zeta potential, and processing advantages that result from their high surface area and discrete particulate morphology.

Application environments, such as elevated temperatures or aqueous dispersions, are often averse to maintaining the properties associated with the discrete nature of nanoparticles. Therefore, nanoparticle compositions and morphologies were developed to enable nanoparticle applications in targeted, adverse application environments.

1.1 Nanoparticle Production

NTC produces nanocrystalline metal oxides by a new, patent-pending, physical vapor synthesis technique – NanoArc™ Synthesis. The process vaporizes precursors in a plasma, followed by rapid quenching, condensation, and formation of metal oxide nanoparticles. The size of the nanoparticles is controlled by the condensation rate and the particle concentration in the quench zone. The discrete nanoparticles form loose aggregates collected as a dry powder. Nanoparticles produced by the NanoArc™ process are crystalline, equiaxed, non-porous, discrete particles with mean diameters in the 7 – 50 nm range. These nanoparticles have a high zeta potential and can be dispersed to provide stable dispersions. The NanoArc™ process is commercially scaled and produces single- and mixed-metal oxides.

1.2 Surface Treatment and Dispersion

NTC developed proprietary surface treatment and dispersion processes for nanoparticles designed to provide one or more of the following properties:

- dispersion into fluids (aqueous and hydrocarbons),
- prevention of particle agglomeration,
- compatibility with formulation ingredients, and
- enhance nanoparticle surface chemistry.

As produced, the metal oxide powders disperse, to a "natural" pH, in aqueous systems. However applications generally require nanoparticle dispersions to be formulated with other ingredients and/or used at pH values other than the "natural" pH. Thus, to derive economic benefit nanoparticles require surface treatment, especially if the application pH requires adjustment through the isoelectric point of the nanoparticle dispersion.

2 EXPERIMENTAL

Nanoparticle surface area was determined by the BET method for determining specific surface area by nitrogen adsorption. Nanoparticle size is calculated from the surface area and particle density, assuming spherical particle morphology.

OSC is the ability of a material to absorb oxygen in an oxidative atmosphere and desorb oxygen in a substantially inert atmosphere. OSC was quantified on a thermogravimetric analyzer which measures the weight of the oxygen storage material as a function of temperature after the oxygen storage material is subjected to sequential oxidation-reduction cycles. Each oxidation-reduction cycle involves (a) heating the nanoparticles to 600°C under oxygen, (b) reducing the nanoparticles with a hydrogen-nitrogen gas (2%/98%, mole basis) at 600°C, and (c) oxidizing the nanoparticles with oxygen at 600°C. OSC is expressed as moles of oxygen per gram of catalyst.

3 DISCUSSION

Ceria and doped-ceria nanoparticles produced by the NanoArc™ process are listed in Tables 1 and 2. The nanoparticles were characterized by zeta potential, and BET surface area and OSC as a function of thermal treatment. XRD measurement of the doped-ceria based nanoparticles indicates crystalline solid solutions of the indicated compositions were formed – these materials remain stable at the thermal conditions investigated.

The OSC of thermally treated ceria and doped-ceria nanoparticles are given in Table 1. The particle size of all the nanoparticles drops (decrease in BET) when subjected to 1050°C thermal treatment for 12 hours. The static OSC also decreases as a result of nanoparticle sintering resulting from thermal treatment.

The temperature dependence of OSC for thermally treated ceria and doped-ceria nanoparticles is given in Table 2. The temperature dependence of OSC is dependent on nanoparticle composition.

Stable aqueous dispersions of ceria and doped-ceria nanoparticle were prepared at up to 50 wt% solid concentrations and pH values from 4 to 8. The high zeta potentials enable dispersions to remain stable for at least 18 months.

4 CONCLUSIONS

Ceria nanoparticles exhibit decreases in OSC when subjected to increasing calcination temperatures. The degradation in static OSC is attributed to a decrease in active surface area of the nanoparticle at temperatures above the sintering temperature.

Doped ceria nanoparticles have greater OSC and are more thermally stable compared with ceria nanoparticles.

Ceria nanoparticles surface-treated with zirconia have greater and more thermally stable OSCs compared with ceria nanoparticles even though the surface area of zirconia-surface treated ceria decreases above the sintering temperature.

The OSC temperature dependence is a function of the nanoparticle composition. Thus, nanoparticle compositions can be tailored to provide high OSC, or high chemical reactivity, at specific application temperatures to provide high performance nanomaterials to meet demanding application needs.

Nanoparticle	ζ-pot	Static Oxygen Storage Capacity at 600°C μmoles O_2/g (BET)	
		600°C	1050°C
Ceria	43 mV	85 (90 m^2/g)	13 (5 m^2/g)
Doped Ceria			
$Ce_xZr_yO_2$ (12% Zr)	30 mV	300 (85 m^2/g)	240 (21 m^2/g)
$Ce_xZr_yLa_zO_2$ (72:20:8)	38 mV	238 (69 m^2/g)	170 (27 m^2/g)
$Ce_xZr_yPr_zO_2$ (73:20:7)	35 mV	229 (78 m^2/g)	174 (24 m^2/g)
$Ce_ySm_zO_2$ (84:16)	37 mV	163 (100 m^2/g)	32 (10 m^2/g)
Zr-Doped Ceria			
$Ce_xZr_yO_2$ (12% Zr)	32 mV	183 (121 m^2/g)	145 (18 m^2/g)
$Ce_xZr_yO_2$ (35% Zr)	28 mV	300 (85 m^2/g)	240 (21 m^2/g)
$Ce_xZr_yO_2$ (55% Zr)	29 mV	295 (74 m^2/g)	225 (21 m^2/g)
Ceria Surface-Treated with ZrO_2	37 mV	167 (75 m^2/g)	143 (5 m^2/g)

Table 1. Static OSC at 600°C for thermally treated ceria and doped-ceria nanoparticles

Nanoparticle	Calcination	ζ-pot	BET, m^2/g	Static Oxygen Storage Capacity μmoles O_2/g	
				500°C	600°C
Ceria (9-nm)	None	43 mV	90	27	85
	1050°C		5		13
Doped Ceria					
$Ce_xZr_yO_2$ (35% Zr)	None	28 mV	85	13	300
	1050°C	31 mV	21		240
$Ce_xZr_yLa_zO_2$ (72:20:8)	None	38 mV	69		238
	800°C		56	87	
	900°C		49	80	166
	1050°C		24		170
$Ce_xZr_yPr_zO_2$ (73:20:7)	None	35 mV	78		229
	800°C		64	93	191
	900°C		52	87	164
	1050°C		24		174

Table 2. Temperature dependence of OSC for thermally treated ceria and doped-ceria nanoparticles

Realizing Complex Microsystems: A Deterministic Parallel Assembly Approach

J. Randall, G. Hughes, A. Geisberger, K. Tsui, R. Saini, M. Ellis, G. Skidmore

Zyvex Corporation, 1321 North Plano Road, Richardson, TX 75081, USA, ghughes@zyvex.com

ABSTRACT

The push towards miniaturization has created substantial interest in microelectromechanical systems (MEMS). To date, most of the developments regarding MEMS technology have relied on monolithic fabrication and integration. The monolithic approach has successfully produced various miniature technologies; however, most of these miniature technologies are essentially discrete devices or, at best, simple systems such as pressure sensors and accelerometers. Truly complex systems, by definition, are a combination of independent but interrelated elements that, in totality, function as a unified entity. Computer-controlled, parallel assembly of micromachined components promises to drive the miniaturization wave by enabling the manufacture of unprecedented complex microsystems. Described in this paper is a deterministic parallel assembly approach that uses silicon MEMS components, such as end-effectors, connectors, and sockets, integrated with high precision robotic systems.

Keywords: MEMS, microassembly, parallel assembly, microsystems

1 INTRODUCTION

Many of the current, commercially-available MEMS devices such as pressure sensors and accelerometers rely on monolithic fabrication methods. As the drive towards miniaturization continues, more complex devices will be developed which will push the limits of monolithic fabrication methods and may require the heterogeneous assembly of materials. Assembly of micro components will complement monolithic approaches and enable the fabrication of complex microsystems comprised of dissimilar materials and structures fabricated from both MEMS and non-MEMS processes.

Microassembly can be subdivided into two broad categories – stochastic assembly and deterministic assembly [1]. Stochastic assembly methods accomplish the parallel placement of components through self-assembly mediated via global, external forces such as fluid flow [2] or vibratory agitation [3]. Each part randomly moves across the substrate which contains sites, such as etched wells, where the parts are desired to be placed. Both the parts and the placement sites must be designed such that the desired placement and orientation of each part is achieved. In contrast to the random nature of the stochastic approach, deterministic assembly incorporates the direct placement of every part in serial or parallel fashion [4], [5]. Each component is picked up and placed into its desired location so that the placement of each component is known.

Zyvex is developing a deterministic parallel assembly approach that uses silicon MEMS components, such as end-effectors, connectors, and sockets, integrated with high precision robotic systems. The end-effectors are deep reactive ion etched structures that are manipulated via robotic systems and are used to pick the MEMS components from the substrate and place them in their respective sockets (which are also micromachined within the substrate). The sockets are designed for mechanical connectivity and can be metalized to provide electrical connection to the component.

2 MICROASSEMBLY TOOLS

The tools required for microassembly are end-effectors, connectors, and sockets. All of these tools can be precisely manufactured using MEMS processes. End-effectors including grippers and jammers are used to pick up the various components. Incorporated with each of the pickable components are connectors designed to couple with a specific end-effector. Sockets within the substrate are designed to accommodate the connectors associated with each of the pickable components. All of these tools and their associated components are designed with standard MEMS and/or integrated circuit (IC) layout software. The computer-aided design (CAD) layout not only contains the tool and component designs, but also provides the necessary data to determine the precise placement of all of the components. This data can be used to automate the robotic assembly system.

2.1 End-effectors

An example of an end-effector used to pick and place components is the microgripper shown in Figure 1.

300 µm

Figure 1: 50-µm thick, electrothermal microgripper.

The microgripper is fabricated using deep reactive ion etching of silicon and uses electrothermal actuation. The grippers can be fabricated with a range of opening sizes and can be designed to either open or close when power is supplied. All of the end-effectors are designed to couple with connectors, described below, that are built into the MEMS components. Grippers can also be designed for non-MEMS components such as wires and coils.

2.2 Connectors

To facilitate assembly, the MEMS components are designed with built-in compliant connectors that are geometrically and mechanically symmetric which results in self-centering of the assembled parts. Examples of connector structures are shown in Figure 2. The thin, angled beam attached to the top of each part is a breakable tether used to hold the components in place until assembly.

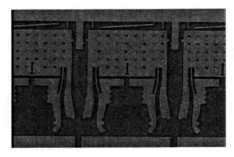

Figure 2: 50-μm thick connector microstructures.

Connectors are designed to accommodate the desired end-effector (gripper, jammer, etc.) which will be used to pick up the attached component and can be metalized for improved electrical connectivity. MEMS components including micromirrors, electrostatic plate deflectors, bent-beam actuators, and thermal bimorphs have all been fabricated with attached connectors.

2.3 Sockets

Once the component attached to a connector is picked up with the appropriate end-effector, the component is then placed into its respective socket located within the substrate. An example socket is shown in Figure 3.

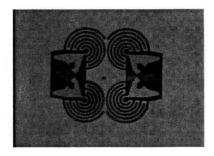

Figure 3: Micromachined socket.

Incorporated with the socket shown above are compliant spring structures which further enhance the self-aligning capability of our microassembly approach. Upon release of the MEMS component within the socket, the connector and socket equilibrate and stabilize to their designed positions due to their compliance. The self-alignment capability allows for precision part placement independent of the initial robotic placement of the part. Metalization of the sockets can be performed to further facilitate electrical connectivity. A final placed part is shown in Figure 4.

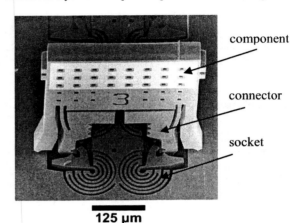

125 μm

Figure 4: A MEMS component placed within a socket.

2.4 Robotics

Our current robotic system consists of a five degree of freedom configuration and is shown in Figure 5.

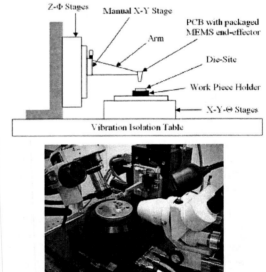

Figure 5: Microassembly robotic system.

Three stages provide for XY and theta motion of the work piece substrate. A Z and phi stage provides motion towards and away from the substrate and rotation about an axis parallel to the XY plane of motion. End-effectors are attached to the end of the arm which can rotate about the phi axis parallel to the XY plane. All five stages are driven by a closed loop control system providing a one micron precision over the range of travel for the linear stages. The system is computer controlled via a GPIB interface and motion sequences can be programmed with a scripting language allowing for automated assembly of micro devices. For instance, the script can be derived from the CAD layout which contains the designs for the MEMS components, connectors, and sockets.

3 PROCESS FLOW AND MICROASSEMBLY

The MEMS components are fabricated on silicon-on-insulator (SOI) wafers. The components are built within the top, thinner layer of silicon (the device layer) and are tethered to fixed structures built within the same layer. The buried oxide is used as a sacrificial layer which is removed during the final process steps. An example of a SOI MEMS fabrication process is shown in Figure 6.

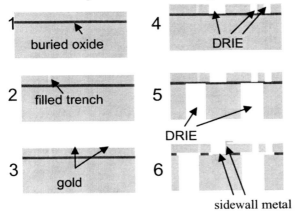

Figure 6: Example SOI MEMS fabrication process.

After the buried oxide is removed and the wafer is diced, the MEMS components to be assembled are de-tethered from the substrate to fully release the components. A simple tether is shown previously in Figure 2. The tethers are broken using an end-effector attached to the robotic arm. Automatic de-tethering can be accomplished by programming the sequence with a script derived from the CAD layout. Each component is then picked up off the substrate by the MEMS end-effector, rotated into position, and inserted into the desired socket. Using the Zyvex-designed MEMulator software, an emulation of the microassembly process is shown in Figure 7.

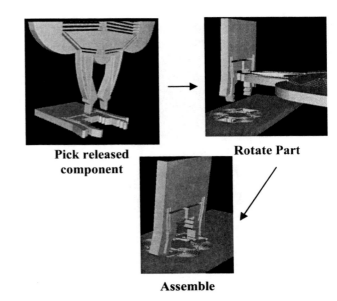

Pick released component   **Rotate Part**

Assemble

Figure 7: Emulation of the microassembly process.

4 PARALLEL ASSEMBLY

The parallel assembly of components has the potential to achieve higher manufacturing throughput over serial assembly. The difference with parallel assembly is the need to array the end-effectors onto the robotic arm. Furthermore, alignment of the connectors and sockets becomes more challenging and requires more complex robotic control schemes. Parallel assembly examples using 1X2 and 1X4 end-effector arrays are shown in Figure 8 and Figure 9, respectively.

Figure 8: Parallel assembly using a 1X2 array.

Figure 9: Parallel assembly using a 1X4 array.

5 MICROASSEMBLED DEVICES

The following are examples of devices that have been assembled using our microassembly tools and robotic system. Figure 10 depicts an assembly of four electrostatic deflector plates which make up a quadrupole structure for charged particle manipulation.

Figure 10: A microassembled quadrupole.

A variable optical attenuator is shown in Figure 11, comprised of three fixed micromirrors and a single, movable micromirror. The movable mirror is placed within a socket attached to electrothermal actuators.

Figure 11: Micromirrors arranged as a variable optical attenuator. Bottom mirror is positioned within a movable socket which rotates the mirror.

Finally, an example of the heterogeneous assembly achievable with our deterministic approach is the variable inductor shown in Figure 12. A copper coil is shown to the left. A silicon micromachined linear stepper motor is located to the right of the coil. Attached to the stepper motor and located within the coil is a nickel-iron core which is manipulated via the stepper motor.

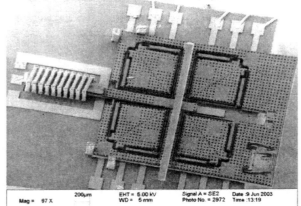

Figure 12: A microassembled variable inductor.

6 CONCLUSION

Several benefits over monolithic fabrication and stochastic assembly are realized when utilizing a deterministic microassembly approach. In comparison to the random placement nature of stochastic assembly, directed placement of each component is achieved with the deterministic approach. Such directed placement enables full tracking of each component and is amenable to "known good die" manufacturing. Another benefit of the deterministic approach is the ability to assemble heterogeneous components within a single microsystem. For instance, components made from various MEMS processes or even different materials can be precisely assembled together to form a truly complex microsystem. Due to the versatility of the deterministic parallel assembly approach described herein, numerous applications of microsystems can be realized including fiber optic components, high frequency devices, portable chemical and biological detection systems, and miniature high performance laboratory and industrial instrumentation.

ACKNOWLEDGMENT

This work was performed under the support of: The U.S. Department of Commerce National Institute of Standards and Technology Advanced Technology Program Cooperative Agreement Number 70NANB1H3021.

REFERENCES

[1] M. B. Cohn, K. F. Böhringer, J. M. Novorolski, A. Singh, C. G. Keller, K. Y. Goldberg, and R. T. Howe, "Microassembly technologies for MEMS," *SPIE Micromachining and Microfabrication*, Santa Clara, CA, (1998).

[2] H. J. Yeh and J. S. Smith, "Fluidic self-assembly of microstructures and its application to the integration of GaAs on Si," *Proceedings IEEE Micro Electro Mechanical Systems*, Oiso, Japan, p. 279-284 (1994).

[3] K. F. Böhringer, M. Cohn, K. Y. Goldberg, R. Howe, and A. Pisano, "Electrostatic self-assembly aided by ultrasonic vibration," presented at the *American Vacuum Society National Symposium on Micro Electro Mechanical System (MEMS)*, San Jose, CA (1997).

[4] G. Yang, J. A. Gaines, and B. J. Nelson, "A supervisory wafer-level 3D microassembly system for hybrid MEMS fabrication," *Journal of Intelligent and Robotic Systems*, vol. 37, p. 43-68 (2003).

[5] M. Ellis, G. Skidmore, A Geisberger, K. Tsui, N. Sarkar, and J. Randall, "Microfabricated silicon mechanical connectors and micro assembly," in *Nanotech 2002*, Houston, TX, American Institute of Aeronautics and Astronautics (2002).

A Novel X-ray Microtomography System with High Resolution and Throughput

Yuxin Wang, Fred Duewer, Shashidhar Kamath, David Scott, and Wenbing Yun

Xradia, Inc., 4075A Sprig Dr., Concord, CA 94520

ABSTRACT

The large penetration depth and rich contrast mechanisms of x rays makes it ideal for non-destructive or non-invasive imaging applications. Projection-type x-ray micro-imaging systems are widely used in micro- and nano-technology industries to study the internal structures of manufactured components such as micro-electro-mechanical (MEM) or semiconductor devices. The resolution of these system is typically determined by the size of the x-ray source. As a consequence, a small x-ray source size and high magnification are required to achieve high resolution. Since both factors reduce the detectable flux, a compromise between the resolution and the throughput must be made. Based on its innovative high-resolution detector, Xradia has developed a new system, the microXCT, that solves this problem by using a unique optical design. It is able to acquire images with 1-micormeter resolution in an exposure time of a few seconds. This instrument includes a fully automated tomographic data acquisition and reconstruction capability for an user to study the 3D structure of a sample at micrometer three-dimensional resolution. This system's unique capability of imaging at high-resolution with minimal compromise in the throughput makes it a valuable tool in non-destructive imaging applications in microtechnology and biotechnology.

1 INTRODUCTION

Since its discovery, the ability of x rays to see through material has been exploited extensively in non-invasive and non-destructive imaging applications in medical and industrial applications. For nano-technology, in particular, x-ray imaging systems provide the ability to non-destructively image the internal structures of devices with tens of nm to a few um resolution. To date, laboratory x-ray microscopes that employ x-ray optical elements have achieved 60 nm resolution [1]. They are typically used in failure analysis applications in semiconductor industries. The more widely used systems are projection-type x-ray microscopes that use no x-ray optics, but only uses a scintillated x-ray detector to record the projected shadow image through a sample (see Fig. 1). A number of manufacturers offer commercial systems with resolution ranging from tens of um to sub-micron. These instruments are widely deployed in a wide spectrum of industrial and scientific applications including semiconductor and electronics testing, biomedical research, archeology, and geology.

Besides commercial system, many such systems have been developed using synchrotron radiation sources. They typically offer much higher throughput to allow real-time study of dynamic events, and continuously tunable x-ray energy, which greatly enhances the material analysis capabilities.

In most commercial system, very high magnification is required to achieve the high resolution. The resolution with these designs is approximately the size of the x-ray source. Many manufacturers have develop nano-focused x-ray sources in order to achieve sub-um resolution. The reduction of source size, combined with the high magnification, have severely limited the throughput of these systems. To overcome this limitation, Xradia has developed a new system using an unique optical design based on its innovative high-resolution detector system. It is able to acquire images with 1-micormeter resolution in an exposure time of a few seconds. This throughput approaches that of synchrotron-based instruments.

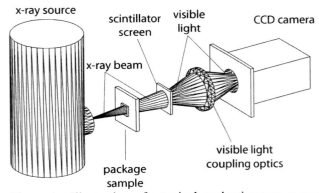

Figure 1. Illustration of a typical projection-type x-ray microscope.

2 IMAGING SYSTEM GEOMETRY

We first look at the imaging properties of the projection type microscopes. As illustrated in Fig. 1, the imaging system consists of a x-ray source, a sample position system, and a detector system, which simply records the shadow of the sample. We first look at the resolution limit of this system. Fig. 2 shows the geometry of such a system with a x-ray source with size s, two point objects separated by distance δ, and their shadow on the detector plane. The source to sample distance is l_s, and the sample to detector distance is l_d. In this geometry, Each point casts a shadow on the detector plane with the size $\left(l_d / l_s \right) s$. This is essentially the point spread function. The magnification is:

$$ M = \frac{l_s + l_d}{l_s} . $$

Ignoring the diffraction effect, we can use the criteria that the two point objects are resolved if the center of the object does not fall into the shadow of the other. That is:

$$ \left(\frac{l_s + l_d}{l_d} \right) \delta \geq \frac{l_d}{l_s} s $$

Therefore the resolution limit determined by the imaging geometry is:

$$ \delta \geq \frac{M - 1}{M} s . $$

A couple of special cases can be observed from this expression:

1. *Contact printing mode*: $M \approx 1$, then $\delta = 0$. That is, infinitely high resolution can be achieved in contact printing mode, where the sample is placed very close to the detector. In practice the system resolution is primarily determined by the detector resolution.

2. *Projection imaging mode*: $l_d \gg l_s$, and $M \geq 1$, then $\delta \approx s$. In this mode, the system resolution is determined primarily the source size. The detector resolution is relaxed because the features are magnified by the diverging x-ray beam. If the detector has sufficiently high resolution to sample the image, the system resolution is approximately that of the source size. Note that the resolution is never worse than the source size.

These are basically the two ways to achieve high resolution: one can use either a very high resolution detector or a x-ray source with very fine spot size. The trade-off is then to build a high resolution detector with high efficiency versus a source with high flux.

How much compromise must be made in each case with practical systems? Non-destructive imaging requires x rays with sufficient energy to penetrate through a complete device with a few mm to centimeter in size. X ray sources with more than 50-150 keV electron bombardment energy are used in most commercial systems. With the projection mode, to achieve 1 um resolution, the detector can be of coarse resolution with good efficiency, but the x-ray source spot size must be kept less than 1 um. The mean range of electrons in a bombardment target can be estimated with these empirical formulae:

$$ \text{lateral} = \frac{0.1 E^{1.5} \left(\text{keV} \right)}{\rho \left(g / cm^3 \right)} $$

$$ \text{depth} = \frac{0.077 E^{1.5} \left(\text{keV} \right)}{\rho \left(g / cm^3 \right)} $$

For example, with 150 keV electrons, the range is about 10 um in tungsten. The dimension of the x-ray generation volume is slightly larger. As illustrated in Fig. 3, the x-ray generation volume of a target is approximately that of the electron focal spot near the surface, but balloons to tens of um deeper into the target. To keep the spot size to 1 um, one must either lower the electron acceleration voltage, or use a thin film target with less than 1 um thickness instead of a solid target. Lowering the acceleration voltage makes the x rays less penetrating and is only acceptable for imaging small or partially destructed samples. Using a thin film target allows x rays to be generated only near the surface and avoid most of the x-ray generation volume. This clearly causes an severe efficiency loss since at 150 keV, no more than a few percent of the x rays are generated near the surface.

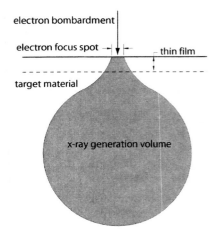

Figure 3. Illustration of the x-ray generation volume of a electron-bombardment target.

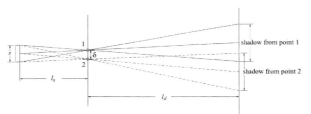

Figure 2. Imaging geometry of projection-type x-ray microscopes.

With contact printing mode, on the other hand, a moderate sized x-ray source can be used to take advantage of its full x-ray generation volume, but a scintillated detector with microscope objective coupling is needed. The x-ray shadow of the sample is converted to a visible light image detected by a single crystal scintillator. The visible light image is then imaged to a CCD camera by an objective lens. To reach 1-um scale resolution, the objective lens must have sufficiently high numerical aperture of at least 0.25. This high numerical aperture restricts the depth of field to tens of um. Therefore, the scintillator must be made thinner than tens of um in order to achieve the desired resolution. This thin scintillator typically absorbs only about 10-20% of the incoming radiation, therefore reducing the efficiency and the throughput. With current designs, the detector efficiency loss with contact printing mode is not as severe as the output loss with the projection mode. As a result, ones using contact mode can provide about tens times better throughput than projection mode systems at 1-um level resolution. If the resolution is relaxed to tens of um, the two system configurations are essentially similar and their throughput becomes comparable. Besides the throughput considerations, the long magnification beam path required in projection systems also have negative consequences on the system's footprint and image artifacts resulting from diffraction effects.

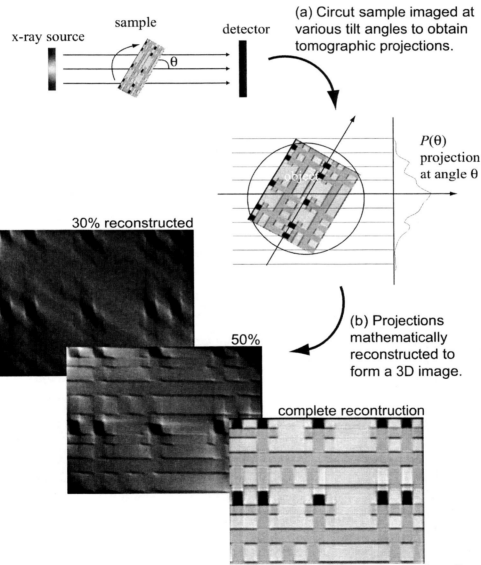

Figure 4. Illustration of the tomographic imaging process: The sample is first imaged at different tilt angles from the x-ray beam to obtain a series of *tomographic projections*. These projections are then recombined mathematically to form a 3D image representing the 3D structure of the sample.

3 THE MICROXCT SYSTEM

The microXCT is a high-performance x-ray imaging system with many innovative conceptual and engineering designs. Its resolution and field of view are adjustable from a few settings according to the sample size and the required resolution. In the high-resolution mode, a resolution of 1 um can be achieved with 1 mm field of view. In the survey mode, it provides 10 mm field of view, with 10 um resolution. An intermediate resolution setting can also be used. The systems is based on a commercial x-ray source with a moderate spot size that makes use of the full x-ray generation volume of the target, and a proprietary detector system developed by Xradia. This system provides very high throughput, particularly for the high resolution modes, where images with 1-um resolution can be acquired with second-scale exposure times. Images can be acquired in real-time in the survey mode. The sample is mounted on a rotation stage to provide full 360 degree sample tilting, thus allowing 3D tomographic data acquisition. The process of tomographic imaging is illustrated in Fig. 4: a series of images are acquired as the sample is rotated to different tilt angles. These images are then assembled mathematically to produce a 3D image of the sample. With this method, the internal structure of the sample is obtained without physical modification [2].

In operation, it offers user-friendly graphical user interface with fully automated data acquisition to acquire the tomographic projections, perform the image processing procedures, reconstruct the 3D structure, and provides the user with a 3D view of the result. Operator attention is needed only for loading the sample and analyzing the data. One to a few hours is required for acquiring and reconstructing a 3D tomographic data set.

Because this is system operates in the contact printing mode, it has a very small footprint of 2.5 ft x 4 ft. It has very modest power consumption and does not require any special facility preparations such as cooling water or special gas connections.

4 PRACTICAL APPLICATIONS

As a non-destructive micro-imaging instrument, the microXCT is a valuable tool in a wide range of scientific and industrial applications. Its strength is imaging the internal structures of samples of 1 mm to tens of mm in size at 1 um to tens of um resolution. Most samples can be mounted in the microscope for imaging without any preparations. We will list a couple of examples of failure analysis with integrated circuits packaging and inspection of MEM devices.

4.1 Semiconductor Packaging Failure Analysis

The feature size of IC packaging have decrease to um-level in the past few years while the complexity have increased drastically. The traditional 2-D x-ray inspection are becoming increasingly inadequate for imaging them because the resolution of these system are typically tens of um, and without quantitative 3D imaging capability, 2D images contain too many overlapping features for the operator to interpret, making fault identification very difficult.

The solution is a high-resolution 3D system with integrated 3D imaging capability that is automated well enough for an operator to use routinely. Fig. 4 shows the image of an AMD Athlon processor packaging taken at an angle normal to the chip surface with the microXCT system. It is clear

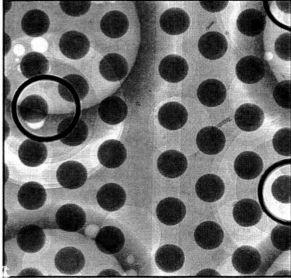

Figure 5. An AMD Athlon chip packaging imaged with the microXCT.

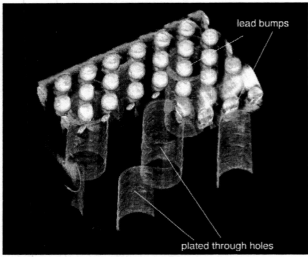

Figure 6. Volume rendering of the reconstructed 3D image of the AMD packaging .

the system has high enough resolution to resolve most features in the sample, but since many features are over lapping, its structure is very difficult to determine from this 2D image. A 3D tomographic reconstruction is able to resolve the features in depth. Figure 6 shows a rendering of the reconstructed 3D image. Features such solder bumps, copper lines, plated through holes are clearly visible. Figure 7 shows one section of the reconstruction in the plane perpendicular to the chip surface which clearly reveals the depth structures of the device. This cross-sectional view is usually obtained by physically cutting the sample. Using the microXCT, however, internal defects of the packaging can be found without the destructive process.

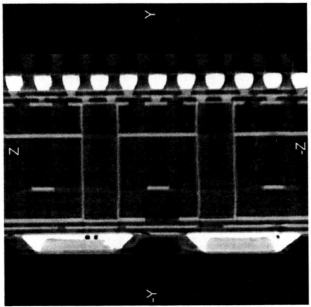

Figure 7. A slice from the reconstructed 3D image of the AMD Athlon package.

4.2 Embedded MEM Structures

Embedded MEM structures can be imaged in a similar way as integrated circuit chip packaging. Figure 8 shows an image of an accelerometer fabricated with MEM technology. Most of today's MEM devices contain embed um-scale features. The microXCT is able to image these features non-destructively. In many cases, the motion and mechanical properties of a MEM device can be studied dynamically as it is in operation.

5 CONCLUSION

We have developed a projection type x-ray microscope based on an unique optical design. This system's capability of imaging at high-resolution with minimal compromise in the throughput makes it a powerful tool in non-destructive imaging applications in microtechnology and biotechnology, for example, integrated circuits packaging

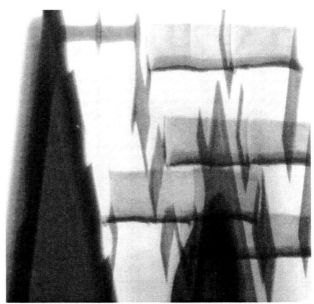

Figure 8. A 2D image of a MEM accelerometer imaged with the microXCT at 60 degree tilt angle..

failure analysis, imaging embedded MEMs structures, material stress failure mode analysis, bone implant interface, etc. Because of its high throughput, quasi-real-time 3D imaging can be performed to study dynamic processes such as formation and propagation of cracks in a bio-mechanical sample or delamination in a integrated circuit packaging.

REFERENCES

[1] Wang, et. al., "A Transmission X-ray Microscope (TXM) for Non-destructive 3D Imaging of ICs at Sub-100 nm Resolution," *ISTFA Proceeding*, ASM International, 2002
[2] Rosenfeld and Kak, Digital Picture Processing, Vol. 1, Academic Press Inc., 1982.

Step and Repeat UV Imprint Process Technology for Wafer-Scale Nano-Manufacturing

Michael Watts*, Van Truskett, Jin Choi, Chris Mackay, Ian McMackin, Philip Schumaker, Daniel Babbs, S.V. Sreenivasan, Norman Schumaker

Molecular Imprints, Inc.
1807-C W. Braker Lane
Austin, Texas 78758, U.S.A.
*mpcwatts@militho.com

The Step and Flash Imprint Lithography (S-FIL™) process is a step and repeat nano-replication technique based on UV curable low viscosity liquids. S-FIL uses field-to-field drop dispensing of the UV curable liquids for the step and repeat patterning. This approach allows for patterning of structures with widely varying pattern densities and complicated structures.

Nano-manufacturing requires the features to be printed reproducibly, aligned precisely and printed with low defect density. In this paper, the latest data on the S-FIL process will be described.

In addition, a brief summary of the overall status of the S-FIL process will also be presented. The presentation will specifically include:

- Full wafer (200 mm) residual thickness control to enable practical etching (thickness variation < 10 nm, 3□)
- Field edge control compatible with 50 um kerf regions
- Field-to-field imprint CD control and line edge roughness
- Overlay alignment results
- Process life and defect data

Molecular Imprints, Inc. (MII) has developed the Imprio™ 100, which is the first commercial step and repeat imprint lithography system with field-to-field alignment (Figure 1). Full wafer step and repeat printing performance is shown in Figure 2. The current status of overlay alignment accuracy is presented in Figure 3. Figure 4 shows the film thickness means and variation data for a full wafer. The data was obtained by at 64 locations per field (field size is 25 mm by 25 mm) over 25 fields on the wafer. Figure 5 shows the edge definition of an imprint field magnified by 1000X. Figure 6 shows process life data for printing pillars and contacts.

Keywords: Step and flash imprint technology, imprinting, nano-manufacturing, wafers, imprint process

Figure 1: Imprio™ 100 from Molecular Imprints, Inc. is a step and repeat imprint lithography system with field-to-field alignment

Figure 2: Full 200 mm step and repeat wafer coverage with lithographically useful residual layer thickness (variation of <20 nm, 3σ) and field edge control compatible with < 500 μm kerf.

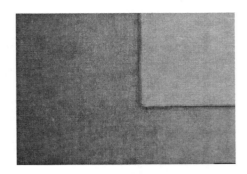

Figure 5: Clearly defined field edge of imprint at 1000 X magnification

X-mean (nm)	Y-mean (nm)	3σ, X (nm)	3σ, Y (nm)	θ (μrad)
-12	20	252	225	6.32

Figure 3: Full wafer overlay alignment accuracy based on three alignment error measurements per field on the wafer.

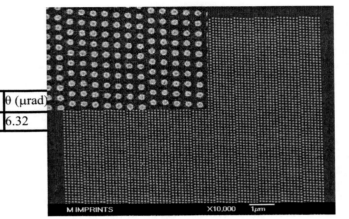

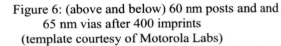

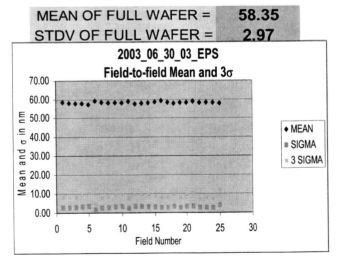

Figure 4: Residual film thickness means and variation data for a full wafer

Figure 6: (above and below) 60 nm posts and and 65 nm vias after 400 imprints (template courtesy of Motorola Labs)

A New Semiconductor-Wafer Market Based on the Deepening of Natural Surface Undulations to Form Strongly Textured Atomic Ridges (STAR) With Pitches from 0.6 to 5.4 nm: Model Demonstrations in the Physical and Life Sciences

D. L. Kendall and M. Kendall

StarMega Corp, Albuquerque, NM, USA, dkendall@starmega.com

ABSTRACT

A broad patent position has been established for semi-conductor wafers with physically-deepened grooves, ridges and dots on particular crystal planes with perfect pitches of 0.9 to 5.4 nm. Optical lithography is used in a standard production facility on top of deepened natural underlying nanotexture to produce quite useful new devices and ICs. See www.starmega.com. The very high value-added com-ponent of these wafers is the business opportunity. One method for producing nanoridges on any crystal surface (including III-Vs) will be discussed. Several applications will be demonstrated with mock-ups: ballistic transport; ultra-low-power MOSFETs; low-cost simultaneous sensing of hundreds of different molecules based on arrays containing nanotubes, DNA, and/or other molecules; high temperature BCS superconductors (not "HTS") with Buckyballs between flexible walls; mega-tip nanoprobes; particle filters with openings down to 0.3 nm for nano-shadow masks and for oxygen enrichment of air.

Keywords: silicon and III-Vs, ballistic transport, nano-mosfets, security sensors, superconducting interconnects

1 PERSPECTIVE

The ability to produce nanometer width grooves and ridges in a single crystal surface opens up a multitude of structural permutations and applications. This is demonstrated in Figure 1, which shows a pair of atomic ridges surrounding an interposed nanometer-width groove. This particular structure is relatively simple to produce over large areas and allows a sort of ballistic transport of electrons in the ridges while a normal inversion layer of electrons exists just below the ridges. The mobility of the highly scattered electrons in such an inversion layer of a MOSFET is typically about 500 cm^2/Vs, which leads to an average drift velocity of about 5×10^5 cm/s in a large transistor of 10 μm length between the source and drain (S-D). By contrast, we estimate that the velocity of the electrons in the narrow ridges can reach a velocity of about 5×10^7 cm/s since they suffer essentially no lateral scattering in the ridges [1].

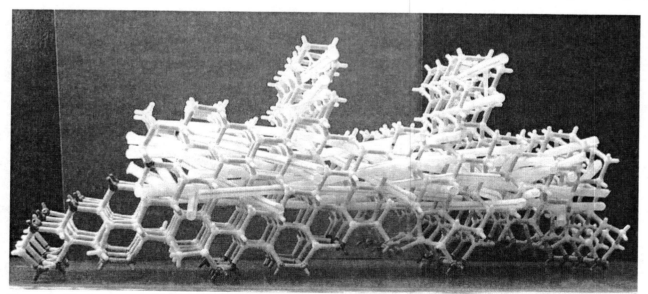

Figure 1: Electron waveguides in parallel with a normal nMOSFET on a ridged (3 3 7) wafer with a 1.57 nm pitch. Source-to-drain current flows into the plane of the figure. The gate conductor and dielectric are omitted for clarity. The (5 5 12) surface with 5.4 nm unit cell is on the bottom of the model. A segment of a {001} surface is shown at the left edge facing up.

We have chosen a large nMOSFET in the example above and a small S-D voltage of only 1V between the source and drain to emphasize one of the strong advantages of such structures for electronic devices, namely the potentially very large increase in the velocity of the electrons in the atomic ridges. The electrons in the inversion layer will have a wide distribution of velocities due to scattering of several types. The plastic straws in the atomic model illustrate the chaotic trajectories of the electrons in the normal inversion layer and the almost un-scattered trajectories in the narrow ridges.

In addition, the "Smallest Meaningful Dimension" (SMD) of only 0.31 nm in the above example allows fabrication of devices and circuits exhibiting reliable and useful quantum effects at room temperature and even far above. Furthermore, it does this using today's optical technology to make otherwise standard-looking MOSFETs piggybacked on STAR wafers. This is in sharp contrast to most of the research devices presently being fabricated using the best modern lithographic methods (e-beam, x-rays, EUV). The latter are generally limited in their SMD to 20 to 50 nm and their research devices must often be operated at very low temperatures of less than 4K. This follows from the fact that the first excited level $\_E_1$ of a deep potential well depends inversely on the square of the width, w, of the well [2], namely:

$$\_E_n = n^2\, K \big/ (w^2\, m_r^*) \;, \tag{1}$$

where $n = 1, 2, 3 \ldots$, w is the width of the well, m_r^* is the conductivity effective mass of the electron relative to the rest mass of the electron (which is about 0.26 for an electron in Si), and K includes well known constants and is 0.38 eV nm^2 when w is in nm. Since the STAR patents [3-6] cover groove and ridge widths for Si from 0.6 to 5.1 nm (and pitches from 0.94 to 5.35 nm), it is easy to show that the first energy level of 0.056 eV at room temperature of 300K (27°C) is active even in the *largest* ridge of 5.1 nm. Since the thermal energy, kT, at 300K is only 0.026 eV, this is not enough to smear out the quantum effects of even this largest nano-ridge. At the smallest attainable stable ridge width of 0.31 nm shown in Figure 1, the first level is at about 15 eV so the devices should operate at *much* higher temperatures reaching in principle above the melting point of Si. The III-V compounds often have smaller effective masses than Si, so ridges and grooves in these materials may operate at even higher temperatures.

We should also point out here that even a very shallow potential well of a depth of just one atom still has bound excited states. The energy levels can be shown to be smaller than the values given for the deep well above by a

factor of π^2 (about one magnitude). Thus, our narrowest ridges and grooves of 0.3 to 0.6 nm can still give robust quantum effects at room temperature at depths as small as one or two atoms.

2 ONE GENERIC TECHNOLOGY

Before discussing several of the many other diverse applications for the STAR structures, we will review one generic technology for obtaining nm-width grooves on Si wafers (which will also be applicable to many other crystalline materials). We start with a wafer of any diameter that has been X-ray aligned in the desired direction, sawed, and chem-mechanically polished (CMP) in a standard Si wafer facility [7]. The wafer is then heated in a UHV system to about 1150°C for a minute at a pressure less than 10^{-10} torr to remove the native oxide and impurities and to allow the surface to restructure. It is then slowly cooled to room temperature, where it can have "pre-assembled" precise values of pitch ranging from 0.94 to 5.35 nm. The *clean* surfaces are the (1 1 4) and the (5 5 12), which produce large regions of perfectly straight atomic ridges with pitches of 1.63 and 5.35 nm, respectively [8]. After reaching room temperature or some other suitable substrate temperature where the arriving atoms will have a sufficiently low surface diffusion coefficient, an etch resistant material is evaporated onto the surface at near grazing incidence so that only the surface atoms that are slightly elevated above the neighboring atomic undulations are coated. The sample is generally rotated during this procedure to avoid the formation of needles pointing to the evaporation source and to form relatively tall vertical etch masks of say 5 to 20 atoms height to serve as a mask against a subsequent chemical or an ion beam etching process [3,4]. The latter etching process is continued until the desired groove depth is obtained, which may only require the removal of a single atom or may proceed for thousands of atoms with highly anisotropic etching processes. An example of the natural undulating surface *before* both the evaporation and the etching processes is apparent from the bottom of Fig. 1 for the (5 5 12) surface which has repeating elevated ridges about 0.3 nm higher than the undulations between the ridges. A directed evaporation at about 1 to 5 degrees from grazing is adequate to insure that very little material is evaporated in the regions between the ridges, and the rotation of the wafer doesn't change the situation to any significant degree [10]. In many cases, this wafer can be submitted to a standard foundry where it is handled exactly like a standard Si wafer to produce decidedly non-standard devices and ICs.

Alternatively, in order to obtain other precise pitches, the high index Si surfaces may have an added evaporation of a very thin layer of a metal such as Au, Ga, or Ag followed by a UHV heat treatment to cause the surface to facet and restructure to particular favored crystal surfaces. *(A "high index surface" is defined as one having at least one of its Miller indices greater than 1. The (1 1 2) qualifies, as does the (5 5 12). The latter is often informally called the "(1 1 2.4)" to make it clear that it is about halfway between the (1 1 2) and the (1 1 3) planes.)* This allows a wide selection of precise pitches at particular values of 0.94, 1.57, 1.63*, 1.91, 2.21, 2.51, 3.14, 3.45, 3.78, and 5.35* nm, where the asterisks denote the clean surfaces mentioned earlier. All of the above high index surfaces are attainable and they can be all be considered as (1 1 X) planes if X is allowed to have non-integer values. For example, the 1.91 nm pitch is the surface unit cell of the (5 5 7) plane, or informally the "(1 1 1.4)" surface.) The ones without asterisks usually require the deposition of a sub-monolayer or monolayer ML thickness of some metal and a UHV heat treatment [4,9].

A good example of the above process involves the smallest attainable pitch with stable groove walls, which is generally the (1 1 2) surface with a ***sub-nanometer*** surface unit cell of 0.94 nm which is produced using a single ML of Ga and a heat treatment at about 450 C. If the Ga layer is not present, the clean (1 1 2) surface after a UHV heat treatment consists of many (111) and (113) facets, and the Ga layer causes it to "unfacet" to form an atomically flat (1 1 2) surface [11]. By contrast, ***an exactly opposite result*** can be obtained by coating an atomically flat clean (5 5 12) wafer (as shown in Fig. 2) with as little as 0.05 ML of Au and heating it to 900C. This sub-ML film of Au causes it to form relatively large facets of the (7 7 15) surface with interspersed (1 1 3) facets (as shown in Fig. 3) [4,9]. Of course, if one ***starts*** with a wafer precisely aligned on the (7 7 15) surface, its initial condition is a faceted surface after a UHV heat treatment until the small amount of Au is added and heated to 800C. After this Au process, it becomes a large atomically flat (7 7 15) surface with ridges with precise and perfectly regular pitch (as in the large facets of Fig. 3).

The ability to attain in a controlled manner such a large number of specific values for the precise pitch in the (partial) list above is a very important aspect of the STAR wafer technology. For example, it allows the fabrication of whole wafers with nanowidth grooves designed to encase a particular diameter of DNA, Buckyball, or nanotube, or a particular spacing between electrons and holes in a new type of surface emitting laser, etc.[3]. Actually, the groove widths have much more variability of choice than do the pitch values. This is because the wall thicknesses can be chosen to be essentially any integer value of the ***"Single Channel Wall Thickness"*** (SCWT) shown in Fig. 1. An example of this is shown in Fig. 4 below where Buckyballs are inserted between two double-SCWTs. They can also be inserted between single SCWTs for greater flexibility [3]. We believe this will be an important factor in the attainment of a good high temperature

superconductor with Bardeen-Cooper-Schreiffer (BCS) character as opposed to the typical HTS type superconductors. The possibility exists that zero-loss interconnects for ICs may be developed that operate above liquid nitrogen temperature. Please note that the Buckballs stack in the flexible walls as if the walls were not present, something like ordered marble layers with thin plastic food wrap interlayers.

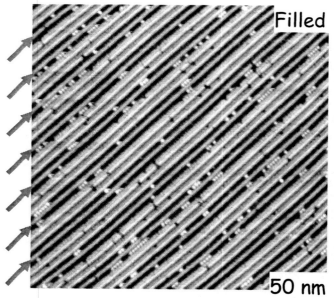

Figure 2: A clean (5 5 12) surface with the arrows showing the regular 5.35 nm spacing [4].

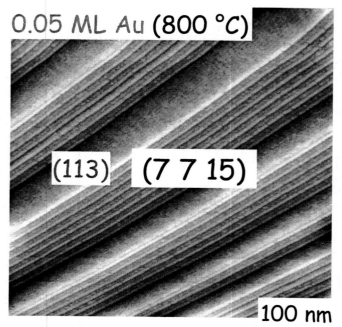

Figure 3: The restructured (5 5 12) surface showing the alternating (7 7 15) and (1 1 3) facets after adding 0.05 ML of Au and UHV heating at 800C. The atomic ridges on the (7 7 15) have a perfect pitch of 3.45 nm [4].

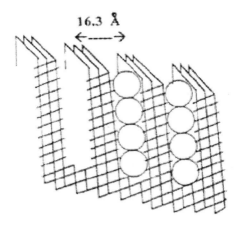

16.3 Å

Figure 4: Buckyballs in a (1 1 4) nano-grooved surface [3].

We invite the interested reader to visit our website at www.starmega.com. where we discuss and have LINKS to all four of the STAR patents. In particular, we discuss the very low power MOSFET operating at an S-D voltage of about 0.5 V that will have a very high electron velocity [1,3]. We also discuss a cantilever mega-tip array on nm-width pitch for SCULPTING true arbitrary nm geometry circuits and other applications [5].

We also propose a radical type of ultra-thin filter based on STAR technology merged with a SIMOX-type process [6]. A detailed fabrication technology is presented which produces openings down to at least 0.3 nm in width on a regular nm-pitch array. These ultra small openings can be made to cover large Si wafers for use as shadow masks for either charged or uncharged particles for many tasks, including a nano-lithography scheme, and a means for passively separating oxygen from normal air. The latter could provide an oxygen deficient patient with a means to produce oxygen enriched air without carrying a heavy oxygen tank.

Finally, we invite the attendees at the Nanotech 2004 in Boston conference to visit our Exhibitors Booth #102 in the Grand Ball Room at the Sheraton Boston on Tuesday and Wednesday, March 8 and 9 to ask questions about the technical and investment possibilities.

2.1 References

[1] Unpublished experiments and calculations.
[2] K. K. Ng, "Complete Guide to Semiconductor Devices" McGraw-Hill, p. 227, 1995.
[3] D. L. Kendall, US Patent 6,667,492, Dec. 23, 2003 "Quantum Ridges and Tips". This is the basic patent (submitted Nov. 9, 1998) underlying the three below.

[4] A. A. Baski and D. L. Kendall, US Patent 6,413,880 B1, Jul..2, 2002, "Stongly Textured Atomic Ridge and Dot Fabrication".
[5] D. L. Kendall, US Patent 6,465,782 B1, Oct. 15, 2002, "Strongly Textured Atomic Ridges and Tip Arrays".
[6]D. L. Kendall and M. J. Guttag, 6,509,619, Jan. 21, 2003, "Stongly Textured Atomic Ridge and Dot MODFETs, Sensors, and Filters.
[7] Virginia Semiconductor Inc., Fredericksburg, VA
[8] A. A. Baski, et al., The Structure of Si Surfaces from (001) to (111), Surf. Sci. 392,69-85 (1997)
[9] K. Jones, K. M. Saoud, and A. A. Baski, in "Cluster and Nanostructure Interfaces" Eds. P. Jena, S. N. Khanna, and B. K. Rao, World Scientific, New York, 2000, p. 49.
[10] K. Robbie, J. C. Sit, M. J. Brett, "Advanced techniques for glancing angle deposition", J. Vac. Sci. Tech. B, 16(3), 1115-1122 (1998).
[11] A.A. Baski, et al. "The Structure of Si (112):Ga-(NX1) Reconstructions", Surf. Sci. Lett. 423, pp. L265-L270 (1999).

Leveraging Mainstream Design and Analysis Tools for MEMS

I. Mirman[*], D. Flanders[**]

[*]SolidWorks Corporation, Concord, Massachusetts, USA, imirman@solidworks.com
[**]Axsun Technologies Inc., Billerica, Massachusetts, USA, dflanders@axsun.com

ABSTRACT

Mainstream CAD/CAE tools are used to rapidly design and commercialize a broad product line of rugged MEMS-based optoelectronic devices for applications including telecom, oil and gas production, process monitoring, medical diagnostics, and scientific discovery. These tools deliver the robust capabilities previously available only with specialized, often expensive, MEMS-focused options.

This paper illustrates some of the considerations in choosing a MEMS CAD/CAE tool, and provides examples of mainstream tools successfully used in MEMS design and analysis.

Keywords: MEMS, CAD, FEA

1 INTRODUCTION

Now that MEMS technology has broken through from research to commercialization, companies around the world are in a race to deliver better, smaller, and cheaper MEMS-based alternatives in virtually every industry and field, including consumer products, automotive, medicine, scientific discovery, process monitoring.

Because time-to-market schedules are now measured in months and quarters, and engineering and manufacturing challenges make up the bulk of the critical path, it is imperative to leverage concurrent engineering practices, and take advantage of tools that help reduce the number of design iterations.

While some MEMS design and analysis challenges are not ready to be tackled by mainstream tools, many are. It is therefore worthwhile to consider the capabilities and advantages mainstream CAD/CAE tools offer:

- one standard CAD tool to design everything
- 3D visualization;
- associative photomask definition;
- sub-micron feature definition;
- design re-use and configuration management; and
- integrated finite element analysis (FEA).

2 3D VISUALIZATION

MEMS designers often use layout software that is inherently two-dimensional. However, solid modeling tools offer MEMS designers better design visualization. 3D gives designers a clear and accurate review of parts (Figure 1) and assemblies (Figure 2) early in the design cycle.

Figure 1: LIGA component model and SEM photo

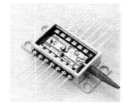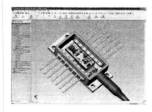

Figure 2: Miniature Spectrometer model and photo

The ease of use offered by mainstream 3D packages is of paramount importance. After all, many MEMS designers are not mechanical engineers or CAD "power users," often driving the CAD system just a few hours per week. CAD/CAE tools built and running native on the ubiquitous Windows operation system leverage standard user interface paradigms such as drag-and-drop and menu structures.

3D visualization provides MEMS designers a first check of design intent, proper operation, collision avoidance, and package stack-up, helping identify potential problems early in the design cycle – before committing to a mask or processing wafers. Modeling and verification of all the components for the high-speed optical interconnect module in Figure 3 within one CAD model provided the confidence to quickly commit to tooling investments for the cooler, package, and assembly equipment.

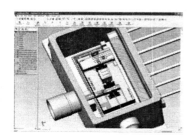

Figure 3: 10 GB/s optical interconnect module

For example, automated checking of clearance design rules can be particularly important for surface micro-machining to ensure proper etching and separation of parts during lift-off processes. Any interference will stop motion between parts that contact and the point of interference can be highlighted.

True motion simulation, integrated into some mainstream CAD packages and long used by machinery designers, can now help the MEMS engineer simulate mechanisms, such as meshing gears, and showing their operation in the assembly. While not a substitute for all simulation (e.g., gear stiction due to Van der Waals forces), many common questions arising during design can be answered immediately.

Illustrated in Figure 4 is a miniature electrostatic jog motor for a lens focusing application, whose rotation is converted to linear actuation through use of a rack-and-pinion assembly. The rotation is gear-reduced first, resulting in very fine positional control. Even at these small scales, mainstream CAD tools can provide kinematic simulation.

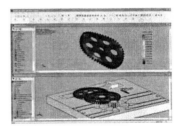

Figure 4: Miniature motor and gear assembly. (Courtesy Schiller Engineering)

3 DESIGNING TO SCALE

Given that dimensions associated with MEMS designs commonly range from fractions of microns to – in the case of package, system design, and assembly equipment – meters, there are advantages to designing to scale, and in 3D.

Not only can features such as optical gratings and micro-gears be designed to scale in 3D, but a MEMS package can be modeled in context of the chip design, and the assembly fixturing and machinery can be modeled in context of the package. Designing in the same scale, using the same CAD tools, it is possible to take advantage of the automatic design propagation that mainstream CAD packages can offer.

In response to needs of MEMS designers, some mainstream CAD packages can support the wide dynamic range of dimensions (for example, exceeding eight orders of magnitude) required by MEMS products and the associated equipment. Figure 5 illustrates a MEMS optoelectronic component, module, along with a model of the manufacturing equipment used on the production line. All of the designs are done to scale, and in the context of

each other; a change in the MEMS component can trigger design changes in the module and related fixturing.

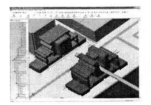
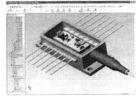

Figure 5: MEMS-based components and equipment

4 ASSOCIATIVE PHOTOMASKS

A common task in MEMS design is the creation of the photomasks associated with the various steps in the MEMS fabrication process. Bi-directional associativity in mainstream CAD packages enables change in one place (e.g., layout sketch for a single cell) to automatically propagate to all related documents (e.g., photomasks drawings), dramatically reducing the time required for engineering design changes, often from days to seconds. Figure 6 illustrates solid models and photomasks of silicon acceleration sensors. Changes in one file automatically propagate to the other.

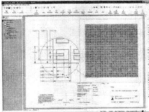

Figure 6: Solid models and photomasks

Masks for positive and negative photoresist can easily be fabricated directly from CAD files (Figure 7).

Figure 7: Photomasks created from drawing files.

For designs of even moderate complexity, generating photomasks for the various layers may be quite challenging with some CAD systems. With today's parametric solid modelers, generating 2D drawings of various cross-sections (e.g., different depths) is automatic. Figure 8 illustrates a MEMS-based variable capacitor solid model (left) and two-dimensional projected and cross-section views (right).

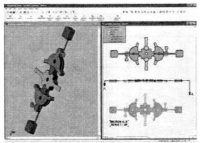

Figure 8: MEMS-based variable capacitor model (Courtesy Microfabrica, Inc.)

5 DESIGN RE-USE

New designs are frequently based on previous ones. Some mainstream CAD tools have rich capabilities to enable the designer to leverage existing designs through 2D and 3D file import.

Today's CAD systems have the ability to create a library of commonly-used features or parts as a powerful tool for designers who have already accumulated a large design database. The engineer can save time by simply dragging required components for a new assembly directly from the library and dropping them into the new design (Figure 9).

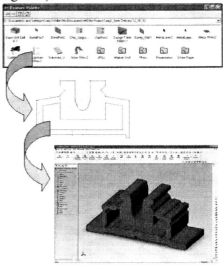

Figure 9: A sketch is dragged from the FeaturePallette in SolidWorks, and used to define a solid model of a LIGA component

In addition to reusing portions of previous designs, it is sometimes advantageous to incorporate design variants, or configurations, within a single file.

Configuration management enables the generation of multiple versions of parts, assemblies, and drawings in a single document with a minimal amount of time and effort. Configurations make use of design tables, derived design data, component properties, relationships, viewing states, and other attributes, storing part and assembly information in one area for greater efficiency. New configurations are easily developed from previously created designs to further speed development and meet market needs for data reuse.

An example application of this capability is illustrated in Figure 10. Several configurations of an optical bench assembly are created within a single SolidWorks model file for simplicity and easy design control. Selecting (activating) configurations requires a single click in the SolidWorks FeatureManager.

"What if" scenarios for different design requirements such as film thickness and modulus can be quickly explored by turning on and off various configurations of a part or assembly. Etched well dimensions and sizing of cutouts can be tied to design data for each chip size. As the chip requirements change, the necessary wafer level dimensions automatically update to reflect the new design.

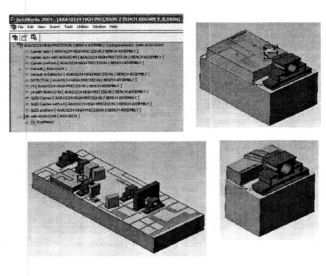

Figure 10: Multiple optical bench configurations

Components involving multistage processing, such as surface micromachining or LIGA, can easily be documented by using multiple configurations of a single part. Configuration management techniques can generate a discrete version of a part or assembly as necessary to reflect a separate version or in-process state. These versions can help compare designs, track expected performance, and develop process plans.

6 INTEGRATED FEA

MEMS products often require a diverse set of analyses, including thermal, stress, deflection, buckling, non-linear, electromagnetic analysis, and computational flow dynamics (CFD).

Increasingly, mainstream finite element analysis tools can be relied on to meet some of these challenges, due to the tight CAD/CAE integration, relatively low cost, ease of use, and the ability to handle small dimensions and large aspect ratios.

Instead of recreating and meshing a model inside a stand-alone FEA package, the CAD solid model can be a starting point for analysis. Component material properties (applied to the product as it is designed in the CAD environment) can be automatically shared with the integrated FEA application. There is also no need to learn a new user interface.

As an example, consider Figure 11, a LIGA component used to precisely position an optical fiber. The COSMOS FEA suite [1], running inside the SolidWorks window, is used to compute the deflections associated with the forces imparted on the LIGA holder through a closed-loop alignment system.

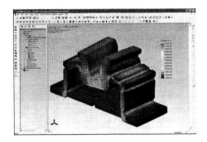

Figure 11: Finite element analysis of LIGA component

Another example is the MEMS-based electro-optic interconnect module in Figure 12. Designing the entire product – from the chip-level through to the entire system – in one CAD package, and analyzing it with just one integrated FEA package, saves not only upfront investment and training, but also the incremental efforts in design optimization.

Figure 12: Thermal analysis of electro-optic interconnect module (Courtesy Schiller Engineering)

7 BUSINESS CONSIDERATIONS

MEMS designers number in the thousands, whereas mechanical designers in the millions. Hence, software tools that serve the mainstream market can amortize their development costs and debugging test cases across many more users. This typically results in software tools that are cheaper, easier to use, and with higher quality.

Because MEMS is a field advanced largely by universities and small firms, the tools' total cost of ownership and learning curve are of paramount concern. With pricing in the USD$4,000-$8,000 range, mainstream tools can offer dramatic savings over more specialized tools that often require proprietary hardware to run.

MEMS designers require powerful functionality from their design and analysis software, but it must be easy to learn and use, as they are not using it all the time. Because of the broad adoption of mainstream 3D CAD and FEA systems worldwide, and their presence in most engineering curricula, it is straightforward to find engineers proficient in these tools.

Another benefit is broad data interoperability. Using the same CAD and FEA tools as other designers within and outside the company will dramatically reduce data interoperability challenges, which typically manifest themselves as design errors and delays associated with remodeling parts from scratch.

Additionally, mainstream tools contain a great deal of useful functionality typically not available in more specialized tools. For example, mainstream CAD tools include extensive collaboration functionality, an increasingly important part of the development process, enabling designers to share designs easily with anyone, anywhere. Some mainstream CAD tools also offer a rich application programming interface (API), enabling the automation and customization of many engineering design tasks.

9 CONCLUSIONS

Advances in mainstream CAD/CAE tools' MEMS functionality, coupled with their power in addressing general design and analysis challenges, broad adoption worldwide, and availability of integrated complementary solutions, make them viable contenders for the MEMS engineer's toolbox.

REFERENCES

[1] SRAC, http://www.cosmosm.com/

UV Laser Micro-Materials Processing Of MEMS, Microfluidics, Sensors, LEDs and Other Miniature Devices

By Jeffrey P. Sercel

JP Sercel Associates Inc. (JPSA)
Hollis, New Hampshire, USA

ABSTRACT

The emerging demands of medical device micro-machining – from drug delivery systems to cell sorting and more – require extremely precise, tight tolerances, high repeatability, and cost-effective processing. Short-wavelength (157-193nm) UV VUV Excimer lasers have proven ideal for such applications particularly with regard to processing difficult materials such as borosilicate glass, quartz, fused silica, and PTFE, exhibiting the ability to execute complex features with large-area and ganged processing capabilities and characteristic smooth cuts in such applications as precise drilling of microscopic apertures. However, the characteristics of short-wavelength UV VUV Excimer laser beam configurations are unique, with drawbacks that balance advantages. For example, short wavelength UV cannot be transmitted through air and must be transmitted through either a vacuum or inert system. In addition, the laser beam itself must be controlled and configured, and this is the role of the VUV Beam Delivery System (BDS). This paper will examine the role and configuration of the BDS in short-wavelength Excimer laser processing, and the specific requirements for a BDS enabling the practical and cost-effective research/commercial application of Excimer lasers to medical device micromachining.

Keywords: excimer, laser, micro-machine, ablation, absorption, beam, shapers.

Emerging challenges in micro-machining, particularly in biomedical applications, include the need to process materials at faster speeds; to process a wider range of materials; and to process them with greater precision, repeatability, and to micro-machine features of increasingly smaller dimensions with ever-tighter tolerances. Many of these applications are beyond the capabilities of mechanical micro-machining, and must be processed with lasers.

Examples of these cutting-edge technologies include lab-on-a-chip, micro-fluidics, drug delivery systems, biosensors, cell sorters, cell trapping, gene sequencing, hemo cytometers, nozzles, MEMS, micro filters, and more. These applications may require complex features, holes, cones, channels, sample chambers etc. of microscopic size, of uniform and consistent size, with certain essential characteristics that may include sharply-defined features, smooth walls, optically clear surfaces and to be produced with high repeatability and at production speeds sufficient to make their production economically feasible. In one example, laser-scribing of glass slides for cell counting replaces manual or mechanical diamond-scribing; in another instance, lasers etch patterns on glass bio-detection sensors that would otherwise be chemically etched - more slowly and at much higher cost per piece.

A wider range of materials – from certain glass types to PTFE, quartz and fused silica – complicate matters. Some cannot be mechanically machined at all to achieve the required features or tolerances; and although they can be micromachined using lasers, some materials cannot be properly or effectively machined by all lasers. For such applications, short-wavelength, Excimer UV lasers are proving to be the best choice due to their unique capabilities and beam characteristics.

Different materials absorb laser energy differently; the greater the absorption of the material, the easier it is to machine it cleanly and consistently. Many materials can be effectively micromachined with longer-wavelength lasers (e.g., Nd:YAG); however many materials such as certain types of glass, cannot tolerate longer wavelengths without cracking; other materials will exhibit rough holes and edges that do not meet the strict requirements of the application. Difficult materials (such as quartz and fused silica) can be effectively processed using short wavelength (157nm) Excimer lasers. Other difficult material can be processed at 193nm such a UV transparent glasses such as borosilicate and sensitive polymers such as nylon, PMMA and PET. They excel at direct write, high-speed, high aspect ration hole drilling, thin film patterning applications and are well suited to many of these cutting-edge medical device micro-machining applications. Due to the high absorptivity of short-wavelength UV by the material, micro-machining is crisp, precise, and repeatable. The process itself is known as photo-ablation.

Excimer lasers are high average power UV laser sources with many significant features and characteristics that make them ideal for high-resolution materials processing. Excimer lasers operate at a variety of user-selectable UV wavelengths from 157nm to 351nm. This allows processes to be optimized based on absorption; e.g., sub-micron layers of materials can be removed with each laser pulse. This characteristic alone makes Excimer lasers remarkably different from other laser types.

Short UV Excimer laser wavelengths can be projected onto material with very high resolution. Even with the use of simple lenses to shape and direct the beam, micron resolution is easily achieved.

PHOTO ABLATION

The method of materials removal with Excimer lasers is unique and a direct function of the laser's characteristic form and energy type. Known as laser (photo) ablation, this occurs when small volumes of materials absorb high peak power laser energy. When matter is exposed to focused excimer light pulses, the energy of the pulse is absorbed in a thin layer of materials, typically <0.1μm thick, due to the short wavelength of deep UV light. The high peak power of an excimer light pulse, when absorbed into this tiny volume, results in strong electronic bond breaking in the material. The molecular fragments that result expand in a plasma plume that carries any thermal energy away from the work piece. As a result, there is little or no damage to the material surrounding the feature produced. Each laser pulse etches a fine sub-micron layer of material; the ejecting material carries the heat away with it. Depth is obtained by repeatedly pulsing the laser; depth control is achieved through overall dosage control.

BEAM DELIVERY SYSTEMS

The Beam Delivery System (BDS) Directs laser energy onto process material; it determines laser power density on the target, determines the size and shape of the beam on the target; and motion control systems are sometimes used for auto-focus, or for articulated beam positioning relative to process material. Excimer laser beams differ from other types of lasers in that the beam covers a generally wide area comparatively, and is characterized by a 'flat-topped' as opposed to Gaussian shape. Excimer laser beams can cover relatively large areas of material with effective processing results.

Excimer laser beams are not perfectly uniform in intensity over the area of the beam and therefore only a portion of the area of the beam is usable for high-uniformity materials processing. In some cases, only the most uniform section or "filet" of the beam will be selected for use, and the non-uniform section of the laser beam will be discarded.

High-quality optics (CaF_2) is part of a VUV BDS; optical techniques are employed to use a larger fraction of the available laser power. Furthermore, due to the premium price associated with UV photons, high beam utilization – known as the Beam Utilization Factor, or BUF – in many cases is a key economic factor, which can qualify or disqualify an otherwise technically feasible application. Beam efficiency enhancers, a.k.a. beam shaping optics and beam homogenizers, can be employed to shape the beam and simultaneously make the laser energy uniform.

Near-field imaging involves the use of a mask to project a pattern of laser light onto a part. The features contained in a pattern are then etched into the target material, at a magnification determined by the relative positioning of the optical elements. Near-field imaging can be used to project a mask image onto a workpiece so that complex features can be patterned. This technique is the basis for excimer laser micromachining in many materials processing systems.

High performance doublet or triplet imaging lenses made from Caf2 are used to improve the image quality over larger fields of view improve the grid distortion as well as the spot size uniformity and minimum spot size achievable. A limitation to just one lens material makes the imaging lens design a significant engineering task. Due to losses in the optics, it is desirable to minimize the number of elements. This places constraints on the accuracy and surface finish of the lenses.

To efficiently use a higher percentage of the available UV Excimer laser beam energy, optical techniques are employed to ensure uniform irradiation over large areas in effectively three dimensions. Imaging optics are used to control feature accuracy over larger fields of view (2-D). Beam shaping or beam homogenizer techniques are used for exposure control (3-D).

The unique nature of Excimer laser beams includes both strengths and weaknesses. The wide nature of the beam allows the beam, through near-field imaging and masks, to be divided up and thus perform multiple tasks simultaneously (such as drilling multiple holes in a part) for maximum BUF; however, beam shapers, homogenizers, and other optical elements of a BDS are required to configure the beam for maximum effectiveness, these being somewhere between the laser and the target.

Short-wavelength UV laser energy's Achilles' heel is air. While longer wavelength laser energy can transmit through air with (depending on the type/wavelength) negligible loss of energy or efficiency, short wavelength – 157nm – UV cannot, and must be transferred through either a vacuum or through an inert gas that will transmit the laser beam without appreciable loss. Indeed, even the presence of 1ppm O_2 in the vacuum or purged BDS will result in degradation and loss of efficiency of the beam, as well as generation of contaminants and ozone that will further absorb the laser energy and reduce its effectiveness dramatically. Therefore, the different modules and elements in the BDS must also comprise a system that is either evacuated or purged with an inert gas (N_2 purification is required) that will facilitate transmission of the UV laser beam.

Large facilities may have the resources to build large and complex evacuated beam delivery systems, but these are not flexible and not economically feasible for the average user or for most commercial use. In addition to being costly to custom manufacture, such systems lack the flexibility to process a high product mix or conduct research. For example, an Excimer BDS may include a movable/changeable mask or series of masks; telescoping parts; field lenses; and turning mirror blocks. Moving parts are not especially compatible with attempts to maintain a tightly sealed vacuum system with the needed <1ppm O_2 required for optimum beam transmission.

BDS CONFIGURATION

A practical Beam Delivery System for 157 and 193nm Excimer laser processing consists of multiple interchangeable modules used to control the intensity and fluence of the laser beam as well as its shape and distribution. These can include variable attenuators; telescopes; field lenses; turning mirror blocks; beam splitters; beam dump; multi-position mask changers; power meter modules; and more. For example, it is known that as an Excimer laser ages, its beam quality changes (not necessarily deterioration, but this is also included as a factor). This fact can affect process control and therefore must be addressed. Beam power meters are used in-line in the BDS to monitor beam shape and strength. In such modules, the beam may be redirected to the meter – either in whole, thus interrupting the beam, or in part (sampling a certain few percent of the beam), by redirecting a slice of it during processing, by one or more mirrors in the module. Thus, a vacuum system is rendered impractical.

A more cost-effective and flexible approach is to employ a purged system. Inert gases such as pure nitrogen and argon are effective transmitters of Excimer laser beams. The BDS is designed such that certain modules have ports for the introduction of slightly pressurized inert gases that flood the system. Controlled leakage at specific points (such as the connection of a movable part such as an actuator to the outside) allows slight egress of gas to the outside at any point where contaminants (ambient air) might otherwise enter the system. The result is a modular, quickly interchangeable and configurable BDS suitable for commercial and research purposes with the needed flexibility to allow frequent beam characteristic changes as well as mask changing and adjustment.

The Excimer laser user has varying options with regard to inerting the BDS. Certainly there is a cost factor involved with the use of high-purity nitrogen ($\leq$1ppm O_2; a requirement) or other gases. High-purity bottled gases are easily obtainable; high-volume users may wish to purchase nitrogen generating equipment and optional high-purity filters to achieve the needed purity levels for the BDS. Of course, higher-volume usage will mean more frequent O_2 filter changes to maintain gas purity, at higher cost with increased frequency of changeouts.

High purity components are a necessary part of the BDS, and by this we mean, more specifically, the lack of contaminants or potential contaminants. A sealed purged system will naturally require gaskets of some sort to maintain the integrity of the inert gas fill and ensure cost-efficient use of gases. However, over millions of laser pulses, repeated discharges cause organic materials within the BDS exposed to UV laser light to degrade under the intense UV exposure that occurs during lasing. These organic materials can be gaskets, lubricants, etc. In the process of degrading, they outgas in the process and bring contaminants into the system. These UV absorbing impurities cause the laser light output to fall over time. Thus, it is important that BDS components are free of grease and lubricants, and that gaskets are shielded from contact with UV laser light, through design.

CONCLUSION

Short-wavelength (157-193nm) UV VUV Excimer lasers are emerging as a precise, cost-effective technology for meeting the emerging demands of medical device micro-machining. The ability to machine complex patterns with extreme accuracy and repeatability, due to the flexibility of the UV VUV laser system and its characteristic clean cut, makes these systems well suited to handling the tight tolerances required, and the ability to process a multitude of ordinary and exotic materials. 157nm systems are currently in production, but these systems require careful attention to multiple details (such as gas composition, etc.) in order to operate repeatably and cost-efficiently. More research is needed, certainly, and process engineering; however, the promise of cost-effective and commercially viable UV VUV laser micromachining is great and will enable the advancement of new medical technologies where older technologies and approaches have already proven to be inadequate or economically unfeasible.

Local Electrode Atom Probes for 3-D Metrology

T. Kelly, T. Gribb, J. Olson, R. Martens, J. Shepard, S. Wiener, T. Kunicki, R. Ulfig, D. Lenz,
E. Strennen, E. Oltman, J. Bunton and D. Strait

Imago Scientific Instruments Corporation, 6300 Enterprise Lane, Madison, WI, USA, tkelly@imago.com

ABSTRACT

Imago Scientific Instruments is developing its Local Electrode Atom Probe (LEAP®) microscope for use in semiconductor, data storage, and other nanotechnology industries to provide 3-D atomic-scale metrology. The LEAP microscope achieves true 3-D atomic-scale analysis by using a high electric field to remove individual atoms from material surfaces and a position-sensitive detector to record information that reveals the atoms' position and identity.

The LEAP microscope represents a new class of tool for atomic-scale characterization whose analytical capabilities surpass those of currently employed metrology tools. The LEAP microscope's atom-by-atom analysis achieves higher spatial resolution and greater sensitivity than TEM, and provides lateral spatial resolution that is not available by SIMS.

Specific industry needs that may be addressed by the LEAP microscope are quantitative analysis of dopant distributions in ultra-shallow junctions, purity/interface structures of ultra-thin dielectrics, and multilayer thin film structure in read/write head and storage media technology.

This paper presents results of LEAP analysis of buried interfaces in electronic materials to show the unique nature of the analytical information provided by the technique and the potential benefit to industry of developing a metrology tool based on LEAP technology.

Keywords: atom probe, characterization, metrology, nanoscale analysis, 3-D imaging

1 THE LEAP MICROSCOPE

Imago's LEAP microscope is an innovative version of a 3-Dimensional Atom Probe (3DAP) microscope. Briefly, 3DAPs analyze materials in the following manner:

- A needle-shaped specimen with a prepared tip (100 nm radius at the apex) is placed into a vacuum chamber.

- A voltage is applied between the specimen and a detector, resulting in an electric field that pulls on the atomic nuclei on the specimen tip. At a high enough field, atoms lose one or more electrons and field evaporate from the tip as positive ions. The voltage is applied in pulses to allow for a known time of departure, and carefully controlled such that atoms are removed one at a time.

- After leaving the specimen, the atom is accelerated by the divergent field toward a position-sensitive detector that records the location and time of impact of the atom. The time of impact is determined by the accelerating voltage and atom mass.

- This approach achieves very high spatial resolution (better than 0.5 nm) and high magnification (10^6X) by projection.

- As the analysis progresses, the entire surface layer of atoms on the specimen is removed, exposing the underlying layer. This process is continued over thousands of atomic layers.

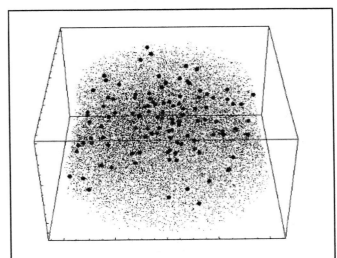

Figure 1. Results of 3DAP analysis of boron-doped silicon. Dots represent individual atoms; Si is shown in gray, boron in blue.

The three-dimensional picture of the atomic structure and composition of the specimen can be re-created, as shown in Figure 1, from the record of position and time of impact of the atoms pulled off the specimen as follows:

Atomic structure: By employing simple geometry, the original position of the atom on the specimen tip can be computed from the position at which the atom collides with the detector, since the top surface of the apex of the specimen maps directly onto the flat surface of the detector. This positional information reveals the 3-D atomic *structure* of the specimen.

Atomic composition: The identity of atoms can be determined from the time of flight from the specimen tip to the detector, equal to the difference in time between the voltage pulse that causes the atom to leave the specimen and time of impact of the specimen atom on the detector.

Time of flight is proportional to the atom's mass-to-charge ratio, and so can be used to identify the detected specimen atoms. This information reveals the 3-D atomic *composition* of the specimen.

2 ANALYTICAL RESULTS

Three-dimensional data from LEAP analysis provides new information and new insights for scientists and engineers developing and manufacturing devices at the nanoscale. An important area of application of 3-D analysis is in the study of interfaces.

Interfaces comprise a major structural component of today's electronic materials, most of which are fabricated as layered structures on planar substrates. The most critical layers are typically on the order of a few nanometers thick. In such thin layers, interfaces become a potentially dominant component of the device. It is therefore critical in device fabrication that both the structure and composition of buried interfaces be well understood. The LEAP microscope excels at such buried interface problems.

2.1 Multilayer Thin Film Stacks

Read sensors in current hard drives are made of multilayer thin film structures that utilize the gigantic magnetoresistance (GMR) effect. Thin metal films on the order of 1 to 3 nm thick are layered in the read sensor to create this effect. Two of the most important layers of a typical structure are made of cobalt or cobalt alloys and pure copper or other non-magnetic spacer elements. The signal magnitude from this type of sensor needs to be maximized in practice and is achieved by controlling two critical parameters: layer thicknesses and interface roughness.

A test structure was fabricated that contains many repeats of these two layers. Figure 2 presents a subset of LEAP microscope analysis results of the multilayer test structure. Six repeats of the layer structure are visible. The figure includes data from 400,000 atoms and only shows the key elements: cobalt, iron, and copper. Low levels of impurities were identified, but are not included here. Cobalt and iron are both shown as blue atoms, and copper is shown as red in the figure.

These materials are known to grow with columnar grains parallel to the growth direction. The location of two suspected grain boundaries is marked in Figure 2. This section of the dataset was chosen to illustrate the presence of the grain boundaries. This dataset provides evidence that that copper diffusion occurs along grain boundaries in these multilayer stacks.

The compositional variations in the structure may be evaluated by an alternate presentation of the analytical data. Evaluation of composition variations at a specific location in the structure is shown in Figure 3. It shows a transverse projection through a thin subvolume of Figure 2. The close-packed planes of the structure ({111} FCC CoFe and {111} FCC Cu) are visible in the image. They serve as an internal length calibration standard. The length scaling in the transverse directions is obtained by assuming uniform atom density in all directions.

The composition profile in Figure 4 is derived from Figure 3. Each datum in the profile corresponds to approximately two atomic layers in Figure 3, and there are 120 atoms/datum. Error bars are not shown in the figure but at each datum, they are dominated by the statistical noise associated with this number of atoms per datum. This composition profile shows a feature that has been observed previously [1,2]: the diffuseness of the copper-on-cobalt interface (following the growth direction) is less than the diffuseness of the cobalt-on-copper interface. This can be seen visually in the subvolume image and in the slopes of the composition profile at the interfaces.

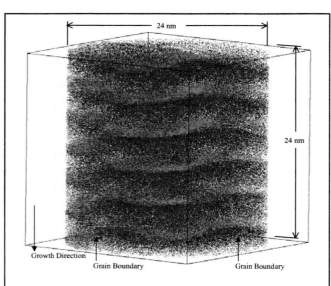

Figure 2. LEAP microscope image of a cobalt-10%iron/copper multilayer test structure. This image shows a 400,000-atom subset of data from a data set that contains information from a total of five-million-atoms. Cobalt and iron atoms are both represented by blue dots, and copper atoms by red dots.

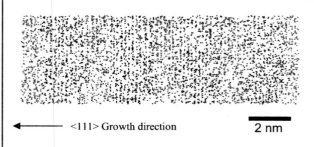

<111> Growth direction   2 nm

Figure 3. A 2-D projection of a thin 3.2 nm by 3.2 nm square cylinder of the image in Figure 1 that shows the close-packed atomic planes ({111}FCC CoFe or {111}FCC Cu). The layer thicknesses are determined from these image data to be 2.4 nm and 1.4 nm for the CoFe.

The interface diffuseness may be evaluated from this image by measuring the distance over which the

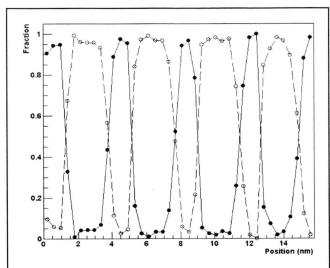

Figure 4. A local composition profile through the subvolume in Figure 2 along the growth direction. The length scale of the profile is calibrated from the spacing of the atomic layers in the image. Filled circles are copper concentration and open circles are cobalt plus iron concentration.

composition changes from 90% to 10% at each interface. This 90-10 interface thickness was determined to be 0.60±0.05 nm for the copper-on-cobalt and 0.90±0.05 nm for the cobalt-on-copper interfaces, in reasonable agreement with previous such measurements [2]. Such observations play a crucial role in the development of these nanoscale devices and have had impact on actual process development in the data storage industry.

2.2 Metal-Oxide-Semiconductor Structure

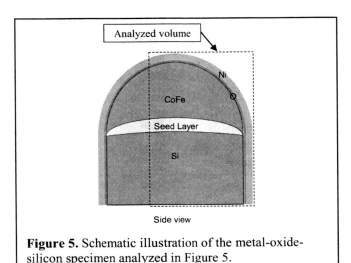

Figure 5. Schematic illustration of the metal-oxide-silicon specimen analyzed in Figure 5.

Specimens of a complex silicon/silicon oxide/metal structure were also prepared for analysis for demonstration purposes. Silicon posts were oxidized in air at room temperature, coated with nickel to form the metal-oxide-

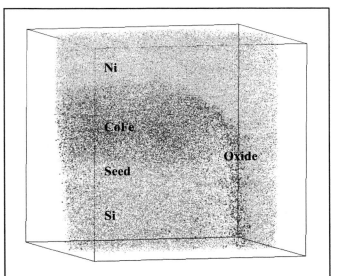

Figure 6. A perspective view of the oxide layer on silicon at a point on the specimen tip below the metal seed layer. Nickel atoms are shown in green, cobalt and iron atoms are blue, silicon is gray, and oxygen is red.

silicon structure and then annealed at 175°C in nitrogen. Figure 5 shows a schematic depiction of the specimen.

Figure 6 shows a section of the atom probe image of the same specimen. The oxide layer on the silicon is clearly visible in the image. This complex specimen mimics the structure in actual MOS devices and demonstrates that it should be possible to analyze the complex structures encountered in semiconductors using the LEAP microscope.

3 METROLOGY APPLICATIONS

Three LEAP instruments have been built by Imago Scientific Instruments. Imago's LEAP microscopes employ innovative technology that greatly improves analysis speed and simplifies specimen preparation relative to other 3DAPs. These developments for the first time make use of atom probe microscopy for metrology applications a potentially workable proposition. The large throughput of the LEAP microscope makes it realistic to consider applications where information is needed in short timeframes like one hour or less.

In process development, there exists a need for feedback about the effect of a given set of operating parameters on a microstructure so that modifications to the process can be made for the next iteration. Often much iteration is required. In some cases, it can take days to get microstructural feedback which means that the process development can take weeks or months. If the feedback information were available in hours, then the process development cycle could be shortened to days. In fast moving industries like the semiconductor industry, where for example, the implantation regimen for dopants in silicon must be developed for each new generation of chips, shortening the development cycle is extremely valuable.

Another application of the LEAP microscope in manufacturing is tool qualification. The first task in the manufacturing process establishment is tool qualification. A tool like a thin film deposition system or ion implantation device must be qualified when it is first commissioned and again after any maintenance takes it out of production. Whether it is film thickness or implant concentration and distribution, in some industries it can take days to iterate and tune a tool to get it ready for production. With the rapid turnaround of a LEAP microscope, this process might be shortened to hours. If so, major savings in process tool redundancy could be realized.

The extreme short-turnaround application is process monitoring. Where process variability is problematic or where feed forward and feedback are utilized in closed loop process control, the LEAP microscope may be able to provide information in a timely manner. One scenario in the semiconductor industry would use monitor wafers to sample the process at sufficient intervals for control purposes. In this type of application, analysis times would need to be on the order of one hour.

A final example of application is in failure analysis. In failure analysis, the objective is to identify defects as quickly as possible. Today's commonly available analytical tools can usually adequately analyze defects that are larger than one micron. Submicron defects can be *located* by currently available techniques, but not *analyzed* easily with these same techniques. The LEAP microscope may solve this problem. In one scenario, a defect review tool (DRT) can be used to rapidly locate small defects, and then the LEAP microscope may be used to rapidly analyze the defect. The entire process may be completed in less than one hour.

4 SUMMARY

Development of the three-dimensional atom probe microscope for metrology will provide industry with a tool that can contribute to shorter development times for new device designs and faster movement along the manufacturing learning curve. Imago Scientific Instruments has demonstrated the viability of this scenario in principle with its innovative 3-D atom probe, the LEAP microscope, and now is working toward practical implementation of the LEAP microscope as the first true 3-D atomic-scale metrology tool.

ACKNOWLEDGMENTS

Peter F. Ladwig and Y. Austin Chang of the University of Wisconsin and David J. Larson, Martin C. Bonsager, Bharat B. Pant, and Allan E. Schultz of Seagate Technology provided the specimens for the multilayer thin film analysis.

REFERENCES

[1] Larson et al., "Atomic-scale analysis of CoFe/Cu and CoFe/NiFe interfaces," Appl. Phys. Lett., **77(5)**, 726-728, 2000.
[2] Larson et al., "Atom probe analysis of roughness and chemical intermixing in CoFe/Cu films," J. of Appl. Phys., **89(11)**, 7517-7521, 2001.

Nanoinformatics:
Emerging Computational Tools in Nano-scale Research

K. Ruping* and B.W. Sherman**

* Massachusetts Institute of Technology, Cambridge, MA, USA, ruping@mit.edu
** Massachusetts Institute of Technology, Cambridge, MA, USA, woody420@mit.edu

ABSTRACT

Over the past several years basic research across several disciplines has revealed the promise of nanotechnology as an applied science in a wide range of industries. This progress has attracted the attention of government granting authorities, commercial researchers, and venture capital investors. Increased funding for nano-scale research initiatives has led to a growing demand for specialized human resource skills, research tools, and laboratory infrastructure.

This paper looks to a new area of computational tools that has emerged to meet the needs of the nano-scale research lab, and predicts the development of the field of nanoinformatics that will enable new advances and new applications in nanotechnology. We start with the historical roots of computer science and informatics. Next we look to the development of informatics into particular areas of science and commercial industries. Finally we document the growing use of computer science, database management tools, and information technology architectures in nano-scale research.

Keywords: informatics, computer science, nanoinformatics.

1. THE ORIGINS OF INFORMATICS

The term "informatics" has taken on various meanings, depending on the time period, geographical location, and field of use. For example, in Russia and parts of Europe the term informatics is closely related with computer science as a whole, and computer programming in particular. [8] Here we adopt a narrower definition of informatics to mean the science of applying computational theories and tools to the gathering, storage, manipulation, and interpretation of digital information. Such a definition still closely couples computer science with informatics, and envisions various applications of computational tools across different industries.

The history of computer science and informatics dates back to the 1830s when Charles Babbage, a Cambridge mathematician, conceived the first mechanical computing machine called a Differential Engine. Having found part of this never-completed device, Howard Aiken built the first electronic computer, the Harvard Mark I in 1944. However, applications for this new class of "calculators" were limited by instruction sets that were either hard-wired or manually inputted into the computational device through an I/O peripheral such as punch cards. This meant that any reprogramming required substantial human effort to re-wire the device or manually input new instructions, both of which were fraught with human error as well as time delays.

The advent of informatics came one year later when John von Neumann proposed that the programming code be incorporated into the electronic data resident on memory.[1, 3] The von Neumann architecture provided a level of abstraction away from the physical hardware, transforming the mundane task of programming into a field of engineering. With computer instructions liberated from the unchanging hardware elements and the continually changing data inputs, informatics emerged as an exciting area of research across university classrooms and industry labs. Government grant organizations, however, did not quickly recognize the strategic importance that informatics would have on other technical fields.

Informatics applications started with the first commercial computer. In 1951 the Census Bureau secured a UNIVAC from the Eckert-Mauchley Division of Remington Rand to tabulate census data. Three years later General Electric and the Metropolitan Life Company were using their UNIVACs to process employee payroll checks and perform other business processing needs. [6, 7] Soon IBM was offering a competing stored memory computers for computational science applications as well as for data management functions. Suddenly a growing number companies such as General Electric, RCA and Honeywell were entering the computer market with large mainframe computers, but with limited software flexibility.

Moving into the Cold War period, government support for emerging technologies became a strategic interest of several granting agencies. The momentum behind informatics as a discipline was not to be overlooked by the early 1960s, and computer science as a whole was increasingly viewed as a strategic differentiator in both economic and military competitive analysis. The first grant institution that included informatics research in its administrative scope was the Information Processing Techniques Office (IPTO), established in 1962 under the Defense Advanced Research Projects Agency (DARPA) and headed by J.C.R. Licklider. In 1967 the National Science Foundation (NSF) united its disparate computer science activities into a single office, the Office of Computing Activities (OCA). The funding strategy was focused on facilities, however, with $11.3 million of the office's $12.8 million total budget going toward institutional support in its first year of operation. [5]

In these early years computing power was seen to come from hardware advances alone. Software typically came bundled with the hardware, with little chance for third-party vendors to develop and sell independent products or extensions to existing software applications. That changed in 1969 when IBM decided to unbundled the pricing of its software in light of antitrust pressures. Into the 1970s a new generation of independent software emerged which enabled the extension of informatics to fields beyond the traditional mathematical, scientific and business processing applications.

2. INFORMATICS GROWS UP

As memory capacity expanded and processor speeds raced forward in what was to become Moore's Law, computer science engineers applied these capabilities to the sophisticated management and manipulation of large databases. These tools became attractive to a group of biologist working on genomic research. The foundation of bioinformatics was set.

2.1 Bioinformatics

Bioinformatics was the result of both the growing capabilities of the hardware systems supporting computational tasks, the increasingly sophisticated software that enabled complex data management, and the needs of research scientists who had growing economic resources with which to either design or purchase high-end informatics software. This new field evolved from the biochemical advances fueling the molecular revolution in biology across the 1970s and 1980s, such as gel electrophoresis, amino acid sequencing, polymerase chain reaction, and gene mapping techniques. [12] Bioinformatics became an industry when government research grants coincided with corporate R&D and venture capital investment into biotechnology. These origins closely map the current state of computational science in nanotechnology, which we will return to shortly.

Throughout the 1960s government support of basic research was broken down into the traditional academic categories, including biology but excluding computer science. In the 1970s new NSF funding was introduced to molecular biology projects. As research programs were formalized, it soon became apparent that new tools were needed to handle data and to provide high-throughput computation of biological data.

Bioinformatics started with government interest in sequencing the human genome, first articulated in 1984 and institutionalized with the Human Genome Project in 1990. The objective was an informatics challenge: to identify the approximately 30,000 genes in human DNA, determine the sequences of the 3 billion chemical base pairs that comprise human DNA, and manage this information in publicly accessible databases. Since the inception of this program, the amount of DNA sequence data has grown exponentially to nearly cover the complete DNA sequence of the human genome. The funding of this project and its related computational demands on the researchers facilitated the development of specialized informatics tools.

As a result, academic research facilities attracted a new generation of multidisciplinary researchers familiar with biology, statistical modeling, and computer science. Advances in statistical methods of data analysis and progress in computational technology enabled a growing community of specialists to effectively deal with the explosion of sequence data. They developed a multidisciplinary research area aimed at organizing, classifying, and parsing the immense richness of sequence data. Soon these tools were to become products for sale to others as the bioinformatics community turned into an industry.

2.2 Cheminformatics

Other fields have emerged from the interface between science and computational engineering. Similar to early genomics, chemistry researchers are now challenged by the complex data structures and the computation power needed to model reactions. Combinatorial chemistry is an increasingly common field of chemistry research where a set of different compounds are reacted in combination with each other to form libraries of resulting substances and their characteristics. [11] High throughput chemical analysis demands process control, data capture, and information processing. Chemical analysis, test equipment, and informational databases are leading the way to the new field of cheminformatics -- the organization of chemical data in a logical form to facilitate the process of understanding and making inferences. [9]

With the vast space of chemical compounds, it is essential to have computational methods and informatics tools to organize this information in a manner that makes research advances possible at an appreciable rate. The value of this organization can be seen in the pride and secrecy that pharmaceutical companies hold related to their chemical databases. These databases typically range between 100,000 and 1 million compounds and obviously require sophisticated data management and computational support. Even with such volumes today, these databases must grow by orders of magnitude to begin to span the chemical specie needed to find optimal drug candidates.

3. NANOINFORMATICS

The early development of nanotechnology is even more controversial than the origins of computer science and bioinformatics, in part because nanotechnology cuts across multiple scientific disciplines. The origins of nanotechnology lie in Richard Feynman's 1959 speech entitled "There's Plenty of Room at the Bottom." [2] In this presentation to the annual meeting of the American Physical Society, Feynman proposed the possibility of manipulating matter at the atomic level. Scientists had few tools with which to enable nanotechnology research, and most of the work at this scale was either theoretical or limited to chemistry.

Some of the most important early advances in nanotechnology have focused on innovative research equipment and new experimental methods applying such equipment. For example, Gerd Binnig and Heinrich Rohrer invented the Scanning Tunneling Microscope (STM) at IBM in 1981, for which they were awarded the Nobel Prize in Physics eight years later. Due to the physical scale of nanotechnology, much of the research equipment performing nano-scale experimentation is computer-driven. These tools have internal operating systems, computational software applications, and data management tools. As such the increasingly complex computer-driven research equipment, producing an ever increasing volume of digital output, were prime candidates for the next generation of informatics, which we call nanoinformatics.

There are two basic characteristics of nanotechnology research, also found in the development of bioinformatics, that has led to the emerging field of nanoinformatics. First is the deluge of data that computer-driven tools generate, particularly in nano-scale experimentation that incorporates a growing set of variables from across an increasing base of scientific knowledge. Second, the need for control -- of the systems themselves and of elements that the researcher hopes to manipulate -- increases in complexity as does the refinement of the research equipment itself. Both data management and system control come together in experimental tasks or industrial processes at a size-scale and time-scale that requires delicate sensing, massive data, complex calculation, and precision movement to translate virtual modeling into actual mechanical movement.

We are at the early stages in the emergence of a nanoinformatics science, and still earlier in the development of a nanoinformatics industry. Progress has been rapid. Early research tools required physical media, such as floppy disks or CDs, to transfer data between

equipment. Data capture and transfer was overcome by today's second generation of equipment networked across the nano-scale research lab. Equipment that was once reliant on an internal system for memory, computational power, and software tools now come with an Ethernet card and a communications interface. More problematic is the growing number of data formats and the lack of interoperability across the laboratory.

4. CONCLUSION

Similar to the liberation of software tools from hardware in the 1970's, the decoupling of laboratory equipment from informatics tools is enabling more dynamic research as well as increased knowledge of the results of such laboratory equipment. We are now at the inflection point of a new era of research management and information tools that will take advantage of computation power at a time when nano-scale data needs are expanding in both scale and complexity.

We believe progress in the application of informatics solutions to nano-scale research challenges will continue. The next generation nanoinformatics tools will focus on computational cooperation and systems integration, at which time the deluge of data will be translated into a windfall of knowledge. These advances will have to move beyond data sharing to task sharing, when a nanotechnology lab experiment touches on the computing power of a set of equipment behind a common user interface. Future progress in this field will move toward intelligent automation, complex data mining, and intuitive visualization of results.

REFERENCES
[1] J. von Neumann, "First Draft of a Report on the EDVAC," 1945.
[2] R. Feynman, "There's Plenty of Room at the Bottom: An Invitation to Enter a New Field of Physics," December 29, 1959, printed in Engineering and Science, February 23, 1960.
[3] W. F. Aspray, "Pioneer Day `82: History of the Stored Program Concept," Ann. Hist. Comp., Vol. 4, No. 4, 1982.
[4] History of the DOE Human Genome Program, http://www.er.doe.gov/production/ober/history.html.
[5] A.L. Norberg and J.E. O'Neill, "A History of the Information Processing Techniques Office of the Defense Advanced Research Projects Agency," Charles Babbage Institute, 1992
[6] Paul Ceruzzi, "A History of Modern Computing," 1998.
[7] M. Campbell-Kelly & W. Aspray, "Computer: A History of the Information Machine," 1996.
[8] V. Kasyanov, "SIMICS: Information System on Informatics History. International Federation of Information Processing," ICEUT 2000

http://www.ifip.or.at/con2000/iceut2000/iceut05-05.pdf

[9] K. Watkins, "Bioinformatics," Chemical & Engineering News, Feb 19, 2001.

[10] D. Baird, A. Shew, "Probing the History of Scanning Tunneling Microscopy," SHOT October 2002, Society for the History of Technology, http://shot.press.jhu.edu

[11] K. Schwall, E. Shanbrom, "Narrowing the Boundaries," Bioinformatics World, Spring 2003

[12] Dibner Institute for the History of Science and Technology, "The History of Bioinformatics," http://hrst.mit.edu

NANOPOLIS: an infrastructure for communication in the nanotech world

Dan BOG [*+] and Florin CIONTU [**]

[*]iMediasoft Group, Grenoble, FRANCE, dan.bog@imediasoft.net
[+]Politehnica University of Bucharest, ROMANIA,
[**]TIMA Labs, Grenoble, FRANCE, florin.ciontu@imag.fr

ABSTRACT

In this paper we present the fundamental concepts that led to the development of Nanopolis as a Distributed Knowledge Network for Nanoscale Science and Engineering. We start with a syncretic picture composed of ideas that have gained momentum in the past decades of cybernetic and sociological research. This approach is destined to provide a formal framework for the co-evolution (self-consistent evolution) of the semantics of communication and the structure of its underlying support. Based on this concept, the link to the more practical sphere of day-to-day actions is made by the fundamental thesis of Nanopolis: a part of the systemic challenges of nanotechnology can be translated into problems related to the need for spontaneous emergence of new communication networks. Finally, we present the current ongoing initiatives of Nanopolis and outline the envisioned future developments.

Keywords: nanotechnology, distributed knowledge network communication, multi-agent systems

1 INTRODUCTION

We are currently starting to foresee the morphing of nanotechnology from a futuristic prediction into the technology of tomorrow. A case for this statement is made by plans for moving the manufacturing of nanostructures from lab to fab and the applications from paper to prototypes. But in the light of the new opportunities we also start foreseeing the transformative effect brought by the drift towards the real-world – we are inevitably entering a phase where the dynamic of nanotech is not dominated anymore by what's happening in a couple of dozens of labs but by the complex dynamic of the research –industry – government triple helix. Research groups are starting to focus on specific details instead of dreaming of opening new paths, industrialists and research groups are entangling their activities more and more and not in the least voices asking for regulatory policies begin to be heard. At this stage we are entitled to take a systemic view of nanotechnology's development and address problems specific to the general dynamic of social systems like those related to the communication networks structuring such systems.

The Nanopolis project (www.nanopolis.net) takes such an approach by focusing on mitigating the impact of some of the challenges characteristic to nanotechnology through the stimulation of new communication patterns between the players in the nanotech world. Under its current manifestation, the project is a distributed knowledge network aiming at transposing today's knowledge on matter exploration at the atomic scale into a virtual representation space. More than 200 research groups are participating in this project coordinated by iMediasoft.

In this paper we present the fundamental ideas lying at the foundation of Nanopolis, its current stage as well as future actions. At first sight, the ideas presented here are abstract and grounded in theoretical developments, but they assemble into a formal framework implemented through a series of real-life actions as shown in the second part of the paper.

The remainder of the paper is structured as follows. In Section 2 we present the ideas on communication in knowledge based systems that sustain the fundamental framework of Nanopolis. The transposition of these ideas into the current actions of the project is presented in Section 3 together with an overview on future actions. We finally end with conclusions in Section 4.

2 FORMAL FRAMEWORK

Communication and the structure of its underlying support co-evolve interdependently driven by the wave of self-organization characterizing social systems and more generally the biosphere. This Section summarizes some ideas supporting this statement and sketches a framework for the development of Nanopolis as a distributed knowledge network.

We proceed with the construction of this framework in two phases:

- At first we take into account self-organization as a general phenomenon in social systems or ecosystems and the role of self-organization in determining the multi-layered structure of these systems.
- These general considerations can be particularized for the case where communication is seen as a semantic layer of a social system

2.1 Self-Organization in Social Systems

Self-organization and entropy decrease have been acknowledged to various degrees as intrinsic characteristics of social systems although their relation with *ab initio* mechanisms at the origin of living systems is still controversial. The former mechanisms have been studied by Ilya Prigogine[1] and his school in a Nobel prize awarded work on entropy decrease in non-equilibrium chemistry. Non-equilibrium systems can export entropy thus evolving to a higher level of order. At critical bifurcation points these systems can either fold back into chaos or evolve to more and more differentiated states. Irrespective of the existence of a link between these molecular level phenomena and the upper levels of the biosphere, Prigogine's work is significant in that it provides a model for a self-organizing system. Moreover, the concepts of entropy and information are rather cybernetic than physical. They can be applied to any system for which the basic assumptions of thermodynamics related to non-differentiability of basic elements hold. As such, Axelrod's model [2] on dissemination of culture is a good example of integrating the concept of order in its entropic sense into the study of social phenomena. The subject is further detailed in [3] through an analysis of the influence of the structure of social networks on the non-equilibrium order-disorder transitions. Finally, self-organization has been tackled in a generic manner in the context of Multi-agent Systems [4] and a sociologic view focused on self-organization in the knowledge society is provided in Ref. [5].

The structure of self-organizing systems is intimately related to the concept of hierarchy. This is unsurprising considering the direct correspondence between hierarchy and entropy reduction. More subtle arguments could also play a role, like the stability of hierarchical systems with respect to fluctuations. The latter concept is allegorically expressed in Simon's parable on Tempus and Hora, the two watch makers [6]. However, purely hierarchical structure can lead to a suboptimal degree of global self-organization of the entire system. In other words, a purely egoistic strategy for an organism can easily lead to a non-optimal result for the species. Thus, hierarchical organization and its variations have been rejected as realistic models for self-organizing systems [7].

This led to the adoption of heterarchies as a more appropriate model for the study of complex multi-level structured systems. As mentioned in [8], heterarchies are structures with "no single governing level", where "various levels exert a determinate influence on each other in some particular respect". Co-evolution is the process describing this kind of self-consistent evolution of a multiple-level structure. The concept has a significant practical value as, beyond theoretical conceptualizations, its existence has been confirmed in various instances like the co-evolution of eukaryotic viruses and their host organisms [9], the study of the interaction of different HIV populations[10] or the relationship between communication networks and social structures[11] which is the focus of this paper.

The conclusion up to this point is that self-organization is a ubiquitous principle governing the dynamics of a broad range of systems with a strong social component. Moreover, heterarchy is it currently considered the most appropriate model for such systems due to its

appropriateness for taking into account the self-consistent evolution on multiple levels in multi-layer structures.

2.2 Communication as a semantic layer

Two ideas will be developed in the next paragraph: (1) how communication can be integrated as a semantic layer within a heterarchy and (2) how communication co-evolves with the other layers of this heterarchy. For communication to be seen as a layer of semantic nature in a heterarchic model of a social system one must derive communication as a partial projection of one of the other layers of the system. Partial projection denominates the operation allowing derivation of a new abstract layer from an existing one through aggregation of elementary agents and interactions. This derivation process is represented in Figure 1.

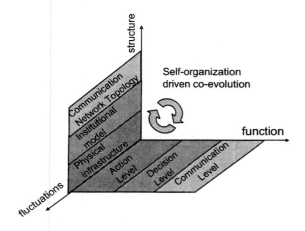

Figure 1. Communication as a partial projection of the decision level in a heterarchy-based representation of the social system

We adopted the function/structure/fluctuation interdependency as defined in [1] in order to be able to distinguish between the co-evolution of function and structure and the co-evolution of layers obtained from successive projections. The fundamental level of an agent-based representation of a social system is structured through analogy with a real-world infrastructure. A first projection allows us to derive the *decision level* by associating to event the corresponding decision at the origin of any action. Furthermore, the correspondence between decision and communication is obtained by selecting three classes of information:

- Communication channels constitute **the co unication network to ology** and are essential to convey information to an agent. Different attributes can be attached to a communication channel.

- *Perception* is essentially a matter of integration within a previously existing referential. How this referential is shaped, can be a matter depending on other communication actions. A fundamental problem of communication is whether they conveyed message has all the attributes allowing it to be integrated into an existing structure.

- *Analysis* is considered a process of almost mechanical nature corresponding to applying a set of operators to a data structure. The set of operators is considered common knowledge to all the agents in a population. The fundamental distinction between perception and analysis is that the former is dependent on the current state of an agent while the latter is considered dependent on knowledge transmissible independently of the individual experience of the agent.

Once communication can be projected as a layer of a heterarchic structure, its co-evolution can be characterized in the multi-level multi-theoretical framework defined in Ref. [11] to study the emergence of new communication networks. The latter are defined in opposition to formal networks which are the ones most commonly defined by our social institutions. Two extreme examples are relevant for formal networks: bureaucracies which are totally hierarchic and "free markets" where every communication action is a kind of bargaining and which are completely anarchic. On the contrary, emergent networks are the ones with connections naturally arising in social networks. According to the same authors, the emergence of communication networks is the result of a complex interplay of basic mechanisms acting at different levels: a single actor, dyad, triad etc. The basic mechanisms can be grouped into several theoretical classes: relating to self-interest, mutual interest and collective action, contagion, cognition, exchange and dependency, homophily proximity etc. The co-evolution of communication and its underlying social substratum can be characterized through the superposition of these basic mechanisms and can be pointed out by observing a critical point in any communication revolution. Thus, after a new communication technology substitutes older ones and respectively enhances communication itself, at a critical point businesses restructure their organizations in order to take full advantage of the new medium(ex: Business models organized around the internet).

In conclusion, communication networks can be seen as a partial projection in a heterarchical structure. Their co-evolution with other components of the global structure can be characterized *ab initio* by taking into account the elementary mechanisms governing the emergence of communication networks.

3 NANOPOLIS: DISTRIBUTED KNOWLEDGE NETWORK

After the previous formal developments we now proceed to explaining the current action lines of Nanopolis. Central to this is the fundamental thesis of Nanopolis which tries to express some of the systemic challenges of nanotechnology in terms of problems related to the need for the emergence of new communications networks. After exemplifying this thesis with two case studies we present the current initiatives. The Section ends with a description of future actions and their envisioned impact.

3.1 Fundamental Thesis

The fundamental philosophy of Nanopolis must be understood by taking into account the emergence of communication networks in the context of new challenges and new opportunities brought in by nanotechnology.

Coping with a part of the systemic challenges specific to nanotechnology requires the activation of new emergent networks mechanisms.

Two examples can be used to illustrate this thesis: the challenge posed by interdisciplinarity and the unbalanced evolution of *the triple helix government-research institutions-industry.*

Interdisciplinarity is becoming the manifesto of all speeches on the challenges of nanotechnology. As mentioned in Section 2.2, decisions to pursue a path are always conditioned by the available communication channels, and the global understanding of the chain of concepts within each topic is integrated (perception and analysis). However, the topology of a communication network is determined of course by physical issues but also by the "distance" between the concepts mastered by individuals. This means that in order to stimulate communication between individuals with different backgrounds, one must counter the absence of proximity in their backgrounds. Once this is countered, and once different agents see a mutual interest, communication will become self-reinforcing. This is a clear case of conceptual distance preventing communication. It is also a case where the enhancement of cognitive mechanisms, would contribute considerably to the emergence of communication mechanisms between agents who don't benefit otherwise from conceptual proximity.

The second example is the unbalanced evolution of education and science. The traditional bureaucratic circuit where new major changes are made at best every 5-7 years is not good enough to keep with the fast pace of discovery. A student coming out of undergraduate school today and wanting to do research in nanotechnology has no means to make an informed choice and most of the time his decision is arbitrary. Lastly, nanotechnology will apparently have most of its impact in traditional fields of the economy where big players are already established but history shows that existing businesses fail to recognize the potential of new technologies.

3.2 Current Initiatives

In its current stage of development, Nanopolis is identifying the critical points in the systemic evolution of nanotechnology and constructing the basic mechanisms allowing the stimulation of the emergence of new communication patterns. The first component is constituted by ongoing sociological research. The second one is currently materialized by two components:

- a software infrastructure allowing the participants in Nanopolis to supply/use knowledge in a distributed e-collaborative environment. Typical transactions supported by these systems are: (1) local processing of knowledge (2) multimedia transposition and the validation protocol and (3) content deployment.
- a considerable amount of scientific content related to matter exploration at the atomic scale, produced through the interactions of the institutions participating in the project. As this content is gathered in real-time, it is subsequently projected in different spheres of interest: industrialists, scientists or the general public. Information was presented via multimedia representations like 2D and 3D computer-generated images and animations.

The resulted informational content is disseminated as an interactive multimedia three-volume encyclopaedia series. The first two volumes are focused on the most frequently used tools for matter exploration at the atomic scale: synchrotron light and neutrons. The third volume is focused on nanotechnology.

"Exploring matter with synchrotron light", the 1st volume, has been developed in partnership with the "European Synchrotron Radiation Facility" and the help of APS (USA), Canadian Light Source, Soleil Synchrotron (Fra) and other synchrotron facilities. The CD-ROM has been awarded the prize for the best scientific multimedia – UNESCO Sept. 01, also recognized of pedagogical interest by the Ministry of Education of France.

The 2nd volume, "Exploring matter with neutrons", was developed in partnership with the Institutes Laue-Langevin (Fra), ISIS (UK), LANSCE (USA), Jülich Forshungszentrum (Ger), LLB-CEA (Fra), GKSS (Ger) and other members of the international neutronics community.

The 3rd volume of the Nanopolis series, "Exploring Nanotechnology" is currently under development and covers the research domains spawned by the emergence of new exploration techniques in the last few years. A first edition is expected to be released in 2004.

In addition to the these 3 tomes encyclopaedia series , Nanopolis is now evolving into a Virtual Communication Pole on the Internet offering precious resources of today's knowledge of the most recent physics, methods and technologies at the atomic scale in direct connection with the head R&D figures from all over the world.

3.3 Future initiatives

Future initiatives will focus on two classes of players: universities and policy-concerned institutes. The iMediaLearn system of iMediasoft will allow universities to integrate the multimedia content available on Nanopolis into their educational curriculum. Policy-concerned institutes, governmental as well as non-governmental will be involved in the project thus clearly mirroring the dynamics of the triple helix research-industry-government and further creating new communication bridges within the nanotech world.

4 CONCLUSIONS

Essential to this paper was our initial observation that nanotechnology reaches a point where its evolution is determined by the complex social factors and at this point we can apply systemic considerations. Our theoretical developments in this direction target to formalize the relationship between communication and the rest of the factors determining the dynamic of the field. The purpose of this approach is the automatic mirroring of this dynamic into Nanopolis as a distributed knowledge network. Furthermore, stimulating elementary mechanisms within this network should generate new self-sustaining patterns of communication. In the end, the emergence of these new communication patterns should provide a solution to some of the systemic challenges of nanotechnology.

REFERENCES

[1] G. Nicolis and I. Prigogine, J. Wiley & Sons, New York, (1977)
[2] R. Axelrod, J. Conflict Res. 41, 203 (1997),
[3] Konstantin, K. et al., Phys. Rev. E 67, 045101(R) (2003)
[4] Van Dyke Parunak, H. and Brueckner, S. , Proceedings of the International Conference on Autonomous Agents, (Agents 2001)m 124-130
[5] Leydesdorff, L., Universal Publishers/ uPUBLISH.com (2001, 2003)
[6] Simon HA (1962), Proc. Am. Phil. Soc. 106: 467--482
[7] Holland, J. H.., MA: Perseus Books., (1998)
[8] Kontopoulos, K. M. (1993). The logics of social structure. New York: Cambridge University Press.
[9] Kaufman R, PNAS, vol 96, no 21, 11693, (1999)
[10] Malins, C., et al., PNAS, Vol. 95, pp. 7637, (1998)
[11] Monge, P. R., & Contractor, N., New York: Oxford University Press, (2003)

Index of Authors

Index of Keywords

swissnanotech™

welcomes you

to the Nanotech 2004
Trade Show & Conference

Switzerland has been at the forefront of the nanotechnology development curve, contributing enormously to the field through inventions such as the Atomic Force Microscope, the Scanning Tunneling Microscope and many other groundbreaking discoveries.

Through the strong leadership, excellent research, and innovative business environment, Switzerland has become a destination for the global nanotechnology market so come visit the Swissnanotech Booth 200.

Exhibitors include:

Nanoworld AG	Swissnanotech
Nanosensors	Swiss Business Hub USA
Nanofair	Consulate of Switzerland, SHARE
Greater Zurich Area	Location:Switzerland
Reinhardt Microtech AG	Basel Area Business Development
	Economic Development Western Switzerland

Experts from Switzerland will also be participating as keynote speakers as well as panelists, including representatives from the Swiss Federal Institute of Technology, Zuerich, M.E. Mueller Institute of Basel, EPFL, University of Neuchatel, and others. More information is available in the official program.

The Swiss delegation at NANOTECH 2004 also includes representatives from Nestlé, UBS Private Banking, NanoDimension, Euresearch, Swiss Re, the Swiss American Business Council, PriceWaterhouseCoopers, CSM Instruments, and Semasopht.

You are also cordially invited to join us for a panel discussion and reception hosted by the Consulate of Switzerland, SHARE on Tuesday, March 9th at 6:15 p.m.

The event will start with short presentations from Prof. Peter Seeberger (ETHZ) and Prof. Andreas Engel (Biozentrum at Uni Basel) and will be followed by a reception gathering a large group of leading nanotech researchers and industry experts from Switzerland as well as many members of the local nanotech community.

SHARE is located at 420 Broadway in Cambridge. For details on the program and directions, please visit our web site at **www.shareboston.org**.

The **swiss**nanotech™ initiative at NANOTECH 2004 is organized, coordinated and sponsored by:

swiss business hub
usa
member of business network switzerland

Location:Switzerland

Swiss House
for Advanced Research and Education
Consulate of Switzerland

Nanotechnology
Investment Opportunities

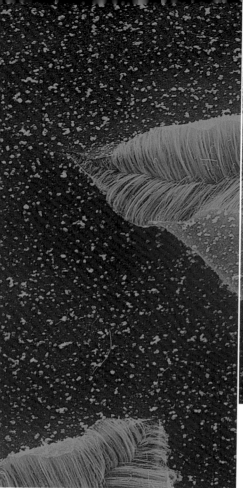

Australian Government

Invest Australia

To find out more about Australian nanotechnology opportunities or set up a meeting with an Investment Advisor, visit the Nanotechnology Australia stand at Nanotech 2004, or contact:

Hill Helen
Invest Australia – New York Office
Phone: **+1 212 351 6575**
Cell Phone: +1 917 608 3430
helen.hill@investaustralia.gov.au

Michael Claessens
Invest Australia – Canberra Office
Phone: **+61 2 6213 6704**
Cell Phone: +61 412 177 526
michael.claessens@investaustralia.gov.au

Australia has an active and rapidly growing nanotechnology capability, with a growing number of specialist nanotechnology firms, MNCs, and foreign government agencies investing in the development and commercialisation of breakthrough Australian technologies.

World-class scientific expertise and R&D facilities, combined with strong government support and a robust economy, have provided investors in Australian nanotechnology with a wide range of commercialisation and partnership opportunities. Practical outcomes are being delivered in manufacturing, materials processing, energy generation, and environmental management.

Products which exemplify this solutions-based philosophy include supercapacitors as light weight power sources in portable electronic devices, low energy filtration membranes, and clear ultraviolet screens for cosmetics, paints, and glass.

Assistance for investors

Invest Australia is the national inward investment agency, set up by the Australian Government to help foreign investors establish or expand a business in Australia. It leads a national approach to investment promotion and facilitation, providing businesses with an initial point of contact for investment enquiries.

Invest Australia understands the challenges of investing in nanotechnology and is committed to attracting new investment which will enable sustainable industry growth and development.

Working in partnership with business, Invest Australia provides organisations with the information and contacts needed to establish or expand your nanotechnology capability into Australia, including:

- expert advice on Australia's nanotechnology capabilities and opportunities for potential and current investors;

- the arrangement of site visits and links to potential joint venture partners;

- information on business costs, skills availability, taxation, R&D and other business related infrastructure;

- advice on investment regulations, streamlining major project approval processes and contacts with key Australian government agencies; and

- the identification of any relevant government industry assistance schemes.

Image courtesy CSIRO Health Sciences and Nutrition: A Nanostar

THE FUTURE IS **here** Nanotechnology Australia